A Virtual Tribute to Quark Confinement and the Hadron Spectrum (vConf21)

Held online

EPJ Web of Conferences Volume 258 (2022)

Stavanger, Norway
2 - 6 August 2021

Editors:

Alexander Rothkopf
Jens O. Andersen
Aleksi Kurkela
Dieter Rohrich
Laura Tolos
Anders Tranberg
Konrad Tywoniuk

ISBN: 978-1-7138-4246-0

Printed from e-media with permission by:

Curran Associates, Inc.
57 Morehouse Lane
Red Hook, NY 12571

Some format issues inherent in the e-media version may also appear in this print version.

For additional information, please contact EDP Sciences – Web of Conferences at the address below.

EDP Sciences – Web of Conferences
17, Avenue du Hoggar
Parc d'Activité de Courtabœuf
BP 112
F-91944 Les Ulis Cedex A
France

Phone: +33 (0) 1 69 18 75 75
Fax: +33 (0) 1 69 28 84 91

contact-edps@webofconferences.org

Additional copies of this publication are available from:

Curran Associates, Inc.
57 Morehouse Lane
Red Hook, NY 12571 USA
Phone: 845-758-0400
Fax: 845-758-2633
Email: curran@proceedings.com
Web: www.proceedings.com

TABLE OF CONTENTS

edp sciences

EPJ Web of Conferences

Statement of Peer review

In submitting conference proceedings to EPJ Web of Conferences, I certify to the Publisher that I adhere to the **Policy on Publishing Integrity** of the journal in order to safeguard good scientific practice in publishing.

1. All articles have been subjected to peer review administered by the proceedings editors.
2. Reviews have been conducted by expert referees, who have been requested to provide unbiased and constructive comments aimed, whenever possible, at improving the work.
3. Proceedings editors have taken all reasonable steps to ensure the quality of the materials they publish and their decision to accept or reject a paper for publication has been based only on the merits of the work and the relevance to the journal.

Title, date and place of the conference

A Virtual Tribute to Quark Confinement and the Hadron Spectrum (vConf21)
August 2nd-6th 2021, virtual (hosted by the University of Stavanger)

Proceedings editor(s):

Alexander Rothkopf (chair), Jens O. Andersen, Aleksi Kurkela, Dieter Röhrich, Laura Tolos, Anders Tranberg, Konrad Tywoniuk

Date and editor's signature

20.12.2021, for the editorial team:

Alexander Rothkopf

17, avenue du Hoggar - PA de Courtabœuf – BP 112 - 91944 Les Ulis Cedex A (France)
Tél. : 33 (0)1 69 18 75 75 - Fax : 33(0)1 69 07 45 17 – www.edpsciences.org

Round table on open quantum systems

Miguel Ángel Escobedo[1,*]

[1]Instituto Galego de Física de Altas Enerxías (IGFAE), Universidade de Santiago de Compostela. E-15782, Galicia, Spain

Abstract. This is a discussion about open quantum systems and its application to the study of hard probes. It reviews some of the things that were said in the corresponding round table. Views are only my own.

1 Introduction

In the last years, many groups have applied the formalism of open quantum systems to study quarkonium suppression [1–14] and jet quenching [15]. In my opinion, one of the reasons is that when you study these problems from the point of view of quantum field theory it is common to find the same type of equations that appear when you study other problems, for example quantum optics, using the open quantum system formalism. In other words, when we study quarkonium's evolution in a medium assuming that we know how the medium behaves what we are effectively doing is making a trace over the environment degrees of freedom.

In the formalism of open quantum system, the *universe* is divided in two pieces. The system of interest and the environment. The state of the system is encoded by the reduced density matrix, which is obtained making a trace over the environment degrees of freedom to the *universe*'s density matrix. The time evolution of the reduced density matrix is ruled by the so-called master equation. When applying this formalism to the study of hard probes in heavy-ion collisions, the challenge is twofold. On one hand, we try to derive the master equation from QCD. On the other hand, it is often difficult to solve the master equation to obtain phenomenological results. I will focus on the application of the open quantum system framework to the study of quarkonium suppression, however, there also other interesting applications in other fields, as for example, jet quenching, cold atoms and cosmology.

2 Challenges

2.1 More general derivations of the master equation

Most derivations of the master equation have used either Hard Thermal Loop perturbation theory [1, 12] or pNRQCD in the limit $\frac{1}{r} \gg T$ [9]. In order to obtain more reliable predictions, we should investigate a wider range of temperature regimes and, at the same time, incorporate more non-perturbative input. Note that we observe many different quarkonium species in heavy-ion collisions (as for example, J/Ψ, $\Upsilon(1S)$, $\Upsilon(2S)$...). Each species has a different radius and, therefore, they probe the medium at different scales. On top of this, the

*e-mail: miguelangel.escobedo@usc.es

medium temperature changes with time as the quark-gluon plasma expands and cools down and, furthermore, depends on the impact parameter of the collision. Therefore, many different types of relations between the radius of quarkonium r and the temperature T are realized nowadays in heavy-ion collisions.

Regarding the combination of Non-relativistic Effective Field Theories (NREFTs) with the open quantum system framework [8–10], the results available at the moment include the derivation of the master equation in the $\frac{1}{r} \gg T$ regime and the application to phenomenology of the more restricted case $\frac{1}{r} \gg T \gg E$. From the point of view of NREFT, the case $\frac{1}{r} \gg T$ is the simpler to analyse. To generalize the nowadays available results to the case $\frac{1}{r} \sim T$ requires some new theoretical developments. The reason is the following. In a non-relativistic system, as heavy quarkonium, different well separated energy scales are present. The mass of the heavy quarks m is much larger than the inverse of the typical radius, $\frac{1}{r}$, and this is much larger than the binding energy E. The use of NREFTs is useful for the study of this kind of systems. Starting from QCD, Non-relativistic QCD (NRQCD) [16, 17] is obtained after integrating out the scale m. If we also integrate out the scale $\frac{1}{r}$ we obtain potential NRQCD (pNRQCD) [18]. In the case $\frac{1}{r} \gg T$, we can perform the matching between NRQCD and pNRQCD ignoring the medium. Then, we can use pNRQCD to compute medium effects and derive the master equation. If, instead, we are considering the case $\frac{1}{r} \sim T$ then we can not ignore the medium when matching NRQCD to pNRQCD. This requires to substantially change the matching procedure.

The computation of thermal effects in real-time requires the use of the Schwinger-Keldysh contour in which one can speak of a doubling of degrees of freedom. We use a path-integral that goes from minus infinity to infinity and then goes back to minus infinity. We can call the fields on the first part of the path integral fields of type 1 (time ordered) and the rest of fields are called fields of type 2 (anti time ordered). In general, when we integrate out thermal effects, we expect to generate terms that mix both types of fields. This is a manifestation of the fact that medium effects can not always be encoded in an effective action. When dissipation is present, we need to use the more general concept of influence functional instead [19]. This is a translation to the path integral language of the observation that not all master equations are of the Liouville-von Neumann type. How does this affects the matching between NRQCD and pNRQCD? That is an open question [1]. In the literature, one can find several studies on how to construct an EFT including dissipation and what are the symmetries of the influence functional (see for example [22, 23]). Many of those studies are focussed on deriving an EFT for hydrodynamics and, perhaps, some of these ideas would be also useful to study quarkonium.

2.2 Incorporating non-perturbative input

There are two types of non-perturbative input that have been used to constraint the master equation of quarkonium. The static potential [24, 25] and transport coefficients, like κ and γ.

Regarding the static potential, it is interesting to discuss if it is possible, and how, to establish that the static potential is useful to the study of heavy quarkonium. At $T = 0$, the static potential coincides with a Wilson coefficient of pNRQCD. The matching between NRQCD and pNRQCD can be performed non-perturbatively in the static limit and, since each term has a well-defined power counting in terms of $\frac{1}{m}$, the Lagrangian so-obtained can be used to describe heavy quarkonium [18]. At finite temperature, an analogous procedure can be followed as long as we only care about time-ordered correlators. The static potential

[1] Note that one thing that we do know is that the doubling of degrees of freedom can be ignored if we are only interested in the time-ordered correlator [20, 21]. Unfortunately, then it is not possible to derive the master equation.

coincides with a Wilson coefficient in pNRQCD that enters the leading-order evolution of quarkonium's time-ordered propagator. Regarding the master equation, less is known. We do not know exactly how the static potential enters the master equation of quarkonium without using perturbative arguments. To understand this we need to construct a version of pNRQCD valid for the derivation of the master equation in the $\frac{1}{r} \sim T$ regime, an issue we discussed previously.

In the limit $\frac{1}{r} \gg T \gg E$ the situation is more clear. The influence of the medium into the master equation can be encoded in two non-perturbative parameters, κ and γ [8, 9]. κ coincides with the heavy quark diffusion parameter [26], useful in the study of open heavy flavour. It is related to the imaginary part of the potential and controls the strength of the dissipative (non-unitary) part of the master equation. On the other hand, γ is related with the real part of the potential and encodes the influence of the medium into the unitary part of the master equation. Physically, it can be related to screening with the caveat that the small r limit is used. We have a lot of information on κ from lattice QCD (see [27] for a recent computation), enough to have an educated guess of the dependence of κ with the temperature. Regarding γ, less information is available. There are determinations of γ using the thermal mass shift of quarkonium [10], which is proportional to $\gamma\langle r^2 \rangle$ assuming the hierarchy $\frac{1}{r} \gg T \gg E$. However, it is not enough at the moment to understand how γ depends on the temperature. An interesting question is whether κ and γ can be used to constraint the master equation beyond the regime $\frac{1}{r} \gg T \gg E$.

There is work that needs to be done in order to find new operators computable on lattice QCD that can bring new information about the master equation when $\frac{1}{r} \sim T$. It is possible that these operators have a complicated time-ordering because of the mixing between the two branches of the Schwinger-Keldysh contour. In fact, this kind of operators emerge on the study of the evolution of a static pair in a plasma [28]. It is not clear how these operators can be computed on lattice QCD, but that does not mean that it can not be done. The issue is that only computations in Euclidean space can directly be performed on lattice QCD, then, using analytic continuation Minkowski space quantities can be obtained. It needs to be understood how to do the analytic continuation in order to compute operators mixing the two branches of the contour.

2.3 Recombination

One of the phenomena that affect the number of quarkonium bound states in a heavy-ion collision is recombination [29, 30], the process in which two heavy quarks meet inside of the medium and form a new bound state. We can distinguish two kinds of recombination, the correlated and the uncorrelated one. In correlated recombination the two heavy quarks were part of a bound state previously, this bound state dissociated but after some time the heavy quarks recombine again. This is the kind of recombination that appears when we study the dilute limit. Uncorrelated recombination means that the heavy quarks that recombine did not belong to the same bound state before or that they might never have been in a bound state. This kind of recombination increases with the number of heavy quarks in the collision and is believed to be a leading-order effect for charmonium at LHC.

Correlated recombination is one of the main ingredients of the open quantum system framework. In fact, correlated recombination compensates the effect of the imaginary part of the potential and make the master equation trace preserving. However, it is challenging to incorporate uncorrelated recombination. Until now, two different strategies have been followed in the literature. The use of semi-classical approximations and the molecular chaos hypothesis.

Using the Langevin equation in the Abelian limit [11] or Langevin-like equations in QCD [12], systems with many heavy particles can be simulated. This is due to the fact that classical equations are much less costly to simulate that quantum master equations. One of the problems identified in QCD is that the applicability region seems to be much narrower than the corresponding QED case, however, there have been recent studies trying to improve this situation [31]. Both in QCD and QED, uncorrelated recombination has been observed, showing that semi-classical equations allow to qualitatively understand this phenomenon. However, it is challenging to understand to which quarkonium state the newly formed bound state correspond. This is because in order to define what is, for example, a 1S state and what is not we need to talk about the wave function, an object that disappears when making the semi-classical approximation.

Another approach is to use the molecular chaos hypothesis [4]. In the context of Boltzmann equations, this hypothesis means that the 2-particle distribution function is assumed to be equal to the product of two 1-particle distribution functions. For the case of quarkonium, this means that the distribution of octets is assumed to be proportional to a product of the uncorrelated distribution of individual heavy quarks. Using this hypothesis, the Boltzmann equation can be solved and a good agreement with experimental data is found. An open question is then to find a scheme in which this hypothesis becomes the leading order result in a controlled approximation. This is still an open question.

At the moment, most of the theoretical computations use the dilute limit. In an effective way, this is equivalent to assuming a system formed by a heavy quark, a heavy antiquark and a medium without other heavy particles. Perhaps, a study considering two pairs of heavy particles in detail would give some useful insight on the phenomenon of recombination from the point of view of open quantum systems.

References

[1] Y. Akamatsu, Phys. Rev. D **91**, 056002 (2015), 1403.5783
[2] A. Rothkopf, Phys. Rept. **858**, 1 (2020), 1912.02253
[3] Y. Akamatsu (2020), 2009.10559
[4] X. Yao, T. Mehen, Phys. Rev. D **99**, 096028 (2019), 1811.07027
[5] X. Yao, B. Müller, Phys. Rev. D **100**, 014008 (2019), 1811.09644
[6] X. Yao, W. Ke, Y. Xu, S.A. Bass, B. Müller, JHEP **01**, 046 (2021), 2004.06746
[7] X. Yao, Int. J. Mod. Phys. A **36**, 2130010 (2021), 2102.01736
[8] N. Brambilla, M.A. Escobedo, J. Soto, A. Vairo, Phys. Rev. D **96**, 034021 (2017), 1612.07248
[9] N. Brambilla, M.A. Escobedo, J. Soto, A. Vairo, Phys. Rev. D **97**, 074009 (2018), 1711.04515
[10] N. Brambilla, M.A. Escobedo, A. Vairo, P. Vander Griend, Phys. Rev. D **100**, 054025 (2019), 1903.08063
[11] J.P. Blaizot, D. De Boni, P. Faccioli, G. Garberoglio, Nucl. Phys. A **946**, 49 (2016), 1503.03857
[12] J.P. Blaizot, M.A. Escobedo, JHEP **06**, 034 (2018), 1711.10812
[13] J.P. Blaizot, M.A. Escobedo, Phys. Rev. D **98**, 074007 (2018), 1803.07996
[14] R. Katz, P.B. Gossiaux, Annals Phys. **368**, 267 (2016), 1504.08087
[15] V. Vaidya, X. Yao, JHEP **10**, 024 (2020), 2004.11403
[16] W.E. Caswell, G.P. Lepage, Phys. Lett. B **167**, 437 (1986)
[17] G.T. Bodwin, E. Braaten, G.P. Lepage, Phys. Rev. D **51**, 1125 (1995), [Erratum: Phys.Rev.D 55, 5853 (1997)], hep-ph/9407339

[18] N. Brambilla, A. Pineda, J. Soto, A. Vairo, Nucl. Phys. B **566**, 275 (2000), hep-ph/9907240
[19] R.P. Feynman, F.L. Vernon, Jr., Annals Phys. **24**, 118 (1963)
[20] N. Brambilla, J. Ghiglieri, A. Vairo, P. Petreczky, Phys. Rev. D **78**, 014017 (2008), 0804.0993
[21] M.A. Escobedo, J. Soto, Phys. Rev. A **78**, 032520 (2008), 0804.0691
[22] A. Baidya, C. Jana, R. Loganayagam, A. Rudra, JHEP **11**, 204 (2017), 1704.08335
[23] M. Crossley, P. Glorioso, H. Liu, JHEP **09**, 095 (2017), 1511.03646
[24] Y. Burnier, O. Kaczmarek, A. Rothkopf, Phys. Rev. Lett. **114**, 082001 (2015), 1410.2546
[25] D. Bala, O. Kaczmarek, R. Larsen, S. Mukherjee, G. Parkar, P. Petreczky, A. Rothkopf, J.H. Weber (2021), 2110.11659
[26] J. Casalderrey-Solana, D. Teaney, Phys. Rev. D **74**, 085012 (2006), hep-ph/0605199
[27] N. Brambilla, V. Leino, P. Petreczky, A. Vairo, Phys. Rev. D **102**, 074503 (2020), 2007.10078
[28] M.A. Escobedo, Phys. Rev. D **103**, 034010 (2021), 2010.10424
[29] P. Braun-Munzinger, J. Stachel, Phys. Lett. B **490**, 196 (2000), nucl-th/0007059
[30] R.L. Thews, M. Schroedter, J. Rafelski, Phys. Rev. C **63**, 054905 (2001), hep-ph/0007323
[31] S. Delorme, also on these proceedings

Particle therapy masterclass

Panagiota Foka[1*], *Aristeidis* Mamaras[2], *Damir* Skrjiel[3], *Joao* Seco[4], *Christian* Graeff[5], *Marco* Pulia[6], *Hans-Peter* Wieser[7] and *Niklas* Wahl[4]
[1]GSI Helmholtzzentrum für Schwerionenforschung, Planckstraße 1, 64291 Darmstadt, Germany
[2]AUTH University Campus, Administration Building, 54124, Thessaloniki, Greece
[3]UNSA Obala Kulina bana 7/II, Sarajevo 71000, Boznia and Herzegovina
[4]DKFZ Im Neuenheimer Feld 280, 69120 Heidelberg, Germany
[5]Department of Electrical Engineering and Information Technology, Technical University, Darmstadt, Germany
[6]CNAO 642W+92 Pavia, Provincia di Pavia, Italy
[7]LMU Geschwister-Scholl-Platz 1, 80539 München, Germany

Abstract. The aim of the new Particle Therapy MasterClass (PTMC) was to develop an educational and training environment in which anyone can learn about fundamental and applied research in particle therapy. The PTMC was recently integrated into the International MasterClass 2021 online programme that attracted 1500 students from 37 institutes in 20 countries, worldwide. The PTMC focuses on the topic of cancer treatment, a particularly sensitive and socially relevant topic. The main idea is to (a) provide a basic understanding of cancer radiation therapy, (b) demonstrate that fundamental properties of particle interactions with matter, which are used for detection in physics experiments, are also the basis for treating cancer tumours; and (c) show that the same accelerator technologies are used in both, research laboratories and therapy centres. For the hands-on session, the open-source professional treatment planning software matRad is used, developed for research and training by the German Cancer Research Center – DKFZ. Ultimately, students are shown "what physics has to do with medicine" and what are the various possibilities that physics and STEM studies may open up for job opportunities in fields that are lacking expert personnel.

1 Introduction

The new Particle Therapy MasterClass, PTMC, developed in 2019 and presented in this paper, aims at highlighting applications of fundamental research for the benefit of society. While the mission and mandate of research institutes, such as CERN, GSI and others, is fundamental research, the developed technologies and acquired knowledge find applications in many different domains. The aim of the PTMC is to enhance awareness of such contributions focusing on the particularly sensitive societal topic of cancer therapy, while also enhancing awareness on advanced possibilities of cancer tumour treatment using particles such as protons and heavier ions.

The concept of the PTMC combines different pedagogical elements (Section 2) giving the opportunity to participants to become scientists for a day. Elaborating on the theme "from particle physics to particle therapy" it was initially developed targeting high-school students within the International MasterClasses, IMC [1], the flagship project of the International Particle Physics Outreach Group IPPOG [2]. Additionally, it can be easily tailored to different needs and education levels. The developed concept and available material via the PTMC web page [3] were already used to complement training of university students, as basis for a full week specialised school and for events such as the CERN Open Days and the Science Days of Montenegro.

Within the IMC project, masterclasses are usually organised as full day events, with a typical agenda, followed by the participating institutes. During 6 weeks before the Easter vacations, 3-5 institutes participate every day, inviting high-school students to immerse in the world of particle physics giving them a taste on "what scientists do". Adapting to an online mode due to the Covid pandemic, students were invited by the institutes to join a zoom session, instead of being invited at the institutes in person. In several cases, this was announced broadly and students from the whole country could join the online sessions. The PTMC specifically shows to students "what physics has to do with medicine", and triggers questions on how particles can be used for cancer therapy.

The PTMC, coordinated by GSI, was included for the first time in the IMC2020 schedule after a first local test at GSI and a pilot one, in early 2019, with the participation of GSI [4], CERN [5] and DKFZ [6] Heidelberg. This core team represents the world leading research institutes in their domains that also provided tangible applications for health: GSI treated the first 450 patients in Europe with carbon ions, CERN hosted the Proton-Ion Medical Machine Study (PIMMS) [7] design team. DKFZ is Germany's national cancer research institute, which is also closely connected to the HIT Heidelberg ion therapy facility [8]. The first PTMC session in IMC2020 was performed in Mexico [9], attracting about

* Corresponding author: yiota.foka@cern.ch

200 students, while the rest of the sessions had to be cancelled. In the IMC2021 season, the PTMC sessions were performed online, and with no limitation on registrations, they attracted more than 1500 students around the world. During the 6 PTMC sessions a total of 37 institutes from 20 countries participated, with about 6 institutes performing the PTMC in the same day.

2 Pedagogical elements

The PTMC agenda integrated and expanded the basic elements of the IMC that include, in the morning, introductory lectures and a visit to an experiment or lab, while in the afternoon, a hands-on session, followed by a common video-conference among all participating institutes. Each PTMC online session started with a video, showing a particle therapy procedure in a virtual therapy centre to trigger curiosity and interest. Further, it provided an overall visual impression, so that students could relate the specific presentations that followed to the broader image and procedures. The lectures covered a wide range of topics, supported by videos to keep the interest of students. Introductory lectures covered the basics of cancer biology and treatment options, the physics and biology of exposure to ionizing radiation, imaging devices and detectors, as well as accelerators used for cancer diagnosis and treatment. Due to the multidisciplinary nature of the topic, the introductory lectures can be tuned to or complemented by lectures on the expertise of the different institutes. This includes, for example, fundamental detector or software developments that are used in medicine for imaging, diagnostics, dosimetry, as they are critical parts of the treatment procedures where the impact of breakthrough developments on detectors for physics experiments is clear. The importance of computing and software developments is also highlighted including machine learning techniques and their possible applications in these fields. The message is always that: **technologies developed for fundamental research find applications for medicine**.

Highlighting the use of accelerators for society, students were surprised to learn that from about 30 000 accelerators that are in operation world-wide, about 6% is used for research, while 1/3 is used for medicine and the rest for different industrial applications. Adapting technologies and methods developed for fundamental particle physics research, sophisticated medical accelerators can deliver particle beams at a desired depth targeting a tumour. Students are also shown that fundamental properties of particle interactions with matter, which are used to detect them in physics experiments, are also the basis for treating cancer tumours. Figure 1a shows, as an example, the Bethe Bloch energy loss used for particle identification in the TPC detector of the ALICE [10] experiment at the CERN LHC, while the 1b plot presents the energy deposition as function of depth in water.

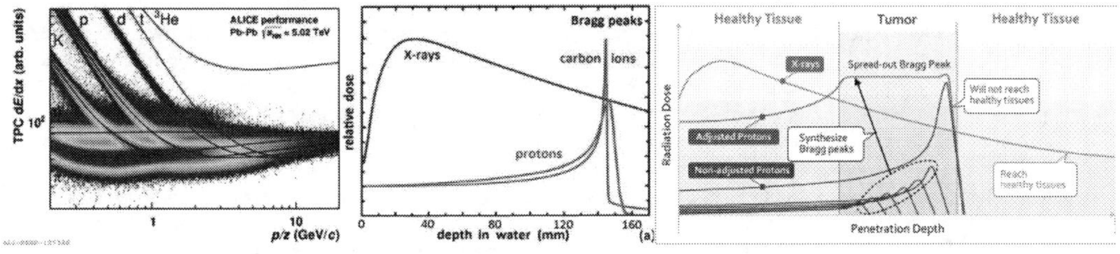

Figure 1: Bethe Bloch energy loss used for particle identification (a); energy deposition as function of depth for photons, protons and carbon ions used for Hadron Therapy (b); and spread-out Bragg peak (c).

Here, protons and carbon ions reveal their dose deposition characteristics, manifesting in the Bragg peak, which is the basic motivation for using "heavy" charged particles radiation therapy. One can clearly observe that photons lose most of their energy entering the tissue and they still deposit energy in healthy tissues behind the tumour while ions deposit most of their energy at the end of their trajectory before they come to a full stop. By tuning the energy of the accelerator and adapting the number of ions that are being accelerated, one can scan the entire tumour homogeneously, what is known as spread-out Bragg peak (SOBP), shown on the last plot.

During the initial pilots in 2019 visits to HIT, CERN and GSI were possible. Since 2020 online visits were offered, due to the pandemic, at the ALICE heavy-ion experiment. They were complemented with online visits during the afternoon common video-conferencing at the GSI experimental room where carbon ion therapy was pioneered in Europe, and in the CNAO [11] ion therapy centre, close to Milano (Figure 2). Experts from these institutes provided an excellent interactive help, discussing with the students the results. These online visits further highlighted the connection between heavy-ion physics and heavy-ion therapy.

In general, the PTMC highlighted that technology and knowledge transfer "from physics to clinics" resulted in the four carbon-ion therapy centres in Europe. Following the PIMMS initiative, accelerator experts of the CERN NIMMS group work currently on the design of components for a next generation ion therapy machine adapting hi-tech technologies for med-ical accelerators. As an example, Figure 3 shows the accelerator layout and architect's concepts (courtesy "Kaprinis architects") for a heavy-ion therapy facility which is proposed to be built in South East Europe (SEE). The PTMC was performed by several SEE institutes that provided material in several local languages, supporting capacity building in the area.

Figure 2: Online visits at GSI heavy-ion research centre, CNAO ion therapy centre and the ALICE heavy-ion experiment.

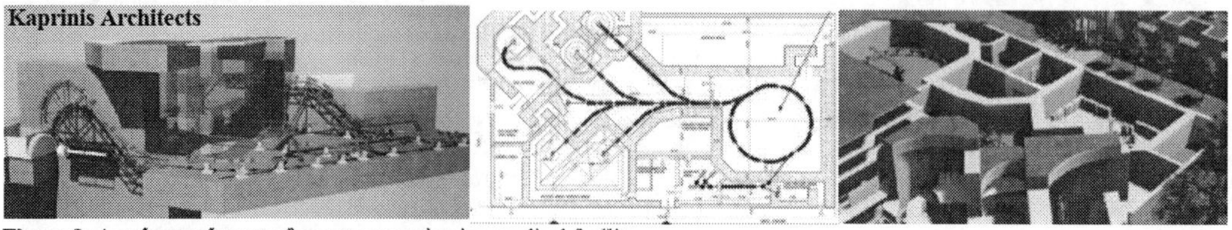

Figure 3: Accelerator layout of a next generation ion medical facility.

Before the PTMC hands-on session, the students were introduced to the concepts of dose calculation, dose optimization and radiotherapy treatment plans. In particular, school students can learn here how elemental mathematical concepts (e.g., learned from calculus) are required and applied to solve actual technical problems. The PTMC hands-on session itself is based on the open-source treatment planning toolkit developed by DKFZ for research and training (Figure 4) [12,13,14]. It is used to simulate and optimize the therapeutic intensity-modulated dose distribution that the accelerating structures have to deliver to the tumour using photons, protons or carbon ions according to "prescriptions" of the students who can appreciate the differences of these radiation modalities and use them optimally according to different cases. While matRad's computations were validated with commercial treatment planning software actually used for therapies, it is not licensed for treatment itself but uses a GPLv3 open-source license. This facilitates matRad's lightweight, flexible, accessible and performant code, useful for research and such educational applications as the PTMC. Further, depending on the level of the student, special educational versions can be provided with reduced treatment planning complexity of the GUI.

The provided data are a TG-119 phantom and anonymous computer tomography (CT) scans of head, liver and prostate tumour cases from the CORT dataset [15]. MatRad includes a graphical user interface (GUI) resembling typical commercial systems to present the data in 3D, provides possibilities to rotate interactively and presents slices in 2D projections. The GUI includes graphical tools to visualize the beam directions and volumes to be irradiated.

For the dose optimisation, one can use multiple objectives and constraints for dose prescription, e.g., define a minimum and/or maximum dose value to be achieved within a structure or organ [14]. The visual result on the display, based on a colour scale, is very intuitive to comprehend and easy to associate with the resulting histograms of the delivered dose to the target tumour and organs at risk to avoid. Students can select different angles for irradiation to minimize the deposited dose in each trajectory while accumulating the required dose in the target volume. They can witness the different characteristics of protons and carbon ions, that, in contrast to photons, deposit very little energy in the tissues before and practically none after the tumour (Figure 4).

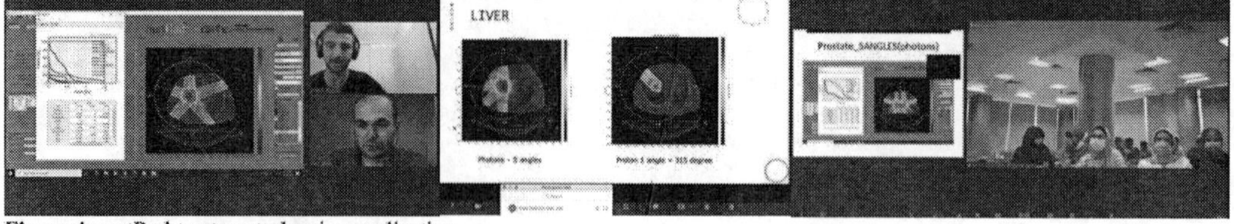

Figure 4: matRad treatment planning application.

The hands-on session is typically divided into three sections. The first section aims at highlighting the difference of photon and charged particle irradation on the TG-119 phantom, showing, for example, that many beams are required with photons to cover the tumour while the sparing of healthy tissue can be challenging, whereas protons achieve at least similar results with already one or two beams.

The second part introduces carbon ions and radio-biological effectiveness on a patient case. The institutes and/or students can choose between the liver and head-and-neck case for this exercise. This introduces versatility to the later videoconference, where students present and exchange their findings.

The last part critically introduces the "downside" of charged particle therapy, that is, its sensitivity to uncertainties that might change the net range in the patient and distort the planned dose distribution when the patient is, e.g., misaligned.

Instructions for installing matRad specific software, prepared by the DKFZ colleagues, are available via the PTMC web pages, including recordings and are sent to participants in advance, although installation can be handled even during the session. In order to support the PTMC local organising teams of each participating institute, and prepare their tutors and moderators, weekly dedicated tutorials were offered during the IMC2021 season but also training sessions at request. The PTMC web pages [3] provide a large variety of material in several languages, including links to the "PTMC in a kit" google drive which contains the necessary material needed to perform the PTMC, including recordings with instructions. In addition, it links to the indico agendas of institutes that performed already the PTMC, including their presentations in their local languages.

3 PTMC tutorial

To enable researchers to become tutors for the PTMC, tutorials are offered as training. These tutorials focus mainly on the hands-on session with matRad to familiarize with the matRad GUI and provide a deeper understanding of the underlying cancer treatment planning concepts.

The tutorials explain the possibilities of the software that is accessible through the GUI (Figure 5), which includes four different tabs: The first "workflow" tab controls the execution of the major steps of the treatment planning process. Thus, it contains buttons for loading the provided patient data as well as DICOM files, data from binary files and already prescribed doses. The "dose calculation" button allows to either compute the dose influence matrix to be used for a subsequent dose optimization of a new treatment plan or recalculate the dose for a different patient geometry (e.g., a shifted patient). The "optimization" button triggers dose optimization of the beam fluence, i.e., the beamlet (named "bixel" in matRad) weights, that yield the best possible dose distribution according to the clinical objectives and constraints underlying the radiation treatment. For photons, a beamlet/bixel is a small photon intensity element used to illustrate a portion of an intensity-modulated radiotherapy (IMRT) beam for the optimization of the intensity distribution or for dose calculations for the treatment planning [16], while for charged particles it refers to an individual spot with distinct lateral position and energy, forming one of the localized Bragg-peaks that compose SOBP.

Next, is the "plan" tab, where the parameters of the plan are inserted. In this tab, the students can input the beamlet/bixel width (or spot spacing for particles), the gantry and couch angles, the type of the radiation method, the number of fractions of the prescribed dose as well as the "isocenter" (the reference point in space around which, the gantry and couch rotate) coordinates.

The plan tab is followed by the "objectives & constraints" tab, which is enabled after inserting the data for each case, showing the organs of interest (e.g target), as well as the organs at risk. Here, the dose optimization can be configured by defining multiple, typical planning objectives and constraints to achieve the optimal dose distribution in different structures or organs.

Last tab is the "visualization" tab, where the user can set visualization properties. For instance, there is a "slice selection" scroll bar, to select and display a specific 2D image out of the 3D voxelized patient volume. In addition, there is the "plane selection" button where the user can switch between axial, sagittal and coronal view. In addition, the "visualize plan/beam lines" button is important, which one can enable for a better understanding of the different beam angles of the irradiation plan.

During the PTMC tutorials, first the TG-119 is demonstrated after the description of important medical physics terms that are also provided in the PTMC webpage. The TG-119 is an acrylic sample that is also known as C-phantom, due to its characteristic shape (Figure 5).

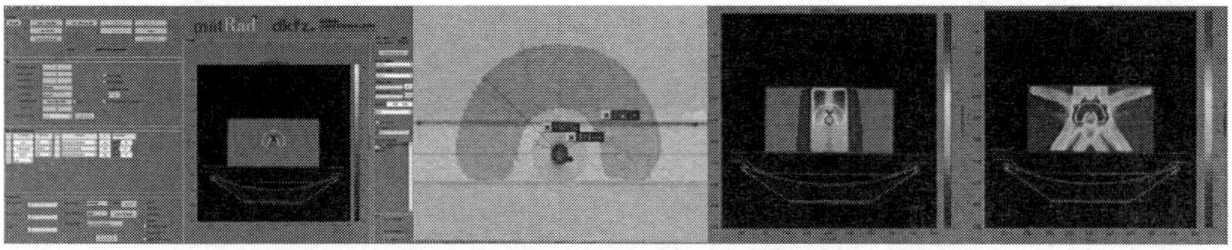

Figure 5: matRad GUI; TG-119 sample; TG-119 irradiation with a single angle of photons and five angles.

This phantom is usually used by professionals for quality assurance to verify the proper function of their equipment. In this case, the aim is to avoid the central bar, which represents the Organ At Risk (OAR, organs more sensitive to irradiation), and irradiate only the C-shape, which is the target where the cancer tumour is located. For this tutorial, two photon treatment plans were designed utilizing a single beam angle and five beam angles to demonstrate characteristics of intensity modulated photon treatment plans. This example demonstrates how the number of beam angles influences the resulting dose distribution. The results can be seen in Figure 5, where the red zones are the areas irradiated the most and the blue ones the least.

Then, for a real-life application, the liver case is typically being demonstrated. In this case, the complexity of the treatment plan is increased due to the sensitive areas that should be taken into consideration during planning. Three radiation methods were used, photons, protons, and carbons. For the photon treatment plan, five different angles were used, focused on the side where the tumor is located. For the proton and carbon treatment plans, a single angle was used. The results can be seen in Figure 6.

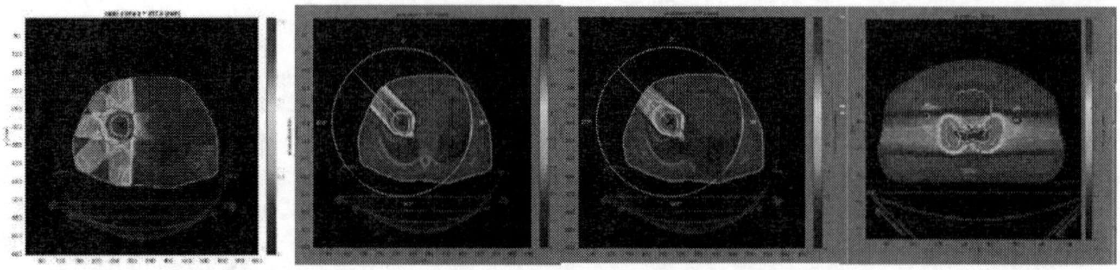

Figure 6: (left to right) Liver case treatment plan with photons, with protons and with carbons. Prostate treatment plan with carbons.

The matRad software also provides Dose-Volume Histograms (DVH) that show how much dose is received by each volume in percent. Ideally, 100% of the prescribed dose should be received by the target structure while 100% of the organ at risk should receive zero dose. DVHs are very useful, especially when comparing quantitatively the effects of the different radiation methods. An example of a DVH can be seen Figure 7, for the liver case.

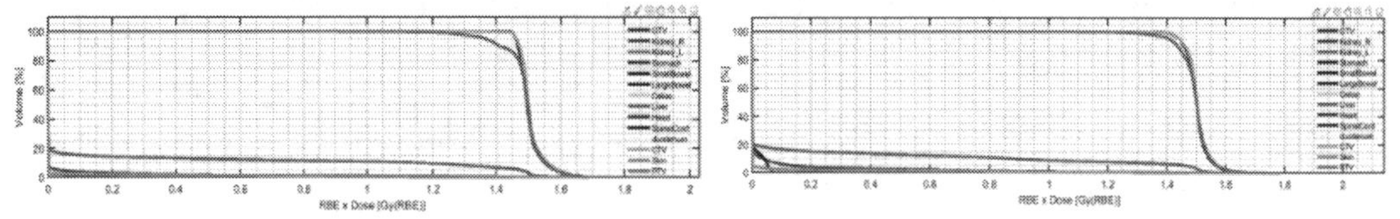

Figure 7: DVH for the liver case with protons (left) and carbons (right).

Through the graphical interpretations of the DVH, the importance of the carbon and proton irradiation is highlighted, since a better and more accurate dose deposition is achieved in the region of importance, avoiding unwanted irradiation to the healthy tissue, thanks to their Bragg peak characteristics and increased radiobiological effectiveness.

The aim of the tutorial is also to demonstrate that there is no a priori "correct" method for treating patients, but the radiation physicist, in collaboration with the medical doctor, must define an individual patient-tailored treatment plan for each case, which in some cases might be the combination of two radiation methods or of different treatment techniques. During the hands-on treatment planning process, the advantages of particle therapy become apparent for the most successful treatment of the patient.

4 Heavy Ion Therapy MasterClass School

Motivated by the received feedback, the format of the PTMC and its pedagogical elements were employed for the full-week Heavy Ion Therapy MasterClass (HITM) school [17], which was organised within the framework of the HITRIplus [19] EU-funded project, from 17-21 May 2021, online, with the support of the PTMC core team of volunteers (Figure 4). It was attended by over 1000 participants, from all over the world, that ranged from undergraduate students to Masters, PhDs, early stage researchers but also professional practitioners, and involved a total of 36 speakers. Participants greatly appreciated the multidisciplinary approach of the school and the presentations from top experts that started from basics and covered most recent developments and future plans in heavy-ion therapy. Recordings are available via the HITM school web page [17].

The worldwide reach of the HITM school demonstrates the increasing interest in heavy-ion therapy but also the power of networks; the participation to the IMC and the HITM school (maps of Figure 8) match almost one-to-one, since the school was also announced through the IPPOG network.

Figure 8: World-wide reach of HITM school and IMC.

An IPPOG colleague from Benha University in Egypt contributed with an interesting initiative by organising the participation of some 100 students as part of the University curriculum, who joined the sessions from an auditorium, and ensuring the support and presence of the dean for the opening session. Figure 4, shows the discussion of their PTMC results for a liver tumour case. Starting from the first pilot in 2019, addressing school children (Figure 9), the PTMC is now adapted to address also university students and has expanded its reach to a new generation of scientists.

Figure 9: First PTMC pilot in 2019.

5 Summary

The PTMC addressing high-school students and the HITM school addressing university students make the young generation aware of the various possibilities that physics, and STEM studies in general, open up for career opportunities in emerging fields where often there is lack of specialised personnel. They also enhance awareness about modern methods for cancer therapy and new research avenues, that are expanding world-wide; and where clearly the development of technology and expertise of research laboratories is important. Participants are impressed performing a treatment planning like professionals and motivated to contribute to the further advancements of this multidisciplinary field, that opens new horizons. Overall, the opportunities for properly trained scientists is a take-home message.

They also come to appreciate the power of networks and importance of international collaborations working on common goals but also the different challenges and approaches of different disciplines (starting from the relative characterisation of "heavy" ions by physicians and physicists) and, hence, the importance of finding common grounds. The PTMC also highlights the importance of close collaboration among research institutes and their coordinated contributions to this multidisciplinary field. It is not a coincidence that the PTMC was developed with the support of GSI, where carbon-ion therapy was pioneered in Europe, CERN, where the PIMMS designs for ion-therapy machines emerged, and DKFZ, that pushes the frontiers of cancer research.

The motivation for the development of the PTMC after the development of the ALICE MasterClasses [18, 21], was to spread the message that heavy-ions and the tools developed for fundamental research, can also be used for cancer therapy. Several institutes choose to perform them back-to-back which enhances the effect. For the developers and all contributors, institute organisers, tutors, moderators, is a great satisfaction to see the impact and experience the feedback of students at all levels; from the dedication of the core team volunteers, the engagement of assistants, to the interest of high-school students. With clear statements they spell out the strong impact it has on them the realisation that what they learn can have tangile impact to society and health.

Acknowledgements

We further thank Mark Bangert for his aid in the initial local test at GSI as well as the Heidelberg Life Science Lab and Kathrin Platzer for supporting the DKFZ masterclasses; the ENLIGHT [20] coordinator, Manjit Dosanjh, for her guidance, introducing the relevant network, and providing material; the students of the Uni of Sarajevo UNSA, and Uni of Thessaloniki AUTH, that contributed setting up the PTMC web pages, dedicated trainings, material in local languages etc; the students of UNAM, Mexico, for setting up the "PTMC in a kit" first in Spanish and then in English; ALICE, GSI, CNAO colleagues for being eager to offer real-time online visits and moderate video-conferences. The help and support of all colleagues that contributed to this endeavour and the taking off of the PTMC should be acknowledged; also, the dedication of the core team of volunteers from AUTH and UNSA for keeping up with the work and challenges that came with the growing interest for the PTMC. Last but not least, the ARIES EU-funded project and the private "Three Physicists Foundation" for supporting the PTMC development and dissemination.

References

1. IMC : https://physicsmasterclasses.org/ publications: https://physicsmasterclasses.org/index.php?cat=papers
2. IPPOG webpage: https://ippog.org
3. Particle Therapy MasterClass webpage: https://indico.cern.ch/event/840212/
4. GSI: https://www.gsi.de/start/aktuelles
5. CERN: https://home.cern/
6. DKFZ: https://www.dkfz.de/en/index.html
7. Bryant, Philip J, *PIMMS : Proton-ion medical machine study,* in CERN Yellow Reports: Monographs, ebook: 10.5170/CERN-2000-006
8. HIT: https://www.klinikum.uni-heidelberg.de/interdisziplinaere-zentren/heidelberger-ionenstrahl-therapiezentrum-hit/
9. PTMC web page, Mexico City : http://epistemia.nucleares.unam.mx/web?name=fisica_y_sociedad_2020
   PTMC webpage, Puebla : https://www.fcfm.buap.mx/ParticulasElementales/seminarios/PT/index.html
10. ALICE collaboration :
    https://alice-collaboration.web.cern.ch/
11. CNAO: https://fondazionecnao.it/en/
12. matRad – an open-source dose calculation and treatment planning toolkit, https://github.com/e0404/matRad https://doi.org/10.5281/zenodo.3879615
13. Cisternas E., Mairani A., Ziegenhein P., Jäkel O., Bangert M. (2015) *matRad - a multi-modality open-source 3D treatment planning toolkit.* In: Jaffray D. (eds) World Congress on Medical Physics and Biomedical Engineering, June 7-12, 2015, Toronto, Canada. IFMBE Proceedings, vol 51. Springer, Cham. https://doi.org/10.1007/978-3-319-19387-8_391
14. Wieser, H.-P., Cisternas, E., Wahl, N., Ulrich, S., Stadler, A., Mescher, H., Müller, L.-R., Klinge, T., Gabrys, H., Burigo, L., Mairani, A., Ecker, S., Ackermann, B., Ellerbrock, M., Parodi, K., Jäkel, O. and Bangert, M. (2017), *Development of the open-source dose calculation and optimization toolkit matRad.* Med. Phys., 44: 2556-2568. https://doi.org/10.1002/mp.122
15. D. Craft, M. Bangert, T. Long, D. Papp, J. Unkelbach, *Shared data for intensity modulated radiation therapy (IMRT) optimization research: the CORT dataset*, GigaScience, Volume 3, Issue 1, December 2014, 2047–217X–3–37, https://doi.org/10.1186/2047-217X-3-37
16. Medical Dictionary: https://medical-dictionary.thefreedictionary.com/bixel
17. HITM school webpage: https://indico.cern.ch/event/1019104/
18. L. Graczykowski, P. Nowakowski, P. Foka, *New developments for ALICE MasterClasses and the new Particle Therapy MasterClass,* in Proceedings of CHEP 2019,
    [2005.02215] New developments for ALICE MasterClasses and the new Particle Therapy MasterClass (arxiv.org)
19. Hitriplus : https://www.hitriplus.eu/
20. ENLIGHT : https://enlight.web.cern.ch/
21. Y. Foka, M. Janik, *ALICE Masterclass on strangeness,* in EPJ Web of Conferences, Volume 71, 2014,
    https://www.epj-conferences.org/articles/epjconf/abs/2014/08/epjconf_icnfp2013_00057/epjconf_icnfp2013_00057.html

EPJ Web of Conferences **258**, 01003 (2022) https://doi.org/10.1051/epjconf/202225801003
A Virtual Tribute to Quark Confinement and the Hadron Spectrum (vConf21)

Direct CP violation and the $\Delta I = 1/2$ rule in $K \to \pi\pi$ decay in the Standard Model

Christopher Kelly[1,*]

[1]Computational Science Institute, Brookhaven National Laboratory,
Upton, New York, USA.

Abstract. We discuss the RBC & UKQCD collaborations' recent [1] lattice calculation of ϵ', the measure of direct CP-violation in kaon decays. This result significantly improves on our previous 2015 calculation, with nearly 4× the statistics and more reliable systematic error estimates. We discuss how our results demonstrate the Standard Model origin of the $\Delta I = 1/2$ rule, and present our plans for future calculations.

1 Introduction

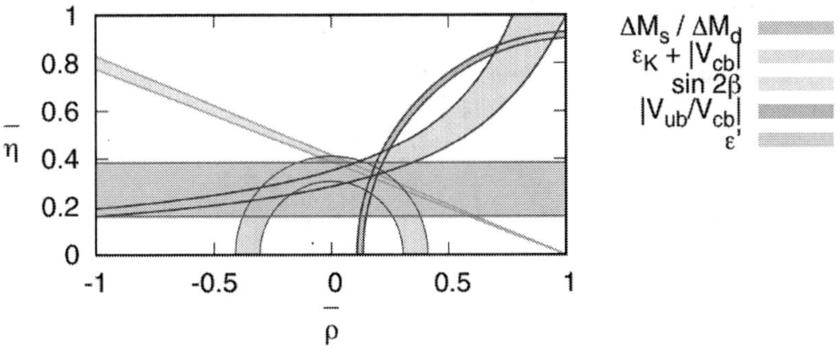

Figure 1. The CKM matrix unitarity triangle in the $\rho - \eta$ plane. The horizontal band is the constraint imposed by our calculation of ϵ'.

Direct CP-violation (CPV) in the decays of neutral kaons into two pions is highly suppressed in the Standard Model, making it an ideal probe for new sources of CPV. Beyond the Standard Model CPV processes appear necessary to explain the predominance of matter over antimatter in the observable Universe assuming that baryogeneses is responsible for this phenomenon. Measurements of these processes also provide a new, horizontal band constraint on the CKM matrix unitarity triangle in the $\rho - \eta$ plane, as illustrated in Fig. 1 where we compare the constraint from our calculation versus those of other sources.

*e-mail: ckelly@bnl.gov

Direct CPV in kaon decays was first observed by the NA48 [2] and KTeV [3] experiments in the late 1990s, and the current world average of the measure of this phenomenon, ϵ', is [4]

$$\mathrm{Re}(\epsilon'/\epsilon) = 16.6(2.3) \times 10^{-4} \tag{1}$$

where ϵ is the measure of indirect CP-violation, itself heavily suppressed: $|\epsilon| = 2.228(11) \times 10^{-3}$. Unfortunately a correspondingly precise Standard Model prediction for this quantity has remained out of reach until recently, due to the fact that the underlying high-energy weak decay processes receive substantial corrections from low-energy hadronic physics that cannot be treated reliably with perturbation theory.

The first *ab initio* calculation of ϵ' in the Standard Model [5] was performed by the RBC & UKQCD collaborations in 2015 using lattice QCD, the only known systematically improvable (i.e. for which all systematic errors can be identified and reduced/eliminated with sufficient computational effort) technique for treating non-perturbative hadronic physics. The calculation was performed in three-flavor, isospin-symmetric QCD where ϵ' emerges from the difference between the complex phases of the decay amplitudes of neutral kaons into two-pions in the $I = 0$ (with amplitude A_0) and $I = 2$ (with amplitude A_2) isospin representations,

$$\epsilon' = \frac{i\omega e^{i(\delta_2-\delta_0)}}{\sqrt{2}} \left(\frac{\mathrm{Im}A_2}{\mathrm{Re}A_2} - \frac{\mathrm{Im}A_0}{\mathrm{Re}A_0} \right), \tag{2}$$

where $\omega = \mathrm{Re}A_2/\mathrm{Re}A_0$ and δ_I are the $\pi\pi$ scattering phase shifts. We obtained the following result,

$$\mathrm{Re}(\epsilon'/\epsilon) = 1.38(5.15)(4.59) \times 10^{-4} \tag{3}$$

where the errors are statistical and systematic respectively. This result exhibits a 2.1σ tension with the experimental number, providing strong motivation for an improved calculation. The calculation also exhibited an unexplained tension of a similar size in the $I = 0$ $\pi\pi$ scattering phase shift that we determine as a necessary component of the $K \to \pi\pi$ measurement, further spurring the desire for a follow-up work. In this document we present the results of an updated calculation [1] published in 2020, where we have reduced our statistical error by a factor of two and dramatically improved our understanding and confidence in the systematic error resulting from the contribution of excited $\pi\pi$ states – thus resolving the phase-shift discrepancy – alongside several other important systematic error sources. We also discuss our ongoing efforts to further refine our calculation over the coming years.

2 Summary of the calculations

We perform an isospin symmetric three-flavor calculation with degenerate light quarks, and a physical strange quark mass. The weak interaction is treated using first order weak effective theory, which for three-flavors is expressed as a Hamiltonian comprising ten four-quark operators Q_i with Wilson coefficients z_i and y_i that encapsulate the high-energy contributions and are computed using perturbation theory:

$$H_W = \frac{G_F}{\sqrt{2}} V_{ud}^* V_{us} \sum_{i=0}^{10} [z_i(\mu) + \tau y_i(\mu)] Q_i(\mu) \tag{4}$$

where $\tau = -\frac{V_{ts}^* V_{td}}{V_{us}^* V_{ud}}$. The $K \to \pi\pi$ amplitudes are then defined as

$$A^I = \langle (\pi\pi)_I | H_W | K^0 \rangle . \tag{5}$$

Both the Wilson coefficients and the operators require renormalization as indicated by the renormalization scale μ in Eq. (4), while their product is renormalization scheme independent. The renormalization is conventionally performed in the $\overline{\text{MS}}$ scheme which is not amenable to a lattice treatment as it involves fractional dimensions. Therefore, for the lattice calculation we first renormalize in an intermediate non-perturbative renormalization (NPR) scheme – the so-called RI-SMOM schemes – at a high energy at which the conversion factors relating the result to $\overline{\text{MS}}$ can be computed perturbatively. Note that the energy at which we perform this matching is bounded by the lattice cutoff and we must stay sufficiently below this scale to avoid discretization errors. This often leads to larger systematic errors arising from the truncation of the perturbative series used for the matching.

The $K \to \pi\pi$ matrix elements are obtained by fitting to three-point functions of the form

$$C_i(t, t_{\text{sep}}^{K\to\text{snk}}) = \langle 0|O_{\text{snk}}^\dagger(t_{\text{sep}}^{K\to\text{snk}})Q_i(t)O_K(0)|0\rangle \tag{6}$$

where O_K is the kaon operator, O_{snk} creates the $\pi\pi$ state and Q_i are the four-quark operators discussed above. The kaon and $\pi\pi$ operators actually generate *all* states with the corresponding quantum numbers, but due to the Euclidean time direction the contribution of each state dies away exponentially in lattice time according to their energy such that at large times only the lowest energy states have significant contributions. A key challenge of the lattice calculation is that with the typical periodic spatial boundary conditions, the O_{snk} operator also generates a state comprising two-pions at rest, hence the unphysical energy-nonconserving amplitude of a kaon of mass ~500 MeV decaying into two pions of total energy ~260 MeV dominates the signal at large times. Multi-state fits to extract the subdominant physical contribution are possible but it is often very difficult to reliably isolate the excited-state terms, particularly with noisy data. Therefore, for these calculations we instead take advantage of the ability to manipulate the spatial boundary conditions (BCs) while incurring only exponentially-suppressed finite-volume errors: In the $I = 2$ case an isospin rotation relates the amplitude of interest to $K^+ \to \pi^+\pi^+$. By replacing the periodic BCs of the down-quark with antiperiodic BCs, the charged pions in the final state also become antiperiodic and as a result their lattice momenta become discretized in odd-integer multiples of π/L where L is the lattice size. We are then able to tune L such that the lowest-energy state comprises two pions with a total energy that matches the kaon mass. Note that this operation explicitly breaks the isospin symmetry, but fortunately any mixing of the final state is prevented by virtue of it being the only allowed doubly-charged state. In the $I = 0$ case this isospin breaking cannot be avoided, therefore we instead utilize G-parity BCs, where a charge-conjugation and isospin rotation are performed at the lattice boundary. Both charged and neutral pions are negative eigenstates of G-parity hence we achieve the effect of imposing antiperiodic BCs on all pions without breaking isospin, at the cost of significantly increasing the computational expense of the lattice calculation[6].

2.1 2015 calculations

The $\Delta I = 3/2$ amplitude A_2 arises from three linear combinations of Q_i that are labeled by their representation under chiral $\text{SU}(3)_L \times \text{SU}(3)_R$. The dominant contribution to $\text{Re}A_2$ arises from the $(27, 1)$ combination, and $\text{Im}A_2$ is dominated by the remaining two $(8, 8)$ operators. The most recent calculation of A_2 [7] was performed in 2015 using physical light and strange quark masses, a large volume to reduce finite-volume systematic errors, and two lattice spacings allowing a continuum limit to be taken. We obtained the following results,

$$\text{Re}A_2 = 1.50(4)(14) \times 10^{-8}\ \text{GeV} \tag{7a}$$

$$\text{Im}A_2 = -6.99(20)(84) \times 10^{-14}\ \text{GeV} \tag{7b}$$

where the errors are statistical and systematic, respectively. A significant achievement of this calculation is obtaining a result with < 1% statistical errors despite having performed a continuum extrapolation. The systematic error is considerably larger and arises mainly from the truncation of the perturbative series used in the matching of our NPR scheme to $\overline{\text{MS}}$ (8%) and in the Wilson coefficients (12%). The value obtained for the real part agrees well with the experimental value of ReA_2 = $1.4787(31) \times 10^{-8}$ GeV from charged kaon decays and $1.570(53) \times 10^{-8}$ GeV from neutral kaon decays.

The $\Delta I = 1/2$ amplitude A_0 receives contributions from all ten effective four-quark operators, with Q_2 the dominant contribution to the real part and Q_6, and to a lesser extent Q_4, dominating the imaginary part. In 2015 we performed the first calculation [5] of this amplitude. This calculation is much more expensive than that of A_2, not just due to the G-parity BCs but also because the $\pi\pi$ final state has vacuum quantum numbers and hence the calculation involves evaluating disconnected diagrams where the two pions are connected to the four-quark operator only through background gluons; these types of diagram are typically much more noisy and require larger statistics, along with more sophisticated methods, to adequately resolve. As a result we performed the calculation on only a single, somewhat coarse lattice spacing which allows us to simulate with a large physical volume, and thus have better control over finite-volume errors, at the cost of incurring larger discretization errors. We used a single $\pi\pi$ operator in which the two constituent pions are generated moving back-to-back with momentum $(\pm 1, \pm 1, \pm 1)\pi/L$. We label this operator $\pi\pi(111)$. For the A_0 amplitude we obtained

$$\text{Re}A_0 = 4.66(1.00)(1.26) \times 10^{-7}\ \text{GeV} \tag{8a}$$

$$\text{Im}A_0 = -1.90(1.23)(1.08) \times 10^{-11}\ \text{GeV}\,, \tag{8b}$$

where the relative error on the imaginary part is much larger than for the real part primarily due to a large cancelation between the dominant Q_4 and Q_6 contributions. Despite the coarse lattice spacing, the systematic error is again dominated by perturbative truncation errors. The real part agrees with the experimental result of ReA_0 = $3.3201(18) \times 10^{-7}$ GeV.

Combining the above results for A_0 and A_2 using Eq. (2), we found

$$\text{Re}(\epsilon'/\epsilon) = 1.38(5.15)(4.59) \times 10^{-4}\,, \tag{9}$$

which is 2.1σ below the experimental value. While the errors on the calculation are large, roughly 3× those of the experiment, this result clearly demonstrates that ϵ' is now accessible to lattice calculation.

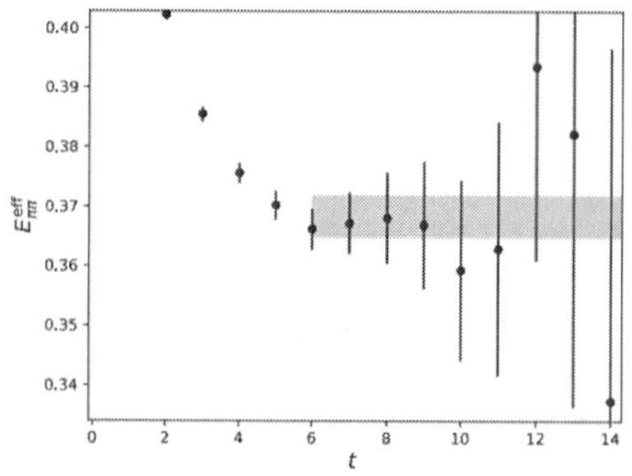

Figure 2. The $\pi\pi$ effective energy obtained using 1438 configurations with the $\pi\pi(111)$ operator.

A puzzle that arose from the $I = 0$ calculation is a $\sim 3\sigma$ disagreement between the value obtained for the $I = 0$ $\pi\pi$ scattering phase shift, $\delta_0(E_{\pi\pi} \approx m_K) = 23.8(4.9)(1.2)^\circ$, and the prediction of $\sim 39^\circ$ obtained by combining the dispersive Roy equations with experimental input [8]. This phase-shift, and the $\pi\pi$ energy from which it is derived, are necessary inputs to the $K \rightarrow \pi\pi$ calculation, hence it is vital that a reliable result is obtained. In order to address this issue, following the 2015 calculation we increased the statistics on the $\pi\pi$ measurement by almost 7× but instead found that the disagreement became worse, with a value obtained of $19.1(2.5)^\circ$ [9]. The corresponding effective mass plot, the plateau value of which indicates the ground-state energy, is shown in Fig. 2. While the plateau appears clear, the rapid reduction of the signal-to-noise ratio makes the most likely explanation for this disagreement the existence of an excited state with an energy close to that of the ground-state, whose contribution is masked by the growth of the error bars.

2.2 Updated calculation of A_0 [1]

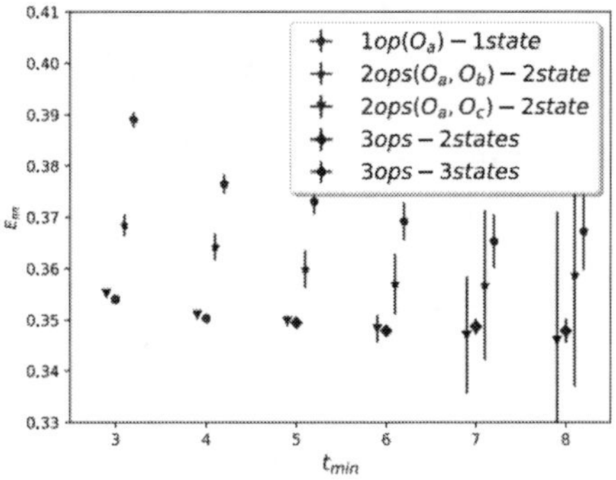

Figure 3. The $I = 0$ $\pi\pi$ energy obtained by varying the lower bound on the fit range $t_{\min}$, the number of operators and number of states. In the legend O_a, O_b, O_c correspond to the $\pi\pi(111)$, $\pi\pi(311)$ and σ operators, respectively. Our best fit is the 3-operator, 2-state result with $t_{\min} = 6$.

In order to address the possibility of excited-state contamination we introduced two additional $\pi\pi$ operators, the $\pi\pi(311)$ operator for which the two pions again move back-to-back but with momenta $(\pm 3, \pm 1, \pm 1)\pi/L$ and permutations thereof, and the $\sigma = (\bar{u}u + \bar{d}d)/\sqrt{2}$ operator, which also has vacuum quantum numbers and hence projects onto the $I = 0$ $\pi\pi$ state. This allows us to perform simultaneous fits to the matrix of $\pi\pi$ correlation functions obtained by varying the source and sink $\pi\pi$ operator to extract the energies and amplitudes of multiple $\pi\pi$ states with considerably higher precision and control. The power of this method is illustrated in Fig. 3 where we show the fitted $\pi\pi$ energy as we vary the number of operators used and the number of states to which we fit. Clearly a significant advantage is gained in both the statistical error and the quality of the plateau. This calculation was performed on 3.4× statistics of our original 2015 calculation. We found [9]

$$\delta_0(471\ \text{MeV}) = 32.3(1.0)(1.4)^\circ\,, \tag{10}$$

which is in much better agreement with the dispersive prediction at this energy of 35.9°. We therefore conclude that excited state contamination was indeed responsible for the disagreement and that it has now been resolved.

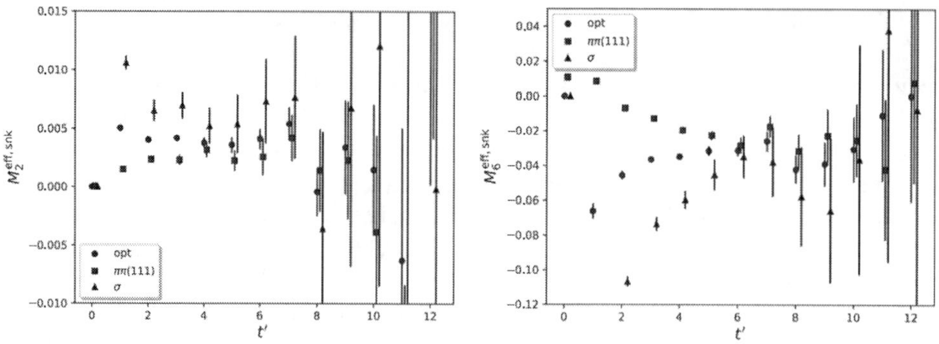

Figure 4. The effective matrix elements of the Q_2 and Q_6 operators for the $\pi\pi(111)$, σ and optimal operators as a function of the time separation t' between the four-quark operator and the $\pi\pi$ sink operator.

The additional operators also had a significant effect on the fits to the $K \to \pi\pi$ matrix elements. In Fig 4 we plot the effective matrix elements defined as,

$$
\begin{aligned}
M_i^{\text{eff,snk}}(t') &= C_i(t, t_{\text{sep}}^{K\to\text{snk}}) \left(\frac{1}{\sqrt{2}} A_K A_{\text{snk}}^0 e^{-m_K t} e^{-E_0(t_{\text{sep}}^{K\to\text{snk}} - t)} \right)^{-1} \\
&= M_i^0 + \sum_j \frac{A_{\text{snk}}^j}{A_{\text{snk}}^0} M_i^j e^{-(E_j - E_0)t'} .
\end{aligned}
\tag{11}
$$

where $t' = t_{\text{sep}}^{K\to\text{snk}} - t$ and C_i are defined in Eq. 6. Here A_K and m_K are the ground-state kaon operator-overlap coefficient and mass, respectively, A_{snk}^j and E_j are the operator-overlap and energy of the $\pi\pi$ ground and excited states, for which the results are obtained from their respective two-point function fits. We apply a uniform cut of $t_{\min} = 6$ to ensure that only the kaon ground state contributes. These effective matrix elements converge to the desired matrix elements M_i^0 at large t'. In the figure we compare the $\pi\pi(111)$ and σ effective matrix elements against the optimal combination of those two operators that best projects onto the ground state. As in the case of the $\pi\pi$ fits, we observe that the additional operators have a significant impact on both the statistical error and the length and quality of the plateau.

In Fig. 5 we plot the fit results for the Q_2 and Q_6 operators as we vary the number of operators, states and the temporal cuts. For Q_2, the dominant contribution to ReA_0, we find good agreement between all of the fits for $t'_{\min} \geq 4$, although we see improved statistical errors with the additional operators. However for Q_6, the dominant contribution to ImA_0, we see a clear pattern of excited state contamination in the one and two-operator fit results, the latter converging later than the former as we would expect. The final best fit for all operators was obtained with three operators and two states with $t'_{\min} = 5$ and $t_{\min} = 6$. The best fit from the 2015 calculation would correspond here to the "1×1 $t_{\min} = 6$" point with $t'_{\min} = 4$, which we see is very different from our new best fit; unfortunately this implies that our $\leq 5\%$ estimate for the excited state contamination in our 2015 calculation was significantly underestimated.

Another notable improvement in this updated calculation is the use of the "step-scaling" technique [10] in the non-perturbative renormalization. Due to the rather coarse lattice spacing, the errors resulting from the truncation of the perturbative series in the RI-SMOM→ $\overline{\text{MS}}$ matching were estimated to be 15% for the 2015 calculation of A_0. Step-scaling is a procedure whereby the matrix parameterizing the non-perturbative running between a low and high

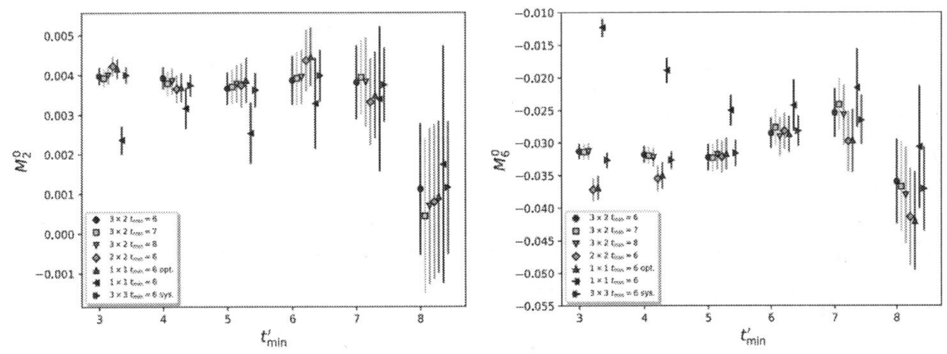

Figure 5. The fitted ground-state matrix elements of the Q_2 and Q_6 operators for various choices of the number of operators, number of states and the cut t_{min} between the kaon and four-quark operators, as a function of the cut t'_{min} between the four-quark operator and the $\pi\pi$ sink. In the legend "$a \times b$" indicates the fit was performed with a operators and b states, and "opt." indicates the optimal operator was used. In the two-operator case we drop the $\pi\pi(311)$ operator and in the one-operator case we further drop the σ. The "sys." results are obtained using three operators and three states for both the $\pi\pi$ and $K \to \pi\pi$ fits and is used in the systematic error estimation. The values are shifted for clarity.

energy scale is computed a fine lattice. The low scale is one that is accessible to the coarse lattice calculation and the high scale is only accessible to the fine lattice. As this running is universal it can be used to run the matrix of renormalization factors determined on the coarse lattice at the low scale to a scale much higher than the lattice cutoff. For the updated calculation we used this procedure to raise the renormalization scale from 1.53 GeV to 4 GeV, which reduced the corresponding systematic error by a factor of 3×. Unfortunately we did not observe a corresponding decrease in the systematic error arising from the use of perturbation theory to compute the Wilson coefficients, which are also evaluated at this 4 GeV scale, due to the fact that perturbation theory is still being applied to match between the 4- and 3-flavor theories at the charm mass scale $m_c \approx 1.3$ GeV.

3 Results of the updated calculation

3.1 The $\Delta I = 1/2$ rule

A longstanding puzzle in $K \to \pi\pi$ physics relates to the experimental observation that the $\Delta I = 1/2$ decay is ~500 more common than the $\Delta I = 3/2$ decay. This $\Delta I = 1/2$ rule is reflected in the ratio

$$\frac{1}{\omega} = \frac{\text{Re}A_0}{\text{Re}A_2} = 22.45(6)\,, \tag{12}$$

the inverse of which appears as a coefficient in the equation for ϵ', Eq. (2). The perturbative running from the weak to the charm scale can be shown to give rise to a factor of two difference in the amplitudes, but there has been no widely accepted explanation as to the origin of the remaining factor of 10. With our lattice calculation we are now able to answer this question: in computing ReA_2 we found a strong, 70%-level cancelation between the two dominant diagrams C_1 and C_2 of the $(27, 1)$ operator amplitude as illustrated in Fig. 6. This is very different from the behavior predicted by naïve color counting which suggests these terms should have the same sign and differ by a factor of three in magnitude. From our lattice

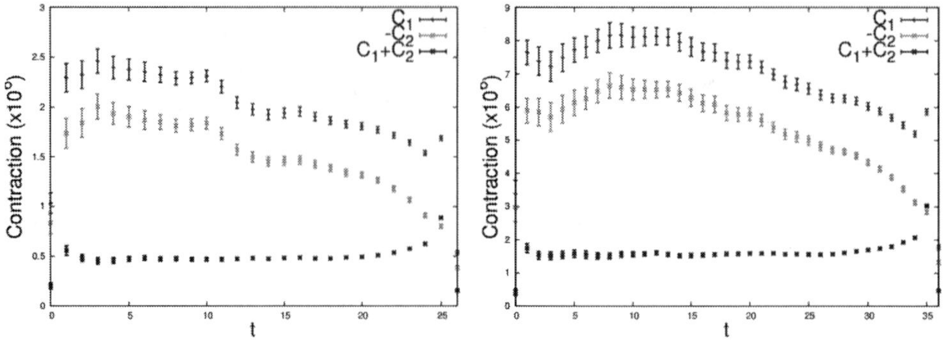

Figure 6. The two dominant contributions C_1 and C_2 to ReA_2 and their sum on the 48I (left) and 64I (right) ensembles, reproduced from Fig. 11 of Ref. [7].

calculation we obtain

$$\frac{1}{\omega} = \frac{\mathrm{Re}A_0}{\mathrm{Re}A_2} = 19.9(5.0) \tag{13}$$

which is completely consistent with the experimental number. We therefore conclude that the $\Delta I = 1/2$ rule is a consequence of low-energy QCD.

3.2 A_0 and ϵ'

For the updated calculation we obtain

$$\mathrm{Re}A_0 = 2.99(0.32)(0.59) \times 10^{-7}\ \mathrm{GeV} \tag{14a}$$

$$\mathrm{Im}A_0 = -6.98(0.62)(1.44) \times 10^{-11}\ \mathrm{GeV} \tag{14b}$$

The former agrees well with the experimental result ReA_0 = $3.3201(18) \times 10^{-7}$ GeV and with our previous result. The value for ImA_0 differs substantially from our previous result however, which gives rise to a similarly large change in ϵ':

$$\mathrm{Re}(\epsilon'/\epsilon) = 21.7(2.6)(6.2)(5.0) \times 10^{-4}\,. \tag{15}$$

Here the first set of parentheses gives the statistical error. The systematic error has been broken into two contributions: the third set of parentheses corresponds to an estimate (23%) of the effects of isospin breaking (IB) and electromagnetism (EM) obtained through next-to-leading order chiral perturbation theory with input from the $1/N_c$ expansion [11]. These effects were not formerly included in the budget as the result was considered a pure-QCD calculation. EM+IB effects typically enter at the percent scale in hadronic quantities, but for $K \to \pi\pi$ their relative effects are enhanced due to the 20× suppression of A_2 by the mechanics of the $\Delta I = 1/2$ rule. The remaining systematic error is given by the second set of parentheses and is dominated by the discretization error (12%) and the perturbative truncation error in the Wilson coefficients (12%). The excited state systematic error is treated as negligible compared to the other sources of error given the control over such effects provided by the multi-operator method demonstrated in the previous section.

This result for Re(ϵ'/ϵ) is now in good agreement with the experimental number.

4 Conclusions and outlook

The RBC & UKQCD collaborations have completed an update [1] to our 2015 calculation [5] of A_0. Among the notable changes are a 3.4× increase in the number of measurements; the addition of two more $\pi\pi$ operators to the data which vastly improves the statistical resolution and the quality of the fit plateaus, and also resolves a discrepancy previously observed between the lattice and dispersive predictions of the $I = 0$ $\pi\pi$ scattering phase shift at the kaon mass scale; and the use of step-scaling to raise the renormalization scale from 1.53 GeV to 4 GeV resulting in a 3× improvement in the formerly-dominant error resulting from the $\overline{\text{MS}}$ matching. By combining the results with our earlier calculation of A_2 we obtain an *ab initio* determination of $\text{Re}A_0/\text{Re}A_2$ that is consistent with experiment and thus demonstrate that the $\Delta I = 1/2$ rule mainly arises from non-perturbative QCD effects. We also obtain a result for $\text{Re}(\epsilon'/\epsilon)$ that is now consistent with the experimental value and has a total error that is ~3.6× that of the experiment.

We believe that ϵ' remains a promising avenue in which to search for new physics, and since the recent publication have commenced work towards further improving our result. The most significant single source of error is due to the missing contributions of isospin breaking and electromagnetism. Unfortunately the lattice treatment of electromagnetism is complicated by the long-range nature of its interactions and so significant theoretical hurdles must be overcome before a first principles calculation is possible. We are presently exploring a novel technique for treating the interactions using a truncated potential in Coulomb gauge where the truncation range is set to confine the interaction within the finite box resulting in finite-volume errors that can be treated analytically using infinite-volume perturbation theory [12].

A second significant source of error arises from the use of perturbation theory to match between the four- and three-flavor effective theories in the Wilson coefficients. While a four-flavor lattice calculation with physical charm quarks is possible, the requirement of large statistics, a fine lattice spacing to control charm discretization effects and a large volume to control finite-volume effects puts this outside of the reach of present and near-future supercomputers. In the interim we are investigating the possibility of computing the $4 \to 3$f matching non-perturbatively [13, 14].

The dramatic improvement in control over excited state effects that we discussed in this document has reopened the question as to whether we can perform multi-state fits to extract the physical $\Delta I = 1/2$ matrix element from a calculation with periodic boundary conditions, forgoing the use of the computationally-expensive G-parity boundary conditions to remove the dominant contribution of the unphysical decay of the kaon to a stationary $\pi\pi$ state. We are presently conducting a preliminary calculation on two different coarse lattices to evaluate this approach [15].

Finally, the largest pure-lattice systematic error is due to the fact that A_0 has only been computed at a single, somewhat coarse lattice spacing. We are presently generating two additional ensembles with G-parity boundary conditions which will allow for the controlled continuum extrapolation of our result for A_0 [16], eliminating this source of error.

References

[1] R. Abbott *et al.* [RBC and UKQCD], Phys. Rev. D **102**, no.5, 054509 (2020) doi:10.1103/PhysRevD.102.054509 [arXiv:2004.09440 [hep-lat]].

[2] J. R. Batley *et al.* [NA48], Phys. Lett. B **544**, 97-112 (2002) doi:10.1016/S0370-2693(02)02476-0 [arXiv:hep-ex/0208009 [hep-ex]].

[3] E. Abouzaid *et al.* [KTeV], Phys. Rev. D **83**, 092001 (2011) doi:10.1103/PhysRevD.83.092001 [arXiv:1011.0127 [hep-ex]].

[4] M. Tanabashi *et al.* [Particle Data Group], Phys. Rev. D **98**, no.3, 030001 (2018) doi:10.1103/PhysRevD.98.030001

[5] Z. Bai *et al.* [RBC and UKQCD], Phys. Rev. Lett. **115**, no.21, 212001 (2015) doi:10.1103/PhysRevLett.115.212001 [arXiv:1505.07863 [hep-lat]].

[6] N. H. Christ, C. Kelly and D. Zhang, Phys. Rev. D **101**, no.1, 014506 (2020) doi:10.1103/PhysRevD.101.014506 [arXiv:1908.08640 [hep-lat]].

[7] T. Blum, P. A. Boyle, N. H. Christ, J. Frison, N. Garron, T. Janowski, C. Jung, C. Kelly, C. Lehner and A. Lytle, *et al.* Phys. Rev. D **91**, no.7, 074502 (2015) doi:10.1103/PhysRevD.91.074502 [arXiv:1502.00263 [hep-lat]].

[8] G. Colangelo, J. Gasser and H. Leutwyler, Nucl. Phys. B **603**, 125-179 (2001) doi:10.1016/S0550-3213(01)00147-X [arXiv:hep-ph/0103088 [hep-ph]].

[9] T. Blum *et al.* [RBC and UKQCD], [arXiv:2103.15131 [hep-lat]].

[10] R. Arthur *et al.* [RBC and UKQCD], Phys. Rev. D **83**, 114511 (2011) doi:10.1103/PhysRevD.83.114511 [arXiv:1006.0422 [hep-lat]].

[11] V. Cirigliano, H. Gisbert, A. Pich and A. Rodríguez-Sánchez, JHEP **02**, 032 (2020) doi:10.1007/JHEP02(2020)032 [arXiv:1911.01359 [hep-ph]].

[12] J. Karpie, PoS **LATTICE2021**, 312 (2021)

[13] M. Tomii, PoS **LATTICE2018**, 216 (2019) doi:10.22323/1.334.0216 [arXiv:1901.04107 [hep-lat]].

[14] M. Tomii, PoS **LATTICE2019**, 174 (2020) doi:10.22323/1.363.0174

[15] M. Tomii, PoS **LATTICE2021**, 394 (2021)

[16] C. Kelly, PoS **LATTICE2021**, 008 (2021)

Data-driven approaches to the evaluation of hadronic contributions to the $(g-2)_\mu$

Gilberto Colangelo[1,*]

[1]Albert Einstein Center for Fundamental Physics, Institute for Theoretical Physics, University of Bern, Sidlerstrasse 5, 3012 Bern, Switzerland

Abstract. In this talk I reviewed the data-driven theoretical calculation of the hadronic contributions to the anomalous magnetic moment of the muon in the Standard Model mainly as it has been presented in the White Paper, but also including the most recent developments. All this is presented in the light of the new measurement of $(g-2)_\mu$ recently released by the Fermilab experiment, which led to an increase of the discrepancy with the Standard Model from 3.7 to 4.2σ.

1 Introduction

In April this year the Fermilab Muon $g-2$ Collaboration has announced their long-awaited first result of its series of measurements of the anomalous magnetic moment of the muon [1]. The comparison between the measurement and the Standard Model (SM) prediction, as it has been presented in the White Paper (WP) [2] showed a discrepancy at the level of 3.7σ with respect to the Brookhaven measurement [3]. The new experimental world average after the Fermilab result differs from the SM by 4.2σ. Table 1 summarizes the various contributions to the SM value for a_μ and shows that, as is well known, the two main sources of uncertainty are both hadronic and are the leading-order contribution, namely the hadronic vacuum polarization (HVP), and the new structure at next-to-leading order, namely hadronic light-by-light (HLbL).

Two further important news happened at the same time as the Fermilab announcement, both coming from the lattice: the BMW calculation of the HVP contribution, the first to reach sub-percent uncertainty, has been published [4] and a second lattice evaluation of the HLbL contribution has appeared on the arXiv [5]. Concerning the former, while the article has been on the arXiv since more than a year, the published version contained a slightly revised result which sits almost exactly in the middle between the experimental and the WP number and has a comparable uncertainty. This is an unsatisfactory situation which needs to be clarified. The second lattice result, on the other hand, agrees very well with both the first lattice calculation of the HLbL contribution as well as the data driven one, whose average is taken as the SM value for HLbL in the WP. Such a confirmation is of course very welcome and makes the perspective of further reductions in the final uncertainty for this contribution even more concrete.

*e-mail: gilberto@itp.unibe.ch

Table 1. Summary of the different contributions to a_μ in the Standard Model [2] and comparison to the present experimental world average.

Contribution	Value $\times 10^{11}$	References
Experiment (E821)	116 592 089(63)	Ref. [3]
Experiment (FNAL)	116 592 040(54)	Ref. [1]
Experiment (World-Average)	116 592 061(41)	
HVP LO (e^+e^-)	6931(40)	Refs. [6–11]
HVP NLO (e^+e^-)	−98.3(7)	Ref. [11]
HVP NNLO (e^+e^-)	12.4(1)	Ref. [12]
HVP LO (lattice, *udsc*)	7116(184)	Refs. [13–21]
HLbL (phenomenology)	92(19)	Refs. [22–34]
HLbL NLO (phenomenology)	2(1)	Ref. [35]
HLbL (lattice, *uds*)	79(35)	Ref. [36]
HLbL (phenomenology + lattice)	90(17)	
QED	116 584 718.931(104)	Refs. [37, 38]
Electroweak	153.6(1.0)	Refs. [39, 40]
HVP (e^+e^-, LO + NLO + NNLO)	6845(40)	
HLbL (phenomenology + lattice + NLO)	92(18)	
Total SM Value	116 591 810(43)	
Difference: $\Delta a_\mu := a_\mu^{\text{exp}} - a_\mu^{\text{SM}}$	251(59)	

2 Hadronic vacuum polarization

The evaluation of the HVP contribution has a long history and mainly relies on the formula first discovered by Bouchiat and Michel sixty years ago [41], which expresses this contribution in terms of the cross section $e^+e^- \to$ hadrons. Many experimental measurements (see [2] for a complete list) have provided essential input for the application of this formula. In recent years mainly two groups have made a systematic and complete evaluation of this contribution and provided regular updates: the latest analyses of DHMZ [10] as well as KNT [11] constitute the basis for the SM number presented in the WP. But there is more than that: other analyses, like the one by Jegerlehner and collaborators [31] have also been considered and critically reviewed. Moreover, analyses of exclusive channels which make use of theoretical arguments (like analyticity and unitarity) to better constrain the data have also been used in the final average [8, 9, 42].

The method adopted to combine these analyses is the following: *1)* central values are obtained by simple averages (for each channel and mass range); *2)* it is always the largest experimental and systematic uncertainty of the analyses considered which is taken; *3)* half of the difference between analyes (or between data by BABAR [43] and KLOE [44–47] in the 2π channel, if this is larger) is added to the uncertainty. This led to the final result reported in Table 1, which has a final relative uncertainty of 0.6%. This is larger than could potentially be achieved in view of the precision of the data: but, as indicated, the combination procedure was aimed to err on the conservative side.

After the WP was published, new results by the SND collaboration [48] as well as an update by the BESIII collaboration [49] have been published. Their precision does not seem to be able to significantly impact the current estimate, but they will motivate updates of the a_μ^{HVP} evaluation. New results by CMD-3 [50] and BABAR are also expected in the near future.

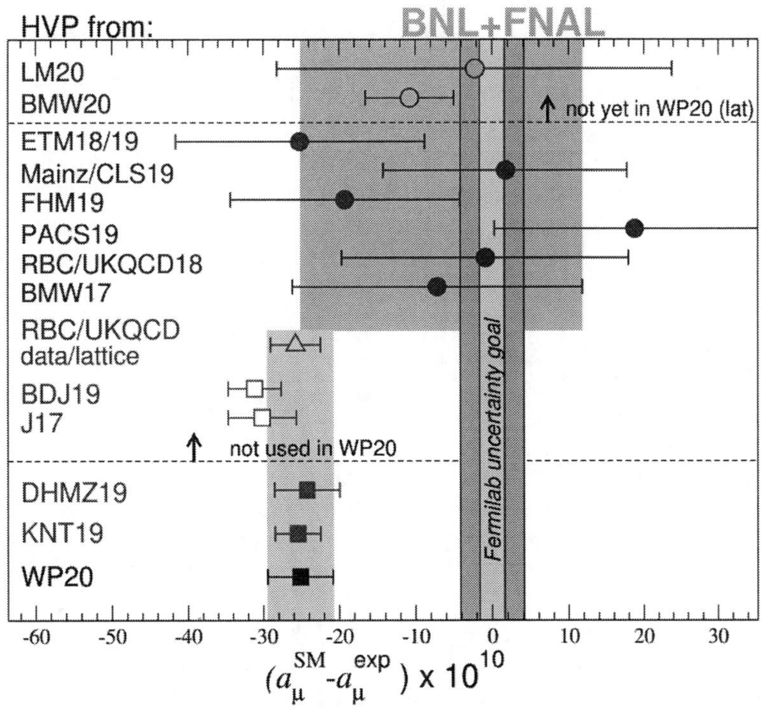

Figure 1. Comparison of different evaluations of the HVP contribution, both data driven (squares) as well as lattice (circles) and mixed (triangle) and of the corresponding total result for $a_\mu^{\text{SM}} - a_\mu^{\text{exp}}$. The evaluations above the lower dashed line have not been included in the WP average (grey band) and the lattice ones above the upper dashed line have not been included in the WP lattice average (light-blue band).

2.1 Lattice

Even though there has been a dedicated talk by Zoltan Fodor on the lattice perspective on the calculation of the HVP contribution to the $(g-2)_\mu$, I have to add a few comments on the most recent lattice results from the data-driven perspective. In particular because the recently published BMW calculation [4]:

$$10^{10} a_\mu^{\text{HVP,LO}}(\text{BMW}) = 707.5(2.3)_{\text{stat}}(5.0)_{\text{syst}} = 707.5(5.5)\,,$$

is higher by about 2.1σ than the data-driven evaluation and, after adding all other contributions summarized in the WP, much closer to the experimental value of a_μ. The situation is illustrated in Fig. 1.

The discrepancy between the data-driven and the BMW result needs to be clarified. In view of the long history of the data-driven approach, the vast experimental database which is used in the calculation and all the independent checks which have been made over the years, I think that it is justified to consider this as the reference SM value for a_μ. The BMW result is the first complete lattice result with a precision comparable to the one of the data-driven approach. Given the relevance of this calculation and of a possible discrepancy with the data-driven approach it is important to first have a consensus lattice result based on different calculations relying on different discretizations and calculational settings. Several collaborations are working hard to reach this goal.

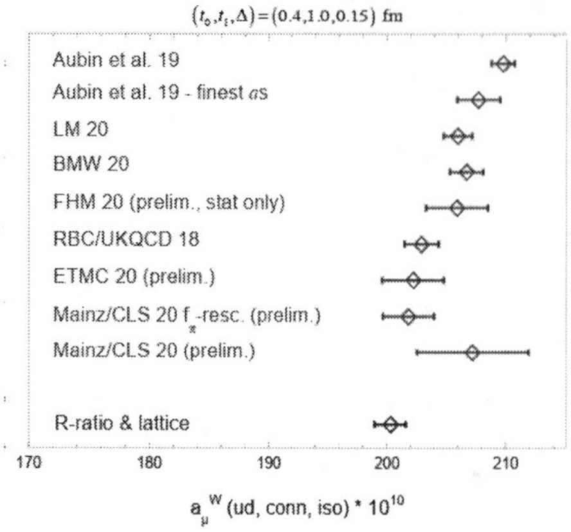

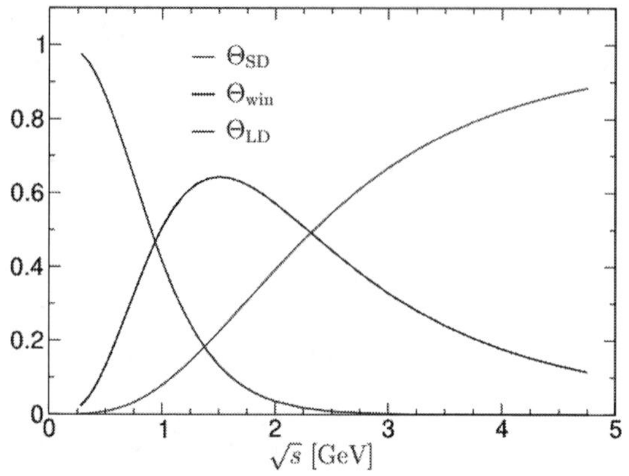

Figure 2. *Left:* Comparison of different lattice evaluations of the intermediate window quantity. Figure courtesy of Davide Giusti [51]. *Right:* Weight functions in s to be used in the time-like integral representation of $a_\mu^{\rm HVP}$ to obtain the window-quantities defined on the lattice. Figure courtesy of Martin Hoferichter.

An important aspect of the BMW result is the potential impact of the higher value of a_μ on the running of $\alpha_{\rm em}$ and on the value of $\alpha_{\rm em}(M_Z)$ [52]: a_μ is determined by an integral over the $e^+e^- \to$ hadrons cross section. The same cross section determines also the running $\alpha_{\rm em}$, though via an integral with a different kernel function: in particular $\alpha_{\rm em}(M_Z)$ is much more sensitive to the high-energy region than to the low-energy one, whereas for a_μ the opposite is true. Any increase in a_μ must be accompanied by an increase in $\Delta\alpha_{\rm em}^{\rm had}(M_Z)$: its size can only be estimated if one knows the energy distribution of the corresponding increase in the hadronic correction. Given the way the lattice calculation is made, this information is not available. Three different analyses [53–55] have estimated the possible impact under different sets of assumptions. They all reached a similar conclusion: unless one accepts to spoil the electroweak fit, the changes in the hadronic cross section have to happen below 2 GeV, the lower the better. Since the region below 1 GeV is dominated by the 2π channel, a specific analysis was dedicated to such a scenario [56]: we investigated the possible form and nature of the discrepancy with the data sets for this channel and pointed out correlations, not only with $\Delta\alpha_{\rm em}^{\rm had}(M_Z)$, but also with the pion charge radius.

Until other lattice collaborations will be able to reach the same level of precision for the complete physical quantity, it is possible to compare the so-called "intermediate window quantity", which is easier to calculate since it is less sensitive to systematic effects. A comparison of different lattice calculations for this window quantity (in the isospin limit) is displayed in Fig. 4 of Ref. [4] and shows a disagreement at the level of $\sim 2\sigma$ among lattice results of similar precision, in particular between the BMW and the one by RBC/UKQCD [36] (whereas BMW agrees well with Aubin et al. [20]). Several lattice collaborations are concentrating on this quantity and aiming to reach a consensus. The current situation has been discussed by Davide Giusti in his talk at this year's Lattice conference [51, 57] and is shown on the left panel of Fig. 2. Reaching a satisfactory consensus for this quantity is the most pressing goal of the lattice community interested in $a_\mu^{\rm HVP}$.

Interestingly, one can evaluate the same window quantity starting from e^+e^- data: the weight function originally defined in terms of Euclidean time for the lattice calculation can be translated in a weight function in s in the timelike region, as illustrated in the right panel of Fig. 2. The outcome of this calculation[1] is also shown in the left panel of Fig. 2: the data-driven result agrees with RBC/UKQCD and a few other preliminary results but is again lower than BMW and Aubin et al. While this may not seem surprising at first, one has to consider that the weight function in s suppresses the contribution from below 1 GeV. Such a disagreement seems to clash with the hypothesis made above that most of the discrepancy for the hadronic cross section occurs below 1 GeV. Together with Martin Hoferichter and Peter Stoffer we have investigated this point more quantitatively: in [56] we have obtained an explicit modification of the $\pi\pi$ cross section below 1 GeV which would give a value of $a_\mu^{\rm HVP}$ in agreement with that of BMW. Evaluating the window quantity with this modified cross section the blue point in Fig. 2 does move slightly to the right but not nearly enough to fill the gap with the BMW20 point. This implies that part of the modification of the hadronic cross section needed to explain the BMW result must happen above 1 GeV. With very minimal and reasonable assumptions about possible distributions of the change in the hadronic cross section we reached the following conclusion: of the $\sim 14 \times 10^{-10}$ units of difference between the data-driven and BMW evaluations of $a_\mu^{\rm HVP}$, at least 5×10^{-10} have to originate from a change in the hadronic cross section above 1 GeV. This is a model-independent lower limit. Note that if all the shift is generated below 1 GeV it amounts to a $\sim$ 2.5% relative change, whereas a shift of 5×10^{-10} generated between 1 and 2 GeV represents a $\sim$ 5% relative change. Moreover the impact on the EW fit will be larger.

3 Hadronic light-by-light

The evaluation of the HLbL contribution in the WP has been much improved with respect to the time of the so-called "Glasgow consensus" [58] (see also [59, 60] for a somewhat different assessment of the situation circa 2009). This is mainly due to the formulation of a dispersive approach for HLbL [24, 61–63] which earlier had been deemed to be impossible. Table 2 illustrates well the improvements, in particular for what concerns the first three rows of the table, added up in the "subtotal" in the fourth row. As one can see by comparing the numbers in that row, the uncertainty reduction has been six- to five-fold with respect to 2009. The remaining rows contain smaller, but still relevant contributions for which the dispersive approach has not yet been applied to its full potential. The reason is that there are conceptual difficulties in including narrow resonances (beyond pseudoscalars) in such an approach: as it has been discussed in Ref. [24] the evaluation of the contribution of single poles in the different channels gives results which depend on the choice of the basis for the HLbL tensor, unless a set of sum rules are satisfied. This is automatically the case for pseudoscalars, but not for any other resonances. A recent discussion of this problem for the case of scalars can be found in Ref. [64], which also shows that progress in this direction is on-going. But it is important to stress that in the four central rows (scalars to short-distance) a superficial comparison of the numbers seems to indicate that uncertainties increased rather than decreased. This just reflects the fact that all possible sources have been accounted for and the explicit goal was to estimate them more conservatively. This is also seen in the way final uncertainties were added for this subset of contributions: linearly in the WP, whereas most previous analyses added them quadratically. For this reason the improvement in the uncertainty looks smaller than it actually is.

Among the remaining more uncertain contributions the axial vectors and the short-distance part are the most relevant ones. For these contributions there is on-going activity and what has been reported in the WP is a snapshot of an evolving situation. The issue of short-distance constraints (SDC) for HLbL has been first pointed out by Melnikov and Vainshtein

[1]Which also implies subtracting contributions from heavier valence quarks as well as isospin-breaking contributions—all done using lattice input—to make the comparison to the lattice meaningful.

Table 2. Comparison of different evaluations of the hadronic light-by-light contribution broken down into its different components identified by the relevant intermediate hadronic state. PdRV(09) is Ref. [58], N/JN(09) Refs. [59, 60] and J(17) Ref. [31].

Contribution	PdRV(09)	N/JN(09)	J(17)	WP(20)
π^0, η, η'-poles	114(13)	99(16)	95.45(12.40)	93.8(4.0)
π, K-loops/boxes	−19(19)	−19(13)	−20(5)	−16.4(2)
S-wave $\pi\pi$ rescattering	−7(7)	−7(2)	−5.98(1.20)	−8(1)
subtotal	88(24)	73(21)	69.5(13.4)	69.4(4.1)
scalars	–	–	–	− 1(3) (scalars + tensors)
tensors	–	–	1.1(1)	
axial vectors	15(10)	22(5)	7.55(2.71)	6(6)
u, d, s-loops / short-distance	–	21(3)	20(4)	15(10)
c-loop	2.3	–	2.3(2)	3(1)
total	105(26)	116(39)	100.4(28.2)	92(19)

(MV) in a seminal paper [22]: besides deriving these constraints, they also proposed a model for how to satisfy them. This essentially consisted in including only the lightest pseudoscalar poles in the HLbL tensor (in the limit of $(g-2)$ kinematics) to satisfy the longitudinal SDC, and the lightest axials to satisfy the SDC of the transverse components. The present precision requirements made it necessary to go beyond this model. Different attempts in this direction have been made recently: in our group we considered a tower of excited pseudoscalars to satisfy the longitudinal SDC [28, 65], whereas two different groups have addressed both the transverse as well as the longitudinal SDC by considering a tower of axial resonances within a model of holographic QCD [66–68]. While it is clear that the axials have to play a prominent role in satisfying the SDC because the contribution of excited pseudoscalars vanishes in the chiral limit, pseudoscalars have the unique advantage that for them the ambiguities mentioned above are absent. This means that each of the two approaches has drawbacks and only represents a step in the direction of a fully satisfactory solution of the SDC. For this reason it is particularly important to compare these two model-dependent solutions. This has been done in great detail in [69] and summarized in the left panel of Fig. 3 which shows the contribution to a_μ of the states responsible for satisfying the longitudinal SDC as a function of a lower cutoff $Q_{\rm min}$ on all three photon momenta. The figure shows that the solution in terms of excited pseudoscalars agrees well with different variants of the holographic model of QCD, and that both give a contribution significantly lower than was predicted by the original MV model: keeping only the pion pole contribution for $g-2$ kinematics is well justified in the high-q^2 region but is a bad approximation at low q^2. Taking into account also the transverse components led to the estimate shown in Table 2 for the contribution of short distance. The figure also shows the curve obtained by Lüdtke and Procura [70] who analyzed the longitudinal SDC on the basis of a set of interpolants and confirmed the estimate in the WP.

Other important developments concern the calculations performed by H. Bijnens and his group of perturbative [71] and non-perturbative [72] corrections to the leading order OPE, which is essentially given by the quark loop [22, 27]. These corrections allow for further reductions of the uncertainties in the evaluation of this contribution, as discussed in [69]. Other theoretical aspects, in particular concerning axial mesons, have been further discussed in [73–76].

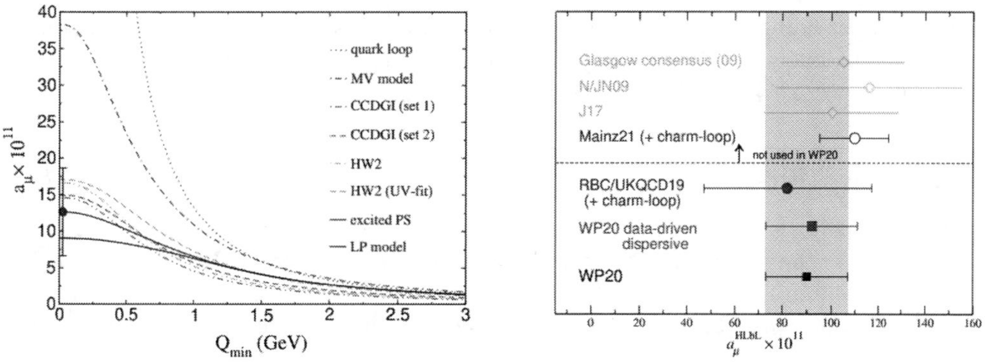

Figure 3. *Left*: contribution to a_μ due to the longitudinal SDC for different solutions to the latter as a function of a lower cutoff on the modulus of photon momenta (from [69]). *Right*: comparison of different evaluations of the HLbL contribution. The two results below the dashed line have been averaged to yield the result presented in the WP. Above the dashed line there is the recent lattice evaluation by the Mainz collaboration and the three other phenomenological ones also shown in Table 2.

3.1 Lattice

Compared to the two-point function which is relevant for HVP, evaluating the four-point function which enters the HLbL contribution is significantly more difficult on the lattice (and not only). In this case, on the other hand, the precision requirements are much less severe. Attempts at calculating the HLbL contribution have started much later than those for HVP, and much of the early (but still recent) work was devoted to developing a calculational strategy [77–81]. These efforts, which were carried out mainly by two lattice groups, RBC/UKQCD and Mainz, have culminated in the first two complete lattice calculations of this contribution: first by the RBC/UKQCD collaboration [36], a result which was early enough to be considered in the WP and which in fact was averaged with the data-driven one and included in the SM prediction for a_μ, see Table 1. The Mainz collaboration first published a result in the SU(3) limit [82], and only very recently completed the calculation for physical quark masses [5], thereby confirming both the RBC/UKQCD result as well as the data-driven one. Both results are shown in Fig. 3.

4 Conclusions

I have briefly reviewed the current status of the Standard Model evaluation of the muon anomalous magnetic moment concentrating in particular on the two main hadronic contributions, HVP and HLbL, and their data-driven as well as lattice evaluations. The comparison with the current experimental world average after the recent Fermilab result shows a 4.2σ discrepancy, which makes the muon $g-2$ one of the most interesting quantities in the search for deviations from the Standard Model and the quest for new physics. The present picture isn't as sharp as it could be because of the lattice calculation of the HVP contribution by the BMW collaboration, which shows a discrepancy with the calculation based on the data-driven approach and, if confirmed, would move the SM value of $(g-2)_\mu$ closer to the measurement. With lattice calculations it is always important to make universality tests, namely to show that different lattice formulations of QCD (in particular for what concerns fermion discretizations) lead to the same result in the continuum limit. This will have to wait until other lattice collaborations will produce results of similar precision as the one by BMW. Work in this direction, in particular for the simpler window quantity discussed above, is on-going and will hopefully soon lead to a full clarification of the situation.

Acknowledgments

I thank the colleagues of the Muon $g-2$ Theory Initiative and in particular of the Steering Committee for the pleasant and fruitful collaboration which produced the White Paper. I am grateful to Martin Hoferichter and Peter Stoffer for the longstanding collaboration on many of the topics described here. This work has been partially supported by the Swiss National Science Foundation.

References

[1] B. Abi et al. (Muon $g-2$), Phys. Rev. Lett. **126**, 141801 (2021), 2104.03281
[2] T. Aoyama et al., Phys. Rept. **887**, 1 (2020), 2006.04822
[3] G.W. Bennett et al. (Muon $g-2$), Phys. Rev. D **73**, 072003 (2006), hep-ex/0602035
[4] S. Borsanyi et al., Nature (2021), 2002.12347
[5] E.H. Chao, R.J. Hudspith, A. Gérardin, J.R. Green, H.B. Meyer, K. Ottnad (2021), 2104.02632
[6] M. Davier, A. Hoecker, B. Malaescu, Z. Zhang, Eur. Phys. J. C **77**, 827 (2017), 1706.09436
[7] A. Keshavarzi, D. Nomura, T. Teubner, Phys. Rev. D **97**, 114025 (2018), 1802.02995
[8] G. Colangelo, M. Hoferichter, P. Stoffer, JHEP **02**, 006 (2019), 1810.00007
[9] M. Hoferichter, B.L. Hoid, B. Kubis, JHEP **08**, 137 (2019), 1907.01556
[10] M. Davier, A. Hoecker, B. Malaescu, Z. Zhang, Eur. Phys. J. C **80**, 241 (2020), [Erratum: Eur. Phys. J. C **80**, 410 (2020)], 1908.00921
[11] A. Keshavarzi, D. Nomura, T. Teubner, Phys. Rev. D **101**, 014029 (2020), 1911.00367
[12] A. Kurz, T. Liu, P. Marquard, M. Steinhauser, Phys. Lett. B **734**, 144 (2014), 1403.6400
[13] B. Chakraborty et al. (Fermilab Lattice, LATTICE-HPQCD, MILC), Phys. Rev. Lett. **120**, 152001 (2018), 1710.11212
[14] S. Borsanyi et al. (Budapest-Marseille-Wuppertal), Phys. Rev. Lett. **121**, 022002 (2018), 1711.04980
[15] T. Blum, P.A. Boyle, V. Gülpers, T. Izubuchi, L. Jin, C. Jung, A. Jüttner, C. Lehner, A. Portelli, J.T. Tsang (RBC, UKQCD), Phys. Rev. Lett. **121**, 022003 (2018), 1801.07224
[16] D. Giusti, V. Lubicz, G. Martinelli, F. Sanfilippo, S. Simula (ETM), Phys. Rev. D **99**, 114502 (2019), 1901.10462
[17] E. Shintani, Y. Kuramashi, Phys. Rev. D **100**, 034517 (2019), 1902.00885
[18] C.T.H. Davies et al. (Fermilab Lattice, LATTICE-HPQCD, MILC), Phys. Rev. D **101**, 034512 (2020), 1902.04223
[19] A. Gérardin, M. Cè, G. von Hippel, B. Hörz, H.B. Meyer, D. Mohler, K. Ottnad, J. Wilhelm, H. Wittig, Phys. Rev. D **100**, 014510 (2019), 1904.03120
[20] C. Aubin, T. Blum, C. Tu, M. Golterman, C. Jung, S. Peris, Phys. Rev. D **101**, 014503 (2020), 1905.09307
[21] D. Giusti, S. Simula, PoS **LATTICE2019**, 104 (2019), 1910.03874
[22] K. Melnikov, A. Vainshtein, Phys. Rev. D **70**, 113006 (2004), hep-ph/0312226

[23] P. Masjuan, P. Sánchez-Puertas, Phys. Rev. D **95**, 054026 (2017), 1701.05829
[24] G. Colangelo, M. Hoferichter, M. Procura, P. Stoffer, JHEP **04**, 161 (2017), 1702.07347
[25] M. Hoferichter, B.L. Hoid, B. Kubis, S. Leupold, S.P. Schneider, JHEP **10**, 141 (2018), 1808.04823
[26] A. Gérardin, H.B. Meyer, A. Nyffeler, Phys. Rev. D **100**, 034520 (2019), 1903.09471
[27] J. Bijnens, N. Hermansson-Truedsson, A. Rodríguez-Sánchez, Phys. Lett. B **798**, 134994 (2019), 1908.03331
[28] G. Colangelo, F. Hagelstein, M. Hoferichter, L. Laub, P. Stoffer, JHEP **03**, 101 (2020), 1910.13432
[29] V. Pauk, M. Vanderhaeghen, Eur. Phys. J. C **74**, 3008 (2014), 1401.0832
[30] I. Danilkin, M. Vanderhaeghen, Phys. Rev. D **95**, 014019 (2017), 1611.04646
[31] F. Jegerlehner, Springer Tracts Mod. Phys. **274**, 1 (2017)
[32] M. Knecht, S. Narison, A. Rabemananjara, D. Rabetiarivony, Phys. Lett. B **787**, 111 (2018), 1808.03848
[33] G. Eichmann, C.S. Fischer, R. Williams, Phys. Rev. D **101**, 054015 (2020), 1910.06795
[34] P. Roig, P. Sánchez-Puertas, Phys. Rev. D **101**, 074019 (2020), 1910.02881
[35] G. Colangelo, M. Hoferichter, A. Nyffeler, M. Passera, P. Stoffer, Phys. Lett. B **735**, 90 (2014), 1403.7512
[36] T. Blum, N. Christ, M. Hayakawa, T. Izubuchi, L. Jin, C. Jung, C. Lehner, Phys. Rev. Lett. **124**, 132002 (2020), 1911.08123
[37] T. Aoyama, M. Hayakawa, T. Kinoshita, M. Nio, Phys. Rev. Lett. **109**, 111808 (2012), 1205.5370
[38] T. Aoyama, T. Kinoshita, M. Nio, Atoms **7**, 28 (2019)
[39] A. Czarnecki, W.J. Marciano, A. Vainshtein, Phys. Rev. D **67**, 073006 (2003), [Erratum: Phys. Rev. D **73**, 119901 (2006)], hep-ph/0212229
[40] C. Gnendiger, D. Stöckinger, H. Stöckinger-Kim, Phys. Rev. D **88**, 053005 (2013), 1306.5546
[41] C. Bouchiat, L. Michel, J. Phys. Radium **22**, 121 (1961)
[42] B.L. Hoid, M. Hoferichter, B. Kubis, Eur. Phys. J. C **80**, 988 (2020), 2007.12696
[43] B. Aubert et al. (BaBar), Phys. Rev. Lett. **103**, 231801 (2009), 0908.3589
[44] F. Ambrosino et al. (KLOE), Phys. Lett. B **670**, 285 (2009), 0809.3950
[45] F. Ambrosino et al. (KLOE), Phys. Lett. B **700**, 102 (2011), 1006.5313
[46] D. Babusci et al. (KLOE), Phys. Lett. B **720**, 336 (2013), 1212.4524
[47] A. Anastasi et al. (KLOE-2), JHEP **03**, 173 (2018), 1711.03085
[48] M.N. Achasov et al. (SND), JHEP **01**, 113 (2021), 2004.00263
[49] M. Ablikim et al. (BESIII) (2020), 2009.05011
[50] F. Ignatov et al. (CMD-3), EPJ Web Conf. **212**, 04001 (2019)
[51] D. Giusti (????)
[52] M. Passera, W. Marciano, A. Sirlin, Phys. Rev. D **78**, 013009 (2008), 0804.1142
[53] A. Crivellin, M. Hoferichter, C.A. Manzari, M. Montull, Phys. Rev. Lett. **125**, 091801 (2020), 2003.04886
[54] A. Keshavarzi, W.J. Marciano, M. Passera, A. Sirlin, Phys. Rev. D **102**, 033002 (2020), 2006.12666
[55] B. Malaescu, M. Schott, Eur. Phys. J. C **81**, 46 (2021), 2008.08107

[56] G. Colangelo, M. Hoferichter, P. Stoffer, Phys. Lett. B **814**, 136073 (2021), 2010.07943

[57] D. Giusti, S. Simula, PoS, **LATTICE2021** p. 189 (????)

[58] J. Prades, E. de Rafael, A. Vainshtein, Adv. Ser. Direct. High Energy Phys. **20**, 303 (2009), 0901.0306

[59] F. Jegerlehner, A. Nyffeler, Phys. Rept. **477**, 1 (2009), 0902.3360

[60] A. Nyffeler, Phys. Rev. D **79**, 073012 (2009), 0901.1172

[61] G. Colangelo, M. Hoferichter, M. Procura, P. Stoffer, JHEP **09**, 091 (2014), 1402.7081

[62] G. Colangelo, M. Hoferichter, B. Kubis, M. Procura, P. Stoffer, Phys. Lett. B **738**, 6 (2014), 1408.2517

[63] G. Colangelo, M. Hoferichter, M. Procura, P. Stoffer, JHEP **09**, 074 (2015), 1506.01386

[64] I. Danilkin, M. Hoferichter, P. Stoffer, Phys. Lett. B **820**, 136502 (2021), 2105.01666

[65] G. Colangelo, F. Hagelstein, M. Hoferichter, L. Laub, P. Stoffer, Phys. Rev. D **101**, 051501 (2020), 1910.11881

[66] J. Leutgeb, A. Rebhan, Phys. Rev. D **101**, 114015 (2020), 1912.01596

[67] L. Cappiello, O. Catà, G. D'Ambrosio, D. Greynat, A. Iyer, Phys. Rev. D **102**, 016009 (2020), 1912.02779

[68] J. Leutgeb, J. Mager, A. Rebhan (2021), 2110.07458

[69] G. Colangelo, F. Hagelstein, M. Hoferichter, L. Laub, P. Stoffer, Eur. Phys. J. C **81**, 702 (2021), 2106.13222

[70] J. Lüdtke, M. Procura, Eur. Phys. J. C **80**, 1108 (2020), 2006.00007

[71] J. Bijnens, N. Hermansson-Truedsson, L. Laub, A. Rodríguez-Sánchez, JHEP **04**, 240 (2021), 2101.09169

[72] J. Bijnens, N. Hermansson-Truedsson, L. Laub, A. Rodríguez-Sánchez, JHEP **10**, 203 (2020), 2008.13487

[73] P. Masjuan, P. Roig, P. Sánchez-Puertas (2020), 2005.11761

[74] M. Knecht, JHEP **08**, 056 (2020), 2005.09929

[75] M. Zanke, M. Hoferichter, B. Kubis, JHEP **07**, 106 (2021), 2103.09829

[76] M. Hoferichter, P. Stoffer, JHEP **05**, 159 (2020), 2004.06127

[77] J. Green, O. Gryniuk, G. von Hippel, H.B. Meyer, V. Pascalutsa, Phys. Rev. Lett. **115**, 222003 (2015), 1507.01577

[78] T. Blum, N. Christ, M. Hayakawa, T. Izubuchi, L. Jin, C. Lehner, Phys. Rev. D **93**, 014503 (2016), 1510.07100

[79] T. Blum, P.A. Boyle, T. Izubuchi, L. Jin, A. Jüttner, C. Lehner, K. Maltman, M. Marinkovic, A. Portelli, M. Spraggs, Phys. Rev. Lett. **116**, 232002 (2016), 1512.09054

[80] T. Blum, N. Christ, M. Hayakawa, T. Izubuchi, L. Jin, C. Jung, C. Lehner, Phys. Rev. Lett. **118**, 022005 (2017), 1610.04603

[81] T. Blum, N. Christ, M. Hayakawa, T. Izubuchi, L. Jin, C. Jung, C. Lehner, Phys. Rev. D **96**, 034515 (2017), 1705.01067

[82] E.H. Chao, A. Gérardin, J.R. Green, R.J. Hudspith, H.B. Meyer, Eur. Phys. J. C **80**, 869 (2020), 2006.16224

Overview of lattice calculations of the x-dependence of PDFs, GPDs and TMDs

Krzysztof Cichy[1,*]

[1]Faculty of Physics, Adam Mickiewicz University,
ul. Uniwersytetu Poznańskiego 2, 61-614 Poznań, Poland

Abstract. For a long time, lattice QCD was unable to address the x-dependence of partonic distributions, direct access to which is impossible in Euclidean spacetime. Recent years have brought a breakthrough for such calculations when it was realized that partonic light-cone correlations can be accessed through spatial correlations computable on the lattice. Appropriately devised observables can be factorized into physical PDFs via a perturbative procedure called matching, analogous to the standard factorization of experimental cross sections. In this short review, aimed at a broader high-energy and nuclear physics community, we discuss the recent highlights of this research program. Key concepts are outlined, followed by a case study illustrating the typical stage of current lattice extractions and by a brief review of the most recent explorations. We finalize with a number of messages for the prospects of lattice determinations of partonic structure.

1 Introduction

Ever since the amazing discovery of nucleon's internal structure, it was one of the most important goals of hadron physics to fully understand and describe this structure. For many years, the focus has been on longitudinal properties, especially for an unpolarized nucleon. Several collider experiments can be well described taking only these properties into account. They are summarized in unpolarized parton distribution functions (PDFs), objects that are well known from multiple global analyses based on thousands of experimental measurements. However, it was realized that the full picture requires also addressing polarized nucleons – thus, also polarized PDFs (helicity and transversity) are important. These functions, in particular transversity, are significantly less constrained by collider data. Moreover, description of some scattering events requires even more – knowledge about the transverse position and motion of partonic constituents of hadrons. Quantification of these aspects necessitates generalizing PDFs to more complicated distributions – generalized parton densities (GPDs) and transverse-momentum-dependent PDFs (TMDs). For these, our knowledge is even more restricted and we are basically at the beginning of the quest to unravel them. This quest needs a coordinated effort of the experimental and theoretical/phenomenological communities. The former will be enabled by the recent or ongoing construction of new experimental facilities, such as the 12 GeV upgrade at Jefferson Lab and the Electron-Ion Collider (EIC), with focus on providing the missing empirical data addressing fundamental questions about the nucleon

*e-mail: kcichy@amu.edu.pl

structure. In turn, the latter will consist in building realistic models, increasing the reliability of perturbative expansions, as well as attempting to access nucleon's properties from first principles. This last possibility is the natural place where lattice QCD (LQCD) can be expected to yield progress, being the only feasible program to quantitatively extract QCD predictions directly from its Lagrangian.

Before calculations of x-dependent distributions became a reality, LQCD had a long history of computing moments of these functions, see e.g. Ref. [1] for a review of this period's achievements. At that stage, around 2013-2014, the common belief was that LQCD was indeed restricted to calculations of moments, quantities accessible from local matrix elements (MEs) in Euclidean spacetime. On the other hand, the full distributions, defined on the light front, were believed to be inaccessible without the Minkowski metric. In principle, such distributions can be reconstructed from a tower of moments, but LQCD was understood to be practically limited to the lowest three, with higher ones suffering from inescapable power-divergent mixings with lower-dimensional operators and exponentially decaying signal-to-noise ratios. Already at that stage, it was known that x-dependence could, in principle, be accessed even in Euclidean space, but the early proposals how to do it [2–4] had not turned into practically feasible programs. In 2013, Ji came up with another idea [5, 6], known as quasi-distributions, to calculate appropriate spatial correlations in a boosted hadron instead of light-cone ones and relate the ensuing new objects with their physical counterparts via perturbation theory. This is possible due to the asymptotic freedom of QCD and the quasi- and light-cone distributions having the same infrared (IR) properties. As their differences emerge only in the ultraviolet (UV), one can use factorization and subtract the UV difference perturbatively, employing the so-called large momentum effective theory (LaMET). This proposal immediately led to first lattice evaluations for the three twist-2 PDFs [7–10] and soon thereafter to a plethora of theoretical and practical studies. The renewed interest in extracting the x-dependence on the lattice led to the revival of the earlier approaches and to the emergence of new ones [11–15], all of which have since achieved significant progress. This dynamical progress is extensively reported in the reviews [16–19].

Since space for this proceeding is limited, it is not possible to discuss here all the progress even of the last year. An interested Reader is, thus, referred to the above mentioned reviews, with the most recent developments covered in Ref. [19]. Instead, we offer here a more general discussion addressed at a wider community, attempting to convey the key messages on what to expect from the lattice in the area of x-dependent distributions.

2 Key concepts

We start by reminding the Reader about the basics of the lattice formulation, to emphasize the implications for the robustness and reliability of lattice-extracted distributions. LQCD is formulated on a discretized grid, with quarks living on lattice sites and gluons on links between them. This grid is finite and discrete, thus IR (the lattice extent, L) and UV (the lattice spacing, a) regulators are naturally introduced. Typical lattice sizes nowadays employed are $L/a \in [32, 128]$ and lattice spacings $a \in [0.03, 0.15]$ fm. As quark masses are free parameters, one can simulate at arbitrary pion masses, typically chosen to be between the physical one and several hundred MeV. The latter, while unphysical, allows one to significantly reduce the computational cost. This cost is still rather huge – the above setup regularizes the formally infinite QCD path integral, but the number of dimensions is still of $\mathcal{O}(10^8 - 10^9)$, necessitating the usage of world's most powerful supercomputers.

Obviously, the ultimate interest is in continuum QCD. Consequently, the regulators need to be removed, as well as effects of other choices like the non-physical pion mass. The key strength of LQCD is that all of this can be done in a controlled manner. However, it also implies that lattice studies on a given subject proceed in stages:

1. Feasibility studies.
2. Qualitative explorations.
3. Advanced studies.
4. Precision studies.

The first stage only aims at establishing whether a given approach can be used in practice. Some methods may prove to be e.g. prohibitively expensive or may suffer from a very unfavorable signal-to-noise ratio. If the feasibility criterion is satisfied, the approach is typically explored qualitatively in a non-physical setup, e.g. at a large pion mass and with only a single lattice spacing. Conclusions from such studies are usually only qualitative, with no quantitative comparison with experiment possible. At this stage, possibities to optimize the method are also realized. Later on, systematic effects are started to be addressed and first quantitatively reliable results can be obtained, although with rather large overall uncertainty. In the last stage, additional optimizations of the method or setup can ensue and the fully quantified errors are reduced to a desired level, depending mostly on the available computer resources and uncertainties external to the lattice, such as truncation effects from a required perturbative input.

Needless to say, the duration of each stage is closely bound to the complexity of the physical question and the lattice method. For the case of nucleon structure, in particular the x-dependent distributions, this complexity is significant and thus, the few years since Ji's breakthrough proposal and the revival/conception of other approaches have not been enough to get to the stage of precision studies. For most approaches and distributions, the studies are somewhere between the first and third stage. It is natural that studies of the simplest quantities, twist-2 PDFs, are most advanced – quantification of systematics is progressing and in a few years, first precision studies may be expected. The 3-dimensional distributions, in turn, lead to additional challenges and thus, are at a more exploratory stage.

Below, we discuss shortly the numerous aspects that need to be addressed to arrive at robust lattice-extracted distributions. To make the arguments more concrete, we concentrate on the two most popular quasi- and pseudo-distribution approaches, in the context of calculating PDFs and GPDs. In both of them, one computes MEs of the form $\mathcal{M}_\Gamma(z,P) = \langle N(P_3)|\overline{\psi}(z)\Gamma\mathcal{A}(z,0)\psi(0)|N(P_3)\rangle$. $|N(P_3)\rangle$ is a boosted hadron state, with boost P_3 taken in the z-direction. A Wilson line of length z in the same direction, $\mathcal{A}(z,0)$, is inserted to maintain gauge invariance. The type of distribution is determined by the Dirac structure Γ. In general, one can consider also momentum transfer, t, with $t = 0$ ($t \neq 0$) for PDFs (GPDs). At $t \neq 0$, it is also necessary to use additional projectors to disentangle the two (four) GPDs in the chiral-even (chiral-odd) case. An important aspect is also that the spacetime locations of the source, the sink and the operator insertion need to be sufficiently separated to isolate the desired ground-state hadron.

The calculation of the above bare MEs is the most expensive part of the calculation. It is important to emphasize that the actual cost may vary even by orders of magnitude, depending on simulation parameters, in particular the nucleon boost and the pion mass. Increasing the nucleon boost leads to an exponentially decaying signal and the problem becomes even more severe when lowering the pion mass. Working with large nucleon boosts (e.g. of $O(3\,\text{GeV})$) at the physical point, the signal is clean only at very low separations between the source and the sink. However, in such a situation, one can hardly extract any ground-state properties, as the correlators receive sizable contributions additionally from tens of excited states, see Ref. [20] for an extensive discussion. The signal is further contaminated with noise when extracting GPDs, due to the nonzero momentum transfer and the necessity to use additional projectors. An important technique to deal with the worsening noise for boosted hadrons is momentum smearing [21]. It does not cure the exponential problem, but moves it to larger momenta. Realistically, the signal-to-noise problem becomes severe when getting to boosts of order 2-2.5 GeV at typically used non-physical pion masses and already around 1.5 GeV at the physical point, see Tab. 1 of Ref. [18] for a comparison of experience of different collaborations. Larger boosts most likely suggest lack of control over excited states and unknown, possibly very large, systematic errors. It should be kept in mind that large boosts also lead to enhanced discretization effects when the boost in lattice units becomes comparable to the inverse lattice spacing, the UV cutoff of the lattice theory.

Apart from the above, the calculation of bare MEs is subject to several standard sources of systematic effects. In early stages of a computational program, these may be largely unexplored. Addressing these systematics is mandatory towards the more mature stages and good control over them is a prerequisite for precision physics. Thus, a fully controlled calculation necessitates the use of at least 3 lattice spacings to perform the continuum limit extrapolation (after removal of divergences), a check of finite volume effects, pion mass effects (if not simulating directly at the physical point) and several other sources of possible contaminations. Depending on the final desired accuracy, one may need to address also subleading effects, e.g. the assumption that up and down quarks are degenerate is totally safe when aiming at 10% precision, but cannot be made if 1% total error is targeted; in such a case, one would need to take into account the different masses and electric charges of the light quarks.

The difficulty of the partonic distributions program consists also in the presence of additional sources of systematic effects beyond the ones for bare MEs. An important issue is the non-perturbative renormalization of the latter. The extra complication with respect to standard lattice hadron structure calculations arises due to the non-locality of inserted operators. In approaches based on the Wilson line, its presence induces a power divergence, in addition to the standard logarithmic one. The most commonly used renormalization strategies are types of RI/MOM renormalization tailored for non-local operators [22–25] and ratio schemes [25, 26]. However, as argued in Ref. [27], contamination from IR effects may ensue at large z. In particular, RI/MOM schemes were shown to evince a residual divergence [28], important numerically at very fine lattice spacings. Proposals to overcome this issue were put forward recently, the so-called hybrid scheme [27] and "self-renormalization" [29].

Further important systematics go beyond bare and renormalized MEs. Since these MEs are Euclidean quantities, it is unsurprising that a delicate step is to "translate" Euclidean MEs to physical x-distributions. This has two separate aspects, related to the change from coordinate to momentum space and from spatial to light-cone correlations. At this stage, also the crucial difference between quasi- and pseudo-distribution approaches emerges. In the former, renormalized MEs are first subjected to x-dependence reconstruction, leading to momentum-space quasi-distributions that are matched to their light-cone counterparts. In turn, matching to physical distributions for the latter is performed in coordinate space, defining light-cone Ioffe-time distributions, then subjected to x-dependence reconstruction. The reconstruction issue, broadly discussed in Ref. [30], originates from the mathematically ill-defined attempt to obtain a continuous distribution from a discrete and truncated set of MEs, which poses a so-called inverse problem. The problem can be tackled by providing an additional input that complements the lattice data, in the form of a physically-justified assumption or a mathematical criterion. As such, the solution of the inverse problem is non-unique and different methods should be used to avoid introducing a bias. The translation from spatial to light-cone correlations (matching) is also subtle. It can be done safely only if MEs were extracted for a sufficiently highly boosted nucleon. In the presence of the exponentially decaying signal, care is needed to estimate effects from a finite boost. These higher-twist effects are power-suppressed, but may be sizable with currently attainable hadron momenta and are expected to be enhanced both at small and large x. Since matching relies on factorization, the involved perturbative expansion is also potentially sensitive to truncation effects. For a long time only 1-loop formulae were known, but recently first calculations at two loops appeared [31–35], leading to conclusions about the size of truncation effects. The latter need to be controlled also at the (separate or concurrent with matching) stage of translating from a lattice-applicable scheme to $\overline{\text{MS}}$.

The above discussion has still not enumerated all possible sources of systematics. More elaborate considerations can be found in the reviews [16–19] and in papers referenced therein. Let us summarize this part with the main message of this discussion – the lattice extraction of partonic distributions is comparatively more complicated with respect to most other lattice calculations and thus, subject to several other sources of systematic effects. At the present stage, there is no complete calculation for any distribution in the sense of robustly quantifying all systematics. As such, the current results are qualitative and no quantitative comparison with available experimental/phenomenological results should yet be attempted. However, it is important to bear in mind that the sources of systematics are known and thus, they can and will be addressed in the rather near future. In the next section, the above will be illustrated more concretely with recent lattice computations.

3 Case study: twist-2 PDFs

Unpolarized twist-2 isovector ($u - d$) PDFs are the simplest nucleon PDFs that can be extracted on the lattice and at the same time, the most well-known from phenomenological global fits. As such, they are the natural benchmark case for LQCD and their accurate reproduction is an important aim to demonstrate the robustness of the lattice approach and good control over the numerous systematic effects. The current state-of-the-art can be well illustrated based on two recent calculations at or near the physical point, from Extended Twisted Mass (ETMC) and HadStruc collaborations, using the pseudo-distribution approach. The ETMC work [36] reused the data of Refs. [37, 38], where unpolarized and helicity PDFs were extracted in the quasi-distribution formalism from a physical point calculation using twisted mass (TM) fermions with a clover term, at a lattice spacing $a = 0.094$ fm and $L/a = 48$. In turn, HadStruc [39] used clover fermions at three pion masses, $m_\pi = 358, 278, 172$ MeV, with the same lattice spacing and $L/a = 32, 64$. The final distributions are shown in Fig. 1, the valence PDF q_v for HadStruc and additionally different combinations with antiquarks for ETMC, $q_{v2s} \equiv q_v + 2\bar{q}$, $q \equiv q_v + \bar{q}$ and $\bar{q}$ alone.

Direct comparison is possible for q_v. Results from both calculations are qualitatively very similar. In HadStruc's work (left part of Fig. 1), there is good quantitative agreement with global fits for $x \lesssim 0.4$ (where the errors are the largest) and a discrepancy with phenomenology observed at larger x. This is the typical situation of present lattice extractions and it has a clear origin in unaccounted for systematics from various sources – the plotted errors are merely statistical and do not reflect that the calculation was done only with a single lattice spacing, using only 1-loop matching etc. Conclusions are similar also for ETMC (right part of Fig. 1) when only statistical errors are considered (purple bands). The tension with global fits is somewhat smaller, which suggests an accidentally more favorable interplay of unquantified systematics. However, in this case, an estimate of these systematics was attempted, taking typical ("plausible") magnitudes (based on other hadron structure studies) of cutoff, finite volume, excited states and truncation effects, as proposed in Ref. [40], with two additional systematic effects assessed more quantitatively. The enlarged total uncertainty from combining statistical and "plausible" systematic effects (cyan band) led to total agreement with global fits for the whole x-range. While such a procedure for systematics cannot be treated

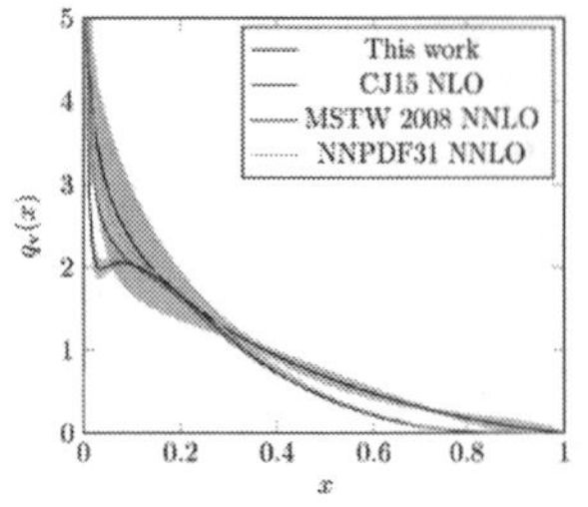

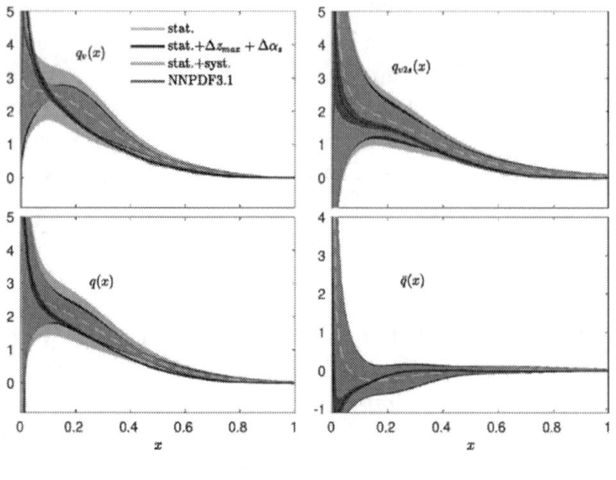

Figure 1: Recent results for the nucleon's unpolarized isovector PDFs. Left: q_v [39]. Right: q_v (upper left), q_{v2s} (upper right), q (lower left), $\bar{q}$ (lower right) [36]. Plots reprinted based on the arXiv distribution license.

as replacement for their careful (and laborious) quantification, it is an important check that systematic effects typically found in other studies may be enough to explain the remaining differences between lattice-extracted PDFs and their phenomenological counterparts.

The same lattice data were also subjected to analyses in the robust NNPDF (Ref. [40] for ETMC and Ref. [41] for HadStruc) and JAM (only for ETMC [42]) global fitting frameworks. We only briefly summarize the main findings. The NNPDF study of ETMC data (in the quasi-PDF formalism) led to conclusions similar to the above, with larger differences with respect to global fits. This is unsurprising, since systematic effects are different at some stages of pseudo- and quasi-PDF extractions, here with their less favorable interplay in the latter. However, again, the "plausible" magnitudes for unquantified systematics brought the final neural-network-reconstructed distributions to agreement with NNPDF3.1. An estimate of systematics was performed also for the HadStruc data combined with the NNPDF framework, leading to better agreement with phenomenology as compared with the case of only statistical errors, but still with some leftover tensions due to incompleteness of systematic estimates at this stage. Finally, the JAM framework as applied to ETMC unpolarized data pointed to similar conclusions, again confirming the need for careful systematic analyses and the need to reduce the overall uncertainties. Interestingly, the JAM work also used lattice data for helicity PDFs. This case is different from the phenomenological point of view, as this kind of PDFs is significantly less constrained by experimental data. The authors concluded that the lattice data can already have some additional constraining power and the tensions with global fits are already smaller than for unpolarized PDFs. There is, however, no fundamental reason why helicity PDFs would have smaller systematic effects than unpolarized ones and these smaller tensions are, again, likely the consequence of a possibly more favorable interplay of the several unaccounted for systematics. It is still fair to assess that LQCD can have an impact on global extractions of helicity PDFs, given the smaller abundance of empirical data for polarized nucleons. This becomes even more obvious for the transversity case (see e.g. Refs. [43, 44]), where LQCD can provide significant constraints that can complement experimental data planned to be obtained in the near future. As argued above, the total uncertainties need to be robustly quantified to achieve this aim (see e.g. Refs. [45, 46] for studies of discretization effects). The unpolarized case can serve as a benchmark for this, given the unlikely prospect of lattice data to contribute to the global extraction of these under the abundance of empirical data taken over tens of years at various experimental facilities.

4 New exploratory directions

In this section, we take a short look at other recent lattice explorations of partonic distributions. For lack of space, only subjectively chosen highlights are mentioned and for a more complete overview of the activities of the last 2 years, we refer to Refs. [18, 19].

- **Flavor-singlet PDFs**. The previous section concentrated on the case of isovector PDFs. Getting flavor-decomposed ones is significantly more challenging, since quark-disconnected diagrams need to be computed. These are more noisy than the quark-connected ones and thus, require more expensive computations and special techniques. The first flavor-singlet PDFs (unpolarized, helicity and transversity) for the light quarks were obtained in Refs. [47, 48] in the quasi-PDF approach. The disconnected diagrams were addressed via techniques optimized earlier for local operators, with explicit low modes treatment and stochastic high modes evaluation using hierarchical probing and the one-end trick. The results were obtained with TM quarks at a non-physical pion mass ($m_\pi = 260\,\text{MeV}$) and are subject to follow-up studies of several sources of systematics. With respect to isovector ones, an additional one is the mixing with gluon PDFs, neglected in this work. However, this mixing can be expected to be a small effect for light quarks (but not for the heavier ones, which appear in the nucleon predominantly from gluon splitting).
- **Gluon PDFs**. Gluonic quantities evince significant noise in LQCD calculations and hence, gluon PDFs are difficult to determine. The recent exploratory pseudo-distribution study by the HadStruc collaboration [49] (clover fermions, $m_\pi = 358\,\text{MeV}$) utilized several special techniques to get a robust signal. Particularly important was the momentum-smeared distillation technique [50], earlier tested for quark PDFs in Ref. [51]. Combining it with the summed GEVP method of extracting the matrix elements, it effectively allows to extract the data at smaller source-sink separations with demonstrable control over excited states. Another important technique, first introduced in Ref. [46], was reconstruction of the x-distribution with Jacobi polynomials. At this stage, mixing with singlet quark PDFs was neglected. The final gluon PDF was found to agree well with phenomenological determinations in the whole x-range. Obviously, this cannot prevent further studies of systematics, leading to robustly quantified total uncertainties.
- **Twist-3 PDFs**. Apart from leading-twist (twist-2) PDFs, full characterization of nucleon's longitudinal structure requires knowledge also of higher-twist distributions. Recently, twist-3 PDFs (chiral-even $g_T(x)$ and chiral-odd $h_L(x)$, $e(x)$) were addressed in a series of papers [52–55]. Two of these papers [52, 54] derived the 1-loop matching for the three twist-3 distributions, with special attention paid to the role of zero-mode contributions for the chiral-odd cases. At this stage of exploratory studies, mixing with quark-gluon-quark operators was neglected (see Refs. [56, 57] for a discussion of taking this mixing into account). The first lattice calculations (TM fermions, $m_\pi = 260\,\text{MeV}$) appeared in the other two articles, for g_T [53] and h_L [55]. The twist-3 functions were compared to their twist-2 counterparts, the helicity and transversity PDFs. It was observed that the twist-2 and twist-3 distributions are similar in magnitude, with a noticeably steeper descent of the latter at small x. In addition, the authors performed tests of Wandzura-Wilczek (WW) approximations [58], wherein the twist-3 functions are fully determined by their twist-2 counterparts. Both for g_T and h_L, the WW relations hold within statistical errors for $x \lesssim 0.5$, with the magnitude of errors such that violations up to around 40% are possible. This still does not take into account several systematic effects, making these conclusions very preliminary.
- **Generalized parton distributions (GPDs)**. These off-forward generalizations of PDFs are one of the crucial distributions for probing nucleon's 3D structure. Their first lattice determination appeared recently from ETMC – for unpolarized (H, E) and helicity ($\tilde{H}$, $\tilde{E}$)

GPDs [47] as well as transversity (H_T, E_T, $\tilde{H}_T$, $\tilde{E}_T$) GPDs, using again TM fermions at $m_\pi = 260\,\text{MeV}$. Results were obtained in the standard Breit frame at zero and non-zero skewness ξ (when momentum transfer has a longitudinal component). Two/four different projectors were applied to disentangle the GPDs for the chiral-even/odd case. It was found that the GPDs are suppressed with respect to PDFs, as expected, particularly at small x. For the non-zero skewness case, qualitatively different regimes were found, the DGLAP ($|x| > \xi$) and ERBL ($|x| < \xi$) regions, with GPDs further suppressed in the latter. First preliminary results were also reported for a twist-3 GPD ($\tilde{G}_2(x)$; its combination with $\tilde{H}(x)$ being the off-forward generalization of $g_T(x)$) in the same setup [59]. The GPDs calculation is significantly more complex than for PDFs and thus, the amount of work needed for their full mapping in terms of the (x, t, ξ) variables and the full quantification of systematics is very large.

- **Transverse-momentum-dependent PDFs (TMDs)**. While GPDs incorporate information about the transverse position of partons, knowledge about their transverse motion is encoded in another class of distributions, the TMD PDFs. In this case, the new theoretical aspect is the need to handle a new kind of divergences, so-called rapidity divergences, originating from gluon radiation and unregulated by dimensional regularization. A rapidity regulator can be incorporated into a (non-perturbative) soft function, which has a rapidity-independent (intrinsic) part and a part responsible for its evolution in rapidity, defining the Collins-Soper (CS) kernel. The soft function can be obtained via LaMET, as shown in Refs. [60, 61]. On the lattice, one can calculate a pseudoscalar meson form factor, factorizable into the intrinsic soft function and a quasi-TMD wave function. The first explorations of this method appeared recently from LPC (clover fermions, $m_\pi = 333\,\text{MeV}$ with heavier valence quarks) [62] and Beijing+ETMC (TM quarks, $m_\pi = 350\,\text{MeV}$, also with heavier valence pion mass) [63], both with tree-level matching. Feasibility of the approach was, thus, established and some first systematics were addressed (e.g. higher-twist contamination, momentum and pion mass dependence). Also the CS kernel calculation was recently undertaken from a ratio of quasi-TMDs at different rapidities [64] or of first Mellin moments of TMDs [65]. Several lattice extractions in different setups [62, 63, 65–67] lead to a qualitatively consistent picture, with the most recent one [67] concluding, unsurprisingly, an important role of power corrections and NLO perturbative matching. Incorporating these is an important next step for the other groups, together with analyses aiming at checks of various systematic effects to establish quantitatively reliable conclusions.
- **Other hadrons**. The above described work concentrated on partonic distributions of the nucleon. Obviously, there is phenomenologically relevant interest also in other hadrons. Recent work for the pion included advanced studies aimed at estimating the systematics of such computations, using e.g. several lattice spacings to check for discretization effects, effects of using chiral fermions, comparing renormalization procedures (including the new hybrid scheme [27]), assessing the role of 2-loop corrections to matching [68–70]. An attempt to reconstruct the x-dependence from 3 accessible low moments was made in Ref. [71]. It was found that adding the 4th moment from phenomenology has a negligible effect and also that the JAM18 pion PDF [72] reconstructed with 3 moments is consistent with the full one with increased uncertainties. Hence, it is clear that higher moments become important only at a higher level of precision. It is worth to mention that studies of pion distributions have been pursued also with other approaches: auxiliary light quark [73], current-current correlators [74], pseudo-PDFs [75] and auxiliary heavy quark (heavy OPE) [76], the latter recently advanced also theoretically [77]. Other than for the pion, distribution amplitudes for K^* and ϕ mesons were recently calculated, phenomenologically important for B/B_s decays wherein tensions with the Standard Model were reported by

LHCb. On the baryonic side, the PDF of the Δ^+ was calculated [78], with its importance consisting in the possibility of shedding light on the sea quark asymmetry in the nucleon.

5 Summary and prospects

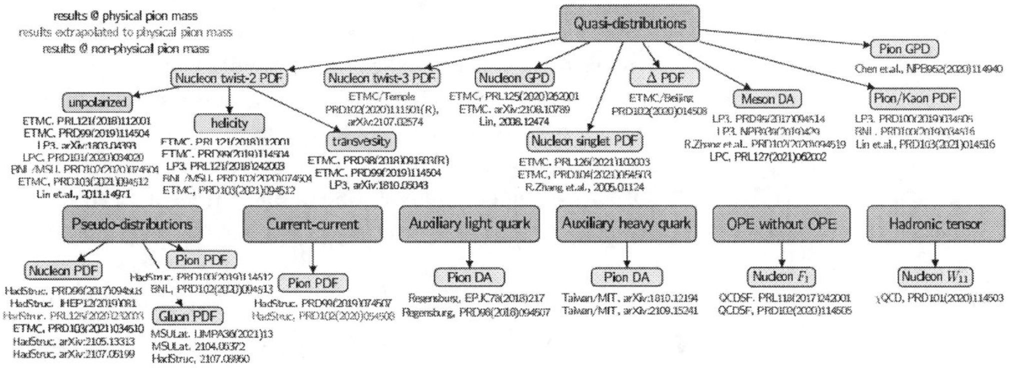

Figure 2: Dynamical progress of lattice PDFs/GPDs, with papers grouped according to the method and topic.

The seminal 2013 paper of Ji [5] sparked intense interest in determining partonic distributions on the lattice, proposing a powerful framework how to access the relevant light-cone correlations with lattice-calculable Euclidean ones and also leading to the revival or conception of alternative methods, all based at some stage on factorization to get to the physical distributions. This dynamical progress is illustrated in Fig. 2, where papers only of the last around 3 years are referenced.

This proceeding has been written with a broader community in mind and below, we summarize the main messages that it tried to convey.

- Extraction of partonic distributions is difficult on the lattice, requiring multiple non-trivial ingredients. It is crucial that control over all these ingredients is possible, but it can be achieved only with dedicated and careful studies of several sources of systematic effects.
- Thus, the road from exploratory to precision studies is long. Most present-day calculations for partonic distributions are at the stage of exploratory or somewhat more advanced studies, but conclusions remain so far qualitative.
- Unpolarized twist-2 PDFs, well-known from global fits to abundant sets of experimental data, are the natural benchmark for the lattice – their reproduction within completely and properly quantified uncertainties will be one of the milestones for the lattice program. At the present stage, there are indications that the size of systematic uncertainties is similar to their magnitude found in other lattice hadron structure studies. Nevertheless, proper quantification of the systematics will require dedicated studies, both on the lattice (additional simulations e.g. at several lattice spacings and volumes) and theoretical (e.g. perturbative corrections at higher order or explicit methods to calculate power corrections).
- For other distributions, lattice can be foreseen to provide constraints that can be incorporated into extractions from experimental data. For several cases where experimental data are basically non-existent, lattice can provide the only available information or one that can be on a similar footing as data that will come from ongoing or planned experiments focused on nucleon's 3D structure (e.g. from the 12 GeV program at JLab or the EIC in a more distant future). The prerequisite for this is, again, the complete quantification of

lattice uncertainties. Under this condition, one can aim at simultaneous fits of different lattice and experimental data, treating the former as "lattice cross sections" in analogy with experimental ones [14, 15].

Overall, this points to a projected very important role for the lattice in understanding hadronic structure. It has been shown that lattice calculations are feasible for a wide range of distributions, with varying level of difficulty, naturally increasing for more complicated distributions due to additional ingredients that need to be taken into account. The remaining task is to make the calculations quantitatively reliable, something that is also feasible with dedicated computational and theoretical efforts. On the lattice side, the task will be augmented by increasing computing power, necessary to handle large boosts for hadrons simulated on large and fine lattices. As the history of lattice computations shows, optimizations in handling the simulations can likely turn out to help as well. This allows us to conclude very optimistic prospects for this rich lattice program and eventually, predict a truly complementary role of LQCD in the quest to understand the details of the structure of the nucleon and other hadrons.

Acknowledgments. I thank the Organizers of the Virtual Tribute to Quark Confinement and the Hadron Spectrum 2021 for the invitation to give this review talk and for the very enjoyable conference. I also thank all my Collaborators in the field of x-dependent partonic distributions. Financial support from the National Science Centre (Poland) is acknowledged, grant SONATA BIS no. 2016/22/E/ST2/00013.

References

[1] M. Constantinou, PoS **LATTICE2014**, 001 (2015), 1411.0078
[2] K.F. Liu, S.J. Dong, Phys. Rev. Lett. **72**, 1790 (1994), hep-ph/9306299
[3] W. Detmold, C.J.D. Lin, Phys. Rev. **D73**, 014501 (2006), hep-lat/0507007
[4] V. Braun, D. Mueller, Eur. Phys. J. **C55**, 349 (2008), 0709.1348
[5] X. Ji, Phys. Rev. Lett. **110**, 262002 (2013), 1305.1539
[6] X. Ji, Sci. China Phys. Mech. Astron. **57**, 1407 (2014), 1404.6680
[7] H.W. Lin, J.W. Chen, S.D. Cohen, X. Ji, Phys. Rev. **D91**, 054510 (2015), 1402.1462
[8] C. Alexandrou, K. Cichy, V. Drach, E. Garcia-Ramos, K. Hadjiyiannakou, K. Jansen, F. Steffens, C. Wiese, Phys. Rev. **D92**, 014502 (2015), 1504.07455
[9] J.W. Chen, S.D. Cohen, X. Ji, H.W. Lin, J.H. Zhang, Nucl. Phys. **B911**, 246 (2016), 1603.06664
[10] C. Alexandrou, K. Cichy, M. Constantinou, K. Hadjiyiannakou, K. Jansen, F. Steffens, C. Wiese, Phys. Rev. **D96**, 014513 (2017), 1610.03689
[11] A.J. Chambers et al., Phys. Rev. Lett. **118**, 242001 (2017), 1703.01153
[12] A.V. Radyushkin, Phys. Rev. **D96**, 034025 (2017), 1705.01488
[13] A. Radyushkin, Int. J. Mod. Phys. A **35**, 2030002 (2020), 1912.04244
[14] Y.Q. Ma, J.W. Qiu, Phys. Rev. **D98**, 074021 (2018), 1404.6860
[15] Y.Q. Ma, J.W. Qiu, Phys. Rev. Lett. **120**, 022003 (2018), 1709.03018
[16] K. Cichy, M. Constantinou, Adv. High Energy Phys. **2019**, 3036904 (2019), 1811.07248
[17] X. Ji, Y.S. Liu, Y. Liu, J.H. Zhang, Y. Zhao, Rev. Mod. Phys. **93**, 035005 (2021), 2004.03543
[18] M. Constantinou, Eur. Phys. J. A **57**, 77 (2021), 2010.02445
[19] K. Cichy, *Progress in x-dependent partonic distributions from lattice QCD*, in *38th International Symposium on Lattice Field Theory* (2021), 2110.07440
[20] J. Green, PoS **LATTICE2018**, 016 (2018), 1812.10574

[21] G.S. Bali, B. Lang, B.U. Musch, A. Schäfer, Phys. Rev. **D93**, 094515 (2016), 1602.05525
[22] M. Constantinou, H. Panagopoulos, Phys. Rev. **D96**, 054506 (2017), 1705.11193
[23] C. Alexandrou, K. Cichy, M. Constantinou, K. Hadjiyiannakou, K. Jansen, H. Panagopoulos, F. Steffens, Nucl. Phys. **B923**, 394 (2017), 1706.00265
[24] I.W. Stewart, Y. Zhao, Phys. Rev. **D97**, 054512 (2018), 1709.04933
[25] T. Izubuchi, X. Ji, L. Jin, I.W. Stewart, Y. Zhao, Phys. Rev. **D98**, 056004 (2018), 1801.03917
[26] K. Orginos, A. Radyushkin, J. Karpie, S. Zafeiropoulos, Phys. Rev. **D96**, 094503 (2017), 1706.05373
[27] X. Ji, Y. Liu, A. Schäfer, W. Wang, Y.B. Yang, J.H. Zhang, Y. Zhao, Nucl. Phys. B **964**, 115311 (2021), 2008.03886
[28] K. Zhang, Y.Y. Li, Y.K. Huo, A. Schäfer, P. Sun, Y.B. Yang (χQCD), Phys. Rev. D **104**, 074501 (2021), 2012.05448
[29] Y.K. Huo et al. (Lattice Parton Collaboration (LPC)), Nucl. Phys. B **969**, 115443 (2021), 2103.02965
[30] J. Karpie, K. Orginos, A. Rothkopf, S. Zafeiropoulos, JHEP **04**, 057 (2019), 1901.05408
[31] V. Braun, K. Chetyrkin, B. Kniehl, JHEP **07**, 161 (2020), 2004.01043
[32] L.B. Chen, W. Wang, R. Zhu, Phys. Rev. D **102**, 011503 (2020), 2005.13757
[33] L.B. Chen, W. Wang, R. Zhu, JHEP **10**, 079 (2020), 2006.10917
[34] L.B. Chen, W. Wang, R. Zhu, Phys. Rev. Lett. **126**, 072002 (2021), 2006.14825
[35] Z.Y. Li, Y.Q. Ma, J.W. Qiu, Phys. Rev. Lett. **126**, 072001 (2021), 2006.12370
[36] M. Bhat, K. Cichy, M. Constantinou, A. Scapellato, Phys. Rev. D **103**, 034510 (2021), 2005.02102
[37] C. Alexandrou, K. Cichy, M. Constantinou, K. Jansen, A. Scapellato, F. Steffens, Phys. Rev. Lett. **121**, 112001 (2018), 1803.02685
[38] C. Alexandrou, K. Cichy, M. Constantinou, K. Hadjiyiannakou, K. Jansen, A. Scapellato, F. Steffens, Phys. Rev. **D99**, 114504 (2019), 1902.00587
[39] B. Joó, J. Karpie, K. Orginos, A.V. Radyushkin, D.G. Richards, S. Zafeiropoulos, Phys. Rev. Lett. **125**, 232003 (2020), 2004.01687
[40] K. Cichy, L. Del Debbio, T. Giani, JHEP **10**, 137 (2019), 1907.06037
[41] L. Del Debbio, T. Giani, J. Karpie, K. Orginos, A. Radyushkin, S. Zafeiropoulos, JHEP **02**, 138 (2021), 2010.03996
[42] J. Bringewatt, N. Sato, W. Melnitchouk, J.W. Qiu, F. Steffens, M. Constantinou, Phys. Rev. D **103**, 016003 (2021), 2010.00548
[43] C. Alexandrou, K. Cichy, M. Constantinou, K. Jansen, A. Scapellato, F. Steffens, Phys. Rev. **D98**, 091503 (2018), 1807.00232
[44] C. Egerer et al. (2021), 2111.01808
[45] C. Alexandrou, K. Cichy, M. Constantinou, J.R. Green, K. Hadjiyiannakou, K. Jansen, F. Manigrasso, A. Scapellato, F. Steffens, Phys. Rev. D **103**, 094512 (2021), 2011.00964
[46] J. Karpie, K. Orginos, A. Radyushkin, S. Zafeiropoulos (2021), 2105.13313
[47] C. Alexandrou, M. Constantinou, K. Hadjiyiannakou, K. Jansen, F. Manigrasso, Phys. Rev. Lett. **126**, 102003 (2021), 2009.13061
[48] C. Alexandrou, M. Constantinou, K. Hadjiyiannakou, K. Jansen, F. Manigrasso, Phys. Rev. D **104**, 054503 (2021), 2106.16065

[49] T. Khan et al. (HadStruc) (2021), 2107.08960
[50] C. Egerer, R.G. Edwards, K. Orginos, D.G. Richards, Phys. Rev. D **103**, 034502 (2021), 2009.10691
[51] C. Egerer, R.G. Edwards, C. Kallidonis, K. Orginos, A.V. Radyushkin, D.G. Richards, E. Romero, S. Zafeiropoulos (2021), 2107.05199
[52] S. Bhattacharya, K. Cichy, M. Constantinou, A. Metz, A. Scapellato, F. Steffens, Phys. Rev. D **102**, 034005 (2020), 2005.10939
[53] S. Bhattacharya, K. Cichy, M. Constantinou, A. Metz, A. Scapellato, F. Steffens, Phys. Rev. D **102**, 111501 (2020), 2004.04130
[54] S. Bhattacharya, K. Cichy, M. Constantinou, A. Metz, A. Scapellato, F. Steffens, Phys. Rev. D **102**, 114025 (2020), 2006.12347
[55] S. Bhattacharya, K. Cichy, M. Constantinou, A. Metz, A. Scapellato, F. Steffens (2021), 2107.02574
[56] V.M. Braun, Y. Ji, A. Vladimirov, JHEP **05**, 086 (2021), 2103.12105
[57] V.M. Braun, Y. Ji, A. Vladimirov (2021), 2108.03065
[58] S. Wandzura, F. Wilczek, Phys. Lett. **72B**, 195 (1977)
[59] S. Bhattacharya, K. Cichy, M. Constantinou, A. Metz, A. Scapellato, F. Steffens, *Twist-3 partonic distributions from lattice QCD*, in *28th International Workshop on Deep Inelastic Scattering and Related Subjects* (2021), 2107.12818
[60] X. Ji, Y. Liu, Y.S. Liu, Nucl. Phys. B **955**, 115054 (2020), 1910.11415
[61] X. Ji, Y. Liu, Y.S. Liu, Phys. Lett. B **811**, 135946 (2020), 1911.03840
[62] Q.A. Zhang et al. (Lattice Parton), Phys. Rev. Lett. **125**, 192001 (2020), 2005.14572
[63] Y. Li et al. (2021), 2106.13027
[64] M.A. Ebert, I.W. Stewart, Y. Zhao, Phys. Rev. **D99**, 034505 (2019), 1811.00026
[65] M. Schlemmer, A. Vladimirov, C. Zimmermann, M. Engelhardt, A. Schäfer, JHEP **08**, 004 (2021), 2103.16991
[66] P. Shanahan, M. Wagman, Y. Zhao, Phys. Rev. D **102**, 014511 (2020), 2003.06063
[67] P. Shanahan, M. Wagman, Y. Zhao (2021), 2107.11930
[68] X. Gao, L. Jin, C. Kallidonis, N. Karthik, S. Mukherjee, P. Petreczky, C. Shugert, S. Syritsyn, Y. Zhao, Phys. Rev. D **102**, 094513 (2020), 2007.06590
[69] X. Gao et al., PoS **LATTICE2021**, 598 (2021)
[70] Y. Zhao et al., PoS **LATTICE2021**, 614 (2021)
[71] C. Alexandrou, S. Bacchio, I. Cloët, M. Constantinou, K. Hadjiyiannakou, G. Koutsou, C. Lauer (ETM), Phys. Rev. D **104**, 054504 (2021), 2104.02247
[72] P. Barry, N. Sato, W. Melnitchouk, C.R. Ji, Phys. Rev. Lett. **121**, 152001 (2018), 1804.01965
[73] G.S. Bali et al., Phys. Rev. **D98**, 094507 (2018), 1807.06671
[74] R.S. Sufian, J. Karpie, C. Egerer, K. Orginos, J.W. Qiu, D.G. Richards, Phys. Rev. **D99**, 074507 (2019), 1901.03921
[75] B. Joó, J. Karpie, K. Orginos, A.V. Radyushkin, D.G. Richards, R.S. Sufian, S. Zafeiropoulos, Phys. Rev. **D100**, 114512 (2019), 1909.08517
[76] W. Detmold, A. Grebe, I. Kanamori, C.J.D. Lin, S. Mondal, R. Perry, Y. Zhao (2021), 2109.15241
[77] W. Detmold, A.V. Grebe, I. Kanamori, C.J.D. Lin, R.J. Perry, Y. Zhao (HOPE), Phys. Rev. D **104**, 074511 (2021), 2103.09529
[78] Y. Chai et al., Phys. Rev. D **102**, 014508 (2020), 2002.12044

Nonequilibrium evolution of quarkonium in medium

Yukinao Akamatsu[1,*] and *Takahiro* Miura[1]

[1]Department of Physics, Osaka University, Toyonaka, Osaka 560-0043, Japan

Abstract. We review recent progress in open quantum system approach to the description of quarkonium in the quark-gluon plasma. A particular emphasis is put on the Lindblad equations for quarkonium and its numerical simulations.

1 Introduction

Relativistic heavy-ion collision experiment provides us with a unique opportunity to study the novel matter composed of quarks and gluons, which are permanently confined inside hadrons at the energy scales of our daily experiences. This novel matter, which is called Quark-Gluon Plasma (QGP), exhibits hydrodynamic collective behavior during the expansion process. The mechanism of early thermalization and the applicability of hydrodynamics to smaller size systems remain fundamental questions in the heavy-ion physics.

In the heavy-ion collisions, a quarkonium acts as an impurity for the QGP and can probe its local color-electric fluctuations[1] [1]. Single heavy quark can also probe the same property of the QGP, but there is a big difference: a quarkonium is made of heavy quark-antiquark pair interacting with each other through attractive force if they are in the color singlet. This attractive force binds the heavy quark pair so that its color is hidden inside the localized wave function. In this sense, a quarkonium is a sensitive probe of QGP at higher temperatures, typically determined by its bound state size. During the last decade, there have been a lot of developments in understanding the non-equilibrium evolution of quarkonium in the QGP as an open quantum system [2–4], which we summarize in this article.

Even though the complexity of heavy-ion collisions is enormous, the first step to describe a quarkonium as an open system is to derive a master equation in the simplest setup, namely the environment is a static and homogeneous QGP and the heavy quarks are the non-relativistic quantum mechanical particles. This simplification implicitly assumes that the (pseudo)critical temperature T_c is much lower than the heavy quark mass M so that one can find a temperature $T_c < T \ll M$. The effects which are yet to be included are as follows:

- Interaction between initially uncorrelated heavy quark pairs.
  – This could be justified for bottomonium because it is rare that two bottom quark pairs are produced simultaneously in a heavy-ion collision. For charmonium, its justification is subtle at RHIC energy scales and probably not given at the LHC. For this reason, our description better applies to bottomonium.

*e-mail: akamatsu@kern.phys.sci.osaka-u.ac.jp

[1]In this article, the word "quarkonium" is used to mean a quantum mechanical state of heavy quark and antiquark, which is not necessarily a bound state or an eigenstate.

- The effects of non-static and inhomogeneous QGP evolution.
  – The master equation contains parameters determined by the temperature, which is assumed to be a constant. Choosing the temperature by the local temperature at (the quantum expectation value of) the center of mass for quarkonium, the above effects can be partially taken into account.
- Heavy quark pair creation in the QGP.
  – This is justified because such process is suppressed by a factor $e^{-2M/T} \ll 1$ with $2M \gg T$.
- Heavy quark pair annihilation in the QGP.
  – This process is suppressed by $1/M^2$ and can be added perturbatively if needed.

The second step is to numerically solve the master equation. Practically, the center-of-mass motion of quarkonium is assumed to follow a free-streaming classical trajectory determined by some model for its initial phase space distribution, and the master equation is solved only for the relative motion. As is the case for the matter evolution, the evolution of quarkonium during the medium thermalization process is not precisely known at present. Therefore, simulation of the master equation starts at the medium thermalization time $\tau_{\text{therm.}}$, which can be estimated to be $\sim$ (0.4-1) fm by the initial time of the hydrodynamic simulations at RHIC [5] and the LHC [6]. To perform the simulation, we need to determine the initial wave function and the freezeout description of quarkonium.

If we ignore the complication from the influences of thermalizing quarks and gluons and color gauge fields, the formation time of quarkonium bound state is estimated by the inverse of its binding energy $\tau_{\text{form.}} \sim 1/E_{\text{bind.}}$, which is $\sim$ (1-2) fm for the bottomonium ground state in the Coulomb potential. Then, the initial wave function of quarkonium at $\tau_{\text{therm.}} \lesssim \tau_{\text{form.}}$ cannot be too different from that at the production, which is approximated by a delta function, up to possible corrections from thermalizing process involving saturation scales. Note that if binding energy is measured from the $B\bar{B}$ threshold, we get $\tau_{\text{form.}} \sim 0.2$ fm and it is more appropriate to evaluate $\tau_{\text{therm.}} \sim \tau_{\text{form.}}$. In any case, the initial wave function of quarkonium at $\tau_{\text{therm.}}$ has not yet been fully investigated and it is customary to choose a singlet or octet Gaussian wave packet (as a proxy for the delta function) or a singlet bound state wave function. Similarly to the initial wave function, the freezeout description of quarkonium at later stage of the heavy-ion collisions is not very much known. As an ad-hoc description, a freezeout temperature $T_f \sim T_c$ is chosen at which we stop the simulation and project the density matrix onto a vacuum state in the singlet.

With the overall picture given as above, we will now review two essential ingredients in the open system descriptions for quarkonium — the master equation (Sec. 2) and its numerical simulation (Sec. 3).

2 Lindblad equations for quarkonium

2.1 Basics of open quantum systems

When a small physical system of our interest is surrounded by environmental degrees of freedom, we can obtain an effective description by integrating out the latter. To be more specific, the total system is given on a Hilbert space $\mathcal{H}_{\text{tot}} = \mathcal{H}_S \otimes \mathcal{H}_E$ with a Hamiltonian $H_{\text{tot}} = H_S \otimes I_E + I_S \otimes H_E + H_I$, and the information of the system quantum states is contained in the reduced density matrix $\rho_S \equiv \text{Tr}_E \rho_{\text{tot}}$. The master equation is an evolution equation of ρ_S. It was found about 45 years ago that if the master equation is Markovian and describes a trace-preserving and completely positive evolution of ρ_S, it must be written in the Lindblad form [7, 8]:

$$\frac{d}{dt}\rho_S(t) = \mathcal{L}(\rho_S) = -i\,[H'_S, \rho_S] + \sum_k \left(L_k \rho_S L_k^\dagger - \frac{1}{2} L_k^\dagger L_k \rho_S - \frac{1}{2} \rho_S L_k^\dagger L_k \right) \tag{1}$$

$$= -i\left(H_{\text{eff}}\rho_S - \rho_S H_{\text{eff}}^\dagger\right) + \sum_k L_k \rho_S L_k^\dagger, \quad H_{\text{eff}} = H'_S - \frac{i}{2}\sum_k L_k^\dagger L_k. \tag{2}$$

Note that the Hamiltonian H'_S in the Lindblad equation is different from the system Hamiltonian H_S due to the coupling with environment. One can also rewrite the Lindblad equation as a sum of non-unitary evolution governed by $H_{\text{eff}} \neq H_{\text{eff}}^\dagger$ and the transition or scattering part $\sum_k L_k \rho_S L_k^\dagger$. The in-medium self-energy is given by $H_{\text{eff}} - H_S$.

When the system-environment coupling is weak, we can obtain the Lindblad equation based on perturbative expansions. Following the standard procedure known as Born-Markov approximation [9], we obtain an evolution equation with similar structure to the Lindblad equation:

$$\frac{d}{dt}\rho_S(t) = \int_0^\infty ds\, \langle V_E(s) V_E(0)\rangle_T \begin{bmatrix} V_S(t-s)\rho_S(t) V_S(t) \\ -\, V_S(t) V_S(t-s)\rho_S(t) \end{bmatrix} + h.c. + O(V^3), \tag{3}$$

where $\langle V_E(s)V_E(0)\rangle_T$ is an environment correlation function in thermal equilibrium. Here the interaction Hamiltonian is $H_I = V_S \otimes V_E$ and the interaction picture is adopted for the time evolution. This form already looks similar to the Lindblad equation, but s-integration must be performed by approximating $V_S(t-s)$ terms. In the regime of quantum Brownian motion [10], it is assumed that the system time scale τ_S is long compared to the environmental correlation time $\tau_E (\ll \tau_S)$, in which $\langle V_E(s)V_E(0)\rangle$ changes substantially[2]. Then, one can approximate as $V_S(t-s) = V_S(t) - is[H_S, V_S(t)] + \cdots$ and obtain after integrating s the Lindblad operator

$$L \propto V_S + \frac{i}{4T}\dot{V}_S + O(V_S^2, \partial_t^2), \tag{4}$$

using the Kubo-Martin-Schwinger (KMS) relations for $\int_{-\infty}^{\infty} e^{i\omega s}\langle V_E(s)V_E(0)\rangle_T$. Note that the derivative expansion in L yields not only the first but also the second order expansion in the master equation, but the precision is ensured only at the first order.

There is a technical remark on (4) here. It must be slightly confusing, but this is the leading order result in the limit $\tau_E/\tau_S \ll 1$ [2]. First, note that the ratio of the first two terms in (4) is $\dot{V}_S/TV_S \sim 1/T\tau_S \ll 1$, but this ratio follows from the KMS relation and does not reflect the smallness of τ_E/τ_S. Furthermore, the second term $i\dot{V}_S/4T$ is an anti-Hermitian correction to the first Hermitian term V_S and can change the dynamical behavior qualitatively because the effective coupling to the environment is now non-Hermitian[3]. Therefore, it is not a simple matter to compare them and estimate their relative importance. So, having both of them can be regarded as a requirement from the fluctuation-dissipation relation of the environment. The true statement about (4) is that this is the Lindblad operator in the leading order in the weak coupling and the leading evaluation in $\tau_E/\tau_S \ll 1$. There are corrections of order τ_E/τ_S to both Hermitian and anti-Hermitian parts of L, which may come from the structure of $\int_0^\infty e^{i\omega s}\langle V_E(s)V_E(0)\rangle_T$ at finite ω and from higher order derivative expansion.

[2]To derive the pre-master equation in the Born-Markov approximation, one needs to assume that τ_E is much smaller than the dynamical time scale $\tau_R (\gg \tau_E)$ of $\rho_S(t)$. The integration range of s can then be extended from $\int_0^t ds$ (if the initial time is 0) to $\int_0^\infty ds$ and the evolution is made Markovian, i.e. dependence on the initial time is lost.

[3]In the (Lindbladized) Caldeira-Leggett model [11–14], $L = x + \frac{ip}{4mT}$ and $L\rho L^\dagger \sim x\rho x$ part describes decoherence and the cross term $L\rho L^\dagger \sim -i(x\rho p - p\rho x)$ describes dissipation.

In the application of open quantum system to quarkonium in the QGP, the validity of the quantum Brownian regime is checked as follows. First, the system time scale τ_S is the intrinsic time scale of quarkonium and is typically given by the inverse of the bound state energy in the Coulomb potential $V(r) = -C_F\alpha_s/r$. For the bottomonium ground state, it is

$$\tau_S \sim \frac{4}{M(C_F\alpha_s)^2} \sim \text{(1-2) fm} \tag{5}$$

for $C_F\alpha_s \sim$ 0.3-0.4 and $M \sim 4.8$ GeV, which we also called $\tau_{\text{form.}}$ in the previous section. Second, the correlation time of the QGP τ_E can be estimated by the inverse of the lowest nonzero Matsubara frequency

$$\tau_E \sim \frac{1}{2\pi T} \sim \frac{0.16 \text{ fm}}{[T/(0.2 \text{ GeV})]} \tag{6}$$

for a strongly coupled QGP with $T/T_c \sim O(1)$. Then, the hierarchy of the scales $\tau_S \gg \tau_E$ is more or less satisfied. Note however that the hierarchy $\tau_S \gg \tau_E$ for bottomonium becomes marginal without the factor 2π in τ_E and the applicability of quantum Brownian method, i.e. the derivative expansion, is subtle. Instead, there is a different approach to derive and solve a classical kinetic theory describing the singlet quarkonia bound states as molecules by interpreting $\tau_S \lesssim \tau_E$ [15–18].

2.2 Quarkonium described by non-relativistic QCD (NRQCD)

When the heavy quark mass is larger than any other scales of QCD and the environment such as Λ_{QCD} and the temperature T, non-relativistic description is available in the heavy quark sector. One of such non-relativistic effective theories for quarkonium is Non-Relativistic QCD (NRQCD). In NRQCD, the heavy quark degrees of freedom are the Pauli spinors for heavy quark ψ and antiquark χ and the effective Lagrangian is written as

$$\mathcal{L}_{\text{NRQCD}} = \mathcal{L}_{\text{QCD}} + \psi^\dagger \left[iD_t + \frac{\vec{D}^2}{2M} \right] \psi + \chi^\dagger \left[iD_t - \frac{\vec{D}^2}{2M} \right] \chi + \cdots, \tag{7}$$

at the leading order of the velocity expansion. For the details of power counting and matching to QCD, see [19]. After neglecting the magnetic interaction, the corresponding Hamiltonian is

$$H_{\text{tot}} = H_{\text{QCD}} + \frac{p_Q^2}{2M} + \frac{p_{Q_c}^2}{2M} + gA_0^a(\vec{x}_Q)t_Q^a - gA_0^a(\vec{x}_{Q_c})t_{Q_c}^{a*} \tag{8}$$

$$= I_S \otimes H_{\text{QCD}} + \left(\frac{p_Q^2}{2M} + \frac{p_{Q_c}^2}{2M} \right) \otimes I_E + \int \frac{d^3k}{(2\pi)^3} \left[e^{i\vec{k}\cdot\vec{x}_Q} t_Q^a - e^{i\vec{k}\cdot\vec{x}_{Q_c}} t_{Q_c}^{a*} \right] \otimes g\tilde{A}_0^a(\vec{k}), \tag{9}$$

where the second line makes it explicit that the heavy quark part is the system and the light sector constitutes the environment. Substituting the formula for the Lindblad operators, we get [20]

$$\frac{d}{dt}\rho_S(t) = -i\left[H'_S, \rho_S\right] + \int \frac{d^3k}{(2\pi)^3} L^a(\vec{k})\rho_S L^{a\dagger}(\vec{k}) - \frac{1}{2}\left\{L^{a\dagger}(\vec{k})L^a(\vec{k}), \rho_S\right\} + O(g^2), \tag{10}$$

$$L^a(\vec{k}) = \sqrt{\tilde{D}(k)} \left[e^{i\vec{k}\cdot\vec{x}_Q} t_Q^a - e^{i\vec{k}\cdot\vec{x}_{Q_c}} t_{Q_c}^{a*} + O(\partial_t) \right], \tag{11}$$

$$H'_S = \frac{p_Q^2}{2M} + \frac{p_{Q_c}^2}{2M} + V(r)t_Q^a t_{Q_c}^{a*} + O(\partial_t), \tag{12}$$

where the real functions $V(r)$ and $D(r) \equiv \int d^3r e^{i\vec{k}\cdot\vec{r}}\tilde{D}(k)$ are defined by the gluon two-point function

$$V(r) + iD(r) = \frac{2i}{N_c^2 - 1}\int_0^\infty dt \langle gA_0^a(t, \vec{r}) gA_0^a(0, \vec{0})\rangle_T. \quad (13)$$

At the length scale of $r \sim 1/gT$, the two-point function receives medium effects from Debye screening mass and Landau damping, which must be resummed to yield the full leading order result:

$$V(r) = -\frac{g^2}{4\pi}e^{-m_D r}, \quad D(r) = g^2 T \int \frac{d^3k}{(2\pi)^3}\frac{\pi m_D^2 e^{i\vec{k}\cdot\vec{r}}}{k(k^2 + m_D^2)^2}. \quad (14)$$

The in-medium self-energy can be obtained from the Lindblad equation as

$$H_S' - \frac{i}{2}\int \frac{d^3k}{(2\pi)^3} L^{a\dagger}(\vec{k}) L^a(\vec{k}) - H_S, \quad (15)$$

whose singlet projection at the leading order in the derivative expansion is the so-called complex potential [21–23]

$$V_{\text{complex}}(r) = C_F[V(r) + i(D(r) - D(0)]. \quad (16)$$

For the further details, such as the next-to-leading terms in the derivative expansion and the Lindblad operators in the projected color singlet and octet basis, see [2, 20]. Also, the master equation at the first order derivative expansion in the strict sense (see Sec. 2.1) is obtained in [24–26].

The derivation here relies on the smallness of g and the condition for the quantum Brownian motion reads

$$\left.\begin{aligned} \tau_E &\sim \frac{1}{gT} && \text{(color electric scale)} \\ \tau_S &\sim \frac{4}{M(C_F\alpha_s)^2} && \text{(Coulomb ground state)} \end{aligned}\right\} \quad \tau_E \ll \tau_S \Leftrightarrow \frac{M}{T} \ll \frac{4g}{(C_F\alpha_s)^2}. \quad (17)$$

The master equation can describe phenomena whose dynamical time scale satisfies the Markovian condition $\tau_R \gg \tau_S$ (see the footnote 2). For example, in order to describe heavy quark kinetic equilibration,

$$\tau_R \sim \frac{M}{g^4 T^2} \quad \text{(equilibration)} \qquad \tau_E \ll \tau_R \Leftrightarrow g^3 \ll \frac{M}{T}, \quad (18)$$

and to describe the decoherence of wave function with size $r \sim 1/m_D$ is

$$\tau_R \sim \frac{1}{D(1/m_D)} \sim \frac{1}{g^2 T} \quad \text{(decoherence at } r \sim 1/m_D) \qquad \tau_E \ll \tau_R \Leftrightarrow g \ll 1. \quad (19)$$

At modestly high temperature $T/T_c \sim O(1)$, QGP is expected to be strongly coupled. In this case, the Lindblad equation derived here is not applicable. Instead, we model the quantum Brownian motion of quarkonia using the same Lindblad equation, but now with model functions

$$C_F V(r) = -\frac{0.3}{r}e^{-2Tr}, \quad C_F D(r) = \frac{T}{\pi}e^{-(Tr)^2}, \quad (20)$$

which is used later in the simulation discussed in Sec. 3.

2.3 Quarkonium described by potential non-relativistic QCD (pNRQCD)

When the heavy quark pair is close to each other, there is another effective description, which describes the pair as a color dipole. The effective field theory is called potential Non-Relativistic QCD (pNRQCD), whose quarkonium degrees of freedom are the singlet $S(t, \vec{R}, \vec{r})$ and the octet $O^a(t, \vec{R}, \vec{r})$ fields. The fields depend on the center of mass ($\vec{R}$) and the relative ($\vec{r}$) coordinates of the quarkonium. The effective Lagrangian at the leading order in $1/M$ and the next-to-leading order in r is

$$\mathcal{L}_{\mathrm{pNRQCD}} = \mathcal{L}_{\mathrm{QCD}} + \int d^3r \mathrm{Tr}\left[\mathrm{S}^\dagger (i\partial_t - V_s(r) + \cdots)\mathrm{S} + \mathrm{O}^\dagger (iD_t - V_o(r) + \cdots)\mathrm{O}\right] \tag{21}$$

$$+ V_A(r)\mathrm{Tr}\left[\mathrm{O}^\dagger \vec{r}\cdot g\vec{E}\mathrm{S} + \mathrm{S}^\dagger \vec{r}\cdot g\vec{E}\mathrm{O}\right] + \frac{V_B(r)}{2}\mathrm{Tr}\left[\mathrm{O}^\dagger \vec{r}\cdot g\vec{E}\mathrm{O} + \mathrm{O}^\dagger \mathrm{O}\vec{r}\cdot g\vec{E}\right] + \cdots,$$

$$\mathrm{S}(t,\vec{R},\vec{r}) \equiv \frac{S(t,\vec{R},\vec{r})}{\sqrt{N_c}}\mathbf{1}, \quad \mathrm{O}(t,\vec{R},\vec{r}) \equiv \sqrt{2}O^a(t,\vec{R},\vec{r})t_F^a, \tag{22}$$

where $V_s(r) = -C_F\alpha_s/r$, $V_o(r) = \alpha_s/2N_c r$, $V_A(r) = V_B(r) = 1$ are the Wilson coefficients at the leading (non-vanishing) order of α_s. Note that the fields in $\mathcal{L}_{\mathrm{QCD}}$ depend on $\vec{R}$. Again, for the details of power counting and matching to NRQCD, see [19]. By redefining the octet field $\mathrm{O}(t, \vec{R}, \vec{r})$ and the color electric field $\vec{E}(t, \vec{R})$ so that the covariant derivative D_t becomes ∂_t while keeping the dipole interaction terms unchanged, the theory becomes simpler and the corresponding Hamiltonian is

$$H = H_{\mathrm{QCD}} + \frac{p^2}{M} - \frac{C_F\alpha_s}{r}|s\rangle\langle s| + \frac{\alpha_s}{2N_c r}|a\rangle\langle a|$$

$$-\left(\sqrt{\frac{1}{2N_c}}(|a\rangle\langle s| + |s\rangle\langle a|) + \frac{1}{2}d^{abc}|b\rangle\langle c|\right)\vec{r}\cdot g\vec{E}^a(\vec{R}), \tag{23}$$

where $|s\rangle$ and $|a\rangle$ ($a = 1, 2, \cdots, N_c^2 - 1$) are color singlet and octet states for a quarkonium. Here, the kinetic term p^2/M for the relative motion is added, which balances with the singlet potential in the bound states. The Lindblad equation in this case reads [27, 28]

$$\frac{d}{dt}\rho_S(t) = -i[H_S', \rho_S] + L_{ai}\rho_S L_{ai}^\dagger - \frac{1}{2}\left\{L_{ai}^\dagger L_{ai}, \rho_S\right\} + O(r^3), \tag{24}$$

$$L_{ai} = \sqrt{\frac{\kappa}{C_F}}\left(\sqrt{\frac{1}{2N_c}}(|a\rangle\langle s| + |s\rangle\langle a|) + \frac{1}{2}d^{abc}|b\rangle\langle c|\right)[r_i + O(\partial_t)], \tag{25}$$

$$H_S' = \frac{p^2}{M} - \frac{C_F\alpha_s}{r}|s\rangle\langle s| + \frac{\alpha_s}{2N_c r}|a\rangle\langle a| + \frac{\gamma}{2C_F}r^2\left(C_F|s\rangle\langle s| + \frac{N_c^2-2}{4N_c}|a\rangle\langle a|\right) + O(\partial_t), \tag{26}$$

where the real coefficients γ and κ are defined by the two-point function of the color electric field:

$$\frac{-\gamma + i\kappa}{C_F} = \frac{2i}{3(N_c^2-1)}\int_0^\infty dt\langle gE_i^a(t,\vec{R})gE_i^a(0,\vec{R})\rangle_T. \tag{27}$$

Note that the two-point function is actually gauge invariant when expressed by the original color electric field. In this derivation, the small-r expansion for the Lindblad equation does not rely on the smallness of the coupling constant g so that the coefficients γ and κ are defined non-perturbatively. The in-medium self-energy for the singlet is $\frac{1}{2}(\gamma - i\kappa)r^2$ at the leading order in the derivative expansion. The coefficient γ causes in-medium mass shift of quarkonium and κ gives in-medium width to the quarkonium spectrum as well as the rate of heavy quark

Table 1. Numerical simulations of Lindblad equation for quarkonium. The entry for "Dissipation" shows whether or not one includes $i\dot{V}_S/4T$ term in Eq. (4).

Description	Setup	Dissipation	Numerical Method
NRQCD	1D, U(1)	no	Stochastic Potential [32, 33]
	3D, U(1)	no	Stochastic Potential [34]
	1D, SU(3)	no	Stochastic Potential [35, 36]
	1D, U(1)	yes	Quantum State Diffusion [37, 38]
	1D, SU(3)	yes	Quantum State Diffusion (this work)
	1D, U(1)	yes	Direct evolution [39]
pNRQCD	1_+D, SU(3)	no	Direct evolution for S and P waves [27, 28]
	3D, SU(3)	no	Quantum Jump [40, 41]

momentum diffusion. Currently available lattice simulations evaluate $\gamma \sim -(0.7\text{-}3.8)T^3$ and $\kappa \sim (0.24\text{-}4.2)T^3$ [29]. For the further details, such as the next-to-leading terms in the derivative expansion and the Lindblad operators in the projected color singlet and octet basis, see [2, 27, 28].

3 Simulation of the Lindblad equations for quarkonium

Having derived the Lindblad equations, the next step is to solve them. There is a challenge here: the density matrix has doubled dimensions of the wave function. So far, several groups have performed numerical simulations of the Lindblad equation for quarkonium in the QGP. Most of them use a class of simulation methods called stochastic unravelling, in which wave functions are evolved stochastically in such a way that their mixed state density matrix is equivalent to the solution of the Lindblad equation. To be explicit, with a solution of the stochastic Schrödinger equation $\psi_i(\vec{x}, t)$ labeled by a sample number $i = 1, 2, 3, \cdots$, the density matrix is given by

$$\rho(\vec{x}, \vec{y}, t) = \lim_{N\to\infty} \frac{1}{N} \sum_{i=1}^{N} \psi_i(\vec{x}, t)\psi_i^*(\vec{y}, t). \tag{28}$$

In this way, one can avoid the huge numerical memory and cost needed to store and evolve the density matrix. Instead, one needs to produce a large ensemble of wave functions to reduce the statistical errors. Among the stochastic unravelling methods, two methods are widely used: Quantum State Diffusion (QSD) [30] and Quantum Jump [31]. The situation of numerical simulation is summarized in the Table 1.

Hereafter, we show the first result of the simulation of Lindblad equation from NRQCD description (with color SU(3) and including the quantum dissipation), which is solved by the QSD method. In the QSD method, in order to simulate (1), one evolves the wave function stochastically by

$$|\psi(t+dt)\rangle = |\psi(t)\rangle - iH'_S|\psi(t)\rangle dt + \sum_k \left(\langle L_k^\dagger \rangle_\psi L_k - \frac{1}{2} L_k^\dagger L_k - \frac{1}{2} \langle L_k^\dagger \rangle_\psi \langle L_k \rangle_\psi \right) |\psi(t)\rangle dt$$
$$+ \frac{1}{\sqrt{2}} \sum_k \left(L_k - \langle L_k \rangle_\psi \right) |\psi(t)\rangle d\xi_k \tag{29}$$

with complex white noises $d\xi_k$ satisfying the following statistical properties

$$\overline{d\xi_k} = \overline{\mathrm{Re}(d\xi_k)\mathrm{Im}(d\xi_\ell)} = 0, \quad \overline{\mathrm{Re}(d\xi_k)\mathrm{Re}(d\xi_\ell)} = \overline{\mathrm{Im}(d\xi_k)\mathrm{Im}(d\xi_\ell)} = \delta_{k\ell} dt. \tag{30}$$

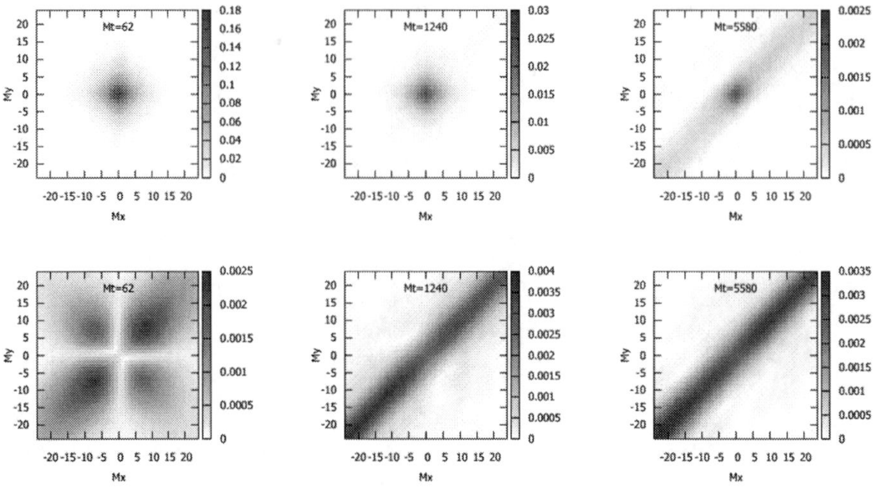

Figure 1. Time evolution of the reduced density matrix in the singlet and octet sectors $|\rho_{s,o}(x, y, t)|^2/M^2$. The figure only shows central domain $-24 \le Mx, My \le 24$. The top figures are the single density matrices and the bottom figures are the octet density matrices.

The quantum Brownian motion of quarkonium is modeled by the NRQCD Lindblad equation (10) with $V(r)$ and $D(r)$ modeled by (20). The simulation is performed on a one-dimensional lattice with $\Delta x = 1/M$, $\Delta t = 0.1M(\Delta x)^2$, and $N_x = 254$ and the QGP temperature is $T = 0.1M$. The origin of the potential $C_F V(r)$ is singular in one dimension and is regulated by

$$C_F V(r) = -\frac{0.3}{\sqrt{r^2 + r_c^2}} e^{-2Tr}, \quad r_c = \frac{1}{M}. \tag{31}$$

As a reference time scale, let us quote the relaxation time τ_{relax} of single heavy quark when the influence of QGP is modeled by $D(r)$:

$$\frac{dp}{dt} = -\eta p, \quad \eta = -\frac{C_F}{MT} \frac{d^2 D(x)}{dx^2}\Big|_0 = \frac{T^2}{\pi M} = \frac{M}{100\pi}, \quad \tau_{\text{relax}} = \frac{100\pi}{M}. \tag{32}$$

3.1 Density matrix

In Fig. 1, time evolution of the density matrix is shown. The top figures are the singlet density matrices and the bottom figures are the octet density matrices. The initial wave function is the singlet ground state. The evolution of the density matrix proceeds by the three steps.

(1) The singlet ground state is excited to octet as a color dipole (left).

(2) The octet density matrix is diagonalized (center).

(3) De-excitation from the octet to the singlet is observed and the system finally reaches a steady state (right).

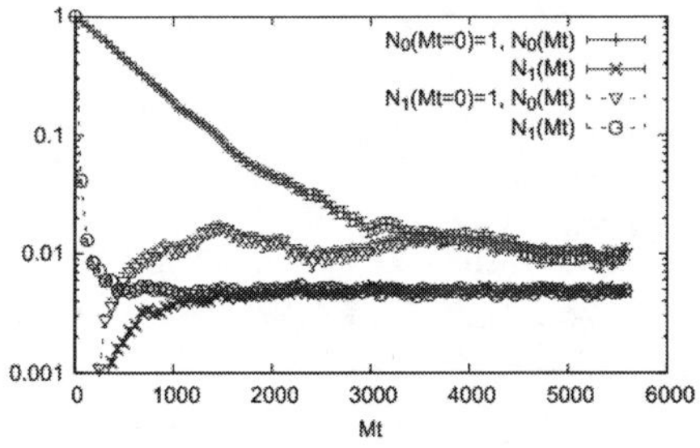

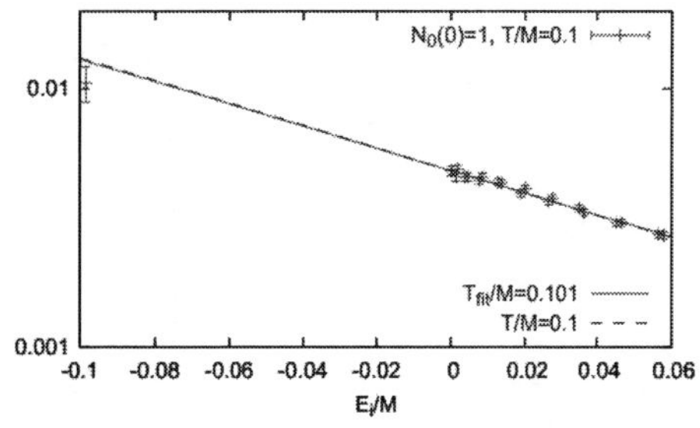

Figure 2. Occupation probabilities for the ground state and the first excited state with different initial conditions (left). Occupation of the lowest 20 levels (singlet) are compared with the Boltzmann distribution (right).

3.2 Equilibration and the effect of dissipation

The equilibration is not necessarily guaranteed in the Lindblad equation, at least it is not proven analytically. Since the Lindblad equation is derived when QGP is in the equilibrium, it is natural to expect that the quarkonium system equilibrates. In order to confirm the equilibration, let us show the occupation probabilities of the eigenstates. In Fig. 2 (left), the occupation probabilities for the ground and the first excited states (both singlet) are plotted as a function of time. Two initial conditions are employed, namely the ground state and the first excited state. At around the last time step $Mt = 5580$, the occupation does not depend on the initial conditions and reaches more or less a steady state. In Fig. 2 (right), the occupation of the lowest 20 levels (singlet) is shown and is nicely fitted by the Boltzmann distribution with a temperature close to the environment $T_{\rm fit} = 0.101M \simeq T$.

Finally, let us deliberately turn off the $i\dot{V}_S/4T$ term in Eq. (4), namely simulate the Lindblad equation (10) derived from NRQCD without $O(\partial_t)$ corrections. Figure 3 compares the results with and without this term when the initial condition is the singlet ground state. As is mentioned in Sec. 2.1, the term $i\dot{V}_S/4T$ can qualitatively change the dynamical property of the Lindblad equation. From the Fig. 3, it is found that the long-time behavior does not show equilibration without this term and the deviation starts even at the very early time. The lack of equilibration is simply because the requirement from the fluctuation-dissipation relation is not met. The deviation at early time can be explained as follows. For small wave function size, the decoherence, described by the first term in $L \propto V_S + i\dot{V}_S/4T$, does not proceed effectively. In such a case, the relative importance of the second term is enhanced [37, 38]. In the same way, the qualitative difference of the late behavior can also be understood. Since the typical coherence length gets shorter and shorter by decoherence, at some stage the importance of the first and the second terms in L balances and the system equilibrates. Without the second term, the decoherence proceeds forever and the system never reaches proper equilibrium.

4 Outlook

- Applicability of the dipole approximation needs to be checked [42]. Since the numerical cost of simulating the Lindblad equation from pNRQCD is much lower and since the Lindblad equation is beyond the weak coupling, it would be very fortunate if the dipole approximation is applicable, at least for finite but long enough time scale for heavy-ion phenomenology.
- At lower temperature $T \sim T_c$, the applicability of the quantum Brownian regime to quarkonium dynamics is subtle. It would be desirable to consider matching to another effec-

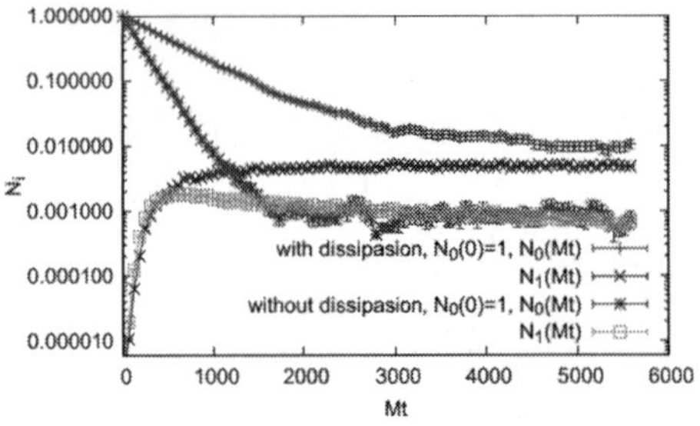

Figure 3. Occupation probabilities for the ground state and the first excited state with and without the dissipative term, i.e. $i\dot{V}_S/4T$ term in Eq. (4).

tive description by classical kinetic theory [15–18], which describes the quarkonium as a molecule with internal bound state levels.

- Initial condition of quarkonium can be elaborated by e.g. simulating under the presence of thermalizing quarks and gluons and gauge fields.

Acknowledgments

The work of Y.A. is supported by JSPS KAKENHI Grant Number JP18K13538.

References

[1] A. Rothkopf, Phys. Rept. **858**, 1 (2020), 1912.02253
[2] Y. Akamatsu (2020), 2009.10559
[3] X. Yao, Int. J. Mod. Phys. A **36**, 2130010 (2021), 2102.01736
[4] R. Sharma, Eur. Phys. J. ST **230**, 697 (2021), 2101.04268
[5] T. Hirano, K. Tsuda, Phys. Rev. C **66**, 054905 (2002), nucl-th/0205043
[6] H. Song, S.A. Bass, U. Heinz, T. Hirano, C. Shen, Phys. Rev. Lett. **106**, 192301 (2011), [Erratum: Phys.Rev.Lett. 109, 139904 (2012)], 1011.2783
[7] V. Gorini, A. Kossakowski, E.C.G. Sudarshan, J. Math. Phys. **17**, 821 (1976)
[8] G. Lindblad, Commun. Math. Phys. **48**, 119 (1976)
[9] H.P. Breuer, F. Petruccione, *The theory of open quantum systems* (Oxford University Press, 2002)
[10] A.O. Caldeira, A.J. Leggett, Physica A **121**, 587 (1983)
[11] L. Diósi, EPL (Europhysics Letters) **22**, 1 (1993)
[12] L. Diósi, Physica A: Statistical Mechanics and its Applications **199**, 517 (1993)
[13] S. Gao, Physical review letters **79**, 3101 (1997)
[14] B. Vacchini, Physical review letters **84**, 1374 (2000)
[15] X. Yao, B. Müller, Phys. Rev. C **97**, 014908 (2018), [Erratum: Phys.Rev.C 97, 049903 (2018)], 1709.03529
[16] X. Yao, T. Mehen, Phys. Rev. D **99**, 096028 (2019), 1811.07027
[17] X. Yao, B. Müller, Phys. Rev. D **100**, 014008 (2019), 1811.09644
[18] X. Yao, W. Ke, Y. Xu, S.A. Bass, B. Müller, JHEP **01**, 046 (2021), 2004.06746
[19] N. Brambilla, A. Pineda, J. Soto, A. Vairo, Reviews of Modern Physics **77**, 1423 (2005)
[20] Y. Akamatsu, Phys. Rev. D **91**, 056002 (2015), 1403.5783

[21] M. Laine, O. Philipsen, P. Romatschke, M. Tassler, JHEP **03**, 054 (2007), hep-ph/0611300
[22] A. Beraudo, J.P. Blaizot, C. Ratti, Nucl. Phys. **A806**, 312 (2008), 0712.4394
[23] N. Brambilla, J. Ghiglieri, A. Vairo, P. Petreczky, Phys. Rev. **D78**, 014017 (2008), 0804.0993
[24] J.P. Blaizot, D. De Boni, P. Faccioli, G. Garberoglio, Nucl. Phys. **A946**, 49 (2016), 1503.03857
[25] J.P. Blaizot, M.A. Escobedo, JHEP **06**, 034 (2018), 1711.10812
[26] J.P. Blaizot, M.A. Escobedo, Phys. Rev. D **98**, 074007 (2018), 1803.07996
[27] N. Brambilla, M.A. Escobedo, J. Soto, A. Vairo, Phys. Rev. **D96**, 034021 (2017), 1612.07248
[28] N. Brambilla, M.A. Escobedo, J. Soto, A. Vairo, Phys. Rev. D **97**, 074009 (2018), 1711.04515
[29] N. Brambilla, M.A. Escobedo, A. Vairo, P. Vander Griend, Phys. Rev. D **100**, 054025 (2019), 1903.08063
[30] N. Gisin, I.C. Percival, Journal of Physics A: Mathematical and General **25**, 5677 (1992)
[31] M.B. Plenio, P.L. Knight, Rev. Mod. Phys. **70**, 101 (1998), quant-ph/9702007
[32] Y. Akamatsu, A. Rothkopf, Phys. Rev. D **85**, 105011 (2012), 1110.1203
[33] S. Kajimoto, Y. Akamatsu, M. Asakawa, A. Rothkopf, Phys. Rev. D **97**, 014003 (2018), 1705.03365
[34] A. Rothkopf, JHEP **04**, 085 (2014), 1312.3246
[35] R. Sharma, A. Tiwari, Phys. Rev. D **101**, 074004 (2020), 1912.07036
[36] Y. Akamatsu, M. Asakawa, S. Kajimoto (2021), 2108.06921
[37] Y. Akamatsu, M. Asakawa, S. Kajimoto, A. Rothkopf, JHEP **07**, 029 (2018), 1805.00167
[38] T. Miura, Y. Akamatsu, M. Asakawa, A. Rothkopf, Phys. Rev. D **101**, 034011 (2020), 1908.06293
[39] O. Ålund, Y. Akamatsu, F. Laurén, T. Miura, J. Nordström, A. Rothkopf, J. Comput. Phys. **425**, 109917 (2021), 2004.04406
[40] N. Brambilla, M.A. Escobedo, M. Strickland, A. Vairo, P. Vander Griend, J.H. Weber, JHEP **05**, 136 (2021), 2012.01240
[41] N. Brambilla, M.A. Escobedo, M. Strickland, A. Vairo, P.V. Griend, J.H. Weber (2021), 2107.06222
[42] T. Miura, Y. Kaida, Y. Akamatsu, M. Asakawa (in preparation)

QCD at high temperature and density: selected highlights

Jon-Ivar Skullerud[1,⋆]

[1] *Department of Theoretical Physics, Maynooth University, Maynooth, Co Kildare, Ireland*

Abstract. I review some of the recent progress in QCD at high temperature and density, with a focus on the nature of the high-temperature transition; cold and dense matter; and hadron properties and transport coefficients at high temperature.

1 Introduction

Understanding the phase diagram of QCD and the properties of hot and dense strongly interacting matter is an endeavour of great scope and interest. Besides the theoretical interest in the exotic forms of matter that may appear under these conditions, a full understanding will cast light on the formation of matter in the early universe, the properties of neutron stars, the formation of heavy elements, and much more. Aligned to this is the experimental programme at RHIC and LHC and, in the near future, at FAIR and NICA.

The theoretical effort that has gone into this field over the past number of years is so large and varied that it will not be possible to do it full justice in one talk, and this overview is therefore necessarily selective. Important topics that will not be covered here include the QCD equation of state at high temperature, QCD in nonzero magnetic field, topology and deconfinement, non-equilibrium dynamics, and much more (see for example [1–5] for complementary overviews). I will instead focus on aspects of the transition(s) at zero chemical potential μ; exploring the chiral critical line at nonzero μ; some progress in understanding matter at large μ and low temperature T; and hadron properties and transport coefficients at large T. The emphasis will be on lattice QCD results, but not exclusively so: in particular, results using other Dyson–Schwinger equations (DSEs), the functional renormalisation group (FRG) and perturbation theory will be referenced where appropriate.

2 Transition(s) at zero chemical potential

2.1 Location and order of the chiral transiton

It is by now well established that the transition from the hadronic phase to the quark–gluon plasma at $\mu = 0$ is a crossover, characterised by a rapid change in the chiral condensate and a peak in the chiral susceptibility. Away from the chiral limit (where a true phase transition is expected) there is therefore no unique definiton of the pseudocritical temperature T_c, but different definitions (e.g., the peak in the susceptibility T_χ or the inflection point of the renormalised chiral condensate T_Δ) give compatible results for physical quark masses. There is also good agreement on the value of T_c at the physical

⋆e-mail: jonivar.skullerud@mu.ie

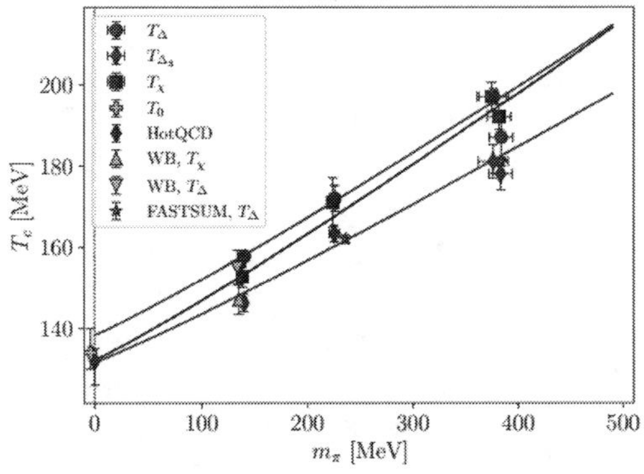

Figure 1. Pseudocritical temperature T_c as a function of the pion mass m_π from different lattice collaborations, from [8].

point between lattice groups using different fermion formulations [6–8] as well as with recent results from the DSE and FRG approaches [9, 10], with all finding $T_\chi = 153 - 158$ MeV. Figure 1 [8] shows the mass dependence of T_c from different lattice calculations [6–8, 11] together with scaling functions, illustrating significant progress on the dependence of T_c on the light quark mass [8, 10, 12], leading to a critical temperature in the chiral limit $T_0 = 130 - 140$ MeV.

The order of the chiral phase transition in the massless 2-flavour limit $m_u, m_d \to 0$ is intimately connected with the fate of the anomalous $U(1)_A$ symmetry. This symmetry is always broken by quantum effects, but if the breaking is small (*effective restoration*) at T_c the chiral transition may be first order or second order in the $U(2)_L \otimes U(2)_R/U(2)_V$ universality class. On the other hand, if the symmetry remains strongly broken, the chiral transition will be second order in the $O(4)$ universality class. Lattice studies have yet to arrive at a consensus on this issue, with [13–17] finding that the $U(1)_A$ symmetry remains broken, while [18–21] finds that it is restored. It is worth noting that of these studies, only [16] has performed a continuum extrapolation. A recent FRG calculation [22] is consistent with $U(1)_A$ restoration and an $O(4)$ universality class.

The chiral transition is expected to be first order in the 3-flavour massless limit, $m_u = m_d = m_s = 0$. An open question is whether the transition is first or second order for physical strange quarks, and if it is first order, what is the critical pion mass m_π^c separating the first order transition region from the crossover region? The evidence from a number of lattice studies is that the first order region shrinks as the continuum limit is approached, with upper limits of $m_\pi^c \lesssim 50$ MeV from highly improved staggered fermions [23] and $m_\pi^c \lesssim 110$ MeV from Wilson–clover fermions [24]. A recent study with Wilson–clover fermions [25] found evidence of a first order transition, while no such evidence was found in a study using staggered fermions [26].

This question has also been addressed by studying systems with $N_f > 3$, where the first order transition in the chiral limit is expected to be stronger and persist to larger quark masses. Ohno *et al* [25] found evidence for a first order transition also for $N_f = 4$ using Wilson–clover fermions. However, Cuteri *et al.* [27], performing a tricritical scaling analysis on staggered fermion data, found that the transition is consistent with second order. A reanalysis of Wilson–clover data gave the same result.

2.2 Some surprises

Recently, some intriguing evidence has emerged suggesting that the phase structure of QCD may be more complicated than previously thought, even at $\mu = 0$. In [28–30] it was found that in the region $T_c < T \lesssim 3T_c$, meson correlators exhibit certain degeneracies that are not present in the non-interacting case. This has been explained in terms of a chiral–spin symmetry $SU(2)_{CS}$ and its flavour extension $SU(2N_f)$, which is a symmetry of the chromoelectric interaction, but is broken by the chromomagnetic interaction. Following this, it has been conjectured that the intermediate-temperature regime is a "stringy fluid" where chiral symmetry is restored and the chromomagnetic interaction is suppressed, but the confining chromoelectric field is still present.

Independently, it was found [31] that a condensate of thermal monopoles forms at roughly the same temperature in full ($N_f = 2+1$) QCD as in pure gauge theory, where this coincides with the deconfinement transition, i.e., at roughly twice the chiral transition temperature. Furthermore, Alexandru and Horvath [32, 33] found an infrared scaling of the Dirac spectral density, $\rho(\lambda) \propto 1/\lambda$, which only set in at $T > T_{\mathrm{IR}} = 200 - 250\,\mathrm{MeV} > T_c$. They conjecture that this scaling is characteristic of a strongly coupled régime which only ends at a much higher (yet to be determined) temperature T_{UV}.

It is worth noting that the temperature scales for the suggested additional transitions are different, and that no connection between the different phenomena has been established. However, these results do point to the need for further investigations into the properties of QCD matter above T_c.

3 Chiral transition line

The shape of the transition line away from $\mu = 0$ has been computed on the lattice using continuation from imaginary μ [7, 34] and Taylor expansion [6, 34] (see also [35] for recent results using a novel reweighting method) as well as from Dyson–Schwinger equations [3], the functional renormalisation group [36] and a combination of the two [9, 37]. The pseudocritical temperature $T_c(\mu_B)$ can be expanded in powers of μ_B,

$$\frac{T_c(\mu_B)}{T_c(0)} = 1 - \kappa_2\left(\frac{\mu_B}{T_c}\right)^2 - \kappa_4\left(\frac{\mu_B}{T_c}\right)^4 + O(\mu_B^6) \tag{1}$$

leaving the parameters κ_2, κ_4 and any higher order parameters to be determined. All the calculations have found that κ_4 is consistent with zero within errors. Results for κ_2 are shown in figure 2. All the lattice results shown have been continuum extrapolated and are consistent with each other. The most recent results from functional methods are also seen to be converging on the same values.

The right panel of figure 2 shows a summary [9] of lattice, functional and experimental data for the chiral transition and freezout curves. In [9] their DSE/FRG calculations were found to be reliable up to $\mu_B/T = 3$, which means that the location of the critical endpoints shown in the figure should only be considered indicative, and the spread between the different results reflect the remaining systematic uncertainties in the approaches.

Fluctuations and correlations of the conserved quantum numbers (baryon number B, electric charge Q and strangeness S) play a crucial rôle in connecting theoretical predictions with experimental results. From the generalised susceptibilities,

$$\chi^{B,Q,S}_{i,j,k} = \frac{\partial^{i+j+k}(p/T^4)}{(\partial\hat{\mu}_B)^i(\partial\hat{\mu}_Q)^i(\partial\hat{\mu}_S)^k}, \qquad \hat{\mu}_a \equiv \frac{\mu_a}{T} \tag{2}$$

we may compute experimentally observable quantities such as the mean $M_B = \chi^B_1$, variance $\sigma^2_B = \chi^B_2$, skewness $S_B = \chi^B_3/(\chi^B_2)^{3/2}$ and kurtosis $\kappa_B = \chi^B_4/(\chi^B_2)^2$ of the baryon number distribution. Cumulants

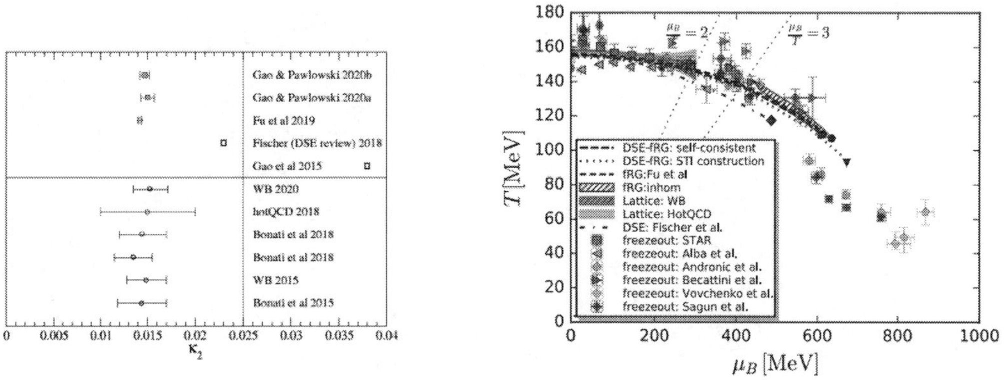

Figure 2. Left: The curvature parameter κ_2 from lattice QCD [6, 7, 34, 38, 39] (bottom panel) and functional methods (blue: DSE, green: FRG) [3, 9, 36, 37, 40] (top panel). Black points have been obtained by analytic continuation from imaginary μ, while red points are from Taylor expansion. Right: Transition line and critical endpoint from lattice and functional methods together with experimental freezeout lines, from [9].

up to 8th order have been computed by the Wuppertal–Budapest collaboration [41] and the HotQCD collaboration [42] and used to determine skewness and kurtosis at nonzero baryon chemical potential as well as preliminary results for the hyper-kurtosis (the sixth moment of the distribution). Results have also been obtained in the FRG framework using a low-energy effective theory [43, 44]. These results may in turn be used to extract freezeout parameters from experimental data and check whether experimental results are consistent with equilibrium QCD.

4 Large density

The methods employed to map out the phase diagram at high T and small μ break down at lower temperature, in particular in the region relevant to nuclear matter and the interior of neutron stars. Recent progress has been made with several alternative methods including complex Langevin simulations, Lefschetz thimble-based methods [45], density of states [46, 47] and strong coupling expansions [48]. In addition, QCD-like theories where the sign problem is absent or weak may serve as a laboratory for generic features of gauge theories at high density as well as provide checks on the validity of other methods. Finally, perturbation theory, valid at asymptotically high densities, can provide constraints on physics at intermediate densities. In the following only a few selected highlights will be presented.

Complex Langevin methods have now reached the point where they have been used to simulate fully dynamical QCD with pion masses down to $\sim$ 500 MeV [49–51]. Figure 3 shows the fermion density as a function of the chemical potential from a simulation with $N_f = 2$ Wilson fermions with $m_\pi \approx 500$ MeV and a lattice spacing $a \approx 0.06$ fm [51]. At low temperature the Silver Blaze phenomenon is clearly seen, as the density remains zero up to $\mu = m_N/3$ with an onset of nuclear matter beyond this point. At very high chemical potential we observe saturation, where all available fermion states on the lattice are occupied.

Recent developments in strong coupling methods [48], based on a character expansion of the gauge field action, have allowed simulations to be carried out for heavy quarks (using a hopping parameter expansion) [53] as well as for light quarks (using a dual formulation) [54, 55]. Recently, a continuous-time formulation for staggered fermions has been developed [52]. Figure 3 shows results

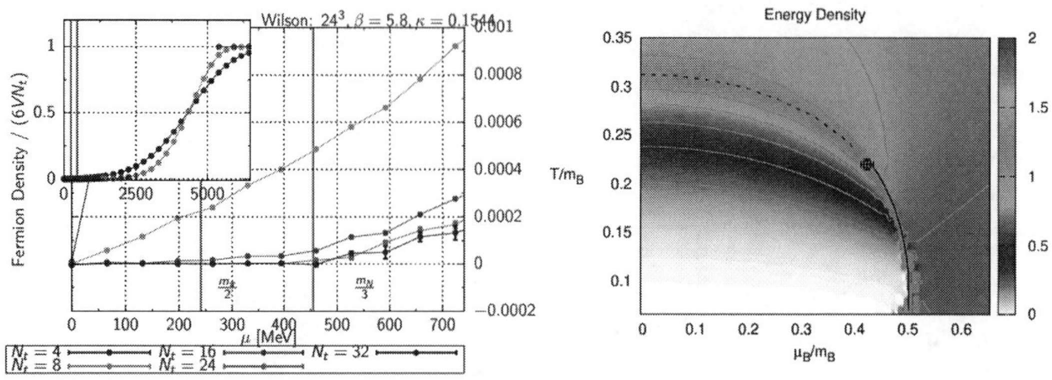

Figure 3. Left: Fermion density as function of chemical potential from complex Langevin with $N_f = 2$ Wilson fermions [51], at temperatures ranging from 100 MeV ($N_t = 32$) to 800 MeV ($N_t = 4$). Right: Energy density as a function of temperature and chemical potential in the strong coupling expansion, from [52]. A first order nuclear gas–liquid transition with a critical endpoint is observed.

using this formulation, exhibiting a first-order nuclear gas–nuclear liquid phase transition including a critical endpoint. The strong coupling expansion has also been applied to large-N_c QCD in the heavy dense limit [56], with findings consistent with the appearance of quarkyonic matter at large N_c.

QCD with nonzero isospin chemical potential μ_I [57, 58] and two-colour QCD (QC_2D) [59–63] do not suffer from the sign problem and are amenable to direct simulations at large density, and may hence provide benchmarks for other methods [64–69]. Both feature a superfluid phase for $\mu_{I,B} > m_\pi$ characterised by condensation of pions and scalar diquarks respectively. The phase diagrams for both theories turn out to be remarkably similar; in particular, the superfluid transition temperature has been found to be almost independent of μ in both cases. In QC_2D simulations with different lattice formulations have found consistent results for the superfluid transition and have also confirmed the existence of a "quarkyonic" phase at large μ, where quarks remain confined but the bulk degrees of freedom are weakly interacting quarks. A BEC régime described by chiral perturbation theory has been found at lower μ for sufficiently light quarks [60]. It remains an open question whether there is a large-μ deconfinement transition at small temperature.

In the absence of reliable, non-perturbative, first-principles information on cold high-density matter, constraints can be placed by combining information on the nuclear matter equation of state with compact star observations. This may be further constrained by requiring the equation of state to match on to perturbation theory at very high density, which has recently been computed to NNLO including NNNLO terms [70]. This approach has been pursued in [70–74]. Specifically, the low-energy equation of state has been computed in a chiral EFT, with the EoS at intermediate density modelled by piecewise polytropes, $p \propto n^\gamma$. Typically, the polytropic index $\gamma \sim 2.5$ for nuclear matter, while high-density perturbation theory corresponds to $\gamma \in [0.5, 0.8]$. Combined with recent observational data including from gravitational waves and the NICER observatory [74] only a subset of all possible equations of state remain. A result of such an analysis [71] is shown in figure 4, where it is concluded that stars with a mass $2M_\odot$ will include a quark matter core, characterised by a small polytropic index, for any EoS that is not ruled out by observations.

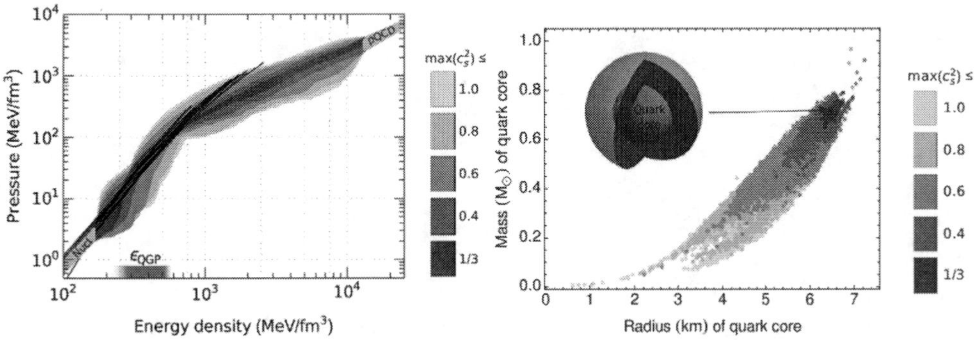

Figure 4. Left: Equation of state for high-density QCD obtained by interpolating between a nuclear eos (bottom left hand corner) and perturbative QCD (top right), constrained by mass and radius data from compact star observations. The kink in the eos suggests a transition from nuclear matter to quark matter. Right: The size of the quark matter core in a $2M_{\odot}$ star, for the different interpolations shown in the left hand plot. Figures taken from [71].

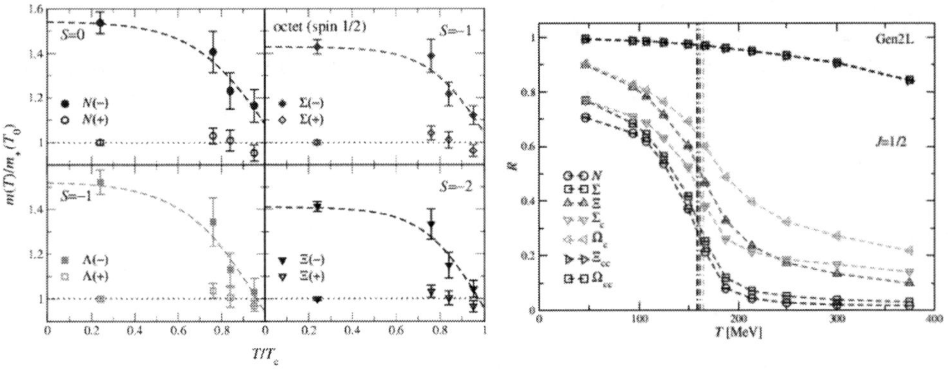

Figure 5. Left: Masses of positive and negative parity octet baryons below the deconfinement transition, from [75]. Right: The parity-doubling R parameter (3) for light, strange and charmed baryons.

5 Hadron properties and transport

5.1 Baryons

The FASTSUM collaboration has studied the properties of baryons at high temperature [11, 75–77], and in particular the emergent degeneracy between positive and negative parity states. Since positive and negative parity are states encoded in the same correlator (as forward and backward moving states respectively), the parameter

$$R = \frac{\sum_{n=0}^{\beta/2-1} R(\tau_n)/\sigma^2(\tau_n)}{\sum_{n=0}^{\beta/2-1} 1/\sigma^2(\tau_n)}\,; \qquad R(\tau) = \frac{G(\tau) - G(\beta-\tau)}{G(\tau) + G(\beta-\tau)} \tag{3}$$

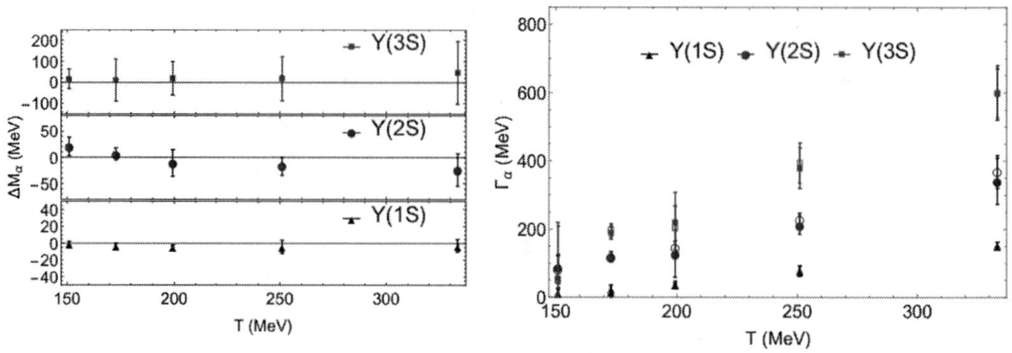

Figure 6. Mass shift (left) and width (right) for the ϒ 1S, 2S and 3S states, from [79].

may serve as a measure of parity doubling, with $R = 0$ when positive and negative parity states are completely degenerate. Figure 5 (left) shows the masses of positive and negative parity octet baryons below the deconfinement transition [77]: we see that while the masses positive-parity baryons remain unchanged, their negative-parity partners become lighter with increasing temperature with degeneracy setting in close to T_c, with potential impact on hadron resonance gas models [75]. The R parameter (**??**) is shown in figure 5 (right). It has an inflection point which is roughly the same for all light, strange and even singly-charmed baryons, and which coincides approximately with the chiral transition temperature [11]. Doubly-charmed baryons, however, do not appear sensitive to the chiral transition.

5.2 Heavy quarkonium

Heavy quarks have long been a focus of attention, both theoretically and experimentally, in the context of the quark–gluon plasma and heavy-ion collisions, see for example the review [78]. In recent years there has been significant progress towards a quantitative understanding of thermal mass shifts and widths of heavy quarkonia, in particular in the beauty sector. In a series of papers [79–81] Larsen *et al.* have employed lattice non-relativistic QCD (NRQCD) on hotQCD ensembles with $N_f = 2 + 1$ and a near-physical pion mass. Using sources optimised to reproduce the zero-temperature spectrum and wavefunctions they have been able to fit the correlators to a gaussian plus low and high energy tails. The Bethe–Salpeter wavefunctions have also been determined [81]. Results for the vector (ϒ) channel are shown in fig. 6, suggesting a very small (negative) mass shift for the ground state and a thermal width increasing with temperature as expected. Similar results are found for the P-wave χ_{b0} (scalar) channel. It may be noted that a small negative mass shift has also been observed by Kim *et al.* [82] and for charmonium by Kelly *et al.* [83], both using the BR method [84].

These results need to be independently confirmed using different approaches. The FASTSUM collaboration has employed a wide range methods to determine the mass shift and width of the ϒ(1S) state [85], including Gaussian fits, moments of correlators, and spectral function reconstruction with linear, Bayesian and machine learning methods. The results are as yet inconclusive, revealing quite different systematics for the different methods, which need to be better understood.

The Bielefeld–CCNU collaboration has conducted a long series of studies of quarkonium in quenched QCD using nonperturbatively improved Wilson fermions. Their most recent results [86] suggest that there is no resonance peak above T_c for J/ψ, while the resonance peak for ϒ persists only up to $1.5T_c$. More work is clearly needed to understand the source of this discrepancy.

5.3 Transport

Transport properties describe the real-time, non-equilibrium properties of the medium, and are as such not directly accessible with the equilibrium methods of euclidean quantum field theory. However, a transport coefficient κ can be related to the spectral functions of the relevant conserved current $J^{\kappa}_{\mu}(x)$ via Kubo relations,

$$\kappa \propto \lim_{\omega\to 0} \frac{\rho^{\kappa}(\omega)}{\omega}\,; \qquad G^{\kappa}_{ij}(\tau) = \int d^3x \left\langle J^{\kappa}_{i}(\tau,\vec{x}) J^{\kappa}_{j}(0,\vec{0})\right\rangle = \frac{1}{2\pi}\int_0^{\infty} K(\omega,\tau;T)\rho^{\kappa}_{ij}(\omega)d\omega\,, \tag{4}$$

where $K(\omega, \tau; T)$ is a known integral kernel. While there has been some progress on shear and bulk viscosity [87, 88] and jet quenching [89, 90], we are still some way from obtaining quantitatively reliable predictions for these quantities, in contrast to the heavy quark diffusion coefficient and the electrical conductivity, which will be reviewed below.

Two methods have been used to compute the heavy quark diffusion coefficient from the lattice. One method [91] uses an effective theory where the heavy quark momentum diffusion coefficient is written in terms of a Polyakov loop correlator with a colour-electric field insertion. The main benefit of this method is that the related spectral function has a smooth behaviour as $\omega \to 0$, greatly reducing any uncertainties related to determining a potentially narrow transport peak. The main drawback is that the correlator is inherently noisy, and noise reduction techniques must be used to get a reasonable signal. Recent continuum extrapolated results from this method have been reported by TUMQCD [92], using a multilevel algorithm which is applicable for pure gauge theories, and by the Bielefeld–Darmstadt collaboration [93, 94] using the Wilson flow, which is also applicable in the presence of dynamical fermions. The results are consistent with each other and with earlier results [95, 96].

The second method uses fermionic vector correlators of charm and beauty quarks, with the benefit that finite quark mass effects are directly included, and the correlators are much less noisy than the gluonic correlators in the EFT method. The main drawback is that it can be very difficult to reliably determine the height of a possibly narrow transport peak. Continuum extrapolated results for both charm and beauty diffusion have recently been presented by the Bielefeld–CCNU collaboration [86].

Figure 7 (left) shows a comparison of lattice results for the diffusion coefficient D together with the NLO perturbative value. The results from each of the two methods are internally consistent, but the results from fermionic correlators are systematically smaller than from the effective field theory.

The electrical conductivity has been computed on the lattice by a number of groups using various approaches (see [98] for a recent review). In addition to earlier results with $O(a)$ improved Wilson fermions and heavier than physical light quarks [99–101], there are now results using staggered fermions with physical up, down and strange quark masses [102, 103]. A summary of these results are shown in figure 7 (right panel). We see that there is good agreement between all the results despite the different methods used, suggesting that we are close to a quantitative, reliable determination of the conductivity of the QGP and its temperature variation. Astrakhantsev *et al.* have also computed the response of the conductivity to an external magnetic field [102] and to a nonzero baryon chemical potential [103]. The latter shows a reduction in the conductivity as μ_B increases, in qualitative agreement with a direct calculation in two-colour QCD [61].

6 Summary

There has been a large amount of progress on understanding the properties of QCD at high temperature and density recently. In particular, we have reached the era of precision physics determining the chiral transition temperature T_c as well as the high-temperature equation of state, and we are within reach of

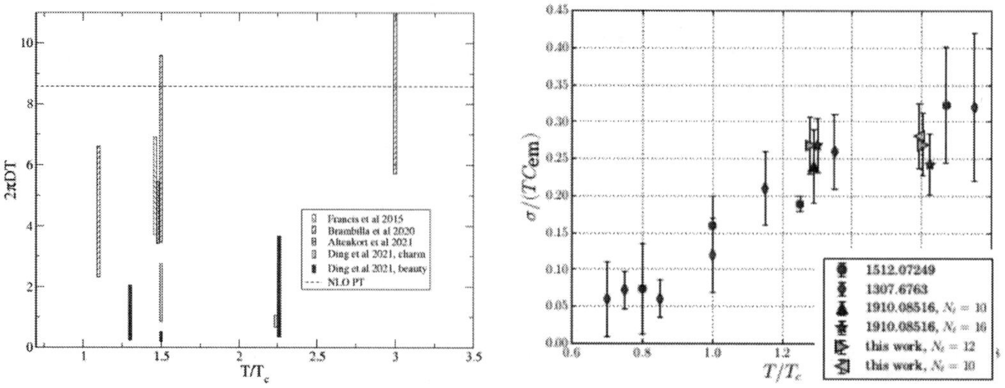

Figure 7. Left: summary of continuum-extrapolated lattice results for the heavy quark diffusion coefficient in pure gauge theory [86, 92–94, 96]. Hatched symbols are obtained using the EFT description; filled symbols are from fermionic operators. Right: Summary of lattice results for the electrical conductivity of QCD matter, from [97].

precision results for fluctuations both at zero and nonzero baryon density. We are also approaching a quantitative understanding of hadron properties in the hot medium as well as some transport properties, in particular the electrical conductivity. At the same time, despite significant progress especially in complex Langevin methods and constraints from compact star observations, we are far from being able to make reliable predictions in the low-temperature, high-density region relevant for neutron stars. First-principles calculations of the shear viscosity beyond theories with gravity duals are also in their infancy. Finally, the strong interaction has given us many surprises in the past and continues to surprise us in the present. It seems likely that there will be more surprises in store also in the future.

References

[1] K. Fukushima, C. Sasaki, Prog. Part. Nucl. Phys. **72**, 99 (2013), 1301.6377
[2] H.T. Ding, PoS **LATTICE2016**, 022 (2017), 1702.00151
[3] C.S. Fischer, Prog. Part. Nucl. Phys. **105**, 1 (2019), 1810.12938
[4] J. Ghiglieri, A. Kurkela, M. Strickland, A. Vuorinen, Phys. Rept. **880**, 1 (2020), 2002.10188
[5] J.N. Guenther, Eur. Phys. J. A **57**, 136 (2021), 2010.15503
[6] A. Bazavov et al. (HotQCD), Phys. Lett. B **795**, 15 (2019), 1812.08235
[7] S. Borsanyi, Z. Fodor, J.N. Guenther, R. Kara, S.D. Katz, P. Parotto, A. Pasztor, C. Ratti, K.K. Szabo, Phys. Rev. Lett. **125**, 052001 (2020), 2002.02821
[8] A.Y. Kotov, M.P. Lombardo, A. Trunin, Phys. Lett. B **823**, 136749 (2021), 2105.09842
[9] F. Gao, J.M. Pawlowski, Phys. Lett. B **820**, 136584 (2021), 2010.13705
[10] J. Braun, W.j. Fu, J.M. Pawlowski, F. Rennecke, D. Rosenblüh, S. Yin, Phys. Rev. D **102**, 056010 (2020), 2003.13112
[11] G. Aarts et al. (2020), 2007.04188
[12] H.T. Ding et al. (HotQCD), Phys. Rev. Lett. **123**, 062002 (2019), 1903.04801
[13] A. Bazavov et al. (HotQCD), Phys. Rev. D **86**, 094503 (2012), 1205.3535
[14] M.I. Buchoff et al., Phys. Rev. D **89**, 054514 (2014), 1309.4149

[15] V. Dick, F. Karsch, E. Laermann, S. Mukherjee, S. Sharma, Phys. Rev. D **91**, 094504 (2015), 1502.06190

[16] H.T. Ding, S.T. Li, S. Mukherjee, A. Tomiya, X.D. Wang, Y. Zhang, Phys. Rev. Lett. **126**, 082001 (2021), 2010.14836

[17] O. Kaczmarek, L. Mazur, S. Sharma (2021), 2102.06136

[18] G. Cossu, S. Aoki, H. Fukaya, S. Hashimoto, T. Kaneko, H. Matsufuru, J.I. Noaki, Phys. Rev. D **87**, 114514 (2013), [Erratum: Phys.Rev.D 88, 019901 (2013)], 1304.6145

[19] B.B. Brandt, A. Francis, H.B. Meyer, O. Philipsen, D. Robaina, H. Wittig, JHEP **12**, 158 (2016), 1608.06882

[20] A. Tomiya, G. Cossu, S. Aoki, H. Fukaya, S. Hashimoto, T. Kaneko, J. Noaki, Phys. Rev. D **96**, 034509 (2017), [Addendum: Phys.Rev.D 96, 079902 (2017)], 1612.01908

[21] S. Aoki, Y. Aoki, G. Cossu, H. Fukaya, S. Hashimoto, T. Kaneko, C. Rohrhofer, K. Suzuki (JLQCD), Phys. Rev. D **103**, 074506 (2021), 2011.01499

[22] J. Braun, M. Leonhardt, J.M. Pawlowski, D. Rosenblüh (QCD) (2020), 2012.06231

[23] A. Bazavov, H.T. Ding, P. Hegde, F. Karsch, E. Laermann, S. Mukherjee, P. Petreczky, C. Schmidt, Phys. Rev. D **95**, 074505 (2017), 1701.03548

[24] Y. Kuramashi, Y. Nakamura, H. Ohno, S. Takeda, Phys. Rev. D **101**, 054509 (2020), 2001.04398

[25] H. Ohno, Y. Kuramashi, Y. Nakamura, S. Takeda, *Critical endpoints in (2+1)- and 4-flavor QCD with Wilson-Clover fermions* (2021), presentation at Lattice 2021

[26] S. Sharma, L. Dini, A. Lahiri, *Chiral phase transition temperature in 3-flavor QCD* (2021), presentation at Lattice 2021

[27] F. Cuteri, O. Philipsen, A. Sciarra (2021), 2107.12739

[28] C. Rohrhofer, Y. Aoki, G. Cossu, H. Fukaya, L.Y. Glozman, S. Hashimoto, C.B. Lang, S. Prelovsek, Phys. Rev. D **96**, 094501 (2017), [Erratum: Phys.Rev.D 99, 039901 (2019)], 1707.01881

[29] C. Rohrhofer, Y. Aoki, G. Cossu, H. Fukaya, C. Gattringer, L.Y. Glozman, S. Hashimoto, C.B. Lang, S. Prelovsek, Phys. Rev. D **100**, 014502 (2019), 1902.03191

[30] C. Rohrhofer, Y. Aoki, L.Y. Glozman, S. Hashimoto, Phys. Lett. B **802**, 135245 (2020), 1909.00927

[31] M. Cardinali, M. D'Elia, A. Pasqui (2021), 2107.02745

[32] A. Alexandru, I. Horváth, Phys. Rev. D **100**, 094507 (2019), 1906.08047

[33] A. Alexandru, I. Horváth, Phys. Rev. Lett. **127**, 052303 (2021), 2103.05607

[34] C. Bonati, M. D'Elia, F. Negro, F. Sanfilippo, K. Zambello, Phys. Rev. D **98**, 054510 (2018), 1805.02960

[35] S. Borsanyi, Z. Fodor, M. Giordano, S.D. Katz, D. Nogradi, A. Pasztor, C.H. Wong (2021), 2108.09213

[36] W.j. Fu, J.M. Pawlowski, F. Rennecke, Phys. Rev. D **101**, 054032 (2020), 1909.02991

[37] F. Gao, J.M. Pawlowski, Phys. Rev. D **102**, 034027 (2020), 2002.07500

[38] C. Bonati, M. D'Elia, M. Mariti, M. Mesiti, F. Negro, F. Sanfilippo, Phys. Rev. D **92**, 054503 (2015), 1507.03571

[39] R. Bellwied, S. Borsanyi, Z. Fodor, J. Günther, S.D. Katz, C. Ratti, K.K. Szabo, Phys. Lett. B **751**, 559 (2015), 1507.07510

[40] F. Gao, Y.x. Liu, Phys. Rev. D **94**, 076009 (2016), 1607.01675

[41] S. Borsanyi, Z. Fodor, J.N. Guenther, S.K. Katz, K.K. Szabo, A. Pasztor, I. Portillo, C. Ratti, JHEP **10**, 205 (2018), 1805.04445
[42] A. Bazavov et al., Phys. Rev. D **101**, 074502 (2020), 2001.08530
[43] W.j. Fu, J.M. Pawlowski, F. Rennecke, B.J. Schaefer, Phys. Rev. D **94**, 116020 (2016), 1608.04302
[44] W.j. Fu, X. Luo, J.M. Pawlowski, F. Rennecke, R. Wen, S. Yin (2021), 2101.06035
[45] A. Alexandru, G. Basar, P.F. Bedaque, N.C. Warrington (2020), 2007.05436
[46] K. Langfeld, PoS **LATTICE2016**, 010 (2017), 1610.09856
[47] C. Gattringer, M. Mandl, P. Törek, Phys. Rev. D **100**, 114517 (2019), 1911.05320
[48] O. Philipsen, *Strong coupling methods in QCD thermodynamics* (2021), 2104.03696
[49] F. Attanasio, B. Jäger, F.P.G. Ziegler, Eur. Phys. J. A **56**, 251 (2020), 2006.00476
[50] Y. Ito, H. Matsufuru, Y. Namekawa, J. Nishimura, S. Shimasaki, A. Tsuchiya, S. Tsutsui, JHEP **10**, 144 (2020), 2007.08778
[51] F. Attanasio, B. Jäger, F.P.G. Ziegler, *With complex Langevin towards the QCD phase diagram*, in *38th International Symposium on Lattice Field Theory* (2021), 2111.02241
[52] M. Klegrewe, W. Unger, Phys. Rev. D **102**, 034505 (2020), 2005.10813
[53] J. Glesaaen, M. Neuman, O. Philipsen, JHEP **03**, 100 (2016), 1512.05195
[54] G. Gagliardi, J. Kim, W. Unger, EPJ Web Conf. **175**, 07047 (2018), 1710.07564
[55] G. Gagliardi, W. Unger, Phys. Rev. D **101**, 034509 (2020), 1911.08389
[56] O. Philipsen, J. Scheunert, JHEP **11**, 022 (2019), 1908.03136
[57] B.B. Brandt, G. Endrődi, S. Schmalzbauer, Phys. Rev. **D97**, 054514 (2018), 1712.08190
[58] B.B. Brandt, F. Cuteri, G. Endrődi, S. Schmalzbauer, Particles **3**, 80 (2020), 1912.07451
[59] T. Boz, P. Giudice, S. Hands, J.I. Skullerud, Phys. Rev. D **101**, 074506 (2020), 1912.10975
[60] N. Astrakhantsev, V.V. Braguta, E.M. Ilgenfritz, A.Y. Kotov, A.A. Nikolaev, Phys. Rev. D **102**, 074507 (2020), 2007.07640
[61] P.V. Buividovich, D. Smith, L. von Smekal, Phys. Rev. D **102**, 094510 (2020), 2007.05639
[62] K. Iida, E. Itou, T.G. Lee (2019), 1910.07872
[63] J. Wilhelm, L. Holicki, D. Smith, B. Wellegehausen, L. von Smekal (2019), 1910.04495
[64] J.O. Andersen, T. Brauner, W. Naylor, Phys. Rev. **D92**, 114504 (2015), 1505.05925
[65] J. Chao, Chin. Phys. C **44**, 034108 (2020), 1808.01928
[66] P. Adhikari, J.O. Andersen, Phys. Lett. B **804**, 135352 (2020), 1909.01131
[67] T.G. Khunjua, K.G. Klimenko, R.N. Zhokhov, JHEP **06**, 148 (2020), 2003.10562
[68] T. Furusawa, Y. Tanizaki, E. Itou, Phys. Rev. Res. **2**, 033253 (2020), 2005.13822
[69] R. Contant, M.Q. Huber (2019), 1909.12796
[70] T. Gorda, A. Kurkela, R. Paatelainen, S. Säppi, A. Vuorinen, Phys. Rev. Lett. **127**, 162003 (2021), 2103.05658
[71] E. Annala, T. Gorda, A. Kurkela, J. Nättilä, A. Vuorinen, Nature Phys. **16**, 907 (2020), 1903.09121
[72] E. Annala, T. Gorda, E. Katerini, A. Kurkela, J. Nättilä, V. Paschalidis, A. Vuorinen (2021), 2105.05132
[73] S.K. Greif, K. Hebeler, J.M. Lattimer, C.J. Pethick, A. Schwenk, Astrophys. J. **901**, 155 (2020), 2005.14164
[74] G. Raaijmakers, S.K. Greif, K. Hebeler, T. Hinderer, S. Nissanke, A. Schwenk, T.E. Riley, A.L. Watts, J.M. Lattimer, W.C.G. Ho, Astrophys. J. Lett. **918**, L29 (2021), 2105.06981
[75] G. Aarts, C. Allton, D. De Boni, B. Jäger, Phys. Rev. **D99**, 074503 (2019), 1812.07393

[76] G. Aarts, C. Allton, S. Hands, B. Jäger, C. Praki, J.I. Skullerud (2015), 1502.03603
[77] G. Aarts, C. Allton, D. De Boni, S. Hands, B. Jäger, C. Praki, J.I. Skullerud, JHEP **06**, 034 (2017), 1703.09246
[78] A. Rothkopf, Phys. Rept. **858**, 1 (2020), 1912.02253
[79] R. Larsen, S. Meinel, S. Mukherjee, P. Petreczky, Phys. Lett. B **800**, 135119 (2020), 1910.07374
[80] R. Larsen, S. Meinel, S. Mukherjee, P. Petreczky, Phys. Rev. D **100**, 074506 (2019), 1908.08437
[81] R. Larsen, S. Meinel, S. Mukherjee, P. Petreczky, Phys. Rev. D **102**, 114508 (2020), 2008.00100
[82] S. Kim, P. Petreczky, A. Rothkopf, JHEP **11**, 088 (2018), 1808.08781
[83] A. Kelly, A. Rothkopf, J.I. Skullerud, Phys. Rev. **D97**, 114509 (2018), 1802.00667
[84] Y. Burnier, A. Rothkopf, Phys. Rev. Lett. **111**, 182003 (2013), 1307.6106
[85] T. Spriggs, S. Offler, B. Page et al. *A comparison of spectral reconstruction methods applied to non-zero temperature NRQCD meson correlation functions* (2021), these proceedings.
[86] H.T. Ding, O. Kaczmarek, A.L. Lorenz, H. Ohno, H. Sandmeyer, H.T. Shu (2021), 2108.13693
[87] N.Y. Astrakhantsev, V.V. Braguta, A.Y. Kotov, Phys. Rev. D **98**, 054515 (2018), 1804.02382
[88] E. Itou, Y. Nagai, *QCD viscosity by combining the gradient flow and sparse model-ing methods* (2021), presentation at Lattice 2021.
[89] M. Panero, K. Rummukainen, A. Schäfer, Phys. Rev. Lett. **112**, 162001 (2014), 1307.5850
[90] A. Kumar, A. Majumder, J.H. Weber (2020), 2010.14463
[91] S. Caron-Huot, M. Laine, G.D. Moore, JHEP **04**, 053 (2009), 0901.1195
[92] N. Brambilla, V. Leino, P. Petreczky, A. Vairo, Phys. Rev. D **102**, 074503 (2020), 2007.10078
[93] L. Altenkort, A.M. Eller, O. Kaczmarek, L. Mazur, G.D. Moore, H.T. Shu, Phys. Rev. D **103**, 014511 (2021), 2009.13553
[94] L. Altenkort, A.M. Eller, O. Kaczmarek, L. Mazur, G.D. Moore, H.T. Shu, *Spectral reconstruction details of a gradient-flowed color-electric correlator*, in *19th International Conference on Strangeness in Quark Matter* (2021), 2109.11303
[95] D. Banerjee, S. Datta, R. Gavai, P. Majumdar, Phys. Rev. D **85**, 014510 (2012), 1109.5738
[96] A. Francis, O. Kaczmarek, M. Laine, T. Neuhaus, H. Ohno, Phys. Rev. D **92**, 116003 (2015), 1508.04543
[97] A. Trunin et al., *Electromagnetic conductivity of quark-gluon plasma at non-zero baryon density* (2021), presentation at Lattice 2021
[98] G. Aarts, A. Nikolaev, Eur. Phys. J. A **57**, 118 (2021), 2008.12326
[99] A. Amato, G. Aarts, C. Allton, P. Giudice, S. Hands, J.I. Skullerud, Phys.Rev.Lett. **111**, 172001 (2013), 1307.6763
[100] G. Aarts, C. Allton, A. Amato, P. Giudice, S. Hands, J.I. Skullerud, JHEP **1502**, 186 (2015), 1412.6411
[101] B.B. Brandt, A. Francis, B. Jäger, H.B. Meyer, Phys. Rev. D **93**, 054510 (2016), 1512.07249
[102] N. Astrakhantsev, V.V. Braguta, M. D'Elia, A.Y. Kotov, A.A. Nikolaev, F. Sanfilippo, Phys. Rev. D **102**, 054516 (2020), 1910.08516
[103] N. Astrakhantsev, V.V. Braguta, M. Cardinali, M. D'Elia, L. Maio, F. Sanfilippo, A. Trunin, A. Vasiliev, *Electromagnetic conductivity of quark-gluon plasma at non-zero baryon density*, in *38th International Symposium on Lattice Field Theory* (2021), 2110.10727

Excited states of isolated fermions in the Higgs phase of gauge Higgs theories

Jeff Greensite[1,*] and *Kazue* Matsuyama[1,**]

[1]Physics and Astronomy Dept., San Francisco State University, San Francisco CA 94132 USA

Abstract. We present evidence that seemingly elementary and non-composite objects, namely isolated static fermions in certain gauge Higgs theories, have a mass spectrum corresponding to localized excitations of the surrounding gauge and Higgs fields.

1 Introduction

Composite systems (molecules, atoms, nuclei, hadrons...) generally have a spectrum of excitations. What about non-composite systems: charged "elementary" particles like quarks and leptons? If the particle is charged, then by Gauss's Law it is accompanied by a surrounding gauge (and possibly other) fields. If these surrounding fields interact with themselves, could they not also exhibit a spectrum of excitations? This would look like a mass spectrum of the isolated elementary particle. That doesn't happen in pure QED where the gauge field does not self-interact, and any energy eigenstate containing a static ± charge pair is just the Coulomb field plus some number of photons. Gauge Higgs theories could be different.

2 Pseudomatter operators

One often reads that physical states are gauge invariant. This statement is almost, but not quite, true. The Gauss law constraint only requires invariance under infinitesimal gauge transformations. In QED, in an infinite volume, a physical state containing a single static charge transforms under a global subgroup of the gauge group. The ground state of pure QED containing a single static electric charge at point $\mathbf{x}$ is the Dirac state

$$\begin{aligned} |\Psi_{\mathbf{x}}\rangle &= \overline{\psi}^{+}(\mathbf{x})\rho_C(\mathbf{x};A)|\Psi_0\rangle \\ \rho_C(\mathbf{x};A) &= \exp\left[-i\frac{e}{4\pi}\int d^3z\, A_i(\mathbf{z})\frac{\partial}{\partial z_i}\frac{1}{|\mathbf{x}-\mathbf{z}|}\right]. \end{aligned} \quad (1)$$

It is easy to check that $|\Psi_{\mathbf{x}}\rangle$ satisfies the Gauss Law. However, let $g(x) = e^{i\theta(x)}$ be an arbitrary U(1) gauge transformation, and we separate out the zero mode $\theta(x) = \theta_0 + \tilde{\theta}(x)$. Then $\psi(\mathbf{x}) \to e^{i\theta(\mathbf{x})}\psi(\mathbf{x})$, but $\rho_C(\mathbf{x};A) \to e^{i\tilde{\theta}(x)}\rho_C(\mathbf{x};A)$. It follows that $|\Psi_{\mathbf{x}}\rangle \to e^{-i\theta_0}|\Psi_{\mathbf{x}}\rangle$ tranforms under a global subgroup of the local gauge symmetry group. Local symmetries cannot break spontaneously, of course (Elitzur). But global symmetries can. The operator $\rho_C(\mathbf{x};A)$ is a first example of what we refer to as a "pseudomatter" field [1]. It is responsible for dressing the bare charge with a Coulomb electric field.

In general, we define a pseudomatter field $\rho(\mathbf{x};A)$ to be a non-local functional of the gauge field which transforms like a matter field in the fundamental representation of the gauge group, except under the global center subgroup of the gauge group. In fact any SU(N) gauge transformation $g_F(\mathbf{x};A)$ to a physical gauge $F(A) = 0$ can be decomposed into N pseudomatter fields $\{\rho_n\}$, and vice-versa: $\rho_n^a(\mathbf{x};A) = g_F^{\dagger an}(\mathbf{x};A)$. In particular, the operator $\rho_C^*(\mathbf{x};A)$ defined earlier is precisely the gauge transformation to Coulomb gauge in an abelian theory. Another example, of particular relevance to this work, is that any eigenstate $\xi_n(\mathbf{x};U)$ of the covariant Laplacian operator $-D^2\xi_n = \kappa_n\xi_n$ in a lattice SU(N) gauge theory is a pseudomatter field, where

$$\begin{aligned} (-D^2)^{ab}_{\mathbf{xy}} &= \sum_{k=1}^{3}\Big[2\delta^{ab}\delta_{\mathbf{xy}} - U_k^{ab}(\mathbf{x})\delta_{\mathbf{y},\mathbf{x}+\hat{k}} \\ &\qquad -U_k^{\dagger ab}(\mathbf{x}-\hat{k})\delta_{\mathbf{y},\mathbf{x}-\hat{k}}\Big] \end{aligned} \quad (2)$$

is the covariant Laplacian. This is the idea behind the Laplacian gauges of Vink and Wiese [2].

We may combine the Higgs field and pseudomatter fields with matter field creation operators to create physical states in gauge Higgs theories.

3 Excited states of "elementary" particles

For a static quark-antiquark pair in a pure gauge theory there is a tower of metastable states

$$\Psi_n(R) = \overline{q}(\mathbf{x})V_n(\mathbf{x},\mathbf{y};U)q(\mathbf{y})\Psi_0 \quad (3)$$

correpsonding to string excitations. This has been observed in computer simulations [3, 4]. For light quarks, the excited states lie on linear Regge trajectories, and these should also exist in the confinement region of a gauge

*e-mail: greensit@sfsu.edu
**e-mail: kazuem@sfsu.edu

Higgs theory. The question is whether, in the spin glass (aka Higgs) phase, there a similar tower of metastable states of the form

$$\Psi_n(R) = \overline{q}^a(\mathbf{x}) \left[\sum_m c_m^{(n)} \rho_m^a(\mathbf{x}) \rho_m^{\dagger b}(\mathbf{y}) \right] q^b(\mathbf{y}) \Psi_0 \,, \quad (4)$$

where the $\{\rho_m(\mathbf{x})\}$ are pseudomatter fields. We have investigated this question in several gauge Higgs models with results that we discuss briefly here. In the study of static fermion excitations, we find that each model has its own special features which must be taken into account.

There are various strategies that can be used to compute the spectrum of a static fermion-antifermion pair. For example, if $|\Psi(R)\rangle$ is some arbitrary physical state containing a static fermion-antifermion pair with separation R, and $E_1(R)$ is the lowest energy of such states above the vacuum energy $\mathcal{E}_0$, then on general grounds, as $T \to \infty$

$$\langle \Psi | \mathcal{T}^T | \Psi \rangle = \sum_n c_n e^{-E_n(R)T} \to c_1 e^{-E_1(R)T} \,, \quad (5)$$

where $\mathcal{T}$ is the transfer matrix multiplied by $e^{\mathcal{E}_0}$. The drawback of this method is that while it is easy to extract the ground state energy, it is not so simple to extract the energies of excited states. Alternatively, let $\{|\Phi_\alpha(R)\rangle\}$ span a subspace of the full Hilbert space with the two static charges. Then one could get an approximate spectrum by diagonalizing $\mathcal{T}$ in this subspace. This approach is followed in some lattice QCD spectrum calculations. Here the drawback is that this requires a pretty big set ~ hundreds of states. This is not practical for our purposes, where it is computationally expensive to generate the $|\Phi_\alpha(R)\rangle$.

Our approach is as follows: Generate a small set of states $\{|\Phi_\alpha(R)\rangle\}$, and diagonalize either $\mathcal{T}$ or $\mathcal{T}^p$ in the small subspace spanned by these states. The hope is that one or more of the eigenstates $|\Psi_n\rangle$ in the subspace will be orthogonal (nearly) to the true ground state. If $|\Psi\rangle$ is such a state, then

$$\langle \Psi | \mathcal{T}^T | \Psi \rangle = \sum_n c_n e^{-E_n(R)T} \to c_{ex} e^{-E_{ex}(R)T} \text{ at large } T \,, \quad (6)$$

where E_{ex} is the energy of an excited state. Of course there are no guarantees this will work, it just has to be tried.

4 Four Models

4.1 SU(3) gauge Higgs theory

We begin with SU(3) gauge Higgs theory with a unimodular Higgs field in the fundamental color representation. Our calculations are carried out on a $14^3 \times 32$ lattice volume, at $\beta = 5.5$ with $\gamma = 0.5$ and $\gamma = 3.5$, in the confinement and Higgs phases respectively [5]. The action is

$$\begin{aligned} S &= -\frac{\beta}{3} \sum_{plaq} \mathrm{ReTr}[U_\mu(x) U_\nu(x+\hat{\mu}) U_\mu^\dagger(x+\hat{\nu}) U_\nu^\dagger(x)] \\ &\quad -\gamma \sum_{x,\mu} \mathrm{Re}[\phi^\dagger(x) U_\mu(x) \phi(x+\widehat{\mu})] \,, \end{aligned} \quad (7)$$

and the goal is to find excited states of static fermions in this system, if such states exist. Here (and in all gauge Higgs models discussed below) we impose the unimodular constraint $\phi^\dagger \phi = 1$.

Let ξ_n denote the eigenstates of the lattice Laplacian operator (no time derivatives) $-D^2 \xi_n = \kappa_n \xi_n$, let $\mathcal{T} = e^{-(H-\mathcal{E}_0)}$ be the (rescaled) transfer matrix, and consider at each quark separation $R = |\mathbf{x}-\mathbf{y}|$ the 4-dimensional subspace spanned by three quark-pseudomatter states, and one quark-scalar state

$$\begin{aligned} \Phi_{n=1,2,3}(R) &= [\overline{q}^a(\mathbf{x}) \xi_n^a(\mathbf{x})] \times [\xi_n^{\dagger b}(\mathbf{y}) q^b(\mathbf{y})] \Psi_0 \\ \Phi_4(R) &= [\overline{q}^a(\mathbf{x}) \phi^a(\mathbf{x})] \times [\phi^{\dagger b}(\mathbf{y}) q^b(\mathbf{y})] \Psi_0 \,. \end{aligned} \quad (8)$$

We calculate numerically the matrix elements and overlaps

$$\mathcal{T}_{mn}(R) = \langle \Phi_m | \mathcal{T} | \Phi_n \rangle \quad , \quad O_{mn}(R) = \langle \Phi_m | \Phi_n \rangle \,. \quad (9)$$

The eigenvalues of $\mathcal{T}$ in the subspace are obtained by solving the generalized eigenvalue problem $[\mathcal{T}]\vec{v}_n = \lambda_n [O] \vec{v}^{(n)}$, and we have eigenstates of $\mathcal{T}$ in the subspace

$$|\Psi_n(R)\rangle = \sum_{i=1}^4 v_i^{(n)} |\Phi_i(R)\rangle \,. \quad (10)$$

Likewise we consider evolving states for Euclidean time T, and compute

$$\begin{aligned} \mathcal{T}_{nn}^T(R) &= \langle \Psi_n | \mathcal{T}^T | \Psi_n \rangle = v_i^{(n)*} \langle \Phi_i | \mathcal{T}^T | \Phi_j \rangle v_j^{(n)} \\ E_n(R,T) &= -\log \left[\frac{\mathcal{T}_{nn}^T(R)}{\mathcal{T}_{nn}^{T-1}(R)} \right] \,. \end{aligned} \quad (11)$$

Integrating out the massive (i.e. static) fermion fields generates a pair of Wilson lines. The numerical computation of $\langle \Phi_i | \mathcal{T}^T | \Phi_j \rangle$ involves the expectation values of products of Wilson lines, terminated by matter or pseudomatter fields.

We have computed $E_{1,2}(R,T)$ in this model on a $14^3 \times 32$ lattice volume, at $\beta = 5.5$ with $\gamma = 0.5$ and $\gamma = 3.5$, in the confinement and Higgs phases respectively. The results for $E_n(R,T)$ and the overlap for $\Psi_1(R), \Psi_2(R)$ in the Higgs phase, in the range $T = 4 - 12$, are shown in Fig. 1. It is seen that the data has converged in this range of Euclidean time T.

There seems to be clear evidence of an excited state of a static fermion-antifermion pair, orthogonal to the ground state, whose energy is only weakly dependent on separation R. At large R the fermions are understood to be isolated from one another. Because the energy gap is far smaller than the threshold for vector boson creation in this model, this excited state is stable.

4.2 The $q = 2$ abelian Higgs model

This is a version of the abelian Higgs model in which the scalar field (like Cooper pairs) carries two units of electric charge:

$$\begin{aligned} S &= -\beta \sum_{plaq} \mathrm{Re}[U_\mu(x) U_\nu(x+\hat{\mu}) U_\mu^*(x+\hat{\nu}) U_\nu^*(x)] \\ &\quad -\gamma \sum_{x,\mu} \mathrm{Re}[\phi^*(x) U_\mu^2(x) \phi(x+\widehat{\mu})] \,. \end{aligned} \quad (12)$$

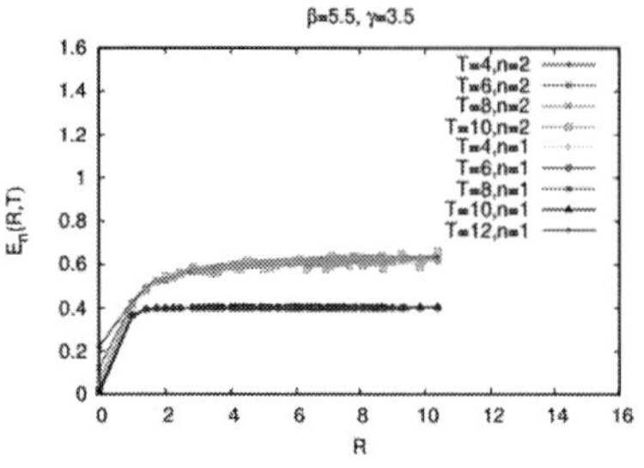

(a) Energies

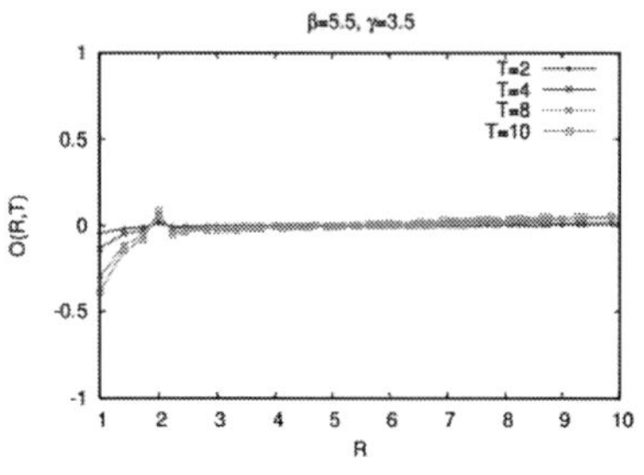

(b) Overlap

Figure 1. (a) Energies $E_n(R,T)$ of states Ψ_1, Ψ_2 after evolution for a period of $(T-1)/2$ units of Euclidean time, at lattice couplings $\beta = 5.5, \gamma = 3.5$. Note the energy gap, which persists out to the largest T values shown, of $E_2(R,T) - E_1(R,T) \approx 0.2$ in lattice units. (b) The overlap $O(R,T)$ between states $\Psi_1(R)$ and $\Psi_2(R)$ which are evolved (and then normalized) for $T/2$ units of Euclidean time. $O(R,T) = 0$ by construction at $T = 0, 1$, but the Euclidean time-evolved states are seen to remain approximately orthogonal for $R > 2$ at larger T.

This theory has a phase diagram consisting of massless, Higgs, and confining phases, in this case separated by thermodynamic transition lines. The search for excited static fermion states was carried out in ref. [6] just inside the Higgs phase, at $\beta = 3.0, \gamma = 0.5$, and in this investigation the fermions were taken to have the same $q = \pm 2$ charges as the scalar field. An example of the relevant data is shown in Fig. 2(a). A surprise is that, following along the same lines of diagonalizing $\mathcal{T}$ in a small subspace, and judging by the fit to $\mathcal{T}_{11}(R,T)$ and $T_{22}(R,T)$, the lowest two states $\Psi_{1,2}$ seem to be very nearly exact eigenstates of the system, even at small T.

The ground and excited state energies, and also a rough estimate of the next excited state energy, vs. fermion separation are shown in Fig. 2(b). Once again, the first excitation is well below the threshold, and is therefore stable against massive photon emission. The second excited state, lying so near the threshold, most likely consists of the ground state plus a massive photon.

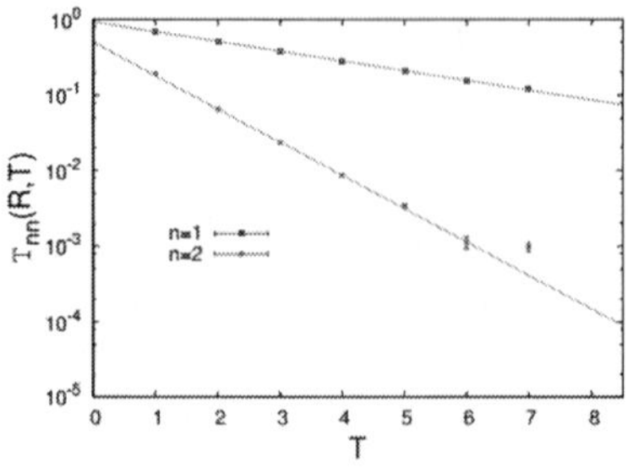

(a)

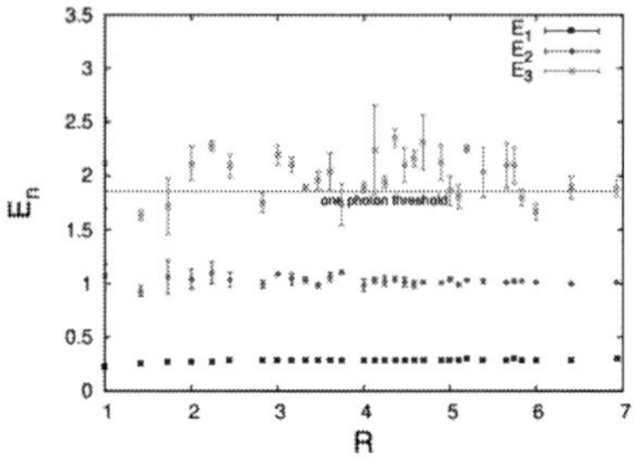

(b)

Figure 2. (a) An example of fits at $R = 6.93$ on a 16^4 lattice in the abelian Higgs model, with couplings $\beta = 3, \gamma = 0.5$. (b) Lowest, first excited, and second excited state energies vs. fermion separation.

4.3 Effective Landau-Ginzburg model

What about real superconductors?[1] The effective Landau-Ginzburg model for ordinary superconductivity is a non-relativistic $q = 2$ abelian Higgs model of this form:

$$\begin{aligned} S &= -\beta \sum_{plaq} \mathrm{Re}[UUU^*U^*] \\ &\quad -\gamma \sum_x \sum_{k=1}^{3} \phi^*(x) U_k^2(x) \phi(x+\hat{k}) \\ &\quad -\frac{\gamma}{v^2} \sum_x \phi^*(x) U_0^2(x) \phi(x+\hat{t}) \,. \end{aligned} \tag{13}$$

where $v \sim 10^{-2}$ in natural units, is on the order of the Fermi velocity in a metal, and $\beta = 1/e^2 \approx 10.9$ for ordinary electrodynamics. Go to unitary gauge, so that $U_0(x) \approx \pm 1$. We then compute the excitations around a pair of static $q = \pm 1$ (e) charges, having electrons, ions, and holes in mind. The parameters γ, β determine the photon mass (inverse to the penetration depth) in lattice units, so for a given γ the penetration depth fixes the lattice spacing in physical units. But this time things are not so simple, and diagonalizing $\mathcal{T}$ in a small subspace doesn't work. Eigenstates in the subspace flow in Euclidean time to the ground state.

What we do instead, at each separation R, is to diagonalize a *power* of the transfer matrix, $\mathcal{T}^{2t_0}$, in the basis Φ_α,

[1] This is work in progress, so we display only preliminary results.

so that

$$\langle \Psi_m | \mathcal{T}^{2t_0} | \Psi_n \rangle = \lambda_n(t_0)\delta_{mn} \,, \tag{14}$$

and define $\Psi_n(t) = \mathcal{T}^t \Psi_n$. Suppose, after evolving Ψ_1 by t_0 units of Euclidean time, that $\Psi_1(t_0)$ is approximately the true ground state in the full Hilbert space. It follows that all $\Psi_{n>1}(t_0)$ are orthogonal to the ground state, simply because $\langle \Psi_m(t_0)|\Psi_n(t_0)\rangle \propto \delta_{mn}$, and therefore, at large $T > 2t_0$

$$\begin{aligned} \mathcal{T}_{22}(R,T) &= \langle \Psi_2 | \mathcal{T}^T | \Psi_2 \rangle \\ &= \langle \Psi_2(t_0) | \mathcal{T}^{T-2t_0} | \Psi_2(t_0) \rangle \\ &\to \text{const} \times e^{-E_{ex}T} \quad \text{where} \quad E_{ex} > E_1 \,. \end{aligned} \tag{15}$$

For this calculation we choose $2t_0 = 9$ and $\gamma = 0.25$. Sample results at separation $R = 5.385$ are shown in Fig. 4. We

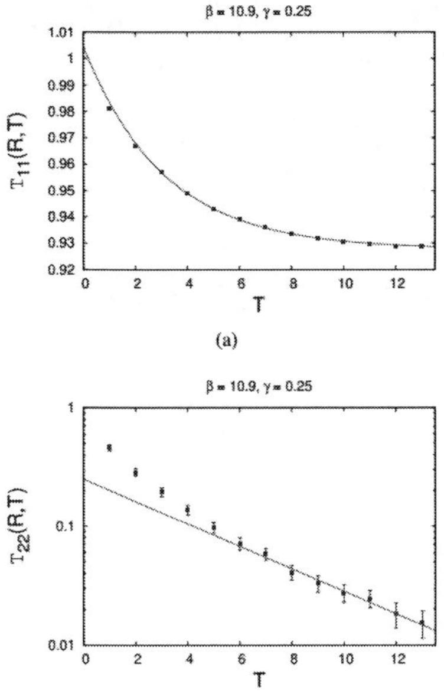

(a)

(b)

Figure 3. $\mathcal{T}_{nn}(R,T)$ at $R = 5.385$ vs T in the effective Landau-Ginzburg model of superconductivity.

fit $\mathcal{T}_{11}$ to

$$f_1(T) = a_1 \exp(-b_1 T) + c_1 \,. \tag{16}$$

The fact that $c_1 \neq 0$ means that the ground state energy $E_1 \approx 0$, while b_1 gives an excited state energy. Then we fit T_{22} in the range $T > 6$ to a single exponential $f_2(T) = a_2 \exp(-b_2)$, and the coefficient $b_2 < b_1$ gives another excitation energy. The data at $R < 4.0$ are rather noisy, with large χ^2. The results for $R > 4$ are shown in Fig. 4. Once again, the first excited state is stable. The next excited state, which is right on the threshold, is presumably the ground state plus a massive photon. The question of whether such new states could be detected experimentally, e.g. by ARPES, is of course of great interest.

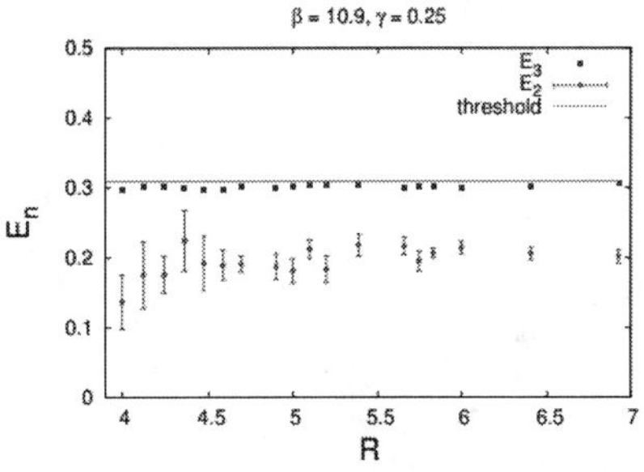

Figure 4. Ground state and excitation energies in the Landau-Ginzburg model. The ground state energy is $E_1 \approx 0$.

5 Excitations in chiral gauge theories

The electroweak sector of the standard model is a chiral gauge theory, and unfortunately there is no known lattice formulation of chiral non-abelian gauge theories with a continuum limit. There is a formulation for U(1) gauge theories due to Lüscher, involving overlap fermions. It is, however, difficult to implement numerically. So in this exploratory work, we chose a simpler option: only static fermions in a quenched formulation of a U(1) lattice gauge theory, with oppositely charged right and left-handed fermions, using a version of the old Smit-Swift lattice action [7, 8]. Lattice doublers restore chiral symmetry, so the original idea was to use a Wilson-style non-local mass term so that the mass of the doublers is infinite in the continuum limit. It is known that the continuum limit doesn't work [9], and the Smit-Swift formulation is not a true chiral gauge theory. Moreover, the positivity of the transfer matrix is unproven. But at least, at finite lattice spacing, there is a mass asymmetry between the desired states and the doublers in part of the phase diagram. So this is a model that can be studied with our methods.

The calculation is a little more involved than in the previous three models, and instead of a Wilson line there is a more complicated covariant object which connects to the pseudomatter fields. We leave the details of this computation to ref. [10], and here only display the results. The numerical simulation is carried out on a $14^3 \times 32$ lattice volume at $\beta = 3.0, \gamma = 1.0$ using $n_{ev} = 4$ Laplacian eigenstates, along with combinations of the $q = 2$ Higgs field and pseudomatter fields. Results are shown in Fig. 5, where $\lambda_n(R,T) \equiv \mathcal{T}_{nn}(R,T)$. In these figures, E_1 is derived from λ_1, E_2 is derived from either λ_3 or λ_4 in the large $T \geq 4$ range, and E_3 is derived from λ_2 in the small $T \lesssim 5$ range. Energies $E_{1,2,3}$ vs. R are displayed in Fig. 6.

6 Conclusions

We have found that in all four gauge Higgs models that we have investigated, the gauge+Higgs fields surrounded a charged static fermion have a spectrum of localized excitations, which cannot be interpreted as simply the ground state plus some massive bosons. This means that charged

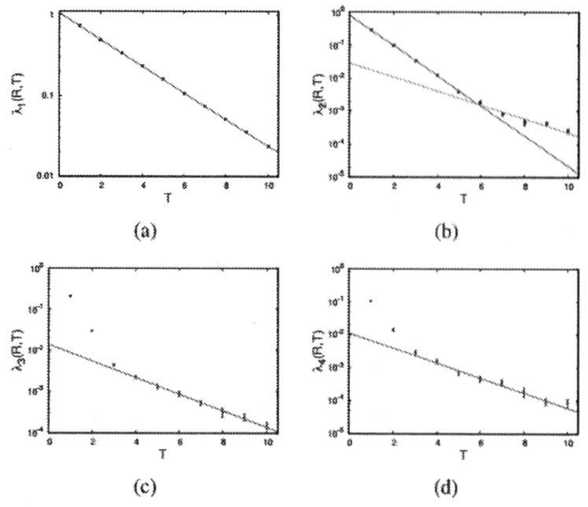

Figure 5. $\lambda_n(R,T) = \mathcal{T}_{nn}(R,T)$ vs. T at $R = 4.243$ in a chiral gauge theory at finite lattice spacing with Smit-Swift regularization.

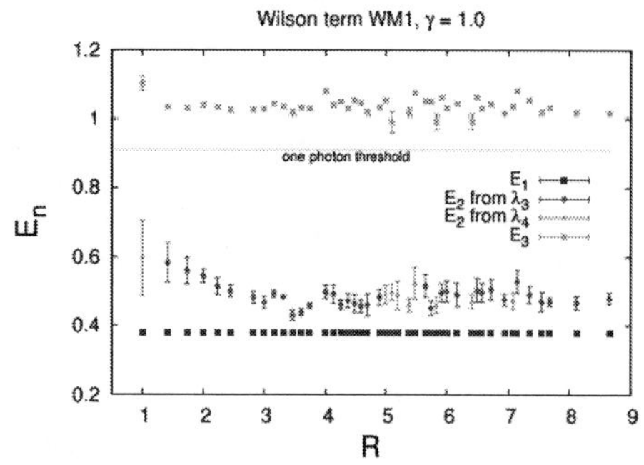

Figure 6. Ground and excitation energies of static fermions in the chiral gauge model at $\beta = 3, \gamma = 1$, shown together with the threshold for creation of a massive photon above the ground state energy.

"elementary" particles, which are not usually thought of as composite objects, can have a mass spectrum in gauge Higgs theories. This conclusion seems robust. We see it in SU(3) gauge Higgs theory, the abelian Higgs and Landau-Ginzburg models, and also in chiral U(1) models. The question now is whether such excitations can be observed experimentally in the gauge Higgs models that we know about, i.e. superconductivity and the electroweak sector of the standard model. We reserve this interesting question for future investigation.

References

[1] J. Greensite, K. Matsuyama, Phys. Rev. **D96**, 094510 (2017), 1708.08979
[2] J.C. Vink, U.J. Wiese, Phys. Lett. **B289**, 122 (1992), hep-lat/9206006
[3] K.J. Juge, J. Kuti, C. Morningstar, Phys. Rev. Lett. **90**, 161601 (2003), hep-lat/0207004
[4] B.B. Brandt, M. Meineri, Int. J. Mod. Phys. A **31**, 1643001 (2016), 1603.06969
[5] J. Greensite, Phys. Rev. D **102**, 054504 (2020), 2007.11616
[6] K. Matsuyama, Phys. Rev. D **103**, 074508 (2021), 2012.13991
[7] J. Smit, Acta Phys. Polon. B **17**, 531 (1986)
[8] P.V.D. Swift, Phys. Lett. B **145**, 256 (1984)
[9] M.F.L. Golterman, D.N. Petcher, J. Smit, Nucl. Phys. B **370**, 51 (1992)
[10] J. Greensite, Phys. Rev. D **104**, 034508 (2021)

Renormalizability of center-vortex sectors in continuum Yang-Mills theory

D. Fiorentini[1], *D.* R. Junior[1,2,*], *L.* E. Oxman[1], and *R.* F. Sobreiro[1]

[1]Instituto de Física, Universidade Federal Fluminense,
Av. Litorânea s/n, 24210-346, Niterói, RJ, Brasil.
[2]Institut für Theoretische Physik, Universität Tübingen,
Auf der Morgenstelle 14, 72076, Tübingen, Deutschland.

Abstract. Recently, a novel approach to quantize SU(N) Yang-Mills theory was proposed, where the configuration space $\{A_\mu\}$ is split into sectors labeled by topological defects, and then the gauge is fixed by a sector dependent condition. As the procedure is local in $\{A_\mu\}$, it could be free from Gribov copies. In this work, we review the renormalizability of sectors labeled by an arbitrary number of elementary center vortices.

Because of the Gribov problem, the understanding of continuum Yang-Mills theory in the infrared regime is certainly not within reach of the usual Fadeev-Popov method. This obstruction was established in Refs. [1, 2], where it was shown that there is no continuous, global gauge fixing condition in configuration space $\{A_\mu\}$, when the gauge group is $SU(N)$. One possible approach to circumvent this problem is to restrict the path integral to the first Gribov region, which was followed in Refs. [3–9].

An alternative approach was proposed by Singer in his seminal paper, which relies on a locally finite open covering $\{\vartheta_\alpha\}$ of the space of all gauge fields $\{A_\mu\}$, with a subordinate partition of unity

$$\sum_\alpha \rho_\alpha(A_\mu) = 1 \quad , \quad \forall A_\mu \in \{A_\mu\} , \tag{1}$$

with the support of the function $\rho_\alpha(A_\mu)$ being on ϑ_α. The idea is to insert this identity in the Yang-Mills partition function, as follows

$$Z_{YM} = \sum_\alpha Z_{(\alpha)} \quad , \quad Z_{(\alpha)} = \int_{\vartheta_\alpha} [DA]\rho_\alpha(A)e^{-S_{YM}(A)} , \tag{2}$$

where the sets $\{\vartheta_\alpha\}$ are to be chosen so as to guarantee the existence of local gauge-fixing conditions $f_\alpha(A) = 0$, $A \in \vartheta_\alpha$, which are free from Gribov copies. Recently, a concrete implementation of this procedure was proposed for the case where the sets ϑ_α are disjoint, thus forming a partition of $\{A_\mu\}$ [10]. The first step is to introduce auxiliary adjoint scalar fields ψ_I with an auxiliary action $S_{aux}(\psi, A)$ in the theory by means of an identity. Each gauge field is then correlated with the $\zeta_I(A)$ that minimizes $S_{aux}(A,\zeta)$. The next step is to perform a generalized polar decomposition on the tuple $\zeta_I(A)$, which allow us to write $\zeta_I(A) = S(A)q_I(A)S^{-1}(A)$, $S \in SU(N)$. The tuple q_I is the rotated ζ_I which is closest to a given

*Presenter; e-mail: davidjunior@id.uff.br.

element of $\mathcal{M}$, the vacuum manifold of S_{aux}. We choose this element to be vT_I, so that q_I satisfies the generalized pure modulus condition $[q_I, T_I] = 0$. One important fact is that the phase S will not always be regular. In general, we may write $S = US_0$, with U regular. We then introduce an equivalence class, where two phases are said to be equivalent if they differ by a regular gauge transformation, i.e.

$$[S] = [S'] \iff S = US' \,, \tag{3}$$

for some regular U. Then, it is natural to split the configuration space $\{A_\mu\}$ into sectors $\mathcal{V}_{S_0}$, where $A_\mu \in \mathcal{V}_{S_0} \iff [S] = [S_0]$. Because of the structure of S_{aux}, the labels S_0 are given by center vortex configurations. The gauge is then fixed by the sector dependent gauge condition

$$f_{S_0}(\psi(A)) = [\psi_I(A), S_0 T_I S_0^{-1}] = 0 \,. \tag{4}$$

In this case, Eq. (2) becomes

$$Z_{YM} = \sum_{S_0} Z_{YM}^{S_0} \quad , \quad Z_{YM}^{S_0} = \int_{\mathcal{V}(S_0)} [DA] e^{-S_{YM}} \delta(f_{S_0}(A)) \mathrm{Det} \frac{\delta f_{S_0}(A^U)}{\delta U}\bigg|_{U=I} \,. \tag{5}$$

Observables may be similarly evaluated by a sum over the partial contributions of the sectors, i.e.

$$\langle O \rangle = \sum_{S_0} \frac{Z_{YM}^{S_0}}{Z_{YM}} \langle O \rangle_{S_0} \,,$$

$$\langle O \rangle_{S_0} = \int_{\mathcal{V}(S_0)} [DA] O(A) e^{-S_{YM}} \delta(f_{S_0}(A)) \mathrm{Det} \frac{\delta \mathrm{f}_{S_0}(\mathrm{A}^{\mathrm{U}})}{\delta \mathrm{U}}\bigg|_{\mathrm{U=I}} \,. \tag{6}$$

Eq. (6) resembles the type of expressions which are considered in phenomenological approaches involving center vortex ensembles (see the reviews [11, 12], and references therein). In these approaches, averages of the Wilson Loop are represented by a sum over center vortex configurations with a weight that is chosen on phenomenological grounds. Therefore, this quantization procedure not only has the potential to avoid the Gribov problem, but also provides a glimpse of a path from Yang-Mills theory to an ensemble of center vortices. An important line of investigation is then to compute the partial contribution of each vortex sector from first principles in order to derive and/or improve the phenomenological approach. A relevant first step in this direction is to establish the renormalizability of the partial contributions Z_{S_0}. This was firstly studied for the vortex-free sector in Ref. [13], and then generalized to a sector containing a general oriented elementary center-vortex configuration in Ref. [14].

In section 1 we present the sector-dependent gauge fixed action that implements the conditions of Eq. (4). In section 2 we review the renormalizability of the vortex-free sector. In section 3 we review the generalization of this result to a sector labeled by an arbitrary number of elementary oriented center vortices. In section 4 we present our conclusions.

1 The gauge-fixed action

From Eq. (5) it is clear that the action will be a sector dependent functional of the fields, as the gauge fixing condition depends on $\mathcal{V}(S_0)$. In order to obtain a concrete expression for it, we introduce in each $Z_{YM}^{S_0}$ the identity

$$1 = \int [D\zeta_I] \delta\left(\frac{\delta S_{aux}}{\delta \zeta_I}\right) \det\left(\frac{\delta^2 S_{aux}}{\delta \zeta_I \delta \zeta_J}\right) \tag{7}$$

which correlates each gauge field A_μ with the tuple $\zeta_I(A)$ that minimizes S_{aux} in the presence of A_μ. As discussed in Ref. [15], it is important to choose S_{aux} carefully in order to obtain a well-defined mapping $A_\mu \to \zeta_I(A)$. In particular, this action should contain Spontaneous Symmetry Breaking (SSB) in order to avoid problems with the injectivity of the map. We considered the choice

$$S_{aux}(\zeta, A) = \int d^4x \Big(D^{ab}_\mu \zeta^b_I D^{ac}_\mu \zeta^c_I + \mu^2 \zeta^a_I \zeta^a_I + \kappa f_{IJK} f^{abc} \zeta^a_I \zeta^b_J \zeta^c_K + \lambda \gamma^{abcd}_{IJKL} \zeta^a_I \zeta^b_J \zeta^c_K \zeta^d_L \Big) , \tag{8}$$

the tensor γ being a general combination of structure constants and deltas. The delta functionals and determinants in Eq. (7) are exponentiated using flavored Lie algebra valued ghost $(\bar{c}_I, c_I)$ and Lagrange multiplier (b_I) fields. Similarly, the gauge-fixing terms in Eq. (5) are exponentiated by $\bar{c}, c$ and b. The resulting additional action is given by

$$\begin{aligned}\Sigma(S_0) = \int d^4x \Big(& D^{ab}_\mu b^b_I D^{ac}_\mu \zeta^c_I + D^{ab}_\mu \bar{c}^b_I D^{ac}_\mu c^c_I + \mu^2 \left(\bar{c}^a_I c^a_I + b^a_I \zeta^b_I \right) + \\ & \kappa f_{IJK} f^{abc} (b^a_I \zeta^b_J \zeta^c_K - 2\bar{c}^a_I \zeta^b_K c^c_J) + \lambda \gamma^{abcd}_{IJKL} \left(b^a_I \zeta^b_J \zeta^c_K \zeta^d_L + 3\bar{c}^a_I c^b_J \zeta^c_K \zeta^d_L \right) \\ & - U^2 \bar{c}^a_I \zeta^a_I - \Lambda f^{abc} f^{cde} \bar{c}^a_I \zeta^b_J \zeta^d_I \zeta^e_J - \mathcal{K} f^{IJK} f^{abc} \bar{c}^a_I \zeta^b_J \zeta^c_K \\ & + i f^{abc} \left(b^a \eta^b_I \zeta^c_I + \bar{c}^a \eta^b_I c^c_I \right) + f^{ecd} f^{eba} \bar{c}^a \eta^b_I \zeta^c_I c^d \Big) ,\end{aligned} \tag{9}$$

where the background field $\eta_I = v S_0 T_I S_0^{-1}$ implements the sector-dependent condition of Eq. (4). It is possible to show that this action has a BRST symmetry, given by

$$\begin{aligned}
sA^a_\mu &= \frac{i}{g} D^{ab}_\mu c^c , & sc^a &= -\frac{i}{g} f^{abc} c^b c^c , \\
s\zeta^a_I &= i f^{abc} \zeta^b_I c^c + c^a_I , & sc^a_I &= -i f^{abc} c^b_I c^c , \\
s\bar{c}^a_I &= -i f^{abc} \bar{c}^b_I c^c - b^a_I , & sb^a_I &= i f^{abc} b^b_I c^c \\
s\bar{c}^a &= -b^a , & sb^a &= 0 , \\
s\mu^2 &= U^2 , & sU^2 &= 0 , \\
s\kappa &= \mathcal{K} , & s\mathcal{K} &= 0 , \\
s\lambda &= \Lambda , & s\Lambda &= 0 .
\end{aligned} \tag{10}$$

The transformations of the gauge fixing parameters μ^2, κ, λ are introduced to ensure that all observables will be independent of them [16, 17].

2 The vortex-free sector

In Ref. [13] we have studied the renormalizability of this sector using the algebraic method. In this approach, the Ward Identities of the action are used to restrict the possible counter terms. In this regard, it is important to introduce external sources coupled with the non-linear BRST variations of the fields, i.e.

$$\begin{aligned}\Sigma^{(1)}_{ext} = \int d^4x \Big(& \frac{i}{g} K^a_\mu (D^{ab}_\mu c^b) - \frac{1}{2} i \bar{C}^a f^{abc} c^b c^c - i f^{abc} \bar{L}^a_I c^b_I c^c + Q^a_I (i f^{abc} \zeta^b_I c^c + c^a_I) \\ & + i f^{abc} B^a_I b^b_I c^c - L^a_I (i f^{abc} \bar{c}^b_I c^c + b^a_I) \Big) .\end{aligned} \tag{11}$$

Moreover, an additional pair of external sources (M_I^{ab}, N_I^{ab}) is introduced by means of the BRST exact term

$$\Sigma^{(2)}_{ext} = s \int d^4x\, M_I^{ab} \bar{c}^a \zeta_I^b \,. \tag{12}$$

They are introduced as a doublet, i.e. $sM_I^{ab} = N_I^{ab}$, $sN_I^{ab} = 0$. To analyze the renormalizability, we employ the algebraic method [17]. In this regard, we write the quantum action Γ at first order as

$$\Gamma^{(1)} = S_{vf} + \hbar \Sigma_C^{vf} \,, \tag{13}$$

Σ_C^{vf} being the counter term, and impose on it all the Ward Identities satisfied by the tree level action S_{vf}, given by

$$S_{\mathrm{vf}} = S_{YM} + \Sigma(S_0) + \Sigma^{(1)}_{\mathrm{ext}} + \Sigma^{(2)}_{\mathrm{ext}} \,. \tag{14}$$

2.1 Ward Identities and renormalization

Besides the Slavnov-Taylor, the action S_{vf} satisfies 6 other Ward Identities. These are:

- The gauge-fixing equation,

$$\frac{\delta S_{\mathrm{vf}}}{\delta b^a} = i f^{abc} \eta_I^b \zeta_I^c - M_I^{ab} \zeta_I^b \,, \tag{15}$$

- The anti-ghost equation,

$$\bar{\mathcal{G}}^a S_{\mathrm{vf}} = \left(\frac{\delta}{\delta \bar{c}^a} - i f^{abc} \eta_I^b \frac{\delta}{\delta Q_I^c} + M_I^{ab} \frac{\delta}{\delta Q_I^b} \right) S_{\mathrm{vf}} = 0 \,. \tag{16}$$

- The ghost number equation,

$$\begin{aligned} \mathcal{N}_{\mathrm{gh}} S_{\mathrm{vf}} = \int d^4x \Big(& c_I^a \frac{\delta}{\delta c_I^a} - \bar{c}_I^a \frac{\delta}{\delta \bar{c}_I^a} + c^a \frac{\delta}{\delta c^a} - \bar{c}^a \frac{\delta}{\delta \bar{c}^a} + \Lambda \frac{\delta}{\delta \Lambda} + \mathcal{K} \frac{\delta}{\delta \mathcal{K}} + \\ & + U^2 \frac{\delta}{\delta U^2} - K_\mu^a \frac{\delta}{\delta K_\mu^a} - Q_I^a \frac{\delta}{\delta Q_I^a} - B_I^a \frac{\delta}{\delta B_I^a} - 2\bar{C}^a \frac{\delta}{\delta \bar{C}^a} - 2\bar{L}_I^a \frac{\delta}{\delta \bar{L}_I^a} \\ & + N_I^{ab} \frac{\delta}{\delta N_I^{ab}} - \xi_I^a \frac{\delta}{\delta \xi_I^a} + m_{IJ}^{ab} \frac{\delta}{\delta m_{IJ}^{ab}} \Big) S_{\mathrm{vf}} = 0 \,. \end{aligned} \tag{17}$$

- The global flavor symmetry,

$$\begin{aligned} Q S_{\mathrm{vf}} = \Big(& \zeta_I^a \frac{\delta}{\delta \zeta_I^a} - b_I^a \frac{\delta}{\delta b_I^a} + c_I^a \frac{\delta}{\delta c_I^a} - \bar{c}_I^a \frac{\delta}{\delta \bar{c}_I^a} - \eta_I^a \frac{\delta}{\delta \eta_I^a} - Q_I^a \frac{\delta}{\delta Q_I^a} \\ & + B_I^a \frac{\delta}{\delta B_I^a} + L_I^a \frac{\delta}{\delta L_I^a} - \bar{L}_I^a \frac{\delta}{\delta \bar{L}_I^a} - \kappa \frac{\delta}{\delta \kappa} - 2\lambda \frac{\delta}{\delta \lambda} - \mathcal{K} \frac{\delta}{\delta \mathcal{K}} - 2\Lambda \frac{\delta}{\delta \Lambda} \\ & - N_I^{ab} \frac{\delta}{\delta N_I^{ab}} - M_I^{ab} \frac{\delta}{\delta M_I^{ab}} \Big) S_{\mathrm{vf}} = 0 \,. \end{aligned} \tag{18}$$

- The linearly broken rigid symmetry,

$$\mathcal{R}S_{\text{vf}} = \left(\zeta_I^a \frac{\delta}{\delta c_I^a} + \bar{c}_I^a \frac{\delta}{\delta b_I^a} - i f^{abc} \eta_I^a \frac{\delta}{\delta N_I^{bc}} - B_I^a \frac{\delta}{\delta \bar{L}_I^a} + \bar{L}_I^a \frac{\delta}{\delta Q_I^a} - \kappa \frac{\delta}{\delta \mathcal{K}} - 2\lambda \frac{\delta}{\delta \Lambda} \right.$$
$$\left. - M_I^{ab} \frac{\delta}{\delta N_I^{ab}} \right) S_{\text{vf}} = \bar{L}_I^a c_I^a + L_I^a \bar{c}_I^a - \zeta_I^a Q_I^a \,. \tag{19}$$

- The ghost equation,

$$\mathcal{G}^a S_{\text{vf}} = \left(\frac{\delta}{\delta c^a} + (f^{abc} f^{cnm} \eta_I^n + i f^{abn} M_I^{mn}) \frac{\delta}{\delta N_I^{mb}} \right) S_{\text{vf}}$$
$$= i f^{abc} \left(\bar{C}^b c^c + Q_I^b \zeta_I^c + \bar{L}_I^b c_I^c + L_I^b \bar{c}_I^c + B_I^b b_I^c \right) + \frac{i}{g} D_\mu^{ab} K_\mu^b \,. \tag{20}$$

These symmetries restricted the possible counter terms dramatically. Indeed, we showed that the most general Σ_C^{vf} is given by

$$\Sigma_C^{\text{vf}} = \int d^4x \Big(a_0 S_{YM}(A) + a_1 (\partial_\mu b_I^a \partial_\mu \zeta_I^a + g f^{abc} b_I^a \partial_\mu \zeta_I^b A_\mu^c + g f^{abc} \partial_\mu b_I^a A_\mu^b \zeta_I^c +$$
$$+ g^2 f^{abe} f^{cde} A_\mu^a b_I^b A_\mu^c \zeta_I^d + \partial_\mu \bar{c}_I^a \partial_\mu c_I^a + g f^{abc} \bar{c}_I^a + g f^{abc} \partial_\mu \bar{c}_I^a A_\mu^b c_I^c \partial_\mu c_I^b A_\mu^c$$
$$+ g^2 f^{abe} f^{cde} A_\mu^a \bar{c}_I^b A_\mu^c c_I^d) + a_2 f^{IJK} f^{abc} (\mathcal{K} \bar{c}_I^a \zeta_J^b \zeta_K^c - 2\kappa \bar{c}_I^a c_J^b \zeta_K^c - \kappa b_I^a \zeta_J^b \zeta_K^c) +$$
$$+ a_{3,IJKL}^{abcd} (\Lambda \bar{c}_I^a \zeta_J^b \zeta_K^c \zeta_L^d - 3\lambda \bar{c}_I^a c_J^b \zeta_K^c \zeta_L^d - \lambda b_I^a \zeta_J^b \zeta_K^c \zeta_L^d) +$$
$$+ a_4 (U^2 \bar{c}_I^a \zeta_I^a - \mu^2 \bar{c}_I^a c_I^a - \mu^2 b_I^a \zeta_I^a) \Big) \,. \tag{21}$$

Then, we showed that it may be absorbed by a redefinition of the parameters, fields and sources of S_{vf}, with appropriate renormalization factors.

3 Sectors labeled by center vortices

In Ref. [14], we considered a sector labeled by n center vortices with guiding centers at the closed worldsurfaces $\Omega_1, \ldots, \Omega_n$. When they carry the same fundamental weight β, the associated defect may be written as $S_0 = e^{i\chi \beta \cdot T}$, where χ is a multivalued angle with respect to $\Omega_1, \ldots, \Omega_n$. Here we are using the notation $\beta \cdot T = \beta|_q T_q$, T_q being the generators of the Cartan subalgebra of $su(N)$. In these sectors, because of the gauge condition ..., the gauge-fixed configurations of auxiliary fields will be of the form $\zeta_I = S_0 q_I S_0^{-1}$, with $[q_I, T_I] = 0$. This means that the components of ζ_I that rotate under S_0 must vanish at $\Omega = \Omega_1 \cup \cdots \cup \Omega_n$. These are the components along the off-diagonal generators $T_\alpha, T_{\bar{\alpha}}$, with $\alpha \cdot \beta \neq 0$, α being the roots of $su(N)$. To impose these boundary conditions, we followed the method of Ref. [18], where the delta function that implements the boundary conditions is exponentiated as follows

$$\prod_\gamma \delta_\Omega(\zeta_I^\gamma) \delta_\Omega(\zeta_I^{\bar{\gamma}}) =$$
$$\int [D\lambda]\, e^{i \sum_\gamma \int d\sigma_1 d\sigma_2 \sqrt{g(\sigma_1,\sigma_2)} \left(\lambda_I^\gamma(\sigma_1,\sigma_2) \zeta_I^\gamma(x(\sigma_1,\sigma_2)) + \lambda_I^{\bar{\gamma}}(\sigma_1,\sigma_2) \zeta_I^{\bar{\gamma}}(x(\sigma_1,\sigma_2)) \right)} \,. \tag{22}$$

Moreover, in order to account for the BRST invariance of these boundary conditions, we should also impose $s\zeta_I^\gamma = s\zeta_I^{\bar{\gamma}} = 0$ on Ω [19]. To this purpose, we represented this expression in an alternative, more symmetric form, by means of the Symanzik method [20], where a generic Lie algebra valued source $J(x)$ was introduced. Then

$$\prod_\gamma \delta_\Omega(\zeta_I^\gamma)\delta_\Omega(\zeta_I^{\bar{\gamma}})\delta_\Omega(s\zeta_I^\gamma)\delta_\Omega(s\zeta_I^{\bar{\gamma}}) \to \int [D\lambda][D\xi]\, e^{-\Sigma^{\text{b.c.}}(\lambda,\xi,J)} \,, \tag{23}$$

with the additional fields satisfying

$$\begin{aligned} s\xi_I^a &= \lambda_I^a \,, & s\lambda_I^a &= 0 \\ sJ &= 0 \,. \end{aligned} \tag{24}$$

The expression of interest is recovered when J is set to take its physical values (see Ref. [14]), and automatically takes into account the BRST invariance of the boundary conditions. Then, it is convenient to introduce external sources m_{IJ}^{ab}, n_{IJ}^{ab} in order to preserve the symmetries of the vortex-free sector. In order to guarantee that these do not alter the physical theory, they are introduced as a BRST doublet, i.e.

$$sm_{IJ}^{ab} = -n_{IJ}^{ab} \,, \qquad sn_{IJ}^{ab} = 0 \,. \tag{25}$$

Therefore, the full action in these sectors is given by

$$\Sigma^{\text{vortex}} = \Sigma(S_0) + S_{\text{ext}}^{(1)} + \Sigma_{\text{ext}}^{(2)} + \Sigma_{\text{ext}}^{(3)} + \Sigma^{\text{b.c.}} \,. \tag{26}$$

To analyze the renormalizability of Σ^{vortex}, we proceeded similarly as in the vortex-free case. Namely, we imposed the Ward Identities in the quantum action to restrict the possible counterterms, and studied the stability of the action.

3.1 Ward Identities and renormalization

Three new Ward Identities are satisfied in the center vortex sectors. These involve the new fields $\lambda_I, \xi_I, J, m_{IJ}, n_{IJ}$, and are given by

- The J equation,

$$\mathcal{J}^a\Sigma^{\text{vortex}} = \left(\frac{\delta}{\delta J^a} - f^{abc}\delta_{IJ}\frac{\delta}{\delta m_{IJ}^{bc}}\right)\Sigma^{\text{vortex}} = 0 \,. \tag{27}$$

- Global symmetry in the boundary-conditions sector,

$$\begin{aligned} \mathcal{F}\Sigma^{\text{vortex}} &= \left(\xi_I^a\frac{\delta}{\delta\xi_I^a} + \lambda_I^a\frac{\delta}{\delta\lambda_I^a} - J^a\frac{\delta}{\delta J^a} - n_{IJ}^{ab}\frac{\delta}{\delta n_{IJ}^{ab}} - m_{IJ}^{ab}\frac{\delta}{\delta m_{IJ}^{ab}}\right)\Sigma^{\text{vortex}} \\ &= 0 \,. \end{aligned} \tag{28}$$

- The linearly broken λ equation,

$$\Lambda_I^a\Sigma^{\text{vortex}} = \frac{\delta\Sigma^{\text{vortex}}}{\delta\lambda_I^a} = f^{abc}\zeta_I^b J^c \,. \tag{29}$$

In Ref. [14], we showed that $\Sigma_C^{\text{vortex}} = \Sigma_C^{\text{vf}}$, and that the action is stable in the vortex sectors.

4 Discussion

Recently, a new way to quantize Yang-Mills theory based on a local gauge fixing procedure, so as to avoid the presence of Gribov copies, was proposed. The path integral was written as a sum over inequivalent sectors labeled by center vortices, and the gauge is fixed with a sector dependent gauge condition. Within this procedure, observables of Yang-Mills theory are evaluated as a sum over partial contributions of each possible center vortex sector, providing a glimpse of a path from Yang-Mills theory to phenomenological center-vortex ensembles. In Ref. [13], we established the renormalizability of the vortex-free sector by using the algebraic method, and then, in Ref. [14], this was generalized to a sector containing an arbitrary number of elementary center vortices. The BRST structure guaranteed the independence of observables from the gauge fixing parameters, and the introduction of appropriate external sources allowed us to to establish a rich set of Ward Identities. These restrictive symmetries implied that the counter-term is the same in all the analyzed sectors, and so is the number of independent renormalization parameters. These results show that this first principles Yang-Mills ensemble is indeed calculable.

Acknowledgments

The Coordenação de Aperfeiçoamento de Pessoal de Nível Superior - Brasil (CAPES) – Finance Code 001, the Deutscher Akademischer Austauschdienst (DAAD) and the Conselho Nacional de Desenvolvimento Científico e Tecnológico (CNPQ) are acknowledged for the financial support.

References

[1] I. M. Singer, *Commun. Math. Phys.* **60** (1978) 7–12.
[2] V. N. Gribov, *Nucl. Phys.* **B139** (1978) 1.
[3] D. Zwanziger, *Nucl. Phys.* **B323** (1989) 513–544.
[4] D. Zwanziger, *Nucl. Phys.* **B399** (1993) 477–513.
[5] D. Dudal, R. F. Sobreiro, S. P. Sorella, and H. Verschelde, *Phys. Rev.* **D72** (2005) 014016.
[6] D. Dudal, J. A. Gracey, S. P. Sorella, N. Vandersickel, and H. Verschelde, *Phys. Rev.* **D78** (2008) 065047.
[7] D. Dudal, S. P. Sorella, and N. Vandersickel, *Phys. Rev.* **D84** (2011) 065039.
[8] M. A. L. Capri, D. Dudal, D. Fiorentini, M. S. Guimaraes, I. F. Justo, A. D. Pereira, B. W. Mintz, L. F. Palhares, R. F. Sobreiro, and S. P. Sorella, *Phys. Rev. D* **92** no. 4, (2015) 045039.
[9] M. A. L. Capri, D. Dudal, D. Fiorentini, M. S. Guimaraes, I. F. Justo, A. D. Pereira, B. W. Mintz, L. F. Palhares, R. F. Sobreiro, and S. P. Sorella, *Phys. Rev.* **D94** no. 2, (2016) 025035.
[10] L. E. Oxman and G. C. Santos-Rosa, *Phys. Rev. D* **92** (2015) 125025.
[11] Greensite, J. *An Introduction to the Confinement Problem*, 2nd ed.; Springer Nature Switzerland: Cham, Switzerland, 2020.
[12] D. R. Junior, L. E. Oxman, G. M. Simões, *Universe* **7(8)**, (2021) 253
[13] D. Fiorentini, D. R. Junior, L. E. Oxman, and R. F. Sobreiro, *Phys. Rev. D* **101** no. 8, (2020) 085007.
[14] D. Fiorentini, D. R. Junior, L. E. Oxman, R. F. Sobreiro, arXiv:2108.11361

[15] D. Fiorentini, D. R. Junior, L. E. Oxman, G. M. Simões, and R. F. Sobreiro, *Phys. Rev. D* **103** (2021) 114010.

[16] O. Piguet and K. S., *Nucl. Phys.* **B253** (1985) 517–540.

[17] O. Piguet and S. P. Sorella, "Algebraic renormalization: Perturbative renormalization, symmetries and anomalies". *Lect. Notes Phys. Monogr.* **28** (1995) 1–134.

[18] R. Golestanian and M. Kardar, *Phys. Rev. A* **58** (1998) 1713–1722.

[19] I. G. Moss and P. J. Silva,*Phys. Rev. D* **55** (1997) 1072–1078.

[20] K. Symanzik, *Commun. Math. Phys.* **16** (1970) 48–80.

EPJ Web of Conferences **258**, 02003 (2022) https://doi.org/10.1051/epjconf/202225802003
A Virtual Tribute to Quark Confinement and the Hadron Spectrum (vConf21)

Gauge-fixed Lattice QCD and the dispersion relation of Wilson fermions

Giuseppe Burgio[1,*] and *Hannes* Vogt[1,**]

[1]Institut für Theoretische Physik, Auf der Morgenstelle 14, 72076 Tübingen (Germany)

Abstract. We show that, when investigating Wilson-fermions correlation functions on the lattice, one is bound to encounter major difficulties in defining their dispersion relation, even at tree level. The problem is indeed quite general and, although we stumbled upon it while studying Coulomb-gauge applications, it also affects gauge fixed studies in covariant gauges, including their most popular version, Landau gauge. In this paper we will discuss a solution to this problems based on a redefinition of the kinematic momentum of the fermion.

1 Introduction

The investigation of the IR-properties of non abelian gauge theories is still one of the most interesting and fruitful chapters of particle physics. In particular, lattice studies have contributed a great deal to our understanding of the gauge invariant aspects of confinement, chiral symmetry breaking, hadron spectrum etc. However, both from a theoretical and from a practical point of view, there exists a broad interest in analysing whether and how such properties are encoded in gauge-fixed Green's functions, mainly since most applicable continuum methods still heavily rely on them. The corresponding first studies for QCD date back to the '70s, see e.g. [1–5], and many analytic and numerical investigations have since followed, both in covariant [6–31] and non-covariant gauges [32–64]. A consistent picture emerges from this corpus, supporting Gribov's original idea [2]: restricting the functional integral to the first Gribov region indeed generates a dynamical mass scale $O(1\,\text{GeV})$. Coulomb-gauge in particular, being the closest we have to a physical description [65, 66], offers many advantages; e.g., one can show that the self energy of the gluon can in fact be described with Gribov's original formula $\omega_A = \sqrt{|\vec{p}|^2 + M^4/|\vec{p}|^2}$, again with $M = O(1\,\text{GeV})$ [46].

A lesson one constantly learns from all lattice investigations at fixed gauge is that, to be able to extract meaningful informations, one must take particular care in treating discretization artifacts, which often cloud the sought results. For example, the gluon self-energy above could only be obtained after devising a proper scheme to remove spurious contributions caused by the discretized time [46]. Similarly, extracting the Coulomb string tension directly from the Coulomb-kernel, which one naively would consider the most straightforward prescription, turns out to be a quite non trivial task; alternative, less intuitive definitions turn out to be more efficient [59]. This holds of course also for quark-quark correlation functions, as the extraction of the chiral mass and of the self energy for staggered fermions, necessary to detect any evidence of chiral symmetry breaking and confinement, clearly exemplifies [55].

*e-mail: giuseppe.burgio@uni-tuebingen.de
**e-mail: hannes@vogt.de

In this paper we will show that for Wilson fermions extra care needs to be taken when calculating fermionic Green's functions. Indeed, the problem lies in the very definition of a suitable kinematic momentum against which to plot fermionic correlation functions, independently of the gauge, and thus affects all gauge fixed investigation, e.g. also in Landau gauge [14]. Of course, one could argue that other lattice fermions could be used for QCD instead, since they should not suffer from such problems. However, the recent surge in investigations of beyond the standard model scenarios, with many lattice simulations performed for different models, has increased the interest in studying the non perturbative properties of correlation functions in QCD-like theories; depending on the model chosen, one will want to keep the freedom of using the most convenient discretization for the fermionic fields. Indeed, that's how we noticed the issue in the first place - when trying to apply the analysis of [55] to SU(2) + (adjoint) Wilson fermions configurations generated by the Edinburgh-Swansea-Odense collaboration as a model for minimal walking Technicolor [67–70].

2 The fermion self energy

The most general structure for the expectation value of the Dirac operator, both in covariant and Coulomb gauges, reads [14, 55]:

$$S^{-1}(\vec{p}, p_4) = i\,\vec{p}\!\!\!/ A_s(\vec{p}, p_4) + i\,p\!\!\!/_4 A_t(\vec{p}, p_4) + i\,p_i\gamma_i p_4\gamma_4 A_d(\vec{p}, p_4) + B(\vec{p}, p_4)\mathbb{1}\,. \qquad (1)$$

A_d can be shown to vanish in general, while of course $A_s(p) = A_t(p)$ for covariant gauges [14, 55]. For Coulomb gauge, moreover, A_s, A_t and B can be shown to be p_4 independent [55]. Factorizing $A_s = Z^{-1}$ out, Eq. (1) can be rewritten as [14, 55]:

$$S^{-1}(\vec{p}, p_4) = Z^{-1}(\vec{p})\left[i\,\vec{p}\!\!\!/ + i\,p\!\!\!/_4\,\alpha(\vec{p}) + M(\vec{p})\mathbb{1}\right] \quad \left(\text{cov. } S^{-1}(p) = Z^{-1}(p)\left[i\,p\!\!\!/ + M(p)\mathbb{1}\right]\right), \qquad (2)$$

where $\alpha = A_t/A_s$ and $M = B/A_s$. The inverse Dirac propagator above can be used to define the static quark propagator $S(\vec{p}) = \int dp_4\, S(\vec{p}, p_4)$, which, in analogy to the free case, corresponds up to a factor $1/2$ to the Hamiltonian operator H. Contrary to covariant gauges, this is straightforward to calculate in Coulomb gauge, yielding [55]:

$$H(\vec{p}) = \frac{Z(\vec{p})}{\alpha(\vec{p})}\frac{\sqrt{\vec{p}^2 + M^2(|\vec{p}|)}}{i\,\vec{p}\!\!\!/ + M(\vec{p})\mathbb{1}} = \frac{Z(\vec{p})}{\alpha(\vec{p})}\frac{-i\,\vec{p}\!\!\!/ + M(\vec{p})\mathbb{1}}{\sqrt{\vec{p}^2 + M^2(|\vec{p}|)}}\,. \qquad (3)$$

The dispersion relation of the fermion can therefore be directly read off from Eq. (3), since the inverse coefficient of $-i\,\vec{p}\!\!\!/ + M(\vec{p})$ corresponds to the eigenvalue of H:

$$\omega(|\vec{p}|) = \frac{\alpha(|\vec{p}|)}{Z(|\vec{p}|)}\sqrt{\vec{p}^2 + M^2(|\vec{p}|)}\,. \qquad (4)$$

As shown in [55], chiral symmetry breaking is encoded in $M(|\vec{p}|)$, while for a confining theory both ω_A and ω should be IR-divergent. Non-perturbative Coulomb gauge therefore nicely decouples the chiral symmetry and confinement properties of QCD-like theories, making it an interesting tool in the analysis of beyond the standard model scenarios [71].

3 The kinematic momentum for Wilson fermions

If the self-energy in Eq. (4) had been obtained in the continuum, there would be nothing one should be careful about. On the lattice, on the other hand, the discrete momenta

$k_i = \frac{2\pi}{aN_i}n_i$, $k_4 = \frac{2\pi}{aN_t}\left(n_4 + \frac{1}{2}\right)$ do not define suitable kinematic variables that can be related to the continuum ones. For staggered and overlap fermions there is however an easy prescription: calculate first, either analytically or numerically, the quark propagator in the free case, $S^{(0)}(k) = i\left(\sum_\mu C_\mu(k_\mu)\gamma_\mu\right) + m\,\mathbb{1}$. Isolating the coefficients of the (euclidean) γ-matrices defines now the kinematic momenta: $p_\mu(k_\mu) = C_\mu(k_\mu)$. The reason why such procedure works is of course that the $\mathbb{1}$-coefficient is k independent. For Wilson fermions, on the other hand, the doubling problem is solved by modifying exactly such coefficient, introducing an explicit k-dependence:

$$S^{(0)}(k) = \frac{i}{a}\sum_{\mu=1}^{4}\gamma_\mu\tilde{p}_\mu(k) + \left(m + \frac{a}{4}\sum_{\mu=1}^{4}p_\mu^2(k)\right)\mathbb{1}\,, \tag{5}$$

with $\tilde{p}_\mu(k_\mu) = \frac{1}{a}\sin(ak_\mu)$ and $p_\mu(k_\mu) = \frac{2}{a}\sin(ak_\mu/2)$; a definition of the kinematic momenta as for staggered/overlap fermions is therefore not possible. Previous studies in Landau gauge [14] have ignored the problem and used p_μ instead; this however does not reproduce the correct dispersion relation even in the free case, in contrast e.g. to ω_A, which is a "natural" function of $p_\mu(k_\mu)$ [13].

To illustrate our proposal, start by considering the free static propagator:

$$\int_{-\infty}^{\infty}\frac{dp_4}{2\pi}S_0(\vec{p}, p_4) = \frac{-i\,\vec{p\!\!/} + m\,\mathbb{1}}{2\sqrt{\vec{p}^2 + m^2}} \equiv \frac{-i\,\vec{p\!\!/} + m\,\mathbb{1}}{2\omega_0}\,, \tag{6}$$

where ω_0 is the free dispersion relation, i.e. the eigenvalue of the free Hamiltonian, the p are all functions of k and:

$$S_0(\vec{p}, p_4) = \frac{-i\,p\!\!/_4 - i\,\vec{p\!\!/} + m\,\mathbb{1}}{p_4^2 + \vec{p}^2 + m^2}\,. \tag{7}$$

Notice now that the free dispersion relation can be directly obtained from the free propagator:

$$\omega_0^{-1}(k) = \frac{1}{\sqrt{\vec{p}^2(k) + m^2}} = \int_{-\infty}^{\infty}\frac{dp_4}{2\pi}\left|S_0(\vec{p}(k), p_4)\right|^2\,, \tag{8}$$

where

$$|S_0(p)|^2 = S_0^\dagger(p)S_0(p) = \frac{1}{p_4^2 + \vec{p}^2 + m^2}\,, \tag{9}$$

The kinematic momentum can be thus "defined" via S_0, calculating $\omega_0(k)$ through Eq. (8) and setting $|\vec{p}(k)| = \sqrt{\omega_0^2(k) - m^2}$, very much in the spirit of the gluonic "tree-level correction" of [13]. Since in lattice calculations, to avoid discretization errors arising from the breaking of rotational invariance, one only considers diagonal momenta of the form $\vec{k} = (k, k, k)$ [13], such prescription is sufficient for all practical purposes. Explicitly (for the sake of readability the lattice spacing will be set to $a = 1$ in the following):

$$S_0^{\rm lat}(k) = \frac{-i\gamma_4\sin(k_4) - i\sum_i\gamma_i\sin(k_i) + (m + \sum_\mu(1 - \cos(k_\mu)))\,\mathbb{1}}{\sin^2(k_4) + \sum_i\sin^2(k_i) + \left[m + \sum_\mu(1 - \cos(k_\mu))\right]^2}\,, \tag{10}$$

$$\left|S_0^{\rm lat}(k)\right|^2 = \frac{1}{\sin^2(k_4) + \sum_i\sin^2(k_i) + \left[m + \sum_\mu(1 - \cos(k_\mu))\right]^2}\,. \tag{11}$$

Replacing $\int \frac{dp_4}{2\pi}$ with $\frac{1}{2\pi}\frac{4\pi}{N_t}\sum_{k_4}$ in Eq. (8) we thus get:

$$\omega_{0,m}^{-1}(k) = \frac{1}{\sqrt{\vec{p}^2 + m^2}} \equiv \frac{2}{N_t}\sum_{k_4}\left|S_0^{\text{lat}}(k)\right|^2 , \tag{12}$$

$$|\vec{p}(k)| = \sqrt{\frac{1}{\left(\frac{2}{N_t}\sum_{k_4}\left|S_0^{\text{lat}}(k)\right|^2\right)^2} - m^2} . \tag{13}$$

Setting now $m = 0$ we get, as operative definition of the kinematic momentum:

$$\begin{aligned} |\vec{p}(k)|^{-1} &= \frac{2}{N_t}\sum_{k_4}\left|S_0^{\text{lat}}(k)\right|^2 \\ &= \frac{2}{N_t}\sum_{n_4=-N_t/2+1}^{N_t/2} \frac{1}{\sin^2(k_4) + \sum_i \sin^2(k_i) + \left[\sum_\mu (1-\cos(k_\mu))\right]^2} . \end{aligned} \tag{14}$$

The values arising from Eq. (14) can be calculated once and for all for each desired lattice size and stored to be used whenever needed. Figure 1 shows the comparison between the "naive" fermionic kinematic momentum $p(k) = 2\sqrt{\sum_i \sin^2(k_i/2)}$, as used e.g. in [14], and our definition Eq. (14) for a 128^4 lattice. As one can see, the differences are quite striking throughout the whole Brillouin-zone. Figure 2 compares the (massive) free fermionic dis-

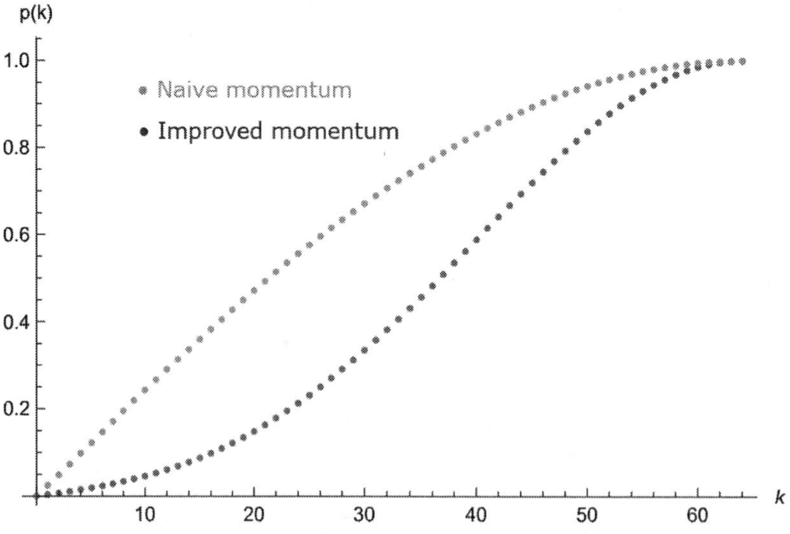

Figure 1. Comparison between $p(k) = 2\sqrt{\sum_i \sin^2(k_i/2)}$ and Eq. (14) for a 128^4 lattice ($\vec{k} = (k,k,k)$).

persion relations plotted against the naive momentum and the one obtained from Eq. (14), with an arbitrarily chosen bare mass. While the former shows a dependence which is physically meaningless for all momenta and is bound to distort any IR results, the latter gives (by definition!) exactly the expected physical behaviour throughout the whole Brillouin-zone.

The above definition will suffice for most applications. Should one however think that mass contributions in the definition of the kinematic momenta might play a role in the application at hand, and therefore wish to extend the prescription to $m \neq 0$, a bit of care is

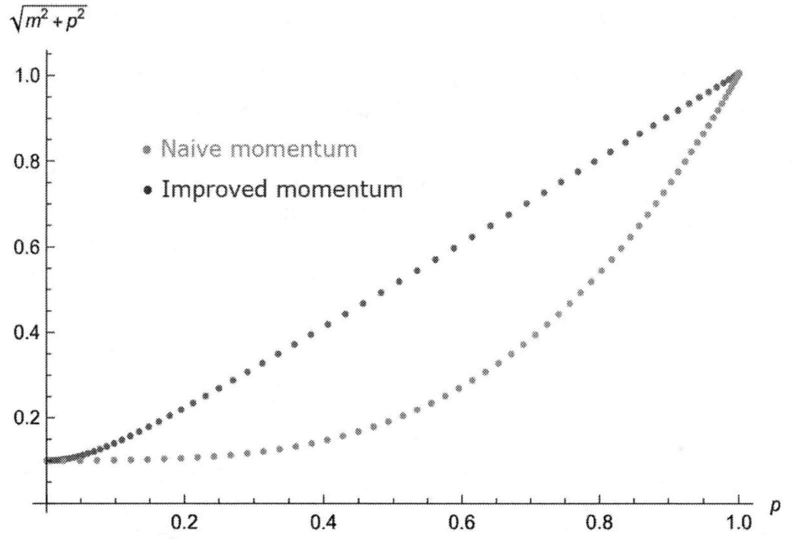

Figure 2. Massive fermionic dispersion relations vs. naive and improved momentum from Eq. (14).

needed. As one can immediately verify, for $m \neq 0$ Eq. (12) shifts the pole from m to $\frac{1}{2}m^2 + m$, so that from Eq. (13) one gets $\vec{p}(k = 0) \neq 0$, moreover with the wrong slope at the origin, $\left.\frac{dp}{dk}\right|_{k=0} \neq 1$. A "mass-improved" prescription should be therefore defined as follows. Expand first $\omega^2_{0,m}(k)$ from Eq. (12) around $k = 0$ up to 2nd order for diagonal momenta, obtaining:

$$\omega^2_{0,m}(k) = \left[\frac{N_t}{2}\frac{1}{\sum_{k_4}\frac{1}{\sin^2(k_4)+3\sin^2(k)+\left[m+2\sin^2\left(\frac{k_4}{2}\right)+6\sin^2\left(\frac{k}{2}\right)\right]^2}}\right]^2 = \omega^2_{0,m}(0)+\frac{1}{2}\left.\frac{\partial^2\omega^2_{0,m}(k)}{\partial k^2}\right|_{k=0} k^2+O(k^4)\,. \tag{15}$$

From this on can then directly define:

$$|\vec{p}(k)|^2 = \left[\omega^2_{0,m}(k) - \omega^2_{0,m}(0)\right]\left[\frac{1}{2}\left.\frac{\partial^2\omega^2_{0,m}(k)}{\partial k^2}\right|_{k=0}\right]^{-1}, \tag{16}$$

with:

$$\omega^2_{0,m}(0) = \left[\frac{N_t}{2\sum_{k_4}\frac{1}{S_m(k_4)}}\right]^2, \qquad \left.\frac{\partial^2\omega^2_{0,m}(k)}{\partial k^2}\right|_{k=0} = \frac{N_t^2\sum_{k_4}\frac{6+6\left(m+2\sin^2\left(\frac{k_4}{2}\right)\right)}{S^2_m(k_4)}}{2\left[\sum_{k_4}\frac{1}{S_m(k_4)}\right]^3},$$

$$S_m(k_4) = \sin^2(k_4) + \left[m + 2\sin^2\left(\frac{k_4}{2}\right)\right]^2 = m^2 + 4(m+1)\sin^2\left(\frac{k_4}{2}\right). \tag{17}$$

4 Conclusions and outlook

In this paper we have given an improved definition of the kinematic momentum for Wilson fermions, which we believe to be crucial if one wishes to extract any information from the IR-behaviour of correlation functions. Such definition, adapted accordingly, should be used in any gauge-fixed analysis involving Wilson-fermions, e.g. also in Landau gauge. Concerning our original motivation, i.e. the application of the analysis of [55] to the configurations used in [67–70], two further problem still need to be solved. First, for Wilson fermions the coefficient B in Eq. (1) is k_4 dependent, both at tree level and in the interacting case. Normalizing B to its tree level expression, i.e. the $\mathbb{1}$ coefficient in Eq. (5), as proposed in [72] for Landau gauge, could only work if the ratio for different energies at fixed $\vec{k}$, $B_{\text{lat}}(\vec{k}, k_4^1)/B_{\text{lat}}(\vec{k}, k_4^2)$ with $k_4^1 \neq k_4^2$, would exactly match their equivalent ratios at tree-level, which one can immediately verify not to be the case. Thus a different prescription needs to be developed, a possible one being to average the normalized ratios over k_4:

$$B(\vec{k}) = m\frac{1}{N_t}\sum_{k_4}\frac{B_{\text{lat}}(\vec{k}, k_4)}{B^0_{\text{lat}}(\vec{k}, k_4)}, \tag{18}$$

where the factor m is needed to guarantee $B(0) = m$. The second problem comes from the fact that for adjoint Wilson fermions the massless limit at $\beta = 2.25$ is approached for a negative bare mass, $-am_0 \approx 1.20$. However this is neither a meaningful choice in Eq. (18) nor does it provide a reasonable tree-level correction for B, as already pointed out for Landau gauge [72]. The proposal there was to use the PCAC mass instead, which might be a good choice in the continuum limit, but can still be the source of other uncertainty at finite lattice spacing. Further work will definitely be needed to settle the issue.

References

[1] A. Casher, J.B. Kogut, L. Susskind, Phys. Rev. D **10**, 732 (1974)
[2] V.N. Gribov, Nucl. Phys. **B139**, 1 (1978)
[3] S. Mandelstam, Phys. Rev. D **20**, 3223 (1979)
[4] T. Kugo, I. Ojima, Prog.Theor.Phys.Suppl. **66**, 1 (1979)
[5] J.M. Cornwall, Phys. Rev. D **26**, 1453 (1982)
[6] D. Zwanziger, Nucl. Phys. **B364**, 127 (1991)
[7] D. Zwanziger, Nucl. Phys. **B399**, 477 (1993)
[8] D. Zwanziger, Nucl. Phys. **B412**, 657 (1994)
[9] C.D. Roberts, A.G. Williams, Prog. Part. Nucl. Phys. **33**, 477 (1994), hep-ph/9403224
[10] A. Cucchieri, T. Mendes, Nucl. Phys. **B471**, 263 (1996), hep-lat/9511020
[11] A. Cucchieri, Nucl. Phys. **B508**, 353 (1997), hep-lat/9705005
[12] A. Cucchieri, Nucl. Phys. **B521**, 365 (1998), hep-lat/9711024
[13] D.B. Leinweber, J.I. Skullerud, A.G. Williams, C. Parrinello (UKQCD), Phys. Rev. **D60**, 094507 (1999), [Erratum: Phys. Rev.D61,079901(2000)], hep-lat/9811027
[14] J.I. Skullerud, A.G. Williams, Phys. Rev. **D63**, 054508 (2001), hep-lat/0007028
[15] R. Alkofer, L. von Smekal, Phys. Rept. **353**, 281 (2001), hep-ph/0007355
[16] P. Boucaud et al., JHEP **04**, 006 (2000), hep-ph/0003020
[17] F.D. Bonnet, P.O. Bowman, D.B. Leinweber, A.G. Williams, J.M. Zanotti, Phys.Rev. **D64**, 034501 (2001), hep-lat/0101013
[18] P.O. Bowman, U.M. Heller, A.G. Williams, Nucl.Phys.Proc.Suppl. **106**, 820 (2002), hep-lat/0110081
[19] F.D.R. Bonnet, P.O. Bowman, D.B. Leinweber, A.G. Williams, J.b. Zhang (CSSM Lattice), Phys. Rev. **D65**, 114503 (2002), hep-lat/0202003
[20] P.O. Bowman, U.M. Heller, A.G. Williams, Phys. Rev. D **66**, 014505 (2002), hep-lat/0203001

[21] D. Zwanziger, Phys. Rev. **D69**, 016002 (2004), hep-ph/0303028
[22] J.C.R. Bloch, A. Cucchieri, K. Langfeld, T. Mendes, Nucl. Phys. **B687**, 76 (2004), hep-lat/0312036
[23] P.O. Bowman, U.M. Heller, D.B. Leinweber, M.B. Parappilly, A.G. Williams, Phys.Rev. **D70**, 034509 (2004), hep-lat/0402032
[24] J. Gattnar, K. Langfeld, H. Reinhardt, Phys. Rev. Lett. **93**, 061601 (2004), hep-lat/0403011
[25] A. Cucchieri, T. Mendes, A. Mihara, JHEP **12**, 012 (2004), hep-lat/0408034
[26] I.L. Bogolubsky, G. Burgio, M. Muller-Preussker, V.K. Mitrjushkin, Phys. Rev. D **74**, 034503 (2006), hep-lat/0511056
[27] W. Schleifenbaum, M. Leder, H. Reinhardt, Phys. Rev. **D73**, 125019 (2006), hep-th/0605115
[28] I.L. Bogolubsky, V.G. Bornyakov, G. Burgio, E.M. Ilgenfritz, M. Muller-Preussker, V.K. Mitrjushkin, Phys. Rev. D **77**, 014504 (2008), [Erratum: Phys.Rev.D 77, 039902 (2008)], 0707.3611
[29] D. Dudal, S. Sorella, N. Vandersickel, H. Verschelde, Phys.Rev. **D79**, 121701 (2009), 0904.0641
[30] N. Vandersickel, D. Zwanziger, Phys. Rept. **520**, 175 (2012), 1202.1491
[31] V. Mader, M. Schaden, D. Zwanziger, R. Alkofer, Eur.Phys.J. **C74**, 2881 (2014), 1309.0497
[32] G. Dell'Antonio, D. Zwanziger, Commun. Math. Phys. **138**, 291 (1991)
[33] D. Zwanziger, Nucl. Phys. **B518**, 237 (1998)
[34] A. Cucchieri, D. Zwanziger, Phys. Rev. **D65**, 014001 (2001), hep-lat/0008026
[35] A. Cucchieri, D. Zwanziger, Phys. Rev. **D65**, 014002 (2001), hep-th/0008248
[36] A. Cucchieri, D. Zwanziger, Phys. Lett. **B524**, 123 (2002), hep-lat/0012024
[37] D. Zwanziger, Phys. Rev. Lett. **90**, 102001 (2003), hep-lat/0209105
[38] C. Feuchter, H. Reinhardt, Phys. Rev. **D70**, 105021 (2004), hep-th/0408236
[39] J. Greensite, S. Olejnik, D. Zwanziger, Phys. Rev. **D69**, 074506 (2004), hep-lat/0401003
[40] D. Epple, H. Reinhardt, W. Schleifenbaum, Phys. Rev. **D75**, 045011 (2007), hep-th/0612241
[41] Y. Nakagawa, A. Nakamura, T. Saito, H. Toki, D. Zwanziger, Phys. Rev. **D73**, 094504 (2006), hep-lat/0603010
[42] M. Quandt, G. Burgio, S. Chimchinda, H. Reinhardt, PoS **LAT2007**, 325 (2007), 0710.0549
[43] D. Epple, H. Reinhardt, W. Schleifenbaum, A.P. Szczepaniak, Phys. Rev. **D77**, 085007 (2008), 0712.3694
[44] M. Quandt, G. Burgio, S. Chimchinda, H. Reinhardt, PoS **CONFINEMENT8**, 066 (2008), 0812.3842
[45] H. Reinhardt, Phys. Rev. Lett. **101**, 061602 (2008), 0803.0504
[46] G. Burgio, M. Quandt, H. Reinhardt, Phys. Rev. Lett. **102**, 032002 (2009), 0807.3291
[47] G. Burgio, M. Quandt, H. Reinhardt, Phys. Rev. **D81**, 074502 (2010), 0911.5101
[48] D.R. Campagnari, H. Reinhardt, Phys. Rev. **D82**, 105021 (2010), 1009.4599
[49] M. Leder, J.M. Pawlowski, H. Reinhardt, A. Weber, Phys. Rev. **D83**, 025010 (2011), 1006.5710
[50] M. Quandt, H. Reinhardt, G. Burgio, Phys. Rev. **D81**, 065016 (2010), 1001.3699
[51] H. Reinhardt, M. Quandt, G. Burgio, Phys. Rev. **D85**, 025001 (2012), 1110.2927

[52] H. Reinhardt, D.R. Campagnari, A.P. Szczepaniak, Phys. Rev. **D84**, 045006 (2011), 1107.3389
[53] J. Heffner, H. Reinhardt, D.R. Campagnari, Phys. Rev. **D85**, 125029 (2012), 1206.3936
[54] G. Burgio, M. Quandt, H. Reinhardt, Phys. Rev. **D86**, 045029 (2012), 1205.5674
[55] G. Burgio, M. Schrock, H. Reinhardt, M. Quandt, Phys. Rev. **D86**, 014506 (2012), 1204.0716
[56] M. Pak, H. Reinhardt, Phys. Rev. **D88**, 125021 (2013), 1310.1797
[57] H. Vogt, G. Burgio, M. Quandt, H. Reinhardt, PoS **LATTICE2013**, 363 (2014), 1311.5707
[58] J. Greensite, A.P. Szczepaniak, Phys. Rev. **D91**, 034503 (2015), 1410.3525
[59] G. Burgio, M. Quandt, H. Reinhardt, H. Vogt, Phys. Rev. **D92**, 034518 (2015), 1503.09064
[60] D.R. Campagnari, H. Reinhardt, Phys. Rev. **D92**, 065021 (2015), 1507.01414
[61] J. Heffner, H. Reinhardt, Phys. Rev. **D91**, 085022 (2015), 1501.05858
[62] P. Vastag, H. Reinhardt, D. Campagnari, Phys. Rev. **D93**, 065003 (2016), 1512.06733
[63] G. Burgio, M. Quandt, H. Reinhardt, H. Vogt, Phys. Rev. D **95**, 014503 (2017), 1608.05795
[64] H. Reinhardt, G. Burgio, D. Campagnari, E. Ebadati, J. Heffner, M. Quandt, P. Vastag, H. Vogt, Adv. High Energy Phys. **2018**, 2312498 (2018), 1706.02702
[65] N.H. Christ, T.D. Lee, Phys. Rev. **D22**, 939 (1980), [Phys. Scripta23,970(1981)]
[66] S.L. Adler, A.C. Davis, Nucl. Phys. B **244**, 469 (1984)
[67] L. Del Debbio, A. Patella, C. Pica, Phys. Rev. **D81**, 094503 (2010), 0805.2058
[68] L. Del Debbio, B. Lucini, A. Patella, C. Pica, A. Rago, Phys. Rev. **D82**, 014509 (2010), 1004.3197
[69] L. Del Debbio, B. Lucini, A. Patella, C. Pica, A. Rago, Phys. Rev. **D82**, 014510 (2010), 1004.3206
[70] L. Del Debbio, B. Lucini, A. Patella, C. Pica, A. Rago, Phys. Rev. **D93**, 054505 (2016), 1512.08242
[71] G. Cossu, M. D'Elia, A. Di Giacomo, G. Lacagnina, C. Pica, Phys. Rev. D **77**, 074506 (2008), 0802.1795
[72] D. August, A. Maas, JHEP **07**, 001 (2013), 1304.4423

On boundary corrections of Lüscher-Weisz string

M. N. Khalil[1,2,3,4,*], *A.* Bakry[5], *X.* Chen[5], *M.* Deliyergiyev[6], *A.* Galal[1], *A.* Khalaf[1], and *P. M.* Zhang[7]

[1]Department of Physics, Al Azhar University, Cairo 11651, Egypt
[2]Department of Mathematics, Bergische Universität Wuppertal, Wuppertal 42097, Germany
[3]Computation-based Science Research Center, Cyprus Institute, Nicosia 2121, Cyprus
[4]Department of Physics, University of Ferrara, Ferrara 44121, Italy
[5]Institute of Modern Physics, Chinese Academy of Sciences, Gansu 730000, China
[6]Department of nuclear and particle physics, University of Geneva, CH-211, Switzerland
[7]Department of physics, Sun Zhongshan University, Gaungdong 510275, China.

Abstract. The potential and the density profile of the QCD flux-tube are investigated within the framework of the Lüscher-Weisz (LW) string action with two boundary terms. The Numerical simulations involve 4D SU(3) Yang-Mills LGT at finite temperature. In general, we detect signatures of the two boundary terms considered in the LW string action. Near the end of QCD Plateau, the LW string is yielding a static potential which is in a good agreement with the lattice data for source separations $R \geq 0.3$ fm. However, at $T/T_c = 0.9$, the fit to the potential data improves with a good fit attained at$R \geq 0.7$ fm. The mean-square width of the energy profile at $T/T_c = 0.9$ matches well the width of the LW string over distance scales $R \geq 0.5$ fm.

1 Introduction

The formation of a stringlike flux tube linking the color sources provides a prospective model for the confinement of quarks. In this picture, the nonperturbave properties of the QCD flux-tubes are expected to conform with that of an open effective bosonic strings in the long string limit. The model is expected to precisely predict a set of infrared (IR) observable which can be compared to the outcomes from the numerical lattice data [1].

Nevertheless, the analysis of string's fine structure in the lattice numerical data for the broadening profile revealed substantial mismatch [2] from the free-string Nambu-Goto (NG) model in the intermediate distance region at high temperatures. The excited spectrum [3] and string tension [4] obtained from the partition function of the leading-order Gaussian formulation of NG string action show a similar disagreement with the numerical data [2, 4, 5] for source separation R less than 1.0 fm.

The precise study of the strings effects suggests considering other features beyond the free NG action, in particular, possible stiffness properties [6] and boundary term corrections [7–9]. The Lorentz-invariant boundary corrections [9] to the static $Q\bar{Q}$ potential have been viable in modeling the data extracted either from the Wilson or the Polyakov-loop operators [2, 7, 8].

The boundary corrections to the static quark potential may account for the deviations among predictions of the effective string and numerical outcomes [9]. Moreover, the contributions of the boundary action to the width profile have neither been theoretically calculated

*e-mail: m.nasr@stimulate-ejd.eu

nor confronted with the numerical lattice data. These boundary effects are hoped to disclose the features of the profile of QCD flux-tube beyond the leading-order approximations which is the goal of this paper.

2 Lüscher-Weisz string action

The Lüscher and Weisz [1] action (physical gauge) which up to four-derivative term reads as

$$S^{\rm LW}[X] = \sigma_0 A + \frac{\sigma_0}{2}\int d^2\zeta\Bigg[\left(\frac{\partial X}{\partial\zeta_\alpha}\cdot\frac{\partial X}{\partial\zeta_\alpha}\right) + \kappa_2\left(\frac{\partial X}{\partial\zeta_\alpha}\cdot\frac{\partial X}{\partial\zeta_\alpha}\right)^2 + \kappa_3\left(\frac{\partial X}{\partial\zeta_\alpha}\cdot\frac{\partial X}{\partial\zeta_\beta}\right)^2\Bigg] + \ldots$$
$$+\gamma\int d^2\zeta\,\sqrt{g}\,\mathcal{R} + \alpha\int d^2\zeta\,\sqrt{g}\,\mathcal{K}^2 + S^b, \tag{1}$$

the vector $X^\mu(\zeta_0,\zeta_1)$ maps the region $C \subset \mathbb{R}^2$ into $\mathbb{R}^4$. In the physical gauge $X^1 = \zeta_0$ and $X^4 = \zeta_1$ which restrict the string fluctuations to transverse directions of the world-sheet C. Invariance under party transform would keep only the terms of even number of derivatives. The map g is the two-dimensional induced metric on the world sheet embedded in the background $\mathbb{R}^4$. The world-sheet area A and the parameters σ_0, α and γ are the string tension, the rigidity and the winding number, respectively. The couplings κ_2, κ_3 are effective low-energy parameters. The open-closed duality [3] imposes constraint on the kinematically-dependent couplings κ_2, κ_3 such that $\kappa_2 + \kappa_3 = \frac{-1}{8\sigma_0}$. The above action Eq. (1) encompasses surface/boundary terms S^b such that the symmetry breaking at the string's boundaries is built in. The boundary term S^b is given by

$$S^b = \int_{\partial\Sigma} d\zeta_0\left[b_1\frac{\partial X}{\partial\zeta_1}\cdot\frac{\partial X}{\partial\zeta_1} + b_2\frac{\partial\partial X}{\partial\zeta_1\partial\zeta_0}\cdot\frac{\partial\partial X}{\partial\zeta_1\zeta_0} + b_3\left(\frac{\partial X}{\partial\zeta_1}\cdot\frac{\partial X}{\partial\zeta_1}\right)^2 + b_4\frac{\partial^2\partial X}{\partial\zeta_0^2\partial\zeta_1}\cdot\frac{\partial^2\partial X}{\partial\zeta_0^2\partial\zeta_1}\right]. \tag{2}$$

where b_i are the couplings [3] of the boundary terms. Consistency with the open-closed string duality [3] implies a vanishing value of the first boundary coupling $b_1 = 0$, the leading-order correction due to second boundary terms with the coupling b_2 appears at higher order than the four derivative term in the bulk. The last two geometrical terms are proportional to the extrinsic and intrinsic curvatures, respectively. We have discussed the geometrical and topological implications of these terms elsewhere [10, 11].

The static potential [12] at leading order reads

$$V^{\rm NG}_{\ell o}(R, L_T) = \sigma_0 R + (d-2)L_T\ \log\eta(\tau) + \mu, \tag{3}$$

where μ is a UV-cutoff and η is the Dedekind η function defined on the real axis as

$$\eta(\tau) = q^{\frac{1}{24}}\prod_{n=1}^{\infty}(1-q^n). \tag{4}$$

The static potential of NG string at second loop order is given by [13]

$$V^{\rm NG}_{n\ell o}(R, L_T) = \sigma_0 R + (d-2)L_T\ \log\eta(\tau) - L_T\log\left(1 - \frac{L_T}{R^3}\frac{(d-2)\pi^2}{1152\sigma_0}\left[2E_4(\tau) + (d-4)E_2^2(\tau)\right]\right) + \mu, \tag{5}$$

with E_2 and E_4 are the second and forth-order Eisenstein series.

The boundary term S^b in Lüscher-Weisz action accounts for the symmetry breaking by the cylindrical boundary conditions owing to the Polyakov lines. The modification to the potential received when considering Dirichlet boundary condition are given by [9]

$$
\begin{aligned}
V^B &= V^{b_2} + V^{b_4}, \\
V^{b_2} &= b_2(d-2)\frac{\pi^3 L_T}{60R^4} E_4(q), \\
V^{b_4} &= \frac{-b_4(d-2)\pi^5 L_T}{126R^6} \mathrm{E}_6(\tau).
\end{aligned} \tag{6}
$$

The expectation value [14, 15] of the mean-square width of the free bosonic string theory in two dimensions

$$
W^2_{\ell o}(\zeta_1,\tau) = \frac{d-2}{2\pi\sigma_0}\log\left(\frac{R}{R_0(\zeta_1)}\right) + \frac{d-2}{2\pi\sigma_0}\log\left|\frac{\vartheta_2(\pi\zeta_1/R;\tau)}{\vartheta_1'(0;\tau)}\right|, \tag{7}
$$

where θ are Jacobi elliptic functions and $\tau = L/T$ is the modular paramter of the cylinder. The width due to the self-interaction is modified with the term [15, 16]

$$
\begin{aligned}
W^2_{n\ell o} =& \frac{(d-2)\pi}{12\sigma^2 R^2}\Big\{\tau\Big(q\frac{d}{dq} - \frac{d-2}{12}E_2(\tau)\Big)\left[E_2(2\tau) - E_2(\tau)\right] - \frac{d-2}{8\pi}E_2(\tau)\Big\} \\
&+ \frac{\pi}{12\sigma R^2}\left[E_2(\tau) - 4E_2(2\tau)\right]\left(W^2_{lo} - \frac{d-2}{4\pi\sigma}\right).
\end{aligned} \tag{8}
$$

The boundary term S^b in Lüscher-Weisz action corrects the mean-square width $W^2_{b_2}, W^2_{b_4}$ [17] acoording to

$$
W^2_{b_2} = \frac{-\pi b_2(d-2)}{4R^3\sigma^2}\left(\frac{1}{8} - \frac{1}{24}E_2(2\tau)\right), \tag{9}
$$

and

$$
W^2_{b_4} = \frac{-\pi^3(d-2)b_4}{32R^5\sigma^2}\left(\mathrm{E}_2(\tau) - \frac{5}{4}\right)\left(\frac{11\mathrm{E}_2(\tau/2)}{36} - \frac{5\mathrm{E}_2(\tau)}{9} - \frac{55}{12}\right), \tag{10}
$$

respectively.

3 Effective string lattice gauge theories

3.1 Simulation setup

We choose to perform our analysis with lattices of temporal extents $N_t = 8$, and $N_t = 10$ slices at a coupling value $\beta = 6.00$. The spatial size is 3.6^3fm^3 with lattice spacing $a = 0.1$ fm. The two lattices correspond to temperatures $T/T_c = 0.9$ just before the deconfinement point, and $T/T_c = 0.8$ which is near the end of QCD plateau.

The gauge configurations were generated using the standard Wilson gauge-action employing a pseudo-heatbath (FHKP) [18] updating to the corresponding three SU(2) subgroup elements (Marinari [19]). Each update step/sweep consists of one heatbath and 5 micro-canonical reflections. The gauge configurations are thermalized following 2000 sweeps. The measurements are taken on 500 bins. Each bin consists of 4 measurements separated by 70 sweeps of updates.

3.2 Static quark potential

The Monte-Carlo evaluation of the $Q\bar{Q}$ potential at source separation R and temperature $T = \frac{1}{L_T}$ is calculated through the expectation value of the Polyakov loop correlator

$$\mathcal{P}_{2Q} = \int d[U]\, P(0)\, P^{\dagger}(R) \exp(-S_w),$$
$$= \exp(-V(R,T)/T). \quad (11)$$

The time links is integrated out as detailed in Ref [20]. Making use of the space-time symmetries of the torus, the above correlator is evaluated at each point of the lattice and then averaged. We define the following possibly interesting combinations of LO and NLO solu-

	Fit Interval $R \in I$	Fit Parameters, $T/T_c = 0.8$ $\sigma_0 a^2$	μ(LU)	b_2(LU)	b_4(LU)	χ^2_{dof}
$V^{b_2}_{\ell o}$	(a)					
	[2,11]	0.0422(1)	-0.4767(6)	-0.148(1)	0.0	197.566
	[3,11]	0.0452(2)	-0.489(1)	-0.607(9)	0.0	5.91231
$V^{b_2,b_4}_{\ell o}$	(b)					
	[2,11]	0.0452(3)	-0.496(2)	-0.44(2)	0.10(1)	1.37428
	[3,11]	0.0450(4)	-0.494(5)	-0.3(2)	0.02(0.1)	1.17367

Table 1. The χ^2 values and the fit parameters; string tension and UV cutoff (σ, μ) together with (b_2, b_4) returned from fits of data to boundary corrected potential at the LO of NG string with $T/T_c = 0.8$. The potential $V^{b_2}_{\ell o}$ and $V^{b_2,b_4}_{\ell o}$ are given by Eq. (12) and Eq. (13), respectively.

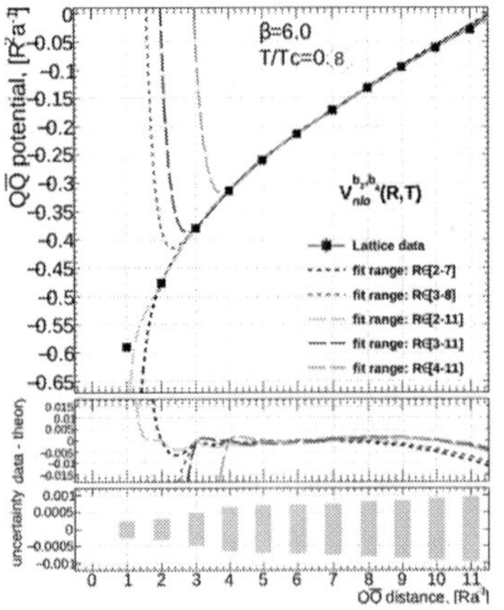

(a)

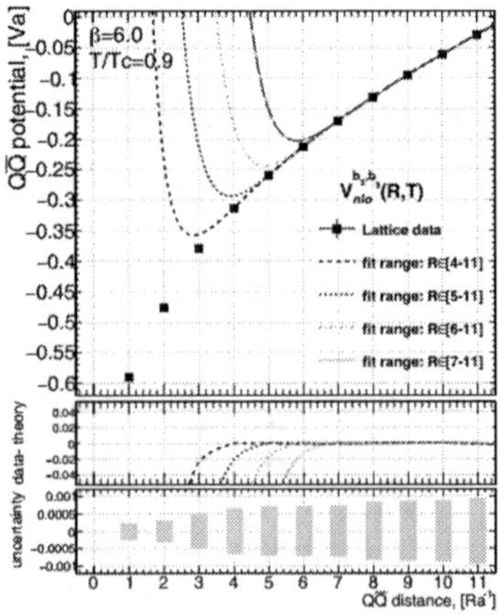

(b)

Figure 1. The $Q\bar{Q}$ potential at (a) $T/T_c = 0.8$ and (b) $T/T_c = 0.9$, the lines correspond to the NLO of NG string with two boundary terms $V^{b_2,b_4}_{n\ell o}$ Eqs. (12) and Eq. (13), respectively.

Fit Interval $R \in I$	Fit Parameters, $T/T_c = 0.9$ $\sigma_0 a^2$	b_2(LU)	b_4(LU)	χ^2
$[R_m, R_M]$				
[6,11]	0.0401(3)	11(4)	0	363.1
[7,11]	0.03954(5)	52(3)	0	99.4
[8,11]	0.0389(2)	205(50)	0	9.4
$[R_m, R_M]$				
[5,11]	0.0400(3)	1185.4 (520.8)	-948.9 (417.7)	336.6
[6,11]	0.0394(5)	23246.0 (14037.0)	-18790.0 (11352.0)	89.2

Table 2. The χ^2 values and the corresponding fit parameters returned from fits to the NLO of NG string potential with boundary terms $V^{b_2}_{n\ell o}$ and $V^{b_2,b_4}_{n\ell o}$ given by Eq. (12) and Eq. (13) at $T/T_c = 0.9$.

tions of Nambu-Goto string for the static potential with the boundary terms,

$$V^{b_2}_{\ell o,n\ell o} = V_{\ell o,n\ell o} + V^{b_2}, \tag{12}$$

$$V^{b_2,b_4}_{\ell o,n\ell o} = V_{\ell o,n\ell o} + V^{b_2} + V^{b_4}, \tag{13}$$

where subscripted $V_{\ell o,n\ell o}$ denotes either LO or NLO of the (NG) potential given according to Eq. (3) and Eq. (5). The $Q\bar{Q}$ potential data are fitted to either one of the boundary terms $V^{b_2}_{n\ell o}$ and $V^{b_2,b_4}_{n\ell o}$ given according to Eq. (12) and Eq. (13), respectively. The corresponding returned values of χ^2 and fit parameters are enlisted in Table 1 at temperature $T/T_c = 0.8$, and Table 2 at $T/T_c = 0.9$, for the depicted fit intervals.

The analysis near the end of QCD plateau can disclose the substantial change in the properties of the confining force with the temperature. At the temperature $T/T_c = 0.8$ where the thermal effects are significantly milder, the values of χ^2 returned from the fits to $V^{b_2}_{\ell o}$ in Eq. (12) indicate minimal residuals of the fits over the interval $R \in [0.3, 1.1]$ fm with good $\chi^2 = 5.9$ taking into account the degree of freedom. Interestingly, the fits to the ansatz $V^{b_2,b_4}_{\ell o}$ with two boundary terms decrease the returned residuals significantly at a shorter interval of $R \in [0.2, 1.1]$ fm

Figure 1(a) illustrates the fitted curves of the potential and the received small residuals over short distances as $R = 0.3$ fm when the boundary term at NLO of NG string potentials. The plot reflects the small value of the fit residuals at all distances when considering $V^{b_2,b_4}_{n\ell o}$ which assumes close values to that enlisted in Table 1. In the contrary to the fits at higher temperatures $T/T_c = 0.9$ shown below the fit ansatz involving the string self-interaction Eq.(5) does not reduce the residuals over source separation $R < 0.5$ fm [21]. Thus we may conclude that the boundary terms becomes the most relevant correction to the free string model with the decrease of the temperature.

The fit to the static potential $V^{b_2}_{n\ell o}$ at $T/T_c = 0.9$ returns values for the parameter b_2 which appear to vary dramatically with the considered range. The values of χ^2 are high when considering the entire intermediate fit-interval $R \in [0.5, 1.1]$ fm. Nevertheless, the returned χ^2 values in Table 2 appear to be much smaller than that obtained by considering merely the pure NG string potential Eqs. (5) (See Ref. [21]). Acceptable value of χ^2 are returned over the fit interval [0.7, 1.2] fm when considering the correction of both the two boundary terms $V^{b_2,b_4}_{\ell o}$. The fitted curves are plotted in Fig. 1(b) for the depicted ranges.

Inspection of Tables 1 and Table 2 indicates that fits to the NLO of the NG string with boundary term $V^{b_2}_{n\ell o}$ produce very close value of $\sigma_0 a^2$ as that of the pure NG string $V_{n\ell o}$. The same observation holds for the fits considering two boundary terms $V^{b_2,b_4}_{n\ell o}$. Despite of the reduction in the minima of χ^2_{dof}, the same returned value [21] of the string tension fit parameter σ_o is consistent with the modular transforms in Eq.(6), where the inverse of the cylinder's modular parameter does not produce terms linear in R and do not contribute to the renormalization of the string tension.

The deviations of the string tension fit parameter from the values returned in Table1 at $T/T_c = 0.8$ is decreased considering other effects such as a possible rigidity property of the confing flux-tube. We reported in Ref [21] that the smoothed string fluctuations by virtue of the rigidity reproduce a close value of $\sigma_0 a^2 \sim 0.44$ returned at $T/T_c = 0.9$.

3.3 Energy-density profile

The corrections provided by the boundary action to static $Q\bar{Q}$ potential seem to eliminate, to some extend, the deviations appearing when constructing the static mesonic states with Polyakov loop correlators. In the following we probe the confining force by profiling the action-density of the Yang-Mills vacuum in the presence of color charges. The action density is evaluated via a three-loop improved lattice field-strength tensor[2]. A scalar field characterization can be defined as

$$C(\vec{\rho};\vec{r}_1,\vec{r}_2) = \frac{\langle \mathcal{P}_{2Q}(\vec{r}_1,\vec{r}_2)\, S(\vec{\rho})\,\rangle}{\langle\, \mathcal{P}_{2Q}(\vec{r}_1,\vec{r}_2)\,\rangle\ \langle S(\vec{\rho})\,\rangle}, \tag{14}$$

with the vector $\vec{\rho}$ referring to the spatial position of the energy probe with respect to some origin, and the bracket $\langle ... \rangle$ stands for averaging over gauge configurations and lattice symmetries.

We implement a number of $n_{sw} = 20$ of improved cooling sweeps to eliminate statistical fluctuations which ought to keep the physical observable intact [2] over source separation upto $R \geq 0.5$ fm. A double Gaussian fit function [2] is implemented to estimate the mean-square width. In analogy with the static $Q\bar{Q}$ potential, we examine same combinations of

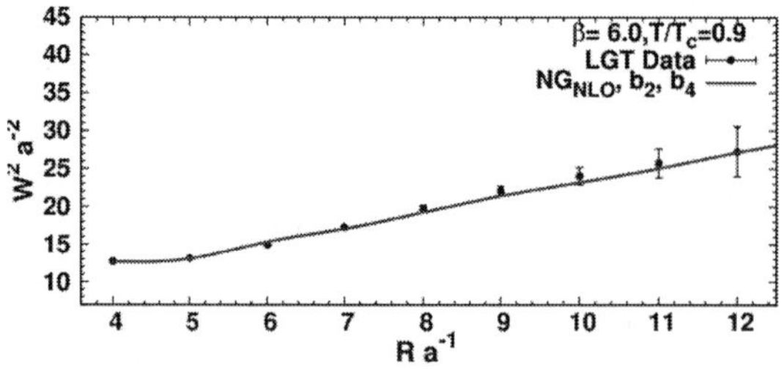

Figure 2. The mean-square width at the middle plane $R/2$ and $T/T_c = 0.9$. The solid line is the square-width of the effective NG string at NLO Eq. (7) and Eq. (8), with two-boundary terms $W^2_{b_2} + W^2_{b_4}$ Eq. (9) and Eq. (10) for fit range $R \in [0.4, 1.2]$ fm.

boundary corrections (Eq. (9) and Eq. (10)) with both the leading and next-to-leading Nambu-Goto string's width defined by (Eq. (7) and Eq. (8)). The resultant χ^2 and fit parameters are enlisted in the Tables. 3 and 4. The returned values in Table 3 indicates better match when including the first boundary correction $W^2_{b_2}$ Eq. (9) compared to the fits along the pure leading (NG) model [2]. The values of χ^2 is further reduced upon switching on the string's self-interaction Eq. (8) together with the boundary corrections.

One can remark that the fits using merely the LO width from the NG string with $W^2_{b_2}$ boundary correction are providing smaller values of residual than that of the corresponding

Plane, NG-order		$R \in [0.5, 1.2]$ fm χ^2	R_0	b_2	$R \in [0.6, 1.2]$ fm χ^2	R_0	b_2
$z=1$	$NG_{\ell o}$	52.6	13.5(3)	12.8(6)	14.98	16.1(5)	22.2(1.6)
	$NG_{n\ell o}$	5.7	13.9(3)	-1.4(3)	2.4	14.8(6)	0.4(1.0)
$z=2$	$NG_{\ell o}$	216.3	25.9	10.9(4)	25.9	10.9(4)	21.7(1.4)
	$NG_{n\ell o}$	27.5	7.1(3)	3.4(2)	7.9	8.9(5)	7.1(9)
$z=\frac{R}{2}$	$NG_{\ell o}$	31.6	3.7(2)	4.0(5)	4.3	6.1(5)	12.2(1.6)
	$NG_{n\ell o}$	9.1	2.5(3)	1.9(2)	0.77	4.0(6)	4.6(1.0)

Table 3. The returned χ^2 and b_2 from the fits of string's width with boundary term $W^2_{b_2}$ Eq. (9).

Plane,NG-order	$R \in [0.5, 1.2]$ fm χ^2	b_2	b_4	$R \in [0.6, 12]$ fm χ^2	b_2	b_4
$z=1, NG_{\ell o}$	52.2	33(31)	-18(27.8)	0.70	-121(38.0)	131(34.8)
$NG_{n\ell o}$	5.66	-11(34.5)	7(28.7)	1.43	-36(36.7)	31(30.8)
$z=2, NG_{\ell o}$	92.5	62(27.2)	-45(24.3)	1.5	-153(35)	161(32.5)
$NG_{n\ell o}$	27.5	4(30.8)	-0.5(25.7)	3.08	-67(34.1)	63(28.7)
$z=\frac{R}{2}, NG_{\ell o}$	24.20	137(29.9)	-119(26.6)	0.29	-59(49.9)	67(46.3)
$NG_{n\ell o}$	6.96	108(35.7)	-89(29.7)	0.07	25(47.5)	-17(40.3)

Table 4. Same as Table 3 for the boundary corrections $W^2_{b_2} + W^2_{b_4}$ given by Eq. (9) and Eq. (10).

pure NG string [2] at NLO. With the consideration of the two boundary corrections $W^2_{b_2} + W^2_{b_4}$ a better match is retrieved over longer string length $R \geq 0.5$ fm (Table. 4). Figure 2 illustrates the fitted curves of the width growth at the middle plane resulting. The width of the string model employing the two boundary corrections $W^2_{b_2} + W^2_{b_4}$ provides good match with the LGT data even up to small source separation distances as $R = 0.4$ fm. Despite the relatively larger uncertainties in the density Eq. (14) compared to $Q\bar{Q}$ potential. However, the successful parameterization of the width profile is an indication of relevance to boundary action of confining flux tube at high energy.

4 Conclusion

In this work, the quark-antiquark potential and energy profile of a static meson is compared to the corresponding theoretical predictions laid down by the Lüscher-Weisz (LW) string with two boundary terms in the action. The static quark-antiquark ($Q\bar{Q}$) potential is calculated using link-integrated Polyakov loop correlators.

Near the end of QCD Plateau region, we detect signatures of the two boundary terms of the Lüscher-Weisz (LW) string in the Monte-Carlo lattice data. The boundary corrected string model is in good agreement with the static $Q\bar{Q}$ potential for color source separation as short as $R = 0.3$ fm. As we approach the deconfinement point $T/T_c = 0.9$, the fits show reduction of the residuals by the inclusion of boundary terms of the Lüscher-Weisz action in the approximation scheme. However, deviations from the returned string tension $\sigma_0 a^2$ persist.

For the energy width profile at $T/T_c = 0.9$, the boundary corrected width with one boundary term for the (LW) action at two-loop order; or two boundary terms with (LW) action at one-loop order displays a good match with the LGT data for color sources separation $R \geq 0.5$ fm.

References

[1] M. Lüscher, P. Weisz, JHEP, **07**, 049 (2002)
[2] A. S. Bakry, M. A. Deliyergiyev, A. A. Galal, A. M. Khalaf, M. N. Khalil, arXiv:hep-lat/2001.02392
[3] M. Lüscher P. Weisz, JHEP, **07**, 014 (2004)
[4] O. Kaczmarek, F. Karsch, E. Laermann, M. Lutgemeier, Phys. Rev. D, **62**, 3 (2000)
[5] KJ. Juge, J. Kuti, C. Morningstar, Phys. Rev. Lett., **90**, 161601 (2003)
[6] H. Kleinert, Phys. Lett., **B174**, 335 (1986)
[7] M. Caselle, M. Panero, R. Pellegrini, D. Vadacchino, JHEP, **01**, 105 (2015)
[8] BB. Brandt, JHEP, **02**, 40 (2011)
[9] M. Billo, M. Caselle, F. Gliozzi, M. Meineri, R. Pellegrini, JHEP, **05**, 130 (2012)
[10] A. S. Bakry, M. A. Deliyergiyev, A. Galal, M. N. Khalil, hep-th/1709.09446.
[11] A. S. Bakry, M. A. Deliyergiyev, A. Galal, M. N. Khalil, hep-lat/2005.04675.
[12] JF. Arvis, Phys. Lett., **127B**, 106 (1983)
[13] K. Dietz and T. Filk Phys. Rev. D, **27**, 2944 (1983)
[14] A. Allais, and M. Caselle, JHEP, **01**, 073 (2009)
[15] F. Gliozzi, M. Pepe, UJ Wiese, Phys. Rev. Lett., **104**, 232001 (2010)
[16] F. Gliozzi, M. Pepe, UJ. Wiese, JHEP, **11**, 053 (2010)
[17] A. S. Bakry, M. A. Deliyergiyev, A. Galal, M. N. Khalil, hep-th/1912.13381
[18] K. Fabricius, O. Haan, Phys. Lett. B, **143**, 459 (1984)
[19] N. Cabibbo and E. Marinari, Phys. Lett. B, **119**, 387 (1982)
[20] P. deForcrand, C. Roiesnel, Phys. Lett., **B151**, 77 (1985)
[21] A. Bakry, M. Deliyergiyev, A. Galal, M. Khalil, hep-lat/2001.04203

EPJ Web of Conferences **258**, 02005 (2022) https://doi.org/10.1051/epjconf/202225802005
A Virtual Tribute to Quark Confinement and the Hadron Spectrum (vConf21)

Monopoles of the Dirac type and color confinement in QCD

Tsuneo Suzuki[1,*], *Atsuki* Hiraguchi[2,**], and *Katsuya* Ishiguro[3,***]

[1]RCNP, Osaka University, Ibaraki Osaka 567-0047, Japan
[2]Institute of Physics, National Yang Ming Chiao Tung University, Hsinchu 30010, Taiwan
[3]Library and Information Technology, Kochi University, Kochi 780-8520, Japan

Abstract. We present results of $SU(3)$ Monte-Carlo studies of a new color confinement scheme proposed recently due to Abelian-like monopoles of the Dirac type corresponding in the continuum limit to violation of the non-Abelian Bianchi identities (VNABI). The simulations are done without any additional gauge-fixing smoothing the vacuum. We get for the first time, in pure $SU(3)$ simulations, (1) the perfect Abelian dominance with respect to the static potentials on $(12 \sim 16)^4$ at $\beta = 5.6 - 5.8$ using the multilevel method, (2) the monopole as well as Abelian dominances with respect to the static potentials by evaluating the Polyakov-loop correlators on $24^3 \times 4$ at $\beta = 5.6$. The Abelian photon part gives zero string tension. The new $SU(3)$ as well as the previous $SU(2)$ results are consistent with the new Abelian picture of color confinement that each one of eight colored electric flux is squeezed by the corresponding colored Abelian-like monopole of the Dirac type corresponding to VNABI.

1 Introduction

Color confinement in quantum chromodynamics (QCD) is still an important unsolved problem. As a picture of color confinement, 't Hooft [1] and Mandelstam [2] conjectured that the QCD vacuum is a kind of a magnetic superconducting state caused by condensation of magnetic monopoles and an effect dual to the Meissner effect works to confine color charges. This conjecture is very interesting, but there are many problems to be unsolved even at the present stage. Especially, to find color magnetic monopoles is not straightforward in QCD.

In gauge theories without scalar fields, it seems necessary to introduce some singularities as shown by Dirac[3] in $U(1)$ quantum electrodynamics. An interesting idea to introduce such a singularity is to project QCD to the Abelian maximal torus group by a partial (but singular) gauge fixing [4]. In $SU(3)$ QCD, the maximal torus group is Abelian $U(1)^2$. Then color magnetic monopoles appear as a topological object at the space-like points corresponding to the singulariry of the gauge-fixing matrix.. Condensation of the monopoles causes the dual Meissner effect with respect to $U(1)^2$. Numerically, an Abelian projection in various gauges such as the maximally Abelian (MA) gauge[5, 6] seems to support the conjecture[7, 8].

*e-mail: tsuneo@rcnp.osaka-u.ac.jp
**e-mail: a.hiraguchi@nycu.edu.tw
***e-mail: ishiguro@kochi-u.ac.jp

Although numerically interesting, the idea of Abelian projection[4] is theoretically very unsatisfactory. 1) There are infinite ways of such a partial gauge-fixing and whether the 't Hooft scheme is gauge independent or not is not known. 2) Especially, consider a Polyakov gauge in which Polyakov loops are diagonalized. In this gauge, the Abelian dual Meissner picture works good [13], where space-like monopole currents play the role of the solenoidal current squeezing the electric field. However, this fact is in contradiction with the 'tHooft idea[10].

In 2010 Bonati et al. [14] found an interesting fact that the violation of non-Abelian Bianchi identity (VNABI) exists behind the Abelian projection scenario in various gauges and hence gauge independence is naturally expected. Along this line, one of the authors (T.S.)[15] found a more general relation that VNABI is just equal to the violation of Abelian-like Bianchi identities corresponding to the existence of Abelian-like monopoles. A partial gauge-fixing is not necessary at all from the beginning. If the non-Abelian Bianchi identity is broken, Abelian-like monopoles necessarily appear due to a line-like singularity leading to a non-commutability with respect to successive partial derivatives This is hence an extension of the Dirac idea of monopoles in QED to non-Abelian QCD.

In this report, (1) the new theoretical scheme for color confinement based on the dual Meissner effect due to the above monopoles is summarized shortly. (2) The first results showing the perfect Abelian dominance and the monopole dominance in pure SU(3 lattice QCD are shown next along with the short review of pure $SU(2)$ results[11]. (3) Finally numerical results showing the continuum limit of the new monopoles[16, 17] are reviewed shortly in the framework of pure $SU(2)$ lattice QCD, since existence of the continuum limit is essentially important for the new confinement scheme.

2 Equivalence of VNABI and Abelian-like monopoles

First of all, we prove that the Jacobi identities of covariant derivatives lead us to conclusion that violation of the non-Abelian Bianchi identities (VNABI) J_μ is nothing but an Abelian-like monopole k_μ defined by violation of the Abelian-like Bianchi identities without gauge-fixing. Define a covariant derivative operator $D_\mu = \partial_\mu - igA_\mu$. The Jacobi identities are expressed as

$$\epsilon_{\mu\nu\rho\sigma}[D_\nu, [D_\rho, D_\sigma]] = 0. \tag{1}$$

By direct calculations, one gets

$$\begin{aligned}[D_\rho, D_\sigma] &= [\partial_\rho - igA_\rho, \partial_\sigma - igA_\sigma] \\ &= -igG_{\rho\sigma} + [\partial_\rho, \partial_\sigma],\end{aligned}$$

where the second commutator term of the partial derivative operators can not be discarded, since gauge fields may contain a line singularity. Actually, it is the origin of the violation of the non-Abelian Bianchi identities (VNABI) as shown in the following. The non-Abelian Bianchi identities and the Abelian-like Bianchi identities are, respectively, $D_\nu G^*_{\mu\nu} = 0$ and $\partial_\nu f^*_{\mu\nu} = 0$. The relation $[D_\nu, G_{\rho\sigma}] = D_\nu G_{\rho\sigma}$ and the Jacobi identities (1) lead us to

$$\begin{aligned}D_\nu G^*_{\mu\nu} &= \frac{1}{2}\epsilon_{\mu\nu\rho\sigma}D_\nu G_{\rho\sigma} = -\frac{i}{2g}\epsilon_{\mu\nu\rho\sigma}[D_\nu, [\partial_\rho, \partial_\sigma]] \\ &= \frac{1}{2}\epsilon_{\mu\nu\rho\sigma}[\partial_\rho, \partial_\sigma]A_\nu = \partial_\nu f^*_{\mu\nu},\end{aligned} \tag{2}$$

where $f_{\mu\nu}$ is defined as $f_{\mu\nu} = \partial_\mu A_\nu - \partial_\nu A_\mu = (\partial_\mu A^a_\nu - \partial_\nu A^a_\mu)\lambda^a/2$. Namely Eq.(2) shows that the violation of the non-Abelian Bianchi identities, if exists, is equivalent to that of the Abelian-like Bianchi identities.

Denote the violation of the non-Abelian Bianchi identities as $J_\mu = D_\nu G^*_{\mu\nu}$. An Abelian-like monopole k_μ without any gauge-fixing is defined as the violation of the Abelian-like Bianchi identities: $k_\mu = \partial_\nu f^*_{\mu\nu}$. Eq.(2) shows that $J_\mu = k_\mu$.

Due to the antisymmetric property of the Abelian-like field strength, we get Abelian-like conservation conditions $\partial_\mu k_\mu = 0$ [18]. The Abelian-like conservation relation gives us eight conserved magnetic charges in the case of color $SU(3)$ and $N^2 - 1$ charges in the case of color $SU(N)$. But these are kinematical relations coming from the derivative with respect to the divergence of an antisymmetric tensor [18]. The number of conserved charges is different from that of the Abelian projection scenario [4], where only $N - 1$ conserved charges exist in the case of color $SU(N)$.

3 Abelian static potentials in $SU(3)$

In $SU(2)$ QCD, perfect Abelian dominance is proved without performing any addtional gauge fixing using the multilevel method[12] in Ref.[9, 11]. Also perfect monopole dominance is proved very beautifully by applying the random gauge transformation as a method of the noise reduction of measuring gauge-variant quantities in the same reference[9, 11]. However, in $SU(3)$ QCD on lattice, it is not straightforward from the beginning. First, to extract Abelian link fields for all eight colors separately from non-Abelian gauge field matrix is not simple, since in $SU(3)$ the non-Abelian gauge field is not expanded by the Lie-Algebra elements in a simple way as in $SU(2)$. We choose the following method to define the Abelian link field by maximizing the following overlap quantity

$$RA = \sum_{s,\mu} \mathrm{Tr}\left(e^{i\theta^a_\mu(s)\lambda^a} U^\dagger_\mu(s)\right), \tag{3}$$

where λ^a is the Gell-Mann matrix and no sum over a is not taken. This choice in $SU(2)$ leads us to the same Abelian link fields adopted in Ref.[9, 11].

For example, we get from the maximization condition of (3) an Abelian link field $\theta_1(s,\mu)$ corresponding to $\lambda_1(SU(3)$ as

$$\theta_1(s,\mu) = \tan^{-1}\frac{Im(U_{12}(s,\mu) + U_{21}(s,\mu))}{Re(U_{11}(s,\mu) + U_{22}(s,\mu))}.$$

Once Abelian link variables are fixed, we can extract Abelian, monopole and photon parts from the Abelian plaquette variable as follows:

$$\theta^a_{\mu\nu}(s) \;=\; \bar{\theta}^a_{\mu\nu}(s) + 2\pi n^a_{\mu\nu}(s) \;\; (|\bar{\theta}^a_{\mu\nu}| < \pi), \tag{4}$$

where $n^a_{\mu\nu}(s)$ is an integer corresponding to the number of the Dirac string. Then an Abelian monopole current is defined by

$$k^a_\mu(s) \;=\; -(1/2)\epsilon_{\mu\alpha\beta\gamma}\partial_\alpha \bar{\theta}^a_{\beta\gamma}(s + \hat{\mu}). \tag{5}$$

This definition (5) satisfies the Abelian conservation condition and takes an integer value which corresponds to the magnetic charge obeying the Dirac quantization condition[19].

3.1 Perfect Abelian dominance in $SU(3)$

Now let us evlaluate Abelian static potentials through Polyakov-loop correlators written by the Abelian Plaquette variable defined above:

$$P_{\mathrm{A}} = \exp[i \sum_{k=0}^{N_t-1} \theta_1(s + k\hat{4}, 4)] \tag{6}$$

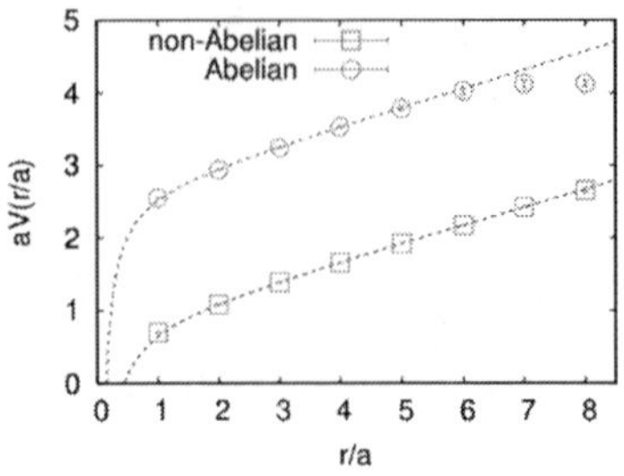

Figure 1. The static-quark potentials from non-Abelian and Abelian PLCF at $\beta = 5.6$ on $16^3 \times 16$ lattice.

Table 1. Simulation parameters for the measurement of static potential using multilevel method. $N_{\rm sub}$ is the sublattice size divided and $N_{\rm iup}$ is the number of internal updates in the multilevel method .

β	$N_s^3 \times N_t$	$a(\beta)$ [fm]	$N_{\rm conf}$	$N_{\rm sub}$	$N_{\rm iup}$
5.60	$16^3 \times 16$	0.2235	6	2	10,000,000
5.70	$12^3 \times 12$	0.17016	6	2	5,000,000
5.80	$12^3 \times 12$	0.13642	6	3	5,000,000

Table 2. Best fitted values of the string tension σa^2, the Coulombic coefficient c, and the constant μa for the potentials $V_{\rm NA}$, $V_{\rm A}$.

$\beta = 5.6, 16^3 \times 16$			
$V_{\rm NA}$	0.239(2)	-0.39(4)	0.79(2)
$V_{\rm A}$	0.25(2)	-0.3(1)	2.6(1)
$\beta = 5.7, 12^3 \times 12$			
$V_{\rm NA}$	0.159(3)	-0.272(8)	0.79(1)
$V_{\rm A}$	0.145(9)	-0.32(2)	2.64(3)
$\beta = 5.8, 12^3 \times 12$			
$V_{\rm NA}$	0.101(3)	-0.28(1)	0.82(1)
$V_{\rm A}$	0.102(9)	-0.27(2)	2.60(3)

Since the above Abelian Polyakov loop operator without any additional gauge-fixing is defined locally, the Poyakov-loop correlators can be evaluated through the multilevel method[12]. Contrary to the $SU2)$ case in Ref.[11], we need much more number of internal updates to get meaningful results. The simulation parameters are shown in Table1. An example of the static potentials from non-Abelian and Abelian Polyakov-loop correlators is whon in Fig.1 The best fitted values of the non-Abelian and Abelian string tensions are plotted in Table2.

3.2 Perfect monopole dominance in $SU(3)$

Without adopting any further gauge fixing smoothing the vacuum, Abelian monopole static potential can reproduce fully the string tension of non-Abelian static potential in $SU(2)$ QCD as shown in Ref.[9, 11]. This is called as perfect monopole dominance of the string tension.

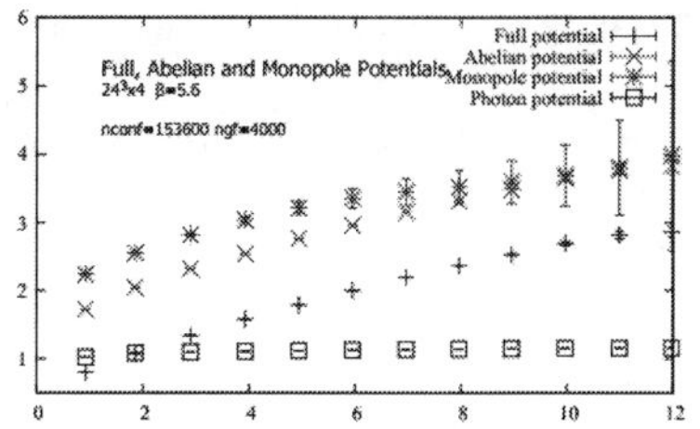

Figure 2. Full, Abelian and Monopole potentials on $24^3 \times 4$ at $\beta = 5.6$

Table 3. Simulation parameters for the measurement of the static potential and the force from P_A, P_{ph} and P_{mon}. N_{RGT} is the number of random gauge transformations.

β	$N_s^3 \times N_t$	$a(\beta)$ [fm]	N_{conf}	N_{RGT}
$SU2, 2.43$	$24^3 \times 8$	0.1029(4)	7,000	4,000
$SU3, 5.6$	$24^3 \times 4$	0.2235	153,600	4,000

Almost perfect monopole dominance was found in $SU(3)$ when MA gauge is adopted in Ref.[20]. However, without such additional gauge-fixing, it is found to be tremendously difficult.

We investigate the monopole contribution to the $SU(3)$ static potential through Polyakov-loop correlators in order to examine the role of monopoles for confinement without any additional gauge-fixing.. The monopole part of the Polyakov loop operator is extracted as follows. Using the lattice Coulomb propagator $D(s - s')$, which satisfies $\partial_\nu \partial'_\nu D(s - s') = -\delta_{ss'}$ with a forward (backward) difference ∂_ν (∂'_ν), the temporal component of the Abelian fields $\theta_4^a(s)$ are written as

$$\theta_4^a(s) = -\sum_{s'} D(s - s')[\partial'_\nu \theta_{\nu 4}^a(s') + \partial_4(\partial'_\nu \theta_\nu^a(s'))] . \tag{7}$$

Inserting Eq. (7) to the Abelian Polyakov loop (6), we obtain

$$\begin{aligned} P_A^a &= P_{ph}^a \cdot P_{mon}^a , \\ P_{ph}^a &= \exp\{-i \sum_{k=0}^{N_t-1} \sum_{s'} D(s + k\hat{4} - s')\partial'_\nu \bar{\theta}_{\nu 4}^a(s')\} , \\ P_{mon}^a &= \exp\{-2\pi i \sum_{k=0}^{N_t-1} \sum_{s'} D(s + k\hat{4} - s')\partial'_\nu n_{\nu 4}^a(s')\} . \end{aligned} \tag{8}$$

We call P_{ph}^a the photon and P_{mon}^a the monopole parts of the Abelian Polyakov loop P^a, respectively [21]. The latter is due to the fact that the Dirac strings $n_{\nu 4}^a(s)$ lead to the monopole currents in Eq. (5) [19]. Note that the second term of Eq. (7) does not contribute to the Abelian Polyakov loop in Eq. (6). We show the simulation parameters and the results in comparison with one typical $SU(2)$ case.

In comparison with those in $SU(2)$ where beautiful Abelian and monopole dominances are observed using reasonable number of vacuum ensembles, we needed much more number of vacuum ensembles even on $24^3 \times 4$ small lattice in $SU(3)$ as shown in Tabel4. Abelian dominance is seen from Abelian-Abelian Polyakov loop correlators. But in the case of

Table 4. Best fitted values of the string tension σa^2, the Coulombic coefficient c, and the constant μa for the potentials V_{NA}, V_{A}, V_{mon} and V_{ph}. V_{FA} (V_{FM}) stands for the potential determined from non-Abelian and Abelian(monopole) Polyakov-loop correlators.

$SU(2)$		σa^2	c	μa	FR(R/a)	χ^2/N_{df}
$24^3 \times 8$	V_{NA}	0.0415(9)	0.47(2)	0.46(8)	4.1 - 7.8	0.99
	V_{A}	0.041(2)	0.47(6)	1.10(3)	4.5 - 8.5	1.00
	V_{mon}	0.043(3)	0.37(4)	1.39(2)	2.1 - 7.5	0.99
	V_{ph}	$-6.0(3) \times 10^{-5}$	0.0059(3)	0.46649(6)	7.7 - 11.5	1.02
$SU(3)$						
$24^3 \times 4$	V_{NA}	0.1707(47)	0.855(107)	1.124(45)	3 - 10	1.170
	V_{A}	0.1771(84)	0.533(54)	3.56(6)	1 - 9	0.81
	V_{FA}	0.1598(67)	1.105(152)	2.203(65)	3 - 10	1.081
	V_{FM}	0.146(17)	0.345(29)	2.477(59)	0 - 8	0.970

monopole-monopole Polyakov loop correlators, we could not get good results. Since Abelian dominance is seen also from non-Abelian and Abelian Polyakov loop correlators as shown in the Tabel4., we try to study non-Abelian and monopole correlators. Since the fit is not good enough, a strong indication of monopole dominance is seen from the hybrid correlators as shown in Fig.2. When we go to larger lattice $40^3 \times 6$ at $\beta = 5.75$ (correponding to a similar temperature), Abelian dominance is seen using non-Abelian and Abelian correlators. But non-Abelian and monopole correlators are much more worse. To improve monopole static potentials for the large separation region, we may study larger T lattice with much more vacuum ensembles. In the case of photon-photon correlators, the string tensions on both cases are almost zero.

4 The continuum limit of the new Abelian-like monopoles

Finally let us discuss an important problem if the above new-type Abelian monopoles have the continuum limit. The studies in the framework of pure $SU(2)$ QCD were done with respect to the monopole density in Ref.[16] and to the effective monopole action in Ref.[17]. Here we review the results shortly.

In both studies, it is inevitable to introduce additional gauge fixings to make the lattice vacuum smooth enough reducing the number of lattice artifact monopoles although the gauge dependence problem appear newly. It is due to the fact that even lattice artifact monopoles contribute to the monopole density or the effective monopole action equally. Hence we adopted four different smooth gauge-fixing methods to check gauge dependence, that is, MCG(maximally center gauge) [22, 23], DLCG(direct Laplacian center gauge)[24], MAWL(maximal Abelian Wilson loop gauge) [25] and MAG+U1(maximal Abelian gauge[5, 6] and $U(1)$ Landau gauge). To make the vacuum smooth, we also introduce a tadpole improved action and the block-spin transformation of monopoles[26, 27]

Original monopoles are defined on a a^3 cube and the n-blocked monopoles are defined on a cube
with a lattice spacing $b = na$ as follows:

$$k_\mu^{(n)}(s_n) = \sum_{i,j,l=0}^{n-1} k_\mu(ns_n + (n-1)\hat{\mu} + i\hat{\nu} + j\hat{\rho} + l\hat{\sigma})$$

We considered $n = 1, 2, 3, 4, 6, 8, 12$ blockings for $\beta = 3.0 \sim 3.9$ on 48^4 lattice.

We evaluated a gauge-invariant density of the n-blocked monopole:

$$\rho(a(\beta), n) = \frac{\sum_{\mu,s_n} \sqrt{\sum_a (k_\mu^{(n)a}(s_n))^2}}{4\sqrt{3}V_n b^3},$$

which is a scale-invariant quantity depending on $a(\beta)$ and n generally. But if we plot ρ versus $b = na(\beta)$, we get a beautiful universal scaling function for all different gauge-fixings as shown in Fig.3. Namely we obtained clear scaling behaviors $\rho(b)$ up to the 12-step blocking transformations for $\beta = 3.0 \sim 3.9$. Hence for fixed b, if we take $n \to \infty$, $a(\beta) \to 0$ that is we go to the continuum limit. The same beautiful scaling behaviors are obtained also in the case of the effective monopole action[17]. In addition to the scaling behaviors, the obtained scaling function is the same for four different gauges. Gauge independence is shown as naturally expected in the continuum limit.

Figure 3. Comparison of the Abelian-like monopole densities versus $b = na(\beta)$ in MCG, AWL, DLCG and MAU1 cases. A uniform scaling acurve is obtained for all gauges.

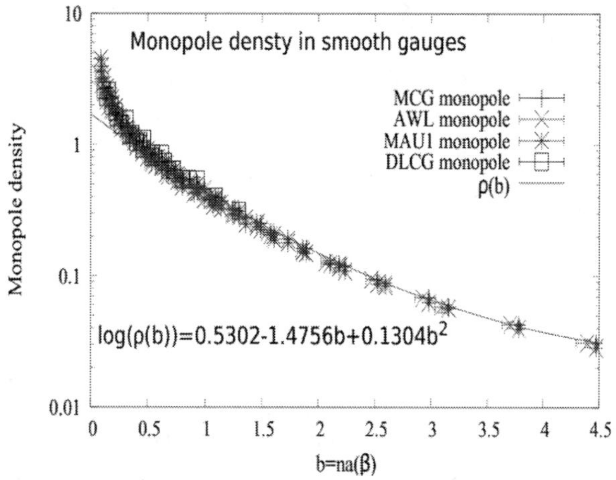

5 ACKNOWLEDGMENTS

The authors would like to thank Y. Koma for his computer code of the multi-level method. The numerical simulations of this work were done using High Performance Computing resources at Cybermedia Center and Research Center for Nuclear Physics of Osaka University, at Cyberscience Center of Tohoku University and at KEK. The authors would like to thank these centers for their support of computer facilities. T.S was finacially supported by JSPS KAKENHI Grant Number JP19K03848.

References

[1] G. 't Hooft, in *Proceedings of the EPS International*, edited by A. Zichichi, p. 1225, 1976.

[2] S. Mandelstam, Phys. Rept. **23**, 245 (1976).

[3] P. Dirac, Proc. Roy. Soc. (London) A 133, 60 (1931).

[4] G. 't Hooft, Nucl. Phys. **B190**, 455 (1981).

[5] A. S. Kronfeld, M. L. Laursen, G. Schierholz, and U. J. Wiese, Phys. Lett. **B198**, 516 (1987).

[6] A. S. Kronfeld, G. Schierholz, and U. J. Wiese, Nucl. Phys. **B293**, 461 (1987).

[7] T. Suzuki, Nucl. Phys. Proc. Suppl. **30**, 176 (1993).

[8] M. N. Chernodub and M. I. Polikarpov, in *"Confinement, Duality and Nonperturbative Aspects of QCD"*, edited by P. van Baal, p. 387, Cambridge, 1997, Plenum Press.

[9] T. Suzuki, K. Ishiguro, Y. Koma and T. Sekido, Phys. Rev. **D77**, 034502 (2008).

[10] M. N. Chernodub, Phys. Rev. **D69**, 094504 (2004),

[11] T. Suzuki, M. Hasegawa, K. Ishiguro, Y. Koma and T. Sekido, Phys. Rev. **D80**, 054504 (2009).

[12] M.Lüscher and P.Weisz, JHEP **09**, 010 (2001)

[13] T. Sekido, K. Ishiguro, Y. Koma, Y. Mori, and T. Suzuki, Phys. Rev. **D76**, 031501 (2007).

[14] C. Bonati, A. Di Giacomo, L. Lepori and F. Pucci, Phys. Rev. **D81**, 085022 (2010).

[15] Tsuneo Suzuki, A new scheme for color confinement due to violation of the non-Abelian Bianchi identities, arXiv:1402.1294

[16] Tsuneo Suzuki, Katsuya Ishiguro and Vitaly Bornyakov, Phys. Rev. **D97**, 034501 (2018). Erratum: Phys. Rev. **D97**, 099905 (2018).

[17] Tsuneo Suzuki, Phys. Rev. **D 97** 034509 (2018).

[18] J. Arafune, P.G.O. Freund and C.J. Goebel, J.Math.Phys. **16**, 433 (1975).

[19] T. A. DeGrand and D. Toussaint, Phys. Rev. **D22**, 2478 (1980).

[20] H. Suganuma and N. Sakumichi, Proceeding of the XIIIth Quark Confinement and Hadron Spectrum, 31 July-6 August 2018, Maynooth University, Maynooth, Ireland, arXiv:1812.06827V2 (hep-lat).

[21] T. Suzuki et al., Phys. Lett. bf B347, 375 (1995).

[22] L. Del Debbio, M. Faber, J. Greensite and S. Olejnik, Phys. Rev. **D55**, 2298 (1997)

[23] L. Del Debbio, M. Faber, J. Giedt, J. Greensite and S. Olejnik, Phys. Rev. **D58**, 094501 (1998)

[24] M. Faber, J. Greensite and S. Olejnik, JHEP **111**, 053 (2001).

[25] T. Suzuki et al., Nucl. Phys. Proc. Suppl. **53**, 531 (1997).

[26] T.L. Ivanenko, A. V. Pochinsky and M.I. Polikarpov, Phys. Lett. **B302**, 458 (1993). ",

[27] H.Shiba and T.Suzuki, Phys. Lett. **B351**, 519 (1995).

Abelian Dirac-type monopoles and the dual Meissner effect in QCD

Atsuki Hiraguchi[1,2,*], *Katsuya* Ishiguro[3,**], and *Tsuneo* Suzuki[4,***]

[1]Institute of Physics, National Yang Ming Chiao Tung University, Hsinchu 30010, Taiwan
[2]Department of Mathematics and Physics, Kochi University, Kochi 780-8520, Japan
[3]Library and Information Technology, Kochi University, Kochi 780-8520, Japan
[4]RCNP, Osaka University, Ibaraki Osaka 567-0047, Japan

Abstract. We investigate the Abelian dual Meissner effect due to violation of the non-Abelian Bianchi identity in $SU(3)$ gauge thoery without gauge fixing. To decide the vacuum type, we evaluate the Ginzburg-Landau parameter from the spatial distribution of color electric fields and squared monopole density. Although the study is done only on 24 $(40)^3 \times 4$ lattice at $\beta = 5.6$, the $SU(3)$ vacuum is found to be of the type 1 near the border of type 1 and type 2. We also confirm the dual Ampère's law directly.

1 Introduction

Quark confinement mechanism is one of the most important problem in particle physics [1]. A dual superconducting picture [2, 3] as an idea of color confinement mechanism, has been investigated for many years. In this picture, quarks are confined by condensation of QCD monopoles. Although the idea of the dual superconducting picture is very interesting in understanding the mechanism of confinement, in constract to SUSY QCD [4] or Georgi-Glashow model [5, 6] with scalar fields, it is not straightforward to define monopoles in QCD. Intoduction of a partial gauge fixing [7] projecting QCD to its maximal Abelian torus group, is one interesting idea to define monopoles in QCD but there are infinte ways of such gauge fixings. Numerically the maximally Abelian gauge [8, 9], which is one of the Abelian projections, seems to support the conjecture [10, 11].

Recently one of the authors (T.S.) found that the violation of the non-Abelian Bianchi identity(VNABI) is equal to Abelian-like monopoles in the continuum theory [12]. This idea is similar to the Dirac monopole [13] in U(1) quantum electrodynamics. It is confirmed that VNABI has the continuum limit in pure $SU(2)$ gauge theory [14, 15]. Therefore showing the continuum limit of VNABI in $SU(3)$ gauge theory is very important. Recent studies about the monopole dominance in $SU(3)$ are presented in another report [16] in this conference. Here, we discuss the Abelian dual Meissner effect due to VNABI in $SU(3)$ gauge theory.

*e-mail: a.hiraguchi@nycu.edu.tw
**e-mail: ishiguro@kochi-u.ac.jp
***e-mail: tsuneo@rcnp.osaka-u.ac.jp

Evaluating the spatial distribution of color electric fields and monopoles has been investigated in $SU(2)$ gauge theory. Especially, in Ref. [17], they investigated the dual Meissner effect in $SU(2)$ gauge theory without any gauge fixing. They confirmed the vacuum type is near the border between type1 and type2 and also the dual Ampère's law is satisfied. The Abelian monopole current squeezes the Abelian color electric field as a solenoidal current. These results agree with the dual superconducting picture. Our purpose of this work is to confirm the dual Meissner effect due to VNABI and to determine the vacuum type in $SU(3)$. In this report, first we explain the lattice definition of monopoles due to VNABI briefly and second show the results of the Abelian dual Meissner effect. Especially, we show the results of the Ginzburg-Landau parameter (GL) and the dual Ampère's law, although at one gauge coupling constant in the present stage.

2 VNABI

We explain the definiton of monopoles due to VNABI in this section briefly. Details of the derivation are shown in Ref.[16].

If gauge fields have a line singularity in the continuum QCD, the non-Abelian Bianchi identiry is violated. Then the violation (VNABI) corresponds to Abelian-like monopoles. In the case of $SU(3)$ gauge theory, VNABI is regarded as eight Abelian-like monopoles in the continuum QCD. The VNABI can be written as

$$D_\nu G^*_{\mu\nu} \;=\; \partial_\nu f^*_{\mu\nu}, \tag{1}$$

where $D_\mu = \partial_\mu - igA_\mu$ is a covariant derivative, $G_{\mu\nu}$ is a non-Abelian field strength and $f_{\mu\nu}$ is defined as $f_{\mu\nu} = \partial_\mu A_\nu - \partial_\nu A_\mu = (\partial_\mu A^a_\nu - \partial_\nu A^a_\mu)\lambda^a/2$. In the case of $SU(3)$, λ^a are the Gell-Mann matrices.

In $SU(3)$ QCD on lattice, it is not straightforward to define the monopole. To extract Abelian link fields for all eight colors separately from non-Abelian gauge field matrix is not simple, since in $SU(3)$ the non-Abelian gauge field is not expanded by the Lie-Algebra elements in a simple way as in $SU(2)$. We use the following method to define the Abelian link field by maximizing the following overlap quantity by taking a functional derivative with respect to an Abelian link field $\theta^a_\mu(s)$:

$$RA = \sum_{s,\mu} \mathrm{Tr}\left(e^{i\theta^a_\mu(s)\lambda^a} U^\dagger_\mu(s)\right), \tag{2}$$

This choice in $SU(2)$ leads us to the same Abelian link fields adopted in Ref. [17].

For example, we get from the maximization condition of (2) an Abelian link field $\theta_1(s,\mu)$ corresponding to σ_1 ($SU(2)$) and λ_1 ($SU(3)$) as

$$\begin{aligned}\theta_1(s,\mu) &= \tan^{-1}\frac{U_1(s,\mu)}{U_0(s,\mu)}, \quad (\mathrm{SU2}: \; U(s,\mu) = U_0(s,\mu) + i\vec{\sigma}\cdot\vec{U}(s,\mu))\\ &= \tan^{-1}\frac{Im(U_{12}(s,\mu)+U_{21}(s,\mu))}{Re(U_{11}(s,\mu)+U_{22}(s,\mu))}, \quad (\mathrm{SU3})\end{aligned}$$

Once Abelian link variables are fixed, we can extract Abelian, monopole and photon parts from the Abelian plaquette variable as follows:

$$\theta^a_{\mu\nu}(s) = \bar{\theta}^a_{\mu\nu}(s) + 2\pi n^a_{\mu\nu}(s) \quad (|\bar{\theta}^a_{\mu\nu}| < \pi), \tag{3}$$

where $n^a_{\mu\nu}(s)$ is an integer corresponding to the number of the Dirac string. Then an Abelian monopole current is defined by

$$k^a_\mu(s) = -(1/2)\epsilon_{\mu\alpha\beta\gamma}\partial_\alpha\bar{\theta}^a_{\beta\gamma}(s+\hat{\mu}) = (1/2)\epsilon_{\mu\alpha\beta\gamma}\partial_\alpha n^a_{\beta\gamma}(s+\hat{\mu}). \tag{4}$$

This definition (4) satisfies the Abelian conservation condition and takes an integer value which corresponds to the magnetic charge obeying the Dirac quantization condition [18].

3 The Abelian dual Meissner effect in $SU(3)$

3.1 Simulation details of the flux-tube profile

In this section, we show the results with respect to the Abelian dual Meissner effect. In the previous work [17] studying the spatial distribution of color electric fields and monopole currents, they used the connected correlations between a non-Abelian Wilson loop and Abelian operators in $SU(2)$ gauge theory without gauge fixing. We apply the same method to $SU(3)$ gauge theory without gauge fixing. Here we employ the standard Wilson action on the 24^3 (40^3) $\times$ 4 lattice with the coupling constant $\beta = 5.6$. We consider a finite temperature system at $T = 0.8T_c$. To improve the signal-to-noise ratio, the APE smearing is applied to the spatial links and the hypercubic blocking is applied to the temporal links. We introduce random gauge transformations to improve the signal to noise ratios of the data concerning the Abelian operators.

To measure the flux-tube profiles, we consider a connected correlation functions as done in [19, 20]:

$$\rho_{conn}(O(r)) = \frac{\left\langle \mathrm{Tr}(P(0)LO(r)L^{\dagger})\mathrm{Tr}P^{\dagger}(d) \right\rangle}{\langle \mathrm{Tr}P(0)\mathrm{Tr}P^{\dagger}(d)\rangle} - \frac{1}{3}\frac{\left\langle \mathrm{Tr}P(0)\mathrm{Tr}P^{\dagger}(d)\mathrm{Tr}O(r) \right\rangle}{\langle \mathrm{Tr}P(0)\mathrm{Tr}P^{\dagger}(d)\rangle}, \tag{5}$$

where P denotes a non-Abelian Polyakov loop, L indicates a Schwinger line, r is a distance from a flux-tube and d is a distance between Polyakov loops. We use the cylindrical coordinate (r, ϕ, z) to parametrize the q-$\bar{q}$ system as shown in Fig 1.

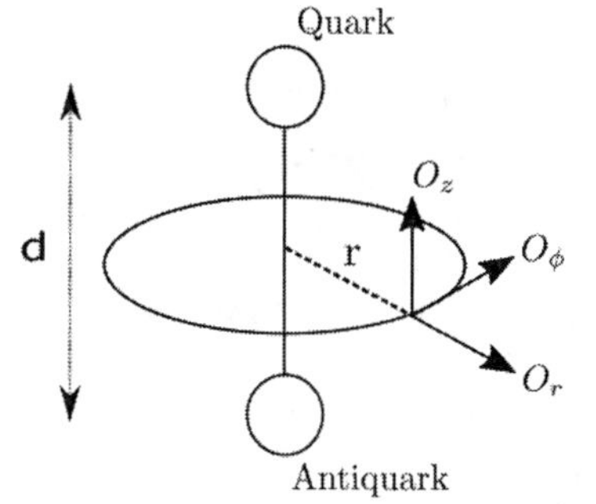

Figure 1. The definition of the cylindrical coordinate (r, ϕ, z) along the q-$\bar{q}$ axis. The d corresponds to the distance between Polyakov loops.

3.2 The spatial distribution of color electric fields

First of all, we show the results of Abelian color electric fields using an Abelian gauge field $\theta_1(s, \mu)$. To evaluate the Abelian color electric field, we adopt the Abelian plaquette as an operator $O(r)$. We calculate a penetration length λ from the Abelian color electric fields for $d = 3, 4, 5, 6$ at $\beta = 5.6$ and check the d dependence of λ. To improve the accuracy of the fitting, we evaluate $O(r)$ at both on-axis and off-axis distances. As a result, we find the Abelian color electric fields E_z^A alone are squeezed as in Fig 2. We fit these results to a fitting function,

$$f(r) = c_1 \exp(-r/\lambda) + c_0. \tag{6}$$

The parameter λ corresponds to the penetration length. We summarize the values of parameters in Table 1. We find the values of the penetration length are almost the same.

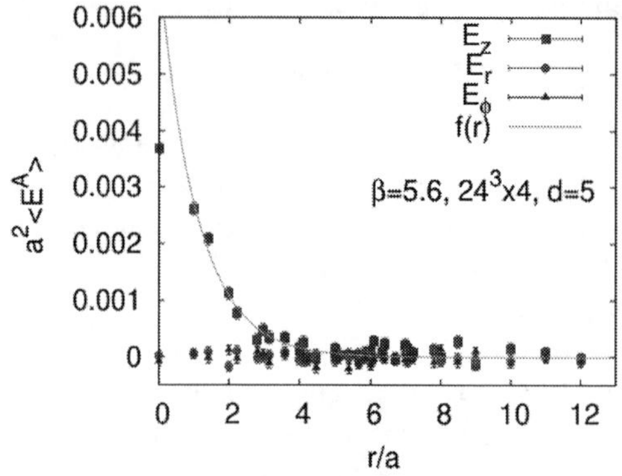

Figure 2. The Abelian color electric field with $d = 5$ at $\beta = 5.6$ on $24^3 \times 4$ lattices.

Table 1. The penetration length λ at $\beta = 5.6$ on $24^3 \times 4$ lattices.

d	λ/a	c_1	c_0	χ^2/N_{df}
3	0.91(1)	0.0100(2)	-0.000002(8)	1.31628
4	1.10(6)	0.0077(4)	-0.00005(4)	0.972703
5	1.09(8)	0.0068(6)	-0.00001(4)	0.995759
6	1.1(1)	0.0055(8)	-0.00008(7)	0.869692

3.3 The spatial distribution of monopole currents

Next we show the result of the spatial distribution of Abelian-like monopole currents. We define the Abelian-like monopole currents on the lattice as in Eq. (4). In this study we evaluate the connected correlation (5) between k^1 and the two non-Abelian Polayakov loops. We use random gauge transformations to evaluate this correlation. As a result, we find the spatial distribution of monopole currents around the flux-tube at $\beta = 5.6$. Only the monopole current in the azimuthal direction, k^1_ϕ, shows the correlation with the two non-Abelian Polyakov loops.

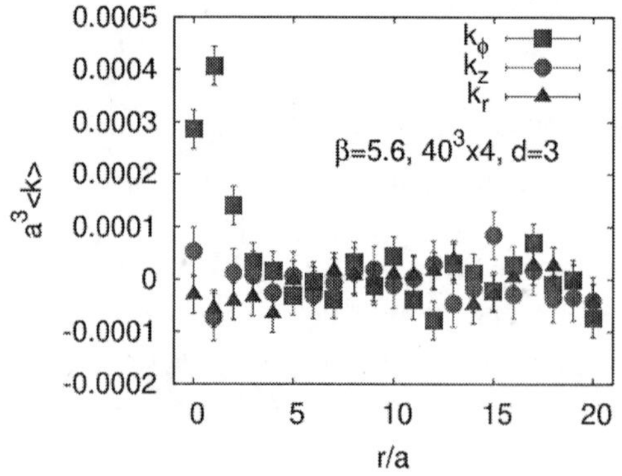

Figure 3. The monopole current at $\beta = 5.6$ on $40^3 \times 4$ lattices.

3.4 The dual Ampère's law

In previous $SU(2)$ researches [17], they investigated the dual Ampère's law to see what squeezes the color-electric field. In the case of $SU(2)$ gauge theory without gauge fixings, they confirmed the dual Ampère's law and the monopole currents squeeze the color-electric fields. In this section we show the results of the dual Ampère's law in the case of $SU(3)$ gauge theory. The definition of monopole currents gives us the following relation,

$$(\mathrm{rot}E^a)_\phi = \partial_t B^a_\phi + 2\pi k^a_\phi, \tag{7}$$

where index a is a color index.

As a results, we confirm that there is no signal of the magnetic displacement current $\partial_t B^a_\phi$ around the flux-tube for $d = 3$ at $\beta = 5.6$ as shown in Fig. 4. It suggests that the Abelian-like monopole current squeezes the Abelian color electric field as a solenoidal current in $SU(3)$ gauge theory without gauge fixing, although more data for larger d are necessary.

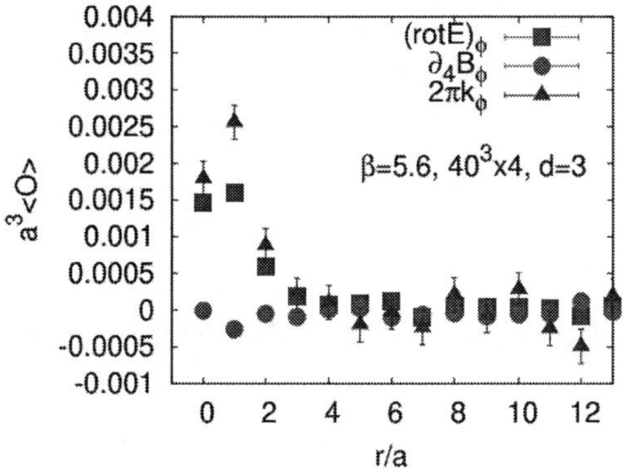

Figure 4. The dual Ampère's law at $\beta = 5.6$ on $40^3 \times 4$ lattices.

3.5 The vacuum type in $SU(3)$ gauge theory without gauge fixing

Finally, we evaluate the Ginzburg-Landau parameter, which characterizes the type of the (dual) superconducting vacuum. In the previous result [17], they found that the vacuum type is near the border between type 1 and type 2 dual superconductors by using the $SU(2)$ gauge theory without gauge fixing. We apply the same method to $SU(3)$ gauge theory.

To evaluate the coherence length, we measure the correlation between the squared monopole density and two non-Abelian Polyakov loops by using the disconnected correlation function [17, 21],

$$\left\langle k^2(r)\right\rangle_{q\bar{q}} = \frac{\left\langle \mathrm{Tr}P(0)\mathrm{Tr}P^\dagger(d)\sum_{\mu,a} k^a_\mu(r)k^a_\mu(r)\right\rangle}{\langle \mathrm{Tr}P(0)\mathrm{Tr}P^\dagger(d)\rangle} - \left\langle \sum_{\mu,a} k^a_\mu(r)k^a_\mu(r)\right\rangle. \tag{8}$$

We fit the profiles to the function,

$$g(r) = c_1' \exp\left(-\frac{\sqrt{2}r}{\xi}\right) + c_0', \tag{9}$$

where the parameter ξ corresponds to the coherence length. We plot the profiles of $\langle k^2(r)\rangle_{q\bar{q}}$ in Fig 5. As a result, we could evaluate the coherence length ξ for $d = 3, 4, 5, 6$ at $\beta = 5.6$ and find almost the same values of $\xi/\sqrt{2}$ for each d as shown in Table 2. Using these parameters λ and ξ, we could evaluate the Ginzburg-Landau parameter. The GL parameter $\kappa = \lambda/\xi$ can be defined as the ratio of the penetration length and the coherence length. If $\sqrt{2}\kappa < 1$, the vacuum type is of the type 1 and if $\sqrt{2}\kappa > 1$, the vacuum is of the type 2. We show the GL parameters in $SU(3)$ gauge theory in Table 3. We find that the vacuum is of the type 1 near the border between type 1 and type 2, although the study is done at one gauge coupling constant $\beta = 5.6$. This is the first result of the vacuum type in pure $SU(3)$ gauge theory without gauge fixing, although different β data are necessary to show the continuum limit.

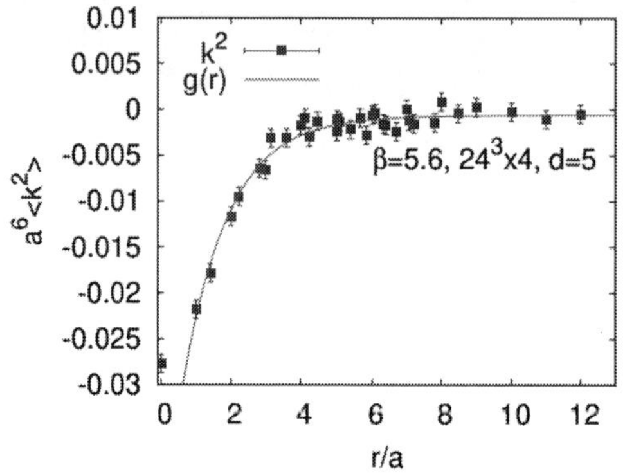

Figure 5. The squared monopole density with $d = 5$ at $\beta = 5.6$ on $24^3 \times 4$ lattices.

Table 2. The coherence length $\xi/\sqrt{2}$ at $\beta = 5.6$ on $24^3 \times 4$ lattices.

d	$\xi/\sqrt{2}a$	c_1'	c_0'	χ^2/N_{df}
3	1.04(6)	-0.050(3)	0.0001(2)	0.997362
4	1.17(7)	-0.052(3)	-0.0003(2)	1.01499
5	1.3(1)	-0.047(3)	-0.0006(3)	0.99758
6	1.1(1)	-0.052(8)	-0.0013(5)	1.12869

Table 3. The Ginzburg-Landau parameters at $\beta = 5.6$ on $24^3 \times 4$ lattice.

d	$\sqrt{2}\kappa$
3	0.87(5)
4	0.93(7)
5	0.83(9)
6	0.9(2)

4 ACKNOWLEDGMENTS

The numerical simulations of this work were done using High Performance Computing resources at Cybermedia Center and Research Center for Nuclear Physics of Osaka University, at Cyberscience Center of Tohoku University and at KEK. The authors would like to thank these centers for their support of computer facilities. T.S was finacially supported by JSPS KAKENHI Grant Number JP19K03848.

References

[1] K. Devlin, *The millennium problems : the seven greatest unsolved mathematical puzzles of our time*,Basic Books, New York (2002).
[2] G. 't Hooft, in *Proceedings of the EPS International*, edited by A. Zichichi, p. 1225, 1976.
[3] S. Mandelstam, Phys. Rept. **23**, 245 (1976).
[4] N. Seiberg and E. Witten, Nucl. Phys. **B426**, 19 (1994).
[5] G. 't Hooft, Nucl. Phys. **B79**, 276 (1974).
[6] A. M. Polyakov, Nucl. Phys. **B120**, 429 (1977).
[7] G. 't Hooft, Nucl. Phys. **B190**, 455 (1981).
[8] A. S. Kronfeld, M. L. Laursen, G. Schierholz, and U. J. Wiese, Phys. Lett. **B198**, 516 (1987).
[9] A. S. Kronfeld, G. Schierholz, and U. J. Wiese, Nucl. Phys. **B293**, 461 (1987).
[10] T. Suzuki, Nucl. Phys. Proc. Suppl. **30**, 176 (1993).
[11] M. N. Chernodub and M. I. Polikarpov, in *"Confinement, Duality and Nonperturbative Aspects of QCD"*, edited by P. van Baal, p. 387, Cambridge, 1997, Plenum Press.
[12] Tsuneo Suzuki, A new scheme for color confinement due to violation of the non-Abelian Bianchi identities, arXiv:1402.1294
[13] P. Dirac, Proc. Roy. Soc. (London) A 133, 60 (1931).
[14] Tsuneo Suzuki, Katsuya Ishiguro and Vitaly Bornyakov, Phys. Rev. **D97**, 034501 (2018). Erratum: Phys. Rev. **D97**, 099905 (2018).
[15] Tsuneo Suzuki, Phys. Rev. **D 97** 034509 (2018).
[16] Tsuneo Suzuki, Atsuki Hiraguchi and Katsuya Ishiguro, "Monopoles of the Dirac type and color confinement in QCD" in Proceeding of vConf21.
[17] T. Suzuki, M. Hasegawa, K. Ishiguro, Y. Koma and T. Sekido, Phys. Rev. **D80**, 054504 (2009).
[18] T. A. DeGrand and D. Toussaint, Phys. Rev. **D22**, 2478 (1980).
[19] Paolo Cea, Leonardo Cosmai, Francesca Cuteri, and Alessandro Papa, Journal of High Energy Physics, 2016, 6 (2016).
[20] Paolo Cea, Leonardo Cosmai, Francesca Cuteri, and Alessandro Papa, EPJ Web Conf., 175, 12006 (2018).
[21] M. N. Chernodub, Katsuya Ishiguro, Yoshihiro Mori, Yoshifumi Nakamura, M. I. Polikarpov, Toru Sekido, Tsuneo Suzuki, and V. I. Zakharov, Phys. Rev. **D72**, 074505 (2005).

Instanton-induced effects in interquark forces, light-front wave functions and formfactors

Edward Shuryak[1,*]

[1]Center for Nuclear Theory, Stony Brook University, Stony Brook NY 11790 USA

Abstract. Exclusive processes are traditionally described by perturbative hard blocks and "distribution amplitudes" (DAs), matrix elements of operators of various chiral structure and twist. One paper (with I.Zahed) calculate instanton contribution to hard blocks, which is found comparable to perturbative one in few-GeV^2 Q^2 region of interest. Another paper aims at comprehensive wave functions of mesons, baryons and pentaquarks. The last ones are also included as 5-quark component of the baryons. The calculation, using 't Hooft operator, gives x-dependence and magnitude of the antiquark PDF. It explains long standing issue of strong flavor asymmetry of antiquark sea. The third paper (also with I.Zahed)is semi-review on the instanton-sphaleron processes in QCD and electroweak theories, with emphasis on their possible experimental observation via double diffractive events at LHC and RHIC. Insert your english abstract here.

1 Brief history

The physics of the nonperturbative vacuum of strong interaction started even before QCD. Nambu and Jona-Lasinio (NJL) [1], inspired by BCS theory of superconductivity, have qualitatively explained that strong enough attraction of quarks can break $SU(N_f)_a$ chiral symmetry spontaneously and, among many other effects, create near-massless pions.

Instantons, the basis for semiclassical theory of QCD vacuum and hadrons, has been discovered in 1970's [2], and soon t'Hooft has found their fermionic zero modes and formulated his famous effective Lagrangian [3]. Not only it solved the famous "$U_A(1)$ problem" – by making the η' non-Goldstone and heavy – but it also produces a strong attraction in the σ and π channels. In the framework of the so called instanton liquid model (ILM) [4], it provided a microscopic (QCD-based) basis for chiral symmetry breaking, chiral perturbation theory and the pion properties. Its two parameters

$$\rho = \frac{1}{3}\,\mathrm{fm}, \qquad n_{I+\bar{I}} = \frac{1}{R^4} = 1\,\mathrm{fm}^{-4} \tag{1}$$

play the same role as the coupling and the cutoff in the NJL model. These parameters have withstood the scrutiny of time, and describe rather well the chiral dynamics related to pions, the Euclidean correlation functions in the few femtometers range, interacting ensemble of instantons, and much more, see [5] for a review.

*e-mail: edward.shuryak@stonybrook.edu

The subject of this talk is several other uses of instantons discussed in recent works. We will consider quark pair production and their role in isospin asymmetry of the nucleon "sea", inclusion of $\bar{I}I$ molecules in mesonic formfactors and forces between quarks in hadrons.

2 The topological landscape and sphaleron production processes

There was significant development in understanding the "topological landscape" and relation between instantons and sphaleron production processes, see [6]. Recently interest to it was renewed, due to possible sphaleron production at RHIC/LHC [7]. Let us focus on two main variables, for all static (3-dimensional and purely magnetic) configurations of the lowest energy, consistent with the walue of those coordinates. One of the coordinates is the topological Chern-Simons number, and the other is the size squared of the field

$$N_{CS} \equiv \frac{\epsilon^{\alpha\beta\gamma}}{16\pi^2}\int d^3x\left(A^a_\alpha\partial_\beta A^a_\gamma + \frac{1}{3}\epsilon^{abc}A^a_\alpha A^b_\beta A^c_\gamma\right),\quad \rho^2 \equiv \frac{\int d^3x\vec{x}^2\vec{B}^2}{\int d^3x\vec{B}^2} \tag{2}$$

If those are kept constant, adding Lagrange multiplies to the action one can find the energy and Chern-Simons number in parametric form

$$U_{\min}(\kappa,\rho) = (1-\kappa^2)^2\frac{3\pi^2}{g^2\rho},\quad N_{CS}(\kappa) = \frac{1}{4}\mathrm{sign}(\kappa)(1-|\kappa|)^2(2+|\kappa|) \tag{3}$$

where $\kappa = 0$ corresponds to top of the barrier, known as "sphaleron".

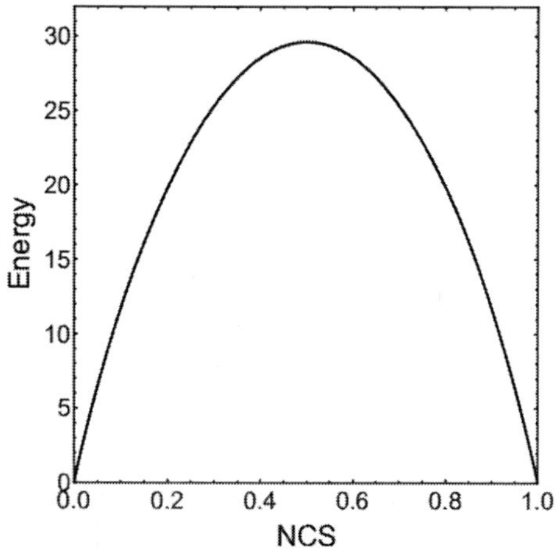

Figure 1. The potential energy $U_{\min}\left(N_{CS},\rho\right)$ (in units of $1/g^2\rho$) versus the Chern-Simons number N_{CS} for the "sphaleron path" between $N_{CS} = 0$ and $N_{CS} = 1$.

Production of sphaleron-path states can be described semiclassically using instanton-antiinstanton "streamline" [8? , 9], and their explosion in Minkowski space-time by [10]. The new discussion is about estimates of how one can produce QCD sphalerons as topologically-charged clusters in double-diffractive events, with a cluster of few GeV mass at the center. Its decay modes into 3 mesons were calculated using t'Hooft Lagrangian.

3 Mesons and baryon light-front wave functions and isospin asymmetry of the nucleon "sea"

High energy processes, like famous electron-nucleon deep inelastic scattering, produced rich "parton phenomenology", in form of parton distribution functions (PDFs), transverse distribution functions (TDFs) etc. Those are certain matrix elements of light front wave functions (LFWFs), which, however, did not obtained much attention. Furthermore, understanding of light front Hamiltonians is only in its initial stage.

In [11] I have calculated LFWFs for several mesons and baryons, including 5-quark component of the latter related to the *isospin asymmetry puzzle*. Canonical process of quark pair production via gluons is "flavor blind", and yet the antiquark sea of the nucleon is very asymmetric, as shown in Fig.2 (right). As noticed in [12], the t'Hooft Lagrangian is strongly flavor asymmetric, say d quark can produce $\bar{u}u$ pair (Fig.2 left) but *not* $\bar{d}d$. If it would be the only process, the $\bar{d}/\bar{u}$ ratio would be 2, as there are two u quarks and only one d in the nucleon: not far from the data at certain x.

Calculating admixture of 5-q states to the nucleon LFWF, I was able to quantify this effect and get the magnitude of the asymmetry and shape of first-generation antiquarks in agreement with the data.

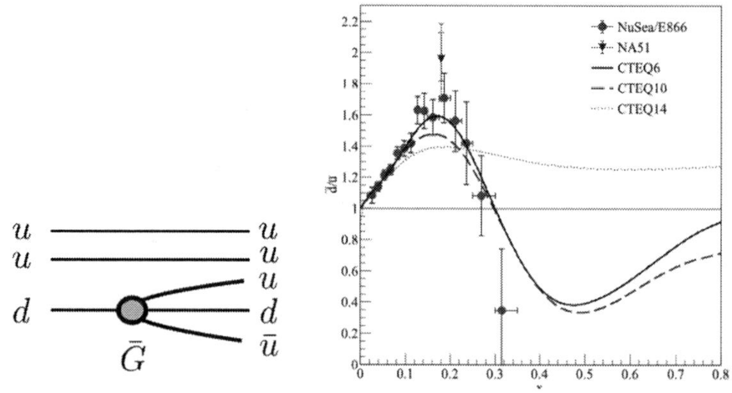

Figure 2. Sea generation via t'Hooft Lagrangian (left) and the data on the $\bar{d}/\bar{u}$ ratio versus x, right.

4 "Dense instanton ensemble"

The original instanton liquid model focused on quark condensate and therefore relatively stand-alone instantons, whose zero modes gets collectivized. But instanton ensemble also includes close instanton-antiinstanton pairs, as it is also observed on the lattice. The stand-alone instantons are seen via "deep cooling" , during which the instanton-antiinstanton molecules get annihilated. In so far, the application of the "molecular component" of the semiclassical ensemble was made only in connection to phase transitions in hot/dense matter. Indeed, this component is the only one which survives at temperatures $T > T_c$, where chiral symmetry is restored. Account for both components together started with [13]. The "molecular component" was also shown to be important at high baryonic densities, where it contributes to quark pairing and color superconductivity [14]. Close $\bar{I}I$ pairs has been qualitatively studied in recent work in [15] which studied their evolution during cooling, see Fig.3. Extrapolated to zero cooling (left side of the plots) one sees that while the instanton size fits previous expectations (1), the density seems to actually be significantly larger.

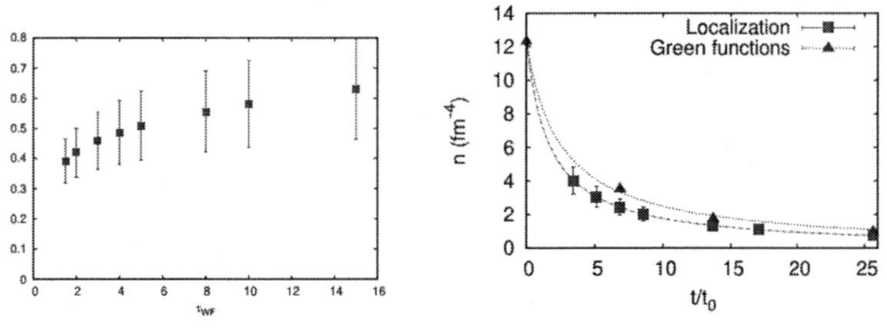

Figure 3. The dependence of the mean instanton sizes (upper plot) and density (lower plot) on the gradient flow cooling time τ (arbitrary units). The quantum vacuum of course corresponds to an extrapolation to $\tau \to 0$.

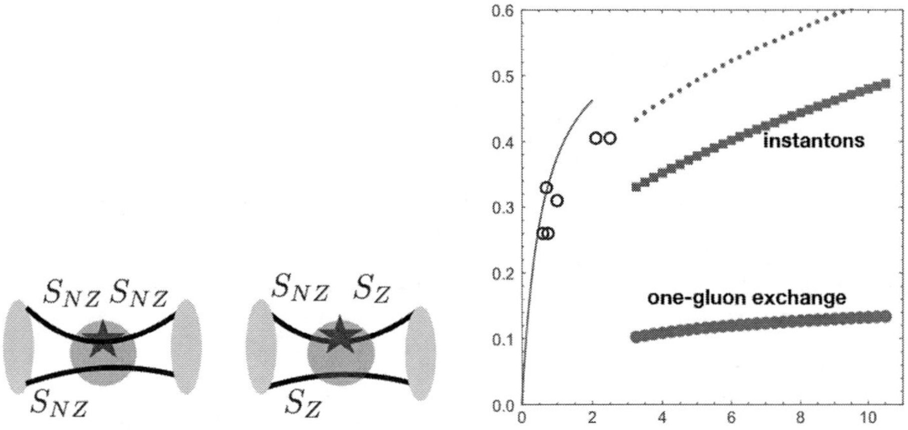

Figure 4. Left: two diagrams with quark propagators in the instanton fields. Right: vector formfactor of the pion $Q^2 F_\pi(Q^2)$. Open points correspond to some experimental data points, thin curve on the left is standard dipole fit. The dotted line on the top is the sum of perturbative and instanton contributions

5 Formfactors: including instantons in hard block

More recently, we have explored non-perturbative contributions to the "hard block" of mesonic form factors [16]. We calculated photon, scalar, graviton and dilaton FFs for pion, rho and scalar a_0 (brother of η?). The field at the instanton center is rather strong, with $\rho = 1/3\, fm$ is large

$$G_{rms} \equiv \sqrt{(G^a_{\mu\nu}(0))^2} = \sqrt{192}/\rho^2 \approx 5\,\mathrm{GeV}^2,$$

comparable to Q^2 is the so called semi-hard region studied.

One example, which was studied a lot experimentally, is electromagnetic pion formfactor, shown in Fig.4 right. It was found that they are mostly dominated by the diagram with three propagator made of non-zero Dirac modes (the left one on the left), not the one with Dirac zero modes. Therefore we included contributions of "dense instanton liquid" , taking the diluteness parameter $\kappa = \pi^2 n_{I+\bar{I}} \rho^4$ to one, in plots like Fig. 4 right. As one can see, in this version the sum of perturbative and instanton-induced formfactor reproduce the data.

The expressions for perturbative, non-zero-mode and zero mode parts are

$$V_a^\pi(Q^2) = \epsilon_\mu(q)(p^\mu + p'^\mu)(e_u + e_{\bar{d}})\Big[\Big(\frac{2C_F\pi\alpha_s f_\pi^2}{N_c Q^2}\Big) \times \int dx_1 dx_2 \Big(\frac{1}{\bar{x}_1\bar{x}_2 + m^2_{\rm gluon}/Q^2}\Big)\Big(\varphi_\pi(x_1)\varphi_\pi(x_2) + 2\frac{\chi_\pi^2}{Q^2}\Big(\varphi_\pi^P(x_1)\varphi_\pi^P(x_2)\Big(\frac{1}{\bar{x}_2 + E_\perp^2/Q^2} - 1\Big) + \frac{1}{6}\varphi_\pi^P(x_1)\varphi_\pi'^T(x_2)\Big(\frac{1}{\bar{x}_2 + E_\perp^2/Q^2} + 1\Big)\Big)\Big)\Big] \tag{4}$$

$$V_c^\pi = \epsilon_\mu(q)(p^\mu + p'^\mu)(e_u + e_{\bar{d}})\Big[\frac{\kappa\pi^2 f_\pi^2\chi_\pi^2}{N_c M_Q^2}\langle\rho^2 \mathbb{G}_V(Q\rho)\rangle \times \int dx_1 dx_2 \bar{x}_1\Big(\varphi_\pi^P(x_1)\varphi_\pi^P(x_2) - \frac{1}{36}\varphi_\pi'^T(x_1)\varphi_\pi'^T(x_2)\Big)\Big] \tag{5}$$

$$V_d^\pi = -\epsilon_\mu(q)(p^\mu + p'^\mu)(e_u + e_{\bar{d}}) \times\Big[\Big(\frac{1}{N_c^2(N_c+1)}\Big)\frac{4\kappa\pi^2 f_\pi^2\chi_\pi^2}{3M_Q^2}\Big\langle\rho^2\frac{K_1(Q\rho)}{Q\rho}\Big\rangle\int dx_1 dx_2 \varphi_\pi^P(x_1)\varphi_\pi'^T(x_2)\Big] \tag{6}$$

Next-twist distribution amplitudes $\varphi_\pi^P, \varphi_\pi^T$ are associated with pseudoscalar γ_5 and tensor $\sigma^{\mu\nu}$ ones, the leading twist φ_π is standard matrix element of the axial current $\gamma_\mu\gamma_5$. For the plot all three are taken to be just constant, independent on x. (Therefore all contributions of the tensor DA, appearing as a derivative over x, becomes actually vanish.)

6 Instanton-induced inter-quark forces

Historically, hadronic spectroscopy got to solid foundation in 1970's, with discoveries of nonrelativistic quarkonia made of heavy c, b quarks. In the first approximation, those are well described by simple Cornell potential

$$V_{\rm Cornell}(r) = -\frac{4\alpha_s}{3}\frac{1}{r} + \sigma_T r \tag{7}$$

which correctly attributes the short-distance potential to perturbative gluon exchange, and its large distance $O(r)$ contribution to the tension of a confining flux tube (the QCD string). One of the issues to be discussed in this paper is the nonperturbative origin of the inter-quark interaction at *intermediate* distances $r \sim 0.2 - 0.5$ fm.

Later developments [17, 18] connected interquark potential to correlator of Wilson lines (central)

$$e^{-V_C(r)T} = \langle W(\vec{x}_1)W^\dagger(\vec{x}_2)\rangle \tag{8}$$

Spin-dependent forces were related to such correlators with two magnetic fields (V_{SS}, V_{tensor}), or magnetic and electric fields for spin-orbit one. To evaluate such nonlocal quantities one needs to use lattice simulation, or rely on certain model of the vacuum fields. In Fig.5 (left) we show "dense instanton model" to the central potential, compared to linear potential and its version from [19] including string quantum vibrations (resummed "Lusher terms"). One can see that instantons can comlpement the flux tube at intermediate distances.

One problem with electric flux tube model is that it does not provide magnetic fields, while instantons are self-dual and have B needed for spin forces. We calculated the instanton contributions to spin forces, see $V_S S$ comparison of lattice potential, perturbative and

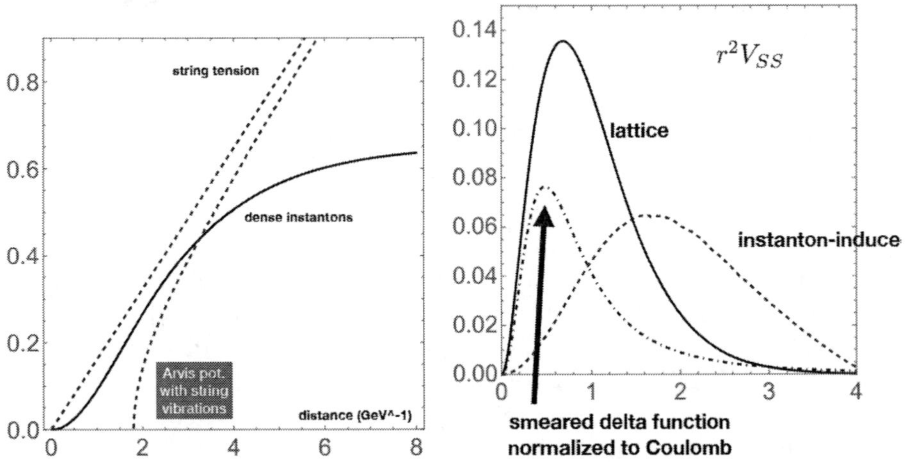

flavors	$M_\Upsilon - M_{\eta_b}$	$M_{J/\psi} - M_{\eta_c}$	$M(D^*) - M(D)$	$M(K^*) - M(K)$	$M(\rho) - M(\pi)$
Exp	61.	116.	137.	398.	636.
$\langle V_{SS}^{lat}/3M_1M_2\rangle$	46.	108.	98.	170.	
$\langle \vec{\nabla}^2 V_C/3M_1M_2\rangle$	28.	58.	48.	82.	
$\langle \vec{\nabla}^2 V_{inst}/3M_1M_2\rangle$	7.	30.	48.	90.	

Figure 5. The central potential V_c (left) and spin-spin $r^2V_{SS}(r)$ (right) versus distance $r, (GeV^{-1})$. The lattice result is one of the parameterization given in [20], the perturbative is Laplacian of (regulated) Coulomb term.

instanton-induced in Fig.5 (right). Note that the area below the perturbative and instanton-induced terms is comparable. The corresponding matrix elements (using Cornell wave functions with proper quark masses) is shown in the table. One can see that e.g. for charmonium the magntitude of spin-spin term is in agreement with lattice and the level splitting. We also considered $L = 1$ families of mesons, from heavy to light, and considered other spin-dependent potentials. We also discussed $\bar{I}I$ molecules, which provide somewhat different potentials due to different pictures of their fields.

However for heavy-light and light-light cases this V_{SS} is not enough: the missing part is then attributed to part of the quark propagators containing zero modes (t'Hooft Lagrangian). It works well for heavy-light and the pions, as expected.

We undertake this study of inter-quark forces in [21], preparing to calculate the light-front Hamiltonian and corresponding light-front wave functions [22].

References

[1] Y. Nambu and G. Jona-Lasinio, Phys. Rev. **122** (1961), 345-358 doi:10.1103/PhysRev.122.345

[2] A. A. Belavin, A. M. Polyakov, A. S. Schwartz and Y. S. Tyupkin, Phys. Lett. B **59** (1975), 85-87 doi:10.1016/0370-2693(75)90163-X

[3] G. 't Hooft, Phys. Rev. D **14** (1976), 3432-3450 [erratum: Phys. Rev. D **18** (1978), 2199] doi:10.1103/PhysRevD.14.3432

[4] E. V. Shuryak, Nucl. Phys. B **203** (1982), 140-156 doi:10.1016/0550-3213(82)90480-1

[5] T. Schäfer and E. V. Shuryak, Rev. Mod. Phys. **70** (1998), 323-426 doi:10.1103/RevModPhys.70.323 [arXiv:hep-ph/9610451 [hep-ph]].
[6] D. M. Ostrovsky, G. W. Carter and E. V. Shuryak, Phys. Rev. D **66** (2002), 036004 doi:10.1103/PhysRevD.66.036004 [arXiv:hep-ph/0204224 [hep-ph]].
[7] E. Shuryak and I. Zahed, [arXiv:2102.00256 [hep-ph]].
[8] J. J. M. Verbaarschot, Nucl. Phys. B **362** (1991), 33-53 [erratum: Nucl. Phys. B **386** (1992), 236-236] doi:10.1016/0550-3213(91)90554-B
[9] V. V. Khoze and A. Ringwald, CERN-TH-6082-91.
[10] E. Shuryak and I. Zahed, Phys. Rev. D **67** (2003), 014006 doi:10.1103/PhysRevD.67.014006 [arXiv:hep-ph/0206022 [hep-ph]].
[11] E. Shuryak, Phys. Rev. D **100** (2019) no.11, 114018 doi:10.1103/PhysRevD.100.114018 [arXiv:1908.10270 [hep-ph]].
[12] A. E. Dorokhov and N. I. Kochelev, Phys. Lett. B **304** (1993), 167-175 doi:10.1016/0370-2693(93)91417-L
[13] E. M. Ilgenfritz and E. V. Shuryak, Nucl. Phys. B **319** (1989), 511-520 doi:10.1016/0550-3213(89)90617-2
[14] R. Rapp, T. Schäfer, E. V. Shuryak and M. Velkovsky, Phys. Rev. Lett. **81** (1998), 53-56 doi:10.1103/PhysRevLett.81.53 [arXiv:hep-ph/9711396 [hep-ph]].
[15] A. Athenodorou, P. Boucaud, F. De Soto, J. Rodríguez-Quintero and S. Zafeiropoulos, JHEP **02** (2018), 140 doi:10.1007/JHEP02(2018)140 [arXiv:1801.10155 [hep-lat]].
[16] E. Shuryak and I. Zahed, Phys. Rev. D **103** (2021) no.5, 054028 doi:10.1103/PhysRevD.103.054028 [arXiv:2008.06169 [hep-ph]].
[17] C. G. Callan, Jr., R. F. Dashen, D. J. Gross, F. Wilczek and A. Zee, Phys. Rev. D **18** (1978), 4684 doi:10.1103/PhysRevD.18.4684
[18] E. Eichten and F. Feinberg, Phys. Rev. D **23** (1981), 2724 doi:10.1103/PhysRevD.23.2724
[19] J. F. Arvis, Phys. Lett. B **127** (1983), 106-108 doi:10.1016/0370-2693(83)91640-4
[20] T. Kawanai and S. Sasaki, Phys. Rev. D **92** (2015) no.9, 094503 doi:10.1103/PhysRevD.92.094503 [arXiv:1508.02178 [hep-lat]].
[21] E. Shuryak and I. Zahed, [arXiv:2110.15927 [hep-ph]].
[22] E. Shuryak and I. Zahed, [arXiv:2111.01775 [hep-ph]].

Another look at the Landau gauge three-gluon vertex

Guilherme T. R. Catumba[1,2,*], *Orlando* Oliveira[2,**], and *Paulo J.* Silva[2,***]

[1]IFIC - University of Valencia - Spain
[2]CFisUC, Department of Physics, University of Coimbra, 3004-516 Coimbra, Portugal

Abstract. We revisit the computation of the three-gluon vertex in the Landau gauge using lattice QCD simulations with large physical volumes of $\sim (6.5\ \text{fm})^4$ and $\sim (8\ \text{fm})^4$ and large statistical ensembles. For the kinematical configuration analysed, that is described by a unique form factor, an evaluation of the lattice artefacts is also performed. Particular attention is given to the low energy behaviour of vertex and its connection with evidence (or lack of it) of infrared ghost dominance.

1 Introduction

The amputated three-gluon correlation function, herein also named vertex, is a fundamental QCD correlation function that allows for the extraction of a strong coupling constant and the computation of a static potential between color charges. Our aim is to revisit a previous calculation [1, 2] using an improved statistics for the larger volume and estimate the corresponding lattice artefacts.

Lattice simulations measure the correlation function $G^{a_1a_2a_3}_{\mu_1\mu_2\mu_3}(p_1, p_2, p_3)$ via the computation of

$$\langle A^{a_1}_{\mu_1}(p_1)\, A^{a_2}_{\mu_2}(p_2)\, A^{a_3}_{\mu_3}(p_3)\rangle = V\,\delta(p_1 + p_2 + p_3)\, G^{a_1a_2a_3}_{\mu_1\mu_2\mu_3}(p_1, p_2, p_3)\,. \tag{1}$$

In terms of the gluon propagator D and of the one-particle irreducible three-gluon diagram (1PI) Γ the correlation function is given by

$$G^{a_1a_2a_3}_{\mu_1\mu_2\mu_3}(p_1, p_2, p_3) = D^{a_1b_1}_{\mu_1\nu_1}(p_1)\, D^{a_2b_2}_{\mu_2\nu_2}(p_2)\, D^{a_3b_3}_{\mu_3\nu_3}(p_3)\, \Gamma^{b_1b_2b_3}_{\nu_1\nu_2\nu_3}(p_1, p_2, p_3). \tag{2}$$

The three-gluon vertex reads

$$\Gamma^{a_1a_2a_3}_{\mu_1\mu_2\mu_3}(p_1, p_2, p_3) = f_{a_1a_2a_3}\,\Gamma_{\mu_1\mu_2\mu_3}(p_1, p_2, p_3) \tag{3}$$

and given that Bose symmetry requires the 1PI function to be symmetric under permutations of any pair (p_i, a_i, μ_i) it follows that $\Gamma_{\mu_1\mu_2\mu_3}(p_1, p_2, p_3)$ has to be antisymmetric under the interchange of any pair (p_i, μ_i). The description of $\Gamma_{\mu_1\mu_2\mu_3}(p_1, p_2, p_3)$ in the continuum formulation requires six Lorentz invariant form factors. Two of the form factors are associated with the transverse component of the vertex, while the remaining define its longitudinal component [3].

*e-mail: gtelo@ific.uv.es
**e-mail: orlando@uc.pt
***Speaker, e-mail: psilva@uc.pt

Here we consider the evaluation of the vertex for the asymmetric momentum configuration that is defined by $p_2 = 0$, as in [1, 2, 4]. Then, the three point connected Green function is given by

$$G_{\mu_1\mu_2\mu_3}(p, 0, -p) = V\frac{N_c(N_c^2-1)}{4}\left[D(p^2)\right]^2 D(0)\frac{\Gamma(p^2)}{3}\; p_{\mu_2}\, T_{\mu_1\mu_3}(p) \tag{4}$$

and, therefore,

$$G_{\mu\alpha\mu}(p, 0, -p)\, p_\alpha = V\frac{N_c(N_c^2-1)}{4}\left[D(p^2)\right]^2 D(0)\;\Gamma(p^2)\; p^2. \tag{5}$$

It follows that the form factor $\Gamma(p^2)$ can be measured by computing the ratio

$$G_{\mu\alpha\mu}(p, 0, -p)p_\alpha / \left[D(p^2)\right]^2 D(0)\,. \tag{6}$$

For large momentum, the gluon propagator becomes small and its fluctuations induce large variations for $\Gamma(p^2)$ that prevent a precise measurement of this form factor in the UV regime. Indeed, such large fluctuations have been observed in previous simulations; see e.g. [1] and references therein.

2 The lattice setup

In the following we show results from a simulation using a 64^4 lattice with an ensemble of 2000 configurations already studied in [1], together with a simulation using a 80^4 lattice and an ensemble of 1800 configurations. The ensembles were generated with the Wilson gauge action at $\beta = 6.0$ using Chroma library [5]. The gauge configurations have been rotated to the Landau gauge using the Fourier accelerated Steepest Descent method [6]. Fast Fourier transforms were implemented using the PFFT library [7]. Our definition for the gluon field being

$$ag_0 A_\mu(x + a\hat{e}_\mu) = \frac{U_\mu(x) - U^\dagger(x)}{2ig_0} - \frac{\mathrm{Tr}\left[U_\mu(x) - U^\dagger(x)\right]}{6ig_0} \tag{7}$$

with its corresponding field in momentum space being given by

$$A_\mu(\hat{p}) = \sum_x e^{-iq(x+a\hat{e}_\mu)} A_\mu(x + a\hat{e}_\mu)\;,\; q_\mu = \frac{2\pi\, n_\mu}{a\, L_\mu}. \tag{8}$$

An improved definition for lattice momenta, motivated by lattice perturbation theory, is

$$p_\mu = \frac{2}{a}\sin\left(\frac{\pi n_\mu}{L_\mu}\right). \tag{9}$$

3 Handling of noise and lattice artefacts

We use two approaches to handle the problem of the large statistical fluctuations at high momenta already discussed in [8, 9], namely

- explore the ambiguity on the scale setting that allow us to bin the momentum data — the momentum are grouped in bins and for each bin the lattice data is replaced by its weighted average, using as weight the inverse of the statistical error of the bin data points;

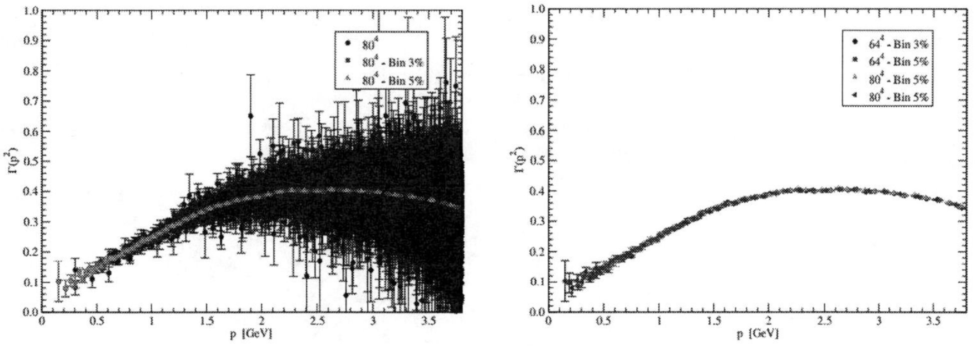

Figure 1. Left plot: original and binned data for $\Gamma(p^2)$, 80^4 lattice. Right plot: Comparison of binned data for $\Gamma(p^2)$ and both lattice volumes.

- perform a $H(4)$ extrapolation of the lattice data [10, 11] — such procedure uses the naive lattice momentum q_μ and explores the lattice $H(4)$ symmetry group that is associated with an hypercube. Accordingly, any lattice scalar function F depends on the $H(4)$ invariants

$$q^2 = q^{[2]} = \sum_\mu q_\mu^2, \quad q^{[4]} = \sum_\mu q_\mu^4, \quad q^{[6]} = \sum_\mu q_\mu^6, \quad q^{[8]} = \sum_\mu q_\mu^8,$$

and one writes $F_{Lat} = F(q^{[2]}, q^{[4]}, q^{[6]}, q^{[8]})$. The corresponding function in the continuum limit is given by $F = F(q^{[2]}, 0, 0, 0)$ up to corrections $O(a^2)$. The procedure requires several data points with the same q^2 but different $q^{[4]}$, $q^{[6]}$, $q^{[8]}$. F can be computed via an extrapolation of F_{Lat} if one assumes that F_{Lat} can be written as a power series of the H(4) invariants. Herein, we ignore contributions due to $q^{[6]}$ and $q^{[8]}$ and consider a linear extrapolation in $q^{[4]}$.

4 Results

The data for $\Gamma(p^2)$ both binned and unbinned is reported in the left plot of Figure 1 for the largest lattice. Clearly, by binning the data the large statistical errors in the high momentum region are suppressed resulting in a well defined and smooth curve. On the right plot of Figure 1, the binned data for the two lattice volumes is compared. The good agreement within errors of both sets of data suggest that the finite volume effects are under control and are small.

The linear H(4) extrapolation of the 64^4 lattice data, including the binned data, is compared with the original binned data in the left plot of Figure 2. Up to momentum $p \sim 2.5$ GeV both the H(4) extrapolation and the binned data are in good agreement. However, for large p the H(4) extrapolation overestimates Γ in comparison with the binned lattice data. As seen in the right plot of Figure 2, in the low momentum regime the two sets of data agree within errors.

As can be observed in all the Figures the form factor does not seem to change sign in the infrared region. Recall that the change of sign is, in general, understood as an indication of ghost dominance in the infrared.

5 Infrared behaviour of $\Gamma(p^2)$

The lack of observing a change of sign in the form factor $\Gamma(p^2)$ can be an indication that either it does not happen and there is no ghost dominance in the infrared region or the change of sign

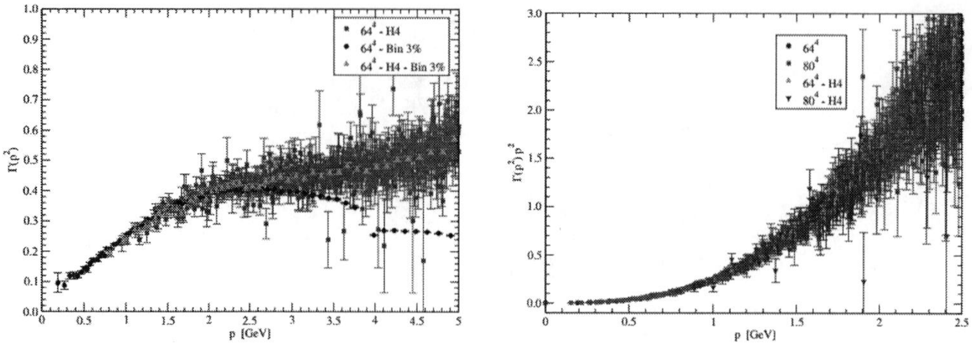

Figure 2. Left plot: results of the H(4) extrapolation of $\Gamma(p^2)$ for the 64^4 lattice volume. Right plot: original and H(4) data for $p^2\Gamma(p^2)$ and for both lattice volumes in the infrared region.

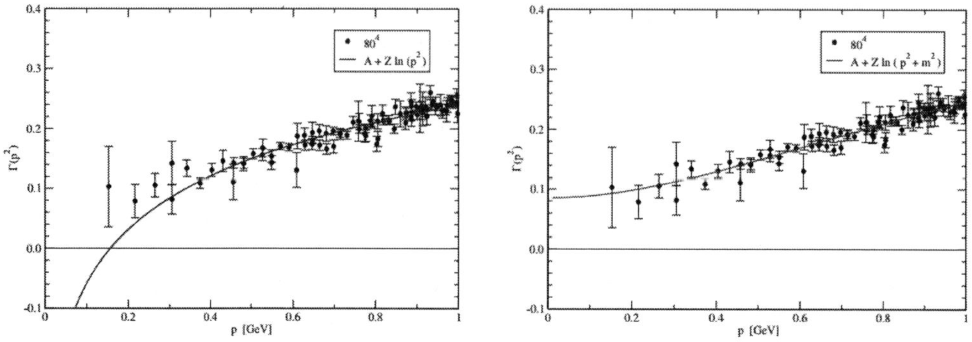

Figure 3. Infrared 80^4 lattice data for $\Gamma(p^2)$ together with some fitting functions. Left plot: $\Gamma(p^2) = A + Z\ln(p^2)$; right plot: $\Gamma(p^2) = A + Z\ln(p^2 + m^2)$.

occurs at momenta that are smaller than those accessed in the simulations. Here, we explore further the infrared behavior of $\Gamma(p^2)$ by fitting the lattice data to various functional forms. In the following, only the 80^4 lattice data for momenta below 1GeV will be considered.

The best fits of the lattice data to $\Gamma_1(p^2) = A + Z\ln(p^2)$ and $\Gamma_2(p^2) = A + Z\ln(p^2 + m^2)$ are reported in Figure 3. $\Gamma_1(p^2)$ was used in recent studies of the infrared three-gluon vertex, see [9] for details, and $\Gamma_2(p^2)$ is a regularized version of $\Gamma_1(p^2)$ that is, in principle, finite over the full range of p^2. The best fit to $\Gamma_1(p^2)$ results in a $\chi^2/d.o.f. = 1.23$, $A = 0.2395(16)$ and $Z = 0.0646(21)$. Hence, the zero crossing should occur at $p_o = 157$MeV. The best fit for $\Gamma_2(p^2)$ gives $A = 0.208(24)$, $Z = 0.124(27)$ and $m = 0.61(15)$ GeV, with a $\chi^2/d.o.f. = 0.95$. No change of sign can be associated with this function; see the right plot of Figure 3.

In order to help clarifying the infrared behavior of $\Gamma(p^2)$, we also considered a power law, $\Gamma_3(p^2) = 1 + c\,p^{-d}$ with the best fit resulting in $c = -0.7621(15)$, $d = 0.1558(49)$ with a $\chi^2/d.o.f. = 1.35$. Accordingly, the change of sign occurs at $p_o = 175$MeV; see Figure 4, left plot. Finally we considered the quadratic function $\Gamma_4(p^2) = a + bp^2 + cp^4$ whose optimal parameters are $a = 0.0978(60)$, $b = 0.218(22)$, and $c = -0.070(18)$, with a $\chi^2/d.o.f. = 0.98$. Similarly to $\Gamma_2(p^2)$, the change of sign of the form factor is also not observed; see Figure 4, right plot.

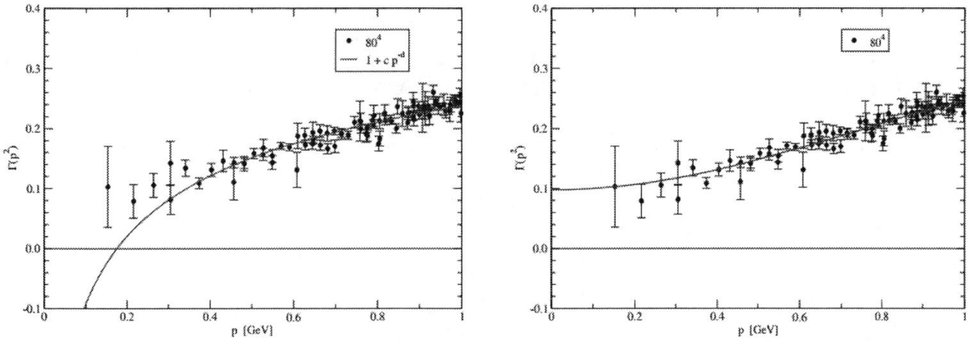

Figure 4. Infrared 80^4 lattice data for $\Gamma(p^2)$ together with some fitting functions. Left plot: $\Gamma(p^2) = 1 + c\,p^{-d}$; right plot: $\Gamma(p^2) = a + bp^2 + cp^4$.

6 Conclusions and outlook

An improved calculation of the three gluon vertex on the lattice and for the asymmetric momentum configuration is described. Two different lattice volumes of (6.5 fm)^4 and (8.2 fm)^4, with a common lattice spacing of $a = 0.102$ fm, have been investigated. In general our approach to handle the fluctuations and the lattice spacing effects seems to be able to produce compatible smooth curves over a wide range of momenta that extends up to $\sim$ 2.5 GeV. Our investigation of the deep infrared region using the lattice data and relying on fits to functional forms are inclusive in what concerns a change of sign of $\Gamma(p^2)$.

This work was granted access to the HPC resources of the PDC Center for High Performance Computing at the KTH Royal Institute of Technology, Sweden, made available within the Distributed European Computing Initiative by the PRACE-2IP, receiving funding from the European Communitys Seventh Framework Programme (FP7/2007-2013) under Grant agreement no. RI-283493. The use of Lindgren has been provided under DECI-9 project COIMBRALATT. We acknowledge that the results of this research have been achieved using the PRACE-3IP project (FP7 RI312763) resource Sisu based in Finland at CSC. The use of Sisu has been provided under DECI-12 project COIMBRALATT2. We acknowledge the Laboratory for Advanced Computing at the University of Coimbra [12] for providing access to the HPC resource Navigator. This work was produced with the support of INCD [13] funded by Fundação para a Ciência e a Tecnologia, I. P. (FCT) and FEDER under the project 01/SAICT/2016 n° 022153. The authors acknowledge Minho Advanced Computing Center [14] for providing HPC resources that have contributed to the research results reported within this paper. This work was produced with the support of MACC and it was funded by FCT under the Advanced Computing Project CPCA/A2/6816/2020, platform Bob. Work supported by national funds from FCT, within the Projects UIDB/04564/2020, UIDP/04564/2020, and CERN/FIS-COM/0029/2017. P. J. S. acknowledges financial support from FCT under Contract CEECIND/00488/2017. G. T. R. C. acknowledges financial support from FCT under Project UIDB/04564/2020, and also from the Generalitat Valenciana (genT program CIDEGENT/2019/040) and Ministerio de Ciencia e Innovacion PID2020-113644GB-I00.

References

[1] A. G. Duarte, O. Oliveira, P. J. Silva, Phys. Rev. **D94**, 074502 (2016).
[2] A. G. Duarte, O. Oliveira, P. J. Silva, PoS(LATTICE2016)351.
[3] J. S. Ball, T.-W. Chiu, Phys. Rev. **D22**, 2550 (1980).
[4] B. Allés, D. S. Henty, H. Panagopoulos, C. Parrinello, C. Pittori, D. G. Richards, Nucl. Phys. **B502**, 325 (1997).

[5] R. G. Edwards, B. Joo, Nucl. Phys. Proc. Suppl. **140**, 832 (2005).

[6] C. T. H. Davies, G. G. Batrouni, G. R. Katz, A. S. Kronfeld, G. P. Lepage, K. G. Wilson, P. Rossi, B. Svetitsky, Phys. Rev. **D37**, 1581 (1988).

[7] M. Pippig, SIAM J. Sci. Comput. **35**, C213 (2013).

[8] G. T. R. Catumba, O. Oliveira, P. J. Silva, Phys. Rev. **D103**, 074501 (2021).

[9] Guilherme Telo Rodrigues Catumba, Master Thesis, University of Coimbra, 2020, arXiv:2101.06074 [hep-lat].

[10] D. Becirevic, Ph. Boucaud, J. P. Leroy, J. Micheli, O. Pène, J. Rodríguez–Quintero, C. Roiesnel, Phys. Rev. **D60**, 094509 (1999).

[11] F. de Soto and C. Roiesnel, JHEP **09**, 007 (2007).

[12] https://www.uc.pt/lca

[13] https://www.incd.pt

[14] https://macc.fccn.pt

Decomposition of the static potential in the Maximal Abelian gauge

Vitaly Bornyakov[1,2,3,*], *Vladimir* Goy[2,4], *Ilya* Kudrov[3], and *Roman* Rogalyov[1]

[1]Institute for High Energy Physics NRC Kurchatov Institute, 142281 Protvino, Russia
[2]Pacific Quantum Center, Far Eastern Federal University, 690950 Vladivostok, Russia
[3]Institute of Theoretical and Experimental Physics NRC Kurchatov Institute, 117218 Moscow, Russia
[4]Institut Denis Poisson CNRS/UMR 7013, Université de Tours, 37200 France

Abstract. Decomposition of SU(2) gauge field into monopole and monopoleless components is studied in $SU(2)$ gluodynamics and in QC_2D with zero and nonzero quark chemical potential after fixing MA gauge. For both components we calculate respective static potential and compare their sum with the nonabelian static potential. We demonstrate good agreement in the confinement phase and discuss the implications of our results.

1 Introduction

We study decomposition of the nonabelian gauge field in the Maximal Abelian gauge (MAG) [1, 2] into the sum of the monopole component and the monopoleless component. We use the lattice gauge theory with gauge group $SU(2)$ for this purpose. In terms of the lattice gauge field $U_\mu(x)$ the decomposition has the following form

$$U_\mu(x) = U_\mu^{mod}(x) u_\mu^{mon}(x) \quad (1)$$

where $u_\mu^{mon}(x)$ is the monopole component and $U_\mu^{mod}(x)$ is the monopoleless component called below also a modified gauge field.

The Abelian dominance for the string tension after fixing MAG [1, 2] is known [3–7] since long (for a review see e.g. [8, 9]). Moreover, it was found [5, 10, 11] that the properly determined monopole component of the gauge field produces the string tension close to its exact value in agreement with conjecture that the monopole degrees of freedom are responsible for confinement [12, 13].

In Refs. [14, 15] it was shown that the topological charge, chiral condensate and effects of chiral symmetry breaking in quenched light hadron spectrum disappear after removal of the monopole contribution from the relevant operators. Similar computations were made within the scope of the Z_2 projection studies [16]. In particular it was shown that after removal of P-vortices the confinement property disappear. We do a similar removal with monopoles. We consider the following types of the static potential: $V_{mod}(r)$ obtained from the Wilson loops of the modified gauge field $U_\mu^{mod}(x)$, $V_{mon}(r)$ obtained from the Wilson loops of the monopole gauge field $u_\mu^{mon}(x)$ and the sum of these two static potentials.

*e-mail: vitaly.bornyakov@ihep.ru

It was demonstrated in [17] for one value of the lattice spacing that $V_{mod}(r)$ could be well fitted by purely Coulomb fit function and the sum $V_{mod}(r)+V_{mon}(r)$ was a good approximation of the original nonabelian static potential, $V(r)$, at all distances.

Here we study this phenomenon at three lattice spacings using the Wilson lattice gauge field action and thus we can make conclusions about the continuum limit. We also present the results for one lattice spacing obtained with the improved lattice field action thus checking the universality. Furthermore, we present results for the $SU(2)$ theory with dynamical quarks, i.e. for QC_2D. These results were partially presented in [18].

2 Details of simulations

We study the $SU(2)$ lattice gauge theory. To fix MAG we used the simulated annealing algorithm [5] with one gauge copy per configuration. Usually after fixing MAG the following decomposition of the nonabelian lattice gauge field $U_\mu(x)$ is made

$$U_\mu(x) = C_\mu(x)u_\mu(x) \quad , \tag{2}$$

where $u_\mu(x)$ is the Abelian field and $C_\mu(x)$ is the nonabelian coset field. The Abelian gauge field $u_\mu(x)$ is further decomposed [19] into the monopole (singular) part $u_\mu^{mon}(x)$ and the photon (regular) part $u_\mu^{ph}(x)$:

$$u_\mu(x) = u_\mu^{mon}(x)u_\mu^{ph}(x) \,. \tag{3}$$

Then it follows from eq. (1) that

$$U_\mu^{mod}(x) = C_\mu(x)u_\mu^{ph}(x). \tag{4}$$

Note that $u^{ph}(x)$ is the Abelian projection of $U_\mu^{mod}(x)$ and involves no monopoles.

We need to compute the usual Wilson loops

$$W(C) = \frac{1}{2}\mathrm{Tr}\mathcal{W}(C), \quad \mathcal{W}(C) = \left(\prod_{l\in C} U(l)\right) \quad , \tag{5}$$

the monopole Wilson loops

$$W_{mon}(C) = \frac{1}{2}\mathrm{Tr}\left(\prod_{l\in C} u^{mon}(l)\right) \quad , \tag{6}$$

and the nonabelian Wilson loops with removed monopole contribution

$$W_{mod}(C) = \frac{1}{2}\mathrm{Tr}\mathcal{W}_{mod}(C), \quad \mathcal{W}_{mod}(C) = \left(\prod_{l\in C} U^{mod}(l)\right) \quad , \tag{7}$$

It is known that MAG fixing leaves $U(1)$ gauge symmetry unbroken. The monopole Wilson loop $W_{mon}(C)$ is invariant under respective residual gauge transformations. This is not true for $W_{mod}(C)$ [17]. Thus we need to fix the Landau $U(1)$ gauge using the gauge condition

$$\max_{\omega} \sum_{x,\mu} Re(\omega(x)u_\mu(x)\omega^\dagger(x+\mu)) \,, \tag{8}$$

where $\omega \in U(1)$ is the gauge transformation. To imrove the noise to signal ratio for the static potential we used the APE smearing [20] in computations of the Wilson loops.

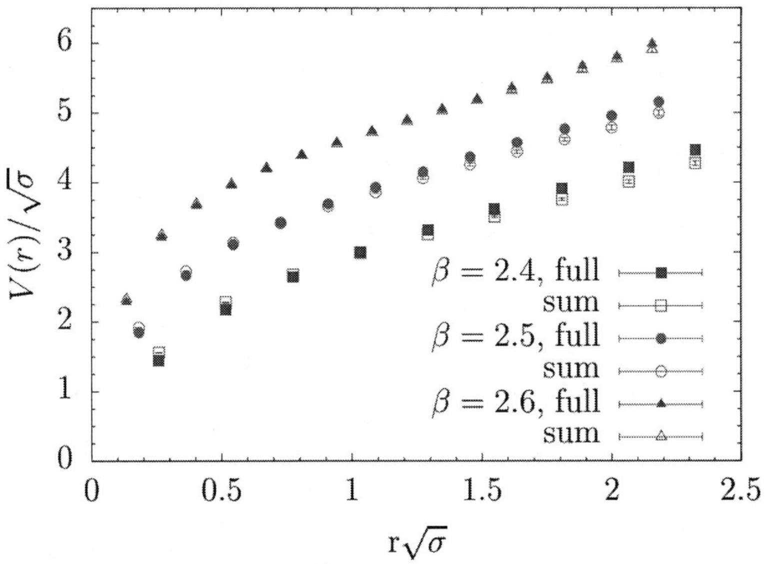

Figure 1. Comparison of the nonabelian static potential $V(r)$ (filled symbols) with the sum $V_{mon}(r) + V_{mod}(r)$ (empty symbols) for Wilson action at $\beta = 2.4$ (squares), $\beta = 2.5$ (circles), $\beta = 2.6$ (triangles).

We generated 100 statistically independent gauge field configurations with the Wilson lattice action at $\beta = 2.4, 2.5$ on 24^4 lattices and at $\beta = 2.6$ on 32^4 lattices. As a check of universality the computations were done with the tadpole improved action at $\beta = 3.4$ on 24^4 lattices. Additionally we present our results obtained in QC_2D on 32^4 lattice at zero and nonzero quark chemical potential [21].

It is worth to note that another decomposition, namely eq. (2) was investigated in Ref. [7] in case if $SU(3)$ gluodynamics. Good agreement between the static potenial $V(r)$ and the sum $V_{abel}(r) + V_{off}(r)$ was found. We believe that this decomposition also deserves further study.

3 Results

In Figure 1 we show the usual nonabelian static potential $V(r)$ denoted as 'full' and compare it with the sum $V_{sum}(r) = V_{mon}(r) + V_{mod}(r)$. One can see that approximate equality

$$V(r) \approx V_{sum}(r) \tag{9}$$

is satisfied for all three lattice spacings and the approximation improves toward the continuum limit. To give further support to this statement we plot in Figure 2 the relative deviation $\Delta(r)$ defined as

$$\Delta(r) = \frac{V(r) - V_{sum}(r)}{V(r)} . \tag{10}$$

It is clear that $\Delta(r)$ decreases with decreasing lattice spacing.

We also studied the question of universality of the decomposition for the static potential eq. (9). The simulations were made with the tadpole improved action at $\beta = 3.4$ with lattice spacing approximately equal to that of the Wilson action at $\beta = 2.5$. The results of our computations are presented in the Figure 3. One can see that agreement between $V(r)$ and $V_{sum}(r)$ is nearly as good as in Figure 1 for $\beta = 2.5$.

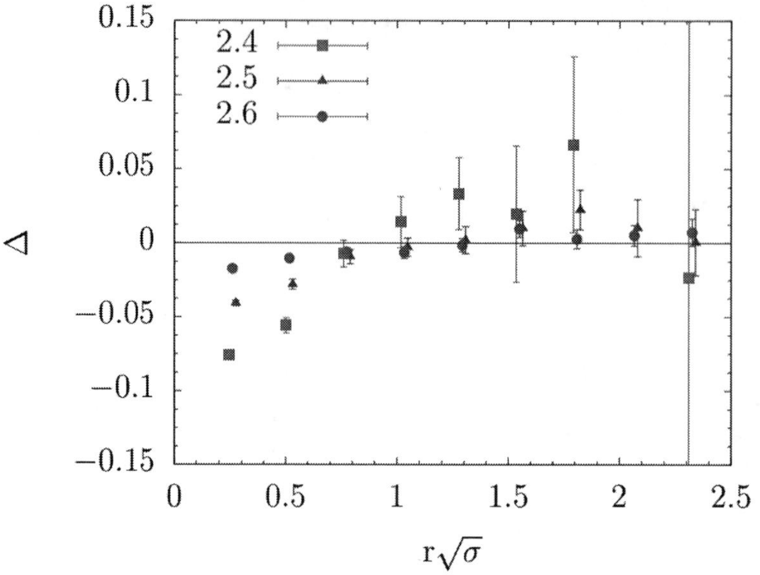

Figure 2. The relative deviation $\Delta(r)$ vs. distance r for three values of the lattice spacing.

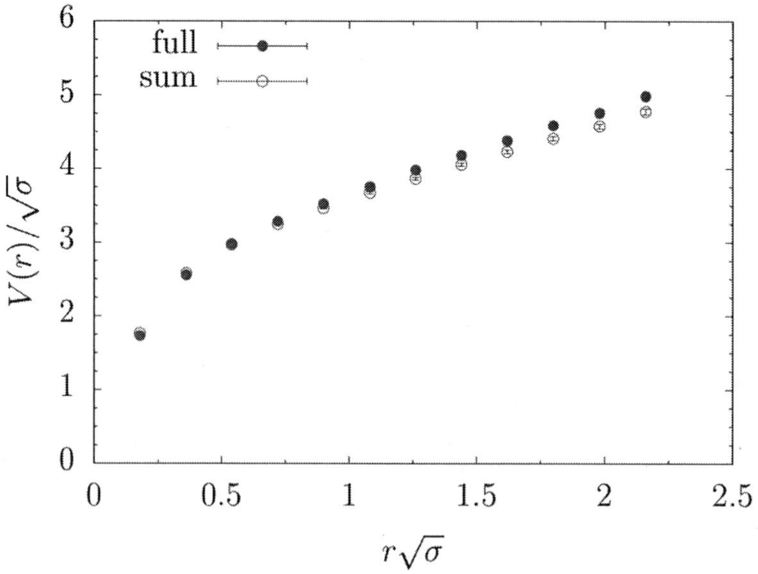

Figure 3. Comparison of the nonabelian static potential $V(r)$ (filled symbols) with the sum of the monopole and modified potentials $V_{sum}(r)$ (empty symbols) for improved action at $\beta = 3.4$.

Furthermore, we completed the same computations in QC_2D on 32^4 lattices with small lattice spacing at zero and nonzero quark chemical potential μ_q (for details of simulations in QC_2D see, e.g. [21]). The results of these computations are presented in Figure 4 for $\mu_q = 0$ and in Figure 5 for $a\mu_q = 0.19$. It can be seen clearly that approximate decomposition (9) is fulfilled with rather high both at zero and nonzero μ_q.

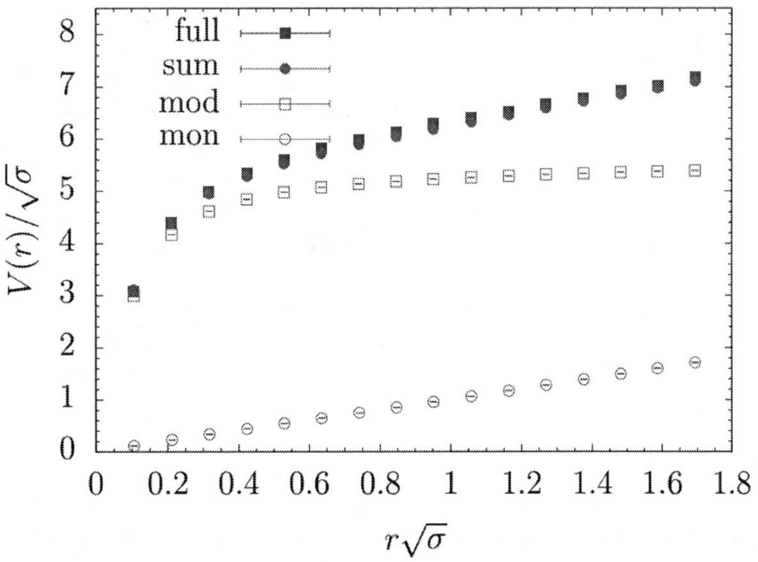

Figure 4. The nonabelian static potential $V(r)$ (filled squares) is compared with the sum of the monopole and modified potentials $V_{sum}(r)$ (filled circles) for QC_2D at zero chemical potential. The potentials $V_{mon}(r)$ (empty circles) and $V_{mod}(r)$ (empty squares) are also shown.

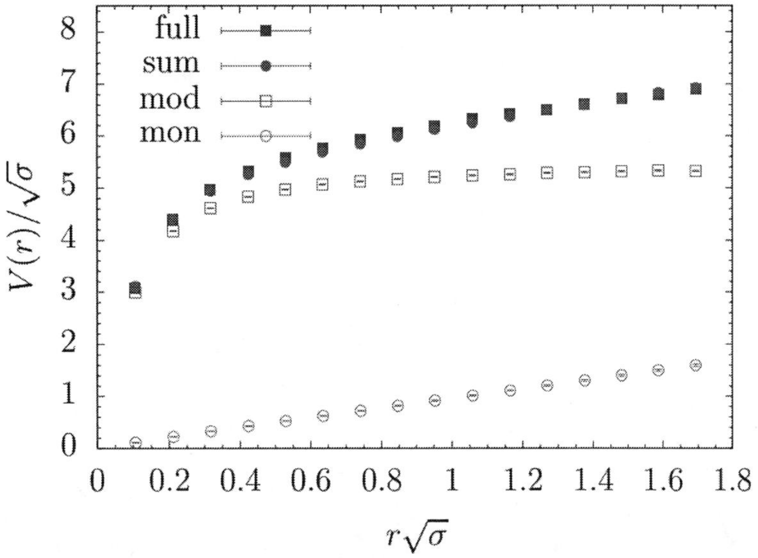

Figure 5. The nonabelian static potential $V(r)$ (filled squares) is compared with the sum of the monopole and modified potentials $V_{sum}(r)$ (filled circles) for QC_2D at nonzero chemical potential $a\mu_q = 0.19$. The potentials $V_{mon}(r)$ (empty circles) and $V_{mod}(r)$ (empty squares) are also shown.

4 Conclusions

We studied the decomposition of the static potential into the linear term produced by the monopole (Abelian) gauge field $U_{mon}(x)$ and the Coulomb term produced by the monopole-less nonabelian gauge field $U_{mod}(x)$. In case of the Wilson action we presented results (Figures 1 and 2) for three lattice spacings demonstrated that the agreement becomes better with decreasing lattice spacing. Thus our results imply that the relation (1) becomes exact in the continuum limit. Further work is needed to provide more evidence for this conclusion. Next, we demonstrated that the decomposition (1) is valid also in the case of improved lattice action (see Figure 3). Furthermore, we found that it works also in QC_2D for both zero and nonzero quark chemical potential. Let us note that in Ref. [18] we also presented results for the adjoint static potential and found that the decomposition (1) works quite well in this case although not as well as in the case of the fundamental representation. It is of course interesting to verify how decomposition (1) works for other observables, e.g. action density and energy density of the flux tube. Also the decomposition should be checked in the case of SU(3) gauge group.

One can draw the following conclusions from the decomposition (1). The monopole part of the gauge field $U_{mon}(x)$ is responsible for the classical part of the hadronic string energy while the monopoleless part $U_{mod}(x)$ produces the fluctuating part of that energy, i.e. while at small distances $U_{mod}(x)$ should reproduce the perturbative results at large distances it contributes to the nonperturbative physics.

Acknowledgments. Computer simulations were performed on the Central Linux Cluster of the NRC "Kurchatov Institute" - IHEP (Protvino) and Linux Cluster of the NRC "Kurchatov Institute" - ITEP (Moscow). This work was supported by the Russian Foundation for Basic Research, grant no.20-02-00737 A. The authors are grateful to G. Schierholz, T. Suzuki, S. Syritsyn, V. Braguta, A. Nikolaev for participation at the early stages of this work and for useful discussions.

References

[1] A. S. Kronfeld, M. L. Laursen, G. Schierholz and U. J. Wiese, Phys. Lett. B **198**, 516 (1987).

[2] G. 't Hooft, Nucl. Phys. B **190** (1981), 455-478 doi:10.1016/0550-3213(81)90442-9

[3] T. Suzuki and I. Yotsuyanagi, Phys. Rev. D **42** (1990) 4257.

[4] S. Hioki, S. Kitahara, S. Kiura, Y. Matsubara, O. Miyamura, S. Ohno and T. Suzuki, Phys. Lett. B **272**, 326 (1991) [Erratum-ibid. B **281**, 416 (1992)].

[5] G. S. Bali, V. Bornyakov, M. Muller-Preussker and K. Schilling, Phys. Rev. D **54** (1996) 2863.

[6] V. Bornyakov and M. Muller-Preussker, Nucl. Phys. Proc. Suppl. **106**, 646 (2002).

[7] N. Sakumichi and H. Suganuma, Phys. Rev. D **90** (2014) no.11, 111501 doi:10.1103/PhysRevD.90.111501 [arXiv:1406.2215 [hep-lat]].

[8] M. N. Chernodub and M. I. Polikarpov, [arXiv:hep-th/9710205 [hep-th]].

[9] R. W. Haymaker, Phys. Rept. **315** (1999), 153-173 [arXiv:hep-lat/9809094 [hep-lat]].

[10] H. Shiba and T. Suzuki, Phys. Lett. B **333**, 461 (1994).

[11] J. D. Stack, S. D. Neiman and R. J. Wensley, Phys. Rev. D **50**, 3399 (1994).

[12] G. 't Hooft, in *Proceedings of the EPS International*, edited by A. Zichichi, p. 1225, 1976.

[13] S. Mandelstam, Phys. Rept. **23** (1976), 245-249

[14] O. Miyamura, Phys. Lett. B **353**, 91 (1995).

[15] S. Kitahara, O. Miyamura, T. Okude, F. Shoji and T. Suzuki, Nucl. Phys. B **533**, 576 (1998)

[16] P. de Forcrand and M. D'Elia, Phys. Rev. Lett. **82**, 4582 (1999) [arXiv:hep-lat/9901020].

[17] V. G. Bornyakov, M. I. Polikarpov, G. Schierholz, T. Suzuki and S. N. Syritsyn, Nucl. Phys. B Proc. Suppl. **153**, 25-32 (2006) [arXiv:hep-lat/0512003 [hep-lat]].

[18] V. G. Bornyakov, I. Kudrov and R. N. Rogalyov, [arXiv:2101.04196 [hep-lat]].

[19] J. Smit and A. van der Sijs, Nucl. Phys. B **355**, 603 (1991).

[20] M. Albanese *et al.* [APE Collaboration], Phys. Lett. B **192**, 163 (1987).

[21] V. G. Bornyakov, V. V. Braguta, E. M. Ilgenfritz, A. Y. Kotov, A. V. Molochkov and A. A. Nikolaev, JHEP **03** (2018), 161 doi:10.1007/JHEP03(2018)161 [arXiv:1711.01869 [hep-lat]].

Correlations and Critical Behavior in Lattice Gluodynamics.

Vitaly Bornyakov[1,2,3,*], *Vladimir* Goy[3,4,**], *Evgeny* Kozlovsky[1,***], *Valentin* Mitrjushkin[5,****], and *Roman* Rogalyov[1,†]

[1]Institute for High Energy Physics of the NRC "Kurchatov Institute", 142281 Protvino, Russia
[2]Institute of Theoretical and Experimental Physics of the NRC "Kurchatov Institute", 117259 Moscow, Russia
[3]Pacific Quantum Center, Far Eastern Federal University, Sukhanova 8, 690950 Vladivostok, Russia
[4]Institut Denis Poisson CNRS/UMR 7013, Université de Tours, 37200 Tours, France
[5]Joint Institute for Nuclear Research, 141980 Dubna, Russia

Abstract. In the Landau-gauge lattice gluodynamics we find that, both in the SU(2) and SU(3) theory, a correlation of the Polyakov loop with the asymmetry of the A^2 gluon condensate as well as with the longitudinal propagator makes it possible to determine the critical behavior of these quantities. We discuss finite-volume corrections and reveal that they can be reduced by the use of regression analysis. We also analyze the temperature dependence of low-momenta propagators in different Polyakov-loop sectors.

1 Introduction

In last decades, the deconfinement transition of strong-interacting matter at finite temperature has received considerable theoretical and experimental study. At physical values of the parameters it is a crossover transition, which probably coincides with the chiral transition.

In the limit of infinitely-heavy quarks, the fermion degrees of freedom can be neglected and the crossover transition goes over into the first-order transition in conventional QCD and into the second-order transition in two-color QCD. For this reason, gluodynamics (that is, pure-gauge theory) is widely employed as a testing tool for the studies of the deconfinement transition.

The temperature and volume dependence of the Green's functions of gauge fields in the vicinity of the critical temperature T_c is of particular interest. The critical behavior of the gluon and ghost propagators was considered, in particular, in [1–3]. In Refs. [4, 5] it was found that, close to criticality, the gluon propagator behaves differently in different Polykov-loop sectors.

A interesting example of critical behavior is provided by the chromoelectric-chromomagnetic asymmetry of the dimension-two gluon condensate [6]. Motivation for the

*e-mail: bornvit@gmail.com
**e-mail: vovagoy@gmail.com
***e-mail: Evgeny.Kozlovsky@ihep.ru
****e-mail: vmitr@theor.jinr.ru
†e-mail: rnr@ihep.ru

studies of the asymmetry and gluon propagators was also discussed in [3, 7, 8] and references therein.

Recently, it was demonstrated that correlations between the Polyakov loop and the zero-momentum longitudinal propagator considerably facilitate the analysis of its critical behavior both in the SU(2) [9, 10] and SU(3) [11] gluodynamics. This made it possible to describe critical behavior of the asymmetry and the propagator with unprecendented precision. Here we discuss the assumptions made in Refs.[9–11] and the role of finite-volume effects.

The paper is organized as follows. In the next Section we introduce the definition and describe the details of our numerical simulations. The correlation between the asymmetry $\mathcal{A}$ and the Polyakov loop $\mathcal{P}$ forms the subject of Section 3. In Section 4 we consider the finite-volume effects and discuss how the Polyakov loop determined the behavior of the asymmetry and the propagator. The propagators at nonzero momenta are considered in Section 5. In Conclusions we summarize our findings.

2 Designations and lattice settings

We study SU(2) and SU(3) lattice gauge theories with the standard Wilson action in the Landau gauge.

Our calculations are performed on $N_t \times N_s^3$ lattices ($N_t = 8$, $N_s = 32, 48, 72, 80$ in the SU(2) case and $N_t = 8$, $N_s = 24$ in the SU(3) case). The temperature T is given by $T = 1/aN_t$ where a is the lattice spacing; $L = aN_s$ is the lattice size. We use the parameter $\tau = \dfrac{T - T_c}{T_c}$.

In the SU(3) case we employ the scale fixing procedure proposed in [12] and use the value of the Sommer parameter $r_0 = 0.5$ fm as in [13]. Making use of $\beta_c = 6.06$ and $\dfrac{T_c}{\sqrt{\sigma}} = 0.63$, Ref. [14] gives $T_c = 294$ MeV and $\sqrt{\sigma} = 0.47$ GeV. We study the range of temperatures $0.904T_c \le T \le 1.104T_c$, the corresponding lattice spacings $0.076 \text{ fm} \le a \le 0.093$ fm. Some 200 independent Monte Carlo gauge-field configurations are generated for each of the sectors of the Polyakov loop $\mathcal{P}$ (also referred to as the center sectors):

$$\begin{array}{llcll} (I) & -\frac{\pi}{3} < & \arg\mathcal{P} & < \frac{\pi}{3} & \text{referred to as} \quad \mathrm{Re}\mathcal{P} > 0 \\ (II) & \frac{\pi}{3} < & \arg\mathcal{P} & < \pi & \text{referred to as} \quad \mathrm{Re}\mathcal{P} < 0 \\ (III) & -\pi < & \arg\mathcal{P} & < -\frac{\pi}{3}\,. & \text{referred to as} \quad \mathrm{Re}\mathcal{P} < 0 \end{array} \tag{1}$$

We make no difference between sectors II and III becuse the quantities under study are independent of Im$\mathcal{P}$ as was shown in [11].

In the SU(2) case we use the string tension $\sigma = 440$ MeV, which gives [15] $T_c = 297$ MeV. We study the range $0.9987T_c \le T \le 1.0083T_c$, the corresponding lattice spacings $a \approx 0.083$ fm. From 400 to 2200 gauge-field configurations are simulated for each set of the parameters. In the case of SU(2) we make no difference between $\mathcal{P}$ and Re$\mathcal{P}$.

Consecutive configurations are separated by $200 \div 400$ sweeps, each sweep includes one local heatbath update followed by $N_s/2$ microcanonical updates.

Definitions of the chromo-electric-magnetic asymmetry $\mathcal{A}$ can be found e.g. in Refs.[6, 9]; the definitions of vector potentials $A_\mu(x)$, and the description of gauge-fixing procedure as well as the expressions for the longitudinal $D_L(p)$ and transverse $D_T(p)$ gluon propagators — in Refs.[3, 11, 13]. Here we only present the expression for the asymmetry in terms of the propagators

$$\mathcal{A} = \frac{2(N_c^2-1)N_t}{\beta a^2 N_s^3}\left[3(D_L(0)-D_T(0)) + \sum_{p\neq 0}\left(\frac{3|\vec{p}|^2 - p_4^2}{p^2}D_L(p) - 2D_T(p)\right)\right]. \tag{2}$$

Here we do not consider details of the approach to the continuum limit and renormalization considering that the lattices with $N_t = 8$ (corresponding to spacing $a \simeq 0.08$ fm at $T \sim T_c$) are sufficiently fine.

3 Correlations of the longitudinal propagator and the asymmetry with the Polyakov loop

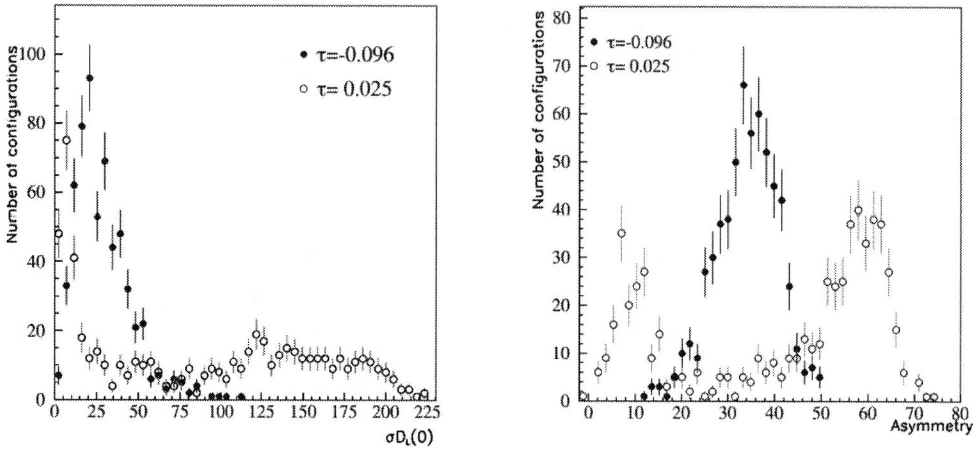

Figure 1: Distributions in the zero-momentum longitudinal propagator (left panel) and the asymmetry (right panel) at $T < T_c$ (filled circles) and $T > T_c$ (empty circles).

It is convenient to begin the analysis of the critical behavior of the longitudinal $D_L(p)$ and transverse $D_T(p)$ propagators by considering the temperature dependence of the distributions of gauge-field configurations in $D_L(0)$.

First, we observe that the distribution in $D_L(0)$ depends strongly on the temperature when $T \sim T_c$ (see Fig. 1). This behavior gets even more complicated due to finite-volume effects, which are considered to be substantial at $T \sim T_c$.

From the plots in Fig. 1 it is clearly seen that not only the average value but also other quantities characterizing the distribution of confugurations in $D_L(0)$ should be thorougly analyzed in order to gain an insight to the critical behavior of the propagators and the asymmetry $\mathcal{A}$. To characterize the propagator distribution, it is instructive to consider its correlation with the Polyakov loop, whose temperature and volume dependence is well understood [16, 17]. That is, we use the asymmetry as the reference.

Thus we consider the conditional distribution $\mathcal{F}(\mathbf{D}|\mathcal{P})$ of $\mathbf{D} = \log[\sigma D_{L,T}(0)]$ at a particular value of the Polyakov loop $\mathcal{P}$ and perform regression analysis based on the linear regession model

$$\langle \mathbf{D}(\tau)\rangle_{\mathcal{P}} \simeq \mathcal{D}_0^{(L,T)}(\tau) + \mathcal{D}_1^{(L,T)}(\tau)\,\mathrm{Re}\,\mathcal{P}(\tau) + \mathcal{D}_2^{(L,T)}(\tau)(\,\mathrm{Re}\,\mathcal{P}(\tau))^2\,, \tag{3}$$

where $\langle \mathbf{D}(\tau) \rangle_{\mathcal{P}}$ is the conditional average of the logarithm of the respective normalized propagator at a given value of the Polyakov loop $\mathcal{P}$. The results for the longitudinal propagator are shown in Fig. 2.

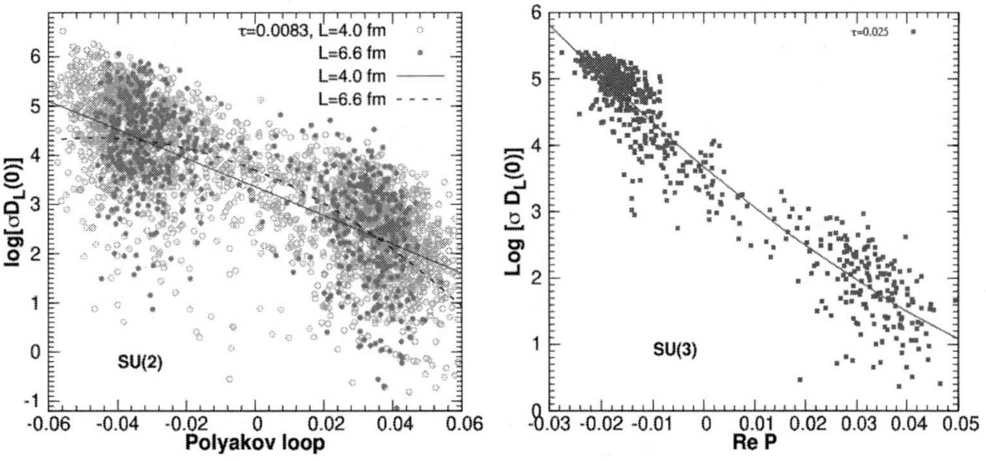

Figure 2: Scatter plots illustrating correlation between the zero-momentum longitudinal propagator and the Polyakov loop. On the left panel data are shown at two different volumes and the same temperature. Shown lines are the regression curves based on the linear model (3).

A similar procedure is performed for the asymmetry,

$$\langle \mathcal{A} \rangle_{\mathcal{P}} \simeq \mathcal{A}_0(\tau) + \mathcal{A}_1(\tau)\,\mathrm{Re}\,\mathcal{P}(\tau) + \mathcal{A}_2(\tau)(\mathrm{Re}\,\mathcal{P}(\tau))^2\,, \tag{4}$$

where the parameters $\mathcal{A}_0$, $\mathcal{A}_1$, and $\mathcal{A}_2$ are also determined from the fit to data. The results are shown in Fig. 3.

4 Finite-volume effects and properties of the correlations

The right-hand side of the formula (4) is useful for an evaluation of the average value of the asymmetry in the infinite-volume limit at the temprature τ, where $\mathcal{P}(\tau)$ is the respective infinite-volume value of the Polyakov loop. Firstly, the sample average value of the asymmetry in the infinite-volume limit coincides with the conditional average at $\mathcal{P}(\tau)$ because the variance of the Polyakov-loop distribution tends to zero as $L \to \infty$. Secondly, finite-volume effects for the predicted value (4) are much less than the sample average as is shown in Fig. 4. Infinite-volume extrapolation of the results shown on the right panel must coincide with those on the left panel. Thus a comparison of panels illustrates huge finite-volume effects due to an exclusion of one center sector, which is common practice at $0 < \tau < 1$. The left panel illustrates that the regression analysis gives much better precision for the finite-volume expectation values of the asymmetry than the conventional use of the sample average. It is also seen that an inclusion of both center sectors results not only in substantial decrease of the finite-volume effects, but also in some decrease of precision.

It is seen on the left panel of Fig. 3 that its correlation with $\mathcal{P}$ becomes more pronounced with an increase of the volume. To study the dependence of the correlation on the volume, we use the Fraction of Variance Unexplained (FVU) designated by S as a measure of the

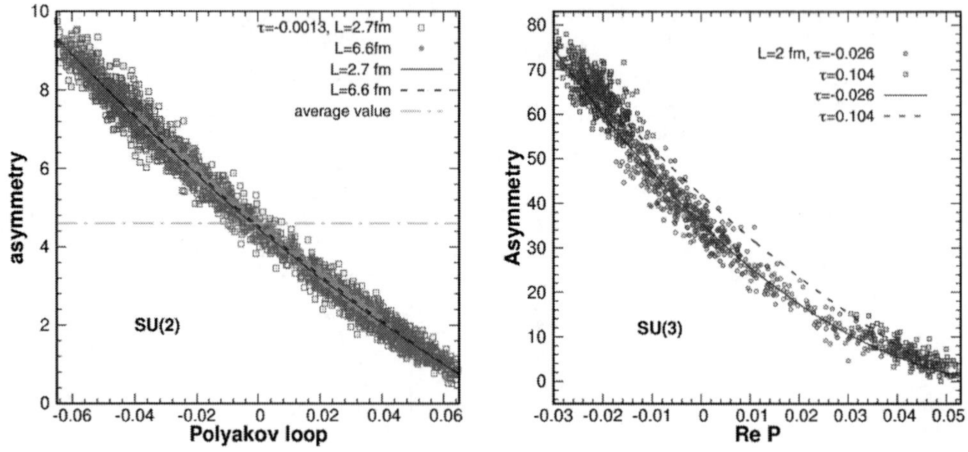

Figure 3: Scatter plots illustrating correlation between the asymmetry and the Polyakov loop. Data are shown at two different lattice sizes and the same temperature in the SU(2) case (left panel) and at two different temperatures and the same lattice size in the SU(3) case (right panel). Shown lines are the regression curves based on the linear model (4).

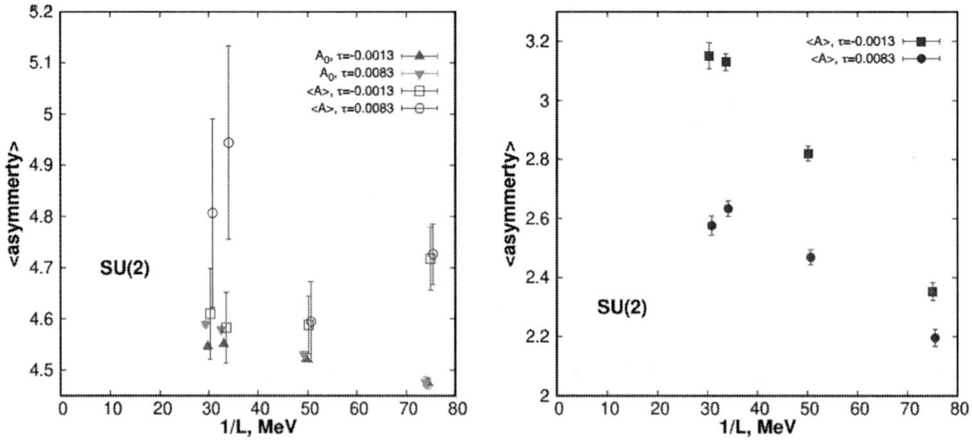

Figure 4: Finite-volume effects for the asymmetry, $\mathcal{A}_0$ $\langle\mathcal{A}\rangle$ is the sample average. Left panel: both center sectors are taken into consideration. Right panel: results for the sector $\mathcal{P} > 0$.

degree of correlation. It represents the ratio of the variance of the residuals $e_A(n) = \mathcal{A}_n - \mathcal{A}_0 - \mathcal{A}_1 \operatorname{Re}\mathcal{P}_n - \mathcal{A}_2(\operatorname{Re}\mathcal{P}_n)^2$ and the sample variance $M_2(\mathcal{A})$,

$$S = \frac{Q}{M_2(\mathcal{A})}, \quad \text{where} \quad Q = \sum_{n=1}^{N_{data}} e_A^2(n), \quad M_2(\mathcal{A}) = \sum_{n=1}^{N_{data}} |\mathcal{A}_n - \langle\mathcal{A}\rangle|^2 . \tag{5}$$

In the general case, $0 \le S \le 1$; $S = 0$ implies that the asymmetry is nothing but a function of the Polyakov loop; $S = 1$ means that they are independent of each other. The results are shown in Fig. 5. A similar estimate of S is performed also for the propagator.

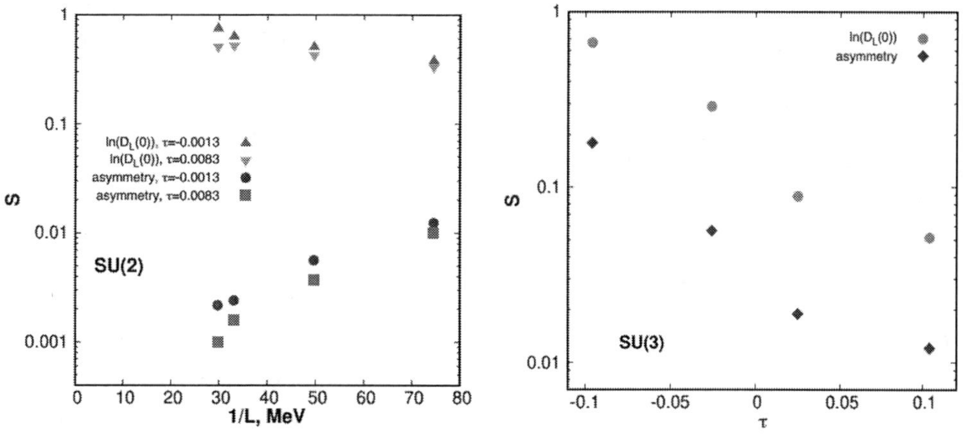

Figure 5: Fraction of variance unexplained: the dependence on lattice size is shown in the SU(2) case (left panel), the dependence on the temperature – for the SU(3) case (right panel) both for $\mathcal{A}$ and $\mathbf{D}$

We see that the FVU for the asymmetry tends to zero as $L \to \infty$, whereas for the propagator it remains on the order of unity. The temperature dependence of FVU is explained by taking all center sectors into account at $T > T_c$ providing a wide range of Polyakov-loop variation. We take all center sectors into consideration because in a finite volume transitions between them are not negligible even at $T > T_c$ when $\tau < 1$.

Our data give some evidence that, in the infinite-volume limit, the zero-momentum longitudinal propagator is not a proper function of the Polyakov loop, it remains a random variable correlated with the Polyakov loop, whereas the asymmetry becomes a proper function of the Polyakov loop.

Nevertheless, our conclusion of the critical behavior of the propagator presenrted in Refs.[9–11] holds true because it is based on the assumption of the smooth dependence of the conditional average $\langle \mathbf{D}(\tau)\rangle_{\mathcal{P}}$ on $\mathcal{P}$, no assumptions on its variance are made. Therewith, the regression curve can be extracted from the data only with a limited precision, therefore, the concept of smoothness should be formulated for the corridor of errors rather than for the proper function.

In this context, smoothness of the regression function $f(\mathcal{P}) \equiv \langle \mathbf{D}(\tau)\rangle_{\mathcal{P}}$ at $\mathcal{P} = \mathcal{P}_s$ means that the value $f(\mathcal{P}_s)$ as well as the derivative $\left.\frac{df(\mathcal{P})}{d\mathcal{P}}\right|_{\mathcal{P}=\mathcal{P}_s}$ evaluated over the range $\mathcal{P} < \mathcal{P}_s$ coincide within statistical error with those evaluated over the range $\mathcal{P} > \mathcal{P}_s$. Our data fulfil this criterion, however, the assumption of smooth dependence should be tested on a larger statistics. Smooth dependence of the conditional averages $\langle \mathcal{A}\rangle_{\mathcal{P}}$ and $\langle \mathbf{D}\rangle_{\mathcal{P}}$ on $\mathcal{P}$ at $\mathcal{P} = 0$ implies that these quantities and, therefore, the infinte-volume limits of the respective sample averages considered as the functions of T have the same singularity/nonanalyticity at $T = T_c$ that the Polyakov loop.

5 Propagators at nonzero momenta

The mentioned test on smoothness can also be motivated by our observation that the temperature dependence of the longitudinal propagator at nonzero momenta differs significantly from that at zero momentum, which is shown in Fig. 6. It is clearly seen that the temperature dependence of the zero-momentum propagator is in a good agreement with that found in Refs. [4, 5].

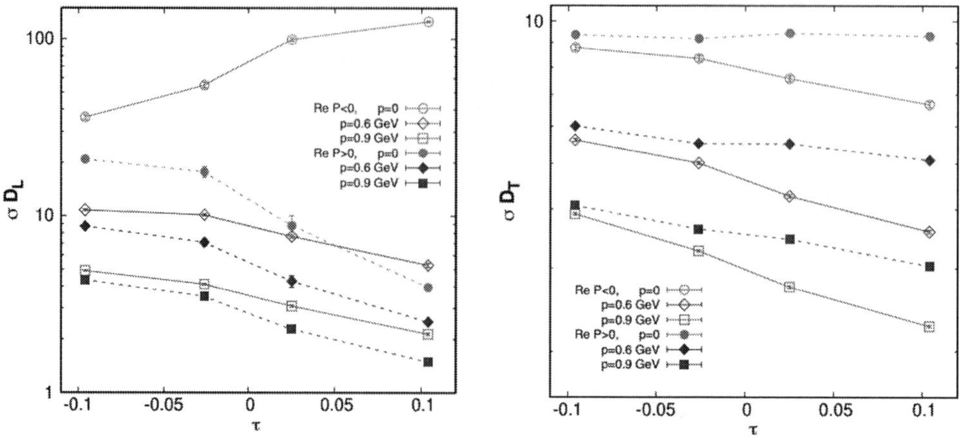

Figure 6: Temperature dependence of the longitudinal (left panel) and transverse (right panel) gluon propagators at various values of momenta in the SU(3) case. Lines are shown just to guide the eye, also notice the logarithmic scale on the ordinate axis.

However, we find that the zero-momentum longitudinal gluon propagator decreases with the temperature in the sector $\mathrm{Re}\mathcal{P} > 0$ and rapidly increases in the sectors $\mathrm{Re}\mathcal{P} < 0$, whereas longitudinal gluon propagator at $p \sim 0.5 \div 1$ GeV decreases with temperature in all Polyakov-loop sectors[1]. Thus a sharp peak of the longitudinal gluon propagator in the sectors $\mathrm{Re}\mathcal{P} < 0$ appears in the deep infrared at $T \sim T_c$. Physical interpretation of such peak may be as follows: a formation of a large bubble of the phase with $\mathrm{Re}\mathcal{P} < 0$ in gluon matter at $T > T_c$ in the phase with $\mathrm{Re}\mathcal{P} > 0$ gives rise to huge fluctuations of chromoelectric fields which should emerge when the size of such bubble exceeds the scale determined by the width of the above-mentioned peak.

Temperature dependence of the transverse propagator at nonzero momenta is shown in the right panel of Fig. 6. It is interesting to remark that it substantially decreases with temperature in the sector $\mathrm{Re}\mathcal{P}$. However, this observation calls for further investigation.

6 Conclusions

We have studied numerically the asymmetry and the longitudinal gluon propagator in the Landau-gauge SU(2) and SU(3) gluodynamics close to criticality. Our findings can be summarized as follows:

- The correlations between $\mathcal{A}$ and $\mathcal{P}$ between $\mathbf{D}$ and $\mathcal{P}$ indicate that the conditional averages $\langle\mathcal{A}\rangle_{\mathcal{P}}$ and $\langle\mathbf{D}\rangle_{\mathcal{P}}$ are smooth functions of $\mathcal{P}$ at $\mathcal{P} = 0$.
- In the infinite-volume limit, the asymmetry is completely determined by the Polyakov loop, whereas the propagators are only partially determined (that is, the gluon propagator at a given value of $\mathcal{P}$ represents a random variable even in the infinite-volume limit.)

[1] Temperature dependence of $D_L(p)$ at $p \sim 0.5 \div 1$ GeV is poorly seen on the plots in Refs. [4, 5]: one can only conclude that it does not increase.

- The use of regression analysis substantially reduces finite-volume effects.
- Finite-volume effects in different center sectors partially cancel each other.
- The temperature dependence of the longitudinal gluon propagator at $p < 1$ GeV and $T \sim T_c$ in the Polyakov-loop sectors with Re$\mathcal{P} < 0$ is qualitatively different at $p = 0$ and $p = p_{min} \sim 600$ MeV.

Acknowledgments. Computer simulations were performed on the IHEP (Protvino) Central Linux Cluster and ITEP(Moscow) Linux Cluster. This work was supported in part by the Russian Foundation for Basic Research, grant no.20-02-00737 A.

References

[1] C.S. Fischer, A. Maas, J.A. Muller, Eur.Phys.J. **C68**, 165 (2010), 1003.1960
[2] A. Maas, J.M. Pawlowski, L. von Smekal, D. Spielmann, Phys. Rev. **D85**, 034037 (2012), 1110.6340
[3] R. Aouane, V. Bornyakov, E. Ilgenfritz, V. Mitrjushkin, M. Muller-Preussker et al., Phys.Rev. **D85**, 034501 (2012), 1108.1735
[4] O. Oliveira, P.J. Silva, PoS **LATTICE2014**, 355 (2014), 1411.0133
[5] P.J. Silva, O. Oliveira, Phys. Rev. D **93**, 114509 (2016), 1601.01594
[6] M.N. Chernodub, E.M. Ilgenfritz, Phys. Rev. **D78**, 034036 (2008), 0805.3714
[7] A. Maas, Phys. Rept. **524**, 203 (2013), 1106.3942
[8] D. Vercauteren, H. Verschelde, Phys. Rev. **D82**, 085026 (2010), 1007.2789
[9] V.G. Bornyakov, V.K. Mitrjushkin, R.N. Rogalyov, Phys. Rev. D **100**, 094505 (2019), 1609.05145
[10] V. Bornyakov, V. Bryzgalov, V. Mitrjushkin, R. Rogalyov, Int. J. Mod. Phys. A **33**, 1850151 (2018), 1801.02584
[11] V.G. Bornyakov, V.A. Goy, V.K. Mitrjushkin, R.N. Rogalyov, Phys. Rev. D **104**, 074508 (2021), 2101.03605
[12] S. Necco, R. Sommer, Nucl. Phys. B **622**, 328 (2002), hep-lat/0108008
[13] V.G. Bornyakov, V.K. Mitrjushkin (2011), 1103.0442
[14] G. Boyd, J. Engels, F. Karsch, E. Laermann, C. Legeland, M. Lutgemeier, B. Petersson, Nucl. Phys. B **469**, 419 (1996), hep-lat/9602007
[15] J.C.R. Bloch, A. Cucchieri, K. Langfeld, T. Mendes, Nucl. Phys. B Proc. Suppl. **119**, 736 (2003), hep-lat/0209040
[16] C. Gattringer, Phys. Lett. B **690**, 179 (2010), 1004.2200
[17] C. Gattringer, A. Schmidt, JHEP **01**, 051 (2011), 1011.2329

Quenched glueball spectrum from functional equations

Markus Q. Huber[1,*], *Christian S.* Fischer[1,2,**], and *Hèlios* Sanchis-Alepuz[3,***]

[1]Institut für Theoretische Physik, Justus-Liebig-Universität Giessen, Heinrich-Buff-Ring 16, 35392 Giessen, Germany
[2]Helmholtz Forschungsakademie Hessen für FAIR (HFHF), GSI Helmholtzzentrum für Schwerionenforschung, Campus Gießen, 35392 Gießen, Germany
[3]Silicon Austria Labs GmbH, Inffeldgasse 33, 8010 Graz, Austria

Abstract. We give an overview of results for the quenched glueball spectrum from two-body bound state equations based on the 3PI effective action. The setup, which uses self-consistently calculated two- and three-point functions as input, is completely self-contained and does not have any free parameters except for the coupling. The results for $J^{PC} = 0^{\pm+}, 2^{\pm+}, 3^{\pm+}, 4^{\pm+}$ are in good agreement with recent lattice results where available. For the pseudoscalar glueball, we present first results from a two-loop complete calculation, rendering also the bound state calculation fully self-consistent.

1 Introduction

The spectrum of bound states in quantum chromodynamics (QCD) is both experimentally and theoretically not completely known yet. Although measured and calculated for many decades by now, there are still some surprises, as, e.g., the experimentally established existence of pentaquarks or the appearance of the XYZ states, but also some remaining gaps that need to be filled. Glueballs, bound states of gluons, belong to the latter class. They can be investigated by ongoing and future experiments like ATLAS, BESIII, PANDA or Glue-X.

On the theory side, their masses were calculated with several methods, but a consensus only exists for the case of quenched QCD, viz., QCD with infinitely heavy quarks. In particular, lattice results for the lightest glueballs have been the benchmark for over twenty years now [1]. These results were confirmed and extended by subsequent lattice computations [2, 3] and are also supported by other methods, e.g., [4–11]. For an overview, we refer to the reviews [12–16].

When quarks are included dynamically, on the other hand, the situation is not as clear for lattice calculations. The reasons are manifold, among them a poor signal-to-noise ratio, incomplete operator bases and a challenging continuum extrapolation [17]. Another new feature is that glueballs can mix with pure quark states which can have the same quantum numbers. This also aggravates the interpretation of experimental results. A recent analysis of radiative J/Ψ-decays of BESIII data, however, identified the scalar glueball [18] with a mass very close to the prediction of quenched QCD.

*e-mail: markus.huber@physik.jlug.de
**e-mail: christian.fischer@theo.physik.uni-giessen.de
***e-mail: helios.sanchis-alepuz@silicon-austria.com

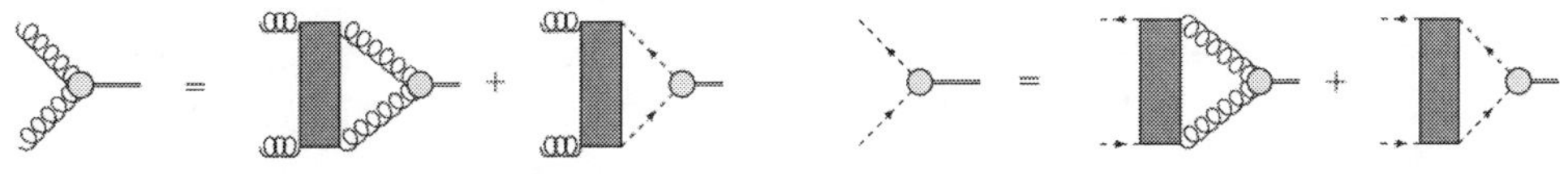

Figure 1. The coupled set of BSEs for a glueball made from two gluons and a pair of Faddeev-Popov (anti-)ghosts. Wiggly lines denote dressed gluon propagators, dashed lines denote dressed ghost propagators. The gray boxes represent interaction kernels given in Fig. 2. The Bethe-Salpeter amplitudes of the glueball are denoted by gray disks.

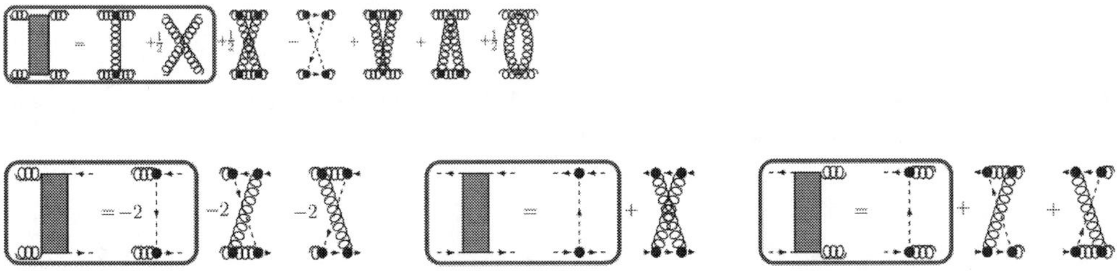

Figure 2. Interaction kernels from the three-loop 3PI effective action. All propagators are dressed; black disks represent dressed vertices. In our calculation, we include the diagrams inside the red rectangles.

Using functional methods, the challenges are completely different than for lattice methods. A crucial point is the input in form of two- and three-point functions that is required. Initially, models were used as in [19–22]. In [10], we used for the first time self-consistent input calculated from Dyson-Schwinger equations [23]. As it turned out, this was decisive to obtain a description of the pseudoscalar and the scalar glueballs at the same time which was not possible before [20]. As an additional benefit, the complete calculation is now self-contained and does not depend on any model parameters. In a subsequent calculation, we extended the setup to nonzero spin [11]. In addition, we present here first results for a fully self-consistent calculation of the pseudoscalar glueball by including also the two-loop diagrams.

In the following, we present the setup of our calculations in Sec. 2 and the results in Sec. 3. We close with a short summary and outlook.

2 Setup

Glueballs can be described by gauge-invariant operators. For example, $O(x) = F_{\mu\nu}(x)F^{\mu\nu}(x)$ has the quantum numbers of the scalar glueball and its mass can be extracted from the pole of the corresponding two-point function $\langle 0|O(x)O(y)|0\rangle$. As opposed to the case of typically used mesonic operators, this function consists of parts with two, three and four gluons. The calculation of the complete two-point function is a complicated task. However, it can be sufficient to consider single pieces for the determination of masses. The reason is that if a pole present is in one piece, this corresponds to the mass of the bound state, as long as no cancellation between different pieces occurs. This is the rationale why we can use a two-body bound state equation to determine the mass of a complicated object like a glueball. More details are provided in [11]. The resulting Bethe-Salpeter equation (BSE) is given in Fig. 1. In general, as we necessarily work in a gauge-fixed setting, the BSE also contains ghost fields and a corresponding amplitude which we call the ghostball-part.

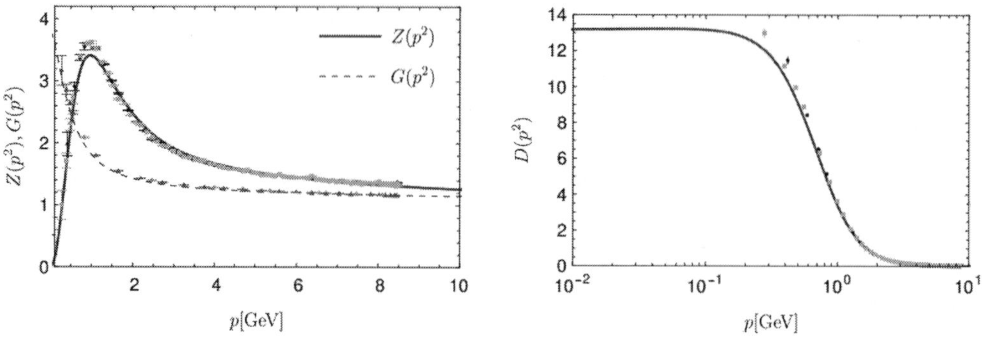

Figure 3. Gluon and ghost dressing functions $Z(p^2)$ and $G(p^2)$, respectively, (left) and gluon propagator $D(p^2)$ (right) in comparison to lattice data [40].

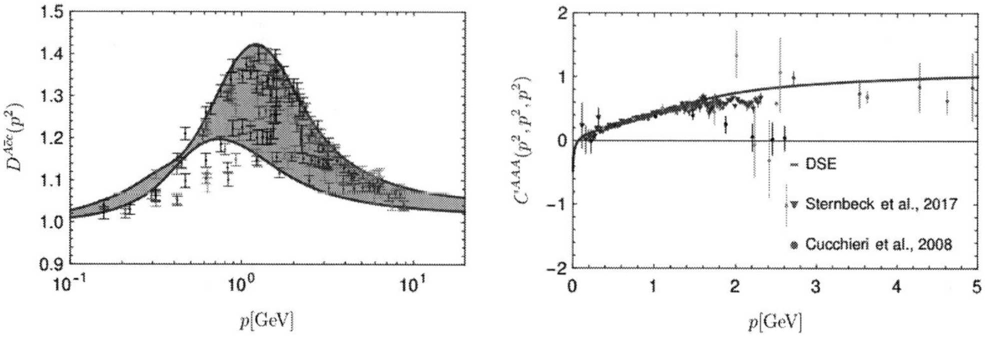

Figure 4. Left: Ghost-gluon vertex dressing function (full kinematic dependence) in comparison to $SU(2)$ lattice data [26]. Right: Three-gluon vertex dressing function at the symmetric point in comparison to lattice data [24, 25]), see Refs. [27, 28] for similar results.

For the calculation, the interaction kernels need to be specified. In analogy to the equations used to calculate the input correlation functions, see below, we derived them from a three-loop truncated 3PI effective action. This yields the diagrams shown in Fig. 2. However, initially we only take into account the diagrams leading to one-loop diagrams, the reason being that the two-loop diagrams are computationally much more expensive. Due to the good agreement of our results with lattice results, it seems that the two-loop contributions are subleading. For the pseudoscalar glueball, which is the computationally least expensive, we performed also a calculation with the two-loop diagrams included. In that case, the truncation is fully self-consistent. For the mass determination, the calculation of the two-loop diagrams was computed with lower precision than the one-loop part. However, for single eigenvalues we increased the precision and checked that our conclusion on the effect of two-loop diagrams remains valid.

In this truncation, the BSE depends on the ghost and gluon propagators and the ghost-gluon and three-gluon vertices. As mentioned above, we take them from a self-consistent, self-contained calculation [23]. In Figs. 3 and 4, we show a selection of the input in comparison with lattice results. We want to stress that this corresponds to one of several possible solutions which vary in their infrared behavior, see, e.g., [29–36]. For the spin-0 case, we tested that the masses agree for different solutions and we thus continued with only one solution for higher spins.

The BSE is solved as an eigenvalue equation. By varying the total momentum P, curves for the eigenvalues λ_i are determined. Physical solutions correspond to eigenvalues $\lambda_i = 1$ and the corresponding value of P determines the mass: $M^2 = -P^2$. For time-like P, the propagators in the integrand are probed at complex values. However, the input we use is only available for Euclidean momenta. Corresponding calculations in the complex plane are currently only available for much simpler truncations and restricted to the propagators [37]. Thus, we can calculate the eigenvalue curves only for space-like P. From these results, we extrapolate to the time-like side using Schlessinger's continued fraction method [38, 39]. We tested this method in a model calculation [10] for which it turned out to be very reliable at least up to approximately 2 GeV. Beyond that, errors start to increase and need to be interpreted carefully. For our highest mass results, we assign some level of uncertainty as indicated in Tab. 1 and Fig. 5. Errors estimated from the extrapolation are obtained by averaging over extrapolations from different subsets of eigenvalues. Details can be found in [10].

To calculate the masses for different spins and parities, appropriate bases need to be constructed. The basis for spin J consists for the ghostball- and glueball-parts of all tensors with J and $J+2$ Lorentz tensors, respectively, which fulfill certain constraints. For the former, they need to be traceless, symmetric and transverse with respect to the total momentum. This leaves only one tensor. For the glueball-part, however, these constraints only apply to the J Lorentz indices for spin, but there are two additional indices from the gluon legs. This increases the number of tensors, but as an additional constraint we need to ensure that we have a basis that is transverse in the gluon legs due to the transversality of the gluon propagator in the Landau gauge employed here. If we included nontransverse parts, this would lead to inconsistencies in the eigenvalue equation that distorts the results. Finally, the parity of the tensors needs to be taken into account. The remaining tensor of the ghostball-part has positive parity which means that there is no ghostball-part for negative parity. Following these considerations, one ends up with at most five tensors. This is a rather small basis, but the expressions become quite lengthy with increasing J due to the symmetrization and the required tracelessness. The detailed expressions for the bases are given in [11].

3 Results

We summarize the results for the glueball masses in Tab. 1 and Fig. 5. We also list results from lattice calculations. To enable a meaningful comparison, we rescaled all results to a common value for the Sommer scale, $r_0 = 0.472(5)\,\text{fm}$, which was used in [3]. The original scale of the functional results was determined via the gluon propagator and a comparison with the lattice results of [40]. The indicated errors for the functional results are estimates from the extrapolation only and should be considered as a lower bound. For higher spins, we reduced the numeric precision due to computation time. From tests with low spin, we know that the ground state is not affected by that but excited states can vanish. This is why we do not see excited states for all quantum numbers. In general, the results are more uncertain for heavier masses due to the extrapolation deeper into the time-like region. We indicate this by a star. The results for $J^{\text{PC}} = 3^{++}$ and 4^{++} are somewhat exceptional, as the extrapolations lead to extremely small errors. With regard to the stability of these states, we want make clear that the employed setup does not contain channels for decays. Thus, all states are stable as long as this possibility is not included, e.g., via dedicated additional diagrams [41]. An interesting finding concerns spin one. Often the Landau-Yang theorem is invoked to explain why spin-1 glueballs cannot be built from two gluons. However, as explained in more detail in [11], in

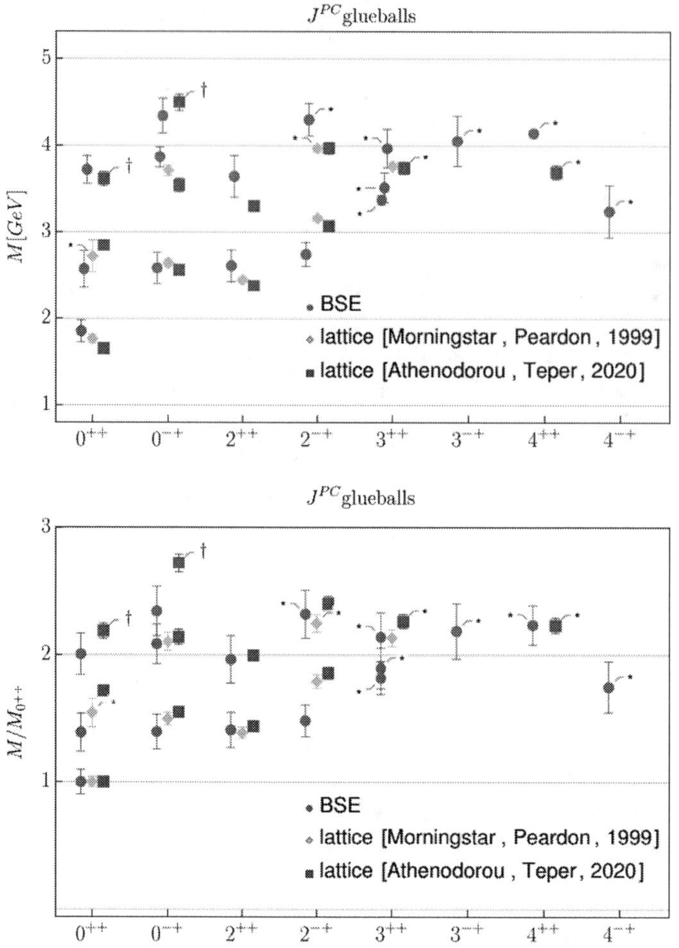

Figure 5. Results for glueball ground states and excited states for the indicated quantum numbers from lattice simulations [1, 3] and functional equations. In the upper plot, we display the glueball masses on an absolute scale set by $r_0 = 1/418(5)\,\text{MeV})$. In the lower plot, we display the spectrum relative to the ground state. Masses with † are conjectured to be the second excited states. Masses with * come with some uncertainty in their identification in the lattice case or in the trustworthiness of the extrapolated value in the BSE case.

our approach it does not apply because the gluons are not on-shell. Nevertheless, we did not find solutions for $J = 1$.

For the pseudoscalar glueball we also performed a calculation with the two-loop diagrams included. This glueball is technically the simplest one for two reasons: There is no ghostball-part and it has a one-dimensional basis. Nevertheless, the calculation of the two-loop diagrams is very costly and we used only a low precision. That calculation showed a very small change in the eigenvalues of less than $0.1\,‰$. More importantly, although in principle conceivable due to the sensitivity of the extrapolation to small changes in the input data, this did not lead to a measurable change in the mass. Finally, we increased the precision further for a few eigenvalues and found that the agreement between the eigenvalues improves

Table 1. Ground and excited state masses M of glueballs for various quantum numbers. Compared are lattice results from [1, 3] with the results of this work and [10]. For [1], the errors are the combined errors from statistics and the use of an anisotropic lattices. For [3], the error is statistical only. In our results, the error comes from the extrapolation method and should be considered a lower bound on errors. All results use the same value for $r_0 = 1/(418(5)\,\mathrm{MeV})$. The related error is not included in the table. Masses with † are conjectured to be the second excited states. Masses with * come with some uncertainty in their identification in the lattice case or in the trustworthiness of the extrapolated value in the BSE case.

	[1]		[3]		This work	
State	M [MeV]	$M/M_{0^{++}}$	M [MeV]	$M/M_{0^{++}}$	M [MeV]	$M/M_{0^{++}}$
0^{++}	1760(50)	1(0.04)	1651(23)	1(0.02)	1850(130)	1(0.1)
0^{*++}	2720(180)*	1.54(0.11)*	2840(40)	1.72(0.034)	2570(210)	1.39(0.15)
0^{**++}	–	–	3650(60)† 3580(150)†	2.21(0.05)† 2.17(0.1)†	3720(160)	2.01(0.16)
0^{-+}	2640(40)	1.50(0.05)	2600(40)	1.574(0.032)	2580(180)	1.39(0.14)
0^{*-+}	3710(60)	2.10(0.07)	3540(80)	2.14(0.06)	3870(120)	2.09(0.16)
0^{**-+}	–	–	4450(140)† 4540(120)†	2.7(0.09)† 2.75(0.08)†	4340(200)	2.34(0.19)
2^{++}	2447(25)	1.39(0.04)	2376(32)	1.439(0.028)	2610(180)	1.41(0.14)
2^{*++}	–	–	3300(50)	2(0.04)	3640(240)	1.96(0.19)
2^{-+}	3160(31)	1.79(0.05)	3070(60)	1.86(0.04)	2740(140)	1.48(0.13)
2^{*-+}	3970(40)*	2.25(0.07)*	3970(70)	2.4(0.05)	4300(190)	2.32(0.19)
3^{++}	3760(40)	2.13(0.07)	3740(70)*	2.27(0.05)*	3370(50)*	1.82(0.13)*
3^{*++}	–	–	–	–	3510(170)*	1.89(0.16)*
3^{**++}	–	–	–		3970(220)*	2.14(0.19)*
3^{-+}	–	–	–	–	4050(290)*	2.19(0.22)*
4^{++}	–	–	3690(80)*	2.24(0.06)*	4140(30)*	2.23(0.15)*
4^{-+}	–	–	–	–	3240(300)*	1.75(0.2)*

even further. Thus we conclude that for the pseudoscalar glueball the two-loop diagrams are not only subleading but totally irrelevant.

4 Summary and outlook

We presented results for the glueball spectrum of pure Yang-Mills theory obtained in a functional framework with Bethe-Salpeter and Dyson-Schwinger equations. The results are in good quantitative agreement with lattice results. We also added some new states, e.g., second excited states of (pseudo)scalar glueballs. The setup is completely free of any parameters to be tuned and thus our results constitute genuine predictions from a first principles calculation.

There are still some improvements that can be made. One is the inclusion of two-loop diagrams which is necessary to confirm if they are indeed as subleading as the comparison with lattice results indicates. This step is computationally demanding but conceptually straightforward. We explored that for the technically simplest glueball, the pseudoscalar one, and found that the two-loop diagrams do not lead to any measurable effect on the mass or the amplitude. This is the first case of a fully self-consistent calculation of a glueball from the three-loop truncated 3PI effective action. More challenging is the circumvention of the extrapolation by working directly with time-like total momentum. In terms of truncations, presently existing results [37] would need to be improved considerably as currently they are lacking two-loop diagrams in the propagators and dynamical vertices. Beyond improvements and tests of the current setup, the inclusion of quarks constitutes the most interesting next step, as this would finally make contact to experiment.

Acknowledgments

This work was supported by the DFG (German Research Foundation) grant FI 970/11-1 and by the BMBF under contracts No. 05P18RGFP1 and 05P21RGFP3. This work has also been supported by Silicon Austria Labs (SAL), owned by the Republic of Austria, the Styrian Business Promotion Agency (SFG), the federal state of Carinthia, the Upper Austrian Research (UAR), and the Austrian Association for the Electric and Electronics Industry (FEEI).

References

[1] C.J. Morningstar, M.J. Peardon, Phys. Rev. **D60**, 034509 (1999), hep-lat/9901004
[2] Y. Chen et al., Phys. Rev. **D73**, 014516 (2006), hep-lat/0510074
[3] A. Athenodorou, M. Teper (2020), 2007.06422
[4] A. Szczepaniak, E.S. Swanson, C.R. Ji, S.R. Cotanch, Phys. Rev. Lett. **76**, 2011 (1996), hep-ph/9511422
[5] A.P. Szczepaniak, E.S. Swanson, Phys. Lett. **B577**, 61 (2003), hep-ph/0308268
[6] D. Dudal, M.S. Guimaraes, S.P. Sorella, Phys. Rev. Lett. **106**, 062003 (2011), 1010.3638
[7] S. Janowski, D. Parganlija, F. Giacosa, D.H. Rischke, Phys. Rev. **D84**, 054007 (2011), 1103.3238
[8] W.I. Eshraim, S. Janowski, F. Giacosa, D.H. Rischke, Phys. Rev. **D87**, 054036 (2013), 1208.6474
[9] D. Dudal, M.S. Guimaraes, S.P. Sorella, Phys. Lett. **B732**, 247 (2014), 1310.2016
[10] M.Q. Huber, C.S. Fischer, H. Sanchis-Alepuz (2020), 2004.00415
[11] M.Q. Huber, C.S. Fischer, H. Sanchis-Alepuz (2021), 2110.09180
[12] E. Klempt, A. Zaitsev, Phys. Rept. **454**, 1 (2007), 0708.4016
[13] V. Crede, C.A. Meyer, Prog. Part. Nucl. Phys. **63**, 74 (2009), 0812.0600
[14] V. Mathieu, N. Kochelev, V. Vento, Int. J. Mod. Phys. E **18**, 1 (2009), 0810.4453
[15] W. Ochs, J. Phys. **G40**, 043001 (2013), 1301.5183
[16] F.J. Llanes-Estrada (2021), 2101.05366
[17] E. Gregory, A. Irving, B. Lucini, C. McNeile, A. Rago, C. Richards, E. Rinaldi, JHEP **10**, 170 (2012), 1208.1858
[18] A.V. Sarantsev, I. Denisenko, U. Thoma, E. Klempt, Phys. Lett. B **816**, 136227 (2021), 2103.09680
[19] J. Meyers, E.S. Swanson, Phys. Rev. **D87**, 036009 (2013), 1211.4648
[20] H. Sanchis-Alepuz, C.S. Fischer, C. Kellermann, L. von Smekal, Phys. Rev. **D92**, 034001 (2015), 1503.06051
[21] E.V. Souza, M.N. Ferreira, A.C. Aguilar, J. Papavassiliou, C.D. Roberts, S.S. Xu, Eur. Phys. J. **A56**, 25 (2020), 1909.05875
[22] L. Kaptari, B. Kämpfer, Few Body Syst. **61**, 28 (2020), 2004.06523
[23] M.Q. Huber, Phys. Rev. D **101**, 11 (2020), 2003.13703

[24] A. Cucchieri, A. Maas, T. Mendes, Phys. Rev. **D77**, 094510 (2008), 0803.1798
[25] A. Sternbeck, P.H. Balduf, A. Kızılersu, O. Oliveira, P.J. Silva, J.I. Skullerud, A.G. Williams, PoS **LATTICE2016**, 349 (2017), 1702.00612
[26] A. Maas (2019), 1907.10435
[27] A. Athenodorou, D. Binosi, P. Boucaud, F. De Soto, J. Papavassiliou, J. Rodriguez-Quintero, S. Zafeiropoulos, Phys. Lett. **B761**, 444 (2016), 1607.01278
[28] P. Boucaud, F. De Soto, J. Rodríguez-Quintero, S. Zafeiropoulos, Phys. Rev. **D95**, 114503 (2017), 1701.07390
[29] P. Boucaud et al., JHEP **06**, 012 (2008), 0801.2721
[30] C.S. Fischer, A. Maas, J.M. Pawlowski, Annals Phys. **324**, 2408 (2009), 0810.1987
[31] R. Alkofer, M.Q. Huber, K. Schwenzer, Phys. Rev. **D81**, 105010 (2010), 0801.2762
[32] A. Maas, Phys. Lett. **B689**, 107 (2010), 0907.5185
[33] A. Maas, Phys.Rept. **524**, 203 (2013), 1106.3942
[34] A. Sternbeck, M. Müller-Preussker, Phys.Lett. **B726**, 396 (2013), 1211.3057
[35] M.Q. Huber, Phys. Rept. **879**, 1 (2020), 1808.05227
[36] G. Eichmann, J.M. Pawlowski, J.a.M. Silva (2021), 2107.05352
[37] C.S. Fischer, M.Q. Huber, Phys. Rev. D **102**, 094005 (2020), 2007.11505
[38] L. Schlessinger, Phys. Rev. **167**, 1411 (1968)
[39] R.A. Tripolt, P. Gubler, M. Ulybyshev, L. Von Smekal, Comput. Phys. Commun. **237**, 129 (2019), 1801.10348
[40] A. Sternbeck (2006), phD thesis, Humboldt-Universität zu Berlin, hep-lat/0609016
[41] C.S. Fischer, D. Nickel, J. Wambach, Phys. Rev. **D76**, 094009 (2007), 0705.4407

Single-spin asymmetry $A_T^{\sin(2\phi-\phi_S)}$ in $\pi^- p$ Drell-Yan process within TMD factorization

Hui Li[1], *Xiaoyu* Wang[2,*], and *Zhun* Lu[3,**]

[1]School of Physics, Southeast University, Nanjing 211189, China
[2]School of Physics and Engineering, Zhengzhou University, Zhengzhou, Henan 450001, China
[3]School of Physics, Southeast University, Nanjing 211189, China

Abstract. We study the single-spin asymmetry $A_T^{\sin(2\phi-\phi_S)}$ in the pion-induced Drell-Yan process within the transverse momentum dependent factorization (TMD factorization). The asymmetry can be expressed as the convolution of the Boer-Mulders function and the transversity function. We numerically estimate the asymmetry $A_T^{\sin(2\phi-\phi_S)}$ at the COMPASS kinematics with the model results for the pion meson distributions from the light-cone wave function approach and the available parametrization for the proton distributions. We also include the TMD evolution formalism both proton and pion parton distribution functions by using two different parametrizations on nonperturbative Sudakov form factor. We find that the asymmetry $A_T^{\sin(2\phi-\phi_S)}$ as functions of x_p, x_π, x_F and $q_\perp$ is qualitatively consistent with the recent COMPASS measurement.

1 Introduction

The T-odd Boer-Mulders function $h_1^\perp$ represents the transversely polarization asymmetry of quarks inside an unpolarized hadron, which manifests the novel structure of hadrons. In particular, a chiral-odd Boer-Mulders function convoluted with the transversity distribution h_1 can be used to describe the asymmetry $A_T^{\sin(2\phi-\phi_S)}$ with ϕ_S the azimuthal angle of target transverse spin. Recently, the first measurement on the $\sin(2\phi-\phi_S)$ asymmetry has been performed by the COMPASS [1], which adopted a pion beam to collide on the transversely polarized nucleon target. Although no clear tendency is observed on the $\sin(2\phi-\phi_S)$ asymmetry due to relatively large statistical uncertainties, it indeed indicates that the asymmetry has a negative sign and substantial size.

In this work, we numerically estimate the asymmetry $A_T^{\sin(2\phi-\phi_S)}$ at the COMPASS kinematics within the TMD factorization [2]. In the calculation, we adopt the model result for the distribution of the pion meson from the light-cone wave function approach, and we choose the available parametrization for the transversity distribution of the proton. Furthermore, we take into account the scale dependence of the pion Boer-Mulders function and the proton transversity function, for which we consider two different parameterizations on the nonperturbative Sudakov form factor.

*e-mail: xiaoyuwang@zzu.edu.cn
**e-mail: zhunlu@seu.edu.cn

2 The TMD evolution of distribution functions

In the $\boldsymbol{b}_\perp$ space, the evolution of the TMD distribution function $\tilde{F}(x, b; \mu, \zeta_F)$ is encoded in Collins-Soper (CS) equation [2] and the renormalization group equation. The general solution for the energy dependence of $\tilde{F}$ can be obtained by solving those equations ($b = |\boldsymbol{b}_\perp|$)

$$\tilde{F}(x, b, Q) = \mathcal{F} \times e^{-S(Q,b)} \times \tilde{F}(x, b, \mu_i), \tag{1}$$

where $\mathcal{F}$ is the factor related to the hard scattering, $S(Q, b)$ is the Sudakov form factor. Hereafter, we set $\mu = \sqrt{\zeta_F} = Q$ and express $\tilde{F}(x, b; \mu = Q, \zeta_F = Q^2)$ as $\tilde{F}(x, b; Q)$ for simplicity. Eq. (1) demonstrates that the distribution $\tilde{F}$ at an arbitrary scale Q can be determined by the same distribution at an initial scale μ_i through the evolution encoded by the exponential form $\exp(-S(Q, b))$.

In order to access the information of $\tilde{F}(x, b; \mu, \zeta_F)$ in the whole b region, we adopt a convenient way to take into account the evolution behavior of $\tilde{F}(x, b; Q)$ in the large b region by including a nonperturbative Sudakov-like form factor $S_{\rm NP}$. Combining the perturbative part and the nonperturbative part, one has the complete result for the Sudakov form factor appearing in Eq. (1)

$$S(Q, b) = S_{\rm P}(Q, b) + S_{\rm NP}(Q, b). \tag{2}$$

with the boundary of the two parts set by the $b_{\max}$ to guarantee b_* is always at the perturbative region. In literature there are different prescription [3, 4, 8] for b_*. In the perturbative region, TMD PDFs at a fixed scale $\tilde{F}(x, b, \mu)$ can be expressed as the convolution of the perturbatively calculable coefficients C and the collinear counterparts of TMD PDFs $\tilde{F}_{q/H}(x, b; \mu) = \sum_i \int_x^1 \frac{d\xi}{\xi} C_{q\leftarrow i}(x/\xi, b; \mu) F_{i/H}(\xi, \mu)$. $F_{i/H}(\xi, \mu)$ is the corresponding collinear distribution at the dynamic scale μ which can be related to b_* by $\mu = c/b_*$, with $c = 2e^{-\gamma_E}$ and the Euler Constant $\gamma_E \approx 0.577$. $C_{q\leftarrow i}(x/\xi, b; \mu) = \sum_{n=0}^{\infty} C_{q\leftarrow i}^{(n)}(\alpha_s/\pi)^n$ is the perturatively calculable coefficient function. Therefore, in the small b region, we can also express the Boer-Mulders function of the pion at a fixed energy scale μ as $\widetilde{h}_{1,q/\pi}^{\alpha\perp}(x, b; \mu) = (\frac{-ib_\perp^\alpha}{2}) T_{q/\pi,F}^{(\sigma)}(x, x; \mu)$, where the hard coefficients are calculated up to LO, and the collinear function $T_{q/\pi,F}^{(\sigma)}(x, x; \mu)$ is the chiral-odd twist-3 quark-gluon-quark correlation function and is related to the first transverse moment of the Boer-Mulders function $h_{1,q/\pi}^{\perp(1)}$ by $T_{q/\pi,F}^{(\sigma)}(x, x; \mu) = 2M_\pi h_{1,q/\pi}^{\perp(1)}$. Note that the perturbative part $S_{\rm P}(Q, b)$ can be expanded as α_s/π series, for which we take up to the accuracy of next-to-leading-logarithmic (NLL) order.

The nonperturbative part of the Sudakov form factor S_{NP} can not be calculated perturbatively, it is usually extracted from experimental data. There are several different parametrizations on S_{NP}, in this paper, we will discuss two of them. One is the SIYY parametrization [5],

$$S_{\rm NP}^{f_{1,q/p}}(Q, b) = \frac{g_1}{2} b^2 + \frac{g_2}{2} \ln \frac{b}{b_*} \ln \frac{Q}{Q_0}, \tag{3}$$

$$S_{\rm NP}^{f_{1,q/\pi}} = g_1^\pi b^2 + g_2^\pi \ln\frac{b}{b_*} \ln\frac{Q}{Q_0}, \tag{4}$$

where $S_{\rm NP}^{f_{1,q/p}}$ and $S_{\rm NP}^{f_{1,q/\pi}}$ correspond to the nonperturbative Sudakov form factors of the unpolarized TMD distribution of the proton and the pion, respectively. Here, b_* follows the choice in Refs. [4] and the parameters are fitted from the NN and $\pi^- N$ Drell-Yan process datas at the initial energy scale $Q_0^2 = 2.4$ GeV^2 yielding $g_1 = 0.212^{+0.006}_{-0.007}$, $g_2 = 0.84^{+0.040}_{-0.035}$, $g_1^\pi = 0.082 \pm 0.022$, $g_2^\pi = 0.394 \pm 0.103$ [6]. Note that the nonperturbative Sudakov form factor $S_{\rm NP}$ for quarks from one proton and antiquarks from another proton satisfies [7]

$$S_{\rm NP}^q(Q, b) + S_{\rm NP}^{\bar{q}}(Q, b) = S_{\rm NP}(Q, b). \tag{5}$$

The other parametrization is the Bacchetta-Delcarro-Pisano-Radici-Signori (BDPRS) parametrization [8],

$$\tilde{f}_1^a(x, b^2; Q^2) = f_1^a(x; \mu^2) e^{-S(\mu^2, Q^2)} e^{\frac{1}{2} g_K(b) \ln(Q^2/Q_0^2)} \tilde{f}_{1\text{NP}}^a(x, b^2), \tag{6}$$

where $g_K = -g_2 b^2/2$. $\tilde{f}_{1\text{NP}}^a(x, b^2)$ is the intrinsic nonperturbative part of the PDFs, which is parameterized as

$$\tilde{f}_{1\text{NP}}^a(x, b^2) = \frac{1}{2\pi} e^{-g_1 \frac{b^2}{4}} (1 - \frac{\lambda g_1^2}{1 + \lambda g_1} \frac{b^2}{4}), \tag{7}$$

with

$$g_1(x) = N_1 \frac{(1-x)^\alpha x^\sigma}{(1-\hat{x})^\alpha \hat{x}^\sigma}. \tag{8}$$

Here, the fitted results of the parameters are $\alpha = 2.95$, $\sigma = 0.17$, $\lambda = 0.86\ \text{GeV}^{-2}$, $g_1(\hat{x}) = 0.28\ \text{GeV}^2$, with $N_1 \equiv g_1(\hat{x})$, $\hat{x} = 0.1$. The b_* prescription in this parametrization follow the choice in Refs. [8].

2.1 Formalism of the asymmetry $A_T^{\sin(2\phi-\phi_S)}$ in Drell-Yan process

In the pion-induced Drell-Yan process $\pi^-(P_\pi) + p^\uparrow(P_p) \longrightarrow \gamma^*(q) + X \longrightarrow l^+(\ell) + l^-(\ell') + X$, the asymmetry $A_T^{\sin(2\phi-\phi_S)}$ is defined as [9]:

$$A_T^{\sin(2\phi-\phi_S)}(x_1, x_2, Q) = \frac{F_T^{\sin(2\phi-\phi_S)}(x_1, x_2, Q)}{F_U^1(x_1, x_2, Q)}. \tag{9}$$

The denominator can expressed as the convolution of the unpolarized distribution functions from each hadron $F_U^1 = \mathcal{C}[f_{1,q/\pi} f_{1,\bar{q}/p}]$, while the numerator is expressed as the convolution of the pion Boer-Mulders distribution and the proton transversity distribution $F_T^{\sin(2\phi-\phi_S)} = -\mathcal{C}[\frac{\boldsymbol{h}\cdot\boldsymbol{k}_{a\perp}}{M_\pi} h_{1,q/\pi}^\perp h_{1,\bar{q}/p}]$ with $\boldsymbol{h} = \hat{\boldsymbol{q}} \equiv \boldsymbol{q}_\perp/|\boldsymbol{q}_\perp|$. After dealing with the convolution and performing the Fourier transformation, we can get the following expressions for the structure functions [10]

$$F_T^{\sin(2\phi-\phi_S)} = -\frac{1}{N_c} \sum_q e_q^2 \int_0^\infty \frac{db}{4\pi} b^2 J_1(q_\perp b) h_{1,q/p}(x_p, \mu_b) T_{\bar{q}/\pi,F}^{(\sigma)}(x_\pi, x_\pi, \mu_b) e^{-\left(S_{\text{NP}}^{f_{1,q/p}} + S_{\text{NP}}^{f_{1,q/\pi}} + S_{\text{P}}\right)} + (q \leftrightarrow \bar{q}),$$
$$F_U^1 = \frac{1}{N_c} \sum_q e_q^2 \int_0^\infty \frac{bdb}{2\pi} J_0(q_\perp b) f_{1,q/\pi}(x_\pi, \mu_b) f_{1,\bar{q}/p}(x_p, \mu_b) e^{-\left(S_{\text{NP}}^{f_{1,q/p}} + S_{\text{NP}}^{f_{1,q/\pi}} + S_{\text{P}}\right)} + (q \leftrightarrow \bar{q}). \tag{10}$$

As there is no extraction on the Boer-Mulders function of the pion meson, we apply a model result based on the light-cone wave function of the pion meson from Ref. [11] at the model scale $\mu_0^2 = 0.25\text{GeV}^2$. For consistency, we apply the unpolarized distribution function of the pion meson $f_{1\pi}(x)$ using the same model. We also apply the result from the light-cone constituent quark model in Ref. [12] for comparison.

For the collinear distributions of the proton, we resort to existing parameterizations, i.e., we adopt the NLO set of the CT10 parametrization (central PDF set) for the unpolarized distribution function $f_1(x)$ of the proton, and we choose the transversity distribution extracted from SIDIS data via the TMD evolution formalism [4].

Furthermore, we apply the QCDNUM evolution package [13] to perform the evolution of $f_{1,q/\pi}$ from the model scale μ_0 to another energy. As for the energy evolution of the twist-3

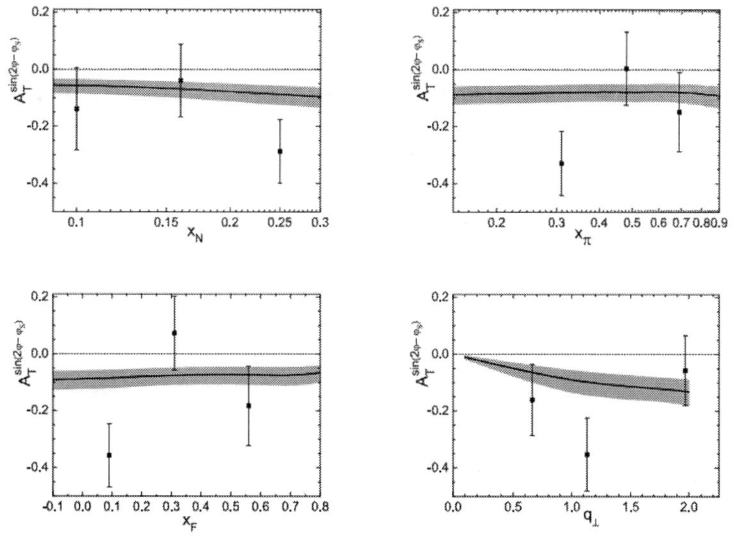

Figure 1. The asymmetry $A_T^{\sin(2\phi-\phi_S)}$ in the $\pi^- N^\uparrow$ Drell-Yan process calculated from the SIYY parametrization on the nonperturbative form factor. The four figures correspond to the asymmetries as functions of x_N (upper left), x_π (upper right), x_F (lower left) and $q_\perp$ (lower right). The solid lines show the results from the central values of the parameters, while the shaded area show the uncertainty bands determined by the uncertainties of the parameters. The solid squares represent the COMPASS data for comparison.

collinear correlation function $T_{q,F}^{(\sigma)}$, we follow the same choice in Ref. [11]. Similarly, we also include the kernel in Ref. [4] to evolve the transversity distribution of the proton.

With all the ingredients above, we calculate the asymmetry $A_T^{\sin(2\phi-\phi_S)}$ in the pion-induced transversely polarized Drell-Yan process at the COMPASS kinematics and compare with data [1]. In Fig. 1, we plot our numerical results of the asymmetry $A_T^{\sin(2\phi-\phi_S)}$ using the SIYY parametrization [Eqs. (3) and (4)] for the nonperturbative Sudakov form factor and the pion Boer-Mulders function from Ref. [11]. The solid curves correspond to the results calculated from the central values of the parameters, while the shaded area shows the uncertainty band determined by the uncertainties of the parameters. To make the TMD factorization valid in the kinematic region, the integration over the transverse momentum $q_\perp$ is performed in the region of $0 < q_\perp < 2$ GeV [14]. In the figure the solid squares show the experimental data measured by the COMPASS Collaboration [15], with the error bars corresponding to the sum of the systematic error and the statistical error.

As shown in Fig. 1, in all the cases the asymmetry $A_T^{\sin(2\phi-\phi_S)}$ in the $\pi^- P$ Drell-Yan from our calculation is negative, in agreement with most of the data from COMPASS. Our estimate also shows that the asymmetry changes slightly with the change of x_N, x_π, or x_F, and the magnitude of the x_N-, x_π-, and x_F-dependent asymmetries is around 0.05 to 0.10. For the $q_\perp$ asymmetry, we find that its magnitude is about 0.05 to 0.15 and moderately increases with increasing $q_\perp$ in the region $q_\perp < 2$ GeV.

To study the impact of different parameterizations of the nonperturbative part on the asymmetry, in Fig. 2, we present the result using the BDPRS evolution formalism [Eqs. (6) and (7)] by the dashed lines. The solid lines show the results in Fig. 1 (central results) for comparison. We find that, in the case of $q_\perp$-dependent asymmetry, the result from the BD-

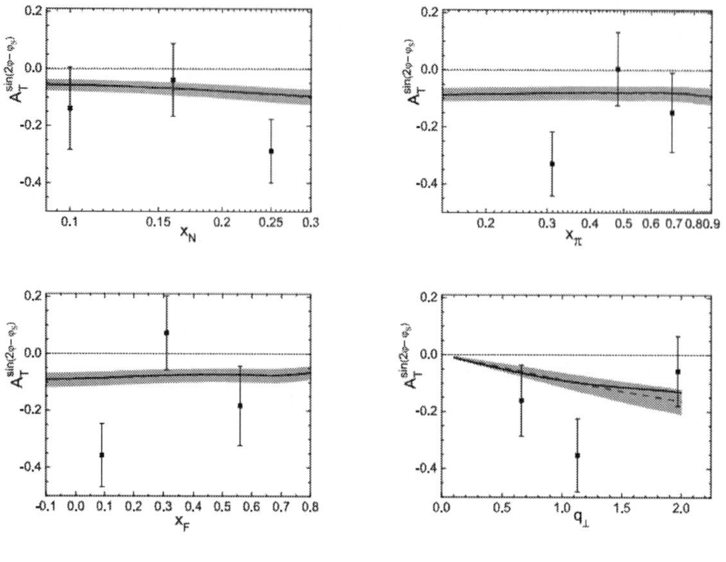

Figure 2. Similar to Fig. 1, but the asymmetry calculated from the BDPRS parametrization on the nonperturbative form factor. The dashed lines plot the central results, while the solid lines are the central results in Fig. 1 for comparison.

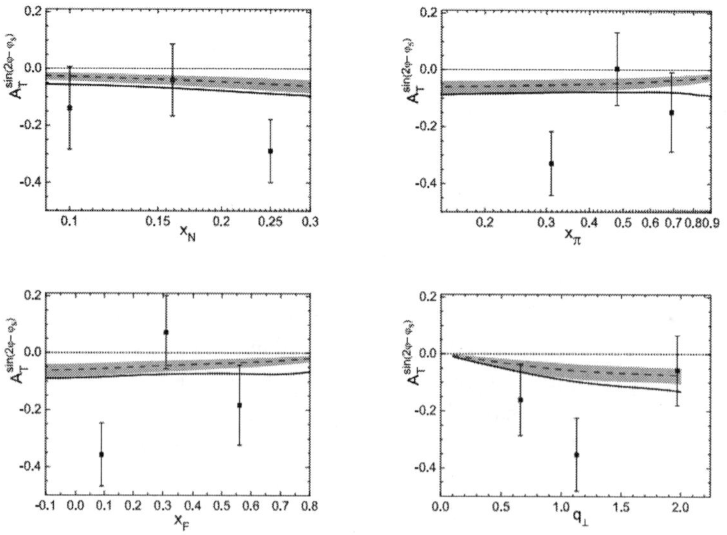

Figure 3. Similar to Fig. 1, but the asymmetry calculated from the Boer-Mulders function of the pion in a light-cone constituent model [12]. The dashed lines plot the central results, while the solid lines are the central results in Fig. 1 for comparison.

PRS parametrization is qualitatively different from the result from the SIYY parametrization, particularly in the region $q_\perp \in [1.5, 2]$ GeV; while for the x_N-, x_π-, and x_F-dependent asymmetries the results from the two evolution formalisms are consistent.

To study the effect of different pion distribution functions on the numerical calculation of $A_T^{\sin(2\phi-\phi_S)}$, in Fig. 3 (the dashed lines), we adopt the pion Boer-Mulders function obtained from the light-front constituent quark model [12] and the SIYY parametrization to perform the calculation. The solid lines correspond to the results in Fig. 1 (central results) for comparison. We find that the sign of the asymmetries from the model results for pion distributions in Ref. [12] are still negative, while their magnitudes are generally smaller than those in Fig. 1.

2.2 Conclusion

In this work, we applied the TMD factorization approach to study the asymmetry $A_T^{\sin(2\phi-\phi_S)}$ in $\pi^- p$ Drell-Yan process at the kinematical region of the COMPASS. The asymmetry arises from the convolution of the pion Boer-Mulders function and the proton transversity distribution. In the calculation, we applied two different approaches (SIYY and BDPRS parametrizations) to perform the TMD evolution of pion and proton distributions for comparison. Their main difference is the treatment on the nonperturbative part of evolution, while the perturbative part is the same and has been kept at NLL accuracy in this work. As the nonperturbative part associated with the pion Boer-Mulders function is still unknown, we assume that it has the same form as that the unpolarized distribution function. The hard coefficients associated with the corresponding collinear functions in the TMD evolution formalism are kept at leading-order accuracy. We then calculated the asymmetry $A_T^{\sin(2\phi-\phi_S)}$ with the available parametrization for the proton distribution and two different model results from the light-cone wave function approach for pion meson distribution. we find that the asymmetry is sensitive to the choice of the pion distribution function, while different choice of the TMD evolution formalism will only on the nonperturbative TMD evolution only affect the shape of the $q_\perp$-dependent of the asymmetry. We also find that the corresponding asymmetry $A_T^{\sin(2\phi-\phi_S)}$ is consistent with the COMPASS measurement in sign and magnitude. Furthermore, our study may provide a framework to access the Boer-Mulders function of the pion and the corresponding nonperturbative Sudakov form factor through transversely polarized $\pi^- p$ data.

References

[1] M. Aghasyan *et al.* (COMPASS Collaboration), Phys. Rev. Lett. **119**, 112002 (2017).
[2] J. C. Collins and D. E. Soper, Nucl. Phys. **B193**, 381 (1981), Erratum: [Nucl. Phys. **B213**, 545 (1983)].
[3] J. C. Collins, D. E. Soper and G. F. Sterman, Nucl. Phys. B **250**, 199-224 (1985).
[4] Z. B. Kang, A. Prokudin, P. Sun and F. Yuan, Phys. Rev. D **93**, 014009 (2016).
[5] P. Sun, J. Isaacson, C. P. Yuan and F. Yuan, Int. J. Mod. Phys. A **33**, 1841006 (2018).
[6] X. Wang, Z. Lu and I. Schmidt, J. High Energy Phys. **08**, 137 (2017).
[7] A. Prokudin, P. Sun and F. Yuan, Phys. Lett. B **750**, 533 (2015).
[8] A. Bacchetta, F. Delcarro, C. Pisano, M. Radici and A. Signori, JHEP **06**, 081 (2017) [erratum: JHEP **06**, 051 (2019)].
[9] F. Gautheron *et al.* [COMPASS], SPSC-P-340.
[10] H. Li, X. Wang and Z. Lu, Phys. Rev. D **101**, 054013 (2020).
[11] Z. Wang, X. Wang and Z. Lu, Phys. Rev. D **95**, 094004 (2017).
[12] B. Pasquini and P. Schweitzer, Phys. Rev. D **90**, 014050 (2014).
[13] M. Botje, Comput. Phys. Commun. **182**, 490 (2011).

[14] P. Sun and F. Yuan, Phys. Rev. D **88**, 114012 (2013).
[15] M. Aghasyan *et al.* [COMPASS], Phys. Rev. Lett. **119**, 112002 (2017).

Improved calculation of the $\gamma^*\gamma \to \pi$ process at low Q^2 using LCSR's and renormalization-group summation*

Sergey Mikhailov[1,*,**], *Alexandr* Pimikov[1,***], and *N. G.* Stefanis[2,****]

[1]Bogoliubov Laboratory of Theoretical Physics, JINR, 141980 Dubna, Russia
[2]Institut für Theoretische Physik II, Ruhr-Universität Bochum, D-44780 Bochum, Germany

Abstract. We study two versions of lightcone sum rules to calculate the $\gamma^*\gamma \to \pi^0$ transition form factor (TFF) within QCD. While the standard version is based on fixed-order perturbation theory by means of a power-series expansion in the strong coupling, the new method incorporates radiative corrections by renormalization-group summation and generates an expansion within a generalized fractional analytic perturbation theory involving only analytic couplings. Using this scheme, we determine the relative nonperturbative parameters and the first two Gegenbauer coefficients of the pion distribution amplitude (DA) to obtain TFF predictions in good agreement with the preliminary BESIII data, while the best-fit pion DA satisfies the most recent lattice constraints on the second moment of the pion DA at the three-loop level.

1 Introduction

In this paper we present our recent work on the calculation of the two-photon process $\gamma^*(Q^2)\gamma(q^2 \sim 0) \to \pi^0$ which contains in the form of a convolution the distribution amplitude of the pion [1]—the simplest bound state in QCD. Our analysis uses the method of lightcone sum rules (LCSR)'s [2, 3] in the extended form developed in [4]. This scheme includes the QCD radiative corrections by means of renormalization-group (RG) summation, ultimately amounting to a generalized version of fractional analytic perturbation theory (FAPT), invented in [5, 6] following [7], and reviewed in [8, 9].

We present the key elements of this formalism and apply it to the recently released preliminary data of the BESIII Collaboration [10, 11], which extend the range of measurements of the $\pi-\gamma$ transition form factor (TFF) to very low $Q^2 \ll 1$ GeV^2 values with an unprecedented precision. At such momenta the conventional LCSR method, based on fixed-order perturbation theory (FOPT), cannot describe this transition process with sufficient accuracy, though it works very well at high Q^2 [12, 13]. Very recently, the full QCD calculation of the two-loop coefficient function of leading twist to the TFF $\gamma^*\gamma \to \pi^0$ was carried out analytically by two different groups using different methods and obtaining coinciding results [14, 15]. This level of computational accuracy of the NNLO radiative correction is required by the expected precision of forthcoming data of the Belle II experiment. In the higher Q^2 region, the measured

*This work is dedicated to the memory of Maxim Polyakov, a deeply admired colleague and friend of ours.
**e-mail: mikhs@theor.jinr.ru
***e-mail: pimikov@mail.ru
****e-mail: stefanis@tp2.ruhr-uni-bochum.de

values of $Q^2F(Q^2)$ reported by the Belle Collaboration [16] agree with the theoretical expectations, while the previous *BABAR* results [17] show above $Q^2 \gtrsim 9$ GeV^2 a rapid growth with Q^2. It is expected that Belle II will collect a large data sample with a much higher accuracy to resolve this discrepancy allowing to test the onset of factorization and identify the scale where asymptotic scaling sets in [18] or establish the existence of scaling violations.

The main ingredients of both LCSR methods are shown in Fig. 1.

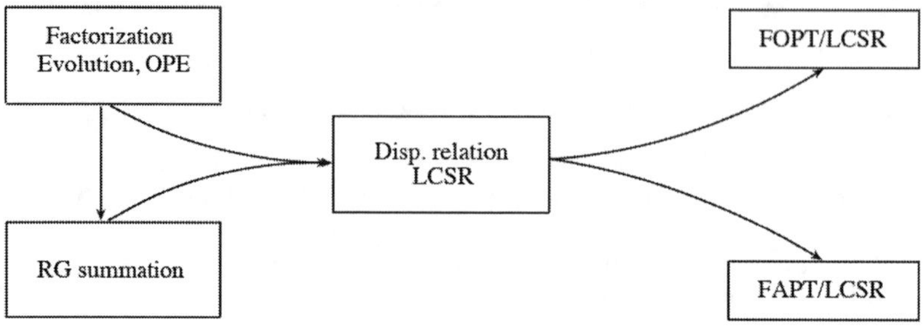

Figure 1. Schematic derivation of the FOPT and FAPT versions of LCSR's to calculate the TFF.

The paper is organized as follows. In Sec. 2 we present a brief description of the current status of the pion-photon transition form factor (TFF) calculation using LCSR's in FOPT and point out its limitations at $Q^2 \lesssim 1$ GeV^2. In the subsequent section Sec. 3 we discuss the extended version of the LCSR method which incorporates the summation of radiative corrections via the renormalization group. While the first version is sufficient to adequately describe this observable at high-Q^2, see, [13], the second version is more suitable at low momenta, where the radiative corrections, together with the higher twists (twist-four and twist-six) contributions, are particularly important.

2 Radiative corrections in FOPT/LCSR

In this section we briefly address the basic results for the TFF using the method of LCSR's within FOPT in QCD. Applying collinear factorization, the TFF for the hard exclusive process $\gamma^*(Q^2)\gamma(q^2) \to \pi^0$ ($Q^2 \gg q^2 \gg \Lambda^2_{\rm QCD}$) in leading twist two can be written in convolution form as a power series expansion in the strong coupling $a_s = \alpha_s/4\pi$ to get

$$F^{(\text{tw-2})}_{\rm FOPT}(Q^2, q^2) = N_{\rm T}\left(T_{\rm LO} + a_s T_{\rm NLO} + a_s^2 T_{\rm NNLO} + \ldots\right) \otimes \varphi_\pi^{(2)}, \tag{1}$$

where $N_{\rm T} = \sqrt{2}f_\pi/3$. The nonperturbative input of the pion structure is encoded in the lightcone distribution amplitude (DA) of twist two (tw-2), $\varphi_\pi^{(2)}$, which describes the partition of the longitudinal momentum between its two valence quarks with fractions $x_q = x = (k^0 + k^3)/(P^0 + P^3) = k^+/P^+$ and $x_{\bar{q}} = 1 - x \equiv \bar{x}$. The hard coefficient functions are given by the following expressions [1]

$$T_{\rm LO} = a_s^0(\mu_{\rm F}^2)T_0(y) \equiv 1/\left(q^2\bar{y} + Q^2 y\right), \tag{2a}$$

$$a_s T_{\rm NLO} = a_s^1(\mu_{\rm F}^2)T_0(y) \otimes \left[\mathcal{T}^{(1)} + \underline{L\, V_0}\right](y, x), \tag{2b}$$

$$a_s^2 T_{\rm NNLO} = a_s^2(\mu_{\rm F}^2)T_0(y) \otimes \Big[\mathcal{T}_\beta^{(2)} - \underline{L\mathcal{T}^{(1)}\beta_0} + \underline{L\mathcal{T}^{(1)} \otimes V_0} - \underline{(L^2/2)\beta_0 V_0} + \underline{(L^2/2)V_0 \otimes V_0} + \underline{\underline{LV_1}}\Big](y, x), \tag{2c}$$

where $L = L(y) = \ln[(q^2\bar{y} + Q^2 y)/\mu_{\rm F}^2]$ and using the following abbreviations: LO (leading order), NLO (next-to-leading order) and NNLO (next-to-next-to-leading order). Here the terms $\mathcal{T}^{(1)}, \mathcal{T}_{\beta}^{(2)}$ and $\mathcal{T}^{(2)}$ represent corrections due to parton subprocesses, while the singly and doubly underlined terms are due to $\bar{a}_s(y)$ and Efremov-Radyushkin-Brodsky-Lepage (ERBL) evolution [19, 20] at one loop (V_0 kernel) and two loops (V_1 kernel), respectively. On the other hand, the evolution of the pion DA is taken into account in terms of the conformal expansion

$$\varphi_{\pi}^{(2)}(z, \mu^2) = \psi_0(x) + \sum_{n=2,4,\ldots}^{\infty} b_n(\mu^2)\psi_n(x), \tag{3}$$

where $\varphi_{\pi}^{\rm asy} = \psi_0(x) = 6x(1-x) \equiv 6x\bar{x}$ is the asymptotic pion DA.

3 Radiative corrections using RG summation: FAPT/LCSR

To implement the RG summation in $F^{(\rm tw\text{-}2)}$, we collect all underlined evolution terms into the running coupling $a_s(\mu^2) \to \bar{a}_s(y) \equiv \bar{a}_s\left(q^2\bar{y} + Q^2 y\right)$ and the ERBL factor [4] to obtain

$$\begin{aligned} F_n^{(\rm tw\text{-}2)}(Q^2, q^2) &= N_{\rm T} T_0(y) \underset{y}{\otimes} \left\{ \left[\mathbb{1} + \bar{a}_s(y)\mathcal{T}^{(1)}(y,x) + \bar{a}_s^2(y)\mathcal{T}^{(2)}(y,x) + \ldots \right] \right. \\ & \left. \underset{x}{\otimes} \exp\left[-\int_{a_s}^{\bar{a}_s(y)} \frac{V(\alpha; x, z)}{\beta(\alpha)} d\alpha \right] \right\} \underset{z}{\otimes} \varphi_{\pi}^{(2)}(z, \mu^2). \end{aligned} \tag{4}$$

Employing in Eq. (4) the Gegenbauer expansion given in Eq. (3), we get in leading logarithmic approximation (LLA)

$$F_n^{(\rm tw\text{-}2)}(Q^2, q^2) \overset{\text{1-loop}}{\longrightarrow} F_{(1l)n}^{(\rm tw=2)} = N_{\rm T} T_0(y) \underset{y}{\otimes} \left[\mathbb{1} + \bar{a}_s(y)\mathcal{T}^{(1)}(y,x) \right] \left(\frac{\bar{a}_s(y)}{a_s(\mu^2)} \right)^{\nu_n} \underset{x}{\otimes} \psi_n(x). \tag{5}$$

One realizes that for $q^2 = 0,\ y \ll 1$, the summation of the evolution terms via $\bar{a}_s(y)$ becomes inapplicable, even if Q^2 is large [1].

We now show that this deficit is amended when a dispersion relation for the TFF is involved. The key elements of this procedure can be summarized as follows.

1. Impose factorization and the twist expansion.

2. Use a dispersive form of the TFF

$$\left[F(Q^2, q^2)\right]_{\rm an} = \int_{m^2}^{\infty} \frac{\rho_F(Q^2, s)}{s + q^2 - i\epsilon} ds,\ \rho_F(s) = \frac{\mathbf{Im}}{\pi}\left[F(Q^2, -s)\right] \tag{6}$$

that inevitably leads to FAPT [5, 6] with analytic couplings $\mathcal{A}_\nu$ (Euclidean space) and $\mathfrak{A}_\nu$ (Minkowski space) in a nonpower series expansion.

3. Dissect the spectral density using the perturbative expansion in Eqs. (4), (5), and employing ρ_n for each of the ψ_n harmonics perform the twist expansion:

$$\rho(Q^2, x) = \sum_{0,2,4,\ldots} a_n(Q^2)\rho_n(Q^2, x) + \rho_{\rm tw\text{-}4}(Q^2, x) + \rho_{\rm tw\text{-}6}(Q^2, x) + \ldots, \tag{7}$$

where the integration variable in the spectral density has been replaced by $s \to x = s/(Q^2 + s)$ and $x_s = s_0/(Q^2 + s_0)$. Then, one obtains the LCSR based on Eq. (6), whereas the FAPT results are encapsulated in the analytic couplings $\mathcal{A}_\nu, \mathfrak{A}_\nu, \mathcal{I}_\nu$. The key point is that the imaginary parts in (6) stem solely from the $a_s^\nu(y)$ factors in Eq. (5). All details of this derivation are discussed in [1].

4. To preserve the perturbative asymptotic limit of the TFF $Q^2F(Q^2 \to \infty) = \sqrt{2}f_\pi$, the calibration condition $\mathcal{A}_\nu(0) = \mathfrak{A}_\nu(0) = 0$ for $0 < \nu \leq 1$ has to be imposed [4]. This yields generalized two-parameter FAPT couplings $\mathcal{I}_\nu$ with explicit expressions given in [1].

Restricting for simplicity our attention to the NNLO$_\beta$ approximation of the partial form factors F_n within FAPT, we derive in the limits $q^2 \to 0, Q(y) \to yQ^2$ the following expression

$$Q^2F^{(\text{tw-2})}_{\text{FAPT;n}}(Q^2) \approx \frac{N_\text{T}}{[a_s(\mu^2)]^{\nu_n}[1 + c_1a_s(\mu^2)]^{\omega_n}}\left\{\underline{\frac{\mathbb{A}_{\nu_n}(m^2,x)}{x} + \left(\frac{\mathbb{A}_{1+\nu_n}(m^2,y)}{y}\right)\underset{y}{\otimes}\mathcal{T}^{(1)}(y,x)}\right.$$
$$+\omega_nc_1\left[\frac{\mathbb{A}_{1+\nu_n}(m^2,x)}{x} + \frac{\mathbb{A}_{2+\nu_n}(m^2,x)}{x}\frac{c_1(\omega_n-1)}{2} + \left(\frac{\mathbb{A}_{2+\nu_n}(m^2,y)}{y}\right)\underset{y}{\otimes}\mathcal{T}^{(1)}(y,x)\right]$$
$$\left.+\underline{\underline{\left(\frac{\mathbb{A}_{2+\nu_n}(m^2,y)}{y}\right)\underset{y}{\otimes}\mathcal{T}^{(2)}(y,x)}}\right\}\underset{x}{\otimes}\psi_n(x), \tag{8}$$

where the terms contributing to the TFF in LLA are underlined. The couplings $a_s^\nu(\mu^2)$ and

$$\mathbb{A}_\nu(m^2,y) = \theta(y \geqslant y_m)\left[\mathcal{A}_\nu(Q(y)) - \mathfrak{A}_\nu(0)\right] + \theta(y < y_m)\left[\mathcal{I}_\nu(m(y),Q(y)) - \mathfrak{A}_\nu(m(y))\right]$$

have to be evaluated with a two-loop running, while $c_1 = \beta_1/\beta_0$ and $\omega_n = [\gamma_1(n)\beta_0 - \gamma_0(n)\beta_1]/[2\beta_0\beta_1]$. The only surviving term in the next-to-leading logarithmic approximation (NLLA) for the numerically important case of the zero-harmonic ($\omega_{n=0} = 0$) is the doubly underlined term. For this reason, the effect of the two-loop evolution in the second line is neglected, i.e., $c_1 = 0$ (for further details, see [1]).

4 FAPT/LCSR for the twist-4 pion TFF

In the previous section we included the RG summation only in the twist-two part of the TFF. To carry out a comprehensive analysis of the experimental data in the low-Q^2 regime $0.35 \leq Q^2 \leq 3.1$ GeV^2, covered by the CELLO [21], CLEO [22], and the preliminary BESIII [10, 11] data, we have to extend the RG procedure to the twist-four term stemming from the contribution of the two-particle DA. To this end, we make the assumption that the three-particle DA term is modified after the RG summation in the same way as the twist-two part and evaluate the complete twist-four contribution in an analogous way.

Following [3], we recast the TFF in the following form

$$F^{\gamma\pi,\text{tw-4}}_{\text{FAPT}}\left(Q^2\right) = \frac{\sqrt{2}f_\pi}{3Q^2}\left[H^{\text{tw-4}}_{\text{FAPT}}(Q^2) + \frac{Q^2}{m_\rho^2}k(M^2)V^{\text{tw-4}}_{\text{FAPT}}(Q^2,M^2)\right], \tag{9}$$

$$V^{\text{tw-4}}_{\text{FAPT}}(Q^2) = -\frac{\delta^2_{\text{tw-4}}(\mu_0^2)}{M^2}\int_{\bar{x}_s}^1 dx\,\frac{\Delta_\nu(s_0,\bar{x})}{(a_s(\mu_0^2))^\nu}\frac{\varphi^{(4)}(\bar{x})}{x^2}\exp\left(\frac{m_\rho^2 - Q^2\bar{x}/x}{M^2}\right), \tag{10}$$

$$H^{\text{tw-4}}_{\text{FAPT}}(Q^2) = -\frac{\delta^2_{\text{tw-4}}(\mu_0^2)}{(a_s(\mu_0^2))^\nu}Q^2\mathbb{A}_\nu(s_0;u)\underset{u}{\otimes}\frac{\varphi^{(4)}(u)}{Q^2(u)}, \tag{11}$$

where we use the same notations and definitions as in [1]. The expression for the function $\varphi^{(4)}(x)$ below has two terms: the first term in the parenthesis is the two-particle DA contribution while the second term is due to the three-particle DA:

$$\varphi^{(4)}(x) = \left(\frac{50}{3} + 10\right)x^2(1-x)^2. \tag{12}$$

Note that the part coming from the three-particle DA is included here only heuristically. Due to the RG summation, the three-particle contribution to twist-four should be considered more rigorously, but this lies beyond the scope of this work. For completion, we also quote the value of the twist-four coupling parameter δ^2 together with its FOPT evolution,

$$\delta^2_{\text{tw-4}}(\mu_0^2) = 0.95\,\lambda_q^2/2 = 0.19\ \text{GeV}^2\,,$$
$$\delta^2_{\text{tw-4}}(Q^2) = \left[\frac{a_s(Q^2)}{a_s(\mu_0^2)}\right]^{\nu} \delta^2_{\text{tw-4}}(\mu_0^2)\,,\, \nu = \gamma_{T4}/\beta_0\,,\gamma_{T4} = 32/9\ [23]\,. \tag{13}$$

The results for $F^{\gamma\pi,\text{tw-4}}\left(Q^2\right)$ in FAPT compared to FOPT are displayed in Fig. 2.

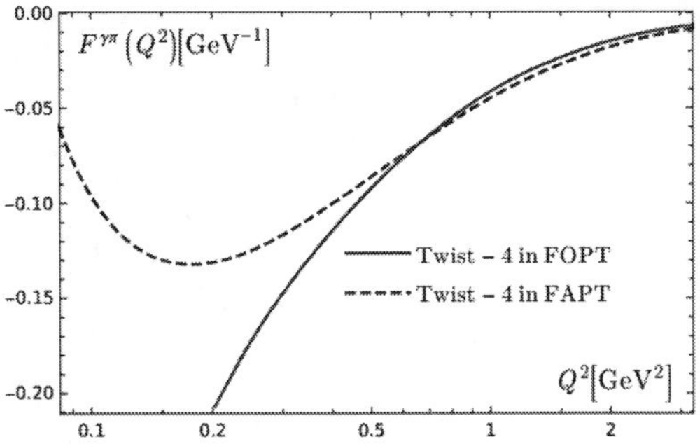

Figure 2. Comparison of the twist-four contribution obtained with FOPT/LCSR's (red solid line) and FAPT/LCSR's (blue dashed line) using in Eq. (9) the single delta-function resonance model with $k(M^2) = 1$.

One may conclude that the FAPT modification of the twist-four contribution provides protection against drastic changes relative to the FOPT behavior, extending this way the applicability domain of the FAPT TFF calculation to lower Q^2 values. On the other hand, both curves come close to each other at $Q^2 > 0.35\ \text{GeV}^2$ so that the estimated ratio (twist-4)/(twist-2) becomes approximately $1/3$ near the lowest BESIII data point, see [10, 11] and Table I in [1].

5 Phenomenological analysis

In this section we present the pivotal phenomenological results of our analysis based on FAPT/LCSR's. This is done for simplicity in terms of $F^{\gamma\pi}_{\text{LCSR}}(Q^2)$ for the harmonics ψ_n, $F^{\gamma\pi}_{\text{LCSR;n}}$, and predictions are shown in Fig. 3 for some selected pion DA's including the best-fit one determined in [1]. Various data from different experiments in the momentum range $0.35 \leq Q^2 \leq 10\ \text{GeV}^2$ are also shown to effect the quality of the fitting procedure at low Q^2.

To this end, we determine the following nonperturbative parameters:

1. b_2, b_4 (conformal coefficients of the twist-2 pion DA) using the following constraints at the scale $\mu_0^2 = 1\ \text{GeV}^2$: $b_2(\mu_0^2) = [0.146, 0.272]$, $b_4(\mu_0^2) = [-0.23, -0.049]$ (domain of twist-2 BMS DA's) [23, 24].

2. $\delta^2_{\text{tw-4}}(\mu_0^2) = 0.19 \pm 0.04\ \text{GeV}^2$ (twist-4 coupling parameter) [23].

3. $\delta^2_{\text{tw-6}}(\mu_0^2) = (1.61 \pm 0.26) \times 10^{-4}\ \text{GeV}^6$ (twist-6 coupling parameter) extracted from the data in [1].

A best fit to the 1σ and 2σ error ellipses of the data up to 3.1 GeV^2 is given in [1] with $\chi^2_{\rm ndf} = 0.38$. It provides agreement with the N^{3}LO lattice constraints from [25] and yields $b_2(\mu_0^2) = 0.159$, which corresponds to $b_4(\mu_0^2) = -0.098$ if depicted within the BMS domain. Note that the platykurtic (pk) range of pion DA's [26, 27] lies entirely within the 1σ error ellipse of this data set and is close to the NNLO lattice strip [25].

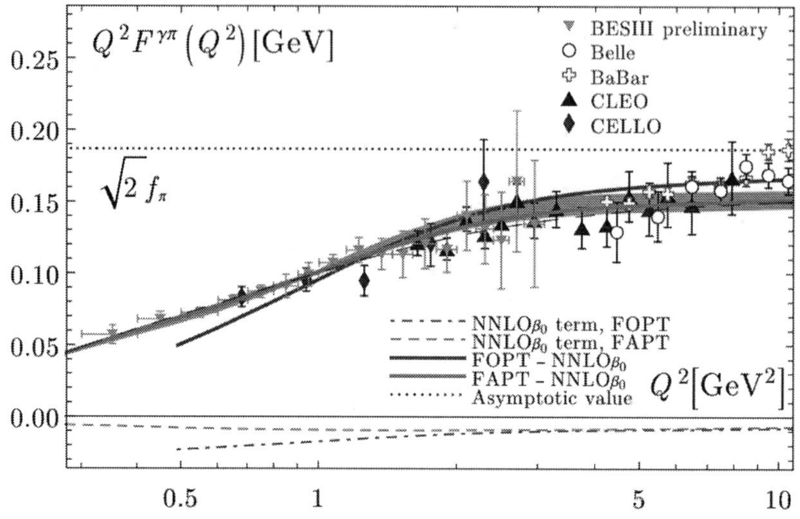

Figure 3. This figure serves to demonstrate that the TFF calculated with FAPT/LCSR's (using $N_f = 3$) compares well with the data up to momenta where it is expected to enter the scaling regime predicted by perturbative QCD, though with reduced accuracy compared to the FOPT/LCSR result, see [13]. The displayed curves are explained in the text.

The TFF predictions calculated this way, are given by

$$F^{\gamma\pi}_{\rm LCSR}\left(Q^2\right) = F^{\gamma\pi}_{\rm LCSR;0}\left(Q^2\right) + \sum_{n=2,4} b_n(\mu^2) F^{\gamma\pi}_{{\rm LCSR};n}\left(Q^2\right) + \delta^2_{\rm tw\text{-}4}(\mu^2) F^{\gamma\pi}_{\rm tw\text{-}4}\left(Q^2\right) + \delta^2_{\rm tw\text{-}6}(\mu^2) F^{\gamma\pi}_{\rm tw\text{-}6}\left(Q^2\right) \tag{14}$$

and are displayed in Fig. 3 for the scaled TFF in comparison with various data.

The green strip shows the theoretical uncertainties related to the BMS DAs calculated with QCD sum rules with nonlocal condensates [24]. The induced uncertainty in the low-Q^2 tail is comparable with the errors of the BESIII experiment, while at higher momenta it is smaller than the data errors. The best-fit results, obtained with the pion DA determined in [1], are denoted by the grey solid line (FAPT/LCSR) and the red solid line (FOPT/LCSR in NNLO), respectively. One notices that the analogous prediction obtained with the platykurtic pion DA [26] within FAPT/LCSR (black dashed line) coincides with the outcome based on the best-fit at the level of $\chi^2_{\rm ndf} = 0.57$. This is significant because this DA amalgamates the key features of NLC's parameterized by the vacuum quark virtuality $\lambda_q^2 = 0.45$ GeV^2 (entailing endpoint suppression) with unimodality—in contrast to the bimodal DA's in the BMS domain with $\lambda_q^2 = 0.40$ GeV^2. We remind that unimodality is welcome because this is a prominent characteristic of pion DA's induced by dynamical chiral symmetry breaking and the emergent generation of hadron mass, see for a recent review [28]. However, such DA's have enhanced tails and fail to reproduce the data with a good accuracy [13].

Finally we note that the dashed curves at the bottom of Fig. 3 representing the dominating contribution NNLO$_\beta$ in FOPT (dashed-dotted line) and FAPT (dashed line) yield above

$Q^2 > 2$ GeV^2 comparable results. This makes it apparent that below $Q^2 \lesssim 1$ GeV^2, the RG summation of the radiative corrections prevents the overestimation of the NNLO contribution in the FOPT/LCSR scheme. A detailed comparison with the total NNLO contribution [14, 15] within our scheme would be useful.

6 Conclusions

In this work we have given a brief presentation of our recent detailed analysis in [1] dealing with the extension of the method of LCSR's provided by the implementation of RG summation of QCD radiative corrections to the $\gamma^*\gamma \to \pi^0$ TFF. We showed that this new scheme inevitably leads to a different perturbative expansion. Instead of a power series expansion in terms of the strong coupling, as in FOPT, on which the standard LCSR method is based, the new perturbation theory employs only analytic couplings amounting to an improved version of FAPT [4]. We used the new FAPT/LCSR scheme [1] to calculate the $F^{\gamma\pi}(Q^2)$ TFF and determined the employed nonperturbative parameters $b_2, b_4, \delta^2_{\text{tw-4}}, \delta^2_{\text{tw-6}}$. To this end, we employed the new preliminary BESIII data [10, 11] in combination with the most advanced lattice calculation of the second moment of the pion DA in three loops [25] and found a bimodal DA which provides a good compromise for all these constraints. The obtained TFF predictions presented above agree with reasonable accuracy also with other data up to $Q^2 = 5$ GeV^2 and beyond with an accuracy that supersedes that of calculations within FOPT at $Q^2 < 1$ GeV^2. Remarkably, the TFF predictions in the FAPT/LCSR scheme obtained with the new best-fit pion DA (Fig. 3) almost coincide with those calculated with the platykurtic pion DA [26], while at the same time both of them are within the uncertainty range (shaded green strip) determined by QCD sum rules with nonlocal condensates [24]. This is important because this DA has a unimodal profile that finds support by the findings of other investigations [28–30] which, however, cannot reproduce the Belle data [16] with the same high accuracy because in contrast to the pk DA, the derived pion DAs are endpoint enhanced [13]. The expected high-precision measurements by the Belle-II Collaboration may provide adjudicative constraints to select the most appropriate pion DA.

Acknowledgments

S. V. M. acknowledges support from the Heisenberg-Landau Program 2021.

References

[1] S.V. Mikhailov, A.V. Pimikov, N.G. Stefanis, Phys. Rev. D **103**, 096003 (2021), 2101.12661
[2] I.I. Balitsky, V.M. Braun, A.V. Kolesnichenko, Nucl. Phys. **B312**, 509 (1989)
[3] A. Khodjamirian, Eur. Phys. J. **C6**, 477 (1999), hep-ph/9712451
[4] C. Ayala, S.V. Mikhailov, N.G. Stefanis, Phys. Rev. D **98**, 096017 (2018), [Erratum: Phys. Rev. D 101, 059901 (2020)], 1806.07790
[5] A.P. Bakulev, S.V. Mikhailov, N.G. Stefanis, Phys. Rev. **D72**, 074014 (2005), [Erratum: Phys. Rev. D72, 119908 (2005)], hep-ph/0506311
[6] A.P. Bakulev, S.V. Mikhailov, N.G. Stefanis, Phys. Rev. **D75**, 056005 (2007), [Erratum: Phys. Rev. D77, 079901 (2008)], hep-ph/0607040
[7] A.I. Karanikas, N.G. Stefanis, Phys. Lett. **B504**, 225 (2001), [Erratum: Phys. Lett. B636, no.6, 330 (2006)], hep-ph/0101031

[8] A.P. Bakulev, Phys. Part. Nucl. **40**, 715 (2009), 0805.0829
[9] N.G. Stefanis, Phys. Part. Nucl. **44**, 494 (2013), 0902.4805
[10] C.F. Redmer (BESIII), *Measurement of meson transition form factors at BESIII*, in *13th Conference on the Intersections of Particle and Nuclear Physics (CIPANP 2018) Palm Springs, California, USA, May 29-June 3, 2018* (2018), 1810.00654
[11] M. Ablikim et al., Chin. Phys. C **44**, 040001 (2020), 1912.05983
[12] S.V. Mikhailov, A.V. Pimikov, N.G. Stefanis, Phys. Rev. **D93**, 114018 (2016), 1604.06391
[13] N.G. Stefanis, Phys. Rev. D **102**, 034022 (2020), 2006.10576
[14] J. Gao, T. Huber, Y. Ji, Y.M. Wang (2021), 2106.01390
[15] V.M. Braun, A.N. Manashov, S. Moch, J. Schoenleber (2021), 2106.01437
[16] S. Uehara et al. (Belle), Phys. Rev. **D86**, 092007 (2012), 1205.3249
[17] B. Aubert et al. (BaBar), Phys. Rev. **D80**, 052002 (2009), 0905.4778
[18] N.G. Stefanis, *Pion-photon transition form factor in QCD. Theoretical predictions and topology-based data analysis* (2019), 1904.02631
[19] A.V. Efremov, A.V. Radyushkin, Theor. Math. Phys. **42**, 97 (1980)
[20] G.P. Lepage, S.J. Brodsky, Phys. Rev. **D22**, 2157 (1980)
[21] H.J. Behrend et al. (CELLO), Z. Phys. **C49**, 401 (1991)
[22] J. Gronberg et al. (CLEO), Phys. Rev. **D57**, 33 (1998), hep-ex/9707031
[23] A.P. Bakulev, S.V. Mikhailov, N.G. Stefanis, Phys. Rev. **D67**, 074012 (2003), hep-ph/0212250
[24] A.P. Bakulev, S.V. Mikhailov, N.G. Stefanis, Phys. Lett. **B508**, 279 (2001), [Erratum: Phys. Lett. B590, 309 (2004)], hep-ph/0103119
[25] G.S. Bali, V.M. Braun, S. Bürger, M. Göckeler, M. Gruber, F. Hutzler, P. Korcyl, A. Schäfer, A. Sternbeck, P. Wein, JHEP **08**, 065 (2019), [Addendum: JHEP 11, 037 (2020)], 1903.08038
[26] N.G. Stefanis, Phys. Lett. **B738**, 483 (2014), 1405.0959
[27] N.G. Stefanis, A.V. Pimikov, Nucl. Phys. **A945**, 248 (2016), 1506.01302
[28] C.D. Roberts, D.G. Richards, T. Horn, L. Chang, Prog. Part. Nucl. Phys. **120**, 103883 (2021), 2102.01765
[29] J.H. Zhang, J.W. Chen, X. Ji, L. Jin, H.W. Lin, Phys. Rev. **D95**, 094514 (2017), 1702.00008
[30] X. Ji, Y.S. Liu, Y. Liu, J.H. Zhang, Y. Zhao, Rev. Mod. Phys. **93**, 035005 (2021), 2004.03543

Spectroscopic Properties of Light Baryon

Chandni Menapara* and *Ajay Kumar* Rai

Department of Physics, Sardar Vallabhbhai National Institute of Technology, Surat-395007, Gujarat, India.

Abstract. Hadron Spectroscopy provides a realm to study the internal quark dynamics within the hadrons through phenomenological, theoretical as well as experimental approaches. In the present article, an attempt has been made to exploit the nucleon N resonances using a non-relativistic hypercentral Constituent Quark Model (hCQM). The properties are studied based on the linear nature of confining part of the potential. The 1S-5S, 1P-3P, 1D-2D and 1F states mostly with four star labelled resonances are explored again with the separation of charge states using different constituent quark masses. Also, Regge trajectories for some obtained states are plotted for examining the linear nature.

1 Introduction

Hadronic Physics aims at understanding the hadrons as bound state of quarks and gluons interaction alongwith the excited spectrum. Baryon spectroscopy particularly attempts to study the resonance states of all the possible spin-parity as well as exploring other properties like magnetic moment, decay, etc [1]. Even after the knowledge of confinement and asymptotic freedom, the strong interaction still poses different challenges and puzzles. Experimental facilities worldwide have tried to produce hadrons through hadronization process in order to reveal yet hidden properties. The light sector of baryons even after so many years of findings still provide us with challenges to certainly understand the interaction inside the composite systems. Also, the decay of many heavy unstable baryons lead to the production of many light, strange baryons in the output.

The N baryons with u and d quark combinations with mixed symmetry wavefunction have been observed experimentally with large number of established resonances [2].
$I = \frac{1}{2}$ P (uud, $I_3 = \frac{1}{2}$), N (udd, $I_3 = -\frac{1}{2}$) with $S = \frac{1}{2}$

N and Δs have been studied through photoproduction decays by ELSA[3]. N has been at focus of the facilties at JLab, CLAS, MAMI [4, 5] and others. The upcoming facilities like PANDA at FAIR-GSI is highly aimed for light, strange baryon studies [6, 7]. Theoretical and phenomenological approaches study N such as Isgur-Karl model [8] Goldstone-boson exchange model [10], relativistic approach [9], quark-diquark system along with Gursey-Radicati exchange intercation [11, 12]. Lately, varied approaches based on QCD SUM Rules [16], basis light-front model [17] and light-front relativistic [18], Lattice QCD [19] and covariant Faddeev approach [20] and others based on n and J^P values and trajectories.

*e-mail: chandni.menaparagmail.com

[21, 23], Skyrme model, AdS/QCD, etc.

In the present article, an attempt has been made to apply hypercentral Constituent Quark Model (hCQM) for both the isospin states of N which is based on varied the ground state 1S. The results have been compared with other models as well as our earlier work where the same 1S state parameters i.e. u and d considered with same constituent mass. Also, later Regge trajectory has been plotted for some states for the reference of linear nature of (n,M^2) and (J,M^2).

2 Theoretical Formalism

The non-relativistic formalism is used to explore the resonance states of baryons. The hypercentral Constituent Quark Model (hCQM) is based on parametrizing all the effects within a baryon in the form of constituent quark masses [24]. The three body interaction is taken care of using Jacobi coordinates, ultimately reducing to hyperradius x and hyperangle ξ. The potential consisting of a Coulomb-like part and another confining part depends solely on the hyperradius x here. Jacobi coordinates takes care of the three body dynamics as given below.

$$\rho = \frac{1}{\sqrt{2}}(\mathbf{r_1} - \mathbf{r_2}); \;\; \lambda = \frac{1}{\sqrt{6}}(\mathbf{r_1} + \mathbf{r_2} - 2\mathbf{r_3}) \tag{1}$$

$$x = \sqrt{\rho^2 + \lambda^2}; \;\; \xi = arctan(\frac{\rho}{\lambda}) \tag{2}$$

The Hamiltonian appears to be,

$$H = \frac{P^2}{2m} + V(x) + V_{SD}(x) + V^1(x) \tag{3}$$

Thus the hyper-radial part of the wave-function as determined by hypercentral Schrodinger equation is

$$\left[\frac{d^2}{dx^2} + \frac{5}{x}\frac{d}{dx} - \frac{\gamma(\gamma+4)}{x^2}\right]\psi(x) = -2m[E - V(x)]\psi(x) \tag{4}$$

The confinement realm of the potential is taken to be of linear nature as earlier explored in various studies.

$$V(x) = -\frac{\tau}{x} + \alpha x \tag{5}$$

Also, the splitting of a given state into all possible angular momentum quantum number J has been considered by adding the spin-dependent terms to the above potential,

$$V_{SD}(x) = V_{SS}(x)(\mathbf{S}_\rho \cdot \mathbf{S}_\lambda) + V_{\gamma S}(x)(\gamma \cdot \mathbf{S}) + V_T \times [S^2 - \frac{3(\mathbf{S}\cdot\mathbf{x})(\mathbf{S}\cdot\mathbf{x})}{x^2}] \tag{6}$$

Here, $V_{SS}(x)$, $V_{\gamma S}(x)$ and $V_T(x)$ are spin-spin, spin-orbit and tensor terms respectively [25].

To further refining the resonance masses, a first order correction term with $\frac{1}{m}$ dependence has been included.

$$V^1(x) = -C_F C_A \frac{\alpha_s^2}{4x^2} \tag{7}$$

where C_F and C_A are Casimir elements of fundamental and adjoint representation. α_s is the running coupling constant. The statis potential with leading order contribution followed by relativistic corrections in the power of $\frac{1}{m}$ term has been noted for quarkonia systems [26].

Also, similar study has been undertaken for heavy-light systems as well [27] which has been a key to include the correction term for light systems.

The Schrodinger equation with the hyper-radial part is numerically solved for calculating the excited state masses [28]. Details of the model can be found in our other articles [31–36].

3 Results and Discussions

Using the above potential model, the masses are computed for 1S-5S, 1P-3P, 1D-2D and 1F states. Earlier N^* resonances have been studied on the footings of the same constituent quark mass i.e. 0.290 GeV for both u and d quarks, employing same value for 1S state for all the members [32]. However, current study segregates all the isospin partners using the separate constituent quark mass as $m_u = 0.290$ and $m_d = 0.300$. The excited states are recalculated for both the isospin states and compared with various results in the table [1] below.

The ground state parameters lead to the value of 938 MeV for P and 948 MeV for N, whereas earlier both the masses were 939 MeV such that excited states cannot be separated out. The next four star status N(1440) with $J^P = \frac{1}{2}^+$ is the 2S state and present results 1419 and 1427 are well within the PDG range and other approaches too. Similarly the 3S and 4S states are found to be with good agreement with experimental results.

The very first state in negative parity is N(1520) with $J^P = \frac{3}{2}^-$ is reproduced as 1494 and 1502 respectively for P and N. Also, this state is lower than its spin partner i.e. $J^P = \frac{1}{2}^-$ N(1535) is also consistent with our results as the model is predicting lower mass for higher spin state. Here, it is noteworthy that N(1650) $J^P = \frac{1}{2}^-$ doesn't appear in the present dat as the $\frac{5}{2}^-$ is just 25 MeV away which falls under 1P partners. In case of 2P states, PDG as well as results from BGR collaboration [30] are compared.

N(1720)$\frac{3}{2}^+$ with four star label is in coherence with most of the references given in comparison. 1D$\frac{1}{2}^+$ appears as a higher state than $\frac{3}{2}^+$ and $\frac{5}{2}^+$. Also, states in 2D are found to vary within 100 MeV difference compared to all.

All the negative parity states of 1F have been observed and assigned three and four star status. In the present results, higher spin state with $J^P = \frac{9}{2}^-$ is slightly under-predicted compared to other results and PDG. The two states which doesn't appear in earlier study have been calculated here. The N(2220) 1G $\frac{9}{2}^+$ obtained to be 2371 and 2376 respectively which is differing by 70 MeV from PDG range. Also, the N(2600) is assigned 1H $\frac{11}{2}^-$ is very well reproduced in the current results.

The other models mentioned for the comparison include relativistic interacting quark-diquark model with Gursey Radicati-inspired exchange interaction [12]. The another very recent study of mass spectra of all light strange baryon is given by N. Amiri et. al. using Bethe Ansatz introducing U(7) algebra [13]. A. V. Anisovich et. al. has reproduced the N and Δ spectrum using the multichannel partial wave analysis of pion and photo-induced reactions [14]. The later are quark model using hyperfine interactions due to two-gluon exchange between quarks [22], semi-relativistic constituent quark model [15], chiral quark model [29], mass formula [23] and dynamical chirally improved quarks [30].

Table 1: Masses of P and N Baryons (in MeV)

State	J^P	P	N	[32]	PDG[2]	Status	[12]	[13]	[14]	[29]	[22]	[15]	[23]	[30]
1S	$\frac{1}{2}^+$	938	948	939	938	****	939	939	960	938	939	938	938	1000 ± 18
2S	$\frac{1}{2}^+$	1419	1427	1425	1410-1470	****	1511	1450.8	1430	1444	1462	1492	1440	
3S	$\frac{1}{2}^+$	1683	1691	1721	1680-1740	***	1776	1699	1710	1832	1748	1763	1710	
4S	$\frac{1}{2}^+$	2028	2035	2089	2050-2150	***								
5S	$\frac{1}{2}^+$	2436	2441	2515										
1P	$\frac{1}{2}^-$	1520	1523	1565	1515-1545	****	1537	1536.1	1501	1567	1497	1511	1538	1539 ± 69
1P	$\frac{3}{2}^-$	1494	1502	1535	1510-1520	****	1537	1550.3	1517	1567	1548	1511	1523	1634 ± 44
1P	$\frac{5}{2}^-$	1461	1469	1495	1665-1680	****					1655		1678	
2P	$\frac{1}{2}^-$	1838	1842	1898	1880-1910	**	1888		1895				2090	1895 ± 128
2P	$\frac{3}{2}^-$	1810	1816	1865	1850-1920	***			1880					1982 ± 128
2P	$\frac{5}{2}^-$	1773	1779	1820										
3P	$\frac{1}{2}^-$	2254	2260	2288										
3P	$\frac{3}{2}^-$	2222	2226	2251	2150				2150			2080		
3P	$\frac{5}{2}^-$	2175	2181	2202										
1D	$\frac{1}{2}^+$	1794	1800	1849	1830-1930	***	1890		1870	1887				1848 ± 120
1D	$\frac{3}{2}^+$	1763	1768	1815	1660-1750	****	1648	1682.7	1690		1734	1725	1700	1773 ± 91
1D	$\frac{5}{2}^+$	1722	1729	1769	1680-1690	****	1799	1704.2	1689	1689.8	1738	1735	1683	
1D	$\frac{7}{2}^+$	1671	1678	1712							1943		1990	
2D	$\frac{1}{2}^+$	2211	2213	2244							2060			1998 ± 59
2D	$\frac{3}{2}^+$	2174	2178	2204	1890-1950	****					2058		1900	2298 ± 191
2D	$\frac{5}{2}^+$	2123	2128	2150	1950-2150				2090		2007		2000	
2D	$\frac{7}{2}^+$	2061	2067	2083	1950-2100				2060				1990	
1F	$\frac{3}{2}^-$	2145	2139	2167	2060-2160	***							2080	2296 ± 129
1F	$\frac{5}{2}^-$	2090	2089	2112	2030-2200	***							2220	
1F	$\frac{7}{2}^-$	2022	2027	2045	2140-2220	****		2135	2180				2150	
1F	$\frac{9}{2}^-$	1940	1951	1963	2250-2320	****		2270	2280	2232.4		2240		
1G	$\frac{9}{2}^+$	2371	2376		2200-2300	****		2273	2200	2174.3			2245	
1H	$\frac{11}{2}^-$	2766	2769		2550-2750	***		2620		2534.5			2650	

4 Regge Trajectory

Regge trajectories have been one of the useful tools in spectroscopic studies. The plot of total angular momentum J and principle quantum number n against the square of resonance mass M^2 have been in the form of linear equation.

$$J = aM^2 + a_0 \tag{8a}$$

$$n = bM^2 + b_0 \tag{8b}$$

The non-intersecting and linearly fitted lines have been in accordance with theoretical and experimental data in many studies. These plots might be helpful in predicting the correct spin-parity assignment of a given state. For the present data, we have plotted (n,M^2) and (J,M^2) with natural parity and have been found to follow the linear trend.

Figure 1. Regge trajectory N for $n \rightarrow M^2$

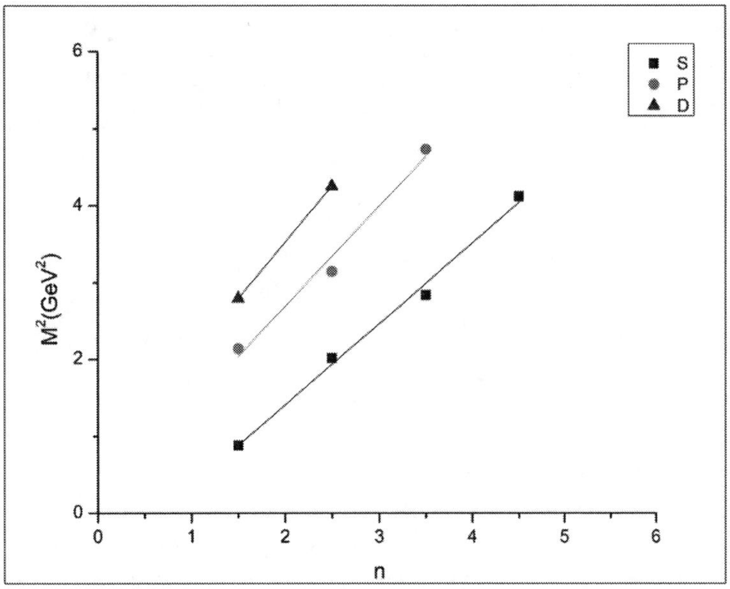

5 Magnetic Moment

Baryon magnetic moment and transition magnetic moment plays a crucial role in providing information regarding the structure and shape of baryon. Magnetic moment shall have contribution from many other effects within the baryon as sea quark, valence quark, orbital etc [37]. Various models have contributed to obtaining the octet and decuplet baryons' magnetic moment [38, 39].

$$\mu_B = \sum_q \left\langle \phi_{sf} | \mu_{qz} | \phi_{sf} \right\rangle \tag{9}$$

where ϕ_{sf} is the spin-flavour wave function.

$$\mu_{qz} = \frac{e_q}{2m_q^{eff}} \sigma_{qz} \tag{10}$$

Figure 2. Regge trajectory N for $J \rightarrow M^2$

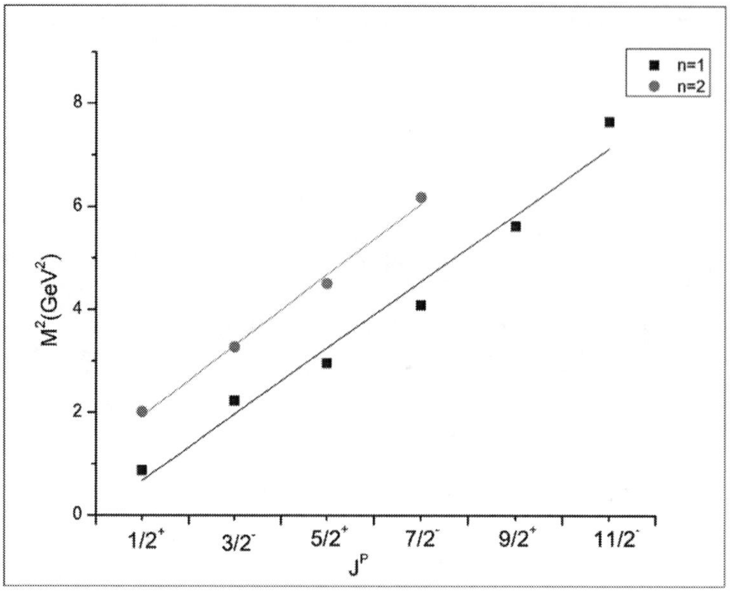

The effective mass of quark m_q^{eff} is calculated using

$$m_q^{eff} = m_q(1 + \frac{\langle H \rangle}{\sum_q m_q}) \tag{11}$$

Using the above equations, magnetic moment have been calculated for P and N with effective quark mass, other than the constituent quark mass. For proton, $\mu_B = 3.03\mu_N$ as compared to $2.79\mu_N$ experimental value. For N, $\mu_B = -1.97\mu_N$ compared to -1.91μ_N experimental value.

6 Conclusion

The present study has employed a non-relativistic, hypercentral Constituent Quark model (hCQM) with potential with linear term, spin-dependent terms as well as first order correction term. It has been attempted to segregate the isospin states of nucleons by introducing a different constituent quark mass for u and d quarks which provided with 938 MeV and 948 MeV respectively. Similarly all the excited states are reproduced for both and comparison shows an overall good agreement with experimental results and various approaches. One state each for 1G and 1H has been obtained as well. The Regge trajectories show a well fitted linear curves for (n,M^2) and (J, M^2) where natural parities upto $\frac{9}{2}$ and $\frac{11}{2}$ are included. In addition, the magnetic moment values have been revised for P and N using the effective quark mass which are quite similar to PDG values.

Acknowledgement

Ms. Chandni Menapara would like to acknowledge the support from the Department of Science and Technology (DST) under INSPIRE-FELLOWSHIP scheme for pursuing this work.
Authors are thankful to the organisers of a Virtual Tribute to Quark Confinement and Hadron Spectrum, 2021 for giving us the opportunity to participate and present our work.

References

[1] V. Crede and W. Roberts, Rept. Prog. Phys., **76**: 076301 (2013), arXiv:1302.7299 [hep-ph]

[2] P.A. Zyla *et al.*, (Particle Data Group), Prog. Theor. Exp. Phys. 2020, 083C01 (2020) and 2021 update

[3] U. Thoma *et al.* (CB-ELSA Collaboration), (2007), arXiv: 0707.3592v4 [hep-ph]

[4] I. G. Aznauryan et al, Int. J. Mod. Phys. E, **22** 1330015 (2013), arXiv:1212.4891 [hep-ph]

[5] V. I. Mokeev, V. D. Burkert, D. S. Carman, L. Elouadrhiri et al, Phys. Rev. C, **93** 025206 (2016)

[6] B. Singh et al. [PANDA Collaboration], J. Phys. G: Nucl. Part. Phys. **46**, 045001 (2019), Phys. Rev. D **95**, 3, 032003 (2017), Eur. Phys. J. A **52** 10, 325 (2016); Eur. Phys. J. A **51**, 107 (2015); Nucl. Phys. A **954**, 323-340 (2016)

[7] G. Barruca at al., [PANDA Collaboration], G. Barucca et al. (PANDA Collaboration), Eur. Phys. J. A **57**, 1-36 (2021); Eur. Phys. J. A **57**, 149 (2021); Eur. Phys. J. A **57**, 1-22 (2021); Eur. Phys. J. A **55**, 42 (2019);

[8] N. Isgur and G. Karl, Phys. Rev. D. **18**, 4187 (1978)

[9] S. Capstick and N. Isgur, Phys. Rev. D **34**, 2809 (1986)

[10] L. Ya. Glozman, Z. Papp and W. Plessas, Phys. Lett. B **381**, 311 (1996)

[11] E. Santopinto, Phys. Rev. C **72**, 022201(R) (2005)

[12] E. Santopinto and J. Ferretti, Phys. Rev. C **92**, 025202 (2015)

[13] N. Amiri, M. Ghapanvari and M. A. Jafarizadeh, Eur. Phys. J. Plus **141**, 136 (2021)

[14] A.V. Anisovich, R. Beck, E. Klempt, V.A. Nikonov, A.V. Sarantsev and U. Thoma, Eur. Phys. J. A **48**, 15 (2012)

[15] M. Aslanzadeh and A. A. Rajabi, Int.J.Mod.Phys. **E26**, 1750042 (2017)

[16] K. Azizi, Y. Sarac and H. Sundu, J.Phys.G **47**, 9, 095005 (2020) arXiv:1909.03323v2 [hep-ph]

[17] J.P.Vary *et al.*, Few Body Syst. **59**, 56 (2018), arXiv:1804.07865 [nucl-th]

[18] I. G. Aznauryan and V. D. Burkert, Phys. Rev. C 85, 055202 (2012), arXiv:1201.5759 [hep-ph]

[19] R. G. Edwards, J. J. Dudek, D. G. Richards, and S. J. Wallace Phys. Rev. D **84**, 074508 (2011)

[20] H. Sanchis-Alepuz and C. S. Fischer, Phys. Rev. D 90, no. 9, 096001 (2014)

[21] Y. Chen, B.Q. Ma, Chin. Phys. Lett. **25**, 3920 (2008)

[22] Y. Chen, B.-Q. Ma, Nucl. Phys. A **831**, 1 (2009)

[23] E. Klempt, arXiv:nucl-ex/0203002 (2002)

[24] M. M. Giannini, E. Santopinto, Chin. J. Phys., **53**, 020301 (2015)

[25] M. B. Voloshin, Prog. Part. Nucl. Phys. **61**, 455-511 (2008)

[26] Y. Koma, M. Koma and H. Wittig, Phys. Rev. Lett. **97**, 122003 (2006)

[27] Z. Shah, K. Thakkar, A. K. Rai and P. C. Vinodkumar, Eur. Phys. J. A **52** 313 (2016); Chin. Phys. C **40**, 12, 123102 (2016)

[28] W. Lucha, F. schoberls, Int. J. Modern Phys. C. **10** 607 (1997)

[29] Z. Ghalenovi and M. Moazzen, Eur. Phys. J. Plus **132**, 354 (2017)

[30] G. P. Engel, C. B. Lang, D. Mohler and A. Schaefer [BGR Collaboration] Phys. Rev. D. **87**, 074504 (2013)

[31] K. Gandhi and A. K. Rai, Eur. Phys. J. Plus, **135**, 213 (2020)

[32] Z. Shah, K. Gandhi and A. K. Rai, Chin. Phys. C **43**, 034102 (2019), arXiv:1812.04858 [hep-ph]
[33] C. Menapara, Z. Shah and A. K. Rai, Chin. Phys. C **45**, 023102 (2021)
[34] C. Menapara and A. K. Rai, Chin. Phys. C **45**, 063108 (2021)
[35] C. Menapara, Z. Shah and A. K. Rai, AIP Conf.Proc. **2220**, 140014 (2020)
[36] C. Menapara and A. K. Rai, DAE Symp.Nucl.Phys. **64** 673-674 (2020)
[37] H. Dahiya and M. Gupta, Phys. Rev. D **67**, (2003) 114015
[38] K. Thakkar, B. Patel, A. Majethiya and P. C. Vinodkumar, PRAMANA J of Physics **77**, (2011) 1053-1067
[39] Fayyazuddin and M. J. Aslam, arXiv:2011.06750 [hep-ph] (2020)

Heavy-quark mesons in the diabatic approach: string breaking and the quarkoniumlike spectrum

Roberto Bruschini[1,*]

[1]Unidad Teórica, Instituto de Física Corpuscular (Universidad de Valencia–CSIC), E-46980 Paterna (Valencia), Spain

Abstract. The Born-Oppenheimer approximation provides a description of heavy-quark mesons firmly based on lattice QCD, but its validity is limited to the lightest states lying far below the first open-flavour meson-meson threshold. This limitation can be overcome in the diabatic framework, a formalism first introduced in molecular physics, where the dynamics is encoded in a potential matrix whose elements can be derived from unquenched lattice QCD studies of string breaking. The off-diagonal elements of the potential matrix provide interaction between heavy quark-antiquark and meson-meson pairs, from which the mixing of quarkonium states with molecular components and the OZI-allowed strong decay widths are directly calculated. This allows for a QCD-based unified description of conventional quarkonium and unconventional mesons containing quark-antiquark and meson-meson components, what has proved to be successful for charmoniumlike and bottomoniumlike resonances.

1 Introduction

For many years the spectrum of heavy-quark mesons has been successfully described by the quark model, in its nonrelativistic [1, 2] and semi-relativistic [3] realizations, from the bound states of a heavy quark-antiquark ($Q\bar{Q}$) pair. This is until the discovery of $X(3872)$ [4], which ushered heavy-quark meson spectroscopy in a new era. In fact this state, and many others discovered afterwards [5], have mass and decay properties that clearly defy the quark model predictions. Finding a unified description of the conventional and unconventional states from QCD, the theory of strong interactions, has been a long standing problem in heavy-quark meson spectroscopy ever since.

A general feature of the unconventional heavy-quark meson states is that their mass lies close to some open-flavor meson-meson threshold with the same quantum numbers. This fact seems to indicate that a proper description of them must at least incorporate meson-meson configurations on top of the quark-antiquark component. Actually, the effect of open flavor meson-meson thresholds on the quarkonium spectrum has been extensively studied through phenomenological models such as the Cornell or $^3P_0$ models [1, 2, 6–18], however these approaches lack a direct connection with QCD.

A connection with QCD can be found using the Born-Oppenheimer (B-O) approximation [19–21]. More precisely, because the mass of the heavy quarks is much larger than the QCD

*e-mail: roberto.bruschini@ific.uv.es

energy scale associated to the gluon field, heavy-quark meson states can be associated to solutions of a Schrödinger-like equation with effective potentials. These potentials correspond to the energy levels of a stationary gluon field in presence of static Q and $\bar{Q}$ sources, and can be calculated directly in lattice QCD. In this way conventional quarkonium $Q\bar{Q}$ states can be identified with the solutions of the ground state (Cornell) potential, and quarkonium hybrids $Q\bar{Q}g$ (where g stands for a gluon) are described through the solutions of the excited state potentials.

However, it is customary to neglect configuration mixing when applying the B-O approximation in QCD, which limits its validity when light quark flavors are included in the calculation of the static potentials [21]. Concretely, lattice QCD calculations of the energy levels of stationary gluon *and light quark* fields in presence of static $Q\bar{Q}$ sources have shown that as the energy of a quarkonium $Q\bar{Q}$ configuration approaches the mass of an open-flavor meson-meson pair a sizable mixing between the two of them (string breaking) takes place [22, 23]. Therefore, the validity of the B-O approximation without configuration mixing is limited to states whose mass lies far below any open-flavor meson-meson threshold.

In this talk we shall overcome this limitation using the diabatic approach, first developed in molecular physics [24], to obtain a QCD-based description of heavy-quark mesons taking into account the configuration mixing between $Q\bar{Q}$ and meson-meson. In this approach the $Q\bar{Q}$–meson-meson mixing induced by string breaking is incorporated together with the confining Q-$\bar{Q}$ interaction in a potential matrix that is directly connected to the static energy levels calculated in lattice QCD. The solution of the multichannel Schrödinger equation with this potential matrix gives the spectrum of both conventional (containing mainly $Q\bar{Q}$) and unconventional (with significant meson-meson component) heavy-quark meson states as well as their OZI-allowed decay widths. This framework is then applied to study charmoniumlike [25, 26] and bottomoniumlike [27] resonances.

The contents of the talk are organized as follows. In Sec. 2 we recap the main steps in the derivation of the B-O approximation. In Sec. 3 we introduce the diabatic approach in QCD and use it to explore the $Q\bar{Q}$–meson-meson mixing. The results of a diabatic study of the charmoniumlike and bottomoniumlike spectra are recollected in Sec. 4, and the main conclusions are summarized in Sec. 5.

2 Born-Oppenheimer approximation

The B-O approximation, first developed for the description of molecules [28], is particularly useful to describe physical systems where there is a clear separation between "heavy" and "light" degrees of freedom (d.o.f.). In molecules this division is provided by the electrons being several thousands times lighter than the nuclei, while in heavy-quark meson the distinction is provided by the QCD energy scale Λ_{QCD} associated to the gluon fields. So heavy quark flavors such as charm (c) and bottom (b), whose mass is much bigger than Λ_{QCD}, are going to be treated as "heavy" d.o.f. whereas gluons (g) and light quark flavors up (u), down (d), and strange (s) are going to be treated as "light" fields.

The huge separation of the energy scales of the light and heavy fields is crucial to the assumption underlying the B-O approximation: that the evolution of the heavy d.o.f. is so slow with respect to that of the light fields that the dynamics of the latter can be solved for *static* heavy quark sources. So in the B-O approximation a heavy-quark meson state $|\psi\rangle$ (containing both heavy and light fields) is a solution of the characteristic equation

$$H\,|\psi\rangle = E\,|\psi\rangle \tag{1}$$

with E the energy of the state and H the Hamiltonian

$$H = K_{Q\bar{Q}} + H^{\mathrm{lf}}_{Q\bar{Q}} \tag{2}$$

where $K_{Q\bar{Q}}$ is the kinetic energy of the heavy quark-antiquark pair and $H^{\rm lf}_{Q\bar{Q}}$ a Hamiltonian operator for the light fields that in general depends on the $Q\bar{Q}$ coordinates but does not contain any derivative with respect to them.

The first step in the solution of (1) is to solve the dynamics of the light fields for static heavy quarks. Formally, this corresponds to putting $K_{Q\bar{Q}}$ to zero, so that the coordinates of the $Q\bar{Q}$ cease being a dynamical variable and become instead a parameter of the residual Hamiltonian for the light fields: $H^{\rm lf}_{Q\bar{Q}} \to H^{\rm lf}_{\rm static}(r)$ where r is the relative $Q\bar{Q}$ distance. The solution of the secular equation for $H^{\rm lf}_{\rm static}(r)$,

$$H^{\rm lf}_{\rm static}(r)\,|\zeta_i(r)\rangle = V_i(r)\,|\zeta_i(r)\rangle \tag{3}$$

with $i = 0, 1, \ldots$ labeling the ground and excited state solutions, yields the light fields states $|\zeta_i(r)\rangle$ and the corresponding *static energy levels* $V_i(r)$, which can be calculated in lattice QCD.

After having solved the light fields for static heavy quarks, the motion of the heavy d.o.f. is reintroduced using the so-called *adiabatic expansion* of the heavy-quark meson state

$$|\psi\rangle = \sum_i \int \mathrm{d}r\, \psi_i(r)\,|r\rangle\,|\zeta_i(r)\rangle \tag{4}$$

where $|r\rangle$ is the quantum state indicating the $Q\bar{Q}$ relative position r and $\psi_i(r)$, the coefficients of the expansion, are called the *adiabatic wave functions* of the heavy-quark meson state. It is important to highlight that the light-field states $|\zeta_i(r)\rangle$ used in the adiabatic expansion (4) are calculated at the same position r of the heavy quarks, what makes it the most natural expansion in the idealized situation where the light fields states adjust instantaneously to the motion of the $Q\bar{Q}$ (hence the name).

Plugging the adiabatic expansion back into (1), and after some manipulations, one can see that the evolution of the adiabatic wave functions $\psi_i(r)$ is determined by the Schrödinger-like equation

$$\sum_j \Big[-\frac{\hbar^2}{2\mu_{Q\bar{Q}}}(\delta_{ij}\nabla + \tau_{ij}(r))^2 + (V_i(r) - E)\delta_{ij}\Big]\psi_i(r) = 0, \tag{5}$$

where the adiabatic potentials $V_i(r)$ are nothing but the static energy levels calculable in lattice QCD. One should realize though that Eq. (5) is not a proper Schrödinger equation because of the non-adiabatic coupling terms $\tau(r)_{ij}$ appearing in the kinetic energy term. These terms, defined as $\tau(r)_{ij} \equiv \langle\zeta_i(r)|\nabla\zeta_j(r)\rangle$, reflect the general nontrivial dependence of the light-field states on the motion of the heavy d.o.f. and couple the various adiabatic wave functions to each other. In B-O it is customary to neglect these terms, what corresponds to the so-called single-channel approximation. Notice that this approximation can be deemed valid only if no significant mixing between the light-field states takes place.

The adiabatic potentials have been first calculated in quenched lattice QCD, this is, without including light quark flavors in the calculation of the energy levels (see for example [19, 20]). They are represented in Fig. 1. The ground state potential shows an attractive Coulomb behavior at short distances and a linear confining shape at large distances, corresponding to the quarkonium $Q\bar{Q}$ configuration. The first excited state on the other hand, while still being confining at large distances, is repulsive at small distances, and is usually associated to the first quarkonium hybrid $Q\bar{Q}g$ configuration.

Thus the B-O approximation can be used to obtain a description of quarkonium and quarkonium hybrid states that is firmly based on quenched lattice QCD, but what happens if light quark flavors are introduced among the light fields? The adiabatic potentials recently

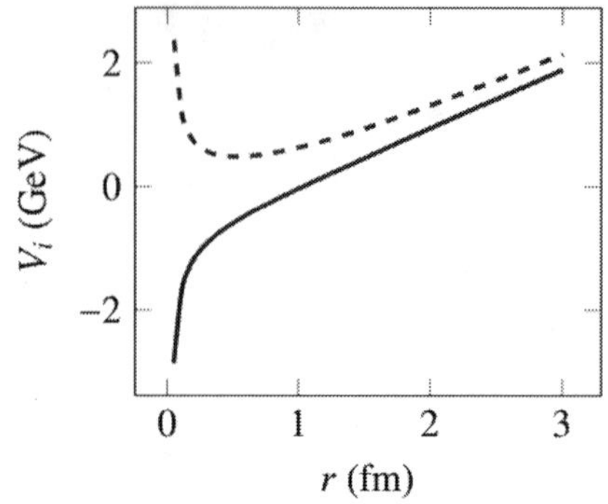

Figure 1. Representation of the adiabatic potentials in quenched lattice QCD. Solid line stands for the ground state (quarkonium) potential and dashed line for the first excited (hybrid) potential.

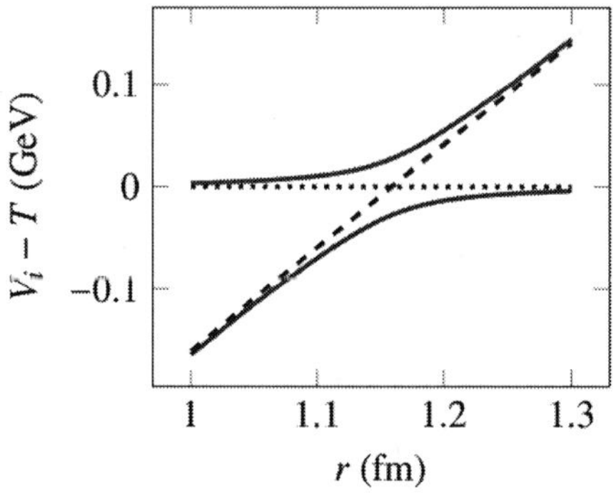

Figure 2. Representetion of the adiabatic potentials in unquenched lattice QCD "normalized" with respect to the threshold mass T. Solid lines stand for the ground state and first excited potentials, dashed line for the quarkonium potential, and dotted line for the threshold energy.

calculated in unquenched (with gluons and light quarks) lattice QCD [22, 23], and represented in Fig. 2, show that the single-channel approximation is untenable when open-flavor meson-meson configurations start to play some role. In fact one can see in Fig. 2 that as r increases the energy of the ground state shifts progressively from linearly increasing, corresponding to the confining $Q\bar{Q}$ potential, to the constant threshold mass value, corresponding to a noninteracting open-flavor meson-meson configuration. At the same time an excited state potential showing the opposite behavior appears above threshold. Notice that these two potentials, unlike the pure $Q\bar{Q}$ and meson-meson energies also represented in Fig. 2, do not cross each other. This phenomenon, referred to as avoided crossing, reflects the creation of a light quark-antiquark pair from the sea by string breaking and their recombination with the heavy quark-antiquark to create an open-flavor meson-meson pair. This configuration mixing between $Q\bar{Q}$ and meson-meson clearly breaks the single-channel approximation and therefore limits its validity to the energy region far below threshold.

3 Diabatic approach

To solve the complications that arise in the B-O approximation when dealing with string breaking, we use a more convenient expansion for the heavy-quark meson state $|\psi\rangle$: the *diabatic expansion*

$$|\psi\rangle = \sum_i \int \mathrm{d}r\, \widetilde{\psi}_i(r, r_0)\, |r\rangle\, |\zeta_i(r_0)\rangle\,, \tag{6}$$

whose crucial difference with respect to the adiabatic expansion (4) is that the light-field states are now calculated for a fixed distance r_0 instead of the $Q\bar{Q}$ distance r. So if we choose r_0 far from the avoided crossing then the expansion coefficients, the diabatic wave functions $\widetilde{\psi}_i(r, r_0)$, correspond to either pure $Q\bar{Q}$ or meson-meson: $\widetilde{\psi}_i(r, r_0) \to \psi_{Q\bar{Q}}(r), \psi_{M\bar{M}}(r)$. In other words, in the diabatic framework the state $|\psi\rangle$ is expanded in $Q\bar{Q}$ and meson-meson components, instead of adiabatic channels that correspond to a mixing of both.

With the diabatic expansion, Eq. (1) becomes the diabatic Schrödinger equation

$$\left[\begin{pmatrix} -\frac{\nabla^2}{2\mu_{Q\bar{Q}}} & 0 \\ 0 & -\frac{\nabla^2}{2\mu_{M\bar{M}}} \end{pmatrix} + \begin{pmatrix} V_{Q\bar{Q}}(r) & V_{\mathrm{mix}}(r) \\ V_{\mathrm{mix}}(r) & T_{M\bar{M}} \end{pmatrix} - E\right]\begin{pmatrix} \psi_{Q\bar{Q}}(r) \\ \psi_{M\bar{M}}(r) \end{pmatrix} = 0, \tag{7}$$

where $\mu_{Q\bar{Q}}$ and $\mu_{M\bar{M}}$ are respectively the $Q\bar{Q}$ and meson-meson reduced masses, $V_{Q\bar{Q}}(r)$ is the $Q\bar{Q}$ potential, $T_{M\bar{M}}$ the open-flavor meson-meson threshold mass, and $V_{\mathrm{mix}}(r)$ the $Q\bar{Q}$–meson-meson mixing potential. Unlike its counterpart Eq. (5), this is a standard multichannel Schrödinger equation with a *diabatic potential matrix* coupling the two channels through the off-diagonal element $V_{\mathrm{mix}}(r)$.

The diabatic potential matrix enjoys a direct connection with the energy levels calculated in lattice QCD, in fact, its eigenvalues are nothing but the unquenched adiabatic potentials [24]. Using this correspondence and the calculated energy levels it can be shown that near the avoided crossing the mixing potential $V_{\mathrm{mix}}(r)$ must be proportional to the energy gap between the adiabatic potentials, whereas far from the avoided crossing its modulus must decrease quite rapidly [25]. The simplest parametrization that respects this requirements is a Gaussian one

$$|V_{\mathrm{mix}}(r)| = \frac{\Delta}{2} e^{-\frac{(V_{Q\bar{Q}}(r) - T_{M\bar{M}})^2}{2\Lambda^2}} \tag{8}$$

where Δ is the energy gap and Λ an energy scale regulating the size of the region where the mixing takes place. Notice that the correspondence between the eigenvalues of the diabatic potential matrix and the adiabatic potentials fixates only the modulus of the mixing potential, not its sign.

The solution of the diabatic Schrödinger equation (7) gives a unified description of heavy-quark mesons composed in general of $Q\bar{Q}$ and open-flavor meson-meson components that is firmly based on lattice QCD, both above and below threshold. Comparison of the resulting spectrum with that of bare quarkonium states shows that the mixing has multiple effects. In general, heavy-quark meson states above threshold acquire decay widths to open-flavor meson-meson pairs, and their mass is shifted due to the coupling with the continuum of meson-meson states. Below threshold, the coupling with off-shell meson-meson configurations makes that all heavy-quark meson states possess some molecular component, and some unconventional state (usually with a significant molecular component) may appear near some threshold.

4 Results

We shall now recollect the results of a diabatic study of the bottomoniumlike [27] and charmoniumlike [25, 26] spectra. Let us notice that the parameters we use in the diabatic potential matrix are not actually fitted on lattice data but are instead a mixture of lattice and phenomenology. There are two reasons for this. First, unquenched lattice QCD energy levels

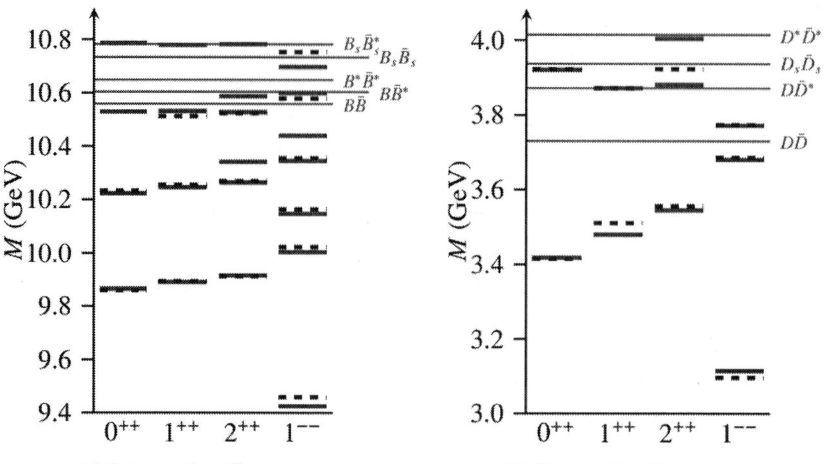

(a) Bottomoniumlike spectrum. (b) Charmoniumlike spectrum.

Figure 3. Spectrum of bottomoniumlike (left) and charmoniumlike (right) meson states. Thick solid lines stand for states calculated in the diabatic approach, thick dashed lines stand for well-established experimental candidates, and long thin horizontal lines stand for open-flavor meson-meson thresholds.

Table 1. Calculated OZI-allowed strong decay widths of bottomoniumlike states above threshold. When available, the experimental total width $\Gamma_{\text{total}}^{\text{Expt}}$ of the corresponding experimental candidate is reported. Masses and widths in MeV units.

J^{PC}	M	$\Gamma_{B\bar{B}}$	$\Gamma_{B\bar{B}^*}$	$\Gamma_{B^*\bar{B}^*}$	$\Gamma_{B_s\bar{B}_s}$	$\Gamma_{\text{total}}^{\text{Theor}}$	$\Gamma_{\text{total}}^{\text{Expt}}$
0^{++}	10785.8	1.6		5.3	0.7	7.6	
1^{++}	10778.9		0.2	1.7		1.9	
2^{++}	10588.4	4.3				4.3	
2^{++}	10782.3	5.4	1.5	21.0	10.4	38.3	
1^{--}	10599.8	21.9				21.9	20.5 ± 2.5
1^{--}	10697.0	2.0	1.0	38.0		41.0	

are only currently available for bottomonium and not for charmonium. Second, a successful description of the low-lying conventional quarkonium states is essential for the description of the unconventional ones, as it depends crucially on the position of the underlying $Q\bar{Q}$ bound states with respect to the open-flavor meson-meson thresholds. Hence, the parameters of the $Q\bar{Q}$ potential are taken from phenomenology, the threshold values are taken as the sum of the corresponding experimental meson masses, and the parameters of the mixing potential are taken from lattice calculations in the bottomoniumlike case, or fitted from the mass of $X(3872)$ in the charmoniumlike case. Notice also that the results presented here have little or no dependence on the particular choice of the parametrization for the mixing potential (8), as long as it respects the conditions of its modulus being approximately maximum near the avoided crossing and vanishing away from it.

The calculated spectra are illustrated in Fig. 3, while the calculated OZI-allowed strong decay widths of quarkoniumlike states above threshold are reported in Tables 1 and 2. From them we can see that the diabatic approach can give a reasonable and unified account of the

Table 2. Calculated OZI-allowed strong decay widths of charmoniumlike states above threshold. When available, the experimental total width $\Gamma^{\rm Expt}_{\rm total}$ of the corresponding experimental candidate is reported. Masses and widths in MeV units.

J^{PC}	M	$\Gamma_{D\bar{D}}$	$\Gamma_{D\bar{D}^*}$	$\Gamma_{D_s\bar{D}_s}$	$\Gamma^{\rm Theor}_{\rm total}$	$\Gamma^{\rm Expt}_{\rm total}$
0^{++}	3920.9	0.6			0.6	
2^{++}	3881.1	49.5	0.4		49.9	35.3 ± 2.8
2^{++}	4003.9	4.8	6.3	3.5	14.5	
1^{--}	3771.7	20.2			20.2	27.2 ± 1.0

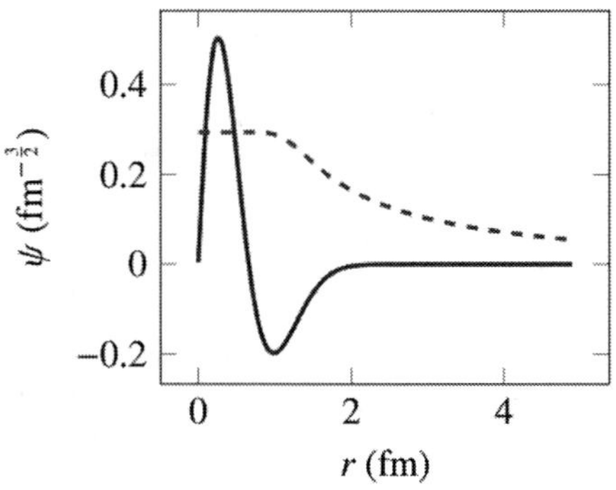

Figure 4. Diabatic wave function of $X(3872)$. The solid and dashed lines stand for the $c\bar{c}$ and $D\bar{D}^*$ components respectively.

spectrum of conventional and unconventional quarkoniumlike states. We see that this is composed of conventional quarkonium states plus a certain number of unconventional ones lying close to some open-flavor meson-meson threshold. The fact that the predicted widths are fairly close to the few well-established experimental values (within a 10% and 30% accuracy for bottomoniumlike and charmoniumlike states respectively) gives further support to the mixing potential being the main mechanism driving both the appearance of unconventional quarkoniumlike states and the OZI-allowed decay of states above threshold.

It may be interesting to examine in more detail the description of $X(3872)$ that emerges in the diabatic approach. From Fig. 4 it can be seen that the diabatic $X(3872)$ at short distance is made of both $c\bar{c}$ and $D\bar{D}^*$ components in comparable amounts. Then as the distance increases the confined $c\bar{c}$ component vanishes and the long $D\bar{D}^*$ tail dominates. This makes the $X(3872)$ mostly a $D\bar{D}^*$ molecule (97% total $D\bar{D}^*$ probability vs. 3% $c\bar{c}$), with a calculated mean root-mean-square radius of near 11 fm.

5 Summary

To summarize, the Born-Oppenheimer approximation provides a description of conventional quarkonia and quarkonium hybrids that is firmly based on lattice QCD. This description takes the energy levels calculated in quenched lattice QCD as the potentials of a Schrödinger equation whose solution provides the quarkoniumlike meson spectrum and wave functions. The diabatic approach allows to include open-flavor meson-meson components in this description. In this approach the dynamics is governed by a diabatic potential matrix whose eigenvalues are directly connected to the unquenched lattice QCD static potentials. This completely non-perturbative description treats the quark-antiquark–meson-meson mixing from string breaking on the same grounds as the confining quark-antiquark interaction. The diabatic approach has been applied to study bottomoniumlike and charmoniumlike mesons, showing that it can give account of both the spectrum and OZI-allowed decay widths of conventional as well as unconventional states.

This work has been supported by Ministerio de Ciencia e Innovación and Agencia Estatal de Investigación of Spain and European Regional Development Fund Grant PID2019-105439 GB-C21 and by EU Horizon 2020 Grant No. 824093 (STRONG-2020). The author acknowledges a FPI fellowship from Ministerio de Ciencia, Innovación y Universidades of Spain under Grant No. BES-2017-079860.

References

[1] E. Eichten, K. Gottfried, T. Kinoshita, K.D. Lane, T.M. Yan, Phys. Rev. D **17**, 3090 (1978), [Erratum: Phys. Rev. D **21**, 313 (1980)]
[2] E.J. Eichten, K. Lane, C. Quigg, Phys. Rev. D **69**, 094019 (2004)
[3] S. Godfrey, N. Isgur, Phys. Rev. D **32**, 189 (1985)
[4] S.K. Choi et al. (Belle Collaboration), Phys. Rev. Lett. **91**, 262001 (2003)
[5] P.A. Zyla et al. (Particle Data Group), Prog. Theor. Exp. Phys. **2020**, 083C01 (2020)
[6] E.J. Eichten, K. Lane, C. Quigg, Phys. Rev. D **73**, 014014 (2006), [Erratum: Phys. Rev. D **73**, 079903 (2006)]
[7] E. van Beveren, G. Rupp, T.A. Rijken, C. Dullemond, Phys. Rev. D **27**, 1527 (1983)
[8] E. van Beveren, G. Rupp, Prog. Part. Nucl. Phys. **117**, 103845 (2021)
[9] K. Heikkilä, N.A. Törnqvist, S. Ono, Phys. Rev. D **29**, 110 (1984)
[10] T. Barnes, S. Godfrey, E.S. Swanson, Phys. Rev. D **72**, 054026 (2005)
[11] T. Barnes, E.S. Swanson, Phys. Rev. C **77**, 055206 (2008)
[12] M.R. Pennington, D.J. Wilson, Phys. Rev. D **76**, 077502 (2007)
[13] I.V. Danilkin, Y.A. Simonov, Phys. Rev. D **81**, 074027 (2010)
[14] I.V. Danilkin, V.D. Orlovsky, Y.A. Simonov, Phys. Rev. D **85**, 034012 (2012)
[15] J. Ferretti, G. Galatà, E. Santopinto, Phys. Rev. C **88**, 015207 (2013)
[16] J. Ferretti, E. Santopinto, Phys. Rev. D **90**, 094022 (2014)
[17] Z.Y. Zhou, Z. Xiao, Eur. Phys. J. A **50**, 165 (2014)
[18] P.G. Ortega, J. Segovia, D.R. Entem, F. Fernández, Phys. Lett. B **778**, 1 (2018)
[19] K.J. Juge, J. Kuti, C.J. Morningstar, Phys. Rev. Lett. **82**, 4400 (1999)
[20] G.S. Bali, Phys. Rep. **343**, 1 (2001)
[21] E. Braaten, C. Langmack, D.H. Smith, Phys. Rev. D **90**, 014044 (2014)
[22] G.S. Bali, H. Neff, T. Düssel, T. Lippert, K. Schilling (SESAM Collaboration), Phys. Rev. D **71**, 114513 (2005)
[23] J. Bulava, B. Hörz, F. Knechtli, V. Koch, G. Moir, C. Morningstar, M. Peardon, Phys. Lett. B **793**, 493 (2019)
[24] M. Baer, *Beyond Born-Oppenheimer: Electronic Nonadiabatic Coupling Terms and Conical Intersections* (John Wiley & Sons, New York, 2006)
[25] R. Bruschini, P. González, Phys. Rev. D **102**, 074002 (2020)
[26] R. Bruschini, P. González, Phys. Rev. D **103**, 074009 (2021)
[27] R. Bruschini, P. González, Phys. Rev. D **103**, 114016 (2021)
[28] M. Born, R. Oppenheimer, Ann. Phys. (Berlin) **389**, 457 (1927)

Towards the NNLO theory prediction for the width difference $\Delta\Gamma_s$

Vladyslav Shtabovenko[1,*]

[1]Institut für Theoretische Teilchenphysik, Karlsruhe Institute of Technology (KIT), 76128 Karlsruhe, Germany

Abstract. The width difference $\Delta\Gamma_s$ that can be extracted from lifetime measurements of the two mass eigenstates of the $B_s^0 - \bar{B}_s^0$ system is one of the key flavor precision observables and has been experimentally measured at per cent level accuracy. The current theory prediction is much less accurate and a sizable reduction of scale uncertainties can only be achieved by means of evaluating the uncalculated 2- and 3-loop QCD corrections. This is precisely the issue addressed in this work where we report on the results that have been obtained so far and explain some of the technical and conceptual challenges that we encountered in the course of our calculations.

1 Introduction

Precision flavor observables constitute an important tool in the task of probing our understanding of the Standard Model (SM) and thus constraining discrepancies between theory and experiments that may be interpreted as manifestations of beyond Standard Model (BSM) physics. The $B_s^0 - \bar{B}_s^0$ system happens to be an ideal laboratory for such investigations, where we can extract three relevant physical quantities that are equally well accessible to theoretical calculations and experimental measurements. These are the oscillation frequency Δm_s, the width difference $\Delta\Gamma_s$ and CP asymmetry in flavor-specific decays a_{fs}^s.

Theoretical investigations of $\Delta\Gamma_s$ allow us to quantify how well our perturbative calculations describe the physics that arises from the SM flavor sector. In principle, light BSM particles that feebly interact with SM model fields could induce a shift in the SM value of the width difference. However, one normally tends to regard $\Delta\Gamma_s$ as a SM precision probe, since its sensitivity to BSM physics (in particular the effects of heavy particles) is rather limited.

The current status quo is that the experimental value [1–3] for $\Delta\Gamma_s$ from [4]

$$\Delta\Gamma_s^{\text{exp}} = (0.085 \pm 0.004)\ \text{ps}^{-1}. \tag{1}$$

is much more precise than the most up-to-date theory prediction [5–10]

$$\Delta\Gamma_s^{\overline{\text{MS}}} = (0.088 \pm 0.011_{\text{pert.}} \pm 0.002_{B,\tilde{B}_S} \pm 0.014_{\Lambda_{\text{QCD}}/m_b})\ \text{ps}^{-1}, \tag{2}$$

$$\Delta\Gamma_s^{\text{pole}} = (0.077 \pm 0.015_{\text{pert.}} \pm 0.002_{B,\tilde{B}_S} \pm 0.017_{\Lambda_{\text{QCD}}/m_b})\ \text{ps}^{-1}, \tag{3}$$

*e-mail: v.shtabovenko@kit.edu

where we would like to attract the attention of the reader to the first uncertainty (denoted as "pert.") that stems from the uncalculated perturbative QCD corrections to the B_s-meson mixing at 2 and 3 loops. Our project aims at calculating these corrections analytically and providing new theory predictions for $\Delta\Gamma_s$ that should hopefully reduce the existing perturbative uncertainties and allow for more meaningful comparisons between experimental and theoretical values.

2 Calculation

Let us begin by explaining the procedure needed to obtain $\Delta\Gamma_s$. The main ingredient is a matching calculation between two effective theories, where on the $|\Delta B| = 1$ side we have [11]

$$\mathcal{H}_{\text{eff}}^{|\Delta B|=1} = \frac{4G_F}{\sqrt{2}}\Bigg[-\lambda_t^s\Big(\sum_{i=1}^{6} C_i Q_i + C_8 Q_8\Big) - \lambda_u^s \sum_{i=1}^{2} C_i (Q_i - Q_i^u) + V_{us}^* V_{cb} \sum_{i=1}^{2} C_i Q_i^{cu} + V_{cs}^* V_{ub} \sum_{i=1}^{2} C_i Q_i^{uc}\Bigg] + \text{h.c.}, \tag{4}$$

with $\lambda_a^s = V_{as}^* V_{ab}$, where V_{ij} are CKM matrix elements. The Wilson coefficients C_i are calculated from matching SM to the $|\Delta B| = 1$ effective Hamiltonian and are known at NNLO accuracy [12]. Eq. (4) does not include evanescent operators $E[Q_i]$ [13, 14], although they are of course properly taken into account in our calculation. Evanescent operators formally vanish in 4 dimensions, as they are of $O(\varepsilon)$, where ε is the dimensional regularization parameter from $d = 4 - 2\varepsilon$. Their presence encodes the ambiguous nature of 4-dimensional identities such as the Chisholm identity

$$\gamma^\mu\gamma^\nu\gamma^\lambda = g^{\mu\nu}\gamma^\lambda + g^{\nu\lambda}\gamma^\mu - g^{\mu\lambda}\gamma^\nu + i\varepsilon^{\mu\nu\lambda\sigma}\gamma_\sigma\gamma^5 \tag{5}$$

or Fierz relations in d dimensions. The fact that a matrix element of an evanescent operator multiplying an ε pole yields something of $O(\varepsilon^0)$ explains why the treatment of such operators in higher order perturbative calculations requires great care. This is also one of the reasons why the present calculation turns out to be highly nontrivial from the conceptual point of view. We refer to our upcoming publication for the complete list of the $|\Delta B| = 1$ operators relevant for our project at 3 loops. The operator basis at 2 loops is given in [15].

As far as the $|\Delta B| = 2$ side of the matching is concerned, the quantity we need to consider is Γ_{12} from the relation $\Delta\Gamma_s \approx 2|\Gamma_{12}|$ which depends on the absorptive part of a bilocal matrix element constructed from time-ordered insertions of two $\mathcal{H}_{\text{eff}}^{|\Delta B|=1}$. Involving Heavy Quark Expansion (HQE) [16–24] one can rewrite Γ_{12} as [8]

$$\Gamma_{12} = -(\lambda_c^s)^2\Gamma_{12}^{cc} - 2\lambda_c^s\lambda_u^s\Gamma_{12}^{uc} - (\lambda_u^s)^2\Gamma_{12}^{uu}, \tag{6}$$

with

$$\Gamma_{12}^{ab} = \frac{G_F^2 m_b^2}{24\pi M_{B_s}}\left[H^{ab}(z)\langle B_s|Q|\bar{B}_s\rangle + \widetilde{H}_S^{ab}(z)\langle B_s|\widetilde{Q}_S|\bar{B}_s\rangle\right] + O(\Lambda_{\text{QCD}}/m_b), \tag{7}$$

where $z \equiv m_c^2/m_b^2$. Our main interest is devoted to the matching coefficients $H^{ab}(z)$ and $\widetilde{H}_S^{ab}(z)$ which we want to determine at NNLO accuracy by calculating contributions from perturbative QCD corrections to the relevant operators on both sides of the matching. The 4-fermion operators Q and $\tilde{Q}_S$ are defined as

$$Q = \bar{s}_i\gamma^\mu(1-\gamma^5)b_i\,\bar{s}_j\gamma_\mu(1-\gamma^5)b_j, \qquad \widetilde{Q}_S = \bar{s}_i(1-\gamma^5)b_j\,\bar{s}_j(1-\gamma^5)b_i, \tag{8}$$

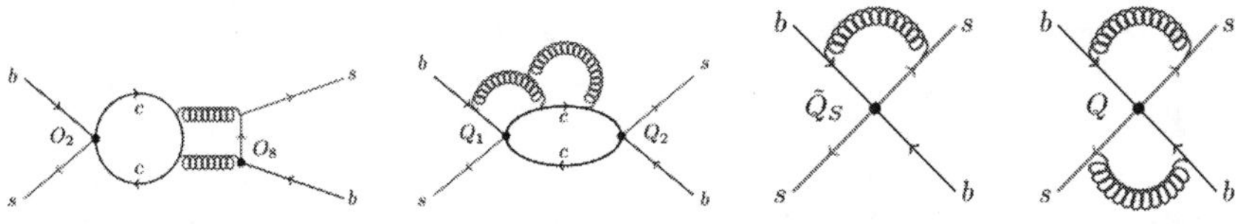

Figure 1. Some of the $|\Delta B| = 1$ and $|\Delta B| = 2$ diagrams entering the $b\bar{s} \to \bar{b}s$ matching calculation. Here (a) and (b) denote $|\Delta B| = 1$ contributions from $Q_2 \times Q_8$ (2 loops) and $Q_1 \times Q_2$ (3 loops), while (c) and (d) represent 1- and 2-loop matrix elements of $|\Delta B| = 2$ operators $\tilde{Q}_s$ and Q.

where i and j denote the fundamental color indices of the b and s quarks. At intermediate stages of our calculation tree-level matrix elements of further $|\Delta B| = 2$ arise so that it is useful to mention them here

$$\widetilde{Q} = \bar{s}_i\gamma^\mu(1-\gamma^5)b_j\, \bar{s}_j\gamma_\mu(1-\gamma^5)b_i \quad Q_S = \bar{s}_i(1-\gamma^5)b_i\, \bar{s}_j(1-\gamma^5)b_j. \tag{9}$$

Apart from that we also must properly account for the $1/m_b$ suppressed operator R_0

$$R_0 = \frac{1}{2}Q + Q_S + \tilde{Q}_S, \tag{10}$$

since, as has already been observed in [5], the unrenormalized matrix element of this operator does not exhibit the $1/m_b$ suppression. We would like to stress that the proper treatment of these contributions from R_0 is crucial to obtain correct matching coefficients at NLO and beyond.

Finally, it is worth mentioning that the $|\Delta B| = 2$ NLO operator basis also comprises a suitable set of evanescent operators which can be found in [15]. The NNLO operator basis will be given in [25].

We choose to carry out our matching calculation on-shell, i. e. with $p_b^2 = m_b^2$ and $p_s^2 = m_s^2$. Since the s quark is taken massless we can choose a kinematics that allows us to set $p_s = 0$, which reduces our calculation to the evaluation of 2-point functions. Even though we do not have external charm legs, m_c still enters our calculation due to internal charm lines induced by $|\Delta B| = 1$ operators and gluon self-energy subloops. From the calculational point of view, the simplest situation arises in the massless charm limit which makes our integrals depend only on a single scale m_b. This is indeed the path we choose at 3 loops, where our matching coefficients are accurate up to $O(z^0)$. At 2 loops we, however, perform an asymptotic expansion in z up to first order.

At 2 loops we consider all possible insertions of $|\Delta B| = 1$ operators, i. e. permutations of $Q_{1,2}$, Q_{3-6} and Q_8 in either of the two vertices. This also comprises the 2-loop double insertion of Q_8 that, strictly speaking, contributes to NNNLO. As far as the 3 loop calculation is concerned, only the double insertion $Q_{1,2}$ (as the dominant contribution) it taken into account. We then match the so-obtained 2- and 3-loop $|\Delta B| = 1$ amplitudes to the 1- and 2-loop $|\Delta B| = 1$ amplitudes respectively, as can be inferred from Fig. 1.

3 Technical details

Let us briefly comment on the calculational setup employed in this computation. All $|\Delta B| = 1$ and $|\Delta B| = 2$ Feynman diagrams are generated with QGRAF [26], while the insertion of Feynman rules and the topology identification are done using our in-house codes Q2E/EXP

[27, 28] or TAPIR [29]. Then, our well-tested CALC framework written in FORM [30] takes care of the amplitude evaluation, asymptotic expansions and further simplifications.

We follow two different strategies to deal with tensor multi-loop integrals appearing in this calculation. In the one case we construct a set of suitable Dirac and color projectors (cf. appendix of [15]) that leave us with scalar loop integrals only. In the other case we calculate all required tensor reduction formulas [31] using FeynCalc [32–34] and Fermat [35] and export them to FORM. We explicitly verify that both approaches lead to the same results, including cases of diagrams with Dirac structures featuring a high number of Dirac matrices (complicated projectors) and diagrams with multiple gluon propagators with a generic gauge parameter (high rank tensor integrals). The implementation of the Feynman rules for various 4-fermion operators was also cross checked (on the level of selected diagrams) using FeynRules [36], FeynArts [37] and FeynCalc. The IBP reduction [38, 39] of the occurring loop integrals is carried out using FIRE [40] and LiteRed [41].

As far as the master integrals are concerned, we mainly need to deal with single scale propagator-type on-shell integrals. Up to 2 loops the corresponding analytic results can be readily extracted from the literature [42, 43]. At 3 loops we encountered many integrals that do not seem to be known in the literature. Luckily, all of them can be evaluated by directly integrating the Feynman parametric representation (derived using FeynCalc) with the aid of HyperInt [44]. To further simplify the HyperInt-results written in terms of Goncharov Polylogarithms (GPLs) [45] containing 6th root of unity, we made use of HyperLogProcedures [46] and PolyLogTools [47]. Finally, every analytic result has been extensively checked numerically with the aid of pySecDec [48–50] and FIESTA [51, 52].

Some new FeynCalc functions that were purposely developed during author's work on this project will become an official part of the upcoming FeynCalc 10 [53] but are already publicly available and documented in the development version of the package. These include `FCFeynmanParametrize` (Feynman parametrization of loop integrals), `FCLoopIntegralToGraph` (graphical representation of loop integrals), `FCLoopFindIntegralMappings` (mappings between master integrals similar to `FindRules` in FIRE) or `FCMatchSolve` (extraction of matching coefficients or renormalization constants).

4 Renormalization and matching

The renormalization of the bare amplitudes on both sides of the matching is rather involved due to the large number of sources of UV-divergences. First of all, we employ the usual QCD renormalization of quark fields, masses, α_s and the gauge parameter ξ, where the latter two are relevant only at 3 loops for the $|\Delta B| = 1$ and at 2 loops for the $|\Delta B| = 2$ amplitudes. In addition to that, we must also take care of the Wilson coefficients of the effective operators that, as it is customary in EFTs, happen to mix under the renormalization. More explicitly, for each of the two effective theories we have

$$(\vec{W}^{\text{bare}}, \vec{W}_E^{\text{bare}}) = (\vec{W}^{\text{ren}}, \vec{W}_E^{\text{ren}})Z \equiv (\vec{W}^{\text{ren}}, \vec{W}_E^{\text{ren}}) \begin{pmatrix} Z_{QQ} & Z_{QE} \\ Z_{EQ} & Z_{EE} \end{pmatrix}, \tag{11}$$

where W_i denote the Wilson coefficients, while the renormalization matrix Z consists of four submatrices describing the mixing of different operator types. In the case of Z_{QQ} and Z_{EE} we have the mixing of physical and evanescent operators among themselves, while Z_{EQ} governs the mixing of evanescent operators into the physical ones. The submatrix Z_{QE} is special in the sense that it contains not only poles but also finite pieces. Those are chosen such, that in the limit $d \to 4$ the contributions of evanescent operators to the renormalized amplitude must vanish.

The $O(\alpha_s^2)$ results for $Z^{|\Delta B|=1}$ can be found in [54]. On the $|\Delta B| = 2$ side of the matching we work with different sets of physical operators and investigate different choices of evanescent operators. Each choice leads to a different 2-loop renormalization matrix, so that we find it more convenient to compute $Z^{|\Delta B|=2}$ in a separate calculation for each relevant operator basis.

While the renormalization procedure renders the bare amplitudes UV-finite, in general they still contain IR divergences. Those can be regularized either dimensionally or by giving the gluon a finite mass. The former makes the evaluation of the amplitudes simpler but complicates the matching procedure. The main reason for this is the appearance of tree level matrix elements of evanescent operators multiplying IR poles, e. g. $\langle E_1^{(1)} \rangle^{(0)} / \varepsilon$, in the renormalized amplitudes. Such quantities are formally of $O(\varepsilon^0)$ so that they do not vanish in the limit $d \longrightarrow 4$. However, if the matching procedure is carried out in a proper way, these terms will cancel in the matching as they should. To this end one has to write down the $|\Delta B| = 2$ matching coefficients as an expansion not only in α_s but also in ε, i. e.

$$H_O = \sum_{i,j \geq 0} H_O^{(i,j)} \varepsilon^j \left(\frac{\alpha_s}{4\pi} \right)^i . \tag{12}$$

Notice that this applies not only to the physical but also to the evanescent operator coefficients. For an NNLO calculation, at LO the matching has to be carried out up to $O(\varepsilon^2)$, while at NLO the $O(\varepsilon^1)$ accuracy is required. Finally, when matching the $O(\alpha_s^2)$ amplitudes, it suffices to determine the Wilson coefficients at $O(\varepsilon^0)$. This procedure is explained in [6] and we have explicitly verified that it works not only at 2 but also at 3 loops.

On the other hand, one might also choose to work with a finite gluon mass, which is unproblematic for diagrams of operator insertions that do not contain 3- and 4-gluon vertices i. e. for most of the 2-loop $|\Delta B| = 1$ and for all 1-loop $|\Delta B| = 2$ diagrams. In this case we need to deal with an additional scale m_g appearing in our master integrals but the matching itself becomes much simpler. This is due to the fact that the UV-renormalized amplitudes are manifestly free of ε poles, so that one can immediately take the limit $d \longrightarrow 4$, in which all matrix elements of evanescent operators vanish. Furthermore, it is worth noting that it is not necessary to compute 2-loop master integrals with full the m_g dependence, as we are only interested in the logarithms of m_g that capture the IR divergences. Here it is fully sufficient to perform an asymptotic expansion in m_g, where we make use of the program ASY.M [55, 56] to reveal the contributing regions of each integral under consideration.

In our calculation we have explicitly verified that for selected 2-loop $|\Delta B| = 1$ correlators the matching with $m_g = 0$ and $m_g \neq 0$ leads to exactly the same matching coefficients. In the 3-loop case we contented ourselves with the dimensional regularization of IR divergences and the fact that all IR poles explicitly cancel in the matching.

5 Results

An overview of the new matching coefficients that were obtained in our calculation is given in Table 1, where the first column denotes relevant operator insertions on the $|\Delta B| = 1$ side of the matching, the second column provides the existing literature result, and the third column highlights our contribution. As one can easily see, in most cases only the so-called n_f-piece, i. e. the fermionic contributions, are currently known in the literature. In this sense our work constitutes an important milestone in the task of obtaining complete predictions that necessarily must incorporate both fermionic and nonfermionic contributions. We would also like to stress that up to $O(z)$ at 2 loops and $O(z^0)$ at 3 loops we explicitly confirm the correctness of all previously computed matching coefficients. This is also true for the n_f-piece of the 3-loop correlator $Q_{1,2} \times Q_{1,2}$ computed in [9], where the check required us to use

Contribution	Literature result	This work
$Q_{1,2} \times Q_{3-6}$	2 loops, z-exact, n_f-part only [10]	2 loops, $O(z)$, full
$Q_{1,2} \times Q_8$	2 loops, z-exact, n_f-part only [10]	2 loops, $O(z)$, full
$Q_{3-6} \times Q_{3-6}$	1 loop, z-exact, full [57]	2 loops, $O(z)$, full
$Q_{3-6} \times Q_8$	1 loop, z-exact, n_f-part only [10]	2 loops, $O(z)$, full
$Q_8 \times Q_8$	1 loop, z-exact, n_f-part only [10]	2 loops, $O(z)$, full
$Q_{1,2} \times Q_{1,2}$	3 loops, $O(\sqrt{z})$, n_f-part only [9]	3 loops, $O(z^0)$, full

Table 1. Overview of the results obtained in the course of this work.

the $|\Delta B| = 1$ basis transformation formulas at $O(\alpha_s^2)$ from [12] in order to convert our result in the operator basis of [11] to the operator basis from [58]. Due to the still ongoing analytic and numerical cross checks of the obtained results, we are not yet in the position to provide an updated theory prediction for $\Delta\Gamma_s$ at NNLO accuracy. The 2-loop matching coefficients for the $Q_{1,2} \times Q_{3-6}$ insertions have been recently made public [15], while the remaining 2- and 3-loop contributions are expected to appear in the near future [25, 59].

In [15] we have shown that the inclusion of nonfermionic contributions to the 2-loop result for $Q_{1,2} \times Q_{3-6}$ induces a sizable shift of $\Delta\Gamma_s$ as compared to the 1-loop result. To this end it is sufficient to consider the ratio between full $\Delta\Gamma_s$ (comprising $Q_{1,2} \times Q_{1,2}$ and $Q_{1,2} \times Q_{3-6}$ contributions up to $O(\alpha_s)$ and $Q_{3-6} \times Q_{3-6}$ contributions up to $O(\alpha_s^0)$) and the $Q_{1,2} \times Q_{3-6}$ piece only. With the 1-loop result for $Q_{1,2} \times Q_{3-6}$ we find

$$\frac{\Delta\Gamma_s^{p,12\times36,\alpha_s^0}}{\Delta\Gamma_s} = 7.6\% \quad \text{(pole)}, \qquad \frac{\Delta\Gamma_s^{p,12\times36,\alpha_s^0}}{\Delta\Gamma_s} = 6.1\% \quad (\overline{\text{MS}}), \tag{13}$$

while when adding the 2-loop $Q_{1,2} \times Q_{3-6}$ piece we obtain

$$\frac{\Delta\Gamma_s^{p,12\times36,\alpha_s}}{\Delta\Gamma_s} = 0.3\% \quad \text{(pole)}, \qquad \frac{\Delta\Gamma_s^{p,12\times36,\alpha_s}}{\Delta\Gamma_s} = 1.4\% \quad (\overline{\text{MS}}). \tag{14}$$

More details on the numerical ingredients of these comparisons can be found in [15]. The captions "$\overline{\text{MS}}$" and "pole" refer to the way how we treat the m_b^2 prefactor in Eq. (7). Here we have the choice between regarding it as an $\overline{\text{MS}}$ or an on-shell mass. We would like to stress that in both cases all masses apart from this m_b^2 prefactor are always treated in the $\overline{\text{MS}}$ scheme.

6 Summary

We reported on the current status of our project which aims to achieve a sizable reduction of the perturbative uncertainties in $\Delta\Gamma_s$. This is needed to match the current experimental precision of the width difference in $B_s^0 - \bar{B}_s^0$ mixing and can be accomplished by extending the available matching calculations between $|\Delta B| = 1$ and $|\Delta B| = 2$ effective theories to 2 and 3 loops for the relevant operator insertions.

At the first stage of this work we have already carried out the corresponding calculations analytically by considering an expansion $z \equiv m_c^2/m_b^2$ up to $O(z)$ and $O(z^0)$ at 2 and 3 loops respectively. While a part of our results has already been published in [15], we still need to do more checks on the remaining contributions, especially at 3 loops. In the near future we plan to publish full analytic results [25, 59] for all of the new matching coefficients obtained in the course of this work and to provide a new update of the theory prediction for $\Delta\Gamma_s$.

Acknowledgments

First of all, the author is very grateful to Marvin Gerlach, Matthias Steinhauser and Ulrich Nierste for the collaboration on this project. We acknowledge Artyom Hovhannisyan for providing to us intermediate results of [9]. We also would like to thank Erik Panzer, Oliver Schnetz, Gudrun Heinrich and Alexander Smirnov for their kind helping with the usage of HyperInt, HyperLogProcedures, pySecDec and FIESTA respectively. This research was supported by the Deutsche Forschungsgemeinschaft (DFG, German Research Foundation) under grant 396021762 — TRR 257 "Particle Physics Phenomenology after the Higgs Discovery". This report has been assigned preprint numbers TTP21-050 and P3H-21-095.

References

[1] R. Aaij et al. (LHCb), Eur. Phys. J. C **79**, 706 (2019), [Erratum: Eur.Phys.J.C 80, 601 (2020)], 1906.08356
[2] A.M. Sirunyan et al. (CMS), Phys. Lett. B **816**, 136188 (2021), 2007.02434
[3] G. Aad et al. (ATLAS), Eur. Phys. J. C **81**, 342 (2021), 2001.07115
[4] Y.S. Amhis et al. (HFLAV), Eur. Phys. J. C **81**, 226 (2021), 1909.12524
[5] M. Beneke, G. Buchalla, C. Greub, A. Lenz, U. Nierste, Phys. Lett. B **459**, 631 (1999), hep-ph/9808385
[6] M. Ciuchini, E. Franco, V. Lubicz, F. Mescia, Nucl. Phys. B **625**, 211 (2002), hep-ph/0110375
[7] M. Ciuchini, E. Franco, V. Lubicz, F. Mescia, C. Tarantino, JHEP **08**, 031 (2003), hep-ph/0308029
[8] A. Lenz, U. Nierste, JHEP **06**, 072 (2007), hep-ph/0612167
[9] H.M. Asatrian, A. Hovhannisyan, U. Nierste, A. Yeghiazaryan, JHEP **10**, 191 (2017), 1709.02160
[10] H.M. Asatrian, H.H. Asatryan, A. Hovhannisyan, U. Nierste, S. Tumasyan, A. Yeghiazaryan, Phys. Rev. D **102**, 033007 (2020), 2006.13227
[11] K.G. Chetyrkin, M. Misiak, M. Munz, Nucl. Phys. B **520**, 279 (1998), hep-ph/9711280
[12] M. Gorbahn, U. Haisch, Nucl. Phys. B **713**, 291 (2005), hep-ph/0411071
[13] M.J. Dugan, B. Grinstein, Phys. Lett. B **256**, 239 (1991)
[14] S. Herrlich, U. Nierste, Nucl. Phys. B **455**, 39 (1995), hep-ph/9412375
[15] M. Gerlach, U. Nierste, V. Shtabovenko, M. Steinhauser, JHEP **07**, 043 (2021), 2106.05979
[16] V.A. Khoze, M.A. Shifman, Sov. Phys. Usp. **26**, 387 (1983)
[17] M.A. Shifman, M.B. Voloshin, Sov. J. Nucl. Phys. **41**, 120 (1985)
[18] V.A. Khoze, M.A. Shifman, N.G. Uraltsev, M.B. Voloshin, Sov. J. Nucl. Phys. **46**, 112 (1987)
[19] J. Chay, H. Georgi, B. Grinstein, Phys. Lett. B **247**, 399 (1990)
[20] I.I.Y. Bigi, N.G. Uraltsev, Phys. Lett. B **280**, 271 (1992)
[21] I.I.Y. Bigi, N.G. Uraltsev, A.I. Vainshtein, Phys. Lett. B **293**, 430 (1992), [Erratum: Phys.Lett.B 297, 477–477 (1992)], hep-ph/9207214
[22] I.I.Y. Bigi, M.A. Shifman, N.G. Uraltsev, A.I. Vainshtein, Phys. Rev. Lett. **71**, 496 (1993), hep-ph/9304225
[23] B. Blok, L. Koyrakh, M.A. Shifman, A.I. Vainshtein, Phys. Rev. D **49**, 3356 (1994), [Erratum: Phys.Rev.D 50, 3572 (1994)], hep-ph/9307247

[24] A.V. Manohar, M.B. Wise, Phys. Rev. D **49**, 1310 (1994), hep-ph/9308246
[25] M. Gerlach, U. Nierste, V. Shtabovenko, M. Steinhauser, in preparation ()
[26] P. Nogueira, J. Comput. Phys. **105**, 279 (1993)
[27] R. Harlander, T. Seidensticker, M. Steinhauser, Phys. Lett. B **426**, 125 (1998), hep-ph/9712228
[28] T. Seidensticker, *Automatic application of successive asymptotic expansions of Feynman diagrams*, in *6th International Workshop on New Computing Techniques in Physics Research: Software Engineering, Artificial Intelligence Neural Nets, Genetic Algorithms, Symbolic Algebra, Automatic Calculation* (1999), hep-ph/9905298
[29] M. Gerlach, F. Herren, in preparation , (),
[30] J. Kuipers, T. Ueda, J.A.M. Vermaseren, J. Vollinga, Comput. Phys. Commun. **184**, 1453 (2013), 1203.6543
[31] A. Pak, J. Phys. Conf. Ser. **368**, 012049 (2012), 1111.0868
[32] R. Mertig, M. Bohm, A. Denner, Comput. Phys. Commun. **64**, 345 (1991)
[33] V. Shtabovenko, R. Mertig, F. Orellana, Comput. Phys. Commun. **207**, 432 (2016), 1601.01167
[34] V. Shtabovenko, R. Mertig, F. Orellana, Comput. Phys. Commun. **256**, 107478 (2020), 2001.04407
[35] R. Lewis, https://home.bway.net/lewis ()
[36] A. Alloul, N.D. Christensen, C. Degrande, C. Duhr, B. Fuks, Comput. Phys. Commun. **185**, 2250 (2014), 1310.1921
[37] T. Hahn, Comput. Phys. Commun. **140**, 418 (2001), hep-ph/0012260
[38] K.G. Chetyrkin, F.V. Tkachov, Nucl. Phys. B **192**, 159 (1981)
[39] F.V. Tkachov, Phys. Lett. B **100**, 65 (1981)
[40] A.V. Smirnov, F.S. Chuharev, Comput. Phys. Commun. **247Â** , 106877 (2020), 1901.07808
[41] R.N. Lee, J. Phys. Conf. Ser. **523**, 012059 (2014), 1310.1145
[42] V.A. Smirnov, *Analytic tools for Feynman integrals*, Vol. 250 (2012)
[43] J. Fleischer, M.Y. Kalmykov, A.V. Kotikov, Phys. Lett. B **462**, 169 (1999), [Erratum: Phys.Lett.B 467, 310–310 (1999)], hep-ph/9905249
[44] E. Panzer, Comput. Phys. Commun. **188**, 148 (2015), 1403.3385
[45] A.B. Goncharov, Math. Res. Lett. **5**, 497 (1998), 1105.2076
[46] O. Schnetz, https://www.math.fau.de/person/oliver-schnetz ()
[47] C. Duhr, F. Dulat, JHEP **08**, 135 (2019), 1904.07279
[48] S. Borowka, G. Heinrich, S. Jahn, S.P. Jones, M. Kerner, J. Schlenk, T. Zirke, Comput. Phys. Commun. **222**, 313 (2018), 1703.09692
[49] S. Borowka, G. Heinrich, S. Jahn, S.P. Jones, M. Kerner, J. Schlenk, Comput. Phys. Commun. **240**, 120 (2019), 1811.11720
[50] G. Heinrich, S. Jahn, S.P. Jones, M. Kerner, F. Langer, V. Magerya, A. Pöldaru, J. Schlenk, E. Villa (2021), 2108.10807
[51] A.V. Smirnov, Comput. Phys. Commun. **204**, 189 (2016), 1511.03614
[52] A.V. Smirnov, N.D. Shapurov, L.I. Vysotsky (2021), 2110.11660
[53] V. Shtabovenko, R. Mertig, F. Orellana, in preparation ()
[54] P. Gambino, M. Gorbahn, U. Haisch, Nucl. Phys. B **673**, 238 (2003), hep-ph/0306079
[55] A. Pak, A. Smirnov, Eur. Phys. J. C **71**, 1626 (2011), 1011.4863
[56] B. Jantzen, A.V. Smirnov, V.A. Smirnov, Eur. Phys. J. C **72**, 2139 (2012), 1206.0546

[57] M. Beneke, G. Buchalla, I. Dunietz, Phys. Rev. D **54**, 4419 (1996), [Erratum: Phys.Rev.D 83, 119902 (2011)], hep-ph/9605259
[58] A.J. Buras, M. Jamin, M.E. Lautenbacher, Nucl. Phys. B **400**, 75 (1993), hep-ph/9211321
[59] M. Gerlach, U. Nierste, V. Shtabovenko, M. Steinhauser, in preparation ()

EPJ Web of Conferences **258**, 04003 (2022) https://doi.org/10.1051/epjconf/202225804003
A Virtual Tribute to Quark Confinement and the Hadron Spectrum (vConf21)

Doubly heavy baryons in Born-Oppenheimer EFT

Jaume Tarrús Castellà[1,2,*]

[1]Department of Physics, Indiana University, Bloomington, Indiana 47401, USA
[2]Center for Exploration of Energy and Matter, Indiana University, Bloomington, Indiana 47408, USA

Abstract. We report on the recent progress on the computation of the doubly heavy baryon spectrum in effective field theory. The effective field theory is built upon the heavy-quark mass and adiabatic expansions. The potentials can be expressed as NRQCD Wilson loops with operator insertions. These are nonperturbative objects and so far only the one corresponding to the static potential has been computed with lattice QCD. We review the proposal for a parametrization of the potentials based in an interpolation between the short- and long-distance regimes. The long-distance description is obtained with a newly proposed Effective String Theory which coincides with the previous ones for pure gluodynamics but it is extended to contain a fermion field. We show the doubly heavy baryon spectrum with hyperfine contributions obtained using these parametrizations for the hyperfine potentials.

1 Introduction

In the past two decades, there have been continuous discoveries at various experimental facilities of hadrons that fall into the category of doubly heavy hadrons. In this category one can find the nearly two dozen exotic quarkonium states discovered so far, such as the double charm tetraquark, T^+_{cc}, recently discovered by the LHCb Collaboration [1], as well as the somewhat older discovery of the pentaquark states [2]. Furthermore, traditional hadrons, in the quark model sense, also belong to this class, such as standard quarkonium and the main subject of this report, doubly heavy baryons, which have been first observed by the LHCb Collaboration [3, 4] in the form of the Ξ^{++}_{cc}.

All doubly heavy hadrons share a set of characteristic scales. The ratios of some of these scales are small and can be used as expansion parameters for the description of the properties of these states. Two of these expansion can be applied to almost all of the cases of interest. First, the heavy-quark mass, m_Q, is larger than the characteristic scale of momentum transfer in hadrons, $\Lambda_{\rm QCD}$, which is the scale of nonperturbative dynamics in QCD. The second is that one can perform an adiabatic expansion between the heavy-quark dynamics and the one of the light degrees of freedom, that is the gluons an light quarks. The characteristic energy scale of the heavy-quark dynamics is their binding energy, which can be written as $m_Q v^2$, with $v \ll 1$ the heavy-quark relative velocity, while the light degrees of freedom dynamics are dominated by nonperturbative effects and therefore characterized by the $\Lambda_{\rm QCD}$ scale.

*e-mail: jtarrus@iu.edu

An Effective Field Theory (EFT) describing any doubly heavy hadron[1] exploiting these two expansion has been proposed in Ref. [5]. Since this EFT reproduces the Born-Oppenheimer (BO) approximation at leading order it is often referred as BOEFT. The EFT was constructed in the single hadron sector up to the heavy-quark spin and angular momentum terms suppressed by $1/m_Q$. Expressions of the potentials as operator insertions in the Wilson loop were obtained by matching the EFT to nonrelativistic QCD (NRQCD) [6–8]. The computation of the Wilson loop with operator insertions cannot be done using perturbative techniques and should be carried out (ideally) in lattice QCD or other nonperturbative approaches.

BOEFT has been applied to doubly heavy baryons in Ref. [9]. In this case, the Wilson loop with light-quark operator insertions corresponding to the static potential, has been obtained in the lattice [10, 11] including several excited states. This lattice data was used in Ref. [9] to obtain the double charm and bottom baryon spectrum at leading order in BOEFT. In the case of the heavy-quark spin and angular-momentum dependent potentials there is no available lattice data. To sidestep this problem, in Ref. [12], a phenomenological parametrization of these potentials was put forward. This parametrization consists of an interpolation between the descriptions of the potentials in the short- and long-distance regimes. In the short-distance regime, $r \ll \Lambda_{\rm QCD}^{-1}$, one can use the multipole expansion to find model-independent parametrizations of the potentials in terms of some nonperturbative constants. On the other hand, for the description of the potentials in the long-distance regime, $r \ll \Lambda_{\rm QCD}^{-1}$, an Effective String Theory (EST) [13] has been developed in Ref. [12]. This EST is an extension of the one which has been used to very successfully describe the standard and hybrid quarkonium potentials [14–17].

2 BOEFT for doubly heavy baryons

Doubly heavy baryons are formed by two distinct components: a heavy quark pair and a light quark. At leading order in NRQCD the heavy quarks are static and the spectrum of the theory is given by the so-called static energies. These are characterized by a set of quantum numbers: the flavor of the light quark, the heavy quark pair relative distance $\boldsymbol{r}$, and the representation of $D_{\infty h}$[2]. The lattice results in Refs. [10, 11] show that the lowest lying static energy corresponds to the representation $(1/2)_g$ followed by three very close states corresponding to the representations $(1/2)_u$, $(3/2)_u$ and $(1/2)'_u$. In the short-distance limit the symmetry group is enlarged from $D_{\infty h}$ to $O(3)$ and the states can be labeled by their spin (κ) and parity (p). Projecting the κ^p states into the heavy quark axis one can obtain states in representations of $D_{\infty h}$. We show the correspondence in Table 1. Here we will consider only doubly heavy baryon states associated to the $\kappa^p = (1/2)^\pm$ light quark states.

$O(3)$	$D_{\infty h}$
$(1/2)^+$	$(1/2)_g$
$(3/2)^-$	$(1/2)_u$, $(3/2)_u$
$(1/2)^-$	$(1/2)'_u$

Table 1. Correspondence between short-distance $O(3)$ representations and $D_{\infty h}$ representations.

[1] Excluding states corresponding to bound states of pairs of singly heavy hadrons.
[2] See Ref. [18] for a detailed discussion on the representations of $D_{\infty h}$.

The Hamiltonian densities associated to the $\kappa^p = (1/2)^\pm$ light-quark states [9] have the following expansion up to $1/m_Q$

$$h_{(1/2)^\pm} = \frac{\boldsymbol{p}^2}{m_Q} + \frac{\boldsymbol{P}^2}{4m_Q} + V^{(0)}_{(1/2)^\pm}(\boldsymbol{r}) + \frac{1}{m_Q} V^{(1)}_{(1/2)^\pm}(\boldsymbol{r}, \boldsymbol{p}) . \tag{1}$$

At leading order we have just the static potential

$$V^{(0)}_{(1/2)^\pm}(\boldsymbol{r}) = V^{(0)}_{(1/2)^\pm}(r) . \tag{2}$$

The heavy-quark spin and angular-momentum dependent operators appear at next-to-leading order and read as

$$V^{(1)}_{(1/2)^\pm \mathrm{SD}}(\boldsymbol{r}) = V^{s1}_{(1/2)^\pm}(r) \boldsymbol{S}_{QQ} \cdot \boldsymbol{S}_{1/2} + V^{s2}_{(1/2)^\pm}(r) \boldsymbol{S}_{QQ} \cdot (\boldsymbol{\mathcal{T}}_2 \cdot \boldsymbol{S}_{1/2}) + V^{l}_{(1/2)^\pm}(r) \left(\boldsymbol{L}_{QQ} \cdot \boldsymbol{S}_{1/2}\right) , \tag{3}$$

with $\mathcal{T}_2^{ij} = \hat{\boldsymbol{r}}^i \hat{\boldsymbol{r}}^j - \delta^{ij}/3$, $\boldsymbol{S}_{1/2} = \boldsymbol{\sigma}/2$ and $2\boldsymbol{S}_{QQ} = \boldsymbol{\sigma}_{QQ} = \boldsymbol{\sigma}_{Q_1} \mathbb{1}_{2\,Q_2} + \mathbb{1}_{2\,Q_1} \boldsymbol{\sigma}_{Q_2}$, where $\boldsymbol{\sigma}$ are the standard Pauli matrices and $\mathbb{1}_2$ is an identity matrix in the heavy-quark spin space for the heavy quark labeled in the subindex.

For the potentials in Eqs. (2) and (3) the matching expressions of Ref. [5] in terms of the Wilson loop with operator insertions reduce to

$$V^{(0)}_{(1/2)^\pm}(\boldsymbol{r}) = \lim_{t\to\infty} \frac{i}{t} \log\left(\mathrm{Tr}\left[\langle 1 \rangle_\Box^{(1/2)^\pm}\right]\right) , \tag{4}$$

and

$$V^{s1}_{(1/2)^\pm}(r) = -c_F \lim_{t\to\infty} \frac{4}{3t} \int_{-t/2}^{t/2} dt' \frac{\mathrm{Tr}\left[\boldsymbol{S}_{1/2} \cdot \langle g\boldsymbol{B}(t', \boldsymbol{x}_1) \rangle_\Box^{(1/2)^\pm}\right]}{\mathrm{Tr}\left[\langle 1 \rangle_\Box^{(1/2)^\pm}\right]} , \tag{5}$$

$$V^{s2}_{(1/2)^\pm}(r) = -c_F \lim_{t\to\infty} \frac{6}{t} \int_{-t/2}^{t/2} dt' \frac{\mathrm{Tr}\left[(\boldsymbol{S}_{1/2} \cdot \boldsymbol{\mathcal{T}}_2) \cdot \langle g\boldsymbol{B}(t', \boldsymbol{x}_1) \rangle_\Box^{(1/2)^\pm}\right]}{\mathrm{Tr}\left[\langle 1 \rangle_\Box^{(1/2)^\pm}\right]} , \tag{6}$$

$$V^{l}_{(1/2)^\pm} = -\lim_{t\to\infty} 2 \int_0^1 ds\, s \frac{\mathrm{Tr}\left[\boldsymbol{S}_{1/2} \cdot \left(\frac{2}{3}\mathbb{1}_2 - \boldsymbol{\mathcal{T}}_2\right) \cdot \langle g\boldsymbol{B}(t/2, z(s)) \rangle_\Box^{(1/2)^\pm}\right]}{\mathrm{Tr}\left[\langle 1 \rangle_\Box^{(1/2)^\pm}\right]} , \tag{7}$$

where $z(s) = \boldsymbol{x}_1 + s(\boldsymbol{R} - \boldsymbol{x}_1)$ and we use the following notation for the Wilson loop averages

$$\langle \dots \rangle_\Box^{(1/2)^\pm} = \langle Q_{(1/2)^\pm}(t/2, \boldsymbol{R}) \dots Q^\dagger_{(1/2)^\pm}(-t/2, \boldsymbol{R}) P\left\{e^{-ig\int_{C_1+C_2} dz^\mu A^\mu(z)}\right\}\rangle , \tag{8}$$

with C_1 and C_2 the upper and lower paths of a rectangular Wilson loop. Note that, unlike the quark-antiquark case, the flow is in the same direction for both paths. The interpolating operators are

$$Q^\alpha_{(1/2)^+}(t, \boldsymbol{x}) = \left[P_+ q^l(t, \boldsymbol{x})\right]^\alpha \underline{T}^l , \tag{9}$$

$$Q^\alpha_{(1/2)^-}(t, \boldsymbol{x}) = \left[P_+ \gamma^5 q^l(t, \boldsymbol{x})\right]^\alpha \underline{T}^l , \tag{10}$$

where $\alpha = -1/2, 1/2$, and we have used the following $\bar{3}$ tensor invariants

$$\underline{T}^l_{ij} = \frac{1}{\sqrt{2}} \epsilon_{lij}, \quad i,\ j,\ l = 1,2,3 . \tag{11}$$

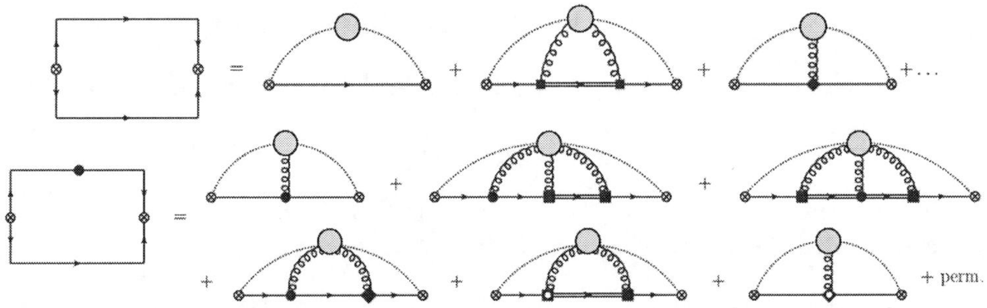

Figure 1. Matching of the Wilson loop with operators insertions with the expansion in weakly-coupled pNRQCD up to next-to-leading order. The single lines represent the antitriplet fields, the double lines the sextet field, the dotted and the curly lines the light-quark and gluon fields respectively. The crossed circles indicate the insertion of a Q operator while the other vertices stand for various multipolar couplings of the heavy-quark pair with chromoelectric or chromomagnetic fields. The upper line corresponds to the static potential while the second and third correspond to the heavy-quark spin-dependent potentials.

3 Short-distance regime

In the short-distance regime $r \ll \Lambda_{\rm QCD}^{-1}$. Since the relative momentum of the heavy quarks scales as $m_Q v \sim r^{-1}$, in this regime it is a perturbative quantity. Therefore, one can integrate it out and build (weakly-coupled) potential NRQCD (pNRQCD) [19–21]. One can then integrate out the $\Lambda_{\rm QCD}$ modes and match pNRQCD to BOEFT. Using this two step matching procedure one finds multipole expanded expressions of the potentials. Other examples of this two-step matching can be found in Refs. [22, 23] for the heavy-quark spin dependent potentials of quarkonium hybrids and in Refs. [24, 25] for the hybrid and standard quarkonium transitions.

The expansion of the static potential in Eq. (4) is given diagrammatically in Fig. 1 and corresponds to the following form

$$V^{(0)}_{(1/2)^\pm}(r) = -\frac{2}{3}\frac{\alpha_s}{r} + \overline{\Lambda}_{(1/2)^\pm} + \overline{\Lambda}^{(1)}_{(1/2)^\pm} r^2 + \dots , \tag{12}$$

with $\overline{\Lambda}_{(1/2)^\pm}$ and $\overline{\Lambda}^{(1)}_{(1/2)^\pm}$ nonperturbative constants. The former can be obtained D or B meson masses from heavy quark-diquark duality [26–29]. One can find analogous expansions for the hyperfine potentials:

$$V^{s1}_{(1/2)^\pm}(r) = c_F\left(\Delta^{(0)}_{(1/2)^\pm} + \Delta^{(1,0)}_{(1/2)^\pm} r^2 + \dots\right) , \tag{13}$$

$$V^{s2}_{(1/2)^\pm}(r) = c_F \Delta^{(1,2)}_{(1/2)^\pm} r^2 + \dots , \tag{14}$$

$$V^{l}_{(1/2)^\pm} = \frac{1}{2}\left[\Delta^{(0)}_{(1/2)^\pm} + \left(\Delta^{(1,0)}_{(1/2)^\pm} - \frac{1}{3}\Delta^{(1,2)}_{(1/2)^\pm}\right) r^2\right] + \dots , \tag{15}$$

with the Δ's corresponding to nonperturbative constants which expression as pNRQCD correlators can be found Ref. [12]. The constants $\Delta^{(0)}_{(1/2)^\pm}$ can be obtained using the heavy quark-diquark duality from the hyperfine splittings of heavy mesons.

4 Long-distance regime

At long distances a flux tube emerges from the heavy quarks joining at the position of the light quark. The dynamics of the flux tube can be described by an EST which coincides with the one for pure gluodynamics with the addition of a fermion field constrained to move on the string.

The action of the gluonic string is just proportional to the area of the string world sheet

$$S_g = -\sigma \int d^2x \sqrt{|\det g_{ab}|}, \tag{16}$$

with σ the string tension, and g_{ab} the induced metric on the string. The action of a four-dimensional Dirac field constrained on a string is given by

$$S_{\text{l.q}} = \int d^2x \sqrt{g}\bar{\psi}(x)\left(i\rho^a \overleftrightarrow{\partial}_a - m_{\text{l.q.}}\right)\psi(x) \quad , \quad \bar{\psi}\rho^a \overleftrightarrow{\partial}_a \psi \equiv \left(\bar{\psi}(\rho^a\partial_a\psi) - (\partial_a\bar{\psi})\rho^a\psi\right)/2 \tag{17}$$

with $\rho^a \equiv \gamma^\mu e^a_\mu$, and $e^\alpha_a \equiv \partial\xi^\alpha/\partial x^a$ the Zweibein. The antisymmetrization of the partial derivative is required by Hermiticity.

Expanding the action in Eq. (16) for small string fluctuations $\partial\xi \sim (r\Lambda_{\text{QCD}})^{-1} \ll 1$ we arrive at

$$S_g = -\sigma \int dtdz\left(1 - \frac{1}{2}\partial^a\xi^l\partial_a\xi^l + \dots\right), \tag{18}$$

and for the case of the fermionic action in Eq. (17) we find

$$S_{\text{l.q}} = \int dtdz\left(\bar{\psi}(t,z)i\gamma^a\overleftrightarrow{\partial}_a\psi(t,z) - m_{\text{l.q}}\bar{\psi}(t,z)\psi(t,z) - \partial^a\xi^l\bar{\psi}(t,z)i\gamma^l\overleftrightarrow{\partial}_a\psi(t,z) + \dots\right), \tag{19}$$

with $l = 1, 2$, and $a = 0, 3$.

The mapping of the chromomagnetic field into string fluctuations can be found in Ref. [14]. However, mappings into string fermion operators are now possible and in fact provide the leading order contribution to the potentials in Eqs. (5)-(7). This mapping is as follows:

$$\boldsymbol{B}^l(t,z) \mapsto \Lambda_f\bar{\psi}(t,z)\frac{\boldsymbol{\Sigma}^l}{2}\psi(t,z), \tag{20}$$

$$\boldsymbol{B}^3(t,z) \mapsto \Lambda'_f\bar{\psi}(t,z)\frac{\boldsymbol{\Sigma}^3}{2}\psi(t,z), \tag{21}$$

with $\boldsymbol{\Sigma} = \text{diag}(\boldsymbol{\sigma}, \boldsymbol{\sigma})$. To convert the two-dimensional spin operators in Eqs. (5) and (6) into four-dimensional spin operators, we have used the prescription $\boldsymbol{S}_{1/2} \mapsto \boldsymbol{\Sigma}/2$.

We compute the potentials in Eqs. (4)-(7) as correlators in the EST using the mapping defined by Eqs. (20)-(21). We find the following result for the static potential:

$$V^{(0)}_{(1/2)^\pm}(\boldsymbol{r}) = \sigma r + E_1, \tag{22}$$

and analogously for the heavy-quark spin and angular-momentum dependent potentials one finds

$$V^{s1}_{(1/2)^\pm}(r) = \frac{c_F}{3r}\left(1 \mp \frac{m_{\text{l.q.}}}{E_1}\right)\left(\Lambda'_f - 2\Lambda_f\right), \tag{23}$$

$$V^{s2}_{(1/2)^\pm}(r) = \frac{c_F}{r}\left(1 \mp \frac{m_{\text{l.q.}}}{E_1}\right)\left(\Lambda'_f + \Lambda_f\right), \tag{24}$$

$$V^{l}_{(1/2)^\pm}(r) = -\frac{1}{2r}\left(1 \mp \frac{4}{\pi^2}\frac{m_{\text{l.q.}}}{E_1}\right)\Lambda_f, \tag{25}$$

with $E_1 = \sqrt{(\pi/r)^2 + m^2_{\text{l.q.}}}$.

Table 2. Global fit of $\kappa^p = (1/2)^+$ $l = 0, 1, 2, 3$ multiplets hyperfine splittings for all the lattice data available for various values of r_0 with $\Delta^{(0)}_{(1/2)^+} = 0.122$ GeV^2 from the B-meson splittings and $\Delta^{(1,0)}_{(1/2)^+} = \Delta^{(1,2)}_{(1/2)^+} = 0$.

r_0 [fm]	Λ_f [GeV]	Λ'_f [GeV]	$\chi^2_{\rm d.o.f}$
0.1	−0.355(10)	−0.265(19)	0.66
0.2	−0.368(13)	−0.264(25)	0.72
0.3	−0.348(19)	−0.270(33)	0.69
0.4	−0.266(27)	−0.286(44)	0.60
0.5	−0.085(41)	−0.314(61)	0.58
0.6	0.224(75)	−0.353(102)	0.83

5 Spectrum with hyperfine contributions

The leading order spectrum for doubly heavy baryons can be obtained by solving the Shrödinger equation with the static potential in Eq. (4) which has been computed on the lattice in Refs. [10, 11]. The results for the leading order spectrum can be found in Ref. [9]. Analogous lattice computations of the hyperfine potentials in Eqs. (5)-(7) are not available. To address this problem a parametrization of these potentials has been proposed in Ref. [12]. The parametrization consists of an interpolation between the short- and long-distance regime descriptions discussed in Sec. 3 and Sec. 4, respectively. The proposed parametrizations of the hyperfine potentials are as follows:

$$V^{s1\,{\rm int}}_{(1/2)^\pm} = c_F\frac{\left(\Delta^{(0)}_{(1/2)^\pm} + \Delta^{(1,0)}_{(1/2)^\pm}r^2\right)r_0^6 + \frac{(\Lambda'_f - 2\Lambda_f)}{3}\left(1 \mp \frac{m_{\rm l.q.}}{E_1}\right)r^5}{r^6 + r_0^6}\,, \tag{26}$$

$$V^{s2\,{\rm int}}_{(1/2)^\pm} = c_F\frac{\Delta^{(1,2)}_{(1/2)^\pm}r^2 r_0^6 + \left(\Lambda'_f + \Lambda_f\right)\left(1 \mp \frac{m_{\rm l.q.}}{E_1}\right)r^5}{r^6 + r_0^6}\,, \tag{27}$$

$$V^{l\,{\rm int}}_{(1/2)^\pm} = \frac{1}{2}\frac{\left[\Delta^{(0)}_{(1/2)^\pm} + \left(\Delta^{(1,0)}_{(1/2)^\pm} - \frac{1}{3}\Delta^{(1,2)}_{(1/2)^\pm}\right)r^2\right]r_0^6 - \Lambda_f\left(1 \mp \frac{4}{\pi^2}\frac{m_{\rm l.q.}}{E_1}\right)r^5}{r^6 + r_0^6}\,. \tag{28}$$

Note that for $r_0 = 0$ we recover the long-distance potentials and for $r_0 \to \infty$ we recover the short-distance potentials.

To obtain the unknown parameters in the interpolated potentials in Eqs. (26)-(28) a χ^2 function was minimized. This function is constructed as the sum of the hyperfine splittings corresponding to the masses of the doubly heavy baryons from the lattice determinations of Refs [30–32, 32–38]. The formulas for the hyperfine contributions to the masses of doubly heavy baryons for the states associated to the static energies $(1/2)_g$ and $(1/2)'_u$ can be found in Ref. [9].

Several fits where performed in Ref. [12]. The results of the fits show that with the current lattice data on the heavy baryon masses it is not possible to determine the values for $\Delta^{(1,0)}_{(1/2)^\pm}$ and $\Delta^{(1,2)}_{(1/2)^\pm}$. Therefore, the preferred fit consists of only the leading order short-distance piece with $\Delta^{(0)}_{(1/2)^\pm}$ fixed by heavy quark-diquark duality. The results of this fit can be found in table 2 for various values of the interpolation parameter r_0. In Fig. 2 we show the spectrum of double charm and bottom baryons including the hyperfine contributions corresponding to the values of the parameters of $r_0 = 0.5$ fm in Table 2.

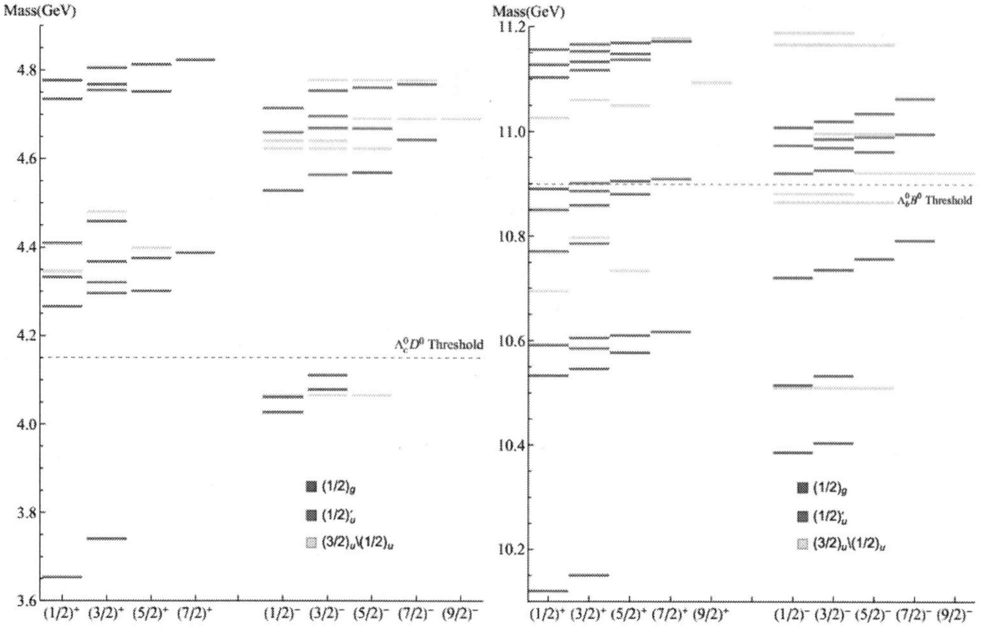

Figure 2. Spectrum of doubly heavy baryons in terms of j^{η_P} states. Each line represents a state. The mixed $(3/2)_u\backslash(1/2)_u$states do not include hyperfine contributions

6 Conclusions

We have reported on the results for the doubly heavy baryon spectrum including hyperfine contributions. The hyperfine potentials, which can be written in terms of the Wilson loop with operator insertions, are nonperturbative quantities that should be, ideally, computed in lattice QCD. However, these computations are often not available. To sidestep this issue, in Ref. [12], a parametrization of the potentials has been proposed. This consists of an interpolation between the short- and long-distance regimes. In the short-distance region the potentials can be computed in terms some unknown constants using weakly-coupled pNRQCD. Similarly, in the long-distance regime an Effective String Theory (EST) can be used. In Sec. 4 we reviewed the EST with a fermion field constrained into the string which is appropriate to compute the potentials for doubly heavy baryons. This procedure to obtain reliable parametrizations of the potentials can be of significant utility in future studies of doubly heavy hadrons.

Acknowledgements

J.T.C. acknowledges financial support by National Science Foundation (PHY-2013184).

References

[1] R. Aaij et al. (LHCb) (2021), 2109.01038
[2] R. Aaij et al. (LHCb), Phys. Rev. Lett. **115**, 072001 (2015), 1507.03414
[3] R. Aaij et al. (LHCb), Phys. Rev. Lett. **119**, 112001 (2017), 1707.01621
[4] R. Aaij et al. (LHCb), Phys. Rev. Lett. **121**, 162002 (2018), 1807.01919

[5] J. Soto, J. Tarrús Castellà, Phys. Rev. D **102**, 014012 (2020), 2005.00552
[6] W.E. Caswell, G.P. Lepage, Phys. Lett. B **167**, 437 (1986)
[7] G.T. Bodwin, E. Braaten, G.P. Lepage, Phys. Rev. D **51**, 1125 (1995), [Erratum: Phys.Rev.D 55, 5853 (1997)], hep-ph/9407339
[8] A.V. Manohar, Phys. Rev. D **56**, 230 (1997), hep-ph/9701294
[9] J. Soto, J. Tarrús Castellà, Phys. Rev. D **102**, 014013 (2020), [Erratum: Phys.Rev.D 104, 059901 (2021)], 2005.00551
[10] J. Najjar, G. Bali, PoS **LAT2009**, 089 (2009), 0910.2824
[11] J. Najjar, Diploma thesis, Universität Regensburg (2009)
[12] J. Soto, J. Tarrús Castellà, Phys. Rev. D **104**, 074027 (2021), 2108.00496
[13] Y. Nambu, Phys. Lett. B **80**, 372 (1979)
[14] G. Perez-Nadal, J. Soto, Phys. Rev. D **79**, 114002 (2009), 0811.2762
[15] N. Brambilla, M. Groher, H.E. Martinez, A. Vairo, Phys. Rev. D **90**, 114032 (2014), 1407.7761
[16] S. Hwang, Ph.D. thesis, Munich, Tech. U. (2018)
[17] R. Oncala, J. Soto, Phys. Rev. D **96**, 014004 (2017), 1702.03900
[18] M. Berwein, N. Brambilla, J. Tarrús Castellà, A. Vairo, Phys. Rev. D **92**, 114019 (2015), 1510.04299
[19] A. Pineda, J. Soto, Nucl. Phys. B Proc. Suppl. **64**, 428 (1998), hep-ph/9707481
[20] N. Brambilla, A. Pineda, J. Soto, A. Vairo, Nucl. Phys. B **566**, 275 (2000), hep-ph/9907240
[21] N. Brambilla, A. Vairo, T. Rosch, Phys. Rev. D **72**, 034021 (2005), hep-ph/0506065
[22] N. Brambilla, W.K. Lai, J. Segovia, J. Tarrús Castellà, A. Vairo, Phys. Rev. D **99**, 014017 (2019), [Erratum: Phys.Rev.D 101, 099902 (2020)], 1805.07713
[23] N. Brambilla, W.K. Lai, J. Segovia, J. Tarrús Castellà, Phys. Rev. D **101**, 054040 (2020), 1908.11699
[24] A. Pineda, J. Tarrús Castellà, Phys. Rev. D **100**, 054021 (2019), 1905.03794
[25] J. Tarrús Castellà, E. Passemar, Phys. Rev. D **104**, 034019 (2021), 2104.03975
[26] M.J. Savage, M.B. Wise, Phys. Lett. B **248**, 177 (1990)
[27] J. Hu, T. Mehen, Phys. Rev. D **73**, 054003 (2006), hep-ph/0511321
[28] T. Mehen, Phys. Rev. D **96**, 094028 (2017), 1708.05020
[29] T.C. Mehen, A. Mohapatra, Phys. Rev. D **100**, 076014 (2019), 1905.06965
[30] R.A. Briceno, H.W. Lin, D.R. Bolton, Phys. Rev. D **86**, 094504 (2012), 1207.3536
[31] Y. Namekawa et al. (PACS-CS), Phys. Rev. D **87**, 094512 (2013), 1301.4743
[32] Z.S. Brown, W. Detmold, S. Meinel, K. Orginos, Phys. Rev. D **90**, 094507 (2014), 1409.0497
[33] C. Alexandrou, V. Drach, K. Jansen, C. Kallidonis, G. Koutsou, Phys. Rev. D **90**, 074501 (2014), 1406.4310
[34] P. Pérez-Rubio, S. Collins, G.S. Bali, Phys. Rev. D **92**, 034504 (2015), 1503.08440
[35] M. Padmanath, R.G. Edwards, N. Mathur, M. Peardon, Phys. Rev. D **91**, 094502 (2015), 1502.01845
[36] C. Alexandrou, C. Kallidonis, Phys. Rev. D **96**, 034511 (2017), 1704.02647
[37] R. Lewis, R.M. Woloshyn, Phys. Rev. D **79**, 014502 (2009), 0806.4783
[38] P. Mohanta, S. Basak, Phys. Rev. D **101**, 094503 (2020), 1911.03741

EPJ Web of Conferences **258**, 04004 (2022) https://doi.org/10.1051/epjconf/202225804004
A Virtual Tribute to Quark Confinement and the Hadron Spectrum (vConf21)

Thermal modification of open heavy-flavor mesons from an effective hadronic theory

Glòria Montaña[1,*]

[1]Departament de Física Quàntica i Astrofísica and Institut de Ciències del Cosmos (ICCUB), Facultat de Física, Universitat de Barcelona, Martí i Franquès 1, 08028 Barcelona, Spain

Abstract. We have developed a self-consistent theoretical approach to study the modification of the properties of heavy mesons in hot mesonic matter which takes into account chiral and heavy-quark spin-flavor symmetries. The heavy-light meson-meson unitarized scattering amplitudes in coupled channels incorporate thermal corrections by using the imaginary-time formalism, as well as the dressing of the heavy mesons with the self-energies. We report our results for the ground-state thermal spectral functions and the implications for the excited mesonic states generated dynamically in the heavy-light molecular model. We have applied these to the calculation of meson Euclidean correlators and transport coefficients for D mesons and summarize here our findings.

1 Introduction

Heavy-ion collision experiments in the Relativistic Heavy Ion Collider (RHIC), at BNL, and in the Large Hadron Collider (LHC), at CERN, recreate in the laboratory the conditions for which the quark-gluon plasma (QGP) was created in the early Universe, short after the Big Bang. Fundamental quantities that permit to characterize this medium are thermal spectral and transport properties.

Heavy quarks, namely the charm and bottom quarks, are produced through the hard-scattering processes at the initial stage of the collisions and interact with the constituents of the QGP. When the temperature of the system goes below the confining transition temperature heavy quarks hadronize into heavy hadrons, mostly into open heavy-flavor (D and $\bar{B}$) mesons for the high energies achieved at RHIC and LHC, where low baryonic densities are reached. The spectral properties of the open heavy-flavor mesons are modified in the hot hadronic medium from their interaction with the light mesons forming the medium.

Finite temperature corrections to the scattering of $D^{(*)}$ and $D_s^{(*)}$ mesons with the Goldstone bosons (π, K, $\bar{K}$ and η mesons) have been implemented within a unitarized thermal approach in [1, 2]. This framework exploits both chiral symmetry and heavy-quark spin-flavor symmetry (HQSFS) to construct the effective Lagrangian that describes the interactions between the heavy and light degrees in vaccuum. HQSFS implies that the interactions are independent of the heavy-quark spin and flavor. The unitarization technique in coupled channels allows us to extend the energy region where the effective theory can be applied, as well as to generate resonances and bound states from the two-body dynamics. In this paper we have studied the thermal modification of these states in the hadronic phase, below the critical temperature T_c.

*e-mail: gmontana@fqa.ub.edu

The finite temperature is incorporated using the imaginary-time formalism (ITF) and the in-medium scattering amplitudes and ground-state self-energies are computed self-consistently.

In Section 2 of this contribution to the proceedings of the Virtual Tribute to Quark Confinement and the Hadron Spectrum 2021 conference we present our latest results on the thermal modification of the properties (masses and widths) of the D and $\bar{B}$ mesons, as well as of the resonances and bound states that are dynamically generated from their scattering with the light mesons. The effective theory results for the thermal scattering amplitudes and the spectral functions were used to compute open-charm meson Euclidean correlators in Ref. [3] and off-shell transport coefficients in Ref. [4]. The main results are summarized in Sections 3 and 4 of this contribution, respectively. Further details can be found in Refs. [1–4].

2 Thermal effective field theory for open heavy-flavor mesons

The interaction in vacuum between the open heavy-flavor mesons ($D^{(*)} = \{D^{(*)}, D_s^{(*)}\}$) and the pseudoscalar Goldstone bosons ($\Phi = \{\pi, K, \bar{K}, \eta\}$) is described in Refs. [1, 2] by means of an effective Lagrangian that is based on chiral symmetry and HQSFS [5–8]. The corresponding tree-level amplitudes of the $D\Phi$ (or $D^*\Phi$) scattering for an incoming channel i and an outgoing channel j read

$$\begin{aligned} V^{ij}(s,t,u) = \frac{1}{f_\pi^2}\Big[&\frac{C_{\rm LO}^{ij}}{4}(s-u) - 4C_0^{ij}h_0 + 2C_1^{ij}h_1 \\ &- 2C_{24}^{ij}\Big(2h_2(p_2\cdot p_4) + h_4((p_1\cdot p_2)(p_3\cdot p_4) + (p_1\cdot p_4)(p_2\cdot p_3))\Big) \\ &+ 2C_{35}^{ij}\Big(h_3(p_2\cdot p_4) + h_5((p_1\cdot p_2)(p_3\cdot p_4) + (p_1\cdot p_4)(p_2\cdot p_3))\Big)\Big], \end{aligned} \tag{1}$$

where $s = (p_1 + p_2)^2$, $t = (p_1 - p_3)^2$ and $u = (p_1 - p_4)^2$ are the usual Mandelstam variables. The values of the low-energy constants (LECs) of the terms at NLO in the chiral expansion, $h_{0,\ldots,5}$, were fitted in [8] to lattice results of scattering lengths and finite-volum spectra. Keeping lowest order in the heavy-quark mass expansion, there are no tree-level diagrams connecting $D\Phi$ and $D^*\Phi$ channels, and the LECs take the same values for these two sectors. The isospin coefficients C_{ij} for the different strangeness and isospin sectors and the values of the LECs are given in Tables III and IV in Ref. [2]. Relying on heavy-quark flavor symmetry (HQFS) the interaction potential of Eq. (1) can be extended to the open-bottom sector upon the replacement $D \to \bar{B}$ and the rescaling of the LECs with the heavy meson masses,

$$\frac{h_i^D}{\hat{m}_D} = \frac{h_i^B}{\hat{m}_B}\,, \quad \text{for } i = \{0,1,2,3\}\,, \quad \text{and} \quad h_{4,5}^D\hat{m}_D = h_{4,5}^B\hat{m}_B\,, \quad \text{for } i = \{4,5\}\,. \tag{2}$$

The s-wave projection of the amplitude in Eq. (1) is unitarized by solving the Bethe-Salpeter equation in coupled channels (Fig. 1a),

$$T_{ij}(s) = V_{ij}(s) + V_{ik}(s)G_kT_{kj}(s)\,, \tag{3}$$

where T is the unitarized amplitude, V is the s-wave interaction kernel and G is the loop function constructed from the two meson propagtors,

$$G^l(s) = i\int\frac{d^4q}{(2\pi)^4}\frac{1}{q^2 - m_D^2 + i\epsilon}\frac{1}{(p-q)^2 - m_\Phi^2 + i\epsilon}\,, \tag{4}$$

which is regularized with a hard momentum cutoff Λ. Equation (3) has an algebraic solution in the on-shell approximation, $T(s) = V(s)[1 - V(s)G(s)]^{-1}$, which is a matrix equation for the case of two or more coupled channels.

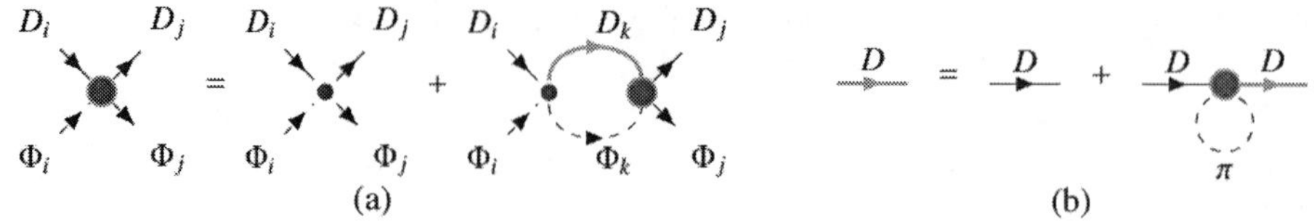

Figure 1: (a) Bethe-Salpeter equation and (b) Dyson-equation for the heavy-meson propagator at finite temperature. Big red circles represent the T-matrix, small blue circles correspond to the interaction kernel and thick orange lines depict dressed propagators. Figure taken from Ref. [1].

At finite temperature the loop function G is calculated through the ITF, in which the time dimension is Wick rotated and compactified in the range $[0, \beta = 1/T]$, with T the temperature of the system, and the energy integral transforms into a summation over discrete Matsubara frequencies, $q^0 \to i2\pi nT$ for bosons [9, 10]. We also introduce the Lehmann representation for the meson propagators

$$\mathcal{D}_M(i\omega_n, \mathbf{q}) = \int d\omega' \frac{S_M(\omega', \mathbf{q})}{i\omega_n - \omega'} , \tag{5}$$

where S_M is the spectral function and M denotes the meson species (D or Φ). After the Matsubara summation and the analytical continuation to real energies, $i\omega_m \to E + i\varepsilon$, the thermal loop function reads

$$G_{D\Phi}(E, \mathbf{p}; T) = \int \frac{d^3q}{(2\pi)^3} \int d\omega \int d\omega' \frac{S_D(\omega, \mathbf{q}; T) S_\Phi(\omega', \mathbf{p} - \mathbf{q}; T)}{E - \omega - \omega' + i\varepsilon} [1 + f(\omega, T) + f(\omega', T)] , \tag{6}$$

where $f(\omega, T) = [\exp(\omega/T) - 1]^{-1}$ is the equilibrium Bose-Einstein distribution function.

The spectral function of the heavy meson is given by

$$S_D(\omega, \mathbf{q}; T) = -\frac{1}{\pi} \mathrm{Im}\, \mathcal{D}_D(\omega, \mathbf{q}; T) = -\frac{1}{\pi} \mathrm{Im} \left(\frac{1}{\omega^2 - \mathbf{q}^2 - m_D^2 - \Pi_D(\omega, \mathbf{q}; T)} \right) , \tag{7}$$

where the self-energy follows from closing the light-meson line in the corresponding T-matrix element (Fig. 1b),

$$\Pi_D(E, \mathbf{p}; T) = \frac{1}{\pi} \int \frac{d^3q}{(2\pi)^3} \int d\Omega \frac{E}{\omega_\Phi} \frac{f(\Omega, T) - f(\omega_\Phi, T)}{E^2 - (\omega_\Phi - \Omega)^2 + i\varepsilon} \mathrm{Im}\, T_{D\Phi}(\Omega, \mathbf{p} + \mathbf{q}; T) . \tag{8}$$

In Refs. [1, 2] we have approximated the spectral function of the light meson to a δ-type spectral function because the in-medium modification of the pions is small. At $T \lesssim 150$ MeV the largest contribution to the self-energy comes from the interaction of the heavy mesons with the pions, as the abundance of heavier light mesons, i.e. kaons, antikaons and eta mesons, is thermally suppressed. This has been discussed in detail in Ref. [2]. In the results below only the pion contribution is considered in Eq. (8) to calculate the heavy-meson self-energy.

Equations (6) to (8) are coupled to each other and require a self-consistent solution. The results obtained for the loop function for $D\Phi$ and $D_s\Phi$ channels are shown in the top and bottom panels of Fig. 2, respectively, for different temperatures. At $T = 0$ (black lines) the imaginary part becomes non-zero in the energy region above threshold ($m_{D/D_s} + m_\Phi$), where the unitary cut develops. At $T > 0$ an additional cut opens up for energies below $m_{D/D_s} - m_\Phi$. This is the so-called Landau cut and it is related to thermal production and absorption processes [9, 10]. From the loop functions in Fig. 2 it is clear that the Landau cut is

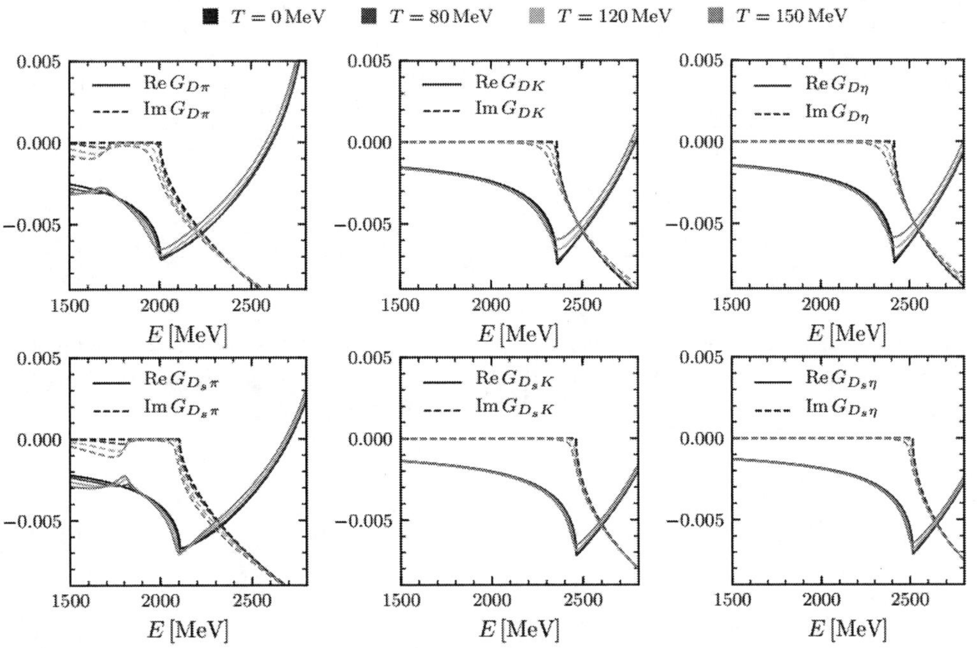

Figure 2: Real and imaginary parts of the two-meson loop function at finite temperature. Top plots: $D\Phi$ case. Bottom plots: $D_s\Phi$ case. Data taken from [2].

much larger for channels involving pions (for K and η mesons it cannot be distinguished with the scale used for the plots), and its importance grows with increasing temperatures because of the larger abundance of thermal pions. In addition, at finite temperature, the boundaries of the cuts are smoothened due to the widening of the heavy-meson spectral function, as discussed in [1, 2].

Figure 3 shows the self-consistent results for the unitarized scattering amplitudes at finite temperature in the sectors in which dynamically generated states with $J^P = 0^+$ appear. In the charm sector with strangeness and isospin $(S, I) = (0, 1/2)$ (first panel) the two resonances are identified with the experimental $D_0^*(2300)$ state, for which a two-pole structure at $T = 0$ was described in vaccuum in [11]. The lower resonance is seen in the $D\pi$ scattering amplitude, while the higher one is visible in the $D_s\bar{K}$ channel. The thermal effects on the generation of these states are moderate and the structures are still clearly seen at $T = 150$ MeV. In the sector with $(S, I) = (1, 0)$ (second panel) the bound state associated to the $D_{s0}^*(2317)$ in vacuum gains a substantial thermal width as the temperature is increased. The bottom counterparts of these states have not been reported experimentally yet. We predict here their thermal modification to be of the same order as in the charm sector, as expected from HQFS (third and fourth panels). The thermal scattering amplitudes for $D^*\Phi$ and $D_s^*\Phi$ and the corresponding resonances with $J^P = 1^+$ are reported in [2].

The thermal modification of the ground states D, D_s, $\bar{B}$ and $\bar{B}_s$ is extracted from their respective spectral functions. Figure 4 displays the evolution of the masses and the widths as a function of temperature in the left and right panels, respectively. The heavy ground states, which are stable degrees of freedom of the model in vacuum, acquire a thermal width due to their interaction with the thermal pions of the bath. Furthermore the masses decrease of the order of $\sim 20 - 50$ MeV at the highest temperature considered ($T = 150$ MeV). The thermal modification of the open-charm vector counterparts is given in [2].

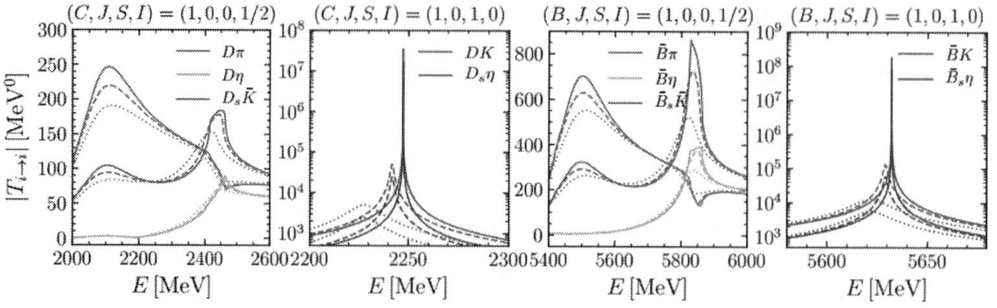

Figure 3: T-matrix at finite temperature for the diagonal transitions in the sectors with charm $C = 1$ (bottom $B = 1$), spin $J = 0$ and strangeness and isospin $(S, I) = \{(0, 1/2), (1, 0)\}$. Figure extracted from [12] with data partially taken from [1].

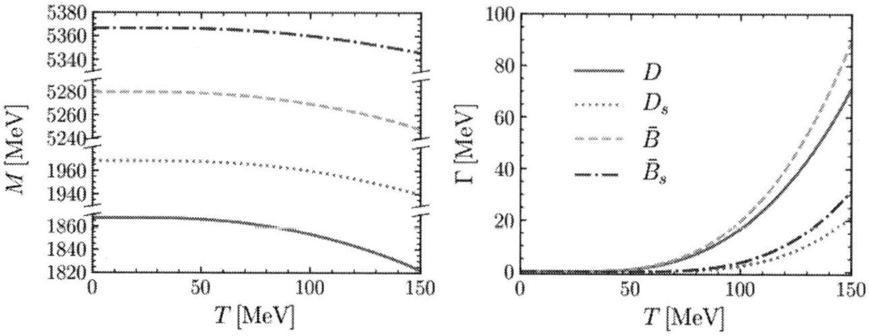

Figure 4: Thermal evolution of the mass (left panel) and the width (right panel) of the open heavy-flavor ground states. Data partially taken from [2].

3 Open-charm Euclidean correlators

In this section we present the results for the Euclidean meson correlators of open-charm mesons that were obtained in our previous work [3]. Temporal Euclidean meson correlators in a hot medium can be computed on the lattice for a discrete number of points in Euclidean time τ. The temperature of the system is related to the temporal extent of the lattice, $T = 1/(aN_\tau)$. The spectral function is related to the Euclidean correlator through the integral relation:

$$G_E(\tau; T) = \int_0^\infty d\omega\, K(\tau, \omega; T)\rho(\omega; T)\,, \tag{9}$$

where the spectral function $\rho(\omega; T)$ is convoluted with a temperature dependent kernel,

$$K(\tau, \omega; T) = \cosh\left[\omega\left(\tau - \frac{1}{2T}\right)\right]\sinh^{-1}\left(\frac{\omega}{2T}\right). \tag{10}$$

The extraction of a continuous spectral function from the lattice QCD correlators requires the inversion of Eq. (9). However such an inversion of a discrete number of data points with statistical errors is an ill-posed problem and the methods that are usually employed rely on specific assumptions for the spectral function. In Ref. [3] we took the opposite approach. In order to compare the effective field theory results with the lattice QCD calculations for open-charm mesons at finite temperature of Ref. [13] we insert the thermal spectral function in Eq. (9) and obtain the corresponding Euclidean correlator.

The full meson spectral function $\rho(\omega; T)$ contains information of the ground state and the additional contribution of excited bound states and the continuum of scattering states. The ground-state spectral function is obtained using Eq. (7) and the continuum is taken into account through a parametrization of the free meson spectral function in the non-interacting limit, while the contribution of the possible excited states is neglected in [3]. Thus, the full spectral function reads

$$\rho(\omega; T) = M_D^4 S_D(\omega; T) + a\rho_{\rm cont}(\omega; T) , \tag{11}$$

where the contribution of the continuum is weighted with respect to the ground-state contribution through the factor a.

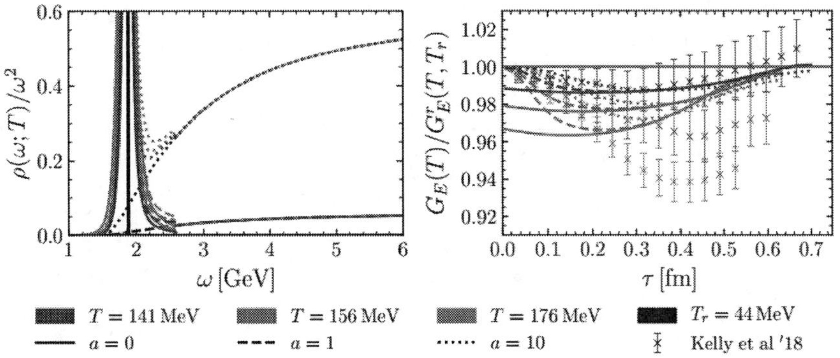

Figure 5: Left panel: Full spectral functions of the D-meson for unphysical meson masses. Right panel: Ratio of the Euclidean correlator and the reconstructed correlator at T_r = 44 MeV. The results are shown for temperatures below T_c = 185 MeV [13] and for various values of the weight parameter a. Figure extracted from [12] with data taken from [3].

The full spectral functions obtained for the D mesons in a thermal medium of pions of mass m_π = 384 MeV, i.e. the unphysical pion mass used in the lattice calculations of Ref. [13], are shown in the left panel of Fig. 5. The results are given for various temperatures below the pseudocritical temperature on the lattice (T_c = 185 MeV) and for the weight parameter $a = \{0, 1, 10\}$ in Eq. (11).

In order to discern the modification of the correlator due to the temperature dependence of the spectral function alone and cancel that of the integration kernel, the ratio with a recostructed correlator is usually computed. This is shown in the right panel of Fig. 5, together with the results of the lattice simulations of Ref. [13]. The inclusion of the continuum improves the behavior of the ratio at small τ and the results are in good agreement with the lattice QCD data within the error bars for the lowest temperatures considered.

4 D-meson transport coefficients

In this section we summarize our latest results on open-charm transport coefficients of Ref. [4], where we implemented the thermal scattering amplitudes and the D-meson spectral functions in order to take into account thermal and off-shell effects. After some justified approximations, the off-shell kinetic equation derived in [4] following the Kadanoff-Baym approach [14] reduces to an off-shell Fokker-Planck equation for the Green's function,

$$\frac{\partial}{\partial t} G_D^<(t,k) = \frac{\partial}{\partial k^i}\left\{\hat{A}(k;T)k^i G_D^<(t,k) + \frac{\partial}{\partial k^j}\left[\hat{B}_0(k;T)\Delta^{ij} + \hat{B}_1(k;T)\frac{k^i k^j}{|\mathbf{k}|^2}\right] G_D^<(t,k)\right\} . \tag{12}$$

In this equation the drag force $\hat{A}$ and the diffusion coefficients $\hat{B}_0$ and $\hat{B}_1$ are defined off shell, as they are calculated using spectral functions, as well as thermal scattering amplitudes. They are defined in [4] as

$$\hat{A}(k^0, \mathbf{k}; T) \equiv \left\langle 1 - \frac{\mathbf{k} \cdot \mathbf{k}_1}{\mathbf{k}^2} \right\rangle, \tag{13}$$

$$\hat{B}_0(k^0, \mathbf{k}; T) \equiv \frac{1}{4}\left\langle \mathbf{k}_1^2 - \frac{(\mathbf{k} \cdot \mathbf{k}_1)^2}{\mathbf{k}^2} \right\rangle, \tag{14}$$

$$\hat{B}_1(k^0, \mathbf{k}; T) \equiv \frac{1}{2}\left\langle \frac{[\mathbf{k} \cdot (\mathbf{k} - \mathbf{k}_1)]^2}{\mathbf{k}^2} \right\rangle, \tag{15}$$

where $\langle \mathcal{F} \rangle$ stands for the average of the quantity $\mathcal{F}$ weighted with the heavy-meson spectral function, the T-matrix and the appropriate Bose-Einstein factors. Full details of the calculation and the results for these transport coefficients are given in [4].

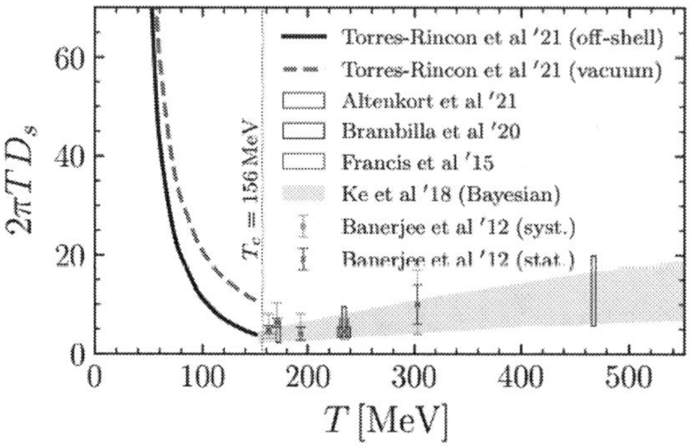

Figure 6. Spatial diffusion coefficient below and above T_c resulting from different approaches (see [4] for details). The empty boxes correspond to a range of values of $2\pi T D_s$ for the central value of T. Figure extracted from [12] with data taken from [4].

The spatial diffusion coefficient D_s is related to $\hat{B}_0$ in the static limit,

$$2\pi T D_s(T) = \lim_{\mathbf{k}\to 0} \frac{2\pi T^3}{\hat{B}_0(E_k, \mathbf{k}; T)}, \tag{16}$$

where E_k is the quasiparticle energy. The result for this transport coefficient is shown in Fig. 6 below T_c for the case of the off-shell thermal calculation (black solid line) and it is compared with the on-shell approximation (green dashed line), in which the spectral function is approximated to a δ and the thermal T-matrix is replaced by the corresponding one in vacuum. The main difference is due to the contribution of the Landau cut when thermal scattering amplitudes are employed. We stress that, when thermal corrections are included, the D-meson spatial diffusion coefficient matches smoothly the recent lattice QCD calculations and Bayesian analyses performed above T_c.

5 Conclusions

In this contribution we have presented our recent results for heavy mesons at finite temperature, including the calculation of Euclidean meson correlators and transport coefficients. We have described the interactions of heavy mesons in a bath of light mesons using a thermal effective field theory based on chiral and heavy quark symmetries. The thermal corrections to the heavy-meson self-energies and the unitarized amplitudes have been obtained in a self-consistent manner. We have shown recent results of the impact that the thermal medium has on the properties of the ground-states (masses and widths) as well as the on the states generated dynamically from the two-meson interaction.

Our results for the Euclidean correlators are compatible with the existing lattice QCD data below T_c. It would be interesting to compare both approaches at smaller temperatures and meson masses closer to the physical values, as well as to apply techniques for the reduction of excited-state contamination from the correlators in the analysis of the lattice QCD data. From the effective theory side, one might need to include the contribution of the thermal kaons to the D-meson self-energy, specially taking into account the reduction of the mass gap between kaons and pions on the lattice.

Finally, we have shown that the use of thermal scattering amplitudes to compute charm transport coefficients entails the appearance of a new term related to the Landau cut of the T-matrix at finite temperature. This contribution is absent if vacuum scattering amplitudes are employed, as commonly done in the past. In particular we have seen that its consideration gives rise to a smooth matching of the transport coefficents with lattice QCD calculations and Bayesian analyses around T_c.

This work has been supported by the projects CEX2019-000918-M (Unidad de Excelencia "María de Maeztu"), PID2019-110165GB-I00 and PID2020-118758GB-I00 financed by MCIN/AEI/10.13039/501100011033. The author also acknowledges support from the FPU17/04910 Doctoral Grant from the Spanish Ministerio de Ciencia Innovación y Universidades.

References

[1] G. Montaña, A. Ramos, L. Tolos, J.M. Torres-Rincon, Phys. Lett. B **806**, 135464 (2020), 2001.11877
[2] G. Montaña, A. Ramos, L. Tolos, J.M. Torres-Rincon, Phys. Rev. D **102**, 096020 (2020), 2007.12601
[3] G. Montaña, O. Kaczmarek, L. Tolos, A. Ramos, Eur. Phys. J. A **56**, 294 (2020), 2007.15690
[4] J.M. Torres-Rincon, G. Montaña, A. Ramos, L. Tolos (2021), 2106.01156
[5] F.K. Guo, C. Hanhart, U.G. Meissner, Eur. Phys. J. **A40**, 171 (2009), 0901.1597
[6] L. Liu, K. Orginos, F.K. Guo, C. Hanhart, U.G. Meissner, Phys. Rev. **D87**, 014508 (2013), 1208.4535
[7] L. Tolos, J.M. Torres-Rincon, Phys. Rev. **D88**, 074019 (2013), 1306.5426
[8] Z.H. Guo, L. Liu, U.G. Meißner, J.A. Oller, A. Rusetsky, Eur. Phys. J. **C79**, 13 (2019), 1811.05585
[9] A.K. Das, *Finite Temperature Field Theory* (World Scientific, New York, 1997), ISBN 9789810228569, 9789814498234
[10] H.A. Weldon, Phys. Rev. **D28**, 2007 (1983)
[11] M. Albaladejo, P. Fernandez-Soler, F.K. Guo, J. Nieves, Phys. Lett. **B767**, 465 (2017), 1610.06727
[12] G. Montana, A. Ramos, L. Tolos, J.M. Torres-Rincon, *Temperature dependence of the properties of open heavy-flavor mesons*, in *19th International Conference on Strangeness in Quark Matter* (2021), 2108.04874
[13] A. Kelly, A. Rothkopf, J.I. Skullerud, Phys. Rev. D **97**, 114509 (2018), 1802.00667
[14] Kadanoff, L.P. and Baym, G., *Quantum statistical mechanics* (New York: WA Benjamin, 1962)

EPJ Web of Conferences **258**, 04005 (2022) https://doi.org/10.1051/epjconf/202225804005
A Virtual Tribute to Quark Confinement and the Hadron Spectrum (vConf21)

Inclusive Hadroproduction of P-wave Heavy Quarkonia in pNRQCD

Hee Sok Chung[1,2,*]

[1]Physik-Department, Technische Universität München, James-Franck-Str. 1, 85748 Garching, Germany
[2]Excellence Cluster ORIGINS, Boltzmannstrasse 2, D-85748 Garching, Germany

Abstract. We compute NRQCD long-distance matrix elements that appear in the inclusive production cross sections of P-wave heavy quarkonia in the framework of potential NRQCD. The formalism developed in this work applies to strongly coupled charmonia and bottomonia. This makes possible the determination of color-octet NRQCD long-distance matrix elements without relying on measured cross section data, which has not been possible so far. We obtain results for inclusive production cross sections of χ_{cJ} and χ_{bJ} at the LHC, which are in good agreement with measurements.

1 Introduction

Hard processes involving heavy quarkonia are considered good probes of perturbative and nonperturbative aspects of QCD [1–4]. Especially, inclusive production cross sections of heavy quarkonia are studied extensively in hadron colliders. A good theoretical understanding of the quarkonium production mechanism is therefore important in deciphering QCD interactions from collider experiments.

The effective field theory nonrelativistic QCD (NRQCD) [5, 6] provides a factorization formalism for hard processes involving heavy quarkonia, which include inclusive production cross sections at large transverse momentum. NRQCD utilizes the separation of the scale of the heavy quark mass m from the scales mv and mv^2, where $v \ll 1$ is the relative velocity of the heavy quark Q and the antiquark $\bar{Q}$ inside the heavy quarkonium. The NRQCD factorization formula for inclusive production rate of a quarkonium $\mathcal{Q}$ can be written as

$$\sigma_{\mathcal{Q}+X} = \sum_n \hat{\sigma}_{Q\bar{Q}(n)+X} \langle \mathcal{O}^{\mathcal{Q}}(n) \rangle, \tag{1}$$

where the $\hat{\sigma}_{Q\bar{Q}(n)+X}$ are short-distance cross sections (SDCSs) for inclusive production of a $Q\bar{Q}$ pair in a specific color and angular momentum state n, while the long-distance matrix elements (LDMEs) $\langle \mathcal{O}^{\mathcal{Q}}(n) \rangle$ correspond to the nonperturbative probabilities for the $Q\bar{Q}$ to evolve into the quarkonium $\mathcal{Q}$. The LDMEs have known scalings in v, so that the sum (1) can be organized in powers of v, and is typically truncated at a desired accuracy in v. While the short-distance cross sections can be computed in perturbative QCD as series in the strong coupling α_s, the LDMEs must be determined nonperturbatively.

*e-mail: heesok.chung@tum.de

While it has been known how to compute the color-singlet LDMEs, determinations of the color-octet LDMEs have usually been done by relying on measured cross section data, because it has not been known so far how to compute them from first principles. This approach led to inconsistent sets of LDME determinations that do not provide a comprehensive description of the important observables involving inclusive production of heavy quarkonia [7]. Therefore, it is much desirable to be able to compute color-octet LDMEs from first principles.

In the strongly coupled regime (characterized by the hierarchy $\Lambda_{\rm QCD} \gg mv^2$), the potential NRQCD (pNRQCD) effective field theory [8–10] which is obtained by further integrating out the scale mv, provides expressions for the LDMEs for decays of heavy quarkonia into light particles in terms of quarkonium wavefunctions and gluonic correlators [11–13]. This greatly enhances the predictive power of the NRQCD factorization formalism, as the wavefunctions can be computed by solving the nonrelativistic Schrödinger equation (see, for example, refs. [14, 15]), and the gluonic correlators are universal quantities that do not depend on the heavy quark flavor or the radial excitation, and in principle can be computed in lattice QCD. It has been anticipated that a similar calculation could be done for the LDMEs for inclusive production cross sections. This formidable task has been accomplished for the first time in ref. [16] for the production of strongly coupled P-wave heavy quarkonia, and a general formalism for strongly coupled quarkonia of arbitrary quantum numbers has been established in ref. [17]. Similarly to the case of decay LDMEs, the results in refs. [16, 17] provide expressions for the production LDMEs in terms of quarkonium wavefunctions and universal gluonic correlators. These results greatly reduce the number of nonperturbative unknowns and enhances the predictive power of the NRQCD factorization formalism for inclusive production of heavy quarkonia.

2 Inclusive production of χ_{cJ} and χ_{bJ} in pNRQCD

The NRQCD factorization formula for the inclusive production rate of χ_{QJ} ($Q = c$ or b) at leading order in v is given by [6]

$$\sigma_{\chi_{QJ}} = (2J+1)\left[\hat{\sigma}_{Q\bar{Q}(^3P_J^{[1]})}\langle \mathcal{O}^{\chi_{Q0}}(^3P_0^{[1]})\rangle + \hat{\sigma}_{Q\bar{Q}(^3S_1^{[8]})}\langle \mathcal{O}^{\chi_{Q0}}(^3S_1^{[8]})\rangle\right], \quad (2)$$

where the LDMEs are defined by [6, 18–20]

$$\langle \mathcal{O}^{\chi_{Q0}}(^3P_0^{[1]})\rangle = \langle\Omega|\frac{1}{3}\chi^\dagger\left(-\frac{i}{2}\overleftrightarrow{\mathbf{D}}\cdot\sigma\right)\psi\mathcal{P}_{\chi_{Q0}(\mathbf{P}=0)}\psi^\dagger\left(-\frac{i}{2}\overleftrightarrow{\mathbf{D}}\cdot\sigma\right)\chi|\Omega\rangle, \quad (3a)$$

$$\langle \mathcal{O}^{\chi_{Q0}}(^3S_1^{[8]})\rangle = \langle\Omega|\chi^\dagger\sigma^i T^a\psi\Phi_\ell^{\dagger ab}\mathcal{P}_{\chi_{Q0}(\mathbf{P}=0)}\Phi_\ell^{bc}\psi^\dagger\sigma^i T^c\chi|\Omega\rangle. \quad (3b)$$

Here, $|\Omega\rangle$ is the QCD vacuum, ψ and χ are Pauli spinor fields that annihilate and create a heavy quark and antiquark, respectively, $\mathbf{D} = \boldsymbol{\nabla} - ig\mathbf{A}$ is the gauge-covariant derivative, $\chi^\dagger\overleftrightarrow{\mathbf{D}}\psi = \chi^\dagger\mathbf{D}\psi - (\mathbf{D}\chi)^\dagger\psi$, $\mathbf{A}$ is the gluon field, σ is a Pauli matrix, T^a is a color matrix, and the operator $\mathcal{P}_{\chi_{Q0}(\mathbf{P})}$ projects onto states that include the quarkonium χ_{Q0} with momentum $\mathbf{P}$. The gauge-completion Wilson line defined by $\Phi_\ell = \mathcal{P}\exp\left[-ig\int_0^\infty d\lambda\ell\cdot A^{\rm adj}(\ell\lambda)\right]$, where $\mathcal{P}$ is path ordering, $A^{\rm adj}$ is the gluon field in the adjoint representation, and ℓ is an arbitrary direction, is inserted to ensure the gauge invariance of the color-octet LDME [18–20]. We have used the heavy-quark spin symmetry to write eq. (2) in terms of the χ_{Q0} LDMEs [6]. An analogous formula holds for the production of h_Q.

The pNRQCD expression for the production LDMEs can be written as [17]

$$\langle\Omega|\mathcal{O}^Q(n)|\Omega\rangle = \frac{1}{\langle\mathbf{P}=\mathbf{0}|\mathbf{P}=\mathbf{0}\rangle}\int d^3x_1 d^3x_2 d^3x_1' d^3x_2'\,\phi_Q^{(0)}(\mathbf{x}_1-\mathbf{x}_2)$$
$$\times\left[-V_{\mathcal{O}(n)}(\mathbf{x}_1,\mathbf{x}_2;\boldsymbol{\nabla}_1,\boldsymbol{\nabla}_2)\delta^{(3)}(\mathbf{x}_1-\mathbf{x}_1')\delta^{(3)}(\mathbf{x}_2-\mathbf{x}_2')\right]\phi_Q^{(0)*}(\mathbf{x}_1'-\mathbf{x}_2'), \quad (4)$$

where $V_{O(n)}$ is a contact term, and $\phi_Q^{(0)}(\mathbf{x}_1 - \mathbf{x}_2)$ is the unit-normalized quarkonium wavefunction at leading order in v. The wavefunction is an eigenfunction of the pNRQCD Hamiltonian, which is given at leading order in v by

$$-\frac{\nabla_1^2}{2m} - \frac{\nabla_2^2}{2m} + V^{(0)}(\mathbf{x}_1 - \mathbf{x}_2), \tag{5}$$

where $V^{(0)}$ is the static potential. The expression in eq. (4) is valid at leading order in v, up to corrections of relative order $1/N_c^2$, where N_c is the number of colors. We refer the readers to ref. [17] for details of the pNRQCD formalism for production LDMEs.

The contact term $V_{O(n)}$ can be computed in expansion in powers of $1/m$. At leading nonvanishing orders in $1/m$, the contact terms for the LDMEs that appear in eq. (2) are given by [17]

$$-V_{O(^3P_0^{[1]})}\big|_{P\text{-wave}} = -\frac{1}{3}\sigma^i \otimes \sigma^j N_c \nabla_{\mathbf{r}}^i \delta^{(3)}(\mathbf{r}) \nabla_{\mathbf{r}}^j, \tag{6a}$$

$$-V_{O(^3S_1^{[8]})}\big|_{P\text{-wave}} = -\sigma^k \otimes \sigma^k N_c \nabla_{\mathbf{r}}^i \delta^{(3)}(\mathbf{r}) \nabla_{\mathbf{r}}^j \frac{\mathcal{E}^{ij}}{N_c^2 m^2}, \tag{6b}$$

where $\mathbf{r} = \mathbf{x}_1 - \mathbf{x}_2$, and we neglect contributions that vanish in (4) when $\phi_Q^{(0)}(\mathbf{x}_1 - \mathbf{x}_2)$ is in a P-wave state. The tensor $\mathcal{E}^{ij}$ is defined by

$$\mathcal{E}^{ij} = \int_0^\infty dt\, t \int_0^\infty dt'\, t' \, \langle \Omega | \Phi_\ell^{\dagger ab} \Phi_0^{\dagger ad}(0;t) g E^{d,i}(t) g E^{e,j}(t') \Phi_0^{ec}(0;t') \Phi_\ell^{bc} | \Omega \rangle, \tag{7}$$

where $E^{i,a}(t) = E^{i,a}(t,\mathbf{0}) = G^{i0,a}(t,\mathbf{0})$ is the chromoelectric field, $G^{\mu\nu,a}T^a = G^{\mu\nu}$ is the gluon field strength tensor, and $\Phi_0(t;t') = \mathcal{P}\exp\left[-ig\int_t^{t'} d\tau\, A_0^{\rm adj}(\tau,\mathbf{0})\right]$ is a Schwinger line.

By using the results for the contact terms $V_{O(n)}$ and eq. (4), we obtain the following expressions for the production LDMEs in pNRQCD [16, 17]

$$\langle O^{\chi_{Q0}}(^3P_0^{[1]})\rangle = \frac{3N_c}{2\pi}|R^{(0)\prime}(0)|^2, \tag{8a}$$

$$\langle O^{\chi_{Q0}}(^3S_1^{[8]})\rangle = \frac{3N_c}{2\pi}|R^{(0)\prime}(0)|^2 \frac{\mathcal{E}}{9N_c m^2}, \tag{8b}$$

where $R^{(0)}(r)$ is the radial wavefunction, and $\mathcal{E} = \frac{3}{N_c}\delta^{ij}\mathcal{E}^{ij}$ is a gluonic correlator. We expect corrections to these expressions to be suppressed by v^2. Since v^2 is comparable to $1/N_c^2$ for bottomonia, while v^2 is larger than $1/N_c^2$ for charmonia, we take the uncertainties in the pNRQCD expressions of the LDMEs to be of relative order v^2.

The correlator $\mathcal{E}$ has a logarithmic scale dependence at one loop level, which is given by

$$\frac{d}{d\log\Lambda}\mathcal{E}(\Lambda) = 12 C_F \frac{\alpha_s}{\pi} + O(\alpha_s^2), \tag{9}$$

where $C_F = (N_c^2 - 1)/(2N_c)$. This implies the following evolution equation for the LDMEs

$$\frac{d}{d\log\Lambda}\langle O^{\chi_{Q0}}(^3S_1^{[8]})\rangle = \frac{4C_F\alpha_s}{3N_c\pi m^2}\langle O^{\chi_{Q0}}(^3P_0^{[1]})\rangle. \tag{10}$$

This agrees with the known one-loop evolution equation derived from perturbative calculations in NRQCD [6]. The consistency with NRQCD factorization at two-loop level has been discussed in ref. [17].

Because the operator definition for $\mathcal{E}$ involves only gluon fields, it is independent of the radial excitation or the flavor of the heavy quark. Hence, the ratio of the color-singlet and color-octet LDMEs at leading order in v given by

$$\frac{m^2\langle \mathcal{O}^{\chi_{Q0}}(^3S_1^{[8]})\rangle}{\langle \mathcal{O}^{\chi_{Q0}}(^3P_0^{[1]})\rangle} = \frac{\mathcal{E}}{9N_c}, \tag{11}$$

is universal for all P-wave quarkonium states. This, in turn, implies that a determination of $\mathcal{E}$ leads to determination of both color-singlet and color-octet production LDMEs for all P-wave charmonium and bottomonium states.

2.1 Inclusive hadroproduction of χ_{cJ}

We now present the phenomenological results for hadroproduction rates of $\chi_{cJ}(1P)$ ($J = 1$, 2) at the LHC. We employ the SDCSs computed at next-to-leading order accuracy in α_s from ref. [21], and take $\Lambda = 1.5$ GeV for the $\overline{\rm MS}$ scale associated with the color-octet LDME.

Since a lattice QCD calculation of $\mathcal{E}$ has not yet been done, we determine $\mathcal{E}$ by comparing the cross sections computed from eq. (2) with the measured cross section ratio $r_{21} = (d\sigma_{\chi_{c2}(1P)}/dp_T)/(d\sigma_{\chi_{c1}(1P)}/dp_T)$ at the LHC, where p_T is the transverse momentum of the χ_{cJ}. Note that the ratio is independent of the radial wavefunction. We obtain [17]

$$\mathcal{E}(\Lambda = 1.5\ \text{GeV})|_{\rm NLO} = 1.17 \pm 0.05. \tag{12}$$

Since the fixed-order calculations of the SDCSs may contain large logarithms of p_T/m, resummation of the logarithms can have significant effects on their shapes in p_T. If we use the SDCSs from ref. [22, 23] which include resummed leading logarithms of p_T/m at leading power (LP) in m/p_T, we obtain [17]

$$\mathcal{E}(\Lambda = 1.5\ \text{GeV})|_{\rm LP+NLO} = 4.48 \pm 0.14. \tag{13}$$

We compare the pNRQCD result for $r_{21} \times B_{\chi_{c2}(1P)}/B_{\chi_{c1}(1P)}$ obtained in ref. [17] as a function of the transverse momentum $p_T^{J/\psi}$ of the J/ψ produced in decays of χ_{cJ}, where $B_{\chi_{cJ}(1P)} = {\rm Br}[\chi_{cJ}(1P) \to J/\psi\gamma] \times {\rm Br}(J/\psi \to \mu^+\mu^-)$, with measurements at the LHC in figure 1.

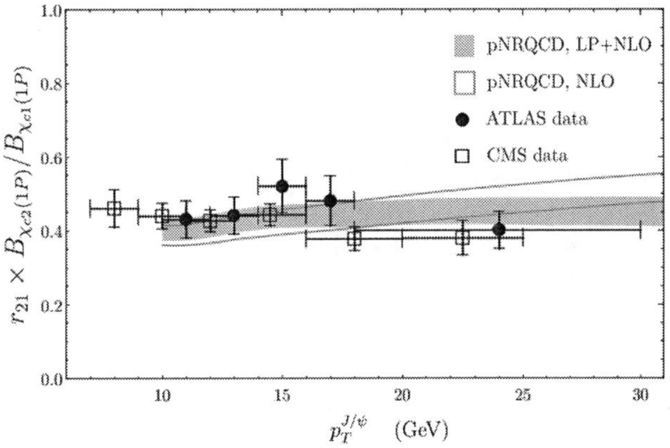

Figure 1. pNRQCD results for the cross section ratio $r_{21} \times B_{\chi_{c2}(1P)}/B_{\chi_{c1}(1P)}$ at the $\sqrt{s} = 7$ TeV LHC in the rapidity region $|y| < 0.75$ compared to CMS [24] and ATLAS [25] measurements. From ref. [17].

Based on the results for $\mathcal{E}$ that we obtain, we can compute absolute cross sections of χ_{cJ} at the LHC. We use $|R^{(0)\prime}(0)|^2 = 0.057$ GeV^5, which is determined in ref. [17] from the two-photon widths of χ_{c1} and χ_{c2} at one-loop level and measurements in ref. [26]. We show the pNRQCD results for the hadroproduction cross sections at the LHC from ref. [17] compared to ATLAS [25] measurements in figure 2.

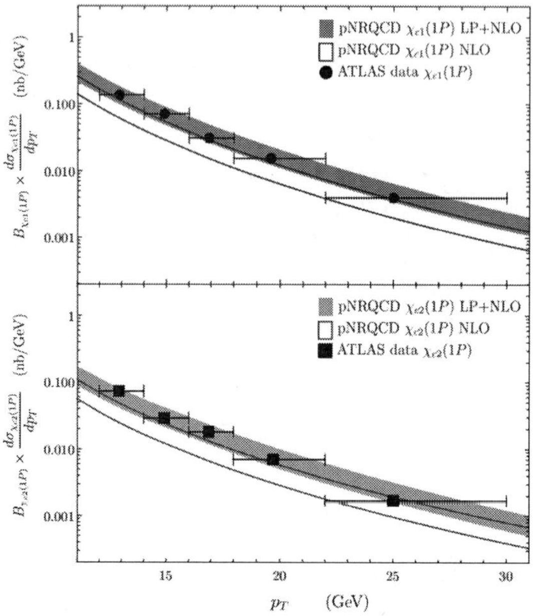

Figure 2. pNRQCD results for the production cross sections of $\chi_{c1}(1P)$ and $\chi_{c2}(1P)$ at the $\sqrt{s} = 7$ TeV LHC in the rapidity region $|y| < 0.75$ compared to ATLAS [25] measurements. From ref. [17].

Finally, we consider the polarization of the J/ψ from decays of $\chi_{c1}(1P)$ and $\chi_{c2}(1P)$. The polarization parameter $\lambda_\theta^{\chi_{cJ}}$ is defined by $\lambda_\theta^{\chi_{cJ}} = (1-3\xi_{\chi_{cJ}})/(1+\xi_{\chi_{cJ}})$, where $\xi_{\chi_{cJ}}$ is the fraction of longitudinally produced J/ψ from decays of $\chi_{cJ}(1P)$. We compare in figure 3 the pNRQCD results for $\lambda_\theta^{\chi_{c1}}$ and $\lambda_\theta^{\chi_{c2}}$ in ref. [17] with the experimental constraints from CMS [27].

2.2 Inclusive hadroproduction of χ_{bJ}

We now compute the hadroproduction rates of $\chi_{bJ}(nP)$ at the LHC, where $J = 1, 2$ and $n = 1$, 2, and 3. Similarly to the case of χ_{cJ}, we employ the SDCSs computed at next-to-leading order accuracy in α_s from ref. [21], and take $\Lambda = 4.75$ GeV for the $\overline{\text{MS}}$ scale associated with the color-octet LDME. We compute $\mathcal{E}(\Lambda = 4.75\text{ GeV})$ from $\mathcal{E}(\Lambda_0 = 1.5\text{ GeV})$ by using the one-loop renormalization-group improved formula

$$\mathcal{E}(\Lambda) = \mathcal{E}(\Lambda_0) + \frac{24C_F}{\beta_0} \log \frac{\alpha_s(\Lambda_0)}{\alpha_s(\Lambda)}, \tag{14}$$

where $\beta_0 = 11N_c/3 - 2n_f/3$ and $n_f = 4$ is the number of active quark flavors. We take the average of eq. (12) and eq. (13) to obtain $\mathcal{E}(\Lambda_0 = 1.5\text{ GeV}) = 2.8 \pm 1.7$. From this we can compute the ratio r_{21} of the $\chi_{b1}(1P)$ and $\chi_{b2}(1P)$, which is independent of the radial wavefunction. We compare the pNRQCD result for the ratio r_{21} from ref. [17] as a function of the transverse momentum $p_T^{\Upsilon(1S)}$ of the $\Upsilon(1S)$ from decays of χ_{bJ} with LHCb [28] and CMS [29] measurements in figure 4.

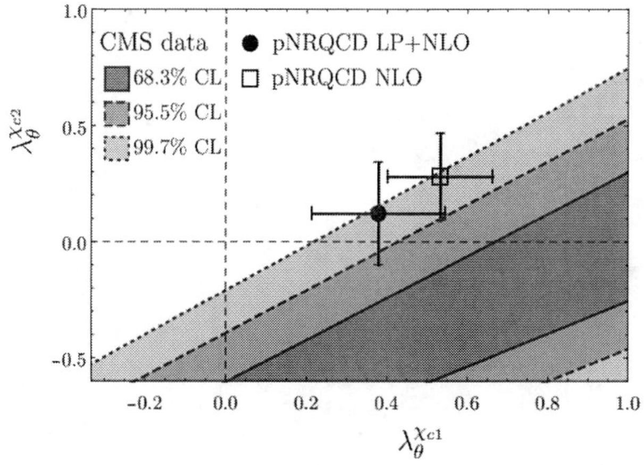

Figure 3. pNRQCD results for the polarization parameters $\lambda_\theta^{\chi_{cJ}}$ for $J = 1$ and 2 at the $\sqrt{s} = 7$ TeV LHC averaged over 8 GeV< $p_T^{J/\psi}$ < 30 GeV compared with the experimental constraints from CMS [27]. From ref. [17].

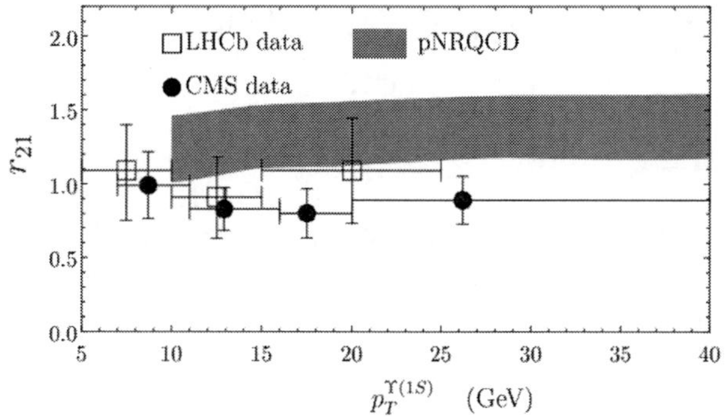

Figure 4. pNRQCD results for the cross section ratio r_{21} for the $\chi_{b1}(1P)$ and $\chi_{b2}(1P)$ states at the $\sqrt{s} = 7$ TeV LHC compared with LHCb and CMS data. From ref. [17].

In order to compute absolute cross sections of $\chi_{bJ}(nP)$, we take the values for $|R^{(0)\prime}(0)|^2$ from the averages of potential-model calculations considered in ref. [13], which are given by $|R^{(0)\prime}_{1P}(0)|^2 = 1.47$ GeV^5, $|R^{(0)\prime}_{2P}(0)|^2 = 1.74$ GeV^5, and $|R^{(0)\prime}_{3P}(0)|^2 = 1.92$ GeV^5. The results obtained in ref. [17] for the absolute cross sections of $\chi_{b1}(nP)$ and $\chi_{b2}(nP)$ for $n = 1, 2$, and 3 are shown in the left panel of figure 5. In order to compare with available measurements from LHCb [30], we compute the feeddown fractions $R^{\chi_b(nP)}_{\Upsilon(n'S)}$ for $n \geq n'$, which are the fractions of $\Upsilon(n'S)$ produced from decays of $\chi_b(nP)$. The fractions $R^{\chi_b(nP)}_{\Upsilon(n'S)}$ are computed from the $\chi_{cJ}(nP)$ production cross sections multiplied by the branching ratios $\text{Br}[\chi_{cJ}(nP) \to \Upsilon(n'S) + \gamma]$, divided by the the inclusive production rate of $\Upsilon(n'S)$. We take the direct $\Upsilon(n'S)$ production cross sections computed in ref. [31], from which we compute the inclusive production rates by adding the feeddown contributions from decays of $\chi_b(nP)$ and $\Upsilon(n''S)$, where $n \geq n'$ and $n'' \geq n' + 1$. We show the results for $R^{\chi_b(nP)}_{\Upsilon(n'S)}$ at the $\sqrt{s} = 7$ TeV LHC from ref. [17] compared with LHCb [30] data in the right panel of figure 5.

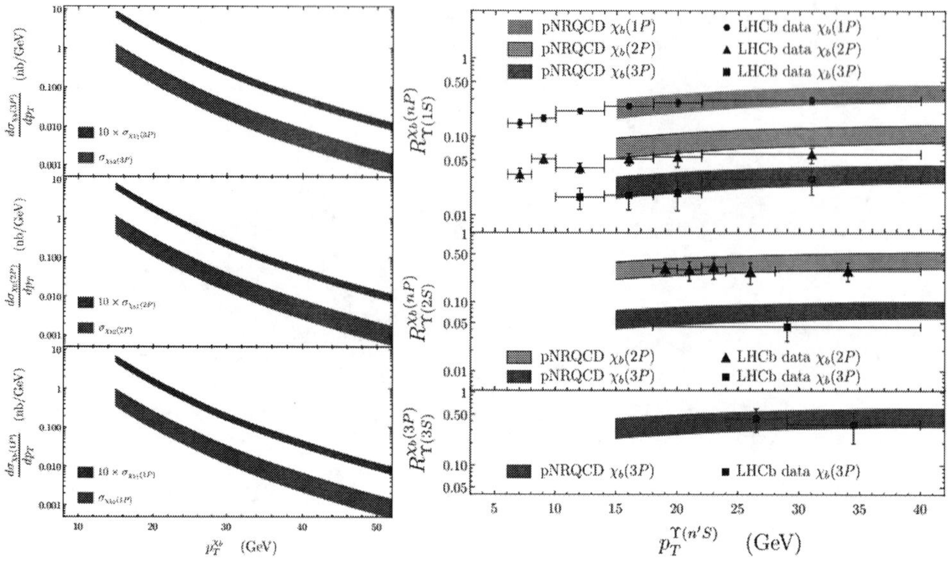

Figure 5. Left panel : pNRQCD results for the inclusive production rates of $\chi_{bJ}(nP)$ (J = 1, 2 and n = 1, 2, and 3) at the $\sqrt{s}$ = 7 TeV LHC in the rapidity range $2 < y < 4.5$. Right panel : pNRQCD results for the feeddown fractions $R^{\chi_b(nP)}_{\Upsilon(n'S)}$ compared to LHCb [30] data. From ref. [17].

3 Summary and outlook

We reviewed recent results for inclusive hadroproduction of P-wave heavy quarkonia in potential NRQCD. By working in the strong coupling regime, expressions for the NRQCD LDMEs for inclusive production of heavy quarkonia have been obtained in terms of quarkonium wavefunctions and universal gluonic correlators. This greatly reduces the number of nonperturbative unknowns, thanks to the universal nature of the gluonic correlators. In the case of P-wave heavy quarkonia, the LDMEs that appear in NRQCD factorization formulas at leading order in v can be determined from the quarkonium wavefunctions and a single gluonic correlator $\mathcal{E} = \frac{3}{N_c}\delta^{ij}\mathcal{E}^{ij}$, where the tensor $\mathcal{E}^{ij}$ is defined in eq. (7). Based on this result, the hadroproduction cross sections of χ_{cJ} and χ_{bJ} have been computed in refs. [16, 17] by using a determination of $\mathcal{E}$ from measurements of the ratio of χ_{c1} and χ_{c2} cross sections at the LHC. The phenomenological results are in reasonable agreement with available LHC data.

The formalism developed in ref. [17] for the LDMEs can be applied to inclusive production rates of any strongly coupled heavy quarkonium state. It would especially be interesting to compute in potential NRQCD the production LDMEs for S-wave quarkonia such as J/ψ, $\psi(2S)$, and Υ, which may shed light on the long-standing puzzle of J/ψ polarization and help understand the η_c production mechanism [32–36].

References

[1] N. Brambilla et al. (Quarkonium Working Group) (2004), hep-ph/0412158

[2] N. Brambilla et al., Eur. Phys. J. C **71**, 1534 (2011), 1010.5827

[3] G.T. Bodwin, E. Braaten, E. Eichten, S.L. Olsen, T.K. Pedlar, J. Russ, *Quarkonium at the Frontiers of High Energy Physics: A Snowmass White Paper*, in *Community Summer Study 2013: Snowmass on the Mississippi* (2013), 1307.7425

[4] N. Brambilla et al., Eur. Phys. J. C **74**, 2981 (2014), 1404.3723
[5] W.E. Caswell, G.P. Lepage, Phys. Lett. **167B**, 437 (1986)
[6] G.T. Bodwin, E. Braaten, G.P. Lepage, Phys. Rev. **D51**, 1125 (1995), [Erratum: Phys. Rev.D55,5853(1997)], hep-ph/9407339
[7] H.S. Chung, PoS **Confinement2018**, 007 (2018), 1811.12098
[8] N. Brambilla, A. Pineda, J. Soto, A. Vairo, Nucl. Phys. **B566**, 275 (2000), hep-ph/9907240
[9] A. Pineda, J. Soto, Nucl. Phys. B Proc. Suppl. **64**, 428 (1998), hep-ph/9707481
[10] N. Brambilla, A. Pineda, J. Soto, A. Vairo, Rev. Mod. Phys. **77**, 1423 (2005), hep-ph/0410047
[11] N. Brambilla, D. Eiras, A. Pineda, J. Soto, A. Vairo, Phys. Rev. Lett. **88**, 012003 (2002), hep-ph/0109130
[12] N. Brambilla, D. Eiras, A. Pineda, J. Soto, A. Vairo, Phys. Rev. D **67**, 034018 (2003), hep-ph/0208019
[13] N. Brambilla, H.S. Chung, D. Müller, A. Vairo, JHEP **04**, 095 (2020), 2002.07462
[14] H.S. Chung, JHEP **12**, 065 (2020), 2007.01737
[15] H.S. Chung, JHEP **09**, 195 (2021), 2106.15514
[16] N. Brambilla, H.S. Chung, A. Vairo, Phys. Rev. Lett. **126**, 082003 (2021), 2007.07613
[17] N. Brambilla, H.S. Chung, A. Vairo, JHEP **09**, 032 (2021), 2106.09417
[18] G.C. Nayak, J.W. Qiu, G.F. Sterman, Phys. Rev. D **72**, 114012 (2005), hep-ph/0509021
[19] G.C. Nayak, J.W. Qiu, G.F. Sterman, Phys. Lett. B **613**, 45 (2005), hep-ph/0501235
[20] G.C. Nayak, J.W. Qiu, G.F. Sterman, Phys. Rev. D **74**, 074007 (2006), hep-ph/0608066
[21] L.P. Wan, J.X. Wang, Comput. Phys. Commun. **185**, 2939 (2014), 1405.2143
[22] G.T. Bodwin, H.S. Chung, U.R. Kim, J. Lee, Phys. Rev. Lett. **113**, 022001 (2014), 1403.3612
[23] G.T. Bodwin, K.T. Chao, H.S. Chung, U.R. Kim, J. Lee, Y.Q. Ma, Phys. Rev. D **93**, 034041 (2016), 1509.07904
[24] S. Chatrchyan et al. (CMS), Eur. Phys. J. C **72**, 2251 (2012), 1210.0875
[25] G. Aad et al. (ATLAS), JHEP **07**, 154 (2014), 1404.7035
[26] M. Ablikim et al. (BESIII), Phys. Rev. D **85**, 112008 (2012), 1205.4284
[27] A.M. Sirunyan et al. (CMS), Phys. Rev. Lett. **124**, 162002 (2020), 1912.07706
[28] R. Aaij et al. (LHCb), JHEP **10**, 088 (2014), 1409.1408
[29] V. Khachatryan et al. (CMS), Phys. Lett. B **743**, 383 (2015), 1409.5761
[30] R. Aaij et al. (LHCb), Eur. Phys. J. C **74**, 3092 (2014), 1407.7734
[31] H. Han, Y.Q. Ma, C. Meng, H.S. Shao, Y.J. Zhang, K.T. Chao, Phys. Rev. D **94**, 014028 (2016), 1410.8537
[32] R. Aaij et al. (LHCb), Eur. Phys. J. C **75**, 311 (2015), 1409.3612
[33] M. Butenschoen, Z.G. He, B.A. Kniehl, Phys. Rev. Lett. **114**, 092004 (2015), 1411.5287
[34] H. Han, Y.Q. Ma, C. Meng, H.S. Shao, K.T. Chao, Phys. Rev. Lett. **114**, 092005 (2015), 1411.7350
[35] H.F. Zhang, Z. Sun, W.L. Sang, R. Li, Phys. Rev. Lett. **114**, 092006 (2015), 1412.0508
[36] R. Aaij et al. (LHCb), Eur. Phys. J. C **80**, 191 (2020), 1911.03326

The dipole picture and the non-relativistic expansion

Miguel Ángel Escobedo[1,*] and *Tuomas* Lappi[2,3,**]

[1]Instituto Galego de Física de Altas Enerxías (IGFAE), Universidade de Santiago de Compostela. E-15782, Galicia, Spain
[2]Department of Physics, P.O. Box 35, FI-40014 University of Jyväskylä, Finland
[3]Helsinki Institute of Physics, P.O. Box 64 00014, University of Helsinki, Finland

Abstract. We study exclusive quarkonium production in the dipole picture at next-to-leading order (NLO) accuracy, using the non-relativistic expansion for the quarkonium wavefunction. The quarkonium light cone wave functions needed in the dipole picture have typically been available only at tree level, either in phenomenological models or in the nonrelativistic limit. Here, we discuss the compatibility of the dipole approach and the non-relativistic expansion and compute NLO relativistic corrections to the quarkonium light-cone wave function in light-cone gauge.

1 Introduction

Exclusive quarkonium production is an ideal observable to study the gluon distribution at low x in Deep Inelastic Scattering and UltraPeripheral collisions. On one hand, exclusive processes depend on the gluon distribution quadratically (as opposed to inclusive ones, which depend on it linearly). On the other hand, we expect heavy quarkonium to be less sensitive to non-perturbative phenomena since the heavy quark mass m is much bigger than Λ_{QCD}.

We are going to study this process using light cone perturbation theory (LCPT). The application of LCPT to the study of exclusive process is called the dipole picture (for a review see [1]). In this framework, at leading order, the cross-section for the process of a virtual photon to interact with a nucleus (without destroying it) and transform into a quarkonium state fulfils the following equation

$$\frac{d\sigma_{T,L}^{\gamma^*+N\to HQ+N}}{dt} = \frac{1}{16\pi}\left|\int d^2r_\perp \int_0^1 \frac{dz}{4\pi}\left(\Psi^*_{HQ}\Psi_{\gamma^*}\right)_{T,L}\sigma_{q\bar{q}}\right|^2 . \tag{1}$$

Here Ψ_{γ^*} is the virtual photon wave function. Its transverse extent is of the order of the inverse of the virtuality Q, which fulfils $Q \gg \Lambda_{QCD}$. Therefore, Ψ_{γ^*} is dominated by perturbative physics. The dipole cross-section $\sigma_{q\bar{q}}$, is the quantity that contains the information about the gluon density in the nucleus. Finally, Ψ_{HQ} is the quarkonium wave function. Note that eq. (1) is just the LO result, at higher orders gluons have to be included in the wave functions of the photon and quarkonium. As a result, it is also needed to take into account the cross-section of the nucleus with more complex partonic configurations, such as $\sigma_{q\bar{q}g}$ and so on.

*e-mail: miguelangel.escobedo@usc.es
**e-mail: tuomas.v.v.lappi@jyu.fi

Our aim is to obtain as much information as possible about the properties of the nucleus from the measurements of quarkonium exclusive production. However, the precision at which this can be done depends on our knowledge of the wave function of quarkonium. Several models have been used in the literature to describe this wave function (see [1] for a discussion of some of them). In this work we are going to study the results that can be obtained by treating quarkonium as a non-relativistic system. There are many reasons to consider this an interesting approach. First, it is generally considered that this is a good approximation. In fact, it is a very common approach in the literature when studying other problems related to quarkonium. Examples of this are the study of inclusive production, spectroscopy and quarkonium decays. Moreover, the non-relativistic limit is a well-defined limit of QCD that is worth studying on its own right. Finally, the non-relativistic limit has been used previously in the dipole picture literature at leading order [2, 3] and we aim to develop a framework that allows to extend these results to higher orders in perturbation theory.

This manuscript is organized as follows, in section 2 the quarkonium wave function in the non-relativistic limit is discussed. In section 3 we describe some cross-checks that this wave function is able to fulfil. In section 4 we discuss the application of these results to the computation of exclusive quarkonium production in the limit $Q \gg m$. Finally, in section 5 we give the conclusions. A more extended discussion of these results can be found in [4].

2 Quarkonium light cone wave function in the non-relativistic limit

At leading order the light cone wave function can be computed only taking into account non-relativistic quarks and their interactions. In LCPT they are defined by having a transverse momentum $p_\perp$ much smaller than m and a momentum fraction close to $\frac{1}{2}$. We allow relativistic degrees of freedom to be part of the wave function of quarkonium. However, they appear as perturbations, that is to say, they appear during small time periods. The mathematical structure of the light cone wave function of quarkonium in our approach is the following. The quarkonium wave function for short distances $\mathbf{x}_\perp \sim \frac{1}{m}$ fulfils

$$\int dz f(z)\Psi^n_{HQ}(z,\mathbf{x}_\perp) = \sum_{m,k}\int dz f(z) C^k_{n\leftarrow m}(z,\mathbf{x}_\perp)\left(\frac{\nabla}{m}\right)^k \int \frac{d\lambda}{4\pi}\phi^m(\lambda,\mathbf{0})\,. \tag{2}$$

In this formula the indices n and m represent the particle content; for higher Fock states the single two-particle relative coordinates $z, \mathbf{x}_\perp$ are replaced by the appropriate variables. The notation Ψ_{HQ} refers to the full light-cone wave function of quarkonium, while the ϕ's are the wavefunctions restricted to the case in which all particles are non-relativistic (therefore $\lambda \equiv z - \frac{1}{2} \ll 1$). Here $f(z)$ is just a test function written to represent the fact that the equality is only valid when integrating over z. Note that each power of $\left(\frac{\nabla}{M}\right)$ acting on ϕ gives a suppression of order v, the non-relativistic velocity of the heavy quarks. At the moment we are interested in the leading order corrections in α_s . In this approximation we only need to take into account terms with $k = 0$. More precisely, we only need to compute the leading order corrections of $C^0_{q\bar{q}\leftarrow q\bar{q}}$ and the leading order contribution to $C^0_{q\bar{q}g\leftarrow q\bar{q}}$. A study of finite k corrections can be found in [5].

3 Cross-checks

The first cross-check we discuss is the computation of the light-cone distribution amplitude. This is a quantity that encodes the properties of hadrons in computations in which the energy of the collision is much bigger than the energy scales ruling the behaviour of the hadron. In the light-cone gauge, it can be obtained from the light cone wave function in the limit $r_\perp = 0$[1]. Moreover, this quantity must fulfil the Efremov-Radyushkin-Brodsky-Lepage (ERBL) equation [6, 7].

$$\frac{\partial D^i(z)}{\partial \log \mu^2} = \frac{\alpha_s C_F}{2\pi} \int_0^1 dz' K_{L,T}(z,z') D^i(z') \,, \tag{3}$$

where $K_{L,T}(z,z')$ is the ERBL kernel. We have checked explicitly in [4], both in longitudinal and transverse polarization, that the distribution amplitudes obtained by our wave functions fulfil this requirement.

Now we move to the discussion of our second cross-check. The cross-section of the process of quarkonium decaying into a pair of leptons can be computed knowing the light cone wave function of quarkonium. At the same time, in the context of NRQCD and related approaches the leading order corrections to this cross-section are well-known [8]. Knowing this, we can deduce that

$$\int_0^1 dz \sum_n \Psi^n_{HQ}(z, \mathbf{0}_\perp) = \left(1 - \frac{2\alpha_s C_F}{\pi}\right) \int d\lambda \phi(\lambda, \mathbf{0}) \,. \tag{4}$$

Note that the previous equation is valid when all divergences are regulated using dimensional regularization. This is not our case since, as it is usually done in LCPT literature, we use a cut-off x_0 to regulate in the + component. This will have consequences in the rest of this subsection.

We focus on the case of longitudinal polarization. Performing this computation we obtain

$$\int_0^1 dz \Psi^{q\bar{q}}_{HQ}(z, \mathbf{0}_\perp) = \left(1 + \frac{\alpha_s C_F}{\pi}\left(\frac{1}{x_0} - 2\right)\right) \int d\lambda \phi(\lambda, \mathbf{0}) \,. \tag{5}$$

The origin of the $\frac{1}{x_0}$ divergence is the Coulomb singularity due to the exchange of gluons with $p_\perp \sim m x_0 \ll m$ and $\lambda \sim x_0 \ll 1$. The computation is consistent if the following equation is fulfilled

$$\frac{d}{dx_0} \int_0^1 dz \Psi^{q\bar{q}}_{HQ}(z, \mathbf{0}_\perp) = -\frac{\alpha_s C_F}{\pi x_0^2} \int d\lambda \phi(\lambda, \mathbf{0}) + \frac{d}{dx_0} \int d\lambda \phi(\lambda, \mathbf{0}) = 0 \,. \tag{6}$$

We have checked that it is the case.

4 Exclusive quarkonium production in the $Q \gg m$ limit

Our final goal is to compute the exclusive quarkonium production in the general case. However, in order to do so we would need the one loop corrections to the photon wave function with massive quarks, which were computed only recently [9]. In this section we focus on the structure of the divergences in the case $Q \gg m$ and for longitudinal polarization[2]. We get that all divergences are correctly cancelled. For this it is crucial to take into account the B-JIMWLK evolution of the target.

The tree level result for the exclusive production of quarkonium has two terms which are sensitive to x_0. One is $\sigma_{q\bar{q}}$, which must fulfil the B-JIMWLK evolution equation. The other is the non-relativistic wave function, with fulfils eq. (6).

[1]In practice this is not so straightforward, to take the limir $r_\perp \to 0$ introduces ultraviolet divergences which we regulate using dimensional regularization

[2]A study of the more general $Q \sim m$ case using the results presented here can be found in [10].

Now we discuss the corrections which only involve the photon wave function. In the limit $Q \gg m$ and in order to study ultraviolet and collinear singularities, we can use the results with massless quarks of [11].

$$\Psi_\gamma(z,r_\perp)\Big|_{NLO} = \Psi_\gamma(z,r_\perp)\Big|_{LO}\left(1 + \delta Z_\gamma(z,r_\perp)\right). \tag{7}$$

In our case we need $Z_\gamma\left(\frac{1}{2}, r_\perp\right)$. We note that this quantity has a dependence on both x_0 and μ.

Now we discuss the corrections which only involve the quarkonium wave function. Regarding the dependence on μ, this is only given by the heavy quark wave function renormalization coefficient Z, which fulfils $\frac{d\delta Z}{d\mu} = \frac{d\delta Z_\gamma(\frac{1}{2},r_\perp)}{d\mu}$. The dependence on x_0 can be divided into two terms. One of these cancels the one given by eq. (6), and the other is equal to the dependence on x_0 given by the photon wave function.

Finally, we discuss the divergences given by diagrams in which a gluon is emitted before the interaction with the target and absorbed after it, in such a way that the amplitude of the process is proportional to $\sigma_{q\bar{q}g}$. The dependence on μ is such that it cancels those given by the photon and quarkonium wave functions. The dependence with x_0 can be divided in two pieces. One is proportional to $\sigma_{q\bar{q}g} - \sigma_{q\bar{q}}$ which cancels the B-JIMWLK evolution of the target. The other is proportional to $\sigma_{q\bar{q}}$ and cancels the divergences given by the photon and quarkonium wave functions, except for the piece which is cancelled by eq. (6). With this we have finished the discussion of the cancellation of all divergences which appear in the calculation.

5 Conclusions

In this work we have discussed the next-to-leading order corrections to the heavy quarkonium light-cone wave function in the non-relativistic limit. As cross-checks, we have seen that the light-cone distribution amplitude that we can obtain from it fulfils the ERBL equation and that using our results we can obtain the one loop corrections to the decay of quarkonium into leptons. To our knowledge, this is the first computation of this process in the light-cone gauge. Regarding the application of our computation to the study of exclusive quarkonium production, which is our final goal, we have checked, by studying the $Q \gg m$ limit, that all divergences are correctly cancelled. A phenomenological application of the results presented here can be found in [10].

Acknowledgements

We are thankful to G. Beuf, V. Guzey and H. Mäntysaari for discussions. The content of this article does not reflect the official opinion of the European Union and responsibility for the information and views expressed therein lies entirely with the authors. This work has been supported by the Academy of Finland, projects No. 267321, 303756 and No. 321840 and by the European Research Council (ERC) under the European Union's Horizon 2020 research and innovation programme (grant agreement No ERC-2015-CoG-681707). The work of M.A.E. was supported by the Academy of Finland project 297058, by *Ministerio de Ciencia e Innovacion* of Spain under project FPA2017-83814-P, Maria de Maetzu Unit of Excellence MDM-2016- 0692, by Xunta de Galicia and FEDER, and by the European Research Council project ERC-2018-ADG-835105 YoctoLHC.

References

[1] H. Kowalski, L. Motyka, G. Watt, Phys. Rev. D **74**, 074016 (2006), hep-ph/0606272

[2] M.G. Ryskin, Z. Phys. C **57**, 89 (1993)
[3] S.J. Brodsky, L. Frankfurt, J.F. Gunion, A.H. Mueller, M. Strikman, Phys. Rev. D **50**, 3134 (1994), hep-ph/9402283
[4] M.A. Escobedo, T. Lappi, Phys. Rev. D **101**, 034030 (2020), 1911.01136
[5] T. Lappi, H. Mäntysaari, J. Penttala, Phys. Rev. D **102**, 054020 (2020), 2006.02830
[6] G.P. Lepage, S.J. Brodsky, Phys. Rev. D **22**, 2157 (1980)
[7] V.L. Chernyak, A.R. Zhitnitsky, Phys. Rept. **112**, 173 (1984)
[8] R. Barbieri, R. Gatto, R. Kogerler, Z. Kunszt, Phys. Lett. B **57**, 455 (1975)
[9] G. Beuf, T. Lappi, R. Paatelainen, Phys. Rev. D **104**, 056032 (2021), 2103.14549
[10] H. Mäntysaari, J. Penttala, Phys. Lett. B **823**, 136723 (2021), 2104.02349
[11] G. Beuf, Phys. Rev. D **96**, 074033 (2017), 1708.06557

EPJ Web of Conferences **258**, 04007 (2022) https://doi.org/10.1051/epjconf/202225804007
A Virtual Tribute to Quark Confinement and the Hadron Spectrum (vConf21)

Molecular interpretation of the LHCb pentaquarks from an analysis of $J/\psi p$ spectrum

Meng-Lin Du[1,2,*]

[1]Instituto de Física Corpuscular (centro mixto CSIC-UV), Institutos de Investigación de Paterna, Apartado 22085, 46071, Valencia, Spain
[2]Helmholtz-Institut für Strahlen- und Kernphysik and Bethe Center for Theoretical Physics, Universität Bonn, D-53115 Bonn, Germany

Abstract. A coupled-channel approach including the $\Lambda_c\bar{D}^{(*)}$ and $\eta_c p$ channels in addition to the $\Sigma_c^{(*)}\bar{D}^{(*)}$ and $J/\psi p$ channels, as required by unitarity and heavy quark spin symmetry, is applied to the hidden-charm pentaquark P_c states, i.e., $P_c(4312)$, $P_c(4440)$ and $P_c(4457)$, discovered by LHCb Collaboration. It is demonstrated that to obtain cutoff independent results, the one-pion exchange potential in the multichannel systems is to be supplemented with next-leading order counter terms responsible for the S-wave-to-D-wave transitions. We show that the experimental data for the $J/\psi p$ mass distributions are fully in line with the $\Sigma_c\bar{D}$ and $\Sigma_c\bar{D}^*$ hadronic molecular interpretation of the $P_c(4312)$ and $P_c(4440)/P_c(4457)$, respectively. A narrow $\Sigma_c^*\bar{D}$ molecule around 4.38 GeV is required by the HQSS with the evidence for its existence seen in the $J/\psi p$ spectrum. Moreover, we predict the line shapes for the elastic and inelastic channels.

1 Introduction

The observation of the $P_c(4380)$ and $P_c(4450)$ [1], as the first well established signals of pentaquarks, in 2015 builds a landmark in searching for the so-called exotic states allowed by quantum chromodynamics (QCD), i.e., the color neutral objects composed rather than a pair of quark and antiquark ($q\bar{q}$) or three quarks (qqq). These exotic structures are observed in the $J/\psi p$ channel in $\Lambda_b^0 \to J/\psi K^- p$ decays, which implies that they contain at least five quarks, $c\bar{c}uud$. The $P_c(4450)$ is confirmed by an updated analysis based on an combined data set collected in Run 1 and Run 2 by LHCb Collaboration in 2019 [2], and it is observed to consist of two narrow overlapping structures, $P_c(4440)$ and $P_c(4457)$. Nevertheless, since the one-dimensional fits to the $m_{J/\psi p}$ distributions alone cannot distinguish broad P_c structures from background varying slowly with $m_{J/\psi p}$, a verification of the broad $P_c(4380)$ observed in Ref. [1] awaits an full amplitude analysis of the new data. At the same time, a new narrow pentaquark state $P_c(4312)$ is discovered in the same channel [2]. The discovery of these pentaquark states immediately spurred many theoretical works and various interpretations on the nature of these pentaquark states were proposed, including hadronic molecules [3–10], compact pentaquark states [11–13], hadrocharmonia [14], and cups effects [15, 16]. An amplitude analysis focusing on the $P_c(4312)$ peak suggests the virtual state origin of

*e-mail: du.menglin@ific.uv.es

the $P_c(4312)$ in Ref. [17]. Among these explanations, the molecular picture is of particular interest since all of the three narrow pentaquark states can be simultaneously interpreted as $\Sigma_c\bar{D}^{(*)}$ bound states with help of heavy quark spin symmetry (HQSS). The charmonium pentaquark with strangeness was predicted in Refs. [18–21] and its signal was first observed by LHCb Collaboration in 2020 [22], i.e., $P_{cs}(4459)$. Its observation has promoted numerous studies of this resonance [23–26]. These studies typically favor a molecular picture with a dominant component of $\Xi_c\bar{D}^*$.

The proximity of the $\Sigma_c\bar{D}^{(*)}$ thresholds to these narrow P_c suggests an important role played by the corresponding two-hadron states in the dynamics of the P_c states. In the common molecular picture, the $P_c(4312)$ is regarded as an S-wave $\Sigma_c\bar{D}$ bound state, while the $P_c(4440)$ and $P_c(4457)$ are bound states of $\Sigma_c\bar{D}^*$ with different spin structures. In order to describe the measured $J/\psi p$ mass distributions, the inelastic channels need to be considered to account for the width of P_c's. In addition to the $J/\psi p$ channel, there are more inelastic channels, most prominently $\Lambda_c\bar{D}^{(*)}$ and $\eta_c p$. The latter is connected to the $J/\psi p$ channel via HQSS. When the $\Lambda_c\bar{D}^{(*)}$ channel is included explicitly, the three-body cut $\Lambda\bar{D}^*\pi$ effects need to be taken into account as the width of $\Sigma_c^{(*)}$ is almost saturated by the decay $\Sigma_c^{(*)} \to \Lambda_c\pi$. In this talk, we will include all the elastic channels $\Sigma_c^{(*)}\bar{D}^{(*)}$ and inelastic channels, $J/\psi p$, $\eta_c p$ and $\Lambda_c\bar{D}^{(*)}$, explicitly, and investigate the role played by the $\Lambda_c\bar{D}^{(*)}$ and the nonperturbative dynamical pions as the proximity of the P_c states to the $\Lambda_c\bar{D}^{(*)}\pi$ thresholds. For the case of only short-distance contact potentials, two solutions were found describing the data almost equally well [9]. The most prominent contribution from the one-pion exchange (OPE) originates from the D-waves, especially the S-wave-to-D-wave transitions [27–29]. Renormalization of the OPE in the energy region of interest requires an inclusion of higher order contact potentials [28, 29]. When the OPE potentials are included together with the necessary next-leading order (NLO) S-D counter terms, only one of the two solutions present in purely contact potentials survives under the requirement of the renormalizability (regulator-independent), which suggests that the quantum numbers of the $P_c(4440)$ and $P_c(4457)$ should be $J^P = \frac{3}{2}^-$ and $\frac{1}{2}^-$, respectively. We also show that there is evidence for the existence of a narrow $\Sigma_c^*\bar{D}$ bound state, required by the HQSS, in the $m_{J/\psi p}$ mass distribution around 4.38 GeV. The width of this $P_c(4380)$ is of the same order of the three narrow P_c states as a consequence of HQSS and thus different from the broad one reported by LHCb in 2015 [1].

2 Framework

2.1 Interactions

The effective interactions contain both short-range contact potentials and long-range OPE. The leading-order (LO) effective Lagrangian for the $\Sigma_c^{(*)}\bar{D}^{(*)}$ and $\Lambda_c\bar{D}^{(*)}$ interactions respecting HQSS reads

$$\begin{aligned}\mathcal{L}_{\rm LO} = & -C_a S^{i\dagger}_{ab} S^{i}_{ba}\langle \bar{H}^\dagger_c \bar{H}_c\rangle - C_b i\epsilon_{jik} S^{j\dagger}_{ab} S^{k}_{ba}\langle \bar{H}^\dagger_c \sigma^i \bar{H}_c\rangle \\ & + C_c\left(S^{i\dagger}_{ab} T_{ca}\langle \bar{H}^\dagger_c \sigma^i \bar{H}_b\rangle - T^\dagger_{ca} S^{i}_{ab}\langle \bar{H}^\dagger_b \sigma^i \bar{H}_c\rangle\right) - C_d T^\dagger_{ab} T_{ba}\langle \bar{H}^\dagger_c \bar{H}_c\rangle, \end{aligned} \tag{1}$$

where $\langle\ldots\rangle$ represents the trace in the spinor space and σ denotes the Pauli matrices. The subindices a, b, c are the flavor indices, and the superindices i, j, k denote the polarization. The heavy quark spin doublets for the ground states (Σ_c, Σ_c^*), $(\bar{D}, \bar{D}^*)$, and Λ_c^+ are grouped in S^i, $\bar{H}$, and T, respectively, [30]

$$S^i = \frac{1}{\sqrt{3}}\sigma^i\Sigma_c + \Sigma_c^{*i}, \qquad \bar{H} = -\bar{D} + \sigma\cdot\bar{D}^*, \tag{2}$$

with

$$S_c^{(*)} = \begin{pmatrix} \Sigma_c^{(*)++} & \frac{1}{\sqrt{2}}\Sigma_c^{(*)+} \\ \frac{1}{\sqrt{2}}\Sigma_c^{(*)+} & \Sigma_c^{(*)0} \end{pmatrix}, \qquad \bar{D}^{(*)} = \begin{pmatrix} \bar{D}^{(*)0} \\ D^{(*)-} \end{pmatrix}, \quad T = \begin{pmatrix} 0 & \frac{1}{\sqrt{2}}\Lambda_c^+ \\ -\frac{1}{\sqrt{2}}\Lambda_c^+ & 0 \end{pmatrix}. \tag{3}$$

The transitions between the $\Sigma_c^{(*)}\bar{D}^{(*)}$ and $J/\psi p$ ($\eta_c p$) can be described by [31]:

$$\mathcal{L}_{\text{ine}} = \frac{g_S}{\sqrt{3}} N^\dagger \sigma^i \bar{H} J^\dagger S^i - \sqrt{3} g_D N^\dagger \sigma^i \bar{H} (\partial^i \partial^j - \frac{1}{3}\delta^{ij}\partial^2) J^\dagger S^j, \tag{4}$$

where N denotes the nucleon doublet, and $J = -\eta_c + \boldsymbol{\sigma} \cdot \boldsymbol{\psi}$ collects the ground state charmonium fields. The LO contact potentials can be derived from a more intuitive way by decomposing the spin structures into the heavy and light degrees of freedom, see e.g. in Refs. [9, 10].

The LO OPE potential can be obtained by the effective Lagrangian for the axial coupling of pions to charmed mesons and baryons [32],

$$\mathcal{L} = \frac{g_1}{4}\langle \boldsymbol{\sigma} \cdot \boldsymbol{u}_{ab} \bar{H}_b \bar{H}_a^\dagger \rangle + i g_2 \epsilon_{ijk} S^{i\dagger}_{ab} u^j_{bc} S^k_{ca} - \frac{1}{\sqrt{2}} g_3 (S^{i\dagger}_{ab} u^i_{bc} T_{ca} + T^\dagger_{ab} u^i_{bc} S^i_{ca}), \tag{5}$$

where $\boldsymbol{u} = -\nabla\Phi/F_\pi$ with $\Phi = \boldsymbol{\tau} \cdot \boldsymbol{\pi}$. Here $F_\pi = 92.1$ MeV is the pion decay constant and $g_1 = 0.57$ is determined from the width of $D^{*+} \to D^0\pi^+$, $g_2 = 0.42$ and $g_3 = 0.71$ are taken from a lattice QCD calculation [33].

2.2 Lippmann-Schwinger equations

In order to describe the $\Lambda_b^0 \to J/\psi p K^-$ decay, we need the bare weak production vertices as well. In this work, we only consider the S-wave $\Sigma_c^{(*)}\bar{D}^{(*)}$ source for the $\Lambda_b^0 \to \Sigma_c^{(*)}\bar{D}^{(*)}K^-$ process since the $\Sigma_c^{(*)}\bar{D}^{(*)}$ thresholds are close to the energy of interest. We introduce seven parameters for the all seven S-wave $\Sigma_c^{(*)}\bar{D}^{(*)}$ sources [9, 10].

With the above ingredients, one can obtain the production amplitude satisfying unitarity by solving the Lippmann-Schwinger equations (LSEs), with a diagrammatic representation shown in Fig. 1,

$$\begin{aligned} U_\alpha^J(E,p) &= P_\alpha^J - \sum_\beta \int \frac{d^3\boldsymbol{q}}{(2\pi)^3} V^J_{\alpha\beta}(E,p,q) G_\beta(E,q) U_\beta^J(E,q), \\ U_i^{(\prime)J}(E,k) &= -\sum_\beta \int \frac{d^3\boldsymbol{q}}{(2\pi)^3} V_{i\beta}^{(\prime)J}(k) G_\beta(E,q) U_\beta^J(E,q), \end{aligned} \tag{6}$$

with U_α^J and P_α denote the production amplitude and the source for the αth elastic channel ($\Lambda_c\bar{D}^{(*)}$ channels as well when they are included) and $U_i^{(\prime)J}$ for the ith inelastic channel $J/\psi p$ ($\eta_c p$). The direct $J/\psi p$ and $\eta_c p$ interactions are neglected since they are Okubo-Zweig-Iizuka suppressed and were found very weak in a recent lattice QCD calculation [34] and a coupled-channel model in Ref. [35]. In addition, we assume that the effect from the interactions of the $\Lambda_c\bar{D}^{(*)}$ with $J/\psi p$ and $\eta_c p$ can be neglected.

To take into account the widths of $\Sigma_c^{(*)}$ which are comparable to those of the P_c states, the two-body propagator is taken as the following form

$$G_\beta(E,\boldsymbol{q}) = \frac{m_{\Sigma_c^{(*)}} m_{D^{(*)}}}{E_{\Sigma_c^{(*)}}(\boldsymbol{q}) E_{D^{(*)}}(\boldsymbol{q})} \frac{1}{E_{\Sigma_c^{(*)}}(\boldsymbol{q}) + E_{D^{(*)}}(\boldsymbol{q}) - E - \tilde{\Sigma}_R(s, m_{\Sigma_c^{(*)}})/(2E_{\Sigma_c^{(*)}}(\boldsymbol{q}))}, \tag{7}$$

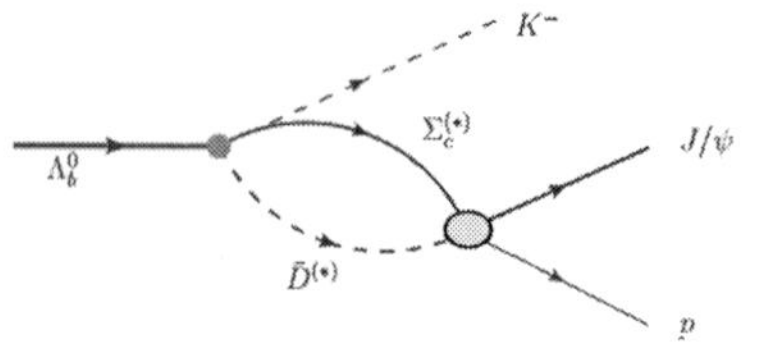

Figure 1. Diagram for the $\Lambda_b^0 \to J/\psi p K^-$ through the $\Lambda_b^0 \to \Sigma_c^{(*)} \bar{D}^{(*)} K^-$.

with $s = (E - E_{D^{(*)}}(\boldsymbol{q}))^2 - \boldsymbol{q}^2$ for the off-shell $\Sigma_c^{(*)}$ and $\tilde{\Sigma}_R(s, m_{\Sigma_c^{(*)}}) = ig^2\lambda^{3/2}(s, m^2_{\Lambda_c}, m^2_\pi)/(8s^{5/2})$ the renormalized self-energy loop diagram. The effect of the $\bar{D}^{(*)}$ width is neglected as it is only several tens of keV. To render the integrals in the LSEs in Eq. (6) well defined we introduce a regulator, with a hard cutoff Λ larger than all typical three-momenta scales to the relevant degrees of freedom.

3 Fit results

To fit the $J/\psi p$ invariant mass distribution, a smooth incoherent background is exploited to model the contributions from misidentified non-Λ_b^0 events, crossed-channel effects and possibly additional broad P_c^+ structures [10]. First we perform fits employing only the contact potentials, referred to as Scheme I. In this circumstance, it is found that the data can be well described without including the $\Lambda_c \bar{D}^{(*)}$ channels and the inclusion of these channels does not improve the fit quality. As in Ref. [9], two solutions, denoted as A and B, are found describing the data almost equally well. Both solutions give seven poles for the $\Sigma_c^{(*)}\bar{D}^{(*)}$ scattering amplitudes, i.e., seven P_c states. In both solutions, the $P_c(4312)$ corresponds to a $\Sigma_c\bar{D}$ bound state with $J^P = \frac{1}{2}^-$. The $P_c(4440)$ and $P_c(4457)$ are bounds states of $\Sigma_c\bar{D}^*$, with quantum numbers $\frac{1}{2}^-$ and $\frac{3}{2}^-$, respectively, in solution A and however interchanged in solution B. Three P_c states dominated by the $\Sigma_c^*\bar{D}^*$ are found in two solutions with $m_{\frac{1}{2}^-} < m_{\frac{3}{2}^-} < m_{\frac{5}{2}^-}$ for solution A and the opposite mass pattern for B. Those states are invisible in the $m_{J/\psi p}$ mass distributions and were suggested to search for in a prompt production at the center-of-mass energy 7 TeV in the pp collision [36]. In particular, in both solutions there is a narrow pole around 4.38 GeV corresponding to a $\Sigma_c^*\bar{D}$ bound state with $J^P = \frac{3}{2}^-$. It is a consequence of the HQSS and different to the broad resonance reported by LHCb in 2015 [1].

As inspired by Scheme I, we assume that the widths of the P_c states are saturated by the $J/\psi p$, $\eta_c p$, the elastic channels (including those via the widths of $\Sigma_c^{(*)}$) and investigate the effects from the OPE for the elastic channels. The importance of the OPE is known as its tensor force mixes the S and D waves and can leave a significant impact on the line shapes. The large typical momenta, $p_{\text{typ}} \sim \sqrt{2\mu\delta} \sim 670$ MeV with δ and μ the largest threshold splitting and the reduced mass of the $\Sigma_c^{(*)}\bar{D}^{(*)}$ system, make the contribution from the tensor force of the OPE even more important than that in the nucleon interaction [27, 28]. The iteration of such a potential within the LSEs was shown to yield a strong regulator dependence for the observable quantities. The cutoff-dependence for the $J/\psi p$ and the $\Sigma_c^{(*)}\bar{D}^{(*)}$ invariant mass distributions are demonstrated in Ref. [10]. In order to cure the cutoff dependence, we include the NLO S-D contact terms for the elastic channels together with the OPE as suggested in Refs. [29]. In what follows, this formulation will be referred to as Scheme II. In this case, there are in principle two best fits corresponding A and B for small cutoffs. The solution B demonstrate only a milder regulator dependence, especially for cutoffs $\Lambda > 1$ GeV which provide a larger separation between the soft and hard scales. On the contrary, the

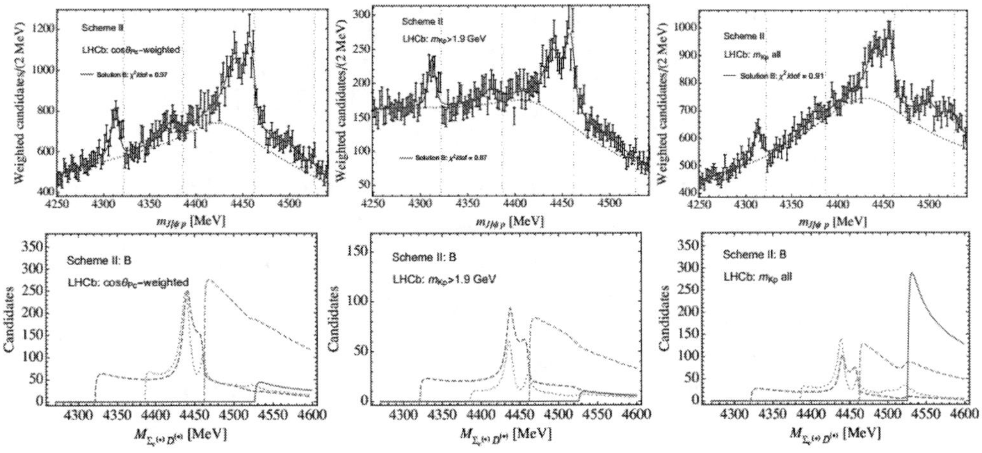

Figure 2. First row: The best fits to the different experimental data [2] for solution B (red solid curves). The corresponding backgrounds are shown as red-dotted curves. Second row: The corresponding predictions for the line shapes of $\Sigma_c\bar{D}$ (blue dashed), $\Sigma_c^*\bar{D}$ (orange dotted), $\Sigma_c\bar{D}^*$ (green dot-dashed), and $\Sigma_c^*\bar{D}^*$ (red solid) mass distributions.

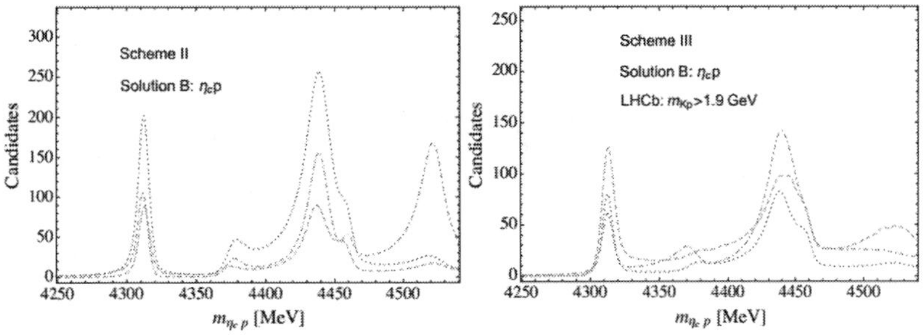

Figure 3. Predictions for the line shapes of the $\eta_c p$ invariant mass distributions based on various fit results for solution B of Scheme II (left) and Scheme III (right), respectively. The blue dotted, orange dashed and green dot-dashed curves correspond to the predictions from the three fit results ($\cos\theta_{P_c}$-weighted, $m_{Kp} > 1.9$ GeV and m_{Kp}-all), in order.

solution A still show a strong cutoff dependence and the $J/\psi p$ line shape can not be well described with cutoffs greater than 1.5 GeV, especially for the peak $P_c(4440)$. Therefore, we discard the solution A and only focus on the results of solution B. More details on the cutoff dependence for two solutions can be found in Ref. [10].

The best fits to the different experimental data [2] are shown in Fig. 2 with corresponding predictions for the line shapes of the $\Sigma_c^{(*)}\bar{D}^{(*)}$ mass distributions. Predictions for the $\eta_c p$ line shape are shown in Fig. 3. The poles lies in the complex energy plane of unphysical Riemann sheets associated to the amplitudes are identified as the P_c resonances, which are collected in Table 1 for those of interest. The $P_c(4312)$ couples dominantly to $\Sigma_c\bar{D}$ with $J^P = \frac{1}{2}^-$ as a (quasi) bound state. The $P_c(4440)$ and $P_c(4457)$ couple dominantly to the $\Sigma_c\bar{D}^*$ with $J^P = \frac{3}{2}^-$ and $J^P = \frac{1}{2}^-$, respectively. In particular, there is a narrow state located around 4.38 GeV where the data show a peak, though less prominent than the well-known three pentaquark states.

Table 1. The pole positions (on the sheets close to the physical region), quantum numbers, the dominant channels (DCs) and their thresholds.

	J^P	DC (threshold [MV])	Pole [MeV]
$P_c(4312)$	$\frac{1}{2}^-$	$\Sigma_c\bar{D}$ (4321.6)	$4313(1) - 3(1)i$
$P_c(4380)$	$\frac{3}{2}^-$	$\Sigma_c^*\bar{D}$ (4386.2)	$4376(1) - 6(2)i$
$P_c(4440)$	$\frac{3}{2}^-$	$\Sigma_c\bar{D}^*$ (4462.1)	$4441(2) - 6(2)i$
$P_c(4457)$	$\frac{1}{2}^-$	$\Sigma_c\bar{D}^*$ (4462.1)	$4461(2) - 5(2)i$
P_c	$\frac{1}{2}^-$	$\Sigma_c^*\bar{D}^*$ (4526.7)	$4525(4) - 9(1)i$
P_c	$\frac{3}{2}^-$	$\Sigma_c^*\bar{D}^*$ (4526.7)	$4520(3) - 12(3)i$
P_c	$\frac{1}{2}^-$	$\Sigma_c^*\bar{D}^*$ (4526.7)	$4500(2) - 9(6)i$

As discussed above, the $J/\psi p$ invariant mass distributions can be well described without including the $\Lambda_c\bar{D}^{(*)}$ channels. It is found that the explicit inclusion of the $\Lambda_c\bar{D}^{(*)}$ as well as the associated $\Lambda_c\bar{D}^{(*)}\pi$ three-body cuts, denoted as Scheme III, on one hand does not improve the fit quality. On the other hand, the fitted inelastic parameters are not effectively constrained as they are very sensitive to the data sets used in the fits and to the background employed. It is hard to discriminate the contributions from the $J/\psi p$ ($\eta_c p$) and $\Lambda_c\bar{D}^{(*)}$ to the width of the P_c states from the $J/\psi p$ mass distribution alone. To illustrate it more clearly, we fits to the data with $m_{Kp} > 1.9$ GeV with three different backgrounds, as shown in Fig. 4. All the fits have values for χ^2 very close to the best fit result. Nevertheless, the fitted parameters are very different, which leads to very different line shapes for the $\Sigma_c^{(*)}\bar{D}^{(*)}$ and $\Lambda_c\bar{D}^{(*)}$ productions, as shown in Fig. 5. In other words, the $J/\psi p$ data alone are not enough to constrain the $\Lambda_c\bar{D}^{(*)}$ interactions. Nevertheless, the predicted line shape for the $\eta_c p$ distribution is consistent with those of Scheme II, see i.e. in Fig. 3, as it is directly related to the $J/\psi p$ channel via HQSS.

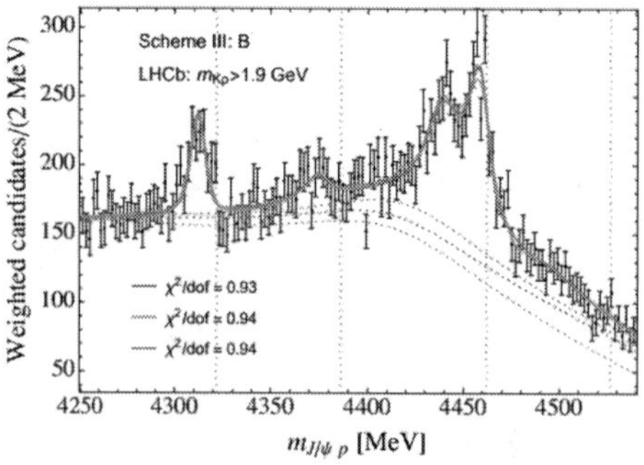

Figure 4. The best fits to the experimental data of $m_{Kp} > 1.9$ GeV [2] with three different fixed backgrounds for Scheme III.

Acknowledgements

This talk is based on the collaboration with Vadim Baru, Feng-Kun Guo, Christoph Hanhart, Ulf-G. Meißner, José A. Oller, and Qian Wang. The project is supported in part by the National Natural Science Foundation of China (NSFC) and the Deutsche Forschungsgemeinschaft (DFG) through the funds provided to the Sino-German Collaborative Research Center TRR110 "Symmetries and the Emergence of Structure in QCD" (NSFC Grant No.

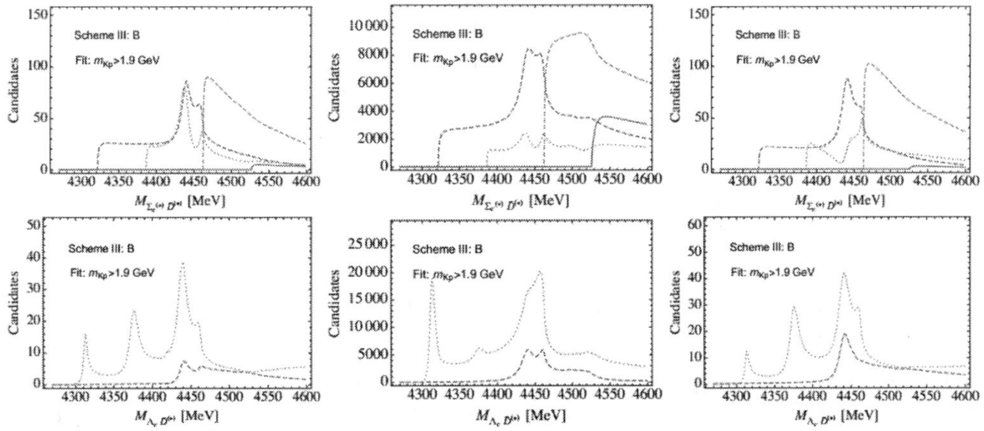

Figure 5. Predictions for the line shapes in the $\Sigma_c^{(*)}\bar{D}^{(*)}$ and $\Lambda_c\bar{D}^{(*)}$ channels based on the fit results of scheme III (solution *B*) for three different backgrounds. The left, middle and right columns show the results that are related to the blue, orange and green lines in Fig. 4, respectively.

12070131001, DFG Project-ID 196253076), by the NSFC under Grants No. 11835015, No. 12047503, No. 11961141012, and No.12035007, by the Chinese Academy of Sciences (CAS) under Grants No. QYZDB-SSW-SYS013 and No. XDB34030000, and by the Munich Institute for Astro- and Particle Physics (MIAPP) of the DFG cluster of excellence "Origin and Structure of the Universe", by the Spanish Ministry of Science and Innovation (MICINN) (Project PID2020-112777GB-I00), by the EU Horizon 2020 research and innovation programme, STRONG-2020 project, under grant agreement No. 824093, by Generalitat Valenciana under contract PROMETEO/2020/023.

References

[1] R. Aaij *et al.* [LHCb], Phys. Rev. Lett. **115**, 072001 (2015)
[2] R. Aaij *et al.* [LHCb], Phys. Rev. Lett. **122**, no.22, 222001 (2019)
[3] H. X. Chen, W. Chen and S. L. Zhu, Phys. Rev. D **100**, no.5, 051501 (2019)
[4] R. Chen, Z. F. Sun, X. Liu and S. L. Zhu, Phys. Rev. D **100**, no.1, 011502 (2019)
[5] M. Z. Liu, Y. W. Pan, F. Z. Peng, M. Sánchez Sánchez, L. S. Geng, A. Hosaka and M. Pavon Valderrama, Phys. Rev. Lett. **122**, no.24, 242001 (2019)
[6] Z. H. Guo and J. A. Oller, Phys. Lett. B **793**, 144-149 (2019)
[7] C. W. Xiao, J. Nieves and E. Oset, Phys. Rev. D **100**, no.1, 014021 (2019)
[8] L. Meng, B. Wang, G. J. Wang and S. L. Zhu, Phys. Rev. D **100**, no.1, 014031 (2019)
[9] M. L. Du, V. Baru, F. K. Guo, C. Hanhart, U. G. Meißner, J. A. Oller and Q. Wang, Phys. Rev. Lett. **124**, no.7, 072001 (2020)
[10] M. L. Du, V. Baru, F. K. Guo, C. Hanhart, U. G. Meißner, J. A. Oller and Q. Wang, JHEP **08**, 157 (2021)
[11] J. B. Cheng and Y. R. Liu, Phys. Rev. D **100**, no.5, 054002 (2019)
[12] A. Ali and A. Y. Parkhomenko, Phys. Lett. B **793**, 365-371 (2019)
[13] R. Zhu, X. Liu, H. Huang and C. F. Qiao, Phys. Lett. B **797**, 134869 (2019)
[14] M. I. Eides, V. Y. Petrov and M. V. Polyakov, Mod. Phys. Lett. A **35**, no.18, 2050151 (2020)

[15] S. Q. Kuang, L. Y. Dai, X. W. Kang and D. L. Yao, Eur. Phys. J. C **80**, no.5, 433 (2020)
[16] S. X. Nakamura, Phys. Rev. D **103**, 111503 (2021)
[17] C. Fernández-Ramírez *et al.* [JPAC], Phys. Rev. Lett. **123**, no.9, 092001 (2019)
[18] J. J. Wu, R. Molina, E. Oset and B. S. Zou, Phys. Rev. Lett. **105**, 232001 (2010)
[19] R. Chen, J. He and X. Liu, Chin. Phys. C **41**, no.10, 103105 (2017)
[20] C. W. Xiao, J. Nieves and E. Oset, Phys. Lett. B **799**, 135051 (2019)
[21] B. Wang, L. Meng and S. L. Zhu, Phys. Rev. D **101**, no.3, 034018 (2020)
[22] R. Aaij *et al.* [LHCb], Sci. Bull. **66**, 1278-1287 (2021)
[23] F. Z. Peng, M. J. Yan, M. Sánchez Sánchez and M. P. Valderrama, Eur. Phys. J. C **81**, 666 (2021)
[24] H. X. Chen, W. Chen, X. Liu and X. H. Liu, Eur. Phys. J. C **81**, no.5, 409 (2021)
[25] J. X. Lu, M. Z. Liu, R. X. Shi and L. S. Geng, Phys. Rev. D **104**, no.3, 034022 (2021)
[26] M. L. Du, Z. H. Guo and J. A. Oller,
[27] V. Baru, E. Epelbaum, A. A. Filin, C. Hanhart, U. G. Meißner and A. V. Nefediev, Phys. Lett. B **763**, 20-28 (2016)
[28] V. Baru, E. Epelbaum, A. A. Filin, C. Hanhart and A. V. Nefediev, JHEP **06**, 158 (2017)
[29] Q. Wang, V. Baru, A. A. Filin, C. Hanhart, A. V. Nefediev and J. L. Wynen, Phys. Rev. D **98**, no.7, 074023 (2018)
[30] A. V. Manohar and M. B. Wise, Camb. Monogr. Part. Phys. Nucl. Phys. Cosmol. **10**, 1-191 (2000)
[31] S. Sakai, H. J. Jing and F. K. Guo, Phys. Rev. D **100**, no.7, 074007 (2019)
[32] T. M. Yan, H. Y. Cheng, C. Y. Cheung, G. L. Lin, Y. C. Lin and H. L. Yu, Phys. Rev. D **46**, 1148-1164 (1992) [erratum: Phys. Rev. D **55**, 5851 (1997)]
[33] W. Detmold, C. J. D. Lin and S. Meinel, Phys. Rev. D **85**, 114508 (2012)
[34] U. Skerbis and S. Prelovsek, Phys. Rev. D **99**, no.9, 094505 (2019)
[35] M. L. Du, V. Baru, F. K. Guo, C. Hanhart, U. G. Meißner, A. Nefediev and I. Strakovsky, Eur. Phys. J. C **80**, no.11, 1053 (2020)
[36] P. Ling, X. H. Dai, M. L. Du and Q. Wang, Eur. Phys. J. C **81**, 819 (2021)

EPJ Web of Conferences **258**, 04008 (2022) https://doi.org/10.1051/epjconf/202225804008
A Virtual Tribute to Quark Confinement and the Hadron Spectrum (vConf21)

Long and short distance behavior of the imaginary part of the heavy-quark potential

Kirill Boguslavski[1,*], *Babak* Kasmaei[2,†], and *Michael* Strickland[3,‡]

[1]Institute for Theoretical Physics, Technische Universität Wien, 1040 Vienna, Austria
[2]Department of Chemistry and Physics, Hood College, Frederick, MD 21701, United States
[3]Department of Physics, Kent State University, Kent, OH 44242, United States

Abstract. The imaginary part of the effective heavy-quark potential can be related to the total in-medium decay width of of heavy quark-antiquark bound states. We extract the static limit of this quantity using classical-statistical simulations of the real-time Yang-Mills dynamics by measuring the temporal decay of Wilson loops. By performing the simulations on finer and larger lattices we are able to show that the nonperturbative results follow the same form as the perturbative ones. For large quark-antiquark separations, we quantify the magnitude of the non-perturbative long-range corrections to the imaginary part of the heavy-quark potential. We present our results for a wide range of temperatures, lattice spacings, and lattice volumes. We also extract approximations for the short-distance behavior of the classical potential.

1 Introduction

The suppression of heavy quark-antiquark bound states, such as bottomonium, is one of the key signatures of the Quark-Gluon Plasma (QGP). Due to the interactions of the heavy quark with the stochastic medium of the soft degrees of freedom, the effective heavy-quark potential possesses an imaginary part which is related to the total in-medium decay width of the states. This imaginary part has been determined using direct quantum field theoretic or effective field theory calculations [1–3].

The imaginary part of the heavy-quark potential has been calculated based on high-temperature quantum chromodynamics (QCD) calculations in the hard thermal loop (HTL) limit [1, 4–10], using effective field theory (pNRQCD) [11–14], finite-temperature lattice QCD [15–24], and real-time classical-statistical solutions of Yang-Mills theory in classical thermal equilibrium [25, 26]. In Ref. [25] the authors presented first results for the imaginary part of the heavy-quark potential using classical-statistical Yang-Mills simulations on spatially 3D lattices of size 12^3 and 16^3.

We extend the previous results to larger lattices up to 252^3 and consider SU(2) and SU(3) gauge theories. Due to the use of rather large lattice sizes, we can now compute the imaginary part of the heavy quark potential at larger values of r/a and reconstruct the functional form of the imaginary part of the heavy-quark potential for a much wider range of distances.

*e-mail: kirill.boguslavski@tuwien.ac.at
†e-mail: kasmaei@hood.edu
‡e-mail: mstrick6@kent.edu

Herein we will present the results for the imaginary part of the heavy-quark potential obtained using classical Yang Mills (CYM) simulations of a thermalized gluonic plasma. The use of CYM simulations is motivated by the fact that in situations where (a) gluonic occupation numbers are large, such as in thermal equilibrium for sufficiently low momenta or in the initial stages of heavy-ion collisions, and (b) the gauge coupling is weak $g^2 \ll 1$, vacuum contributions to observables are suppressed by powers of the gauge coupling.

The system is initialized close to thermal equilibrium using momentum-space initialization and the fields self-thermalize in real time before the extraction of the observables. The advantage of this procedure is that we can run simulations on very large lattices with small lattice spacings a at moderate computational cost.

An issue which can arise when dealing with classical-statistical treatments of gauge theories is that a finite ultraviolet limit does not exist due to the Rayleigh-Jeans divergence [27–32]. For this reason, it is important to identify a suitable manner in which one can scale results in order to extract relevant information. We will use the Debye mass m_D computed in the hard-classical-loops framework to demonstrate that, when plotted as a function of $m_D r$, the imaginary part of the heavy-quark potential is only mildly sensitive to the lattice spacing, or more generally to the simulation parameter $\beta \propto 1/(g^2Ta)$, at small distances $m_D r \lesssim 1$.

1.1 Theory and numerical setup

We considered mainly the pure SU(3) gauge theory with the Yang-Mills classical action. We use a standard real-time lattice discretization approach where fields are discretized on cubic lattices with N^3 sites and lattice spacing a (see, e.g., Refs. [33, 34] and references therein for more details). In this real-time approach, spatial gauge fields are replaced by gauge links $U_j(t,\mathbf{x}) \approx \exp\big(igaA_j(t,\mathbf{x})\big)$ at discrete coordinates $x_k = n_k a$ for $n_k = 0,\ldots,N-1$, while temporal gauge with $A_0 = 0$ is used.

We are interested in extracting the imaginary part of the classical potential $V_{\rm cl}(t,r)$ with $r \equiv |\mathbf{x}|$. Following Refs. [1, 25], it can be calculated using

$$i\partial_t C_{\rm cl}(t,r) = V_{\rm cl}(t,r) C_{\rm cl}(t,r)\,, \tag{1}$$

as the asymptotic temporal slope of $\log[C_{\rm cl}(t,r)]$. The classical thermal Wilson loop $C_{\rm cl}(t,r)$ is defined as

$$C_{\rm cl}(t,r) \equiv \frac{1}{N_c}\,{\rm Tr}\,\langle W[(t_0,\mathbf{x});(t,\mathbf{x})]\,W[(t,\mathbf{x});(t,\mathbf{0})]\,W[(t,\mathbf{0});(t_0,\mathbf{0})]\,W[(t_0,\mathbf{0});(t_0,\mathbf{x})]\rangle\,, \tag{2}$$

with temporal Wilson lines $W[(t_0,\mathbf{x});(t,\mathbf{x})] = \mathbf{1}$ and spatial Wilson lines $W[(t,\mathbf{0});(t,\mathbf{x})] = U_j(t,\mathbf{0})U_j(t,\mathbf{a}_j)U_j(t,2\,\mathbf{a}_j)\cdots U_j(t,\mathbf{x})$ for $\mathbf{x} = \hat{\mathbf{a}}_j r$ and $\hat{\mathbf{a}}_j = \mathbf{a}_j/a$ being a spatial unit vector. Since the classical thermal state is homogeneous, the Wilson loop is additionally averaged over all lattice points by averaging over the reference coordinates $\mathbf{0}$.

In order to extract the imaginary part of $V_{\rm cl}(t,r)$ we compute the time-dependence of $C_{\rm cl}(t,r)$. Due to the imaginary part of the in-medium heavy quark potential, this quantity will decay exponentially at late times with the rate of exponential decay set by the imaginary part of $V_{\rm cl}(t,r)$. We then define the imaginary part of the static classical potential as the late-time limit

$${\rm Im}[V_{\rm cl}(r)] \equiv \lim_{t\to\infty} {\rm Im}[V_{\rm cl}(t,r)]\,. \tag{3}$$

We initialize the fields in a quasi-thermal configuration in momentum-space, and then allow them to self-thermalize dynamically.

In Ref. [1], an expression is given for the imaginary part of the heavy-quark potential to leading-order in the strong coupling constant using the continuum hard thermal loop framework and dimensional regularization. The final result could be expressed compactly as

$$\mathrm{Im}[V^{(2)}(r)] = -\frac{C_F g^2 T}{4\pi}\phi\left(m_D^{\mathrm{HTL}}\, r\right). \tag{4}$$

where $C_F = (N_c^2 - 1)/2N_c$,

$$\phi(x) \equiv 2\int_0^\infty \mathrm{d}z \frac{z}{(1+z^2)^2}\left[1 - \frac{\sin(zx)}{zx}\right], \tag{5}$$

and m_D^{HTL} is the continuum hard thermal loop Deybe mass.

The result for the imaginary part of the classical potential in the infinite volume and infinite time limit ($N \to \infty,\ t \to \infty$) from second-order perturbation theory regularized on a cubic lattice (HCL: Hard Classical Loop) of size $(aN)^3$ is calculated in Ref. [25]. We find that the numerical HCL curves are fit very well by the functional form

$$\mathrm{Im}[V_{\mathrm{cl}}^{(2)}(r)] = g^2 T A_\infty \phi(B\, m_D r). \tag{6}$$

with $\phi(x)$ given in Eq. (5).

2 Results

We performed simulations of the signal, with $g^2T_0 = 0.44$ and $a = \{1,\ 0.5,\ 0.25,\ 0.2\}$ keeping $aN = 12$ fixed and also for $g^2T_0 = 0.45$ and $a = 0.1$ with $N = 252$. The $a = 0.1$ CYM lattice simulation results are compared with the corresponding results (same a and T) obtained from HCL perturbation theory, included as a blue dashed curve. For both our CYM lattice results and the HCL results, $\mathrm{Im}[V_{\mathrm{cl}}(r)]/g^2T$ is plotted as a function of $m_D r$, where we use the leading-order Debye mass m_D^{HCL} calculated in HCL theory. On the right edge of the figure we indicate the asymptotic $r \to \infty$ values obtained using both the dimensionally-regularized continuum HTL result (4) and the extrapolated $N,\beta \to \infty$ HCL result.

As can be seen from this Figure (figure. 1), the CYM lattice results seem to approach a finite large-β form when plotted as a function of $m_D r$, with the CYM results with $\beta = 136$ even overlapping with the corresponding HCL curve at $m_D r \lesssim 1$. At large values of r our CYM simulation results approach the corresponding HCL curve for the respective β value. However, figure. 1 indicates that with increasing β, the CYM potential is approaching the perturbative HCL large-β limit from below.

It should be noted that, while previous studies obtained the potential for different β values as functions of r/a, we found that plotting it as a function of $m_D r$ makes the comparison more intuitive. Most importantly, this allows to study the classical potential even at large β values while incorporating the dominant UV divergence into the mass.

We provide evidence that, at small distances, the classical potential extracted from our lattice simulation data also follows the functional form (6) for different values of β. This is shown in the left panel of figure. 2, where $\mathrm{Im}[V_{\mathrm{cl}}(r)]/g^2T$ for fixed lattice spacing $a = 0.1$ is plotted for different temperatures corresponding to $100 \lesssim \beta \leq 300$, as a function of $m_D\, r$. Fits to each data set using (6) in the considered interval $m_D r \leq 6$ are included as continuous lines.

We can also perform a short-distance expansion of the fitting function for the imaginary part of the potential (6), neglecting terms $O((m_D r)^4)$,

$$\mathrm{Im}[V_{\mathrm{cl}}(r)] \simeq -r^2\,\frac{1}{9}\,|A_\infty| B^2\, g^2 T\, m_D^2\,(4 - 3\gamma - 3\log(B\, m_D\, r)). \tag{7}$$

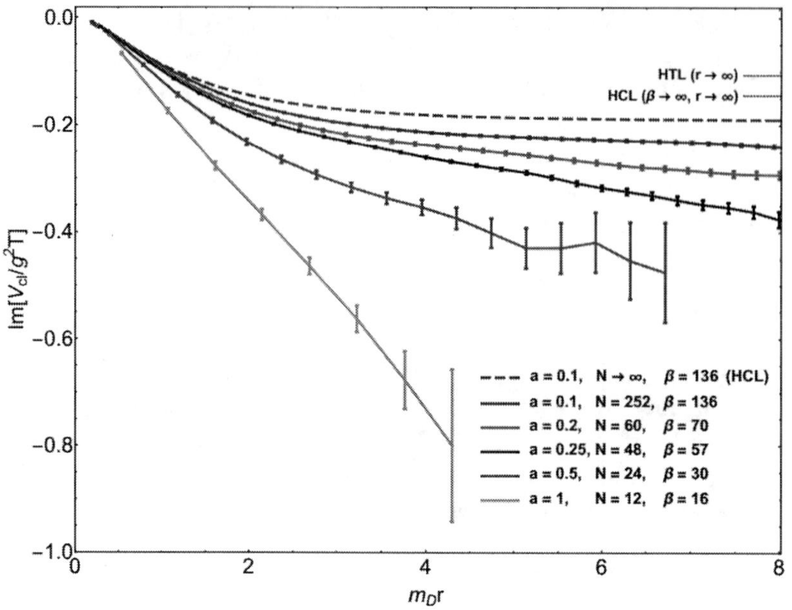

Figure 1. $\mathrm{Im}[V_{\mathrm{cl}}(r)]/g^2T$ for different lattice spacing from nonperturbative Wilson loops (solid lines) and from HCL perturbation theory for $a = 0.1$ (blue dashed line)

The resulting curve is shown as a red line and is observed to agree well with our data points for $m_D\, r \lesssim 0.5$. Thus, for a wide range of β values, the short-distance behavior of our CYM lattice data agrees well with the perturbative functional form (6) and its leading short-distance expansion, which is parametrically given by $|\mathrm{Im}[V_{\mathrm{cl}}(r)]| \sim C_F\, g^2T\ (m_D\, r)^2\ \log(m_D\, r)$.

In the right panel of figure. 2, we combine all of the curves of the left panel and compare them to the function (6). As expected, one observes very good agreement with our data, especially at low distances $m_D r$. We can also perform a short-distance expansion of the fitting function for the imaginary part of the potential (6), neglecting terms $O((m_D r)^4)$,

$$\mathrm{Im}[V_{\mathrm{cl}}(r)] \simeq -r^2\, \frac{1}{9}\, |A_\infty| B^2\, g^2 T\, m_D^2\, (4 - 3\gamma - 3\log(B\, m_D\, r))\,. \tag{8}$$

The resulting curve is shown as a red line and is observed to agree well with our data points for $m_D\, r \lesssim 0.5$. Thus, for a wide range of β values, the short-distance behavior of our CYM lattice data agrees well with the perturbative functional form (6) and its leading short-distance expansion, which is parametrically given by $|\mathrm{Im}[V_{\mathrm{cl}}(r)]| \sim C_F\, g^2T\ (m_D\, r)^2\ \log(m_D\, r)$.

3 Conclusions

We used classical-statistical lattice simulations of the pure Yang-Mills theory to extract the imaginary part of the heavy-quark potential as a function of the quark-antiquark separation. To carry out our simulations on large and fine lattices, we used a self-thermalization scheme to generate thermalized gauge field configurations which relied on initialization of chromo-electric fields in momentum-space followed by a period of self-thermalization.

We extended the previous classical-statistical lattice calculations of $\mathrm{Im}[V_{\mathrm{cl}}]$ by considering rather large lattice sizes. We also found that both the lattice simulation and the HCL results were very well approximated by a functional form which can be obtained from a leading-order hard-thermal loop calculation. Using fits of this form and then expanding the

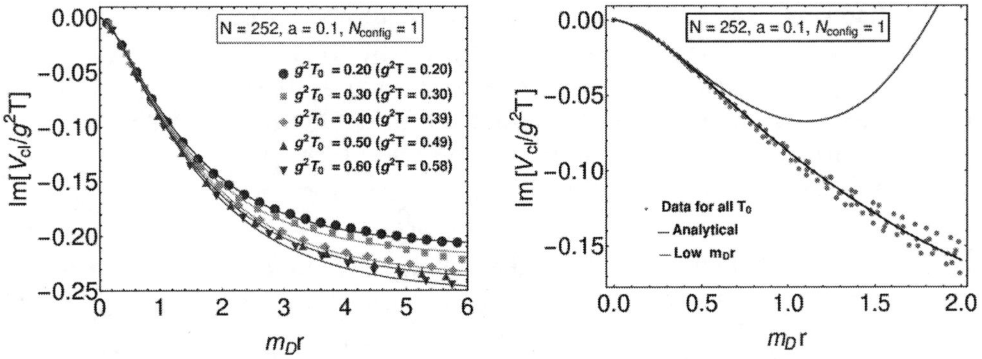

Figure 2. (Left) The extracted classical potential $\text{Im}[V_{\text{cl}}(r)]/g^2T$ as a function of $m_D r$ for fixed lattice spacing $a = 0.1$ but for different temperatures. (Right) The combined data of the left panel, with additional data sets is compared to the analytical form (6).

result at small $m_D r$, we were able to extract small-distance approximations for the imaginary part of the heavy-quark potential. More details about our study can be found in Ref. [35].

References

[1] M. Laine, O. Philipsen, P. Romatschke, M. Tassler, JHEP **03**, 054 (2007), hep-ph/0611300
[2] N. Brambilla, M.A. Escobedo, J. Soto, A. Vairo, Phys. Rev. **D96**, 034021 (2017), 1612.07248
[3] N. Brambilla, M.A. Escobedo, J. Soto, A. Vairo, Phys. Rev. **D97**, 074009 (2018), 1711.04515
[4] A. Dumitru, Y. Guo, M. Strickland, Phys.Lett. **B662**, 37 (2008), 0711.4722
[5] N. Brambilla, J. Ghiglieri, A. Vairo, P. Petreczky, Phys. Rev. **D78**, 014017 (2008), 0804.0993
[6] Y. Burnier, M. Laine, M. Vepsalainen, Phys.Lett. **B678**, 86 (2009), 0903.3467
[7] A. Dumitru, Y. Guo, M. Strickland, Phys.Rev. **D79**, 114003 (2009), 0903.4703
[8] A. Dumitru, Y. Guo, A. Mocsy, M. Strickland, Phys.Rev. **D79**, 054019 (2009), 0901.1998
[9] M. Margotta, K. McCarty, C. McGahan, M. Strickland, D. Yager-Elorriaga, Phys.Rev. **D83**, 105019 (2011), 1101.4651
[10] Y. Guo, L. Dong, J. Pan, M.R. Moldes, Phys. Rev. D **100**, 036011 (2019), 1806.04376
[11] E. Eichten, K. Gottfried, T. Kinoshita, K.D. Lane, T.M. Yan, Phys. Rev. D **21**, 203 (1980)
[12] W. Lucha, F.F. Schoberl, D. Gromes, Phys. Rept. **200**, 127 (1991)
[13] N. Brambilla, A. Pineda, J. Soto, A. Vairo, Rev. Mod. Phys. **77**, 1423 (2005), hep-ph/0410047
[14] N. Brambilla, J. Ghiglieri, P. Petreczky, A. Vairo, Phys. Rev. **D82**, 074019 (2010), 1007.5172
[15] A. Rothkopf, T. Hatsuda, S. Sasaki, PoS **LAT2009**, 162 (2009), 0910.2321
[16] A. Rothkopf, T. Hatsuda, S. Sasaki, Phys. Rev. Lett. **108**, 162001 (2012), 1108.1579
[17] Y. Burnier, A. Rothkopf, Phys. Rev. **D86**, 051503 (2012), 1208.1899
[18] Y. Burnier, A. Rothkopf, Phys. Rev. Lett. **111**, 182003 (2013), 1307.6106

[19] Y. Burnier, A. Rothkopf, Phys. Lett. **B753**, 232 (2016), 1506.08684
[20] Y. Burnier, O. Kaczmarek, A. Rothkopf, JHEP **12**, 101 (2015), 1509.07366
[21] Y. Burnier, A. Rothkopf, Phys. Rev. **D95**, 054511 (2017), 1607.04049
[22] Y. Burnier, O. Kaczmarek, A. Rothkopf, JHEP **10**, 032 (2016), 1606.06211
[23] D. Bala, S. Datta, Phys. Rev. D **101**, 034507 (2020), 1909.10548
[24] D. Bala, S. Datta, Phys. Rev. D **103**, 014512 (2021), 2009.00773
[25] M. Laine, O. Philipsen, M. Tassler, JHEP **09**, 066 (2007), 0707.2458
[26] A. Lehmann, A. Rothkopf, JHEP **07**, 067 (2021), 2012.10089
[27] K. Kajantie, K. Rummukainen, M.E. Shaposhnikov, Nucl. Phys. B **407**, 356 (1993), hep-ph/9305345
[28] J. Ambjorn, A. Krasnitz, Phys. Lett. B **362**, 97 (1995), hep-ph/9508202
[29] P.B. Arnold, D. Son, L.G. Yaffe, Phys. Rev. D **55**, 6264 (1997), hep-ph/9609481
[30] G.D. Moore, K. Rummukainen, Phys. Rev. D **61**, 105008 (2000), hep-ph/9906259
[31] J. Berges, K. Boguslavski, S. Schlichting, R. Venugopalan, JHEP **05**, 054 (2014), 1312.5216
[32] T. Epelbaum, F. Gelis, B. Wu, Phys. Rev. D **90**, 065029 (2014), 1402.0115
[33] K. Boguslavski, A. Kurkela, T. Lappi, J. Peuron, Phys. Rev. D **98**, 014006 (2018), 1804.01966
[34] J. Berges, K. Boguslavski, S. Schlichting, R. Venugopalan, Phys. Rev. D **89**, 114007 (2014), 1311.3005
[35] K. Boguslavski, B.S. Kasmaei, M. Strickland, JHEP **21**, 083 (2020), 2102.12587

EPJ Web of Conferences **258**, 04009 (2022) https://doi.org/10.1051/epjconf/202225804009
A Virtual Tribute to Quark Confinement and the Hadron Spectrum (vConf21)

The static force from generalized Wilson loops using gradient flow

Viljami Leino[1,*], *Nora* Brambilla[1,2,3,**], *Julian* Mayer-Steudte[1,***], and *Antonio* Vairo[1,****]

[1]Physik Department, Technische Universität München,
James-Franck-Strasse 1, 85748 Garching, Germany
[2]Institute for Advanced Study, Technische Universität München,
Lichtenbergstrasse 2 a, 85748 Garching, Germany
[3]Munich Data Science Institute, Technische Universität München,
Walther-von-Dyck-Strasse 10, 85748 Garching, Germany

Abstract. We explore a novel approach to compute the force between a static quark-antiquark pair with the gradient flow algorithm on the lattice. The approach is based on inserting a chromoelectric field in a Wilson loop. The renormalization issues, associated with the finite size of the chromoelectric field on the lattice, can be solved with the use of gradient flow. We compare numerical results for the flowed static potential to our previous measurement of the same observable without a gradient flow.

1 Introduction

The static potential $V(r)$ between a static quark and an antiquark is a quantity that has been studied for a long time in QCD. The perturbative expansion for $V(r)$ is known up to N^3LO accuracy. The perturbative expression of the static potential at short distances allows one to accurately extract the strong coupling, α_s, from lattice QCD computations of the same quantity. This α_s extraction is competitive with other existing determination from different observables [1].

In lattice regularization, the static potential comes with a linear divergence proportional to the inverse lattice spacing $1/a$. This divergence is associated with contributions coming from the self-energy of the static quarks. The self-energy vanishes in dimensional regularization. In dimensional regularization, the perturbative expression for the static potential is, however, affected by a renormalon of order Λ_{QCD} [2, 3]. Both the renormalon and the linear divergence on the lattice can be absorbed into an additive constant. This constant contribution has no physical significance and can be removed by considering the static force $F(r) = \partial_r V(r)$ instead. The static force encodes the shape of $V(r)$ and carries all the physical information needed to extract α_s, while being finite and renormalon free.

A precise determination of the static force from the static potential on the lattice can be challenging. The errors associated with the numerical derivative of the data are directly

*e-mail: viljami.leino@tum.de
**e-mail: nora.brambilla@ph.tum.de
***e-mail: julian.mayer-steudte@tum.de
****e-mail: antonio.vairo@ph.tum.de

proportional to the density of the data points. In quenched QCD simulations, the data can be dense enough for a reliable definition of the numerical derivative [4, 5]. In full QCD simulations, the data at short distances is, however, still too sparse [6]. This problem can be avoided by considering a recently suggested method [7, 8], where the force between a static quark and a static antiquark is measured directly on the lattice by considering a Wilson loop with a chromoelectric field E inserted into one of the temporal Wilson lines. A proof of concept study of this method has recently been carried out in [9, 10].

In the previous study [9, 10], it was found that the insertion of the chromoelectric field induces additional self-energy contributions on the lattice. These self-energy contributions cause a slow convergence towards the continuum limit. Such behavior is expected, as it is well known that operators that involve components of the field strength tensor often come with sizable discretization errors associated with the slow convergence of lattice perturbation theory when expanded with respect to the bare coupling. This issue can be reduced by introducing a multiplicative renormalization factor Z_E. This renormalization factor can be estimated in multiple different ways, see e.g, Refs. [11–16]. In Ref [10], we suggested a method to measure this renormalization factor non-perturbatively by considering the ratio of the static force defined from the Wilson loop with a chromoelectric field insertion and the static force defined from the finite difference of the static potential measured on the lattice.

An interesting alternative approach to this renormalization issue is offered by an algorithm known as the gradient flow [17]. In this proceeding, we study the static force measured on the lattice with the gradient flow and show that the gradient flow reduces sizably discretization effects related to insertion of components of the field strength tensor. The original talk at the conference also summarized the results from Ref. [10]. However, due to page limitations, this summary has been split to a separate proceeding contribution [18].

The proceeding is organized as follows. In Section 2 we introduce our observables. The gradient flow algorithm is explained and applied to our observables in section 3. We discuss the simulation details and present the numerical results in section 4.

2 Definition of the static force

The static potential between a static quark and an antiquark is related to the trace of the Wilson loop with temporal extent T and spatial extent r,

$$\mathrm{Tr}\{\mathrm{P}\, W_{r\times T}\} = \mathrm{Tr}\left\{\mathrm{P}\,\exp\left(ig \oint_{r\times T} dx_\mu\, A_\mu(x)\right)\right\}, \tag{1}$$

from where the static potential can be extracted on the lattice with the following limit

$$V(r,a) = \lim_{T\to\infty} V_{\mathrm{eff}}(r,T,a)\,, \quad V_{\mathrm{eff}}(r,T,a) = -\frac{1}{a}\ln\frac{\langle \mathrm{Tr}\,\{\mathrm{P}\, W_{r\times(T+a)}\}\rangle}{\langle \mathrm{Tr}\,\{\mathrm{P}\, W_{r\times T}\}\rangle}\,. \tag{2}$$

The traditional definition of the static force $F(r)$ via a finite difference is then,

$$F_{\partial V}(r,a) = \frac{V(r+a,a) - V(r-a,a)}{2a}\,. \tag{3}$$

An alternative method for computing the static force was proposed in Refs. [7, 8],

$$F(r) = \lim_{T\to\infty} -i\frac{\langle \mathrm{Tr}\,\{\mathrm{P}\, W_{r\times T}\, \hat{\mathbf{r}}\cdot g\mathbf{E}(\mathbf{r},t^*)\}\rangle}{\langle \mathrm{Tr}\,\{\mathrm{P}\, W_{r\times T}\}\rangle}\,. \tag{4}$$

Here the chromoelectric field $\mathbf{E}(\mathbf{r}, t^*)$ is inserted in one of the temporal Wilson lines at a fixed time t^*. The choice of t^* is arbitrary, because the t^* dependence of $\langle \mathrm{Tr}\{\mathrm{P}\, W_{r\times T}\, gE_j(\mathbf{r},t^*)\}\rangle$

disappears in the limit $T \to \infty$ if t^* is kept constant. In practice, for the purposes of lattice simulations, we set t^* to be in the middle of the temporal Wilson line in order to maximize the distance from the temporal boundaries of the Wilson loop.

The chromoelectric field components are defined in terms of the field strength tensor as $E_j(x) = F_{j0}(x)$. These components have to be discretized on the lattice. We define the chromoelectric field using a so called cloverleaf formulation [13]:

$$gE_j = \frac{\Pi_{j0} - \Pi_{j0}^\dagger}{2ia^2} + O(a^2)\,, \quad \Pi_{j0} = \frac{P_{j,0} + P_{0,-j} + P_{-j,-0} + P_{-0,j}}{4}\,, \tag{5}$$

where $P_{\mu,\nu} = 1 + ia^2 gF_{\mu\nu} = U_\mu(x)U_\nu(x+\hat{\mu})U_\mu^\dagger(x+\hat{\nu})U_\nu^\dagger(x)$ is the plaquette defined as a product of link variables $U_\mu(x) = e^{iagA_\mu(x)}$. We then define an effective force to extract the static force from lattice simulations,

$$F_E(r,a) = \lim_{T\to\infty} F_{E,\mathrm{eff}}(r,T,a) \quad , \quad F_{E,\mathrm{eff}}(r,T,a) = -i\frac{\langle \mathrm{Tr}\,\{\mathrm{P}\, W_{r\times T}\, \hat{\mathbf{r}}\cdot g\mathbf{E}(\mathbf{r},t^*)\}\rangle}{\langle \mathrm{Tr}\,\{\mathrm{P}\, W_{r\times T}\}\rangle}\,. \tag{6}$$

3 Gradient flow

The Yang-Mills gradient flow [17] is a smearing algorithm that evolves a given gauge field configuration towards the classical solution,

$$\partial_{\tau_F} B_{\tau_F,\mu} = -g^2\frac{\delta S_{\mathrm{YM}}}{\delta B_{\tau_F,\mu}} = D_{\tau_F,\nu} G_{\tau_F,\nu\mu} \tag{7}$$

$$G_{\tau_F,\mu\nu} = \partial_\mu B_{\tau_F,\nu} - \partial_\nu B_{\tau_F,\mu} + \left[B_{\tau_F,\mu}, B_{\tau_F,\nu}\right] \tag{8}$$

$$B_{0,\mu} = A_\mu\,, \tag{9}$$

where τ_F is the flow time and S_{YM} is the gauge action. For this study, we choose S_{YM} to be the Lüscher-Weisz action. On the lattice, the gauge fields are replaced with the link variables U_μ and the flow equation can be expressed as,

$$\partial_{\tau_F} U_\mu(\vec{x},\tau_F) = -g_0^2\left(\frac{\partial S_{\mathrm{YM}}[U]}{\partial U_\mu(\vec{x},\tau_F)}\right) U_\mu(\vec{x},\tau_F)\,. \tag{10}$$

This equation smears the gauge links with a flow radius $\sqrt{8\tau_F}$.

In perturbation theory, the flowed fields B can be written in terms of the gluon fields A by iteratively solving (7). This leads to a set of Feynman rules for the flowed fields. At the leading order of perturbation theory, the flow modifies the gluon propagator,

$$D(p,\tau_F) = D(p,\tau_F = 0)e^{-2\tau_F p^2} = \frac{1}{p^2}e^{-2\tau_F p^2}\,. \tag{11}$$

This allows a tree-level definition of the static potential and the static force at a finite flow time,

$$V^{\mathrm{LO}}_{\mathrm{continuum}}(r,\tau_F) = -\frac{\alpha_{\mathrm{s}} C_{\mathrm{F}}}{r}\mathrm{erf}\left(\frac{r}{\sqrt{8\tau_F}}\right) \tag{12}$$

$$F^{\mathrm{LO}}_{\mathrm{continuum}}(r,\tau_F) = \frac{\alpha_{\mathrm{s}} C_{\mathrm{F}}}{r^2}\left[\mathrm{erf}\left(\frac{r}{\sqrt{8\tau_F}}\right) - \frac{r}{\sqrt{2\pi\tau_F}}e^{-\frac{r^2}{8\tau_F}}\right]\,, \tag{13}$$

where erf is the error function. For the purposes of this proceeding, the leading order result is enough. However, we note that the static potential and the static force at finite flow time

β	$N_\sigma \times N_t$	a[fm]	# configurations
6.284	20×40	0.060	1949
6.481	26×56	0.046	1999
6.594	30×60	0.040	1997

Table 1. Gauge link ensembles and their parameters

have been calculated up to one-loop level in Ref. [19]. This calculation will be important for the final extraction of the static force at finite flow time on the lattice.

On the lattice, we use different actions for the simulation itself (Wilson action) and the gradient flow (Lüscher-Weisz action). In this case, the flowed propagator becomes [20],

$$D_{\text{total,lattice}}(p, \tau_F) = e^{-\tau_F D^{-1}_{\text{flow,lattice}}(p,\tau_F=0)} D_{\text{simulation,lattice}}(p, \tau_F = 0) e^{-\tau_F D^{-1}_{\text{flow,lattice}}(p,\tau_F=0)} \tag{14}$$

$$D^{-1}_{*,\text{lattice}} = 4 \sum_{i=1}^{4} \left(\sin^2 \frac{p_i}{2} + c_w \sin^4 \frac{p_i}{2} \right), \tag{15}$$

where $c_w = 0$ for the Wilson action and $c_w = 1/3$ for the Lüscher-Weisz action. Furthermore, the static potential at a finite flow time is then given by

$$V^{\text{LO}}_{\text{lattice}}(r, \tau_F) = -C_\text{F}\, g^2 \int \frac{\mathrm{d}^3 p}{(2\pi^3)} D_{\text{total,lattice}}(p, \tau_F)\,. \tag{16}$$

The static force in the leading order of lattice perturbation theory is then given by a finite difference as defined in Eq. (3) [10].

Having defined the finite flow time static force in both continuum and lattice perturbation theories, we can improve the lattice results for the static force at tree-level by defining an improved force,

$$F^{\text{imp}}_{\text{lattice}}(r, \tau_F, a) = \frac{F^{\text{LO}}_{\text{continuum}}(r, \tau_F)}{F^{\text{LO}}_{\text{lattice}}(r, \tau_F)} F_\text{E}(r, \tau_F, a)\,. \tag{17}$$

Similarly, the static potential can be tree-level improved by taking a ratio of Eqs. (12) and (16).

4 Simulations and results

We discretize the SU(3) Yang-Mills gauge theory using the standard Wilson plaquette action. Furthermore, for the flow equation we use the Lüscher-Weisz discretization. We generate the gauge configurations using the heatbath and overrelaxation algorithms. On these configurations, we then integrate the discretized gradient flow (10) using an optimized adaptive algorithm [21]. We generate three ensembles with parameters and statistics listed in table 1, where the lattice spacing a in physical units is related to the gauge coupling with parameterization from Ref. [4]. We have chosen the set of ensembles to be close to the set used in [10] to ease the comparison with existing results.

First, we test the renormalization property of the gradient flow. It is known that observables involving components of the field strength tensor often exhibit sizable discretization errors at the values of the gauge coupling typically used in numerical simulations. This is due to a slow convergence of lattice perturbation theory, when expanded in the bare coupling [12]. The discretization errors can be reduced with a suitable multiplicative renormalization factor

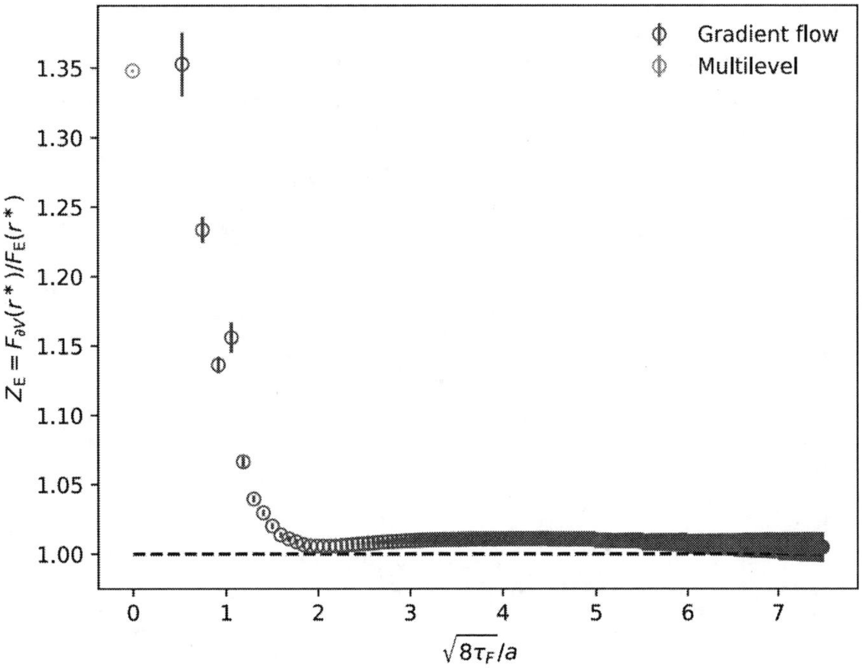

Figure 1. The renormalization factor $Z_E = F_{\partial V}/F_E$ as a function of the flow radius $\sqrt{8\tau_F}$. The point in orange shows the measured Z_E at zero flow time from Ref. [10].

as discussed in Refs. [11, 13–16]. This renormalization factor is the main difference in discretization effects between the two definitions of the static force $F_E(r, a)$ and $F_{\partial V}(r, a)$ given by Eqs. (6) and (3) respectively. Hence, we can define a multiplicative improvement factor Z_E that determines this renormalization,

$$Z_{\mathrm{E}}(a) = \frac{F_{\partial V}(r^*, a)}{F_E(r^*, a)}, \tag{18}$$

where r^* is an arbitrary separation. We use the tree-level improved (17) static force and potential when taking the ratio.

For each flow time τ_F, we vary r^* over all values of r and find a range of intermediate r values, where the data can be described with a constant fit. This constant then defines the renormalization factor Z_E as a function of flow time. The flow time dependent renormalization constant is shown in Fig. 1. We observe that at zero flow time we replicate the result from the previous study [10]. As the flow time is increased, Z_E decreases rapidly until settling to a constant value $Z_E \approx 1$ for $\sqrt{8\tau_F} > 1.6a$. The renormalization factor becoming unity indicates that the gradient flow, indeed, removes the sizable discretization effects caused by the finite discretization of the chromoelectric field. The remaining structure in Z_E for $\sqrt{8\tau_F} > 1.6a$ is most likely due to underestimated systematic errors.

Next, we compare the gradient flow result to an existing zero flow time measurement of the derivative of the static potential. In Ref. [10], we measured the static potential traditionally at zero flow time and performed a fit to the Cornell ansatz. The Cornell potential can then be

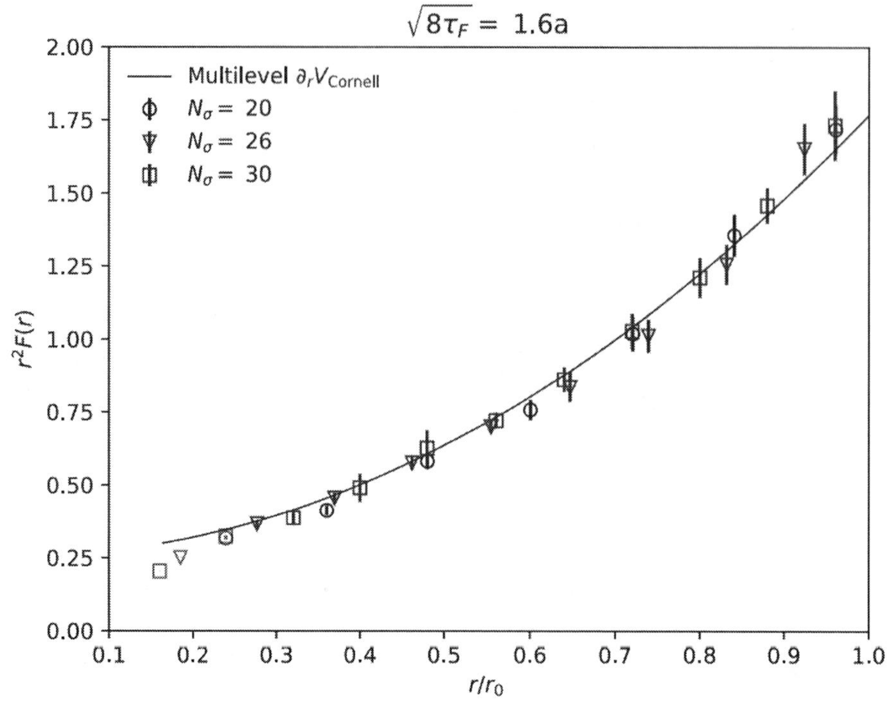

Figure 2. The static force $r^2 F_E$ at flow time $\sqrt{8\tau_F} > 1.6a$ compared to $\partial_r V_{\text{Cronell}}(r)$ from Ref. [10]. The gray points present the data where the flow radius is much larger than the separation r.

differentiated analytically. We use this Cornell fit as our ground truth when testing the new static force formulations.

Ideally, one should perform the zero flow time limit only after the continuum limit has been taken. Unfortunately, we currently lack the required range of lattice spacings to confidently perform the continuum limit. Hence, we cannot perform the zero flow time limit either. Instead, we compare the zero flow time result to the static force measured at the smallest flow time $\sqrt{8\tau_F} > 1.6a$ where $Z_E \sim 1$. This is shown in Fig. 2. We see, that despite the lack of continuum and zero flow time limits, the static force from the gradient flow simulations agrees with the previous result for the derivative of the static potential measured at zero flow time. Since the flow smears the gauge link, we expect to see enhanced discretization effects when the flow radius grows to order of $\sqrt{8\tau_F} \gtrsim r/2$. These points are shown in gray in Fig. 2, where we observe indeed that these points slightly deviate from the expected behavior.

In summary, we have shown preliminary results of the static force measured with gradient flow from the expectation value of a Wilson loop with a chromoelectric field insertion. We observe that at finite flow times the renormalization constant Z_E, which removes the sizable discretization errors associated with the chromoelectric field insertion, becomes consistent with unity. This shows in practice, the well known renormalization property of the gradient flow algorithm. We have also shown an early comparison between our flowed data and the existing result at zero flow time. The continuum and zero flow time limits are left for a later publication. These limits and the strong coupling extraction could be improved further with the recent perturbative calculation of the flowed static force at one loop level [19].

Acknowledgements

This work has been supported by the Deutsche Forschungsgemeinschaft (DFG, German Research Foundation) cluster of excellence ORIGINS that is funded by the DFG under Germany's Excellence Strategy EXC-2094-390783311. The simulations were performed using the publicly available MILC code [22] and were carried out on the computing facilities of the Computational Center for Particle and Astrophysics (C2PAP) of the cluster of excellence ORIGINS.

References

[1] S. Aoki et al. (Flavour Lattice Averaging Group) (2019), 1902.08191
[2] A. Pineda, Phd thesis, University of Barcelona (1998)
[3] A.H. Hoang, M.C. Smith, T. Stelzer, S. Willenbrock, Phys. Rev. D **59**, 114014 (1999), hep-ph/9804227
[4] S. Necco, R. Sommer, Nucl. Phys. **B622**, 328 (2002), hep-lat/0108008
[5] S. Necco, R. Sommer, Phys. Lett. B **523**, 135 (2001), hep-ph/0109093
[6] A. Bazavov, N. Brambilla, X.G. Tormo, I, P. Petreczky, J. Soto, A. Vairo, Phys. Rev. D **90**, 074038 (2014), [Erratum: Phys.Rev.D 101, 119902 (2020)], 1407.8437
[7] A. Vairo, EPJ Web Conf. **126**, 02031 (2016), [EPJ Web Conf.120,07004(2016)], 1512.07571
[8] A. Vairo, Mod. Phys. Lett. **A31**, 1630039 (2016)
[9] N. Brambilla, V. Leino, O. Philipsen, C. Reisinger, A. Vairo, M. Wagner (TUMQCD), PoS **LATTICE2019**, 109 (2019), 1911.03290
[10] N. Brambilla, V. Leino, O. Philipsen, C. Reisinger, A. Vairo, M. Wagner (2021), 2106.01794
[11] A. Huntley, C. Michael, Nucl. Phys. **B286**, 211 (1987)
[12] G.P. Lepage, P.B. Mackenzie, Phys. Rev. D **48**, 2250 (1993), hep-lat/9209022
[13] G.S. Bali, K. Schilling, A. Wachter, Phys. Rev. **D56**, 2566 (1997), hep-lat/9703019
[14] Y. Koma, M. Koma, Nucl. Phys. **B769**, 79 (2007), hep-lat/0609078
[15] D. Guazzini, H.B. Meyer, R. Sommer (ALPHA), JHEP **10**, 081 (2007), 0705.1809
[16] C. Christensen, M. Laine, Phys. Lett. B **755**, 316 (2016), 1601.01573
[17] M. Lüscher, JHEP **08**, 071 (2010), [Erratum: JHEP 03, 092 (2014)], 1006.4518
[18] V. Leino, N. Brambilla, O. Philipsen, C. Reisinger, A. Vairo, M. Wagner (2021), 2111.07916
[19] N. Brambilla, H.S. Chung, A. Vairo, X.P. Wang (2021), 2111.07811
[20] Z. Fodor, K. Holland, J. Kuti, S. Mondal, D. Nogradi, C.H. Wong, JHEP **09**, 018 (2014), 1406.0827
[21] A. Bazavov, T. Chuna (2021), 2101.05320
[22] *http://physics.utah.edu/~detar/milc.html*

Non-perturbative phenomena in jet modification

Guy D. Moore[1,*], *Sören* Schlichting[2,**], *Niels* Schlusser[1,3,4,***], and *Ismail* Soudi[2,5,****]

[1]Institut für Kernphysik, Technische Universität Darmstadt
Schlossgartenstraße 2, D-64289 Darmstadt, Germany
[2]Fakultät für Physik, Universität Bielefeld
D-33615 Bielefeld, Germany
[3]Department of Physics & Helsinki Institute of Physics
P.O. Box 64, FI-00014 University of Helsinki
[4]Biozentrum, Universität Basel
Spitalstrasse 41, 4056 Basel, Switzerland
[5]Department of Physics and Astronomy, Wayne State University
Detroit, MI 48201.

Abstract. The interaction of a jet with the medium created in heavy-ion collisions is not yet fully understood from a QCD perspective. This is mainly due to the non-perturbative nature of this interaction which affects both transverse jet momentum broadening and jet quenching. We discuss how lattice simulations of Electrostatic QCD, can be matched to full, four dimensional QCD, to determine non-perturbative contributions to the momentum broadening kernel. We determine the momentum broadening kernel in impact parameter and momentum space and finally show how these results can be used in phenomenological calculations of in-medium splitting rates.

1 Introduction

One crucial signature of Quark-Gluon-Plasma (QGP) formation in ultra-relativistic heavy-ion collisions is the suppression of highly energetic particles or jets in the final state. Before being detected, the hard partons must traverse the QGP and interact with the medium, leading to a substantial loss of their energy. In addition to the elastic interactions with the medium constituents, multiple soft scatterings between the hard partons and the medium trigger the partons to radiate. Since soft scatterings occur frequenctly, an infinite number of diagrams has to be resummed in order to compute effective $1 \to 2$ in-medium radiation rates. Different formalisms are employed in the literature to obtain the rate of radiation [1–8]. While these formalisms differ on many aspects and approximations, they all rely on the description of the elastic scattering with the medium, obtained using the transverse momentum broadening kernel

$$C(q_\perp) \equiv \frac{(2\pi)^2 \mathrm{d}^3\Gamma}{\mathrm{d}^2 q_\perp \, \mathrm{d}L}, \tag{1}$$

*e-mail: guy.moore@physik.tu-darmstadt.de
**e-mail: sschlichting@physik.uni-bielefeld.de
***e-mail: niels.schlusser@unibas.ch
****e-mail: isma@physik.uni-bielefeld.de

which describes the rate at which the hard partons exchange transverse momentum $\boldsymbol{q}_\perp$ with the medium. To obtain the broadening kernel, one can use various treatments of the medium, e.g., if one treats the medium as many random, static, screened color centers, one obtains $C(q_\perp) \propto \frac{1}{(q_\perp^2+m_\mathrm{D}^2)^2}$, with the Debye screening mass m_D^2. Furthermore, taking the medium to be made up of dynamical moving charges at leading order in perturbation theory [9] leads to $C(q_\perp) \propto \frac{1}{q_\perp^2(q_\perp^2+m_\mathrm{D}^2)}$. At very large parton energies, the interaction with the medium can be approximated as many individually small scatterings, leading to transverse momentum diffusion: $C(b_\perp) \approx \hat{q}\, b_\perp^2/4$, also known as the harmonic oscillator approximation. All these treatments rely on approximations and perturbative expansions, but the medium is quite strongly coupled and there can be large non-perturbative contributions even at high temperatures. It would be important to capture genuine non-perturbative contributions, in order to improve on the treatment of the jet-medium interactions. For that sake, it has been found by Casalderrey-Solana and Teaney [10] that the broadening kernel $C(q_\perp)$ can be related to its position-space ($b_\perp$) version $C(b_\perp)$ via the modified Fourier transform

$$C(b_\perp) \equiv \int \frac{\mathrm{d}^2 q_\perp}{(2\pi)^2} \left(1 - e^{i\boldsymbol{q}_\perp \cdot \boldsymbol{b}_\perp}\right) C(q_\perp) . \tag{2}$$

$C(b_\perp)$, in turn, can be computed from a light-like Wilson-loop operator with real-time techniques. Despite the jet being a high energy observable, its interactions with the medium are dominated by infrared contributions. The requirement of infrared resummation on the one hand and real-time techniques on the other hand pose a serious challenge. This issue can be resolved by treating the correlators to be slightly spacelike, making the infrared part of the calculation amenable to effective field theory (EFT) treatment within the framework of electrostatic QCD (EQCD) [11]. To further support the convergence of the EQCD result, an evaluation on the lattice was pioneered [12] and finally extrapolated to the continuum [13–15].

2 Non-perturbative broadening kernel

2.1 Broadening kernel in position space

EQCD is a long-distance effective treatment of thermal QCD and therefore only reproduces the infrared (IR) regime of QCD well. In order to match the predictions of full QCD in the ultraviolet (UV) regime of $C(q_\perp)$ (see [16]), the incorrect EQCD UV limit has to be removed by subtraction and the correct UV limit from full thermal QCD has to be reinstated, such that following the strategy of [17]

$$\begin{aligned} C_\mathrm{QCD}(b_\perp) &= (C_\mathrm{QCD}(b_\perp) - C_\mathrm{EQCD}(b_\perp)) + C_\mathrm{EQCD}(b_\perp) \\ &= \left(C_\mathrm{QCD}^{O(g^4)}(b_\perp) - C_\mathrm{EQCD}^{O(g^4)}(b_\perp)\right) + C_\mathrm{EQCD}^\mathrm{latt}(b_\perp) + O(g^6) , \end{aligned} \tag{3}$$

with the strong coupling constant g_s. The difference between EQCD and full QCD in brackets is UV dominated and therefore safe to be evaluated perturbatively. The perturbative results are available at $O(g^4)$ in full QCD [16] and EQCD [11, 17]. Continuum-extrapolated lattice data was computed in [15]. Since the lattice data is only available at finite $b_\perp$, we have to rely on analytical information beyond that window of availability.

In the *limit of large impact parameter* $b_\perp$ – small $q_\perp$ in transverse momentum space – the dominant behavior should be linear in $b_\perp$ caused by the Wilson loop following an area law [18]. These contributions are subject to a fit to the large-$b_\perp$ tail of our lattice EQCD data,

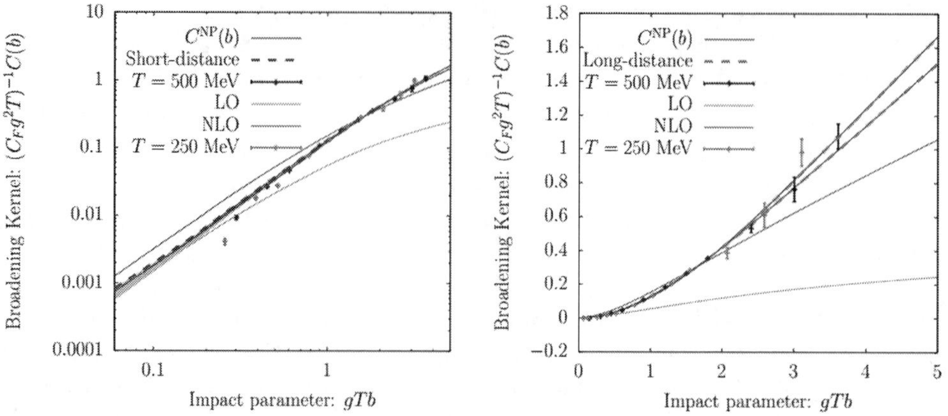

Figure 1. Non-perturbative elastic broadening kernel interpolation spline in the short-distance (left) and large-distance (right) regimes. We compare to both the short-distance limit from Eq. (5) and the long-distance limit from Eq. (4).

yielding the fit constant A and the string tension $\sigma_{\rm EQCD}$ from the literature [19]. The perturbative terms in (3) can only contribute sub-dominantly, in fact, they give rise to a logarithmic term. Thus, the overall large-$b_\perp$ limit reads

$$\frac{C_{\rm QCD}}{g_{\rm 3d}^2}(b_\perp) \xrightarrow{b_\perp \gg 1/g_{\rm 3d}^2} A + \frac{\sigma_{\rm EQCD}}{g_{\rm 3d}^4} g_{\rm 3d}^2 b_\perp + \frac{g_s^4 C_{\rm R}}{\pi}\left[\frac{y}{4}\left(\frac{1}{6} - \frac{1}{\pi^2}\right) + \frac{C_{\rm A}}{8\pi^2 g_s^2}\right]\log\left(g_{\rm 3d}^2 b_\perp\right), \quad (4)$$

in units of the three-dimensional coupling $g_{\rm 3d}^2 \approx g^2 T$, with the dimensionless screening mass ratio $y \equiv \frac{m_{\rm D}^2}{g_{\rm 3d}^4}\Big|_{\mu=g_{\rm 3d}^2}$ and the Casimir operators $C_{\rm R}$ of the representation R of the jet particle and $C_{\rm A}$ the Casimir of the adjoint representation.

The *opposite limit* is governed by the full QCD perturbative part of (3), since lattice EQCD and EQCD perturbation theory agree in the limit of small impact parameter. Dominated by the harmonic oscillator term $C(b_\perp) \approx \hat{q}\, b_\perp^2/4$, subleading effects due to screening occur at $O(b_\perp^2 \ln b_\perp)$. Altogether, the result reads

$$\frac{C_{\rm QCD}}{g_{\rm 3d}^2}(b_\perp) \xrightarrow{b_\perp \ll 1/m_{\rm D}} -\frac{C_{\rm R}}{8\pi}\frac{\zeta(3)}{\zeta(2)}\left(-\frac{1}{2g_s^2} + \frac{3y}{2}\right) g_{\rm 3d}^4 b_\perp^2 \log\left(g_{\rm 3d}^2 b_\perp\right) + \frac{1}{4}\frac{\hat{q}_0}{g_{\rm 3d}^6} g_{\rm 3d}^4 b_\perp^2, \quad (5)$$

with the scale-independent part of the jet-quenching parameter $\hat{q}_0$.

We present in Fig. 1 the fully matched results for $T = 250$MeV and 500MeV in orange and black points, respectively. Using the data points obtained and the limiting behaviors of Eqns. (4-5), we construct a continuous spline for each temperature as discussed in [20] shown as a violet line for $T = 250$MeV and as a blue line for $T = 500$MeV. Overall, we observe that even though the non-perturbative results are obtained at different temperatures and coupling constants, they scale onto very similar broadening kernels when we consider the broadening kernel in units of g^2T as a function of impact-parameter $b_\perp$ in units of $\frac{1}{gT}$. The right panel of Fig. 1 shows how the long-distance limit is approached by the data points and splines obtained from fully matched lattice results at two different temperatures. Moreover, purely perturbative estimates of $C(b_\perp)$ at leading (LO) and next-to-leading order (NLO) are shown as a benchmark. The left panel of Fig. 1 displays how the interpolating curve $C^{\rm NP}(b_\perp)$ approaches the correct small-$b_\perp$ limit from (5) and still matches the fully matched data points

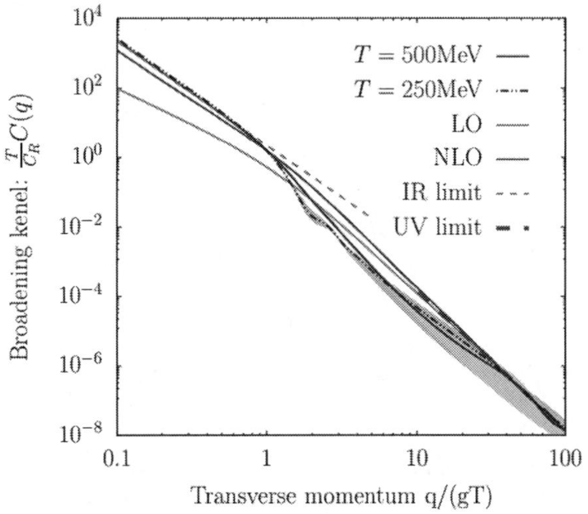

Figure 2. Elastic broadening kernel in momentum space for both $T = 250, 500$MeV, where the blue and red bands represent the error in the choice of the spline for 250MeV and 500MeV respectively. We compare to both the UV and IR limit, given as the Fourier transforms of Eqs. 5 and 4, see [21].

based on lattice EQCD as well as possible. The perturbative results approach the same short-distance behavior at different scales.

2.2 Broadening kernel in momentum space

The lattice EQCD calculation is performed in impact-parameter space ($\boldsymbol{b}_\perp$), which is of direct use for obtaining splitting rates in a medium of infinite size. However, in order to obtain splitting rates in a realistic medium of finite size, it is more convenient to work in momentum space ($q_\perp$). Therefore, we proceed to Fourier transform the resulting fully matched $C(b_\perp)$ to momentum space. The inverse Fourier transform seems to be straightforward to compute, however, since the data points are sparse and the kernel is divergent at large impact parameter, performing the highly oscillatory integrals involved is rather challenging. We can evade these difficulties, if we instead Fourier transform the coordinate space derivative $\frac{\mathrm{d}C(b_\perp)}{\mathrm{d}b_\perp}$ of the momentum broadening kernel. Using Eq. (2), one can write

$$C(q_\perp) = \frac{2\pi}{q_\perp}\int_0^\infty \mathrm{d}b_\perp\, b_\perp J_1(b_\perp q_\perp)\frac{\mathrm{d}C(b_\perp)}{\mathrm{d}b_\perp}\,. \tag{6}$$

Numerical details of the integration are discussed in [21] and we present the resulting broadening kernel in momentum space ($q_\perp$) in Fig. 2 compared to perturbative broadening kernels. We observe how both the non-perturbative and NLO results display similar behavior ($\propto 1/q_\perp^3$) in the IR limit which is the Fourier transformation of the linear long-distance behavior in Eq. (5), however, they come with different prefactors. Conversely, all the different kernels recover the UV behavior ($\propto 1/q_\perp^4$) at large momentum $q_\perp$, which is associated with hard scatterings.

3 Splitting rates in an infinite medium

While there are different formalisms used to compute medium induced radiation in QCD plasma [2, 22–24], here we follow the formalism of Arnold, Moore and Yaffe (AMY) [24],

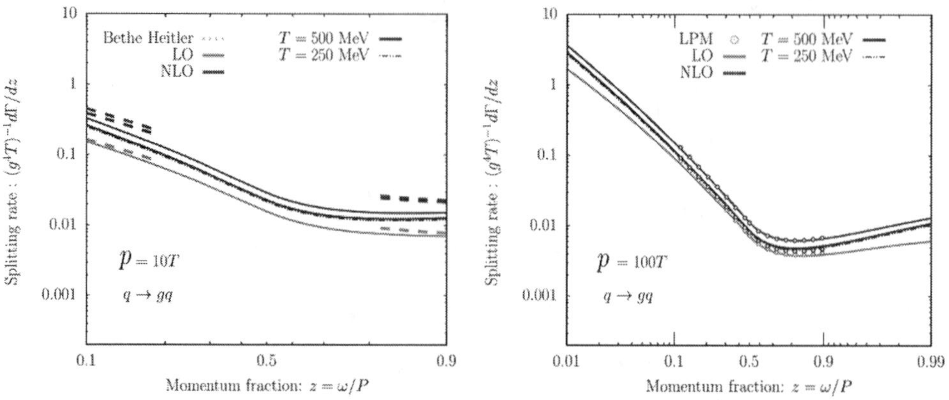

Figure 3. In-medium splitting rate of gluon by a parent quark in a medium of temperature $T = 250$MeV (dashed blue lines) and $T = 500$MeV (full purple lines). Different columns correspond to parent energies $P = 10T$ (left) and $P = 100T$ (right). We compare with rates computed using the perturbative leading order (orange) and next-to-leading order (green) elastic broadening kernels. The Bethe-Heitler rates and LPM rates are shown with dashed lines and circles, respectively, using the color of the corresponding kernel.

which can directly make use of the broadening kernel in impact-parameter space ($\boldsymbol{b}_\perp$) and provides an effective rate $\mathrm{d}\Gamma_{ij}/\mathrm{d}z(P, z)$, which corresponds to the rate at which particle i with energy P radiates particle j with energy $\omega = zP$ in an infinite medium. We refer the reader to [20] for details of the formalism and the numerical procedure.

Our results are presented in Fig. 3, where in addition to the splitting rates obtained using the non-perturbative broadening kernels at $T = 250, 500$MeV, we compare to the rates obtained by employing leading and next-to-leading order results for momentum broadening [20]. Furthermore, we show the Bethe-Heitler rates (dashed lines) and the deep LPM rates (circles) which are semi-analytical approximations of the rate calculation at small typical momentum $Pz(1-z) \ll T$ and large typical momentum $Pz(1-z) \gg T$, respectively (c.f. for details of the derivation [20]). We find that the two temperatures considered for the non-perturbative results do not display remarkable difference when the rate is measured in units of $[g^4T]$, which is a result of the scaling behavior we observed in the broadening kernel. The LPM suppression is recovered at large typical energies, where the splitting rates match the deep LPM rates, while the unsuppressed Bethe-Heitler rates are only reached at low typical momentum beyond the range of momentum fraction shown. When comparing to the rates obtained using perturbative broadening kernels, we observe that at high momentum $Pz(1-z) \gg T$ the rates are closer to the LO results, since this region probes the short-distant behavior of the broadening kernel where the non-perturbative kernel is matched to the LO one. Conversely, at low momentum $z(1-z)E \ll T$, where long-distance region is more relevant, the non-perturbative rate is closer to the NLO rate since they feature similar linear behavior at large impact parameter.

4 Splitting rates in a finite size medium

Using the broadening kernel $C(q_\perp)$ in momentum space, following [8], we are able to obtain the splitting rates of medium-induced radiation in a plasma of finite size. In [21], we provide details of the procedure we employ to obtain the splitting rate $\mathrm{d}\Gamma_{ij}/\mathrm{d}z(P, z, t)$, which describes

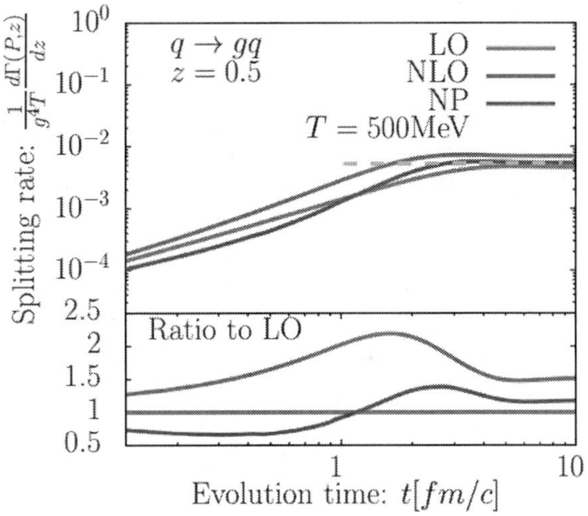

Figure 4. Medium-induced splitting rate of a gluon with momentum fraction $z = 0.5$ from a parent quark with energy $P = 300T$ in an equilibrium medium with temperature $T = 500$MeV as a function of the evolution time t. We compare calculation done using the different collisional broadening kernel as shown in Fig. 2 (the temperature and coupling constant for the perturbative results are matched to the $T = 500$MeV data in Tab. 1 of [20]). The lower panel displays the ratio to the LO results.

the rate at which particle i with energy P radiates particle j with energy $\omega = zP$ after a time t inside the medium. In Fig. 4, we display results of quark with energy $P = 300T$ radiating a gluon with momentum fraction $z = 0.5$ as a function of the evolution time t the quark spends in the medium. We present results at $T = 500$MeV using the obtained non-perturbative broadening kernel (NP) as well as the Leading Order (LO) and Next-to-Leading Order (NLO) perturbative broadening kernels. Dashed gray line represent the infinite medium results obtained in the previous section. We also show in the lower panel the ratio with respect to the LO results. At early times, the splitting rates display a linear behavior and quickly saturate at late times when the rate recovers the infinite medium results. We observe that the non-perturbative rate in fact starts lower than the LO before it evolves in between the LO and NLO.

5 Discussion

In this work, we considered non-perturbative contributions to the momentum broadening kernel $C(q_\perp)$ calculated in lattice EQCD [15]. Since EQCD is a long-distance effective theory of QCD, we supplied our lattice data with the correct short-distance behavior to obtain the non-perturbative momentum broadening kernel in QCD valid at all scales. Employing this kernel in impact parameter space, we computed medium-induced splitting rates by a hard parton traversing a QCD plasma of infinite size. We compared the resulting splitting rates to ones obtained using leading and next-to-leading order perturbative broadening kernels. We find that while the LO results are more relevant for hard momentum ($Pz(1-z) \gg T$) where the short-distance limit is more relevant, the rates display a similar behavior to the NLO for soft momentum ($Pz(1-z) \gg T$). Although the perturbative series for the broadening kernel $C(b_\perp)$ is apparently not convergent [11], we observe that the splitting rates for the non-pertubatively determined $C(b_\perp)$ mostly fall between the LO and NLO results.

Beyond the splitting rates in a medium of infinite size, we were able to transform the non-perturbative momentum broadening kernel $C(q_\perp)$ to momentum space, which is more favorable to compute splitting rates in media of finite extent. Following [8], we computed splitting rates in a finite size QCD medium, and compared our results to calculations with the leading order and next-to-leading order perturbative broadening kernels. We observe that while the rates from NLO display a large difference from the LO rates, non-perturbative results do not deviate beyond a band of 50% around the LO.

Since medium-induced emissions are a dominant mechanism of jet energy loss in QCD plasmas, it will be important to incorporate our results in phenomenological studies of jet quenching. Furthermore, there has been ongoing progress in obtaining non-perturbative contributions to the thermal masses [25] and we envision including these results in future work.

References

[1] M. Gyulassy, P. Levai, I. Vitev, Nucl. Phys. B **571**, 197 (2000), hep-ph/9907461
[2] M. Gyulassy, P. Levai, I. Vitev, Nucl. Phys. B **594**, 371 (2001), nucl-th/0006010
[3] M. Gyulassy, I. Vitev, X.N. Wang, B.W. Zhang (2003), nucl-th/0302077
[4] U.A. Wiedemann, Nucl. Phys. B **588**, 303 (2000), hep-ph/0005129
[5] C.A. Salgado, U.A. Wiedemann, Phys. Rev. D **68**, 014008 (2003), hep-ph/0302184
[6] P.B. Arnold, G.D. Moore, L.G. Yaffe, JHEP **06**, 030 (2002), hep-ph/0204343
[7] M. Djordjevic, U.W. Heinz, Phys. Rev. Lett. **101**, 022302 (2008), 0802.1230
[8] S. Caron-Huot, C. Gale, Phys. Rev. **C82**, 064902 (2010), 1006.2379
[9] P. Aurenche, F. Gelis, H. Zaraket, JHEP **05**, 043 (2002), hep-ph/0204146
[10] J. Casalderrey-Solana, D. Teaney, JHEP **04**, 039 (2007), hep-th/0701123
[11] S. Caron-Huot, Phys. Rev. D **79**, 065039 (2009), 0811.1603
[12] M. Panero, K. Rummukainen, A. Schäfer, Phys. Rev. Lett. **112**, 162001 (2014), 1307.5850
[13] M. D'Onofrio, A. Kurkela, G.D. Moore, JHEP **03**, 125 (2014), 1401.7951
[14] G.D. Moore, N. Schlusser, Phys. Rev. **D100**, 034510 (2019), 1905.09708
[15] G.D. Moore, N. Schlusser, Phys. Rev. **D101**, 014505 (2020), 1911.13127
[16] P.B. Arnold, W. Xiao, Phys. Rev. D **78**, 125008 (2008), 0810.1026
[17] J. Ghiglieri, H. Kim, JHEP **12**, 049 (2018), 1809.01349
[18] M. Laine, Eur. Phys. J. C **72**, 2233 (2012), 1208.5707
[19] M. Laine, Y. Schroder, JHEP **03**, 067 (2005), hep-ph/0503061
[20] G.D. Moore, S. Schlichting, N. Schlusser, I. Soudi, JHEP **10**, 059 (2021), 2105.01679
[21] S. Schlichting, I. Soudi (2021), in preparation
[22] R. Baier, Y.L. Dokshitzer, A.H. Mueller, S. Peigne, D. Schiff, Nucl. Phys. B **483**, 291 (1997), hep-ph/9607355
[23] B.G. Zakharov, JETP Lett. **63**, 952 (1996), hep-ph/9607440
[24] P.B. Arnold, G.D. Moore, L.G. Yaffe, JHEP **12**, 009 (2001), hep-ph/0111107
[25] G.D. Moore, N. Schlusser, Phys. Rev. D **102**, 094512 (2020), 2009.06614

On transverse momentum broadening in real-time lattice simulations of the glasma and in the weak-field limit

Andreas Ipp[1,*], *David I.* Müller[1,**], and *Daniel* Schuh[1,2,***]

[1]Institute for Theoretical Physics, TU Wien,
Wiedner Hauptstr. 8-10, 1040 Vienna, Austria
[2]Speaker and corresponding author

Abstract. In these proceedings, we report on our numerical lattice simulations of partons traversing the boost-invariant, non-perturbative glasma as created at the early stages of collisions at RHIC and LHC. Since these highly energetic partons are produced from hard scatterings during heavy ion collisions, they are already affected by the first stage of the medium's time evolution, the glasma, which is the pre-equilibrium precursor state of the quark-gluon plasma. We find that partons quickly accumulate transverse momentum up to the saturation momentum during the glasma stage. Moreover, we observe an interesting anisotropy in transverse momentum broadening of partons with larger broadening in the rapidity than in the azimuthal direction. Its origin can be related to correlations among the longitudinal color-electric and color-magnetic flux tubes in the initial state of the glasma. We compare these observations to the semi-analytic results obtained by a weak-field approximation, where we also find such an anisotropy in a parton's transverse momentum broadening.

1 Introduction

The quark-gluon plasma (QGP) is an extensively researched state of matter that is created after a heavy-ion collision. It is effectively described by viscous hydrodynamics [1, 2], which allows for the determination of bulk properties, such as particle multiplicities and flow harmonics [3–7], without exact information about the initial conditions of the hydrodynamic evolution because of the existence of a hydrodynamic attractor [8]. However, there are also probes that carry information throughout the whole evolution, and, hence, they are influenced by the other stages of matter that are created after a heavy-ion collision. Some of these probes do not interact strongly, such as photons and dileptons [9–12], while others, such as jets [13–15], do. The seeds of jets (partons) are created by hard scatterings during the collision and are, therefore, already present in the pre-equilibrium precursor state of the QGP, which is the glasma. The glasma is based on an effective theory of high-energy QCD, namely the color-glass condensate (CGC) [16, 17] and consists of expanding chromo-electric and chromo-magnetic flux tubes [18]. This state is highly anisotropic because said flux tubes are initially longitudinal and, at sufficiently high energies, boost invariant.

*e-mail: ipp@hep.itp.tuwien.ac.at
**e-mail: dmueller@hep.itp.tuwien.ac.at
***e-mail: schuh@hep.itp.tuwien.ac.at

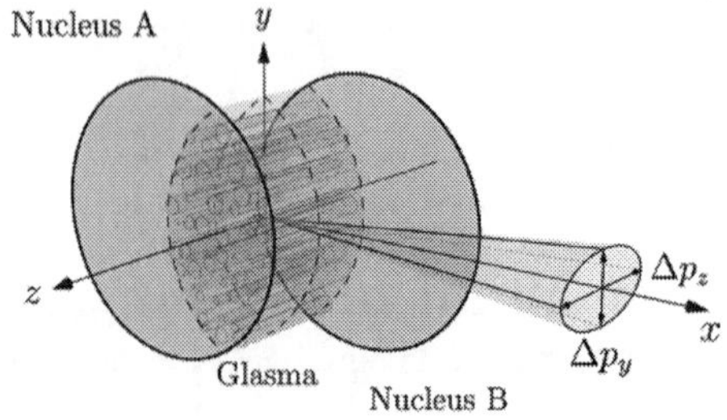

Figure 1. Visualization of a heavy-ion collision, taken from [19]. Two colliding nuclei (gray) produce the glasma, which is depicted by colorful cylinders. During the same collision event, a parton (traveling in x-direction) is produced and experiences momentum broadening transverse to its movement.

In these proceedings, which are based on [19, 20], we discuss the transverse momentum broadening that a parton accumulates during the glasma stage. We do that in a boost-invariant approximation, which causes the system to be effectively 2+1 dimensional, since its rapidity dependence is dropped. The dilute limit [21], in which the color fields are considered to be small, allows for a largely analytic calculation of the transverse momentum broadening. This gives rise to an analytic relation of the momentum broadening anisotropy of an ultra-relativistic test parton to the correlations among the flux tubes in the initial state of the glasma. Recent progress in the dilute limit has been made in 3+1 dimensions [22] as well. We also discuss transverse momentum broadening in a different limit, namely the lattice approximation, which is a real-time lattice field theoretic approach. It permits the treatment of strong fields in the glasma, providing the opportunity to study this more realistic case. The transverse plane is approximated by a square lattice with periodic boundary conditions, and the discretized Yang-Mills equations are solved with finite differences.

These proceedings are organized in the following way: in sec. 2, we will discuss the theoretical framework of describing momentum broadening in a non-Abelian background field. Then, in sec. 3, we will specify the background field to be the glasma and examine two approximations in which we can compute said momentum broadening, namely a lattice approximation (subsec. 3.1) and a weak-field approximation (subsec. 3.2). We will present our results in sec. 4 and conclude in sec. 5.

2 Momentum broadening in a non-Abelian background field

The glasma is produced during the collision of two high-energy nuclei by interactions among their soft partons. We consider an ultra-relativistic parton that is created through hard scatterings during such a collision. We treat it as a test parton and the glasma as a non-Abelian background field, i.e. back-reactions from the parton on the glasma are neglected, but the glasma exerts a non-Abelian Lorentz force on the parton. Since the parton is ultra-relativistic, its trajectory is considered to be unaffected by this force, but it still accumulates momentum along its path. The concrete geometry is shown in fig. 1. The nuclei, depicted in gray, move in positive and negative z-direction, respectively, and collide at the origin. The parton travels along the x-direction ($u^\mu = (1, 1, 0, 0)^\mu$), the cone indicates the broadening in momentum transverse to the particle trajectory and the glasma is visualized by colorful cylinders.

The methods that we use to obtain an expression for transverse momentum broadening for an ultra-relativistic colored particle are largely based on its definition via lightlike Wilson lines [23–25] and the dynamics of classical colored particles [26–30]. We review them in [19] and also derive a formula for said momentum broadening. For a quark, it reads

$$\langle p_i^2(t)\rangle_q = \frac{2g^2}{N_c}\int_0^t \mathrm{d}t' \int_0^t \mathrm{d}t'' \left\langle \mathrm{Tr}\left[U(0,t')F_{i+}(x(t'))U(t',0)U(0,t'')F_{i+}(x(t''))U(t'',0)\right]\right\rangle, \quad (1)$$

where the subscript q denotes that it holds for a quark, g is the Yang-Mills coupling constant, N_c corresponds to the gauge group $\mathrm{SU}(N_c)$ and $i \in \{y, z\}$. The lightlike Wilson line $U(0,t)$ connects the starting position of the particle at its creation at $t_0 = 0$ with its current position at t, and the field strength tensor is denoted by F. Note that the plus sign in the subscript of the field strength tensor refers to the light-cone coordinates $x^\pm = (t \pm x)/\sqrt{2}$. The transverse momentum broadening for a corresponding gluon is related to eq. (1) via Casimir scaling

$$\langle p_i^2\rangle_g = \frac{C_A}{C_F}\langle p_i^2\rangle_q, \quad (2)$$

where C_A and C_F are the Casimirs in the adjoint and fundamental representation, respectively.

3 Momentum broadening in the glasma

The glasma is described in the CGC framework. The hard partons of the incoming nuclei A and B are described by the color charge densities $\rho_{(\mathrm{A,B})}$. Their color currents in covariant gauge $\partial_\mu A^\mu = 0$ are given by $J^\mu_{(\mathrm{A,B})} = \delta^\mu_\pm \rho_{(\mathrm{A,B})}(x^\mp, \mathbf{x})$, with the Kronecker delta $\delta^\mu_\pm$. The light-cone coordinates are defined as $x^\pm = (t \pm z)/\sqrt{2}$, i.e. this definition differs by the one used in eq. (1) by the replacement $x \leftrightarrow z$. This difference is connected to the fact that in the glasma, the transverse coordinates are viewed as transverse to the beam axis (z-axis), while in the context of momentum broadening, the transverse plane is considered to be transverse to the particle direction (x-axis).

The charge densities $\rho_{(\mathrm{A,B})}$ are described by the McLerran-Venugopalan model [31, 32]. In this model, the nuclei are approximated by infinitely thin color sheets $\rho_{(\mathrm{A,B})}(x^\mp, \mathbf{x}) = \delta(x^\mp)\rho_{(\mathrm{A,B})}(\mathbf{x})$ that are infinitely big in the transverse directions. Furthermore, the color charge density ρ is viewed as a random field. Its distribution is given by a Gaussian probability functional $W[\rho]$, which is specified by the charge density correlator

$$\langle \rho^a_{(\mathrm{A,B})}(x)\rho^b_{(\mathrm{A,B})}(y)\rangle = (g\mu)^2\delta^{ab}\delta(x^\mp - y^\mp)\delta(x^\mp)\delta^{(2)}(\mathbf{x}-\mathbf{y}), \quad (3)$$

with the model parameter μ that determines the saturation momentum $Q_s \propto g^2\mu$. Expectation values of observables are obtained by functional integration over ρ, weighted by $W[\rho]$

$$\langle \ldots \rangle = \int \mathcal{D}\rho\, W[\rho]\,(\ldots). \quad (4)$$

The high-momentum (hard) partons act as sources for the low-momentum (soft) partons. The glasma is obtained from solving the classical Yang-Mills equations in the forward light cone of the collision, which we describe in Milne coordinates. These consist of proper time $\tau = \sqrt{2x^+x^-}$, spacetime rapidity $\eta = \ln(2x^+x^-)/2$ and the transverse coordinates x and y. In the boost-invariant approximation, the Yang-Mills equations are source free and physical observables do not depend on η. The glasma initial conditions [21] in temporal gauge $A^\tau = 0$ read

$$A^i(0,\mathbf{x}) = \alpha^i_{\mathrm{A}}(\mathbf{x}) + \alpha^i_{\mathrm{B}}(\mathbf{x}),\ \ A^\eta(0,\mathbf{x}) = \frac{ig}{2}\left[\alpha^i_{\mathrm{A}}(\mathbf{x}), \alpha^i_{\mathrm{B}}(\mathbf{x})\right],\ \ \partial_\tau A^i(0,\mathbf{x}) = 0,\ \ \partial_\tau A^\eta(0,\mathbf{x}) = 0. \quad (5)$$

The fields $\alpha^i_{(\mathrm{A,B})}$ are given by $\alpha^i_{(\mathrm{A,B})}(\mathbf{x}) = \frac{1}{ig}V_{(\mathrm{A,B})}(\mathbf{x})\partial^i V^\dagger_{(\mathrm{A,B})}(\mathbf{x})$, with the asymptotic Wilson lines

$$V^\dagger_{(\mathrm{A,B})}(\mathbf{x}) = \lim_{x^\mp\to\infty} V^\dagger_{(\mathrm{A,B})}(x^\mp,\mathbf{x}) = \lim_{x^\mp\to\infty}\mathcal{P}\exp\left(ig\int_{-\infty}^{\infty}\mathrm{d}x^\mp\,\frac{\rho_{(\mathrm{A,B})}(x^\mp,\mathbf{x})}{\nabla^2 - m^2}\right), \quad (6)$$

where m is an infrared regulator.

Given a solution to the Yang-Mills equations under the glasma initial conditions, we can evaluate the accumulated transverse momentum, given by eq. (1). We obtain

$$\langle p_i^2(\tau)\rangle_q = \frac{g^2}{N_c}\int_0^\tau \mathrm{d}\tau' \int_0^\tau \mathrm{d}\tau'' \,\langle \mathrm{Tr}\left[f^i(\tau')f^i(\tau'')\right]\rangle, \tag{7}$$

with the functions f^i that represent the color-rotated Lorentz force

$$f^y(\tau) = U(\tau)\big(E_y(\tau) - B_z(\tau)\big)\,U^\dagger(\tau), \qquad f^z(\tau) = U(\tau)\big(E_z(\tau) + B_y(\tau)\big)\,U^\dagger(\tau). \tag{8}$$

The lightlike Wilson line U in the fundamental representation along the particle trajectory in temporal gauge is given by

$$U(\tau) = \mathcal{P}\exp\Big(-ig\int_0^\tau \mathrm{d}\tau'\, A_x(\tau')\Big). \tag{9}$$

3.1 Lattice approximation

The lattice approximation [33, 34] is a real-time lattice gauge theoretic method. We approximate the transverse plane by a square lattice and describe each nucleus by N_s color sheets along the longitudinal direction, which accounts for path ordering [35]. The degrees of freedom change to the gauge links in the transverse plane $U_{x,\hat{i}}$, the rapidity component of the gauge field $A_{x,\eta}$ and the conjugate momenta P^i_x and P^η_x. The Yang-Mills equations are replaced by a leapfrog scheme for finite time steps, and the glasma initial conditions are discretized. Details of the derivation of the appropriate lattice version of transverse momentum broadening can be found in [19]. The result reads

$$\langle p_i^2(t_n)\rangle_q \approx \frac{g^2 a_T^2}{N_c}\langle\, \mathrm{Tr}\Big[\big(\sum_{i=0}^{n} f^i(t_n)\big)^2\Big]\rangle, \tag{10}$$

where a_T is the transverse lattice spacing, f^i are discretized versions of eq. (8) and t_n are the times at which the particle position coincides with a lattice site.

3.2 Weak-field approximation

In the weak-field approximation [21], the charge density ρ that describes the nuclei is deemed small. Consequently, path ordering effects do not contribute at the lowest order that is not pure gauge, and the time evolution becomes Abelian at this (fourth) order [36]. The solution of the color-electromagnetic fields can be written as

$$E_z(\tau,\mathbf{x}) = \int \frac{\mathrm{d}^2\mathbf{k}}{(2\pi)^2}\, J_0(k\tau)\tilde{E}_z(0,\mathbf{k})e^{i\mathbf{k}\cdot\mathbf{x}}, \qquad B_z(\tau,\mathbf{x}) = \int \frac{\mathrm{d}^2\mathbf{k}}{(2\pi)^2}\, J_0(k\tau)\tilde{B}_z(0,\mathbf{k})e^{i\mathbf{k}\cdot\mathbf{x}}, \tag{11}$$

$$E_y(\tau,\mathbf{x}) = -\int \frac{\mathrm{d}^2\mathbf{k}}{(2\pi)^2}\frac{ik_x}{k}J_1(k\tau)\tilde{B}_z(0,\mathbf{k})e^{i\mathbf{k}\cdot\mathbf{x}}, \quad B_y(\tau,\mathbf{x}) = \int \frac{\mathrm{d}^2\mathbf{k}}{(2\pi)^2}\frac{ik_x}{k}J_1(k\tau)\tilde{E}_z(0,\mathbf{k})e^{i\mathbf{k}\cdot\mathbf{x}}, \tag{12}$$

with the Bessel functions J_n of the first kind, the norm k of the vector $\mathbf{k}$ and the Fourier components of the initial field strengths $\tilde{E}_z(0,\mathbf{k})$ and $\tilde{B}_z(0,\mathbf{k})$. As can be seen from the order of the Bessel function,[1] the fields are purely longitudinal with respect to the beam axis at $\tau = 0$. The functions f^i, given by eq. (8), turn into

[1] The Bessel function J_0 is positive at zero, while the Bessel function J_1 vanishes, i.e. $J_0(0) = 1$ and $J_1(0) = 0$.

$$f^y(\tau) = -\int \frac{\mathrm{d}^2\mathbf{k}}{(2\pi)^2}\,\gamma(\tau,\mathbf{k})\tilde{B}_z(0,\mathbf{k}), \qquad f^z(\tau) = \int \frac{\mathrm{d}^2\mathbf{k}}{(2\pi)^2}\,\gamma(\tau,\mathbf{k})\tilde{E}_z(0,\mathbf{k}), \tag{13}$$

with

$$\gamma(\tau,\mathbf{k}) = \left(\frac{ik_x}{k}J_1(k\tau) + J_0(k\tau)\right)e^{ik_x\tau}. \tag{14}$$

The Wilson line U does not contribute because it is approximated by a unit matrix at lowest order. The function γ appears in both components of f^i, which means that the time dependent part of their integrands is identical. Inserting eq. (13) into eq. (7) and evaluating the trace leads to

$$\langle p^2_{(y,z)}(\tau)\rangle_q = \int \frac{\mathrm{d}^2\mathbf{k}}{(2\pi)^2}\,g(\tau,\mathbf{k})\,c_{(B,E)}(k,m), \quad g(\tau,\mathbf{k}) = \frac{N_c^2-1}{2}g^8\mu^4\left|\int_0^\tau \mathrm{d}\tau'\,\gamma(\tau,\mathbf{k})\right|^2, \tag{15}$$

with the infrared regulator m. Since γ shows up in both components of f^i, the function g appears in both components of the transverse momentum broadening. They only differ by the initial time correlators c_B and c_E

$$c_B(k,m) = \int \frac{\mathrm{d}^2\mathbf{p}}{(2\pi)^2}\frac{(\mathbf{p}\times\mathbf{k})^2}{(p^2+m^2)^2(|\mathbf{k}-\mathbf{p}|^2+m^2)^2}, \quad c_E(k,m) = \int \frac{\mathrm{d}^2\mathbf{p}}{(2\pi)^2}\frac{(\mathbf{p}\cdot(\mathbf{k}-\mathbf{p}))^2}{(p^2+m^2)^2(|\mathbf{k}-\mathbf{p}|^2+m^2)^2}. \tag{16}$$

This lets us link an anisotropy $\langle p_z^2\rangle/\langle p_y^2\rangle \neq 1$ in the transverse momentum broadening of an ultra-relativistic test parton directly to the difference between the color-electric and color-magnetic correlators at the time of the collision.

4 Results and discussion

Before presenting the results for the physically more relevant lattice simulations that treat strong color fields, we discuss the results of the weak-field limit, in which we have analytic formulas that allow for a better understanding of the mechanisms that lead to momentum broadening. Figure 2(a) displays the accumulated transverse momentum broadening in the dilute glasma. It compares the lattice simulations to the weak-field calculations and shows that they agree for a dilute glasma. Furthermore, it illustrates that the two components of transverse momentum broadening behave differently, which manifests itself by an anisotropy in momentum broadening $\langle p_z^2\rangle/\langle p_y^2\rangle \neq 1$. In eq. (15), we linked such an anisotropy to the difference between the color-electric and color-magnetic correlators during the collision, and Figs. 2(b) and 2(c) show these correlators in momentum space and in coordinate space, respectively. They agree for large momenta $k \gg m$, but they highly differ from one another in the infrared regime $k < m$. While the electric correlator stays positive as the momentum tends toward zero, the magnetic correlator vanishes

$$\lim_{k\to 0} c_E(k,m) = \frac{1}{12\pi^2m^2} > 0, \qquad \lim_{k\to 0} c_B(k,m) = 0. \tag{17}$$

In coordinate space, the difference between the two correlators manifests itself in that the electric correlator is positive everywhere, while the magnetic correlator exhibits a region of anti-correlation around $mr \approx 1$. This anti-correlation is responsible for the flattening of $\langle p_y^2\rangle$. Note that it is the infrared modes of the correlators that differ and are, therefore, responsible for the anisotropy in transverse momentum broadening and that observables that are less sensitive to these modes show a smaller discrimination between electric and magnetic flux tubes. The energy density of electric and magnetic flux tubes, for example, is very similar because there, the infrared modes are suppressed.

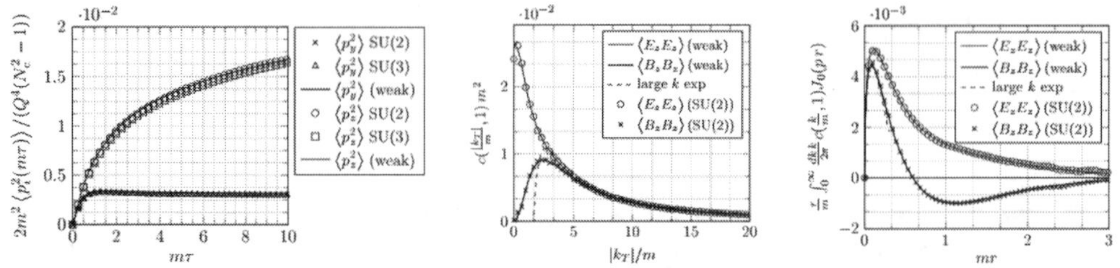

(a) Transverse momentum broadening in the dilute glasma.

(b) Initial correlators of the dilute glasma in momentum space.

(c) Initial correlators of the dilute glasma in coordinate space.

Figure 2. Transverse momentum broadening and initial field correlators of an ultra-relativistic test parton in the dilute glasma, taken from [19]. The lines depict the results of our semi-analytical weak-field approach, and the symbols correspond to our real-time lattice simulations, with SU(2) and SU(3) as gauge groups. Equation (15) provides the relationship between (a) and (b, c).

Figures 3(a) and 3(b) depict the momentum broadening for strong fields $m/(g^2\mu) \ll 1$. The behavior at early times looks qualitatively similar to the dilute glasma, while $\langle p_z^2 \rangle$ differs at later times in that it starts to decrease just below $Q_s\tau = 10$, as does the total transverse momentum broadening $\langle p_\perp^2 \rangle = \langle p_y^2 \rangle + \langle p_z^2 \rangle$. The y-component of transverse momentum broadening depends only weakly on the density of the glasma, whereas the numerical values of the z-component highly differ for varying $m/(g^2\mu)$, so much so that the anisotropy vanishes at very late times for a very strong glasma. Note, however, that at such late times ($Q_s\tau \gtrsim 10$), the glasma is no longer a valid description for the state of matter that is produced in heavy-ion collisions. Figures 3(c) and 3(d) show the initial correlators for strong fields in momentum and coordinate space, respectively. While each of them individually depends on the glasma density numerically, there is also a qualitative difference between them. In momentum space, they differ at low momenta, and the color-electric correlator tends towards a positive value as the momentum approaches zero. The color-magnetic correlator admits a negative value for vanishing momentum, which becomes closer to zero as the glasma becomes more dilute, as one would expect from the weak-field results, where it vanishes for $k = 0$. In coordinate space, the color-magnetic correlator looks qualitatively similar to the corresponding weak-field correlator, with increasing similarity for rising $m/(g^2\mu)$. The color-electric correlator, however, exhibits a negative region for glasmas denser than $m/(g^2\mu) \lesssim 0.1$. This is in contrast to the dilute glasma. Note that despite some similarities in the results of transverse momentum broadening and the initial correlators between the dilute and the dense glasma, we do not have an analytic relation that lets us link the momentum broadening anisotropy to the initial time correlators for a dense glasma. Nevertheless, we expect a relation akin to eq. (15) due to these similarities.

5 Conclusions and outlook

In these proceedings, we discussed transverse momentum broadening of an ultra-relativistic test parton in the glasma, which is the precursor state of the QGP. After revisiting the formalism that provides the basis for our calculations, we presented two approximations with which we were able to compute said momentum broadening and analyzed their similarities and differences. In the weak-field approximation, we found a formula that provides a link between the anisotropy in transverse momentum broadening and the difference in the low momentum modes of the color-electric and color-magnetic correlators at the time of the collision. For

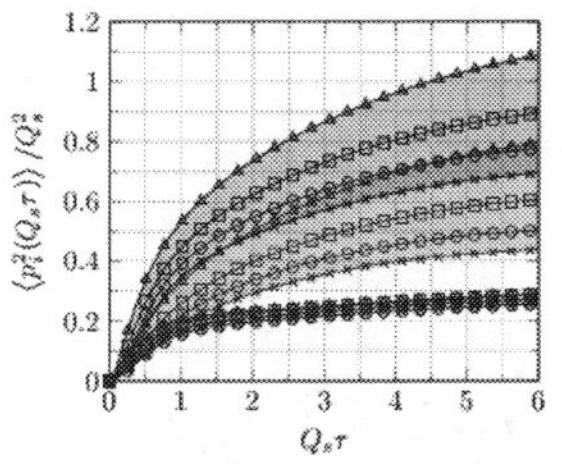

(a) Momentum broadening in the dense glasma at early times.

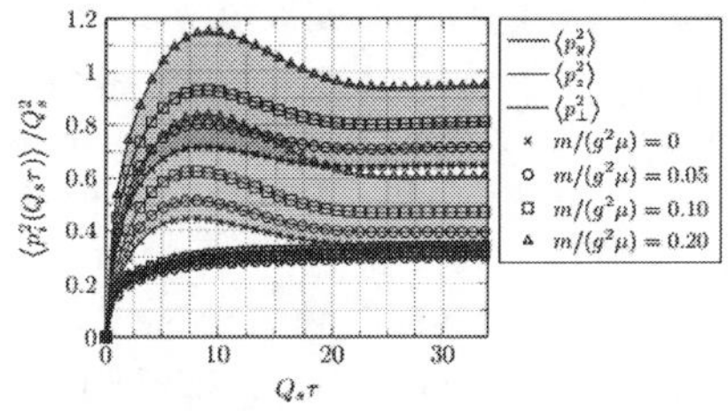

(b) Momentum broadening in the dense glasma up to later times.

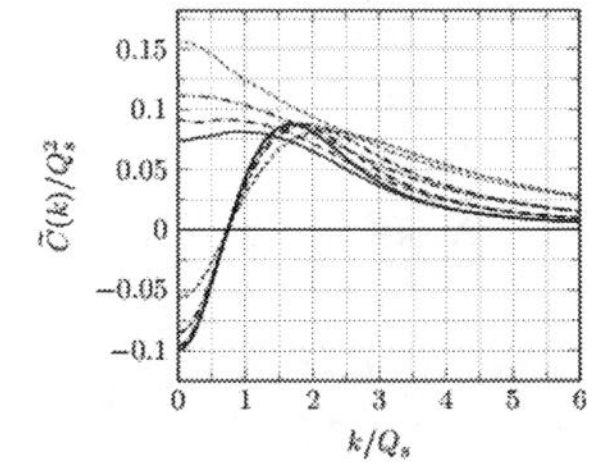

(c) Initial correlators of the dense glasma in momentum space.

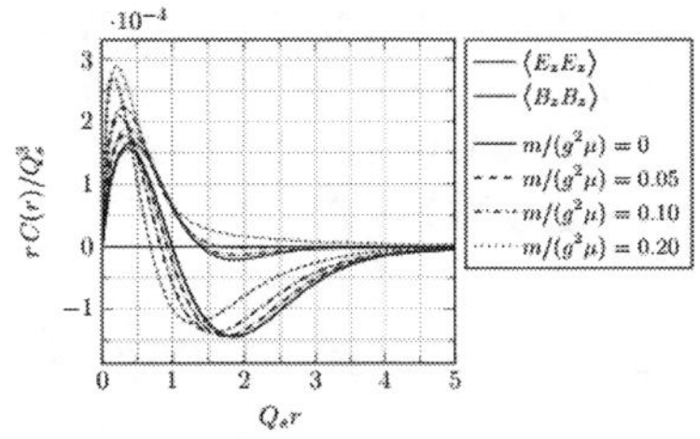

(d) Initial correlators of the dense glasma in coordinate space.

Figure 3. Transverse momentum broadening and initial field correlators of an ultra-relativistic test parton in the dense glasma from SU(3) real-time lattice simulations, taken from [19]. The different symbols correspond to various densities of the glasma, determined by the ratio $m/(g^2\mu)$, and the bands indicate the values that momentum broadening can take for the chosen ratios. The initial color-electric correlator in coordinate space is defined as $C_E(|\mathbf{x}-\mathbf{y}|) = \langle \mathrm{Tr}\left[E_z(\mathbf{x})U_{\mathbf{x}\to\mathbf{y}}E_z(\mathbf{y})U_{\mathbf{y}\to\mathbf{x}}\right]\rangle$, where the Wilson line $U_{\mathbf{x}\to\mathbf{y}}$ connects the lattice points $\mathbf{x}$ and $\mathbf{y}$. The color-magnetic correlator is defined analogously ($E \to B$) and $\widetilde{C}(k)$ are the Hankel transforms of order zero.

realistic fields, we found somewhat similar results in the relevant time scales, which suggests that the weak-field limit can be seen as an interesting approximation in which the underlying mechanisms can be studied analytically.

In order to extend this work, one could use more realistic initial conditions for the glasma or relax the ultra-relativistic test particle approximation. The former contains the treatment of the nuclei as having finite thickness [22], which implies a rapidity dependence of the system, and finite spatial extent in the transverse direction, which allows to study off-central collisions. The latter leads to a deflection of the parton from its straight trajectory and to the inclusion of its back-reaction on the glasma. It would also be interesting to study other observables, such as the energy loss of a test parton traversing the glasma, in order to better understand the early-time behavior of partons that create jets at later stages.

This work has been supported by the Austrian Science Fund FWF No. P32446-N27 and No. P28352. The Titan V GPU used for this research was donated by the NVIDIA Corporation.

References

[1] C. Gale, S. Jeon, B. Schenke, Int. J. Mod. Phys. **A28**, 1340011 (2013), 1301.5893

[2] P. Romatschke, U. Romatschke, *Relativistic Fluid Dynamics In and Out of Equilibrium*, Cambridge Monographs on Mathematical Physics (Cambridge University Press, 2019), ISBN 9781108483681, 9781108750028, 1712.05815

[3] B. Schenke, S. Jeon, C. Gale, Phys. Rev. **C82**, 014903 (2010), 1004.1408
[4] B. Schenke, S. Jeon, C. Gale, Phys. Rev. **C85**, 024901 (2012), 1109.6289
[5] B. Schenke, P. Tribedy, R. Venugopalan, Phys. Rev. **C86**, 034908 (2012), 1206.6805
[6] C. Gale, S. Jeon, B. Schenke, P. Tribedy, R. Venugopalan, Phys. Rev. Lett. **110**, 012302 (2013), 1209.6330
[7] H. Niemi, K.J. Eskola, R. Paatelainen, Phys. Rev. **C93**, 024907 (2016), 1505.02677
[8] J. Berges, K. Boguslavski, S. Schlichting, R. Venugopalan, Phys. Rev. **D89**, 114007 (2014), 1311.3005
[9] M. Martinez, M. Strickland, Phys. Rev. Lett. **100**, 102301 (2008), 0709.3576
[10] A. Ipp, C.H. Keitel, J. Evers, Phys. Rev. Lett. **103**, 152301 (2009), 0904.4503
[11] A. Ipp, P. Somkuti, Phys. Rev. Lett. **109**, 192301 (2012), 1207.0197
[12] G. Vujanovic, J.F. Paquet, G.S. Denicol, M. Luzum, B. Schenke, S. Jeon, C. Gale, Nucl. Phys. **A932**, 230 (2014), 1404.3714
[13] Y. Mehtar-Tani, J.G. Milhano, K. Tywoniuk, Int. J. Mod. Phys. **A28**, 1340013 (2013), 1302.2579
[14] M. Connors, C. Nattrass, R. Reed, S. Salur, Rev. Mod. Phys. **90**, 025005 (2018), 1705.01974
[15] W. Busza, K. Rajagopal, W. van der Schee, Ann. Rev. Nucl. Part. Sci. **68**, 339 (2018), 1802.04801
[16] F. Gelis, E. Iancu, J. Jalilian-Marian, R. Venugopalan, Ann. Rev. Nucl. Part. Sci. **60**, 463 (2010), 1002.0333
[17] F. Gelis, Int. J. Mod. Phys. **A28**, 1330001 (2013), 1211.3327
[18] T. Lappi, L. McLerran, Nucl. Phys. **A772**, 200 (2006), hep-ph/0602189
[19] A. Ipp, D. Müller, D. Schuh, Physical Review D **102** (2020), 2001.10001
[20] A. Ipp, D. Müller, D. Schuh, Physics Letters B **810**, 135810 (2020), 2009.14206
[21] A. Kovner, L.D. McLerran, H. Weigert, Phys. Rev. **D52**, 3809 (1995), hep-ph/9505320
[22] A. Ipp, D.I. Müller, S. Schlichting, P. Singh (2021), 2109.05028
[23] H. Liu, K. Rajagopal, U.A. Wiedemann, Phys. Rev. Lett. **97**, 182301 (2006), hep-ph/0605178
[24] J. Casalderrey-Solana, D. Teaney, JHEP **04**, 039 (2007), hep-th/0701123
[25] M. Panero, K. Rummukainen, A. Schäfer, Phys. Rev. Lett. **112**, 162001 (2014), 1307.5850
[26] Majumder, Abhijit and Müller, Berndt and Mrówczyński, Stanislaw, Phys. Rev. **D80**, 125020 (2009), 0903.3683
[27] M.E. Carrington, S. Mrówczyński, B. Schenke, Phys. Rev. **C95**, 024906 (2017), 1607.02359
[28] S. Mrówczyński, Eur. Phys. J. **A54**, 43 (2018), 1706.03127
[29] M. Ruggieri, S.K. Das, Phys. Rev. D **98**, 094024 (2018), 1805.09617
[30] Y. Sun, G. Coci, S.K. Das, S. Plumari, M. Ruggieri, V. Greco, Phys. Lett. B **798**, 134933 (2019), 1902.06254
[31] L.D. McLerran, R. Venugopalan, Phys. Rev. **D49**, 3352 (1994), hep-ph/9311205
[32] L. McLerran, R. Venugopalan, PRD **49**, 2233 (1994), hep-ph/9309289
[33] A. Krasnitz, R. Venugopalan, Nucl. Phys. **B557**, 237 (1999), hep-ph/9809433
[34] T. Lappi, Phys.Rev. **C67**, 054903 (2003), hep-ph/0303076
[35] K. Fukushima, Phys. Rev. **D77**, 074005 (2008), 0711.2364
[36] H. Fujii, K. Fukushima, Y. Hidaka, Phys. Rev. **C79**, 024909 (2009), 0811.0437

Fermion and gluon spectral functions far from equilibrium

Kirill Boguslavski[1,*], *Tuomas* Lappi[2,3,**], and *Sören* Schlichting[4,***]

[1]Institute for Theoretical Physics, Technische Universität Wien, 1040 Vienna, Austria
[2]Department of Physics, P.O. Box 35, 40014 University of Jyväskylä, Finland
[3]Helsinki Institute of Physics, P.O. Box 64, 00014 University of Helsinki, Finland
[4]Fakultät für Physik, Universität Bielefeld, D-33615 Bielefeld, Germany

Abstract. Motivated by the quark-gluon plasma, we develop a simulation method to obtain the spectral function of (Wilson) fermions non-perturbatively in a non-Abelian gauge theory with large gluon occupation numbers [1]. We apply our method to a non-Abelian plasma close to its non-thermal fixed point, i.e., in a far-from-equilibrium self-similar regime, and find mostly very good agreement with perturbative hard loop (HTL) calculations. For the first time, we extract the full momentum dependence of the damping rate of fermionic collective excitations and compare our results to recent non-perturbative extractions of gluonic spectral functions in two and three spatial dimensions [2, 3].

1 Introduction

Nonperturbatively strong bosonic fields can be encountered in a variety of systems, including heavy-ion collisions at early times [4, 5], cosmological reheating [6], or the infrared sector of gluonic and scalar fields in thermal equilibrium. Understanding their interactions with fermions is important for phenomenological applications including jets (energy loss, jet quenching) or electromagnetic observables, to mention a few in heavy-ion collisions.

For a microscopic description of the dynamics, the calculation of spectral functions $\rho(\omega, p)$ is very useful, since they encode medium interactions and the spectrum of collective excitations.

In this talk we will present a new method to compute the spectral function of Wilson fermions non-perturbatively in a weakly-coupled non-Abelian gauge theory with large gluon occupation numbers $f \gg 1$ (Sec. 2). In this far-from-equilibrium situation, we can use

*Speaker, e-mail: kirill.boguslavski@tuwien.ac.at.
K.B. would like to acknowledge support by the Austrian Science Fund (FWF) under project P 34455-N.
The authors are grateful to A. Kurkela, M. Mace and J. Peuron for collaboration and discussions. The authors wish to acknowledge the National Energy Research Scientific Computing Center, a U.S. Department of Energy Office of Science User Facility supported under Contract No. DE-AC02-05CH11231, the CSC - IT Center for Science, Finland, and the Vienna Scientific Cluster (VSC, Austria) under the project number 71444 for computational resources.
**T.L. is supported by the Academy of Finland, project 321840 and under the European Union's Horizon 2020 research and innovation programme by the European Research Council (ERC, grant agreement No. ERC-2015-CoG-681707) and by the STRONG-2020 project (grant agreement No 824093). The content of this article does not reflect the official opinion of the European Union and responsibility for the information and views expressed therein lies entirely with the authors.
***S.S. is supported by the Deutsche Forschungsgemeinschaft (DFG, German Research Foundation) through the CRC-TR 211 "Strong-interaction matter under extreme conditions" Project number 315477589.

classical-statistical lattice simulations to describe the underlying dynamics [7, 8]. Coupling them to the Dirac equation allows us to extract the fermion spectral function nonperturbatively. Our results are then discussed and compared to perturbative hard-loop (HTL) computations (Sec. 3) and to gluonic spectral functions in different dimensions (Sec. 4, for the comparison see Sec. 5). We conclude in Sec. 5. More details on our method and results for the fermion spectral function can be found in Ref. [1]. Our discussion of the gluon spectral function results is based on Refs. [2, 3].

2 Method & setup

2.1 Theory

We consider SU(N_c) gauge theory on a cubic lattice of size N_s^3 and lattice spacing a_s. The Yang-Mills and Dirac Hamiltonians read in $A_0 = 0$ gauge

$$\begin{aligned} H_{YM} =& \frac{1}{g^2 a_s} \sum_{\mathbf{x},i} \mathrm{Tr}[E_i(t',\mathbf{x})^2] + \frac{1}{2}\sum_{j} \mathrm{ReTr}[1 - U_{ij}(t',\mathbf{x})] \\ \hat{H}_W =& \frac{1}{2}\sum_{\mathbf{x}} \left[\hat{\psi}^\dagger(t',\mathbf{x}), \gamma^0 \left(-i\not{D}_s[U] + m\right)\hat{\psi}(t',\mathbf{x})\right], \end{aligned} \tag{1}$$

with the link fields $U_j \approx \exp\left(ig\, a_s A_j\right)$ replacing gauge fields, plaquettes U_{ij} computed from link fields, and $-i\not{D}_s[U]\hat{\psi}$ being the tree-level improved Wilson Dirac operator [9, 10]. We neglect the backreaction of fermions on the evolution of gauge fields since it is subleading when gluonic occupation numbers are large, as in our case. We will consider $N_c = 2$ in the following to reduce the computational complexity of the calculations.

The fermionic fields are formulated in terms of a mode expansion [11]

$$\hat{\psi}(t',\mathbf{x}) = \frac{1}{\sqrt{V}} \sum_{\lambda,\mathbf{p}} \hat{b}_{\lambda,\mathbf{p}}(t)\, \phi^u_{\lambda,\mathbf{p}}(t',\mathbf{x}) + \hat{d}^\dagger_{\lambda,\mathbf{p}}(t)\, \phi^v_{\lambda,\mathbf{p}}(t',\mathbf{x}), \tag{2}$$

where $\lambda = 1, \cdots, 2N_c$ labels spin and color indices. At the reference time $t' = t$, where t will be the time when we extract the spectral function, the creation and annihilation operators $\hat{b}$, $\hat{d}$ satisfy the usual anti-commutation relations at t and the wave functions are set to plane waves

$$\phi^u_{\lambda,\mathbf{p}}(t',\mathbf{x})\Big|_{t'=t} = u_\lambda(\mathbf{p}) e^{+i\mathbf{p}\cdot\mathbf{x}}, \quad \phi^v_{\lambda,\mathbf{p}}(t',\mathbf{x})\Big|_{t'=t} = v_\lambda(\mathbf{p}) e^{-i\mathbf{p}\cdot\mathbf{x}}. \tag{3}$$

2.2 Fermion spectral function

The spectral function of fermions is defined as

$$\rho^{\alpha\beta}(x,y) = \left\langle \left\{ \hat{\psi}^\alpha(t',\mathbf{x}), \hat{\bar{\psi}}^\beta(t,\mathbf{y}) \right\} \right\rangle, \tag{4}$$

where we have written the Dirac indices α, β explicitly. In our classical-statistical framework, the spectral function can be computed efficiently in momentum space. Plugging in the mode expansion (2) into (4), Fourier transforming with respect to $\mathbf{x} - \mathbf{y}$, and using (3) for time t, one arrives at

$$\begin{aligned} &\rho^{\alpha\beta}(t',t,\mathbf{p}) \\ &= \frac{1}{V}\sum_{\lambda,\mathbf{q}} \left\langle \tilde{\phi}^{u,\alpha}_{\lambda,\mathbf{q}}(t',\mathbf{p}) \left(\tilde{\phi}^{u,\gamma}_{\lambda,\mathbf{q}}(t,\mathbf{p})\right)^* + \tilde{\phi}^{v,\alpha}_{\lambda,\mathbf{q}}(t',\mathbf{p}) \left(\tilde{\phi}^{v,\gamma}_{\lambda,\mathbf{q}}(t,\mathbf{p})\right)^* \right\rangle \gamma_0^{\gamma\beta} \\ &= \frac{1}{V}\sum_{\lambda} \left\langle \tilde{\phi}^{u,\alpha}_{\lambda,\mathbf{p}}(t',\mathbf{p})\, u^{\dagger,\gamma}_\lambda(\mathbf{p}) + \tilde{\phi}^{v,\alpha}_{\lambda,-\mathbf{p}}(t',\mathbf{p})\, v^{\dagger,\gamma}_\lambda(-\mathbf{p}) \right\rangle \gamma_0^{\gamma\beta}, \end{aligned} \tag{5}$$

where $\langle . \rangle$ denotes a classical-statistical average over gluonic configurations. The initialization and the specific form of the initial conditions in Eq. (3), where $\phi^{u/v}_{\lambda,\mathbf{q}}(t,\mathbf{x})$ is given by a plane wave, allows to simplify the expression in the last line above.

2.3 Simulation algorithm

The simulation algorithm can be summarized as follows:

1. Set *initial conditions* for gluons at $t' = 0$, generating a configuration with $\langle E^*_T(t'{=}0,\mathbf{p})E_T(t'{=}0,\mathbf{q})\rangle \propto p\, f(t'{=}0,p)(2\pi)^3\delta(\mathbf{p}-\mathbf{q})$.
2. Solve classical equations of motion (EOMs) for *gauge fields* for $0 \le t' \le t$, set Coulomb-type gauge $\partial^j A_j\big|_t = 0$ at $t' = t$, i.e., at the time when the spectral function is measured.
3. For each momentum mode $\mathbf{p}$ initialize $\phi^{u/v}_{\lambda,\mathbf{p}}$ at $t' = t$ using Eq. (3).
4. *Evolve gauge and fermionic fields* for $t' > t$ using a leap-frog scheme to solve the classical EOMs for gauge and the Dirac equation for fermionic fields.
5. Calculate the fermionic spectral function $\rho(t',t,\mathbf{p})$ for each momentum mode $\mathbf{p}$ according to Eq. (5).

3 Nonperturbatively computed fermion spectral functions

3.1 Non-equilibrium state: self-similar turbulent attractor

We now consider spectral functions in a state of high gluonic occupancy $f_g(t{=}0, p \lesssim Q) \sim \frac{1}{g^2} \gg 1$. Such a system approaches a self-similar attractor with

$$f(t,p) = (Qt)^{\alpha} f_s\left((Qt)^{\beta} p\right), \tag{6}$$

after some short transient time. The scaling exponents and scaling function then become insensitive to details of initial conditions and their universal values in Minkowski spacetime are given by $\beta = -1/7$ and $\alpha = 4\beta$ for a 3+1 dimensional [12–14] and by $\beta = -1/5$ and $\alpha = 3\beta$ for a 2+1D plasma [15]. Since we know the time-evolution of characteristic momentum scales at those attractor states, like the hard scale $\Lambda \sim Q(Qt)^{-\beta}$ and the soft mass scale $m \sim Q(Qt)^{\beta}$, these states are useful assets to study spectral functions in more detail.

3.2 Fermion ρ in 3+1D at an attractor

We run our simulations on a 256^3 lattice with $Qa_s = 0.75$ with almost vanishing fermion mass $m = 0.003125\,Q$ to avoid ambiguities of the Wilson-Dirac equation. The fermion spectral function is extracted at the reference time $Qt = 1500$, i.e., sufficiently late to make sure that we are close to the self-similar regime of Eq. (6). The (nonvanishing) vector components of the spectral function

$$\rho^0_V = \frac{1}{4}\mathrm{Tr}(\rho\gamma^0), \quad \rho_V = -\frac{E_{\mathbf{p}}\, p^j}{4\,p^2}\mathrm{Tr}(\rho\gamma^j), \tag{7}$$

are shown in the left panel of Fig. 1 as functions of $\Delta t \equiv t' - t$ for different momenta. One observes that they exhibit damped oscillations with a well defined frequency (dispersion relation) and damping rate.

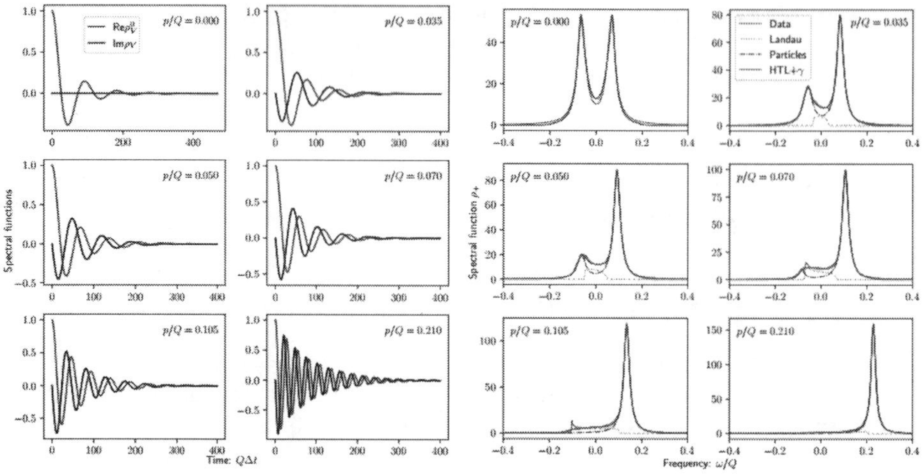

Figure 1. *Left:* Components of the spectral function $\rho(t{+}\Delta t, t, p)$ at different momenta. *Right:* The spectral function $\rho_+(t, \omega, p) \equiv \rho_V^0 + \rho_V$ in frequency space for the same momenta. The values at negative frequency stem from $\rho_+(t, -\omega, p) = \rho_-(t, \omega, p) \equiv \rho_V^0 - \rho_V$. Figures taken from Ref. [1].

This can be made more quantitative after Fourier transforming it with respect to Δt into the frequency domain, as shown in the right panel of Fig. 1. One finds two quasiparticle peaks at $\omega = \pm\omega_\pm(p)$ with respective width (damping rates) $\gamma_\pm(p)$. To extract these values, we recall that the spectral function can be computed perturbatively in HTL as [16]

$$\begin{aligned}\rho_+^{\mathrm{HTL}}(\omega, p) &= 2\pi\beta_+(\omega/p, p)\\ &\quad + 2\pi\left[Z_+(p)\delta(\omega - \omega_+(p)) + Z_-(p)\delta(\omega + \omega_-(p))\right],\end{aligned} \tag{8}$$

with the Landau damping term β_+. We extend this expression to allow for a finite peak width, denoting it HTL+γ,

$$\begin{aligned}\rho_+^{\mathrm{HTL}+\gamma}(\omega, p) &= 2\pi\beta_+(\omega/p, p)\\ &\quad + \frac{2Z_+(p)\gamma_+(p)}{(\omega - \omega_+(p))^2 + \gamma_+^2(p)} + \frac{2Z_-(p)\gamma_-(p)}{(\omega + \omega_-(p))^2 + \gamma_-^2(p)},\end{aligned} \tag{9}$$

and show it in the right panel of Fig. 1, together with its Landau damping and quasiparticle contributions separately. One finds very good agreement with our data. The good agreement with our data thus allows us to extract $\omega_\pm(t, p)$, $Z_\pm(t, p)$ and $\gamma_\pm(t, p)$ by fitting to (9).

These extracted quantities are shown in Fig. 2, together with results from a smaller and coarser 64^3 lattice with $Qa_s = 1$ for comparison. The HTL predictions for the dispersion relations $\omega_\pm^{\mathrm{HTL}}(p)$ and residues $Z_\pm^{\mathrm{HTL}}(p)$ are included as dashed lines. Note that they are completely determined after calculating the HTL fermion mass

$$m_F = \left[C_F \int \frac{\mathrm{d}^3p}{(2\pi)^3}\, \frac{g^2 f_g(p)}{p}\right]^{1/2} \tag{10}$$

whose value is indicated by red arrows. Hence, the nice agreement between our simulation results and leading order HTL expressions shows that these quantities are dominated by perturbative physics.

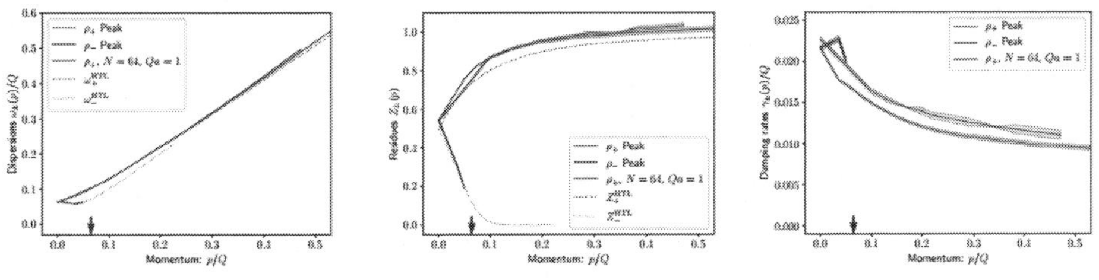

Figure 2. Extracted dispersion relations $\omega_{\pm}(t,p)$, residues $Z_{\pm}(t,p)$ and damping rates $\gamma_{\pm}(t,p)$ as functions of p by fitting to (9). Figures taken from Ref. [1].

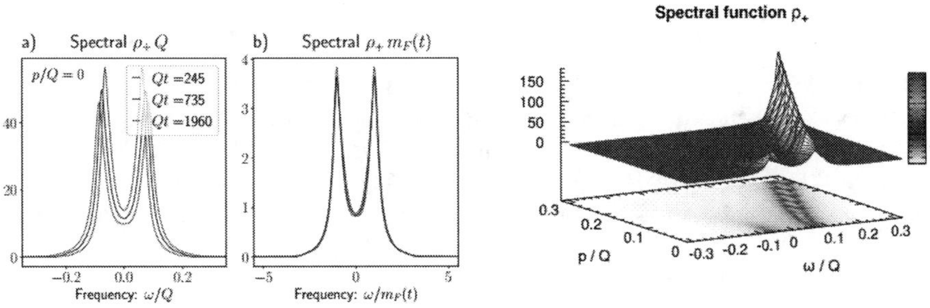

Figure 3. *Left:* Fermion spectral function $\rho_{+}(t,\omega,p{=}0)$ at different times in terms of *a)* the constant scale Q and *b)* the time-dependent fermion mass $m_F(t)$. *Right:* A summary of our results: ρ_{+} as a function of frequency and momentum. Figure taken from Ref. [1].

In addition, our simulations allow for a first-principles insight into the momentum dependence of damping rates $\gamma_{\pm}(p)$, which is hard within a perturbative treatment. In general, we find that $\gamma_{+}(p)$ decrease with momentum and that its values are much smaller than the respective dispersion $\omega_{+}(p)$, validating a quasiparticle picture for (almost) all momentum modes.

From HTL [17] one would expect the damping rate to scale with the effective temperature $\gamma^{\mathrm{HTL}}(t,p{=}0) \propto g^2 T_*(t) \sim Q(Qt)^{-3/7}$. Instead, we find that it rather scales with the fermion mass $\gamma(t,p{=}0) \sim m_F(t) \sim Q(Qt)^{-1/7}$, as we show in the left panel of Fig. 3. By rescaling all dimensionful quantities with $m_F(t) \equiv \omega_{\pm}(t,p{=}0)$, all curves corresponding to different (reference) times t collapse on a single curve. Note that this also confirms a self-similar evolution of the spectral function at the attractor.

4 Gluonic spectral functions

We will now discuss the gluon spectral function $\rho(x',x) = i\left\langle\left[\hat{A}(x'),\hat{A}(x)\right]\right\rangle$ in different dimensions at corresponding self-similar attractors. Its measurement follows a very similar algorithm to the one for fermions and combines classical-statistical simulations with linear response theory. We refer to Refs. [2, 3, 18] for more details. Closely related algorithms were also applied to scalar theories at self-similar attractors in [19, 20].

We will discuss the color averaged transversely polarized (dotted) gluonic spectral functions $\dot{\rho}_T(t',t) = \partial_{t'}\rho_T(t',t)$. One obtains $\dot{\rho}_T(t,\omega) \approx \omega\rho_T$ by Fourier transforming in $\Delta t \equiv t'-t$.

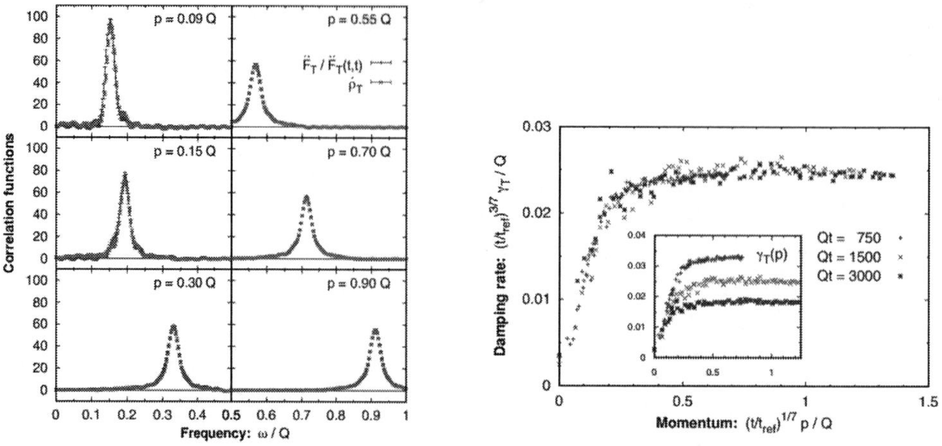

Figure 4. Gluonic simulation results in a **3+1D** plasma. *Left:* The spectral function at different momentum modes as a function of ω. In contrast to the right panel of Fig. 1, we show only the positive frequency part for the gluon spectral function, which agrees with the values at negative frequency due to $\dot{\rho}_T(t,-\omega,p) = \dot{\rho}_T(t,\omega,p)$. *Right:* The extracted damping rate at different times as a function of momentum, rescaled by powers of time (main plot) and without rescaling (inset). Figures taken from Ref. [3].

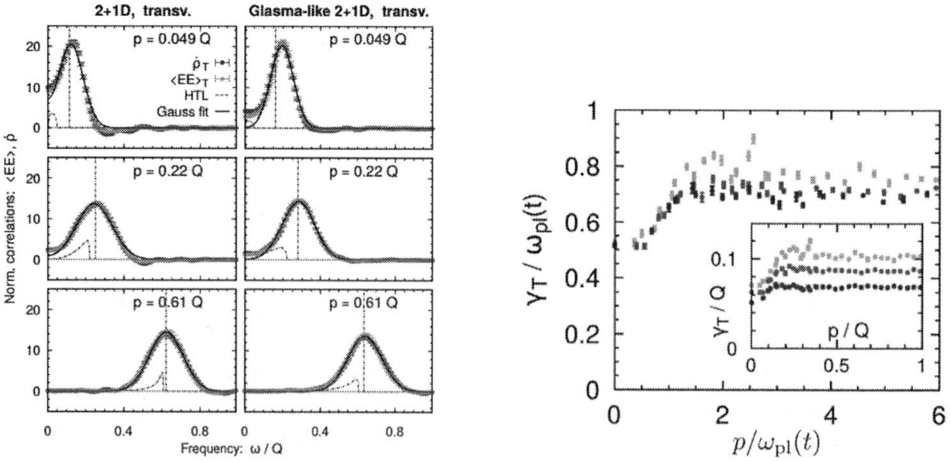

Figure 5. Gluonic simulation results in **2+1D** plasmas. *Left:* The transversely polarized spectral function at different momentum modes as a function of ω for a 2+1D (left) and a 2+1D simulation coupled to an adjoint scalar field (right). *Right:* The extracted damping rate of the genuinely 2+1D plasma at different times as a function of momentum, rescaled by the time-dependent plasmon mass $\omega_{pl}(t)$ (main plot) and without rescaling (inset). Figures taken from Ref. [2].

Apart from the spectral function one can also measure the statistical correlation function $\langle E(t')E(t)\rangle \equiv \ddot{F}(t',t)$. While in thermal equilibrium it is related to the spectral function via the fluctuation-dissipation relation (FDR), far from equilibrium it was found numerically in Refs. [2, 3] that statistical and spectral correlation functions are related via a generalized FDR (also seen in the left panels of Figs. 4 and 5).

4.1 Gluon spectral function in isotropic 3+1D plasmas

For the 3+1D simulations, the same lattice parameters and initial condition are used as in Sec. 3.2 (see Ref. [3] for details on the gluonic simulation results). The gluonic spectral function $\dot{\rho}_T \approx \omega\rho_T$ is shown in the left panel of Fig. 4, extracted at the same reference time $Qt = Qt_{\text{ref}} = 1500$ as we did for the fermion ρ_+. One finds that $\dot{\rho}_T$ is dominated by narrow Lorentzian quasiparticle peaks. The damping rate (peak width) is hence seen to be much smaller than the dispersion (peak position) $\gamma_T(t,p) \ll \omega_T(t,p)$. It is further shown in Ref. [3] that HTL provides an accurate description, with the dispersion relation $\omega_T(p)$, the quasiparticle residue $Z_T(p)$ and the Landau damping contribution for $|\omega| \leq p$ (here suppressed by one power of ω) agreeing well with HTL predictions.

Similarly to the fermionic case in Eq. (9), the damping rate $\gamma_T(t,p)$ is extracted from the gluon spectral function by fitting Lorentzian shapes to the data. The results are shown in the right panel of Fig. 4. The full momentum dependence is obtained and plotted at different times in the inset. Rescaling it as $\gamma_T(t,p)/g^2T_*$, with the effective temperature $g^2T_* \sim Q(t/t_{\text{ref}})^{-3/7}$, and plotting it as a function of $p/\omega_{\text{pl}}(t)$, with the plasmon mass $\omega_{\text{pl}}(t) \equiv \omega_T(t,p{=}0) \sim m_F(t) \sim Q(t/t_{\text{ref}})^{-1/7}$, all the curves are seen to fall on top of each other. Hence, as time grows the ratio $\frac{\gamma_T(t,p)}{\omega(t,p=0)} \sim (Qt)^{-2/7}$ decreases. This shows that the peaks become even more narrow with time and quasiparticle approximations remain valid.

4.2 Gluon spectral function in 2+1D plasmas

In Sec. 4.1 we considered a 3+1D plasma with a gluonic distribution $f(t,p)$ that is isotropic in momentum space. If we now allow for momentum anisotropy with the extreme case being $f(t,p_T,p_z{=}0)$, the system becomes effectively 2+1 dimensional. Here we can distinguish between a genuinely 2+1D plasma consisting of only gluonic excitations and one that is coupled to an adjoint scalar field, which we refer to as Glasma-like. The latter corresponds to the case of extreme momentum anisotropy mentioned above.

We consider these 2+1D systems far from equilibrium and discuss the spectral functions close to their self-similar attractors (see Ref. [2] for details). The numerical results are summarized in Fig. 5. In the left panel $\dot{\rho}_T$ is shown in the frequency domain for different momenta. Different from gluonic and fermionic spectral functions in 3+1D, it exhibits broad non-Lorentzian peaks of approximately Gaussian shape and differs considerably from perturbative HTL predictions that are displayed as green dashed lines. The failure of HTL perturbation theory is not surprising in 2+1D. There the HTL approximation fails because soft momentum modes provide important contributions [15].

The damping rate $\gamma_T(t,p)$ is extracted by performing Gaussian fits to the data and is shown at different times in the right panel of Fig. 5. Rescaled by the time-dependent plasmon mass $\omega_{\text{pl}}(t)$, the damping rate becomes stationary, showing $\gamma(t,p) \sim \omega_{\text{pl}}(t)$ even quantitatively. Hence, the damping rate remains of the order of $\omega_{\text{pl}}(t)$. Since the dispersion is $\omega_T(t,p) \approx \omega_{\text{pl}}(t)$ for modes $p \lesssim \omega_{\text{pl}}(t)$, quasiparticle approximations break down for these soft momenta.

5 Comparison and conclusion

We have presented a new tool to extract the fermion spectral function in highly occupied plasmas nonperturbatively and we have revisited a similar method for gluonic spectral functions. They enable a first-principles determination of damping rates and were applied to classical self-similar attractors in different dimensions.

The extracted fermion and gluon spectral functions share many properties in 3+1D. They exhibit Lorentzian quasiparticle peaks and are well described by perturbative HTL expressions. Their damping rates are in general much smaller than the dispersions, confirming a valid quasiparticle picture. However, there are also interesting differences in the momentum dependence of the damping rates. While the gluonic damping rate γ_T increases with momentum, the fermionic one γ_+ decreases. Moreover, the gluonic damping rate decreases in time faster than the mass, while the fermionic one scales with the mass for $p = 0$.

Interestingly, the scaling $\gamma(t, p{=}0) \sim \omega_{\rm pl}(t)$ has also been observed for the gluonic spectral function in 2+1D systems (see Sec. 4.2). However, there the peaks in the spectral function are generally much broader than in 3+1D and a perturbative HTL description breaks down.

So far, we have applied the methods to systems in Minkowski space-time. In the future, we want to apply our framework to Bjorken expanding (anisotropic) systems and to heavy quarks, which are of phenomenological relevance to heavy-ion collisions. Moreover, non-perturbative properties of the spectral or statistical correlators out of equilibrium can have a sizable impact on transport coefficients, as was shown in [21]. Thus, it will be interesting to study how observables like the photon production rate are affected by our numerical results.

References

[1] K. Boguslavski, T. Lappi, M. Mace, S. Schlichting (2021), 2106.11319
[2] K. Boguslavski, A. Kurkela, T. Lappi, J. Peuron, JHEP **05**, 225 (2021), 2101.02715
[3] K. Boguslavski, A. Kurkela, T. Lappi, J. Peuron, Phys. Rev. **D98**, 014006 (2018), 1804.01966
[4] S. Schlichting, D. Teaney, Ann. Rev. Nucl. Part. Sci. **69**, 447 (2019), 1908.02113
[5] J. Berges, M.P. Heller, A. Mazeliauskas, R. Venugopalan (2020), 2005.12299
[6] R. Allahverdi, R. Brandenberger, F.Y. Cyr-Racine, A. Mazumdar, Ann. Rev. Nucl. Part. Sci. **60**, 27 (2010), 1001.2600
[7] J. Berges, AIP Conf. Proc. **739**, 3 (2004), hep-ph/0409233
[8] J. Berges, K. Boguslavski, S. Schlichting, R. Venugopalan, JHEP **05**, 054 (2014), 1312.5216
[9] M. Mace, N. Mueller, S. Schlichting, S. Sharma, Phys. Rev. D **95**, 036023 (2017), 1612.02477
[10] M. Mace, N. Mueller, S. Schlichting, S. Sharma, Phys. Rev. Lett. **124**, 191604 (2020), 1910.01654
[11] G. Aarts, J. Smit, Nucl. Phys. B **555**, 355 (1999), hep-ph/9812413
[12] A. Kurkela, G.D. Moore, Phys. Rev. D **86**, 056008 (2012), 1207.1663
[13] S. Schlichting, Phys. Rev. D **86**, 065008 (2012), 1207.1450
[14] J. Berges, K. Boguslavski, S. Schlichting, R. Venugopalan, Phys. Rev. D **89**, 114007 (2014), 1311.3005
[15] K. Boguslavski, A. Kurkela, T. Lappi, J. Peuron, Phys. Rev. D **100**, 094022 (2019), 1907.05892
[16] J.P. Blaizot, E. Iancu, Phys. Rept. **359**, 355 (2002), hep-ph/0101103
[17] E. Braaten, R.D. Pisarski, Phys. Rev. D **46**, 1829 (1992)
[18] A. Kurkela, T. Lappi, J. Peuron, Eur. Phys. J. C **76**, 688 (2016), 1610.01355
[19] A. Piñeiro Orioli, J. Berges, Phys. Rev. Lett. **122**, 150401 (2019), 1810.12392
[20] K. Boguslavski, A. Piñeiro Orioli, Phys. Rev. D **101**, 091902 (2020), 1911.04506
[21] K. Boguslavski, A. Kurkela, T. Lappi, J. Peuron, JHEP **09**, 077 (2020), 2005.02418

EPJ Web of Conferences **258**, 05004 (2022) https://doi.org/10.1051/epjconf/202225805004
A Virtual Tribute to Quark Confinement and the Hadron Spectrum (vConf21)

Improving the cold quark-matter pressure via soft interactions at N3LO

Tyler Gorda[1,2*]

[1]Technische Universität Darmstadt, Department of Physics, D–64289 Darmstadt, Germany
[2]Helmholtz Research Academy for FAIR, D–64289 Darmstadt, Germany

Abstract. The propagation of long-wavelength gluons through a dense QCD medium at high baryon chemical potential μ_B is qualitatively modified by the effects of screening, arising from scatterings off the high-momentum quarks in the medium. This same screening phenomenon also impacts gluons occurring in loop corrections to the pressure of cold quark matter, leading to contributions from the parametric scale $\alpha_s^{1/2}\mu_B$, starting at next-to-next-to-leading order (N2LO) in the strong coupling constant α_s. At next-to-next-to-next-to leading order (N3LO), interactions between these long-wavelength gluonic modes contribute to the pressure. These interaction corrections have recently been computed in Ref [1, 2], and the inclusion of these interactions slightly improves the convergence of the equation of state of cold quark matter. In these proceedings, we present these results and provide details summarizing how this lengthy calculation was performed.

1 Introduction

In recent years, the cold (zero-temperature, $T = 0$) quark-matter (QM) equation of state (EoS) has been used as a nontrivial high-density limit [3] to constrain the EoS of neutron-star (NS) matter at much lower baryon densities n_B [4–6]. Typically, this cold-QM EoS can only be used at densities $n_B \gtrsim 40n_0$, with $n_0 \approx 0.16\ \text{fm}^{-3}$ corresponding to nuclear saturation density, where calculations in perturbative Quantum chromodynamics (pQCD) [1, 2, 4, 7] show a small renormalization-scale dependence and are thus under perturbative control.

In calculations at high density, one typically chooses the renormalization scale $\bar{\Lambda}$ to be proportional to the baryon chemical potential

$$\bar{\Lambda} = X\mu_B, \tag{1}$$

and furthermore chooses $X = 2$ as the central value and $X \in [1, 4]$ as the reasonable range quantifying the uncertainty [4, 7]. As one sees from Fig. 1, the uncertainty on the cold QM EoS at next-to-next-to-leading order (N2LO) in the strong coupling α_s (computed in Ref. [7]) quickly explodes below these densities where pQCD calculations show good convergence. Moreover, as of yet, no further prescription can be used to constrain the coefficient X, such as the principle of minimal sensitivity (PMS) [8], as the current functional form of the EoS of cold QM does not exhibit a stationary point as a function of X. However, including higher-order next-to-next-to-next-to-leading order (N3LO) corrections to the cold QM EoS may

*e-mail: tyler.gorda@physik.tu-darmstadt.de

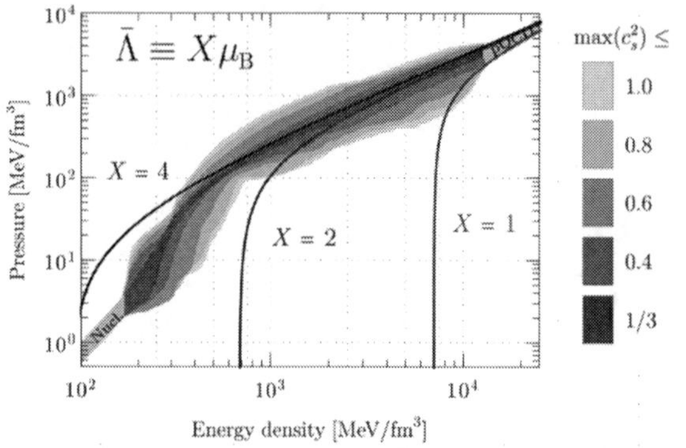

Figure 1. An illustration of the renormalization-scale dependence of the cold-QM EoS at N2LO [7]. The cold-QM EoS is shown for the three different choices of the renormalization parameter $X = 1, 2, 4$, plotted on top of the regions of viable NS-matter EoSs as determined in Ref. [9] using robust theoretical and astrophysical constraints. These regions are color-coded by the maximal value of the speed of sound squared c_s^2 obtained at any point along the EoS, with lower $\max(c_s^2)$ regions plotted on top of higher ones.

change this, as new functional forms will be added to the pressure. If this does happen, it could lead to a drastic improvement in the behavior of this thermodynamic quantity.

Improving the EoS of cold QM from N2LO to N3LO is complicated by the physics of dynamical screening of long-wavelength, low-energy gluonic modes. This phenomena can be understood by examining the dispersion relation for gluons with frequency ω and momenta $\mathbf{k}$ within loop corrections in cold QM:

$$-\omega^2 + |\mathbf{k}|^2 + \Pi(\omega, \mathbf{k}) = 0. \tag{2}$$

Parametrically, $\Pi = O(m_{\rm E}^2) = O(\alpha_s \mu_{\rm B})$, with $m_{\rm E}$ a screening mass, related to the one-loop Debye mass.[1] This holds true even for $\omega, |\mathbf{k}| \ll \mu_{\rm B}$, and so for gluons with $\omega, |\mathbf{k}| \simeq m_{\rm E}$, the self-energy corrections are as large as the free term. For these modes, one cannot treat the self energy as a small perturbation, and instead one must use a resummed gluonic propagator, with the self energy included in the denominator.

At high temperatures, where ω can only take on discrete values given by the Matsubara modes $\omega_n \equiv 2\pi n T$, $n \in \mathbb{Z}$, only the $n = 0$ mode has low enough energy to require resummation, which can be treated within the dimensionally reduced effective field theory (EFT) of electrostatic QCD (EQCD) [10–12]. However, within cold QM at $T = 0$, the discrete Matsubara summation becomes replaced by a normal integral over Euclidean frequencies, and thus there is no longer a clear separation between those energies which are small enough to require resummation and those which are not. This in turn means that there is no clear way to separate the effects from the long-wavelength screened gluons as a distinct EFT within cold QM, and rather the pressure will receive contributions of three different types:

1. *hard* contributions, from short-wavelength modes with $\omega, |\mathbf{k}| \gtrsim \mu_{\rm B}$ and their interactions,
2. *soft* contributions, from long-wavelength modes with $\omega, |\mathbf{k}| \lesssim m_{\rm E}$ and their interactions,
3. *mixed* contributions, sensitive to interactions between the hard and soft modes.

[1] For a single massless quark with quark chemical potential μ_q, $m_{\rm E}^2 \equiv (2/\pi)\alpha_s \mu_q^2$

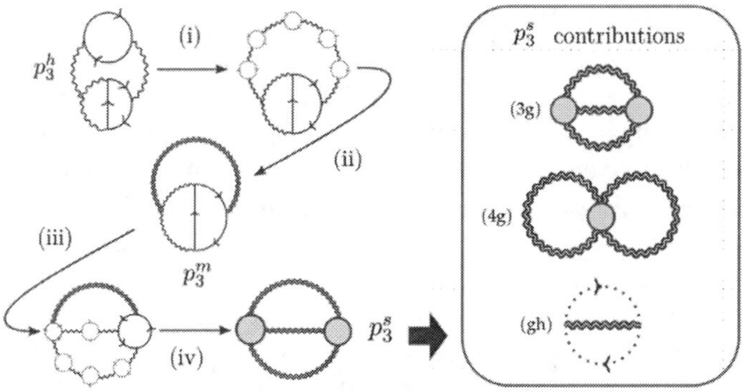

Figure 2. (Left) An illustration of how to arrive at contributions from different momentum scales from a hard diagram. When a gluon momentum in a four-loop diagram becomes soft (i), additional self-energy corrections do not change the order of the diagram. (ii) This soft gluon line can be resummed within the HTL theory, producing the HTL propagator denoted by a thick wavy line and leads to a three-loop 'mixed' contribution if the other gluon momenta are hard. (iii) If the other gluon momenta in the diagram also become soft, then the other gluon propagators and interaction vertices must be dressed with additional loops as well, giving rise to (iv) a fully soft HTL contribution to p_3^s. (Right) All the two-loop HTL diagrams that contribute to p_3^s, derivable in this manner.

These contributions to the pressure p can be organized as follows [1]:

$$\begin{aligned} p = p_0 + \alpha_s p_1^h + \alpha_s^2 p_2^h + \alpha_s^3 p_3^h \quad &\text{(hard modes)} \\ + \alpha_s^2 p_2^s + \alpha_s^3 p_3^s \quad &\text{(soft modes)} \\ + \alpha_s^3 p_3^m \quad &\text{(both).} \end{aligned} \tag{3}$$

The above equation contains all contributions up to N3LO. That the soft modes do not enter until N2LO follows from the phase-space suppression required from the soft kinematics: $\mathrm{d}^4P = O(\alpha_s^2\mu_B^4)$ for soft modes. That the soft interactions remain perturbative, with interaction corrections leading to additional powers of the coupling (unlike at high T—see [13] for a discussion) follows from a more subtle argument discussed in Refs. [1, 14], but the physical reason is simple: at $T = 0$, gluons are not thermally populated by the medium, and thus do not receive an infrared (IR) over-occupation (Bose enhancement) as they do at high T. The correct treatment of the interacting soft scale is given by the hard-thermal-loop (HTL) [or hard Dense loop (HDL)] effective theory [15, 16].

The lack of a clear split between the hard and soft modes leads to an ambiguous *semisoft* range of energies and momenta $m_E \ll \omega, |\mathbf{k}| \ll \mu_B$ in which resummations are not necessary, but one may still use the kinematic simplifications of the HTL theory. This leads to IR divergences in the hard sector which will cancel against ultraviolet (UV) divergences in the soft sector (as well as mixed IR-UV divergences in the mixed sector). These canceling divergences also leads to the $p_i^{s/m/h}$ depending on a factorization mass scale Λ_h, which cancels out when summing over the kinematic sectors at a given order. When performing calculations in dimensional regularization, as done in Refs. [1, 2], this Λ_h appears in the integration measure, as an additional $\overline{\text{MS}}$ scale.

2 Detailed summary of the N3LO soft calculation

At N3LO, the p_3^s contribution in Eq. (3) arises from two-loop HTL diagrams. This can be seen following the analysis in Fig. 2. Note that fermion lines never need to be resummed at high density because the soft fermion modes are Pauli blocked by the medium. The gray blobs at the vertices in this figure indicate the use of one-loop HTL-corrected vertex functions, which are a sum of a bare (Γ_0) and HTL ($\delta\Gamma$) vertices: $\Gamma \equiv \Gamma_0 + \delta\Gamma$. The thick gluon lines correspond to propagators dressed with one-loop HTL self-energies (Π). (We are here suppressing Lorentz structure). In Ref. [1], the calculation of these diagrams was performed for massless quarks in dimensional regularization using the $\overline{\text{MS}}$ scheme, taking $d = 3 - 2\epsilon$, and was performed in Feynman ($\xi = 1$) gauge, though it was checked that the full result is gauge invariant, as also stated in Ref. [17]. Note that this reference also studied these two-loop diagrams at high temperatures, but notably did so by expanding in the ratio m_{E}/T. At $T = 0$, such an expansion is not valid and the entire resummed diagrams must be computed. Below, we provide a summary of key details from Ref. [1], in which the full, unexpanded calculation is performed at $T = 0$.

These three HTL graphs shown in Fig. 2 contribute to the pressure according to

$$\alpha_s^3 p_3^s = g^2 N_c d_A [I_{3g} + I_{4g} + I_{gh}], \tag{4}$$

with the individual expressions given in Ref. [1]. Here, $g \equiv \sqrt{4\pi\alpha_s}$, N_c is the number of quark colors, and $d_A \equiv N_c^2 - 1$ is the number of gluons. One important property of the HTL gluon self energy Π is that it does not depend on the magnitude of the (Euclidean) four momentum of the gluon. This means that each loop integral in the diagram is divergent in the UV, leading to each of the above HTL graphs contributing terms to the pressure at orders $1/\epsilon^2$, $1/\epsilon$, and ϵ^0. Their sum can thus be expressed in the form

$$\alpha_s^3 p_3^s = \frac{g^2 N_c d_A m_{\text{E}}^4}{(2\pi)^6} \left(\frac{m_{\text{E}}}{\Lambda_{\text{h}}}\right)^{-4\epsilon} \left[\frac{p_{-2}}{(2\epsilon)^2} + \frac{p_{-1}}{2\epsilon} + p_0\right], \tag{5}$$

and from the detailed computation [1], one finds the following compact, final expressions:

$$p_{-2} = \frac{11}{6} \int_\Omega \text{tr}\left[\hat{\Pi}^2(\hat{K})\right] = \frac{11\pi^2}{24}, \tag{6}$$

$$\begin{aligned} p_{-1} &= \int_\Omega \left\{ \frac{19 + 11\pi^2}{72} - \frac{11}{6}\text{tr}\left[\hat{\Pi}^2(\hat{K}) \ln[\hat{\Pi}(\hat{K})]\right] + \int_0^{\pi/2} \frac{\mathrm{d}\chi \sin(2\chi)\, \text{tr}\left[\delta\hat{\Gamma}_{3g}^2(\hat{K}\sin\chi, \hat{P}\cos\chi)\right]}{24[1 + \sin(2\chi)\hat{K}\cdot\hat{P}]} \right\} \\ &\approx 8.7432 + 2.5068 + 0.4340(15) = 11.6840(15), \end{aligned} \tag{7}$$

$$p_0 = 17.150(7), \tag{8}$$

where the hatted notation on the HTL functions indicates that the m_{E} has been scaled out, and the trace-and-square notation indicates that the Lorentz indices of the two corresponding HTL functions are fully contracted with one another. The hatted notation on the four-momenta indicates that their magnitudes have been divided out, leaving a unit Euclidean four-vector. The angular integral $\int_\Omega$ and the angle χ will be defined further below. Let us now gives some details from Ref. [1] of how this result is derived.

Within the full expression, the UV-sensitive and UV-finite terms are not clearly separated, since the resummed gluon propagators are a sum of an infinite number of terms, each scaling differently in the UV. Hence, a strategic approach is required to extract the p_{-2} and p_{-1} coefficients from the resummed two-loop integrals. As an observation, we note the following: the logarithmically UV-sensitive terms (those which do not simply vanish in dimensional regularization) are related to pieces of the full expression that scale as m_{E}^4 times a dimensionless integral in the UV, as such a structure allows it to contribute throughout the whole UV tail of the integral. If one could isolate all such terms, while keeping their resummed nature intact, then one would be able to extract the UV divergences clearly.

This motivates the following notation:

$$\Delta_n^{\mu\nu}(K) \equiv (-1)^n \frac{[\Pi(K)^n]^{\mu\alpha}}{(K^2)^n} \Delta_0^{\alpha\nu}(K); \qquad D_n^{\mu\nu}(K) \equiv (-1)^n \frac{[\Pi(K)^n]^{\mu\alpha}}{(K^2)^n} D^{\alpha\nu}(K), \tag{9}$$

where Δ_0 is the bare and D is the HTL-resummed gluon propagator, and $(\Pi^n)^{\mu\alpha} \equiv \Pi^{\mu\nu_1}\Pi^{\nu_1\nu_2}\cdots\Pi^{\nu_{n-1}\alpha}$. The first of these definitions corresponds precisely to the terms occurring in the UV expansion of the HTL propagator. Together, these allow one to peel off bare parts of the resummed propagator without expanding: the relation $D = \sum_{k=0}^{n-1}\Delta_k + D_n$ holds exactly for any $n \geq 1$, and both the bare and resummed pieces scale like $\Delta_n(K) \sim D_n(K) \sim m_{\rm E}^{2n}K^{-2(n+1)}$ in the UV. If one chooses an n large enough ($n = 2$ suffices in all cases here), one can substitute this exact relation for all of the resummed propagators in each contribution $I_{\rm 3g}$, $I_{\rm 4g}$, $I_{\rm gh}$; multiply out the terms; and select those scaling as $m_{\rm E}^4$ in the UV.[2] To do this, one must also take note of the UV scaling of the vertices $\Gamma = \Gamma_0 + \delta\Gamma$: for the three-gluon vertex, $\Gamma_{0,3\rm g} \sim K$ and $\delta\Gamma_{3\rm g} \sim m_{\rm E}^2K^{-1}$, while for the four-gluon vertex, $\Gamma_{0,4\rm g} \sim K^0$ and $\delta\Gamma_{4\rm g} \sim m_{\rm E}^2K^{-2}$.[3] We shall now discuss in detail the manipulations that are performed on the UV-sensitive terms in Ref. [1], leaving a discussion of the finite terms to the end.

In the UV-sensitive terms, one can perform the contractions and arrive at compact expressions, since these terms often involve at least one bare propagator, which directly contracts the vertices. Via these direct contractions, one can make extensive use of the generalized Ward identities for the HTL vertex functions and reduce some four-point functions to three-point functions and some three-point functions to (two-point) self energies. Combining all the UV-sensitive terms, one arrives at the following elegant expression for the UV-sensitive part of $I_{\rm tot} \equiv I_{\rm 3g} + I_{\rm 4g} + I_{\rm gh}$:

$$\begin{aligned}[I_{\rm tot}]^{\rm UV} = \int_{KP}\bigg\{&\frac{2P^2}{R^2}[\hat{P}\cdot D_1(K)\cdot\hat{P}]\mathrm{tr}\,[D_1(P)] - \frac{1}{4}\mathrm{tr}\,[D_1(K)]\,\mathrm{tr}\,[D_1(P)] + \frac{2K^2}{P^2R^2}\mathrm{tr}\,[D_2(K)] \\ &+ \frac{d-1}{R^2}[\hat{P}\cdot D_2(K)\cdot\hat{P}] + \frac{1}{12}\delta\Gamma_{3\rm g}^{\mu\nu\rho}\delta\Gamma_{3\rm g}^{\mu'\nu'\rho'}D^{\mu\mu'}(K)D^{\nu\nu'}(P)D^{\rho\rho'}(R)\bigg\}. \end{aligned}\tag{10}$$

Here, $R \equiv -K - P$, and the measure for a single momentum is defined as $[e^{\gamma_{\rm E}}\Lambda_{\rm h}^2/(4\pi)]^\epsilon \mathrm{d}^{d+1}K/(2\pi)^{d+1}$, with $\gamma_{\rm E}$ the Euler–Mascheroni constant. This expression contains the entirety of the p_{-2} and p_{-1} coefficients, and part of the constant p_0. To proceed further, one may perform the following two steps:

1. Rescale the magnitudes of momenta by $|K| \mapsto m_{\rm E}|K|$ etc., and introduce the hatted notation on the HTL functions. This extracts the $m_{\rm E}$ dependence from all integrals.

2. Change variables in the K and P integrals to write them as integrals over the magnitude of the four vectors $|K|$ and $|P|$ and the remaining angles: namely (Φ_K, Φ_P, θ) with $\tan\Phi_K = |\mathbf{k}|/K_0$, $\tan\Phi_P = |\mathbf{p}|/P_0$, and $\cos\theta = \hat{\mathbf{k}}\cdot\hat{\mathbf{p}}$. Here, the four-vectors are written as $K = (K_0, \mathbf{k})$ and $\hat{\mathbf{k}} \equiv \mathbf{k}/|\mathbf{k}|$.

3. Further transform from magnitudes of the momenta $(|K|, |P|)$ to Euclidean polar (X, χ) coordinates, given by $|K| = X\sin\chi$, $|P| = X\cos\chi$, with $\chi \in [0, \pi/2]$, $X \in [0, \infty]$.

Here, we stress that by moving to the (X, χ) coordinates, one only changes the magnitudes $|K|, |P|$, and so $\hat{K}, \hat{P}$ remain unchanged. After these steps, the integration measure becomes

$$\int_{KP} = C(d)\int_\Omega \int_0^{\pi/2}\mathrm{d}\chi\,\sin^d\chi\cos^d\chi\int_0^\infty \mathrm{d}X\,X^{2d+1}, \tag{11}$$

[2]The propagator manipulations also introduce scaleless integrals, containing only Δ_n terms, which vanish in dimensional regularization. These terms additionally have too few powers of $m_{\rm E}$ to contribute at N3LO.

[3]Note that there are multiple momentum scales for each vertex. These scalings hold for each momentum in the UV.

with the definitions:

$$C(d) \equiv \left(\frac{e^{\gamma_E}\Lambda_h^2}{4\pi m_E^2}\right)^{3-d} \frac{4\pi^{d-\frac{1}{2}}}{(2\pi)^{2d+2}\Gamma\left(\frac{d}{2}\right)\Gamma\left(\frac{d-1}{2}\right)}; \qquad \int_\Omega \equiv \prod_{i\in\{K,P\}} \int_0^\pi \mathrm{d}\Phi_i \sin^{d-1}\Phi_i \int_0^\pi \mathrm{d}\theta \sin^{d-2}\theta. \quad (12)$$

In all cases, the radial X integral can be performed analytically, and in most of the terms of Eq. (10) the angular χ integral can also be performed analytically, even in general d. Physically, the radial X integral generates one $1/\epsilon$ divergence from the UV, and in terms with a $1/\epsilon^2$ contribution, the χ integral generates the second, arising from the boundaries near $\chi = 0$ or $\pi/2$. Translating this back to the original $(|K|, |P|)$ coordinates, one sees these double divergences arise when $|K| \gg |P| \gg 0$ or $|P| \gg |K| \gg 0$, precisely matching the analysis conducted in Ref. [14]. In the final term of Eq. (10), containing $\delta\Gamma_{3g}$, the χ integral cannot be performed analytically, but it is finite.

Before doing the χ integral, one can further simplify the tensor structure of the terms of the form $\hat{P}\cdot D_n(K)\cdot\hat{P}$ in Eq. (10) in general d. This procedure uses certain angular averages performed in Ref. [18]. This removes the products of four-momenta with the self-energies, leaving only traces of powers of the self energies. One may then perform the χ integration analytically, where possible, and expand near $d = 3$ to arrive at the coefficients p_{-2} and p_{-1} above, and part of the p_0 result. Note that one must use the full d-dimensional self energies and three-gluon HTL vertex function in Eq. (10) to obtain the correct results [1].

In the p_{-2} coefficient, one finds that only the terms containing D_2 in Eq. (10) contribute: the terms containing only D_1 cancel each other, and the term containing $\delta\Gamma_{3g}$ only contributes to the p_{-1} because the χ integral is finite. This leads to the dramatically compact expression for the integrand in Eq. (6) for p_{-2}.

The p_{-1} result shown in Eq. (7) contains three terms, the first two of which are straightforward to compute. However, the final term contains the three-gluon HTL vertex function, which requires more work to evaluate. In general, these HTL vertex functions are given by a d-dimensional integral representation. To reduce the numerical complexity of the expressions, one may introduce a modified Feynman parametrization, which allows one to perform these d-dimensional integrals order-by-order in ϵ. This leaves only the one Feynman parameter left to be integrated over for each vertex. One finds that the contribution to p_{-1} from this vertex function contains only a particularly simple contraction between Lorentz indices, since the propagators $D^{\mu\nu}(K) \mapsto \delta^{\mu\nu}/K^2$ in the UV (in Feynman gauge). However, the contractions contributing to p_0 are considerably more complicated, since the full longitudal and transverse structures within $D^{\mu\nu}(K)$ contribute. To deal with these structures, one can repeatedly use the generalized Ward identities to trade spatial contractions for temporal contractions and self-energy terms, both of which are easier to evaluate [1]. With this approach, the contributions to the final result from these vertex functions can be written as six-dimensional integrals (over χ, Φ_K, Φ_P, θ, and two Feynman parameters). These integrals can be computed numerically using Monte Carlo integration provided by the CUBA library [19]. This provides one with the final contribution to the p_{-1} in Eq. (7), and a part of p_0.

This completes the evaluation of the UV-sensitive terms in Eq. (10), and leaves one with only the original UV-finite pieces to compute. These remaining finite pieces contain the full HTL-resummed propagators, as well as the three- and four-gluon HTL vertex functions, and therefore there are no general simplifications that arise from the contractions in these terms. However, the finite nature of these terms makes them simpler to evaluate numerically, as one may directly set $d = 3$ in them. For this reason, they can be evaluation in an automated fashion, by performing any remaining non-Lorentz-invariant tensor contractions using an adapted version of the implementation discussed in Ref. [20] (although in Euclidean space). Properties of the vertex corrections, in particular the generalized Ward identities and their

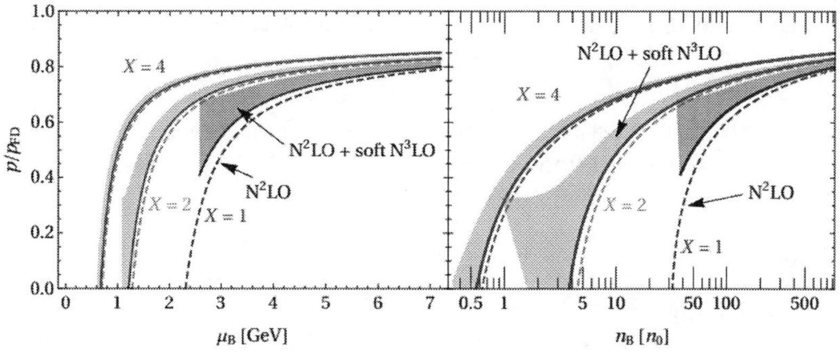

Figure 3. A comparison between the state-of-the-art pressure of cold QM at partial N3LO, including the interactions between soft screened modes, and the corresponding N2LO pressure given as functions of the baryon chemical potential μ_B (left) and baryon number density n_B (right). Both pressures are shown normalized by the free Fermi–Dirac pressure p_{FD}. The three colors correspond to the renormalization-scale variation in the hard sector, and the solid lines in the N3LO result correspond to fixing the factorization scale Λ_h to the PMS value. The filled bands in the N3LO result correspond to varying Λ_h by a factor of two in both directions about the PMS value.

tracelessness, are used extensively to make the resulting expressions as simple as possible. From the simplified expressions, one may perform the same steps 1-3 as presented above, and once again the X integrals can be performed analytically. In the end, one must again perform up to six-dimensional numerical integrals to derive the remaining part of the coefficient p_0 presented in Eq. (8).

3 Results and Discussion

The analytic and numerical results for the soft contribution to the pressure of cold QM at N3LO were already shown above in Eqs. (6)-(8); a plot of these results is shown in Fig. 3. The functional form of the result for the pressure at this order allows one to use the principle of minimal sensitivity to set the factorization scale Λ_h by solving $dp/d\Lambda_h = 0$. One finds $\Lambda_h^{PMS} = \exp[-p_{-1}/(2p_{-2})]m_E$. In Fig. 3, using this value corresponds to the thick lines on the lower boundary of the shaded partial N3LO result. The result also still has a dependence on the UV renormalization scale $\bar{\Lambda}$, which we display over the usual range of $X \in \{1, 2, 4\}$ in different colors. We note here that the PMS value of Λ_h also corresponds to the one which maximizes the $\bar{\Lambda}$ dependence of the result: this means that the PMS value for Λ_h is the most conservative choice within this partial N3LO result on two fronts.

As can be seen from Fig. 3, including the contribution from the interacting soft screened modes at N3LO shifts the pressure to slightly higher values, and thus slightly decreases the UV-renormalization-scale dependence of the result; that is, including these interactions slightly improves the convergence of the pressure of cold QM. Note that this is contrary to the case at high T, where these soft modes limit the convergence of the pressure [21].

We note a further interesting observation, namely that the p_{-2} coefficient seems to be related to the one-loop beta function of pure gauge theory, corresponding to the running of the m_E parameter in the N2LO result. This connection has been investigated even further by the authors of Ref. [22], where an all-orders resummation for the leading and next-to-leading contributions from these soft modes has been conducted.

Finally, we note that the improvements to the pressure presented above and in Ref. [1, 2] may have implications for the NS-matter EoS, and the improved convergence of the results is in agreement with analyses conducted using astrophysical observations to constrain the NS-matter EoS [9]. There it has been found (as also can be seen from Fig. 1 above) that the NS-matter EoS follows the behavior of the pQCD cold-QM EoS to much lower densities than where pQCD calculations are currently converged. This result certainly motivates further improvements to the cold-QM EoS of pQCD matter, which are currently underway.

Acknowledgments

This proceedings is a summary of work conducted in collaboration with Aleksi Kurkela, Risto Paatelainen, Saga Säppi, and Aleksi Vuorinen. TG was in part supported by the Deutsche Forschungsgemeinschaft (DFG, German Research Foundation) – Project-Id 279384907 – SFB 1245.

References

[1] T. Gorda, A. Kurkela, R. Paatelainen, S. Säppi, A. Vuorinen, Phys. Rev. D **104**, 074015 (2021), 2103.07427
[2] T. Gorda, A. Kurkela, R. Paatelainen, S. Säppi, A. Vuorinen, Phys. Rev. Lett. **127**, 162003 (2021), 2103.05658
[3] O. Komoltsev, A. Kurkela (2021), 2111.05350
[4] A. Kurkela, E.S. Fraga, J. Schaffner-Bielich, A. Vuorinen, Astrophys. J. **789**, 127 (2014), 1402.6618
[5] E. Annala, T. Gorda, A. Kurkela, A. Vuorinen, Phys. Rev. Lett. **120**, 172703 (2018), 1711.02644
[6] E.R. Most, L.R. Weih, L. Rezzolla, J. Schaffner-Bielich, Phys. Rev. Lett. **120**, 261103 (2018), 1803.00549
[7] A. Kurkela, P. Romatschke, A. Vuorinen, Phys. Rev. D **81**, 105021 (2010), 0912.1856
[8] P.M. Stevenson, Nucl. Phys. B **231**, 65 (1984)
[9] E. Annala, T. Gorda, A. Kurkela, J. Nättilä, A. Vuorinen, Nature Phys. **16**, 907 (2020), 1903.09121
[10] T. Appelquist, R.D. Pisarski, Phys. Rev. D **23**, 2305 (1981)
[11] K. Kajantie, M. Laine, K. Rummukainen, M.E. Shaposhnikov, Nucl. Phys. B **458**, 90 (1996), hep-ph/9508379
[12] E. Braaten, A. Nieto, Phys. Rev. D **51**, 6990 (1995), hep-ph/9501375
[13] J. Ghiglieri, A. Kurkela, M. Strickland, A. Vuorinen, Phys. Rept. **880**, 1 (2020), 2002.10188
[14] T. Gorda, A. Kurkela, P. Romatschke, M. Säppi, A. Vuorinen, Phys. Rev. Lett. **121**, 202701 (2018), 1807.04120
[15] E. Braaten, R.D. Pisarski, Nucl. Phys. B **337**, 569 (1990)
[16] C. Manuel, Phys. Rev. D **53**, 5866 (1996), hep-ph/9512365
[17] J.O. Andersen, E. Braaten, E. Petitgirard, M. Strickland, Phys. Rev. D **66**, 085016 (2002), hep-ph/0205085
[18] J.H. Ee, D.W. Jung, U.R. Kim, J. Lee, European Journal of Physics **38**, 025801 (2017)
[19] T. Hahn, Comput. Phys. Commun. **168**, 78 (2005), hep-ph/0404043
[20] V. Shtabovenko, R. Mertig, F. Orellana, Comput. Phys. Commun. **207**, 432 (2016), 1601.01167
[21] J.P. Blaizot, E. Iancu, A. Rebhan, Phys. Rev. D **68**, 025011 (2003), hep-ph/0303045
[22] L. Fernandez, J.L. Kneur (2021), 2109.02410

Bottomonium observables in an open quantum system using the quantum trajectories method

Peter Vander Griend[1,*]

[1]Physik-Department, Technische Universität München, James-Franck-Str. 1, 85748 Garching, Germany

Abstract. We solve the Lindblad equation describing the Brownian motion of a Coulombic heavy quark-antiquark pair in a strongly coupled quark gluon plasma using the Monte Carlo wave function method. The Lindblad equation has been derived in the framework of pNRQCD and fully accounts for the quantum and non-Abelian nature of the system. The hydrodynamics of the plasma is realistically implemented through a 3+1D dissipative hydrodynamics code. We compute the bottomonium nuclear modification factor and elliptic flow and compare with the most recent LHC data. The computation does not rely on any free parameter, as it depends on two transport coefficients that have been evaluated independently in lattice QCD. Our final results, which include late-time feed down of excited states, agree well with the available data from LHC 5.02 TeV PbPb collisions.

1 Introduction

A striking feature of the strong nuclear force is its confining nature. At low energies and macroscopic scales, the quarks and gluons charged under the strong interaction only exist in neutral configurations. In the first few instants after the Big Bang, the energy density of the Universe was much higher than today, and the quarks and gluons, now confined in nucleons, may have propagated freely. A recent triumph of experimental nuclear physics, the heavy ion collision experiments at the Large Hadron Collider (LHC) at CERN and the relativistic heavy ion collider (RHIC) at Brookhaven National Laboratory may have achieved sufficiently high energy densities to create a new state of matter, the quark gluon plasma (QGP), in which quarks and gluons are deconfined. This has spurred theorists to investigate observables signaling the formation of a QGP.

One such observable is the nuclear modification factor R_{AA} defined as the ratio of the yield of a heavy quarkonium state in an AA collision to the yield in a pp collision (adjusted for the number of binary collisions). $R_{AA} < 1$ signals heavy quarkonium suppression and has long been theorized to signal the formation of a QGP [1, 2]. It was initially theorized that such a suppression would be caused by Debye screening of the inter-quark potential at distances larger than approximately the inverse of the Debye mass. In the years since these initial predictions, our understanding of the mechanism of heavy quarkonium suppression shifted due to the finding of thermal corrections to the real part of the in-medium potential related to screening and a non-zero imaginary part related to in-medium dissociation due to Landau

*e-mail: vandergriend@tum.de

damping [3]. This was subsequently extended by the inclusion of non-Abelian singlet-octet transitions using the effective field theory (EFT) potential non-relativistic QCD (pNRQCD) [4–7].

Due to the hierarchies of scale inherent to in-medium heavy quarkonium, EFT methods, and specifically pNRQCD, and the formalism of open quantum systems (OQS) are especially useful tools to describe the combined system. In Refs. [8, 9], pNRQCD and OQS methods were applied to in-medium bottomonium, and an in-medium evolution equation taking the form of a Lindblad equation was derived. Refs. [10, 11] made use of the `QTraj` code presented in Ref. [12], which implements the quantum trajectories algorithm, to solve the Lindblad equation and present R_{AA} and the elliptic flow v_2 of the $\Upsilon(1S)$, $\Upsilon(2S)$, and $\Upsilon(3S)$ as functions of centrality and transverse momentum p_T. The remainder of this proceeding is structured as follows: in Sec. 2, we provide the necessary theory background to describe in-medium heavy quarkonium; in Sec. 3, we detail the quantum trajectories algorithm and the `QTraj` code used to solve the Lindblad equation describing the in-medium evolution of heavy quarkonium; in Sec. 4, we present our results; and we conclude in Sec. 5. We emphasize that the method and results presented in this work are fully quantum, non Abelian, heavy quark number conserving, and account for dissociation and recombination.

2 Theory background

In this section, we present the theoretical tools, namely pNRQCD and OQS methods, necessary to describe in-medium heavy quarkonium realizing the hierarchy of scales

$$M \gtrsim 1/a_0 \gg \pi T \sim m_D \gg E, \tag{1}$$

where M is the heavy quark mass, a_0 is the Bohr radius of the bound state, T is the temperature of the medium, $m_D \sim gT$ is the Debye mass, and E is the binding energy of the state. We direct the reader to Refs. [8, 9] for a more detailed discussion.

pNRQCD is an EFT of the strong interaction obtained from full QCD via the sequential integrating out of the hard scale M from full QCD to obtain the EFT non-relativistic QCD (NRQCD) [13, 14]. From NRQCD, further integrating out the soft scale Mv, where $v \ll 1$ is the relative quark-antiquark velocity in a heavy-heavy bound state, gives pNRQCD [15–17]. The degrees of freedom of the resulting theory are heavy-heavy bound states in color singlet and octet configurations and gluons and light quarks at the ultra-soft scale Mv^2. At the Lagrangian level, pNRQCD implements a double expansion in the bound state radius r and the inverse of the heavy quark mass M

$$\mathcal{L}_{\rm pNRQCD} = \mathrm{Tr}\Big[S^\dagger(i\partial_0 - h_s)S + O^\dagger(iD_0 - h_o)O + O^\dagger \mathbf{r}\cdot g\mathbf{E}\, S + S^\dagger \mathbf{r}\cdot g\mathbf{E}\, O + \frac{1}{2}O^\dagger\{\mathbf{r}\cdot g\mathbf{E}\,,\, O\}\Big], \tag{2}$$

where $h_{s,o}$ are the singlet and octet Hamiltonians $h_{s,o} = \frac{\mathbf{p}^2}{M} + V_{s,o}$ with

$$V_s = -\frac{C_f \alpha_s(1/a_0)}{r}, \quad V_o = \frac{\alpha_s(1/a_0)}{2N_c r}, \tag{3}$$

are the attractive singlet Coulombic and repulsive octet Coulombic potentials, respectively. $C_f = (N_c^2-1)/(2N_c)$ where $N_c = 3$ is the number of colors, and $\alpha_s(1/a_0)$ is the strong coupling evaluated at the inverse of the Bohr radius. The potentials encode the information of the integrated-out soft scale.

The OQS formalism allows for the rigorous description of a quantum system of interest coupled to an environment. Within the OQS formalism, three time scales characterize the combined system: the system intrinsic time scale τ_S, the environment correlation time τ_E, and the relaxation time τ_R. For in-medium heavy quarkonium, these scales are given by

$$\tau_S \sim \frac{1}{E}, \quad \tau_E \sim \frac{1}{\pi T}, \quad \tau_R \sim \frac{1}{\Sigma_s} \sim \frac{1}{a_0^2(\pi T)^3}, \tag{4}$$

where Σ_s is the thermal self energy of the quarkonium.

The hierarchy of scales of Eq. (1) implies that $\tau_S, \tau_R \gg \tau_E$ and the system realizes quantum Brownian motion allowing for the use of two simplifying assumptions: the Born and Markov approximations. The Born approximation amounts to a temporal course graining to time scales over which the quarkonium has no effect on the plasma, i.e., time scales over which any excitation caused by the presence of the quarkonium in the medium have relaxed back. At the calculational level, this implies the density matrix of the combined quarkonium-plasma system may be taken to factorize, i.e., $\rho(t) \propto \rho_S(t) \otimes \rho_E$. The Markov approximation implies that only the state of the quarkonium at time t is necessary to describe the evolution at time t. This eliminates a memory integral and gives evolution equations which are local in time.

Refs. [8, 9] use pNRQCD and the OQS formalism to derive the evolution equations of in-medium heavy quarkonium in a strongly coupled medium. The evolution equations take the form of a master equation which in the limit $\pi T \gg E$ takes the form of a Lindblad equation [18, 19]

$$\frac{d\rho(t)}{dt} = -i\,[H, \rho(t)] + \sum_n \left(C_i^n \rho(t) C_i^{n\dagger} - \frac{1}{2}\left\{ C_i^{n\dagger} C_i^n, \rho(t) \right\} \right), \tag{5}$$

where

$$\rho(t) = \begin{pmatrix} \rho_s(t) & 0 \\ 0 & \rho_o(t) \end{pmatrix}, \quad H = \begin{pmatrix} h_s & 0 \\ 0 & h_o \end{pmatrix} + \frac{r^2}{2}\gamma \begin{pmatrix} 1 & 0 \\ 0 & \frac{N_c^2-2}{2(N_c^2-1)} \end{pmatrix}, \tag{6}$$

$$C_i^0 = \sqrt{\kappa} r^i \begin{pmatrix} 0 & \frac{1}{\sqrt{N_c^2-1}} \\ 1 & 0 \end{pmatrix}, \quad C_i^1 = \sqrt{\frac{(N_c^2-4)\kappa}{2(N_c^2-1)}}\, r^i \begin{pmatrix} 0 & 0 \\ 0 & 1 \end{pmatrix}. \tag{7}$$

$\rho_{s,o}(t)$ are the singlet and octet density matrices. κ is the heavy quark momentum diffusion coefficient describing the in-medium diffusion of a single heavy quark [20, 21], and γ is its dispersive counterpart first identified in Refs. [8, 9]. They are defined as

$$\kappa = \frac{g^2}{6N_c}\int_0^\infty ds \left\langle \left\{ \tilde{E}^{a,i}(s,\mathbf{0}), \tilde{E}^{a,i}(0,\mathbf{0}) \right\} \right\rangle, \quad \gamma = -i\frac{g^2}{6N_c}\int_0^\infty ds \left\langle \left[\tilde{E}^{a,i}(s,\mathbf{0}), \tilde{E}^{a,i}(0,\mathbf{0}) \right] \right\rangle, \tag{8}$$

where

$$\tilde{E}^{a,i}(t,\mathbf{0}) = \Omega^\dagger(t) E^{a,i}(t,\mathbf{0}) \Omega(t), \quad \Omega(t) = \exp\left[-ig \int_{-\infty}^{t} dt' A_0(t', \mathbf{0}) \right]. \tag{9}$$

κ and γ are the only free parameters entering the Lindblad equation.

3 Computational methods

In this section, we introduce the `QTraj` code presented in Ref. [12] and used in Refs. [10, 11] to solve the Lindblad equation describing the in-medium evolution of heavy quarkonium.

3.1 Quantum trajectories algorithm

We make use of the quantum trajectories algorithm to solve the Lindblad equation given in Eq. (5); for a general introduction to this method, see Ref. [22]. The central idea of the quantum trajectories algorithm is to split the quantum number conserving diagonal pieces of the evolution specified by the Lindblad equation from the off-diagonal pieces which alter the quantum numbers of the state. The former are collected into an effective Hamiltonian

$$H_{\rm eff} = H - \frac{i}{2}\sum_n C_i^{n\dagger} C_i^n. \tag{10}$$

The `QTraj` code presented in Ref. [12] and used in Refs. [10, 11] implements the quantum trajectories algorithm to solve the Lindblad equation derived in Refs. [8, 9] as follows:

1. Initialize the wave function $|\psi(t_0)\rangle$.
2. Generate a random number $0 < r_1 < 1$ and evolve with $H_{\rm eff}$ until

   $$\| \, e^{-i\int_{t_0}^{t} dt' H_{\rm eff}(t')} |\psi(t_0)\rangle \, \|^2 \leq r_1.$$

   Designate the first time step fulfilling the above inequality the jump time t_j; if $t_j < t_f$, where t_f is the simulation end time, proceed to step 3 to initiate a quantum jump; otherwise, end the simulation at time t_f.
3. Perform a quantum jump:
   (a) If the state is in a singlet configuration, jump to the octet configuration; if the state is in an octet configuration, generate a random number $0 < r_2 < 1$ and jump to the singlet configuration if $r_2 < 2/7$; otherwise, remain in the octet configuration.
   (b) Generate a random number $0 < r_3 < 1$; if $r_3 < l/(2l+1)$, take $l \to l-1$; otherwise, take $l \to l+1$.
   (c) Multiply the wave function by r and normalize.
4. Return to step 2.

Each realization of the above algorithm is a *quantum trajectory*. The average of N quantum trajectories converges to the solution of the Lindblad equation as $N \to \infty$.

3.2 Simulation details

3.2.1 Input parameters

We estimate our systematic uncertainties by varying κ and γ. We utilize three different parametrizations of $\hat{\kappa}(T) = \kappa(T)/T^3$ which we denote $\hat{\kappa}_U(T)$, $\hat{\kappa}_C(T)$, and $\hat{\kappa}_L(T)$ corresponding to the upper, central, and lower "fit" curves of Fig. 13 of Ref. [23]. In contrast to the direct extraction of κ, we utilize three values of $\hat{\gamma} = \gamma/T^3 = \{-3.5, -1.75, 0\}$ extracted indirectly from unquenched lattice measurements of $\delta M(1S)$; this procedure is outlined in Ref. [24]. Recent lattice studies [25, 26] favor $\delta M(1S) \simeq 0$ (and hence $\gamma = 0$) while the studies [27, 28] used in Ref. [24] favor more negative values of $\delta M(1S)$ (and hence of γ).

We take M as the mass of the bottom quark from $M = m_b = m_{\Upsilon(1S)}/2 = 4.73$ GeV with $m_{\Upsilon(1S)}$ from Ref. [29]. We set the value of the strong coupling α_s by solving

$$a_0 = \frac{2}{C_f \alpha_s(1/a_0) m_b}, \tag{11}$$

using the 1-loop, 3-flavor running of α_s with $\Lambda_{\overline{MS}}^{N_f=3} = 332$ MeV [30] giving $\alpha_s = 0.468$.

3.2.2 Lattice parameters

As a numerically tractable discretization of a Dirac delta function, we initialize each quantum trajectory as a Gaussian

$$\psi_\ell(t_0) \propto r^\ell e^{-r^2/(ca_0^2)}, \tag{12}$$

with $r\psi(r)$ normalized to 1 over the 1-dimensional lattice with $c = 0.2$. We utilize a lattice of radial volume L= 80 GeV^{-1} with `NUM`= 4096 lattice sites corresponding to a lattice spacing $a = 0.0195$ GeV^{-1}; we utilize a time step of `dt` = 0.001 GeV^{-1}. For an analysis of the systematic error associated with the temporal and spatial discretization and finite size effects, we direct the reader to Sec. 7 of Ref. [12].

3.2.3 Medium interaction

Interaction with the medium is implemented by coupling to a 3 + 1D dissipative relativistic hydrodynamics code using a realistic equation of state fit to lattice QCD measurements [31]. The hydrodynamics code uses the quasiparticle anisotropic hydrodynamics (aHydroQP) framework [32–34] and in Ref. [35] was tuned to soft hadronic data from 5.02 TeV collisions using smooth optical Glauber initial conditions. In Ref. [10], per centrality bin, approximately 132000 trajectories through the plasma were Monte Carlo sampled and the temperature evolution averaged in each bin. The Lindblad equation was solved using the single average temperature evolution in each centrality bin. In Ref. [11], approximately $7-9\times10^5$ physical trajectories through the plasma were generated by Monte Carlo sampling. For each physical trajectory, the production point was sampled in the transverse plane using the nuclear binary collision overlap profile $N_{AA}^{bin}(x, y, b)$, the initial p_T from an E_T^{-4} spectrum, and the azimuthal angle ϕ uniformly in $[0, 2\pi]$. Approximately $50-100$ quantum trajectories were generated per physical trajectory. The distinct physical trajectories allow for the extraction of differential observables including v_2 and the presentation of results as a function of p_T. The wave function is initialized at $t = 0$ fm and evolved in the vacuum until $t = 0.6$ fm at which time interaction with the medium is initialized; the wave function is evolved in the medium until the local temperature falls below $T_f = 250$ MeV, from which time forward vacuum evolution is again used, to ensure that the hierarchy of scales of Eq. (1) is fulfilled at all times.

3.2.4 Feed down of excited states

Using the quantum trajectories algorithm as implemented in the `QTraj` code [12], one can calculate the survival probability of a quarkonium state, i.e., the probability a state traverses the QGP without dissociating. In order to compare with experiment, we must take into account the probability that an excited state decays into a lower-lying state after exiting the plasma. The experimentally observed production cross section of a state is related to the direct production cross section by $\vec{\sigma}_{\rm exp} = F\vec{\sigma}_{\rm direct}$ where the vectors contain the states and F is a matrix describing the feed down. We take into account the states $\{\Upsilon(1S), \Upsilon(2S), \chi_{b0}(1P), \chi_{b1}(1P), \chi_{b2}(1P), \Upsilon(3S), \chi_{b0}(2P), \chi_{b1}(2P), \chi_{b2}(2P)\}$. F_{ij} where $i < j$ is the branching ratio of the state j to i, the diagonal of F is 1, and $F_{ij} = 0$ for $i > j$. R_{AA} of state i is calculated from the relation

$$R_{AA}^i(c, p_T, \phi) = \frac{(F\cdot S(c, p_T, \phi)\cdot\vec{\sigma}_{\rm direct})^i}{\vec{\sigma}_{\rm exp}}, \tag{13}$$

where $S(c, p_T, \phi)$ is a diagonal matrix containing the survival probabilities and c labels the centrality class. The feed down fractions of F are taken from the Particle Data Group [29].

Our experimental cross sections are $\vec{\sigma}_{\rm exp} = \{57.6, 19, 3.72, 13.69, 16.1, 6.8, 3.27, 12.0, 14.15\}$ nb and are computed using data from the experimental measurements of Refs. [36, 37] as explained in Sec. 6.4 of Ref. [10].

4 Results

In Fig. 1, we present our `QTraj` results for R_{AA} of the $\Upsilon(1S)$, $\Upsilon(2S)$, and $\Upsilon(3S)$ as functions of p_T against experimental measurements of the ALICE [38], ATLAS [39], and CMS [36] collaborations. We observe good agreement with the experimental data and note the generally flat dependence of R_{AA} on p_T of both our theoretical and the experimental results. In Figs. 2 and 3, we present our `QTraj` results for v_2 of the $\Upsilon(1S)$ as a function of centrality and of p_T, respectively, against measurements of the ALICE [40] and CMS [41] collaborations. We note that our results lie within the uncertainty bounds of the experimental measurements.

In Ref. [11], we present results for R_{AA} of the $\Upsilon(1S)$, $\Upsilon(2S)$, and $\Upsilon(3S)$ as functions of centrality. We additionally plot the double ratios of $R_{AA}[\Upsilon(2S)]$ and $R_{AA}[\Upsilon(3S)]$ to $R_{AA}[\Upsilon(1S)]$ against centrality and p_T. We also give the elliptic flow of the $\Upsilon(2S)$ and $\Upsilon(3S)$ against centrality.

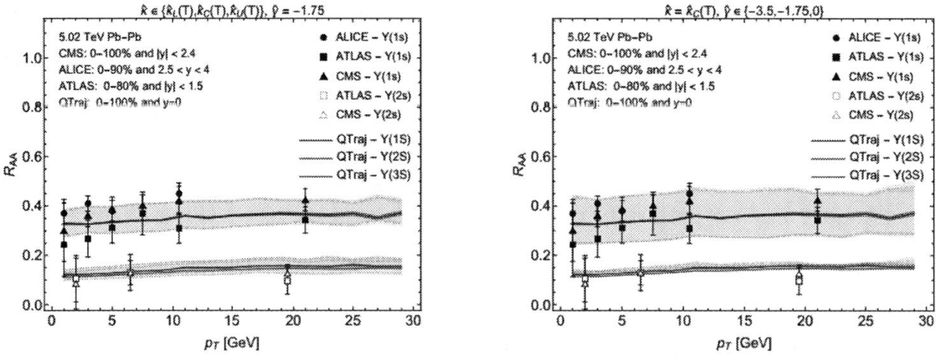

Figure 1. Nuclear modification factor R_{AA} of the $\Upsilon(1S)$, $\Upsilon(2S)$, and $\Upsilon(3S)$ as a function of p_T compared to experimental measurements. The experimental data are taken from the ALICE [38], ATLAS [39], and CMS [36] collaborations. The bands in the theoretical curves indicate variation with respect to $\hat{\kappa}(T)$ (left) and $\hat{\gamma}$ (right). The central curves represent the central values of $\hat{\kappa}(T)$ and $\hat{\gamma}$, and the dashed and dot-dashed lines represent the lower and upper values, respectively, of $\hat{\kappa}(T)$ and $\hat{\gamma}$. Taken from Ref. [11].

5 Conclusions

The hierarchy of scales inherent to in-medium bottomonium make the EFT pNRQCD and the OQS formalism natural tools to describe the system and extract observables of interest. Carrying out this analysis, the authors of Refs. [8, 9] found that the in-medium evolution equations of bottomonium take the form of a master equation which in the limit $\pi T \gg E$ takes the form of a Lindblad equation. In Refs. [10, 11], the `QTraj` code [12] was utilized to solve the Lindblad equation and extract the nuclear modification factor R_{AA} and the elliptic flow v_2 of the $\Upsilon(1S)$, $\Upsilon(2S)$, and $\Upsilon(3S)$ as functions of centrality and transverse momentum p_T. The method is fully quantum, non-abelian, heavy quark number conserving, and accounts for dissociation and quantum recombination; furthermore, the R_{AA} and v_2 results show good agreement with experimental measurements of the ALICE, ATLAS, and CMS collaborations.

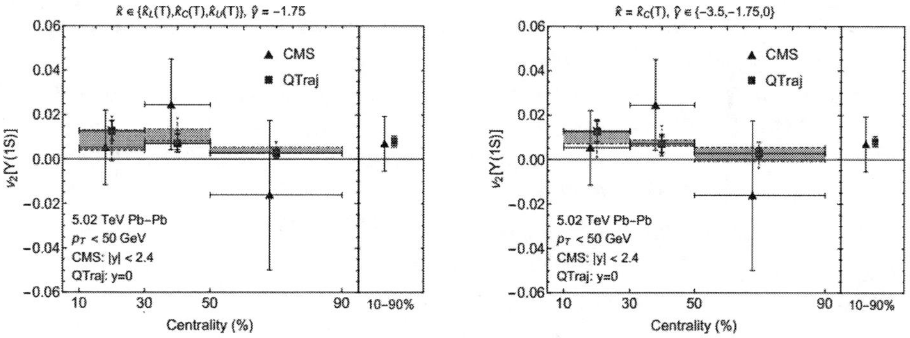

Figure 2. Elliptic flow v_2 of the $\Upsilon(1S)$ as a function of centrality compared to experimental measurements of the CMS [41] collaboration. The bands represent uncertainties as in Fig. 1. Taken from Ref. [11].

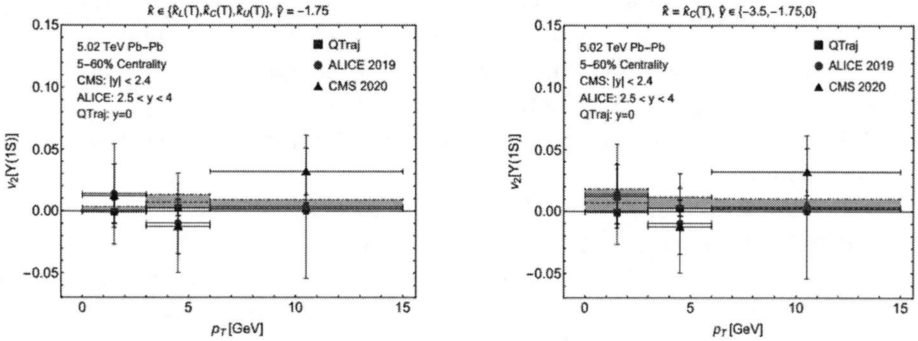

Figure 3. Elliptic flow v_2 of the $\Upsilon(1S)$ as a function of p_T compared to experimental measurements of the ALICE [40] and CMS [41] collaborations. The bands represent uncertainties as in Fig. 1. Taken from Ref. [11].

References

[1] T. Matsui, H. Satz, Phys. Lett. B **178**, 416 (1986)
[2] F. Karsch, M.T. Mehr, H. Satz, Z. Phys. C **37**, 617 (1988)
[3] M. Laine, O. Philipsen, P. Romatschke, M. Tassler, JHEP **03**, 054 (2007), hep-ph/0611300
[4] N. Brambilla, J. Ghiglieri, A. Vairo, P. Petreczky, Phys. Rev. D **78**, 014017 (2008), 0804.0993
[5] M.A. Escobedo, J. Soto, Phys. Rev. A **78**, 032520 (2008), 0804.0691
[6] N. Brambilla, M.A. Escobedo, J. Ghiglieri, J. Soto, A. Vairo, JHEP **09**, 038 (2010), 1007.4156
[7] A. Beraudo, J.P. Blaizot, C. Ratti, Nucl. Phys. A **806**, 312 (2008), 0712.4394
[8] N. Brambilla, M.A. Escobedo, J. Soto, A. Vairo, Phys. Rev. D **96**, 034021 (2017), 1612.07248
[9] N. Brambilla, M.A. Escobedo, J. Soto, A. Vairo, Phys. Rev. D **97**, 074009 (2018), 1711.04515

[10] N. Brambilla, M.A. Escobedo, M. Strickland, A. Vairo, P. Vander Griend, J.H. Weber, JHEP **05**, 136 (2021), 2012.01240
[11] N. Brambilla, M.A. Escobedo, M. Strickland, A. Vairo, P. Vander Griend, J.H. Weber (2021), 2107.06222
[12] H.B. Omar, M.A. Escobedo, A. Islam, M. Strickland, S. Thapa, P. Vander Griend, J.H. Weber (2021), 2107.06147
[13] W.E. Caswell, G.P. Lepage, Phys. Lett. B **167**, 437 (1986)
[14] G.T. Bodwin, E. Braaten, G.P. Lepage, Phys. Rev. D **51**, 1125 (1995), [Erratum: Phys.Rev.D 55, 5853 (1997)], hep-ph/9407339
[15] A. Pineda, J. Soto, Nucl. Phys. B Proc. Suppl. **64**, 428 (1998), hep-ph/9707481
[16] N. Brambilla, A. Pineda, J. Soto, A. Vairo, Nucl. Phys. B **566**, 275 (2000), hep-ph/9907240
[17] N. Brambilla, A. Pineda, J. Soto, A. Vairo, Rev. Mod. Phys. **77**, 1423 (2005), hep-ph/0410047
[18] G. Lindblad, Commun. Math. Phys. **48**, 119 (1976)
[19] V. Gorini, A. Kossakowski, E.C.G. Sudarshan, J. Math. Phys. **17**, 821 (1976)
[20] J. Casalderrey-Solana, D. Teaney, Phys. Rev. D **74**, 085012 (2006), hep-ph/0605199
[21] S. Caron-Huot, G.D. Moore, Phys. Rev. Lett. **100**, 052301 (2008), 0708.4232
[22] A.J. Daley, Adv. Phys. **63**, 77 (2014), 1405.6694
[23] N. Brambilla, V. Leino, P. Petreczky, A. Vairo, Phys. Rev. D **102**, 074503 (2020), 2007.10078
[24] N. Brambilla, M.A. Escobedo, A. Vairo, P. Vander Griend, Phys. Rev. D **100**, 054025 (2019), 1903.08063
[25] R. Larsen, S. Meinel, S. Mukherjee, P. Petreczky, Phys. Rev. D **100**, 074506 (2019), 1908.08437
[26] S. Shi, K. Zhou, J. Zhao, S. Mukherjee, P. Zhuang (2021), 2105.07862
[27] S. Kim, P. Petreczky, A. Rothkopf, JHEP **11**, 088 (2018), 1808.08781
[28] G. Aarts, C. Allton, S. Kim, M.P. Lombardo, M.B. Oktay, S.M. Ryan, D.K. Sinclair, J.I. Skullerud, JHEP **11**, 103 (2011), 1109.4496
[29] P.A. Zyla et al. (Particle Data Group), PTEP **2020**, 083C01 (2020)
[30] P. Petreczky, J.H. Weber (2020), 2012.06193
[31] A. Bazavov, J. Phys. Conf. Ser. **446**, 012011 (2013), 1303.6294
[32] M. Alqahtani, M. Nopoush, M. Strickland, Phys. Rev. C **92**, 054910 (2015), 1509.02913
[33] M. Alqahtani, M. Nopoush, M. Strickland, Phys. Rev. C **95**, 034906 (2017), 1605.02101
[34] M. Alqahtani, M. Nopoush, M. Strickland, Prog. Part. Nucl. Phys. **101**, 204 (2018), 1712.03282
[35] M. Alqahtani, M. Strickland (2020), 2008.07657
[36] A.M. Sirunyan et al. (CMS), Phys. Lett. B **790**, 270 (2019), 1805.09215
[37] R. Aaij et al. (LHCb), Eur. Phys. J. C **74**, 3092 (2014), 1407.7734
[38] S. Acharya et al. (ALICE), Phys. Lett. B **822**, 136579 (2021), 2011.05758
[39] Songkyo Lee (ATLAS Collaboration), *Quarkonium production in Pb+Pb collisions with ATLAS*, Quark Matter 2020 https://indico.cern.ch/event/792436/contributions/3535775/ (2017)
[40] S. Acharya et al. (ALICE), Phys. Rev. Lett. **123**, 192301 (2019), 1907.03169
[41] A.M. Sirunyan et al. (CMS), Phys. Lett. B **819**, 136385 (2021), 2006.07707

EPJ Web of Conferences **258**, 05006 (2022) https://doi.org/10.1051/epjconf/202225805006
A Virtual Tribute to Quark Confinement and the Hadron Spectrum (vConf21)

Medium evolution of a static quark-antiquark pair in the large N_c limit

Miguel Ángel Escobedo[1,*]

[1]Instituto Galego de Física de Altas Enerxías (IGFAE), Universidade de Santiago de Compostela. E-15782, Galicia, Spain

Abstract. We study the transitions between the different color states of a static quark-antiquark pair, singlet and octet, in a thermal medium. This is done non-perturbatively exploiting the infinite mass limit of QCD. This study is interesting because it can be used for future developments within the framework of Effective Field Theories (EFTs) and because it can be combined with other techniques, like lattice QCD or AdS/CFT, to gain non-perturbative information about the evolution of quarkonium in a medium. We also study the obtained expressions in the large N_c limit. This allows us to learn lessons that are useful to simplify phenomenological models of quarkonium in a plasma. More details can be found in [1].

1 Introduction

One of the most interesting probes of the medium created in heavy-ion collisions is heavy quarkonium. There are several features that make this particle interesting in the context of quark-gluon plasma studies. The mass of heavy quarks is large. Therefore, a heavy quark pair can only be created at the beginning of the collision in a perturbative process not affected by the medium. However, the interaction with the medium does modify the probability that a bound state is formed and its lifetime. Therefore, measuring R_{AA} we can extract information about the medium.

There are three mechanism that modify the number of quarkonium states inside of a quark-gluon plasma: colour screening, inelastic collisions with medium particles and recombination. It is desirable to have a formalism in which the three mechanism can be described consistently. Moreover, we need a quantum mechanical description. The reason is that, when thermal modifications are a leading order effect, the question we wish to answer is whether a bound state can exists in such conditions or not.Note that bound state formation (as for example the hydrogen atom) is a problem that requires applying quantum mechanics.

A promising formalism that meets these requirements is that of open quantum systems. In this framework, we consider a *universe* made of heavy quarks and a environment, that is made of light quarks and gluons. The state of this *universe* is determined by its density matrix. This matrix evolves following an unitary evolution, however, it is a very complex object that contains information about all the degrees of freedom of the *universe*. It is convenient to work with a simpler object called the reduced density matrix. It is obtained from the full density matrix after performing a trace over the environment degrees of freedom. The reduced

*e-mail: miguelangel.escobedo@usc.es

density matrix contains all the information needed to determine the state of the heavy quarks, however, its evolution is not unitary any more. In general, the time evolution of the reduced density matrix is ruled by the so-called master equation.

The master equation for the evolution of quarkonium in a medium has been derived in two different limits. In perturbation theory [2, 3] and in the limit $\frac{1}{r} \gg T$ [4]. There are other derivations in the literature that do not discuss the Non-Abelian nature of QCD, which is one of the focus of this work. Once the master equation is known, it is not an easy task to apply it to obtain phenomenological results due to the high computational cost. However, the use of Monte Carlo methods has improved the situation in recent years [5, 6].

Our motivation in this work is to gain information about the master equation in more general settings. More specifically, we ask ourselves if it is possible to know something about the master equation in the case $\frac{1}{r} \sim T$ beyond the perturbative regime. We expect that the needed non-perturbative medium information can be encoded in expectation values of gauge invariant operators. These operators might be computed using lattice QCD or other non-perturbative techniques in the future. Even if this computation is not possible at the moment, to obtain these expectation values could be useful to get non-perturbative insights from the large N_c limit. A first step in this direction is to study the static limit, defined as the limit in which the heavy quark mass is infinite.

2 Non-relativistic effective field theories

In this section we review one of the main tools we are going to use in our investigation, Non-relativistic effective field theories (NREFTs). Heavy quarks have a mass m that is much larger than Λ_{QCD} and the temperature of the medium. Other relevant energy scales, are the inverse of the typical radius $\frac{1}{r} \sim mv$ (with $v \ll 1$) and the binding energy $E \sim mv^2$. Therefore, quarkonium is a system in which several well-separated energy scales are present. In these situations it is useful to apply EFTs. The large separation between energy scales can break the convergence of naive perturbation theory or make lattice computations very costly. Using EFTs we can reorganize the computation in such a way that we only deal with one energy scale at each step, solving the previously mentioned problems. In the case of heavy quarks, we can integrate the physics at the scale m to go from QCD to non-relativistic QCD (NRQCD) [7, 8]. This can be done pertubatively and ignoring medium effects, since $m \gg \Lambda_{QCD}, T$. In NRQCD, heavy quarks are not represented by a bispinor field. Instead, we have a spinor field for heavy quarks and another spinor field for the antiquarks. One can also integrate out the scale $\frac{1}{r}$ to go from NRQCD to potential NRQCD (pNRQCD) [9]. The pNRQCD Lagrangian at $T = 0$ is

$$\mathcal{L}_{pNRQCD} = \int d^3\mathbf{r} Tr\left[S^\dagger (i\partial_0 - h_s) S + O^\dagger (iD_0 - h_o) O\right] + V_A(r) Tr(O^\dagger \mathbf{r} g\mathbf{E} S + S^\dagger \mathbf{r} g\mathbf{E} O) + \frac{V_B(r)}{2} Tr(O^\dagger \mathbf{r} g\mathbf{E} O + O^\dagger O \mathbf{r} g\mathbf{E}) + \mathcal{L}_g + \mathcal{L}_q \,. \quad (1)$$

The degrees of freedom of the heavy quarks and antiquarks are reorganized in a singlet S and octet O field. The pNRQCD Lagrangian is manifestly gauge invariant. This simplifies the connection with quantities that can be computed using lattice QCD. If $\frac{1}{r} \gg T$, we can use the previous Lagrangian as a starting point. If this is not the case, then the matching between NRQCD and pNRQCD has to be done taking into account the presence of the medium.

Let us now remark an interesting property of the matching at $T = 0$. We can match NRQCD to pNRQCD in the static limit. In this limit the matching can be done non-perturbatively since the propagation of a heavy quark is described by a time-like Wilson line. Using this we can relate all Wilson coefficients to expectation values of gauge invariant operators. Once the we have obtained the pNRQCD Lagrangian in the static limit, we can

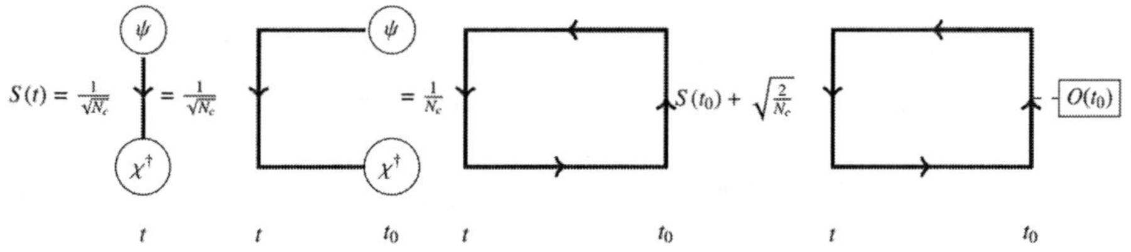

Figure 1. Backwards evolution of the singlet. We are using a birdtrack notation, more details in [1]

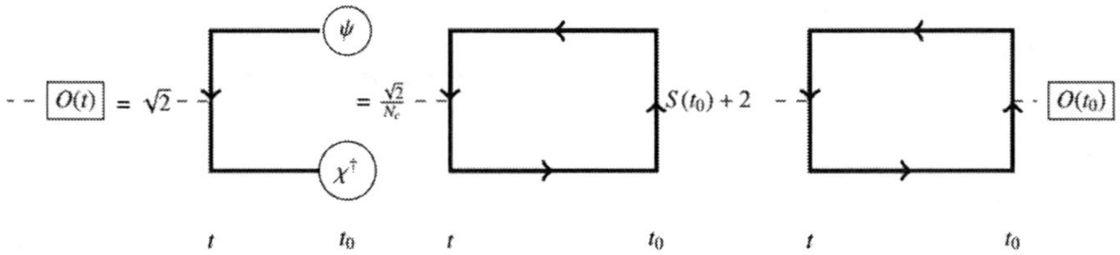

Figure 2. Backwards evolution of the octet.

apply it to study situations in which the mass is not infinite, as for example real quarkonium. The reason is that each term has a well-defined power counting in terms of $\frac{1}{m}$. We expect that this nice property is also true at finite temperature, although more developments are needed to match NRQCD to pNRQCD in this case. Therefore, we move in the next section to study the evolution of a pair of static quarks in the medium.

3 The static limit

We use the NRQCD Lagrangian in the static limit

$$\mathcal{L}_{NRQCD} = \mathcal{L}_{QCD} + \psi^\dagger iD_0\psi + \chi^\dagger iD_0\chi\,. \tag{2}$$

We study the evolution of operators that transform like singlets or octets made of static quark fields, they are called interpolating fields in the pNRQCD literature [9]. We expect that these fields are close to the singlet and octet fields of pNRQCD in the limit of small r, however the precise relation needs to be worked out. From now on we are going to use a slight abuse of language and identify the singlet and octet with their corresponding interpolating field. In order to work out the evolution of a static quark-antiquark pair we need to take into account that the evolution of a static quark is given by a time-like Wilson line and make use of the Fierz identity. The evolution of the singlet is represented in fig. 1 and that of the octet in fig. 2. Using these results we can obtain the probability to find a singlet or octet at time t knowing that there was one singlet or octet at time t_0. In [1] we found the expectation value that correspond to each of the four cases. An unusual characteristic which they share is that they mix fields in different branches of the Schwinger-Keldysh contour. Even though a non-pertubative study of these expectation values is beyond the scope of this work, we can obtain some information from studying them in the large N_c limit.

3.1 Singlet to singlet transition

The probability to find a singlet at time t assuming that there is a singlet at time t_0 is proportional to

$$Tr(W_{SS}^{\dagger}(\mathbf{R}, \mathbf{r}; t, t_0) W_{SS}(\mathbf{R}, \mathbf{r}; t, t_0)\rho_l), \quad (3)$$

where W_{SS} is a Wilson loop of the kind that defines the static potential. In the large N_c limit diagrams mixing gluons attached to W_{SS} and gluons attached to $W_{SS}^{\dagger}$ are suppressed. Therefore,

$$Tr(W_{SS}^{\dagger}(\mathbf{R}, \mathbf{r}; t, t_0) W_{SS}(\mathbf{R}, \mathbf{r}; t, t_0)\rho_l) = |Tr(W_{SS}(\mathbf{R}, \mathbf{r}; t, t_0)\rho_l)|^2 \,. \quad (4)$$

This result suggests that, in the large N_c limit, the survival probability of the singlet can be studied using an effective non-Hermitian Hamiltonian.

3.2 Singlet to octet transition

For this particular transition the analysis of the consequences of the large N_c limit allows to relate it to two different expectation values. One of them is proportional to the singlet to singlet transition while there is another new term appearing in this transition. This structure is reminiscent of the fact that the probability to find an octet is equal to the probability to find anything minus the probability to find a singlet.

3.3 Octet to singlet transition

This observable has a structure similar to the singlet to octet transition. However, there is an important quantitative difference. This transition is suppressed in the large N_c limit by a factor $\frac{1}{N_c^2}$.

3.4 Octet to octet transition

In this case we observe that diagrams that connect the quark with the antiquark are suppressed by a power of $\frac{1}{N_c^2}$. This means that in the large N_c limit the octet behaves like an uncorrelated pair of particles. In this sense, the large N_c limit seems to justify the molecular hypothesis often used in the literature to solve the Boltzmann equation [10], at least in the low density limit in which this study is performed.

3.5 Qualitative picture

We can distinguish two different cases. The first case is the one in which the density of singlets D_s is similar to the density of octets D_o

$$D_s(t) = S\,d(t - t_0) D_s(t_0)\,, \quad (5)$$

$$D_o(t) = (Q(t - t_0) - S\,d(t - t_0)) D_s(t_0) + Qd(t - t_0) D_o(t_0)\,. \quad (6)$$

The singlet decays into octets following an evolution that can be encoded with an effective Hamiltonian. The octet evolves like an uncorrelated pair of heavy quarks sourced by the decay of singlets. This situation is maintained until the density of singlets decreases so much that we can consider that $D_s \sim \frac{1}{N_c^2}$. Then, the power counting changes and the evolution equations at leading order in the large N_c limit are

$$D_s(t) = Sd(t-t_0)D_s(t_0) + \frac{Iq(t-t_0) - Sd(t-t_0)}{N_c^2 - 1} D_o(t_0)\,, \tag{7}$$

$$D_o(t) = Qd(t-t_0)D_o(t_0)\,. \tag{8}$$

Now, in order to compute the density of singlets, it is important to consider the decay of octets into singlets. Regarding the octet, we can neglect the decay of singlets and just consider that the octets evolve like uncorrelated pairs of quarks.

4 Conclusions

We tried to justify that the study of the static limit can be relevant to understand the physics of *real* heavy quarkonium. We have studied all the possible transitions between singlets and octets in the large N_c limit and wrote the corresponding expectation value of gauge invariant operators, which may be computed using non-perturbative techniques. They have the special feature that they involve the two branches of the Schwinger-Keldysh contour.

We studied the consequences that we can obtain from using the large N_c limit. Although we have not discussed it explicitly, these consequences are compatible with previous computations using perturbation theory or pNRQCD in the $\frac{1}{r} \gg T$ limit. The evolution of the singlet can be encoded in an effective Hamiltonian (that is not necessarily Hermitian). The decay of octets into singlets can be ignored, at least as long as the density of singlets is not very low. The octet propagates like an uncorrelated pair of particles. This suggests that the large N_c limit might be used to justify the molecular chaos hypothesis, however, to confirm this would need a study considering large densities of heavy quarks.

Acknowledgements

The work of M.A.E. was supported by *Ministerio de Ciencia e Innovacion* of Spain under project FPA2017-83814-P, Maria de Maetzu Unit of Excellence MDM-2016- 0692, by Xunta de Galicia and FEDER, and by the European Research Council project ERC-2018-ADG-835105 YoctoLHC.

References

[1] M.A. Escobedo, Phys. Rev. D **103**, 034010 (2021), 2010.10424
[2] Y. Akamatsu, Phys. Rev. D **91**, 056002 (2015), 1403.5783
[3] J.P. Blaizot, M.A. Escobedo, Phys. Rev. D **98**, 074007 (2018), 1803.07996
[4] N. Brambilla, M.A. Escobedo, J. Soto, A. Vairo, Phys. Rev. D **97**, 074009 (2018), 1711.04515
[5] Y. Akamatsu, M. Asakawa, S. Kajimoto (2021), 2108.06921
[6] N. Brambilla, M.A. Escobedo, M. Strickland, A. Vairo, P. Vander Griend, J.H. Weber, JHEP **05**, 136 (2021), 2012.01240
[7] W.E. Caswell, G.P. Lepage, Phys. Lett. B **167**, 437 (1986)
[8] G.T. Bodwin, E. Braaten, G.P. Lepage, Phys. Rev. D **51**, 1125 (1995), [Erratum: Phys.Rev.D 55, 5853 (1997)], hep-ph/9407339
[9] N. Brambilla, A. Pineda, J. Soto, A. Vairo, Nucl. Phys. B **566**, 275 (2000), hep-ph/9907240
[10] X. Yao, T. Mehen, Phys. Rev. D **99**, 096028 (2019), 1811.07027

News from the strong interactions program of NA61/SHINE

Wojciech Bryliński[1,*]
for the NA61/SHINE Collaboration

[1]Warsaw University of Technology

Abstract. NA61/SHINE (SPS Heavy Ion and Neutrino Experiment) is a fixed-target experiment operating at the CERN SPS accelerator. The main goal of the strong interactions program of NA61/SHINE is to study the properties of the phase transition between confined matter and quark-gluon plasma by performing a two-dimensional scan in beam momentum and size of collided nuclei. Within this program, collisions of different systems (p+p, p+Pb, Be+Be, Ar+Sc, Xe+La, Pb+Pb) over a wide range of beam momenta (13A-150(8)A GeV/c) have been recorded.
This contribution discusses the latest results of hadron production in p+p, Be+Be, Ar+Sc and Pb+Pb reactions measured by the NA61/SHINE. In particular, the results include charged kaons and pions spectra and higher-order moments of multiplicity and net charge distributions. The presented data are compared with the predictions of different theoretical models as well as the results from other experiments. Finally, the motivation and plans for future NA61/SHINE measurements are discussed.

1 Introduction

SPS Heavy Ion and Neutrino Experiment (SHINE) [1] is a fixed-target experiment operating at the Super Proton Synchrotron (SPS) at the European Organization for Nuclear Research (CERN). The main physics motivation of the NA61/SHINE experiment is to study the properties of the phase transition between hadronic matter and quark-gluon plasma. Within this program, NA61/SHINE performed a two-dimensional scan in collision energy (13A–150(8)A GeV/c) and system size (p+p, Be+Be, Ar+Sc, Xe+La, Pb+Pb). The phase diagram of strongly interacting matter as well as the overall summary of collected data is presented in Figure 1.

The NA61/SHINE detector is a multi-purpose spectrometer optimised to study hadron production in various types of collisions. The schematic picture of the detector is shown in Figure 2. The main subdetectors of the whole setup are the Time Projection Chambers. Two Vertex-TPCs (VTPCs), located in the magnetic field, together with two large volume Main-TPCs (MTPCs) are main tracking devices and are able to register a large number of particle tracks (up to 1500 in central Pb+Pb collisions). Four smaller TPCs: GAP-TPC and 3 Forward-TPCs (FTPCs) are located along the beam axis. Such a setup gives an excellent capabilities in charged particles momenta measurement and allows for the particle identification complemented by the information from the Time-of-Flight (ToF) detectors. The last

*e-mail: wojciech.brylinski@cern.ch

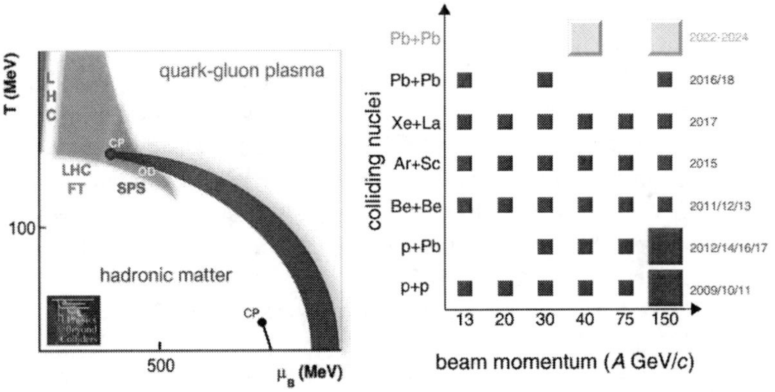

Figure 1. Phase diagram of strongly interacting matter (left). Schematic picture presenting the data collected within system size – beam momentum scan performed by NA61/SHINE (right).

detectors on the beamline are Projectile Spectator Detectors (MPSD and FPSD), which measure the energy of spectators. This information is used to determine the centrality in A+A collisions with very good accuracy. Beam particles are measured by an array of beam detectors. They are used for the beam trajectory measurement as well as the identification of beam particles.

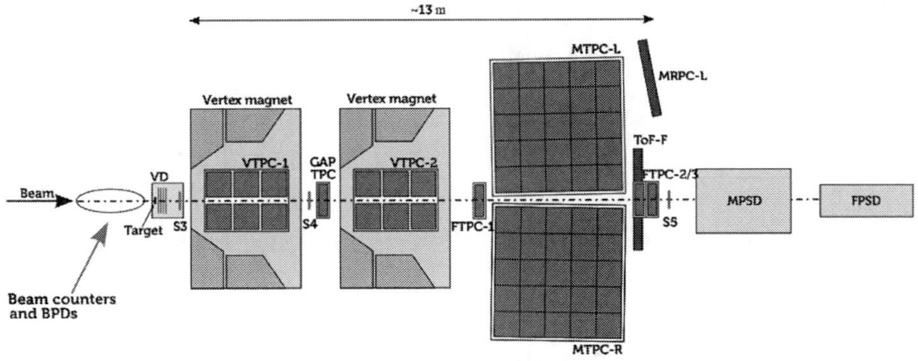

Figure 2. The schematic picture of the NA61/SHINE detector.

2 Study of the onset of deconfinement

The Statistical Model of the Early Stage (SMES) [2] predicts several signatures of the first order phase transition between the hadron gas and the quark-gluon plasma. In the transition region, constant temperature and pressure in the mixed-phase and an increase of the number of internal degrees of freedom are expected.

The energy and system-size dependence of K^+/π^+ ratio at mid-rapidity is presented in Figure 3. The results were obtained using data collected by NA61/SHINE: p+p [3], Be+Be

[4], Ar+Sc (preliminary) as well as NA49: Pb+Pb [5]. The visible structure for Pb+Pb and Au+Au data (so-called horn) in K^+/π^+ ratio was predicted by SMES as the signature of the onset of deconfinement. The presented results were compered with dynamical and statistical models (see right pannel of Figure 3). The dynamical models that do not include the phase transition (EPOS [6], UrQMD [7] and SMASH [8]) describe the data for small systems (p+p and Be+Be), but fail to describe the data for heavier systems (Ar+Sc and Pb+Pb). On the other hand, the PHSD [9] model which includes the phase transition agrees with the data for heavier systems and overestimates the data for smaller systems. Both statistical models which were taken into consideration – with (SMES [10]) and without (HRG [11]) phase transition included – overestimate the K^+/π^+ ratio especially for small systems.

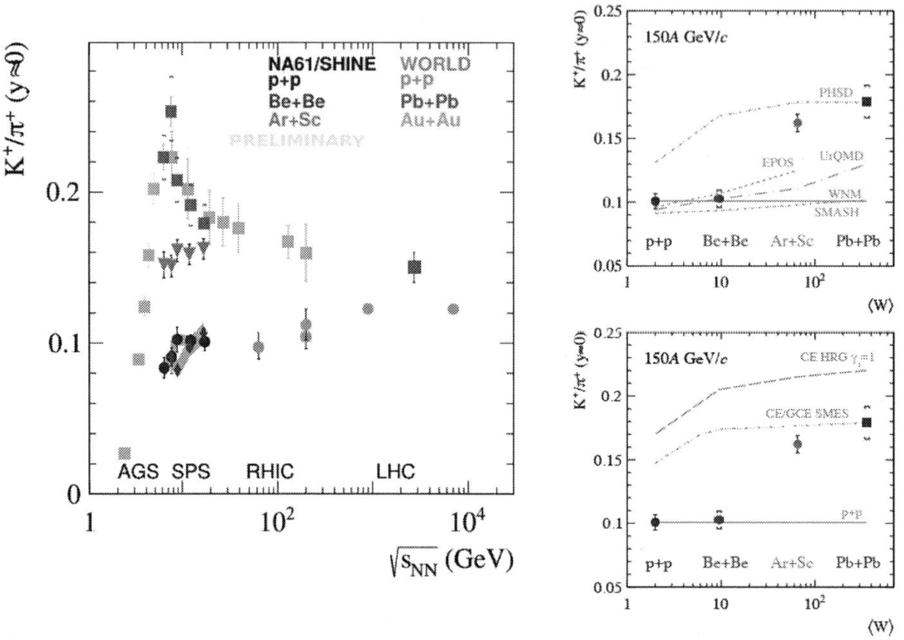

Figure 3. System-size and energy dependence of K^+/π^+ ratio compared with other experiments (left) as well as dynamical and statistical models (right).

The energy dependence of K^+/π^+ ratios as well as inverse slope parameter for p+p collisions are shown in Figure 4. The comparison to the corresponding results for heavier systems reveals similarities in the collision energy dependence – a rapid change of the hadron production properties in the same energy range. This phenomena for p+p collisions occurs very close the energy that corresponds to the phase transition in heavy systems (approximately 8 GeV).

3 Search for the critical point

The expected signature of the critical point (CP) is a non-monotonic dependence of various fluctuations measured in the NA61/SHINE system size – energy scan. A special interest is devoted to the fluctuations of conserved charges (electric, strangeness or baryon number) [12–14]. In order to compare the fluctuations in systems of different sizes, one should use the quantities which are insensitive to the size of the system. Such quantities are called intensive quantities and can be obtained by the division of cumulants κ_i of the measured distribution,

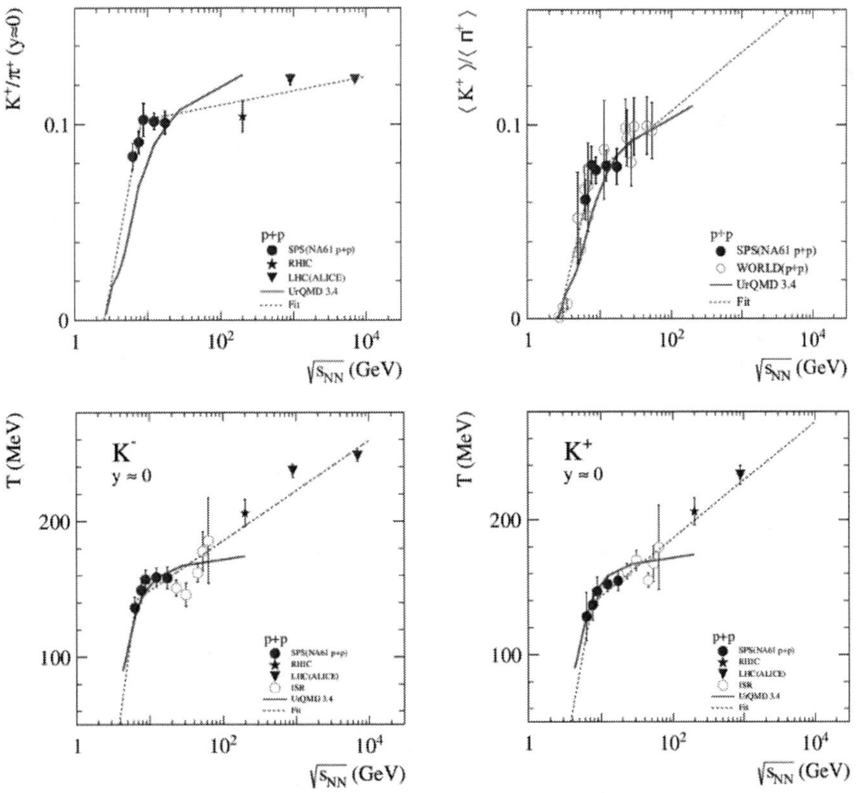

Figure 4. Energy dependence of the K^+/π^+ ratio in inelastic p+p interactions at mid-rapidity (top-left) and in the full phase-space (top-right) as well the inverse slope parameter T of transverse mass spectra at mid-rapidity for K^- (bottom-left) and K^+ (bottom-right) mesons.

where i is the order of the cumulant. For second, third and fourth order cumulants intensive quantities are defined as: κ_2/κ_1, κ_3/κ_2 and κ_4/κ_2. Figure 5 shows the system size and energy dependence of second, third and fourth order cumulant ratio of negatively charged hadrons in p+p, Be+Be and Ar+Sc collisions. So far, there is no clear evidence of the critical point, however more detailed studies are needed.

Proton intermittency is another possible tool that can be used to search for critical point. In the proximity of CP a local power-law fluctuation of the baryon density should be manifested. It can be searched for by studying the scaling behaviour of second factorial moments, $F_2(\delta) = \frac{\langle \frac{1}{M}\sum_{i=1}^{M} n_i(n_i-1)\rangle}{\langle \frac{1}{M}\sum_{i=1}^{M} n_i\rangle^2}$ with the cell size or, equivalently, with the number of cells M in (px, py) space of protons at mid-rapidity [15–17]. NA61/SHINE measures $F_2(M)$ using statistically independent points and cumulative variables. Preliminary results on $F_2(M)$ of mid-rapidity protons measured in 0-20% most central Ar+Sc collisions at 150A GeV/c and 0-10% most central Pb+Pb collisions at 30A GeV/c are presented in Figure 6. The intermittency index φ_2 for a system freezing out at the QCD critical endpoint is expected to be $\varphi_2 = 5/6$ assuming that the latter belongs to the 3-D Ising universality class. $F_2(M)$ of protons for Ar+Sc at 150A GeV/c and Pb+Pb at 30A GeV/c show no indication for power-law increase with a bin size.

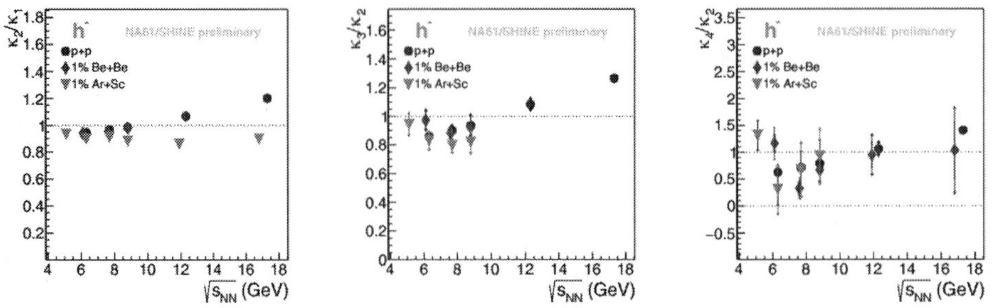

Figure 5. Preliminary results on the energy and system-size dependence of multiplicity fluctuations of negatively charged hadrons.

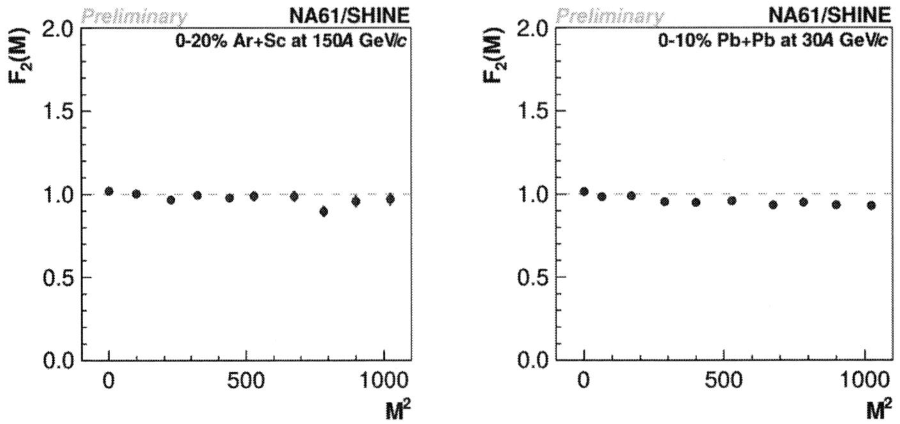

Figure 6. Preliminary results on $F_2(M)$ of mid-rapidity protons measured in 0-20% most central Ar+Sc collisions at 150*A* GeV/*c* (left) and 0-10% most central Pb+Pb collisions at 30*A* GeV/*c* (right).

4 Open charm measurements

The physics program of NA61/SHINE was extended by measurements of open charm production in A+A collisions which is the main goal of NA61/SHINE beyond 2021. It was motivated by three main questions:

- What is the mechanism of open charm production?
- How does the onset of deconfinement impact open charm production?
- How does the formation of quark-gluon plasma impact J/ψ production?

The mean multiplicity of charm quarks $\langle c\bar{c}\rangle$ produced in a full phase space in heavy-ion collisions is not yet known and has to be measured in order to answer these questions. The acceptance of the NA61/SHINE detector is large enough to extrapolate the measurements to the full phase space with relatively small uncertainties. This unique feature makes NA61/SHINE the only experiment which is able to perform such measurement in the near future.

To meet the challenges of the required spatial resolution of primary and secondary vertex reconstruction, the detector was upgraded by a micro vertex detector. A Small-Acceptance version of the Vertex Detector (SAVD) was successfully commissioned in December 2016 and first pilot data were collected for Pb+Pb collisions at a beam momentum of 150*A* GeV/*c*.

This data allowed to validate the measurement concept and to perform the first direct measurement of open charm hadron production in collisions of nuclei at the SPS energies. Figure 7 (left) shows the first indication of a D^0 and $\overline{D^0}$ peak. In October and November 2017, the large statistics datasets for Xe+La collisions at 75*A* and 150*A* GeV/*c* were registered. The distribution of the longitudinal coordinate of the primary vertex is shown in Figure 7 (right).

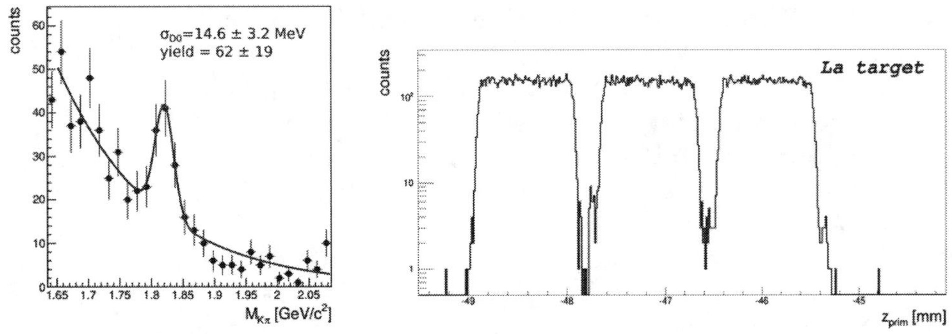

Figure 7. Invariant mass distribution of D^0 and $\overline{D^0}$ candidates in central Pb+Pb collisions at after the background suppression cuts (left). Distribution of longitudinal coordinate of the primary vertex for interactions in the La target, which was composed of three 1 mm plates (right).

5 NA61/SHINE detector upgrades

The NA61/SHINE spectrometer is undergoing a significant upgrade during the Long Shutdown 2 at CERN (2019-2021) [18]. Most of the upgrades are dedicated to open charm physics program which requires the increase of the phase space coverage of Vertex Detector and the tenfold increase of data taking rate to about 1 kHz. The most important points of the upgrade are summarized below:

- 10 fold increase of data taking rate up to 1kHz thanks to the replacement of the TPC readout electronics,
- improvement of acceptance and efficiency of the Vertex Detector,
- improvement of radiation tolerance of the PSD hadron calorimeter,
- introduction of new TOF detector based on mRPC technology,
- replacement of old readout electronics based on CAMAC and FASTBUS standards,
- development of the new trigger and data acquisition system.

The first physics data taking with the upgraded setup is planned for 2022.

Acknowledgments: This work was supported by WUT-IDUB and the National Science Centre, Poland under grant no. 2019/34/H/ST2/00585.

References

[1] N. Abgrall *et al.* [NA61 Collaboration], JINST **9**, P06005 (2014) doi:10.1088/1748-0221/9/06/P06005 [arXiv:1401.4699 [physics.ins-det]].

[2] M. Gazdzicki and M. I. Gorenstein, Acta Phys. Polon. B **30**, 2705 (1999) [hep-ph/9803462].

[3] A. Aduszkiewicz *et al.* [NA61/SHINE], Eur. Phys. J. C **77** (2017) no.10, 671 doi:10.1140/epjc/s10052-017-5260-4 [arXiv:1705.02467 [nucl-ex]].

[4] A. Acharya *et al.* [NA61/SHINE], Eur. Phys. J. C **81** (2021) no.5, 397 doi:10.1140/epjc/s10052-021-09135-3 [arXiv:2101.08494 [hep-ex]].

[5] S. V. Afanasiev *et al.* [NA49], Phys. Rev. C **66** (2002), 054902 doi:10.1103/PhysRevC.66.054902 [arXiv:nucl-ex/0205002 [nucl-ex]].

[6] T. Pierog and K. Werner, Nucl. Phys. B Proc. Suppl. **196** (2009), 102-105 doi:10.1016/j.nuclphysbps.2009.09.017 [arXiv:0905.1198 [hep-ph]].

[7] A. Motornenko *et al.*, Phys. Rev. C99 (2019) 3, 034909 doi:10.1103/PhysRevC.99.034909

[8] J. Mohs, S. Ryu and H. Elfner, J. Phys. G **47** (2020) no.6, 065101 doi:10.1088/1361-6471/ab7bd1 [arXiv:1909.05586 [nucl-th]].

[9] V. Kireyeu, I. Grishmanovskii, V. Kolesnikov, V. Voronyuk and E. Bratkovskaya, Eur. Phys. J. A **56** (2020) no.9, 223 doi:10.1140/epja/s10050-020-00232-7 [arXiv:2006.14739 [hep-ph]].

[10] R. V. Poberezhnyuk, M. Gazdzicki and M. I. Gorenstein, Acta Phys. Polon. B **46** (2015) no.10, 1991 doi:10.5506/APhysPolB.46.1991 [arXiv:1502.05650 [nucl-th]].

[11] A. Motornenko, V. V. Begun, V. Vovchenko, M. I. Gorenstein and H. Stoecker, Phys. Rev. C **99** (2019) no.3, 034909 doi:10.1103/PhysRevC.99.034909 [arXiv:1811.10645 [nucl-th]].

[12] M. A. Stephanov, Phys. Rev. Lett. **102** (2009), 032301 doi:10.1103/PhysRevLett.102.032301 [arXiv:0809.3450 [hep-ph]].

[13] M. Asakawa and M. Kitazawa, Prog. Part. Nucl. Phys. **90** (2016), 299-342 doi:10.1016/j.ppnp.2016.04.002 [arXiv:1512.05038 [nucl-th]].

[14] M. A. Stephanov, K. Rajagopal and E. V. Shuryak, Phys. Rev. Lett. **81** (1998), 4816-4819 doi:10.1103/PhysRevLett.81.4816 [arXiv:hep-ph/9806219 [hep-ph]].

[15] A. Bialas and R. B. Peschanski, Nucl. Phys. B **273** (1986), 703-718 doi:10.1016/0550-3213(86)90386-X

[16] L. Turko, Phys. Lett. B **227** (1989), 149-152 doi:10.1016/0370-2693(89)91298-7

[17] F. K. Diakonos, N. G. Antoniou and G. Mavromanolakis, PoS **CPOD2006** (2006), 010 doi:10.22323/1.029.0010

[18] A. Aduszkiewicz *et al.* [NA61 Collaboration], CERN-SPSC-2018-008, SPSC-P-330-ADD-10

Transport near the chiral critical point

Alexander Soloviev[1,*]

[1]Center for Nuclear Theory, Department of Physics and Astronomy, Stony Brook University, Stony Brook, New York 11794, USA

Abstract. The evolution of a heavy ion collision passes close to the O(4) critical point of QCD, where fluctuations of the order parameter are expected to be enhanced. Using the appropriate stochastic hydrodynamic equations in mean field near the the pseudo-critical point, we compute how these enhanced fluctuations modify the transport coefficients of QCD. Finally, we estimate the expected critical enhancement of soft pion yields, which provides a plausible explanation for the excess seen in experiment relative to ordinary hydrodynamic computations.

1 Introduction

Relativistic viscous hydrodynamics [1] is remarkably successful at describing measurements performed at the Relativistic Heavy Ion Collider (RHIC) and the Large Hadron Collider (LHC). One feature present in heavy ion collisions, yet absent from current hydrodynamic simulations, is chiral symmetry breaking and the associated chiral phase transition. Although chiral symmetry is not an exact symmetry due to the finite quark mass, for small enough quark mass the pattern of chiral symmetry breaking should have some influence on the evolution of the system. This is supported by results from the lattice [2, 3], where certain aspects of QCD thermodynamics, such as chiral susceptibility, could be understood using $O(4)$ scaling functions. As such, chiral symmetry and its breaking should be included to increase the predictive power of hydrodynamics.

The reorganization of hydrodynamics to include chiral physics was presented recently in [4, 5], building on earlier work on $O(4)$ physics going back to Rajagopal and Wilczek [6] almost thirty years ago. The basic idea is that at long distances and high temperature, the appropriate effective theory for QCD is ordinary hydrodynamics, approximately respecting the $SU_L(2) \times SU_R(2) \sim O(4)$ symmetry. As the temperature drops below the critical temperature, the system undergoes the chiral phase transition and develops a chiral condensate, breaking the symmetry down to $SU_V(2)$. Thus, the ordinary hydrodynamic theory, described by energy-momentum and charge conservation, should be expanded to include the fluctuations of the order parameter, e.g. the pions, which are the pseudo-Goldstone modes associated with the broken chiral symmetry. The resulting theory is reminiscent of a non-abelian superfluid [7, 8].

In these proceedings, we review the work of [5], where we computed the contribution of the pseudo-Goldstone modes to the hydrodynamic transport coefficents, namely the shear

*e-mail: alexander.soloviev@stonybrook.edu

viscosity and iso-vector conductivity. We will work in the mean field limit to attain a qualitative understanding of the scaling functions, while noting that recently the real time dynamics of the critical theory near the $O(4)$ critical point have been studied [9].

The organization of the paper is as follows: first, we provide a recap of the theoretical set up in Sec. 2. Then in Sec. 3 we outline the computation of the modification of the transport coefficients (here we discuss the shear viscosity. For the isovector conductivity, see [5]). Finally, we conclude in Sec. 4 by making a few phenomenological remarks on the expected soft pion enhancement compared to hydrodynamic models which do not include chiral effects.

2 Hydrodynamics near the O(4) critical point

We begin with the generating functional

$$W[g_{\mu\nu}, A_\mu, \phi_a] = \int d^3x \left[p(T) + \frac{\chi_0}{4}\mu_{ab}^2 - \frac{1}{2}\Delta^{\mu\nu} D_\mu\phi_a D_\nu\phi_a - V(\phi) \right], \tag{1}$$

where p is the pressure, μ_{ab} is the $O(4)$ chemical potential (e.g. $\mu_{0i} = \mu_A$), the $O(4)$ vector is $\phi_a = (\sigma, \pi_i)$ with the σ denoting the condensate and the π^i are the pions, the gauge covariant derivative is $D_\mu\phi = \partial_\mu\phi - iA_\mu\phi/2$, the potential is $V(\phi_\alpha\phi_\alpha) = m_0^2(t)\phi_a\phi_a + \lambda(\phi_a\phi_a)^2/4 - H_a\phi_a$, with the mass $m_0^2(t) \sim (T - T_c) \sim z$ and H_a is the applied magnetic field or quark mass, i.e. the explicit symmetry breaking term. We define the temperature via

$$T = \frac{1}{\sqrt{-\beta^\mu g_{\mu\nu}\beta^\nu}}, \tag{2}$$

where $\beta^\mu = u^\mu/T$ and u^μ is the normalized four velocity, $u^\mu u_\mu = -1$. Note that a and b denote the $O(4)$ index, Greek indices denote spacetime indices and i, j and k are spatial indices.

We compute the energy momentum tensor and the current via

$$T^{\mu\nu} = \frac{2}{\sqrt{-g}}\frac{\delta W}{\delta g_{\mu\nu}}, \quad J^\rho_{ab} = \frac{2}{\sqrt{-g}}\frac{\delta W}{\delta (A_\rho)_{ab}}. \tag{3}$$

The former is exactly conserved, while the latter is only partially conserved

$$\partial_\mu T^{\mu\nu} = 0, \quad \text{and} \quad \partial_\mu J^\mu_{ab} = \phi_a H_b - \phi_b H_a. \tag{4}$$

At ideal order, we should check that entropy, given by the Gibbs-Duhem relation,

$$s = \frac{\varepsilon + p - \frac{1}{2}\mu \cdot n}{T}, \tag{5}$$

is not being produced. We find that the entropy is conserved up to a term on the right hand side

$$\partial_\mu(su^\mu) = \frac{1}{T}[u\cdot\partial\phi_a + \mu_{ab}\phi_b]\left[\partial_\mu(\Delta^{\mu\nu}\partial_\mu\phi_a) - \frac{1}{\phi^2}\frac{\partial V}{\partial\phi^2}\phi_a + H_a)\right].$$

Requiring ideal entropy production to be zero, we see immediately the Josephson constraint

$$u\cdot\partial\phi_a + \mu_{ab}\phi_b = 0, \tag{6}$$

which encodes the coupling between the pions and the chemical potential.

Next we add dissipation via

$$T^{\mu\nu} = T^{\mu\nu}_{\text{ideal}} + \Pi^{\mu\nu} \tag{7}$$
$$J^{\mu} = J^{\mu}_{\text{ideal}} + q^{\mu} \tag{8}$$
$$u \cdot \partial\phi_a + \mu_{ab}\phi_b = \Xi_a \tag{9}$$

The requirement of positive entropy production leads to dissipative equations [5], namely the relaxation equation for the order parameter

$$u^{\mu}\partial_{\mu}\phi_a + \mu_{ab}\phi_b = \Gamma\left[\partial_{\mu}(\Delta^{\mu\nu}\partial_{\mu}\phi_a) - \frac{1}{\phi^2}\frac{\partial V}{\partial\phi^2}\phi_a + H_a)\right] + \zeta^{(1)}\phi_a\nabla_{\alpha}u^{\alpha} \tag{10}$$

and the partially conserved axial current ($\partial_{\mu}J^{\mu}_{ab} = \phi_a H_b - \phi_b H_a$)

$$J^{\mu}_{ab} = n_{ab}u^{\mu} + \Delta^{\mu\nu}(\phi_b(D_{\nu}\phi)_a - \phi_a(D_{\nu}\phi)_b) - T\sigma\Delta^{\mu\nu}\partial_{\nu}\left(\frac{\mu_{ab}}{T}\right), \tag{11}$$

For completeness, the conserved energy momentum tensor ($\partial_{\mu}T^{\mu\nu} = 0$) reads

$$\begin{aligned} T^{\mu\nu} &= (\varepsilon + p)u^{\mu}u^{\nu} + pg^{\mu\nu} + (\partial_{\mu}\phi_a)^2 - u^{\mu}u^{\nu}(u^{\sigma}\partial_{\sigma}\phi_a)^2 \\ &-\eta\sigma^{\mu\nu} - \Delta^{\mu\nu}\zeta^{(0)}\nabla\cdot u + \Delta^{\mu\nu}\zeta^{(1)}\phi_a\left[\partial_{\mu}(\Delta^{\mu\nu}\partial_{\mu}\phi_a) - \frac{1}{\phi^2}\frac{\partial V}{\partial\phi^2}\phi_a + H_a)\right], \end{aligned} \tag{12}$$

where the first line and second line are the ideal and dissipative contributions, respectively.

As mentioned in the introduction, we will work in the mean field limit with $\phi_a = (\sigma, \mathbf{0})$ and $H_a = (H, \mathbf{0})$, which is found by minimizing the potential

$$0 = \frac{dV}{d\phi} = m_0^2(t)\sigma + \lambda\sigma^3 - H. \tag{13}$$

We will call the solution of the above equation $\bar{\sigma} = \bar{\sigma}(T, H)$. We can compare our scaling function in the mean field to the one computed on the lattice [10, 11] and find that there is close qualitative agreement. Next, we linearize our equations around mean field, $\phi_a \to \phi_a + \delta\phi_a = (\bar{\sigma} + \delta\sigma, \bar{\sigma}\varphi_i)$, which leads to the following set of equations:

$$\partial_t\varphi = -\mu_A + \Gamma(\nabla^2 - m^2)\varphi, \tag{14}$$
$$\partial_t\mu_A = v^2(-\nabla^2 + m^2)\varphi + D_0\nabla^2\mu_A, \tag{15}$$
$$\partial_t\delta\sigma = \Gamma(\nabla^2 - m_{\sigma}^2)\delta\sigma. \tag{16}$$

Parameters here depend on the mean field (13) and temperature

$$v^2 = \bar{\sigma}/\chi, \qquad m^2 = H/\bar{\sigma}, \quad \Gamma = \text{const}, \quad D_0 = \text{const},$$
$$m_{\sigma}^2 = m_c^2\left(z + 3f_G^2(z)\right).$$

We can compute the propagators by going to Fourier space, adding a fictitious source and solving the resulting equation. As an example, the diffusive equation for the condensate reads

$$-i\omega + \Gamma(k^2 + m_{\sigma}^2)\delta\sigma = \Gamma H, \tag{17}$$

where H is the source. Then the retarded propagator is simply

$$G_R^{\sigma\sigma} = \frac{\Gamma}{-i\omega + \Gamma(k^2 + m_{\sigma}^2)}. \tag{18}$$

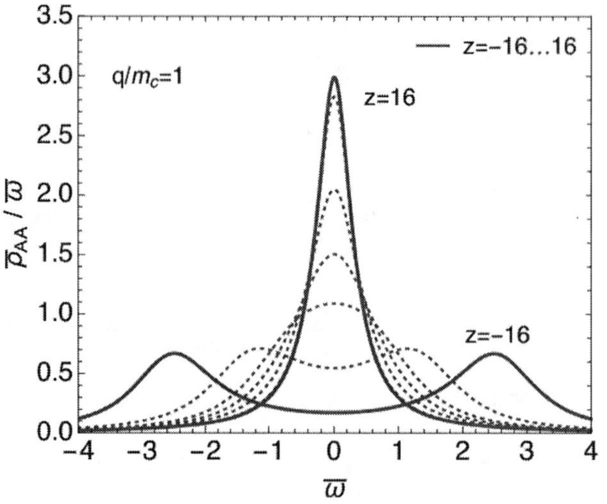

Figure 1. The spectral density as a function of frequency at various values of z above and below the phase transition. For $z \gg 0$, $\rho_{AA}/\omega \propto Dk^2/(\omega^2 + (Dk^2)^2)$, which is the diffusive peak associated with quark diffusion. For $z \ll 0$, $\rho_{AA} \propto \Gamma_k/((-\omega + \omega_k)^2 + (\Gamma_k/2)^2)$, there are two distinct poles, which can be interpreted as a pair of propagating pions. For the choice of units, see [5].

To find the symmetrized propagator, we simply need

$$G_{\mathrm{sym}} = \frac{T}{i\omega}(G_R - G_A), \tag{19}$$

where the advanced propagator is simply the complex conjugate of the retarded propagator, $G_A = (G_R)^*$. The complete set of symmetric propagators is found to be

$$\begin{aligned} G_{\mathrm{sym}} &= \begin{pmatrix} G^{\varphi\varphi}_{\mathrm{sym}} & G^{\varphi\mu}_{\mathrm{sym}} & G^{\varphi\sigma}_{\mathrm{sym}} \\ G^{\mu\varphi}_{\mathrm{sym}} & G^{\mu\mu}_{\mathrm{sym}} & G^{\mu\sigma}_{\mathrm{sym}} \\ G^{\sigma\varphi}_{\mathrm{sym}} & G^{\sigma\mu}_{\mathrm{sym}} & G^{\sigma\sigma}_{\mathrm{sym}} \end{pmatrix} \\ &= \begin{pmatrix} \frac{2T}{\chi_0}\frac{g_1(\omega^2+g_2^2)+g_2^2\omega_k^2}{(-\omega^2+\omega_k^2+g_1g_2)^2+(\omega\Gamma_k)^2} & \frac{2T}{\chi_0}\frac{-i\omega_k\omega\Gamma_k}{(-\omega^2+\omega_k^2+g_1g_2)^2+(\omega\Gamma_k)^2} & 0 \\ \frac{2T}{\chi_0}\frac{i\omega_k\omega\Gamma_k}{(-\omega^2+\omega_k^2+g_1g_2)^2+(\omega\Gamma_k)^2} & \frac{2T}{\chi_0}\frac{g_2(\omega^2+g_1^2)+g_1^2\omega_k^2}{(-\omega^2+\omega_k^2+g_1g_2)^2+(\omega\Gamma_k)^2} & 0 \\ 0 & 0 & \frac{2T\Gamma}{\omega^2+\Gamma^2(k^2+m_\sigma^2)} \end{pmatrix}, \end{aligned} \tag{20}$$

where we used the following shorthand

$$g_1 = \Gamma(k^2 + m^2), \quad g_2 = Dk^2, \quad \Gamma_k = g_1 + g_2, \quad \omega_k = v^2(k^2 + m^2). \tag{21}$$

We can see the behavior of the propagator arising from (14) in Fig. 1, where we plot the spectral density of the axial charge density, which is related to the propagator via

$$(\chi_0\omega_k)^{-2} G^{\varphi\varphi}_{\mathrm{sym}} = \frac{T}{\omega}\rho_{AA}. \tag{22}$$

The figure neatly summarizes the changing degrees of freedom through the chiral phase transition: from diffusive quarks for $z \gg 0$, i.e. $T > T_c$, to soft pions for $z \ll 0$.

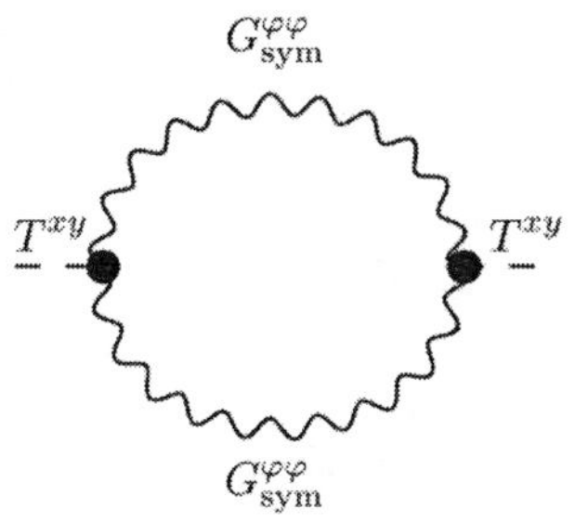

Figure 2. A part of the diagram we need to compute for the shear viscosity.

3 Modification to transport coefficients: focus on shear viscosity

In this section, we outline how to compute the correction to the shear viscosity due to the pion modes. We will compute this using the Kubo formula

$$2T\eta = \int d^4x \left\langle \frac{1}{2}\{T^{xy}(t,\boldsymbol{x}), T^{xy}(0,\boldsymbol{0})\} \right\rangle , \tag{23}$$

where the relevant component of the shear tensor (12) in the linearized regime is given by

$$\Delta T^{xy} = \frac{1}{4}\mathrm{tr}(\partial^x\Sigma\partial^y\Sigma^\dagger + \partial^y\Sigma\partial^x\Sigma^\dagger) = \partial^x\delta\sigma\partial^y\delta\sigma + \bar{\sigma}^2\partial^x\varphi_a\partial^y\varphi_a\,. \tag{24}$$

As we see, there are two contributions, one arising from the condensate and the other from the linearized pions. Thus, it is expedient to break up the computation into two pieces:

$$\Delta\eta \equiv I^{xy}_{\sigma\sigma} + I^{xy}_{\varphi\varphi}$$

Computing the hydrodynamic loop dependent on the φ propagator, $G_{\rm sym}^{\varphi\varphi}$, which is illustrated in Fig. 2, we have

$$I^{xy}_{\varphi\varphi} = 2\int \frac{d^3k}{(2\pi)^3}\frac{d\omega}{(2\pi)}(k^xk^yG^{\varphi\varphi}_{sym})^2 = \frac{2T^2}{30\pi^2\Gamma}\Lambda - \frac{2T^2m_\sigma}{32\pi\Gamma} + \ldots$$

The ellipsis denotes subleading terms in the momentum cut-off, Λ. Likewise, performing the integration dependent on the σ propagator, we find

$$I^{xy}_{\sigma\sigma} = \frac{2T^2}{30\pi^2\Gamma}\int_0^\Lambda \frac{k^6dk}{(k^2+m_\sigma^2)^3} = \frac{T^2\Lambda}{15\pi^2\Gamma} - \frac{T^2m_\sigma}{16\pi\Gamma} + \ldots \tag{25}$$

Thus, we see that we have two pieces: a divergent term which can be absorbed into the definition of the hydrodynamic shear viscosity and a universal finite piece. Thus, we are ready to state one of the main results of [5]: the shear viscosity gets a correction due to pions

$$\eta(z) = \eta_{\rm hydro} + \Delta\eta(z)\,. \tag{26}$$

The correction is plotted in Fig. 3. To understand the plot and the physics behind it, it is helpful to consider the two extreme limits. First, in the unbroken phase at large z, the propagator (20) simplifies dramatically and we have

$$\Delta\eta_\infty = -\frac{Tm_\sigma}{8\pi\Gamma}, \tag{27}$$

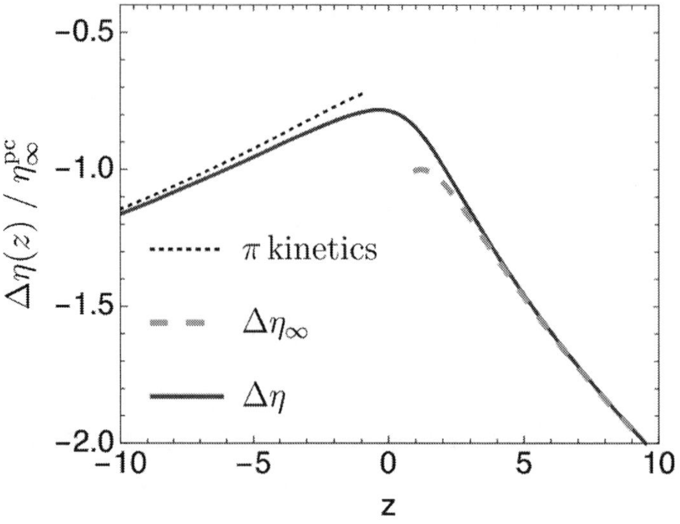

Figure 3. The correction to the shear viscosity, normalized by the value at the pseudocritical point at $z_{\rm pc} = 1.19$, which we estimate to be $4\pi\eta_\infty^{\rm pc}/s = 0.3$ [5].

which corresponds to the orange dotted line in the plot.

In the broken phase for $z \ll 0$, the relevant degrees of freedom are the soft pions. As such, one can compute the correction to the shear viscosity using another method, namely the Boltzmann equation in the relaxed time approximation for the pion distribution, f_π, in the effective fluid metric $G^{\mu\nu} = -u^\mu u^\nu + v^2(x)\Delta^{\mu\nu}$, where v^2 is the pion velocity, i.e.

$$\frac{\partial\mathcal{H}}{\partial q_\mu}\frac{\partial f_\pi}{\partial x^\mu} - \frac{\partial\mathcal{H}}{\partial x^i}\frac{\partial f_\pi}{\partial q_i} = -\Gamma_q(\omega_q f_\pi - T), \tag{28}$$

where the effective Hamiltonian is given by $\mathcal{H} = \frac{1}{2}G^{\mu\nu}q_\mu q_\nu + \frac{1}{2}v^2 m^2$. The contributions of soft pions to the transport coefficients of QCD in the broken phase were previously computed in [4].

4 Outlook to soft pion production

As a look towards the future, in this section we will make an estimate for the soft pion enhancement near the critical point. As the system passes close to the chiral phase transition, the condensate and its phase are generated, resulting in the production of pions. These pions will have small momenta and could possibly imprint a signal of chiral dynamics in the detector.

We estimate the number of soft pions via the Bose-Einstein distribution $n = 1/(e^{\omega(p)/T} - 1)$. As such, we will need to provide the soft pion dispersion curve

$$\omega^2 = v^2(p)p^2 + m_p^2(p), \tag{29}$$

which should interpolate between the two limiting dispersions: for soft pions with $q \ll \pi T$,

$$\omega_{q\ll\pi T}^2 = v_0^2 p^2 + m_{0p}^2, \tag{30}$$

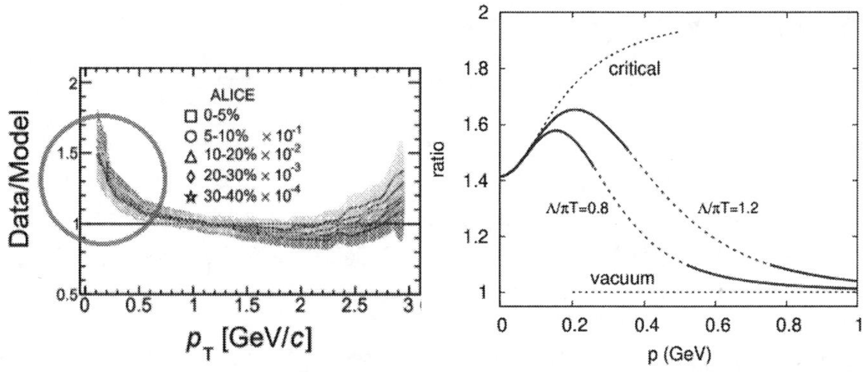

Figure 4. Left: comparison of experimental data to hydrodynamic model of pion yields from [13] with blue circle indicating the deviation for soft pions. Right: our estimate of soft pion yields using the model described in text.

while at larger momenta, we should recover the vacuum dispersion

$$\omega_{\rm vac}^2 = q^2 + m_{\rm vac}^2. \tag{31}$$

The various parameters estimated at the pseudocritical point $z_{\rm pc} = 1.19$ are as follows

$$v_0^2(z_{\rm pc}) \simeq 0.25, \quad m_{0p}(z_{\rm pc}) \simeq 0.1{\rm GeV}, \quad m_{\rm vac}^2 = 0.14{\rm GeV}, \tag{32}$$

We model the interpolation via

$$v^2(p) = c^2(1 - F(p/\Lambda) + v_0^2 F(p/\Lambda), \tag{33}$$

$$v^2(p) = m_{\rm vac}^2(1 - F(p/\Lambda) + m_{0p}^2 F(p/\Lambda), \tag{34}$$

where F is any interpolating function, which is quadratic for small momenta and approaches zero when $p \sim \Lambda$. For our present purposes, we take

$$F\left(\frac{p}{\Lambda}\right) = \frac{1}{1 + \frac{1}{2}\left(\frac{p}{\Lambda}\right)^2 + \left(\frac{p}{\Lambda}\right)^4}. \tag{35}$$

Now, we are ready to compute the ratio of the interpolation to the vacuum form, namely

$${\rm ratio} = \frac{\frac{dN_{\rm crit}}{d^3p}}{\frac{dN_{\rm vac}}{d^3p}} = \frac{e^{\omega_{\rm vac}(p)/T} - 1}{e^{\omega(p)/T} - 1}, \tag{36}$$

which we plot in Fig. 4 for a variety of cut-offs. The results are promising with the enhancement appearing in the same ballpark as the data. Of course, the model we outlined was rather crude and a more robust treatment is required, which would involve e.g. constraining dispersion via lattice, second order hydro, including resonance decays, etc. Moreover, since the parameters in (32) are Euclidean quantities, they can be measured more precisely on the lattice. In the future, there will be the upgrade to Inner Tracking System at ALICE [12], which would allow experimentalists the chance to observe more low p_T particles, especially soft pions.

Acknowledgments

The author is supported by the Austrian Science Fund (FWF), project no. J4406.

References

[1] S. Jeon, U. Heinz, Int. J. Mod. Phys. E **24**, 1530010 (2015), 1503.03931

[2] H.T. Ding et al. (HotQCD), Phys. Rev. Lett. **123**, 062002 (2019), 1903.04801

[3] O. Kaczmarek, F. Karsch, A. Lahiri, L. Mazur, C. Schmidt, *QCD phase transition in the chiral limit* (2020), 2003.07920

[4] E. Grossi, A. Soloviev, D. Teaney, F. Yan, Phys. Rev. D **102**, 014042 (2020), 2005.02885

[5] E. Grossi, A. Soloviev, D. Teaney, F. Yan, Phys. Rev. D **104**, 034025 (2021), 2101.10847

[6] K. Rajagopal, F. Wilczek, Nucl. Phys. B **399**, 395 (1993), hep-ph/9210253

[7] D.T. Son, Phys. Rev. Lett. **84**, 3771 (2000), hep-ph/9912267

[8] D.T. Son, M.A. Stephanov, Phys. Rev. D **66**, 076011 (2002), hep-ph/0204226

[9] A. Florio, E. Grossi, A. Soloviev, D. Teaney (2021), 2111.03640

[10] J. Engels, O. Vogt, Nucl. Phys. B **832**, 538 (2010), 0911.1939

[11] J. Engels, F. Karsch, Phys. Rev. D **85**, 094506 (2012), 1105.0584

[12] (2018)

[13] D. Devetak, A. Dubla, S. Floerchinger, E. Grossi, S. Masciocchi, A. Mazeliauskas, I. Selyuzhenkov, JHEP **06**, 044 (2020), 1909.10485

Quarkonia dynamics in the QGP with open quantum systems

Stéphane Delorme[1,*], *Thierry* Gousset[1], *Roland* Katz[1], and *Pol-Bernard* Gossiaux[1]

[1]SUBATECH, Université de Nantes, IMT Atlantique, IN2P3/CNRS, 4 rue Alfred Kastler, 44307 Nantes cedex 3, France

Abstract. We investigate the real-time dynamics of a correlated heavy quark-antiquark pair inside the Quark-Gluon Plasma using new quantum master equations derived from first QCD principles and based on the work of Blaizot & Escobedo [4]. The full equations are directly numerically solved in one-dimension to reduce computing costs and is used to gain insight on the dynamics in both a static and evolving medium following a Björken-like temperature evolution. The effect of the initial state on the dynamics is also studied.

1 Quarkonium suppression and recombination

A key observable for the study of the Quark-Gluon Plasma (QGP) in nucleus-nucleus (A-A) collisions is the suppression of quarkonia. It is characterized by a lower production of quarkonia in A-A collisions compared to proton-proton (p-p) collisions. To explain quarkonia suppression, several effects are at play. Due to the presence of the QGP, the interaction between the heavy quark and antiquark composing a quarkonium is modified, as numerous other color charges are present. The density of color charges also increases as temperature increases, leading to a stronger screening of the $Q\bar{Q}$ interaction and a melting of the pairs. [1] The bound $Q\bar{Q}$ pair may also scatter on the plasma constituents (light quarks and gluons), which can also lead to the dissociation of the pair. Those two effects are usually described by a complex potential, which depends on the medium temperature and the relative distance between the two quarks. The screening effects are described by the real part V of this potential and the dynamical effects are described by the imaginary part W of this potential. The combination of both screening and collisions leads to quarkonia suppression. This picture is however incomplete: one would expect a stronger suppression as the collision energy increases, however, experimental measurements at high-energy colliders (such as the LHC and the RHIC) showed a more complicated pattern. Indeed, suppression was found to be much smaller than expected for the J/Ψ with respect to some quantities [3]. A possible cause for this smaller suppression is the recombination of uncorrelated charm quarks and antiquarks during the evolution and the hadronization phase. Recombination is therefore crucial for charmonia phenomenology, as the number of c and $\bar{c}$ produced in A-A collisions is very high. In recent years, a lot of effort has been done to provide a geniune quantum treatment of the real-time dynamics of heavy quarkonia in the QGP, using the open quantum systems formalism. In that framework, one can study the dynamics of a quantum system (a quarkonium) in interaction with an environment (the QGP). As we are only interested in the $Q\bar{Q}$

*e-mail: delorme@subatech.in2p3.fr

pair dynamics, we perform a trace on the environment degrees of freedom in order to obtain a quantum master equation describing the quarkonium real-time dynamics.

2 The model

This work is based on the previous work by Blaizot & Escobedo [4]. In their article, they derived coupled quantum master equations in the quantum brownian motion regime (which is valid for temperatures $T >> 200$ MeV) by assuming a weak coupling between the plasma constituants (light quarks and gluons) and the heavy quarks forming the pair. Those equations take into account color degrees of freedom and can be written as:

$$\frac{d}{dt}\begin{pmatrix}\mathcal{D}_s\\ \mathcal{D}_o\end{pmatrix} = \mathcal{L}\begin{pmatrix}\mathcal{D}_s(\mathbf{s},\mathbf{s}',t)\\ \mathcal{D}_o(\mathbf{s},\mathbf{s}',t)\end{pmatrix}, \quad (1)$$

where $\mathcal{D}_s$ and $\mathcal{D}_o$ are respectively the singlet and octet density matrices. $\mathbf{s}$ and $\mathbf{s}'$ are respectively the relative distance between the quark and the antiquark and the conjugated variable. More specifically, one can decompose the $\mathcal{L}$ operator in the following way:

$$\mathcal{L} = \begin{pmatrix}\mathcal{L}_{ss} & \mathcal{L}_{so}\\ \mathcal{L}_{os} & \mathcal{L}_{oo}\end{pmatrix}, \quad (2)$$

with the various $\mathcal{L}_{ss}, \mathcal{L}_{so}, \mathcal{L}_{os}$ and $\mathcal{L}_{oo}$ operators describing transitions between color states. Another way of decomposing $\mathcal{L}$ (and the other color transition operators) is to write $\mathcal{L} = \mathcal{L}_0 + \mathcal{L}_1 + \mathcal{L}_2 + \mathcal{L}_3$. $\mathcal{L}_0$ describes the free motion of the pair, $\mathcal{L}_1$ involves the real part of the imaginary potential V and describes the static screening effects, $\mathcal{L}_2$ is related to the fluctuations in the medium which involves the imaginary part of the potential W) and $\mathcal{L}_3$ describes dissipative effects in the medium, which involves derivatives of W. A key property of well-defined quantum master equations is the conservation over time of the density operator positivity. In their approach, the equations derived by Blaizot & Escobedo break the positivity at the $\mathcal{L}_3$ level. In our work, we extended their equations[1], resulting in the modification of $\mathcal{L}_3$ operator and the presence of a new higher-order operator $\mathcal{L}_4$. With the new equations, positivity is preserved and the master equation is well-defined. We chose to solve the quantum master equations numerically in a one-dimensional case, in order to reduce the computing cost, as it is sufficient to gain insight on the dynamics. The resolution was performed on a bi-dimensional grid of size 20 fm × 20 fm, with a spatial step $\Delta s = 0.04$ fm and a time step $\Delta t = 0.1$ fm/c, using the Crank-Nicolson scheme.

3 One-dimensionnal dynamics of a quarkonium in the QGP

3.1 Color dynamics

In this section, we consider a QGP with a fixed temperature, set to $T = 300$ MeV and consider two cases: 1) a "QED-like" case where color transitions are forbidden and 2) a proper QCD case where color transitions are taken into account. In both cases, we take as initial state the 1S-like eigenstate, obtained from the real part V of the vacuum potential. We then perform at each time the instantaneous projections on the first three eigenstates (as if the freeze-out was instantaneous), defined as $w_n^{QED} = \langle\Psi_n^{QED}|\mathcal{D}|\Psi_n^{QED}\rangle$ in the QED-like case and $w_n^{QCD} = \langle\Psi_n^{QCD}|\mathcal{D}_s|\Psi_n^{QCD}\rangle$ in the QCD case. Figure 1 shows the evolution over time of

[1]Following a procedure described in Appendix B of [4].

the projections on the eigenstates. One can clearly see that the populations of states in the QCD case are smaller than in the QED-like case. This can be explained by the color transitions, which transfer part of the norm to octet color sector. Color transitions are therefore crucial to understand properly the dynamics of a quarkonium in the Quark-Gluon Plasma.

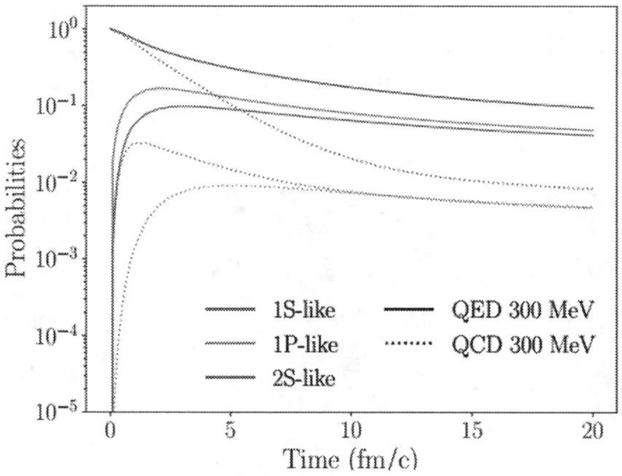

Figure 1. Evolution over time of the vacuum eigenstates populations for a temperature $T = 300$ MeV with color transitions (dashed lines) and without color transitions (solid lines)

3.2 Fixed temperature case

We first consider a medium with fixed temperature and take as the initial state the 1S-like eigenstate obtained from the vacuum potential V. We show on figure 2 the evolution over time of the populations of the first three eigenstates of the vacuum potential for three different medium temperatures $T = 200, 300$ and 400 MeV.

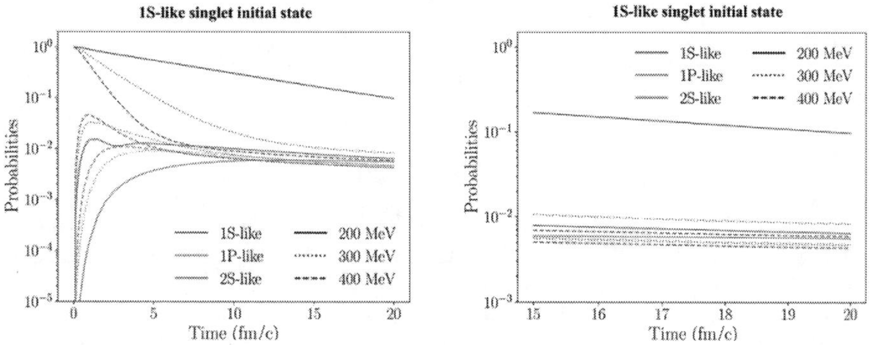

Figure 2. *Left panel:* Evolution over time of the vacuum eigenstates populations for a temperature $T = 200, 300$ and 400 MeV. *Right panel:* Zoom on the late-time evolution.

We can observe at early times a transient phase where transitions from the initial 1S-like state occur, followed by an equilibration phase of the populations. We can note that the 2S-like state is populated faster than the 1P-like state, due to the different nature of the transitions. In the case of the 2S-like state, a direct transition from the 1S-like state (due to the real part of the potential V) is possible, while for the 1P-like state, dipolar transitions must occur. Those transitions however also induce a color state transition, meaning that the transition from a 1S-like singlet state to a 1P-like singlet state requires multiple dipolar transitions, resulting in a slower population of the 1P-like state. At later times, the decay rate is the same for all three eigenstates (except for $T = 200$ MeV) and populations seems to follow a temperature distribution. However, they do not follow a Maxwell-Boltzmann distribution at the considered temperatures. The evolution of the populations is characterized by the interplay between binding, diffusion and transition between states.

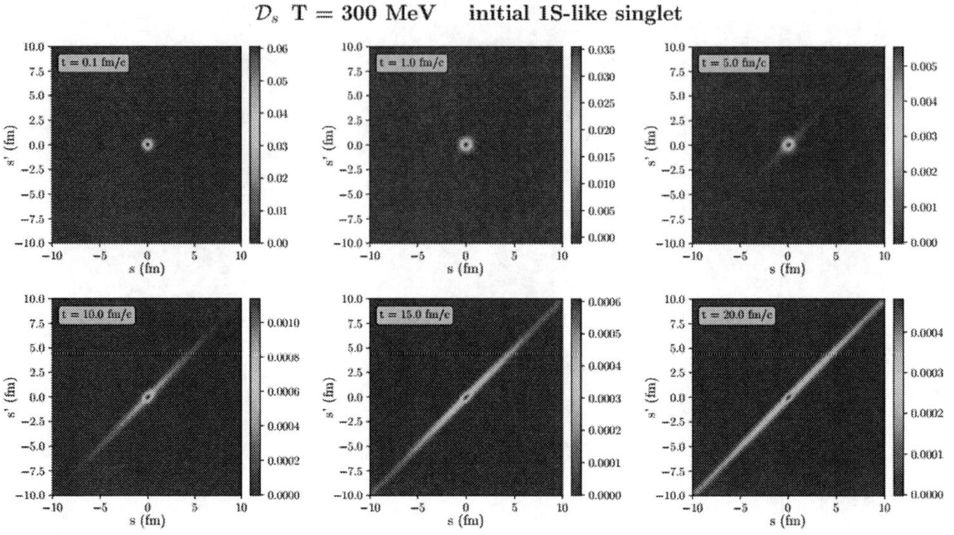

Figure 3. Evolution of the singlet density matrix $\mathcal{D}_s$ over time for a temperature $T = 300$ MeV.

Figure 3 shows the evolution of the singlet density matrix $\mathcal{D}_s$ over time. The initial state progressively delocalizes along the $s = s'$ axis and two components are observed: 1) a remnant of the initial state, present around $s = s' = 0$ fm and 2) a semi-classical-like component on the rest of the $s = s'$ axis. We can observe that the width of the distribution along the $s = -s'$ is small, indicating that the quarkonium evolution is affected by strong decoherence effects induced by the QGP.

We now change the initial state and take a P-like state in the octet color channel. The evolution over time of the state populations for that initial state is shown in Figure 4. We can observe that compared to the singlet initial state, the final populations are greater, due to the transitions from all possible octet channels. The second observation we can make is the formation of bound states due to color transitions at early times. One should however note that the two quarks are already very close to each other at the start of the evolution, which helps the early time formation.

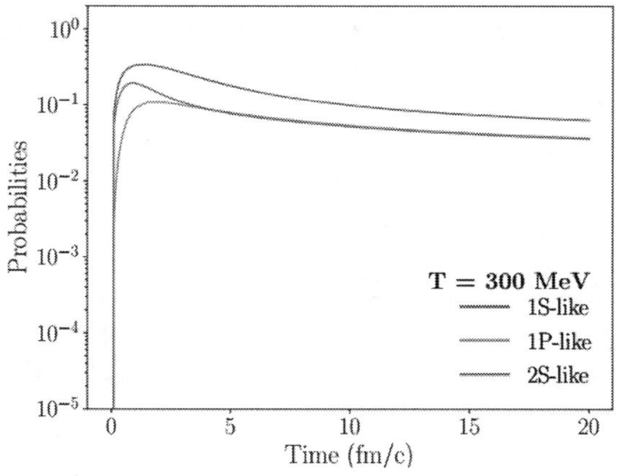

Figure 4. Evolution over time of the vacuum eigenstates populations for a temperature $T = 300$ MeV with an initial P-like octet state.

The evolution of the singlet density matrix $\mathcal{D}_s$ over time is shown in figure 5. We can see that the singlet state formed at early times from color transitions also delocalize progressively along the $s = s'$ axis. The same observations as in the color singlet initial state case can be made, indicating that the limit reached at later times may be independent of the initial state considered.

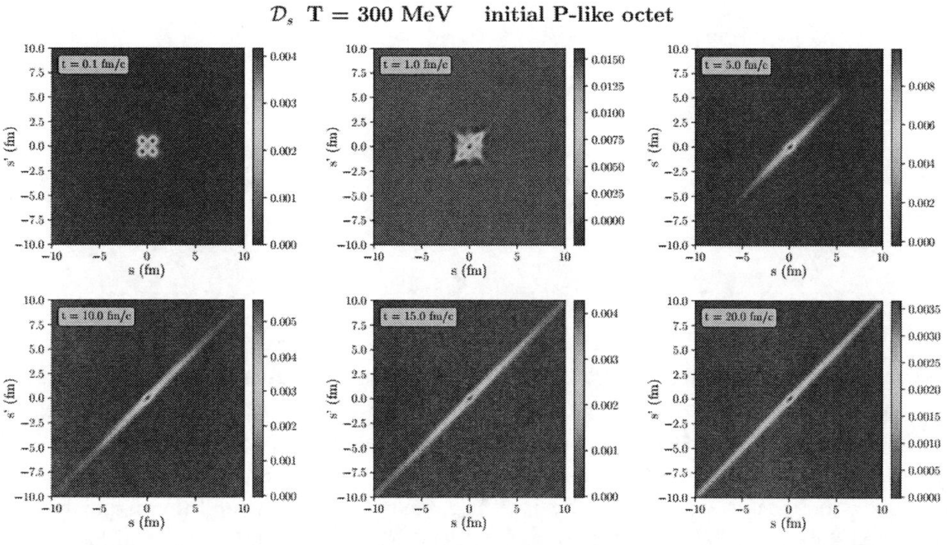

Figure 5. Evolution of the singlet density matrix $\mathcal{D}_s$ over time for a temperature $T = 300$ MeV and an octet P-like initial state.

3.3 Cooling medium case

We now relax the fixed temperature medium approximation used previously and adopt a medium temperature evolution following Björken's model [5], $T(t) = T_0 \left(\frac{t_0}{t_0+t}\right)^{1/3}$, where T_0 is the medium initial temperature and t_0 the initial time taken to be equal to 1 fm/c. We consider a QGP medium with initial temperature $T_0 = 600$ MeV and an initial P-like state in the octet color channel. We perform as before the instantaneous projection of the singlet density matrix $\mathcal{D}_s$ on the vacuum eigenstates. The results obtained are shown in figure 6. We again observe the creation at early times of excited singlet bound states and the progressive re-equilibration of the populations, followed by a decay phase. One can also note that starting from $t \sim 13$ fm/c, the 1S-like eigenstate starts to get repopulated. This can be explained by the progressive cooling of the medium, leading to the binding of excited $c\bar{c}$ pairs and $c\bar{c}$ pairs from the continuum. This effect was not present when the medium temperature was fixed, indicating that a realistic treatment of the medium is required.

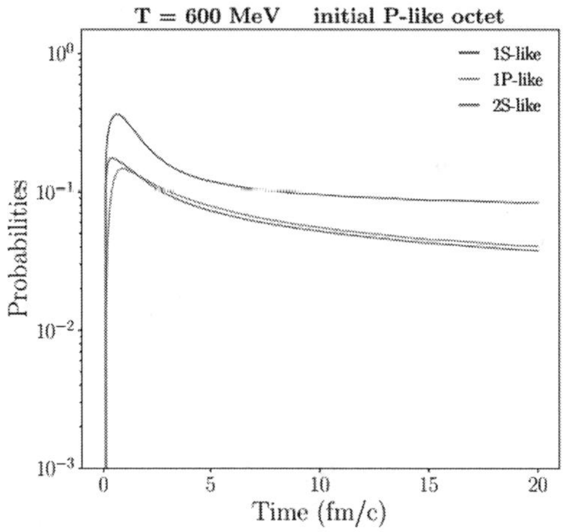

Figure 6. Evolution over time of the vacuum eigenstates populations from an initial temperature $T_0 =$ 600 MeV.

4 Conclusion

We presented results on real-time quarkonium dynamics in the quark-gluon plasma described by new coupled quantum master equations. The equations involve transitions between color states as well as dissipative effects, and were solved numerically in one dimension. The global evolution is the result of the interplay between binding effects, diffusive effects and transitions between states. The analysis of the density matrix showed the delocalization of the initial state along a specific direction in the (s, s') space and exhibited a semi-classical-like behaviour. The analysis of this semi-classical component is under way and updated results will be published in the near future.

Acknowledgements

The authors thank Jean-Paul Blaizot and Alexander Rothkopf for valuable discussions.

References

[1] T. Matsui, H. Satz, Phys. Lett. B **178**, 416 (1986)
[2] D. Lafferty, A. Rothkopf, Phys. Rev. D, **101** (2020)
[3] ALICE collaboration, Phys. Letter. B., **766**, 212-224 (2017)
[4] J.P. Blaizot, M.A. Escobedo, J. High Energ Phys. **2018**, 34 (2018)
[5] J.D. Björken, Phys. Rev. D, **27**, 140-151 (1983)

The ϕ meson production in small collision systems observed by PHENIX

Mariia Mitrankova[1,*], *Alexander* Berdnikov[1], *Yaroslav* Berdnikov[1], *Dmitry* Kotov[1], and *Iurii* Mitrankov[1]

[1]Peter the Great St.Petersburg Polytechnic University (SPbPU), Saint Petersburg, Russia

Abstract. The measurements of light hadron production in small collision systems (such as p+Al, p+Au, d+Au, ^{3}He+Au) may allow to explore the quark-gluon plasma formation and to determine the main hadronization mechanism in the considered collisions. Such research has become particularly crucial with the observation of the light hadrons collective behavior in $p/d/^3$He+Au collisions at $\sqrt{s_{NN}}$ = 200 GeV and in p+Al collisions at the same energy at forward and backward rapidities. Among the large variety of light hadrons, ϕ meson is of particular interest since its production is sensitive to the presence of the quark-gluon plasma. The paper presents the comparison of the obtained experimental results on ϕ meson production to different light hadron production in p+Al and ^{3}He+Au at $\sqrt{s_{NN}}$ = 200 GeV at midrapidity. The comparisons of ϕ meson production in p+Al, p+Au, d+Au, and ^{3}He+Au collisions at $\sqrt{s_{NN}}$ = 200 GeV at midrapidity to theoretical models predictions (PYTHIA model and default and string melting versions of the AMPT model) are also provided. The results suggest that the QGP can be formed in $p/d/^3$He+Au collisions, but the volume and lifetime of the produced medium might be insufficient for observation of strangeness enhancement effect. Conceivably, the main hadronization mechanism of ϕ meson production in p+Al collisions is fragmentation, while in $p/d/^3$He+Au collisions this process occurs via coalescence.

1 Introduction

The quantum chromodynamics (QCD) predicts [1] that at a sufficiently high temperature and / or density a state of deconfined quarks and gluons, so-called quark-gluon plasma (QGP), is formed. According to calculations based on the lattice QCD [2], the phase transition from hadron matter to the QGP occurs at the temperature of 150-200 MeV, which corresponds to the energy density of $\sim$ 1 GeV/fm^3. The experimental study of the phase transition and the QGP properties will help to understand the processes of the Universe evolution, the structure of neutron stars, and to clarify the features of the fundamental theory of strong interactions.

Collider experiments provided the evidence of a strongly interacting QGP formation in the laboratory conditions in relativistic heavy ion collisions (Au+Au, Cu+Cu, etc.) at the energy $\sqrt{s_{NN}}$ = 200 GeV [1, 3]. Recently, the signatures of the QGP existence were also observed in small collision systems such as p+Au, d+Au, and ^{3}He+Au [4]. Nonetheless, the evidences of the QGP in p+Al collisions at $\sqrt{s_{NN}}$ = 200 GeV were obtained only in

*for the PHENIX collaboration, e-mail: mashalario@gmail.com

forward and backward rapidities [5]. The comprehensive study of particle production in small collision systems at midrapidity ($|\eta| < 0.35$) can reveal the main light hadron production mechanism in these systems and the sufficient conditions for the QGP formation.

The study of ϕ meson production has been proven to be a convenient tool for the investigation of the QGP effects [6, 7]. The ϕ meson contains strange quarks ($s\bar{s}$) [8], and whereby its production in relativistic ion collisions is considered as a good probe for the study of strangeness enhancement effect [9] . Additionally, the ϕ meson has a mass comparable to the mass of the proton, and may provide additional study of the baryon enhancement effect [10].

However, not only QGP effects may influence the peculiarities of the ϕ meson production in relativistic ion collisions. Various cold nuclear matter (CNM) effects that reflect the initial state of the collision can also modify the ϕ meson production [11, 12]. Among CNM effects are multiple parton scattering, the initial state energy loss and the modification of the initial parton distribution functions in nuclei and others.

To interpret experimental results on ϕ production in p+Al, p+Au, d+Au, and ^{3}He+Au collisions at $\sqrt{s_{_{NN}}} = 200$ GeV at midrapidity, reveal the possible QGP effects and distinguish them from CNM ones the comparisons with all available experimental data and different theoretical models are provided.

As a non-QGP baselines PYTHIA [13, 14] and default version of AMPT model [15] are used. To describe the QGP effects the experimental data were compared to the predictions of the string melting version of AMPT [16].

2 Data analysis and theoretical models

Data sets used in the current study were collected by the PHENIX experiment at RHIC [17] in p+Al, p+Au, d+Au, and ^{3}He+Au collisions at $\sqrt{s_{_{NN}}} = 200$ GeV at midrapidity ($|\eta| < 0.35$). The ϕ meson production was studied via kaon decay channel.

According to Glauber model [18] in absence of the collective effects, relativistic ion collision can be presented as a superposition of elementary nucleon-nucleon collisions. However, various hot and cold nuclear effects may affect the evolution of colliding system. To study the effects affecting the particle production in ultrarelativistic collisions, nuclear modification factors R_{AB} are used [19] and are calcucated as

$$R_{AB} = \frac{d^2N_{AB}/dydp_T}{N_{coll} \cdot d^2N_{pp}/dydp_T}$$

where $d^2N_{AB}/dydp_T$ ($d^2N_{pp}/dydp_T$) is the per-event yield of particle production in $A + B$ ($p + p$) collisions and N_{coll} is a number of binary nucleon-nucleon collisions. $\langle N_{\rm coll}\rangle$ values were estimated with Glauber model using Monte-Carlo simulation.

To reveal the possible QGP effects the ϕ meson R_{AB} measured in the current analysis were compared to previous results on π^0, $\pi^\pm$, $K^\pm$, K^* and $p(\bar{p})$ production in central p+Al and ^{3}He+Au collisions at midrapidity. The parameters of analysed particles are shown in the Table 1 [8].

To theoretically interpret the obtained experimental results, PYTHIA and AMPT model calculations were used.

The PYTHIA model was developed for the generation of events in high-energy collisions considering that QGP phase is not beeing formed. Any modifications in particle production are scrutinized as a result of various CNM effects.

The flow of PYTHIA program can be roughly divided in three main stages. In the first stage the nature of the event and the physics process are determined. Then all subsequent

Table 1: Particle masses, decay channels, mean lifetimes and branching ratios

Decay	Mass, MeV/c^2	Mean lifetime, fm/c	Br, %
$\pi^0 \to \gamma\gamma$	134.9770 ± 0.0005	(2.54 ± 0.54)·10^7	98.823± 0.034
$K^* \to \pi^\pm K^\mp$	891.66 ± 0.26	4.16 ± 0.05	66.6 ± 0.1
$\phi \to K^+K^-$	1019.455 ± 0.020	46.3 ± 0.4	48.9 ± 0.5
Charged hadrons ($h^\pm$)			
$\pi^\pm$	139.57018 ±0.00035	(7.7763± 0.0001)·10^{15}	-
$K^\pm$	493.677 ± 0.016	(3.6980 ± 0.0006)·10^{15}	-
$p(\bar{p})$	938.2720813± 0.0000058	stable	-

activities on the partonic level are generated, involving initial- and final-state radiation, multiple parton–parton interactions and the structure of beam remnants. As a result of this stage a realistic partonic structure is obtained. Finally, Lund string fragmentation model [20] is used for the hadronization of obtained parton configuration.

For generation of the events in relativistic p+A and A+B collisions a new model PYTHIA/Angatyr was developed [14]. Within the framework of this model, the A + B collision is described as a superposition of elementary nucleon-nucleon collisions of different types (elastic, diffractive, absorbing).

In the current study ϕ meson R_{AB} obtained via PYTHIA model were calculated as a ratio of ϕ meson invariant spectra in A+B collision calculated with PYTHIA/Angatyr to the ϕ meson invariant spectra in p+p collision calculated with PYTHIA 8, normalized to the experimental $\langle N_{\rm coll}\rangle$ values.

The AMPT model comprises of two different configurations - default and string melting versions - and thereby provides a comprehensive study of relativistic ion collision evolution. In AMPT model particle production from initial colliding system is considered to be from either hard or soft process. In the default version of AMPT model only partons from hard processes are involved into the Zhang's Parton Cascade (ZPC) [21], and than Lund string fragmentation model [20] is implied for hadronization. In the final stage the hadron cascade based on A Relativistic Transport model for hadrons (ART) [22] is implemented.

The string melting version of AMPT model differs from the default version in the partons cascade and the hadronization mechanism. In this case, exicited strings from soft processes are melted to the partons and are involved in the ZPC along with partons from hard processes. Therefore, QGP phase is formed. Additionaly, the coalescense from the QGP [23] is used for hadronization.

The ϕ meson R_{AB} based on AMPT model were calculated as a ratio of ϕ meson invariant p_T spectra in A+B collision to the experimental p+p baseline [24] and normalized to the experimental $\langle N_{\rm coll}\rangle$ values.

3 Results

The comparison of the ϕ meson R_{AB} to π^0, $\pi^\pm$, $K^\pm$, K^* and $p(\bar{p})$ R_{AB} is shown in the Fig. 1 for the most central (0-20% centrality class [25]) p+Al and ^{3}He+Au collisions at $\sqrt{s_{NN}}$ = 200 GeV. In p + Al collisions all light hadron production show conformity at midrapidity. Nevertheless, in central ^{3}He+Au collisions protons yields are enhanced relatively to the binary scaled yields in p+p collisions, i.e. baryon enhancement might be observed. At the same time all mesons R_{AB} independently of quark content lie on the same curve, and are equal to unity within uncertainties at high-p_T. These results might draw an assumption that

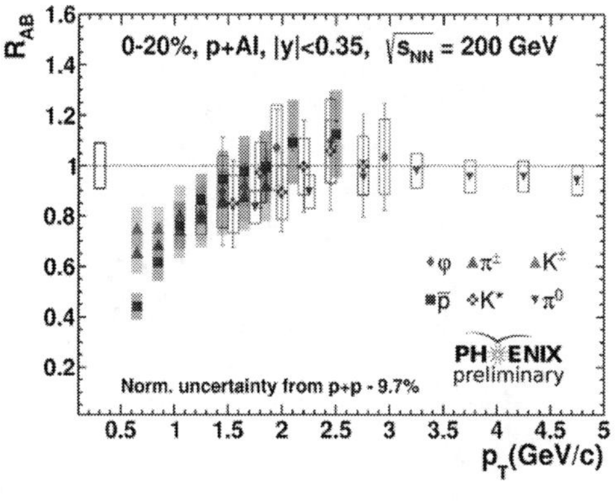

(a)

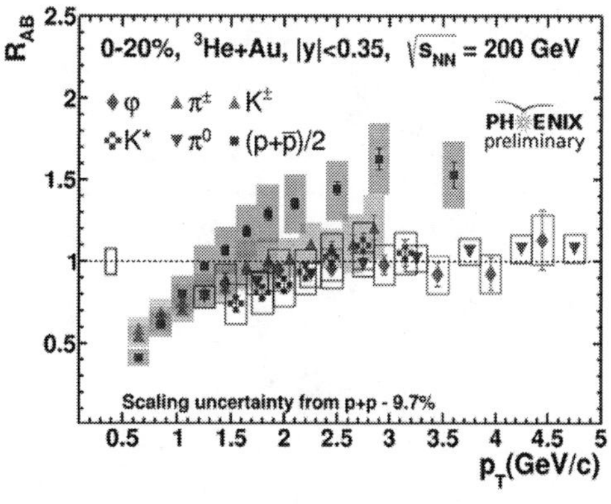

(b)

Figure 1: The comparison of ϕ meson to π^0, $\pi^\pm$, $K^\pm$, K^* and $p(\bar{p})$ nuclear modification factors in (a) p+Al and (b) ^{3}He+Au collisions at $\sqrt{s_{NN}}$ = 200 GeV at midrapidity ($|\eta| < 0.35$). The normalization uncertainty from p+p ($\sim$ 9.7%) is not shown. Here and below vertical bars correspond to statistical errors and rectangles – to systematic ones. The box near unity corresponds to the normalization uncertainty.

strangeness enhancement effect does not reveal itself in p+Al and ^{3}He+Au collisions at $\sqrt{s_{NN}}$ = 200 GeV at midrapidity.

Figure 2 shows the comparison of experimental R_{AB} results on ϕ meson production in p+Al, p+Au, d+Au [7], and ^{3}He+Au at $\sqrt{s_{NN}}$ = 200 GeV to the PYTHIA model predictions. From the figure one can seen that PYTHIA model calculations are in agreement with p+Al experimental data within uncertainties. However PYTHIA calculations fail to predict the $p/d/^3$He+Au data, having smaller ϕ meson R_{AB} values.

Figure 3 shows the comparison of experimental ϕ meson R_{AB} in p+Al, p+Au, d+Au, and ^{3}He+Au collisions at $\sqrt{s_{NN}}$ = 200 GeV to the AMPT model predictions. The ϕ meson R_{AB} in p+Al collisions are overpredicted by the string melting version of the AMPT model calculations, whereas the default version calculations demonstrate more conformity. The ϕ meson R_{AB} in $p/d/^3$He+Au collisions are in agreement with string melting version of the AMPT model calculations within uncertainties and are underestimated by default version of AMPT.

4 Summary

PHENIX has measured ϕ meson production in small collision systems (p+Al, p+Au, d+Au, and ^{3}He+Au) at $\sqrt{s_{NN}}$ = 200 GeV at midrapidity (η < 0.35). The results are compared to different light hadron production in considered collisions. The comparisons of ϕ meson R_{AB} to theoretical models predictions (PYTHIA model, default and string melting versions of the AMPT model) are also provided.

The comparison of ϕ meson R_{AB} to π^0, $\pi^\pm$, $K^\pm$, K^* and $p(\bar{p})$ R_{AB} suggests that baryon enhancement might be observed in ^{3}He+Au collisions and it seems to be absent in p+Al collisions, while strangeness enhancement effect does not reveal itself in both collision systems within uncertainties. The obtained results may indicate that particle production mechanism is different in p+Al and ^{3}He+Au collisions. Moreover it evidences in favour that QGP can

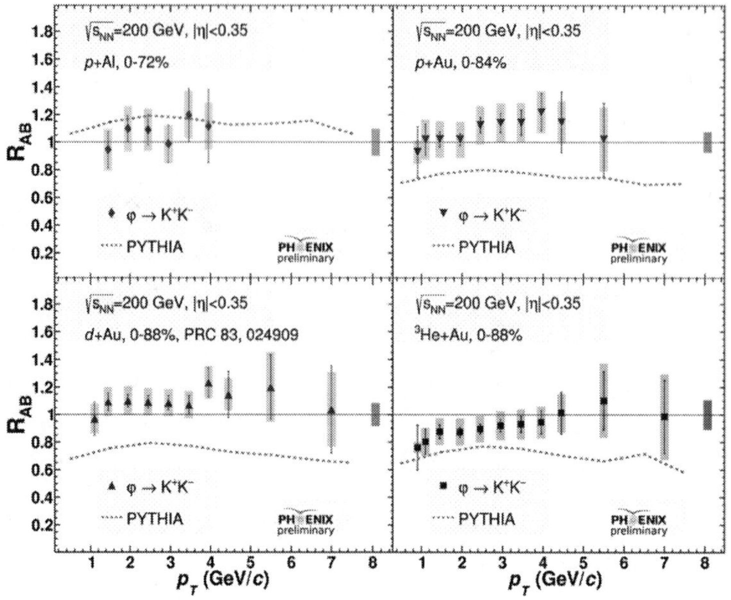

Figure 2: The comparison of ϕ meson R_{AB} in p+Al, p+Au, d+Au, and ^{3}He+Au collisions at $\sqrt{s_{NN}}$ = 200 GeV at midrapidity ($|\eta|$ < 0.35) obtained in the experiment to PYTHIA model predictions.

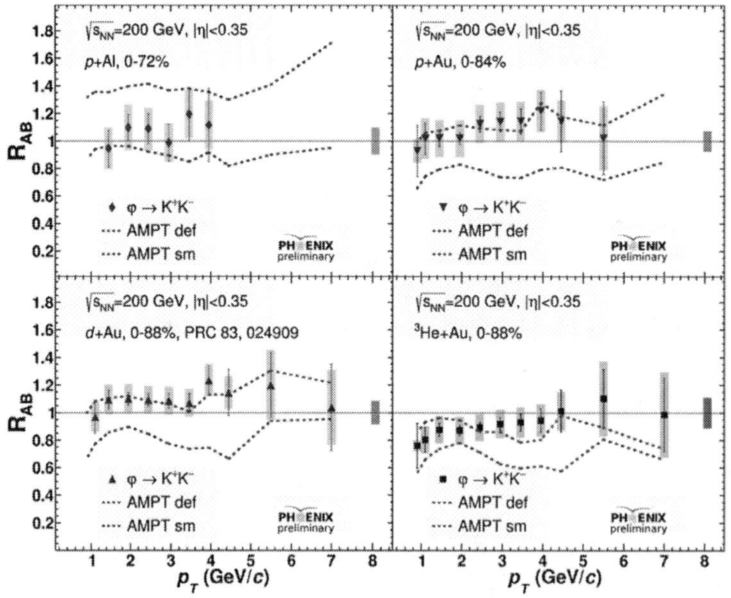

Figure 3: The comparison of ϕ meson R_{AB} in p+Al, p+Au, d+Au, and ^{3}He+Au collisions at $\sqrt{s_{NN}}$ = 200 GeV at midrapidity ($|\eta|$ < 0.35) obtained in the experiment to default (def) and string melting (sm) versions of the AMPT model predictions.

be formed in ^{3}He+Au collisions, but the volume and lifetime of the produced medium might be insufficient for strangeness enhancement effect observation.

The ϕ meson R_{AB} in p+Al collisions at midrapidity are well described with PYTHIA model and default version of AMPT model predictions. This may indicate that the main ϕ meson hadronzation mechanism is fragmentation, while the influence of coalescence mechanism is insignificant.

The ϕ meson R_{AB} in $p/d/^3$He+Au collisions at midrapidity are in agreement with predictions of the string melting version of AMPT model within uncertainties, whereas PYTHIA model and default version of AMPT model predictions underestimates the experimental data. This result evidences in favour that QGP is beeing formed in $p/d/^3$He+Au collisions at $\sqrt{s_{NN}}$ = 200 GeV and main ϕ meson production mechanism in this case might be coalescence.

Acknowledgments

We acknowledge support from Russian Ministry of Education and Science, state assignment for fundamental research (code FSEG-2020-0024).

References

[1] K. Adcox, S. Adler, S. Afanasiev, C. Aidala, N. Ajitanand, Y. Akiba, A. Al-Jamel, J. Alexander, R. Amirikas, K. Aoki et al., Nuclear Physics A **757**, 184–283 (2005)

[2] F. Karsch, Lect. Notes Phys. **583**, 209 (2002), hep-lat/0106019

[3] A. Berdnikov, Y. Berdnikov, D. Kotov, M. Larionova, I. Mitrankov, Bulletin of the Russian Academy of Sciences: Physics **84**, 1275 (2020)

[4] C. Aidala et al. (PHENIX), Nature Phys. **15**, 214 (2019), 1805.02973

[5] C. Aidala, Y. Akiba, M. Alfred, V. Andrieux, N. Apadula, H. Asano, B. Azmoun, V. Babintsev, N. Bandara, K. Barish et al., Physical Review C **101** (2020)

[6] A. Shor, Phys. Rev. Lett. **54**, 1122 (1985)

[7] A. Adare, S. Afanasiev, C.e.a. Aidala (PHENIX Collaboration), Phys. Rev. C **83**, 024909 (2011)

[8] P. Zyla et al. (Particle Data Group), PTEP **2020**, 083C01 (2020)

[9] P. Koch, B. Muller, J. Rafelski, Phys. Rept. **142**, 167 (1986)

[10] A. Adare, S. Afanasiev, C. Aidala, N.N. Ajitanand, Y. Akiba, H. Al-Bataineh, J. Alexander, A. Angerami, K. Aoki, N. Apadula et al., Physical Review C **88**, 024906 (2013)

[11] C.O. Dorso, P.A.G. Molinelli, J.I. Nichols, J.A. López, *Cold nuclear matter* (2013), 1211.5582

[12] A. Adare, C. Aidala, N.N. Ajitanand, e.a. Akiba (PHENIX Collaboration), Phys. Rev. C **92**, 044909 (2015)

[13] T. Sjöstrand, S. Ask, J.R. Christiansen, R. Corke, N. Desai, P. Ilten, S. Mrenna, S. Prestel, C.O. Rasmussen, P.Z. Skands, Computer Physics Communications **191**, 159–177 (2015)

[14] A.V. da Silva, W.M. Serenone, D.D. Chinellato, J. Takahashi, C. Bierlich, *Studies of heavy-ion collisions using pythia angantyr and urqmd* (2020), 2002.10236

[15] Z.W. Lin, C.M. Ko, B.A. Li, B. Zhang, S. Pal, Phys. Rev. C **72**, 064901 (2005)

[16] Z.W. Lin, S. Pal, C. Ko, B.A. Li, B. Zhang, Nuclear Physics A **698**, 375–378 (2002)

[17] K. Adcox et al. (PHENIX), Nucl. Instrum. Meth. A **499**, 469 (2003)

[18] M.L. Miller, K. Reygers, S.J. Sanders, P. Steinberg, Ann. Rev. Nucl. Part. Sci. **57**, 205 (2007), nucl-ex/0701025

[19] M. Mitrankova, Y. Berdnikov, A. Berdnikov, I. Mitrankov, D. Kotov, St. Petersburg State Polytechnical University Journal: Physics and Mathematics **13**, 152 (2020), cited By 1

[20] B. Andersson, G. Gustafson, G. Ingelman, T. Sjostrand, Phys. Rept. **97**, 31 (1983)

[21] B. Zhang, Computer Physics Communications **109**, 193–206 (1998)

[22] B.A. Li, C.M. Ko, Phys. Rev. C **52**, 2037 (1995)

[23] R.J. Fries, B. Muller, C. Nonaka, S.A. Bass, Phys. Rev. Lett. **90**, 202303 (2003), `nucl-th/0301087`

[24] A. Adare et al. (PHENIX), Phys. Rev. D **83**, 052004 (2011), `1005.3674`

[25] M.L. Miller, K. Reygers, S.J. Sanders, P. Steinberg, Annual Review of Nuclear and Particle Science **57**, 205–243 (2007)

A comparison of spectral reconstruction methods applied to non-zero temperature NRQCD meson correlation functions

Thomas Spriggs[1,*], *Gert* Aarts[1,2], *Chris* Allton[1], *Timothy* Burns[1], *Rachel* Horohan D'Arcy[3], *Benjamin* Jäger[4], *Seyong* Kim[5], *Maria-Paola* Lombardo[6], *Sam* Offler[1], *Ben* Page[1], *Sinead* M. Ryan[7], and *Jon-Ivar* Skullerud[3]

[1]Department of Physics, Swansea University, Swansea SA2 8PP, United Kingdom
[2]European Centre for Theoretical Studies in Nuclear Physics and Related Areas (ECT*) & Fondazione Bruno Kessler Strada delle Tabarelle 286, 38123 Villazzano (TN), Italy
[3]Department of Theoretical Physics, National University of Ireland Maynooth, County Kildare, Ireland
[4]CP3-Origins & Danish IAS, Department of Mathematics and Computer Science, University of Southern Denmark, 5230 Odense M, Denmark
[5]Department of Physics, Sejong University, Seoul 143-747, Korea
[6]INFN, Sezione di Firenze, 50019 Sesto Fiorentino (FI), Italy
[7]School of Mathematics, Trinity College, Dublin, Ireland

Abstract. We present results from the FASTSUM collaboration's programme to determine the spectrum of the bottomonium system as a function of temperature. Three different methods of extracting spectral information are discussed: a Maximum Likelihood approach using a Gaussian spectral function for the ground state, the Backus Gilbert method, and the Kernel Ridge Regression machine learning procedure. We employ the FASTSUM anisotropic lattices with 2+1 dynamical quark flavours, with temperatures ranging from 47 to 375 MeV.

1 Introduction

There has been a great deal of interest in onia systems in the context of heavy-ion collision experiments, particularly since the proposal they behave as a proxy for the temperature [1]. While initial work concentrated on the charmonium system, interest turned to bottomonium for several reasons: these mesons are produced copiously in LHC heavy-ion collision experiments, they act as probes of the quark-gluon plasma, and results from the CMS Collaboration indicate sequential suppression in this system [2].

The FASTSUM collaboration has had a long programme of studying the bottomonium spectrum at non-zero temperature using the NRQCD method [3], principally using the Maximum Entropy Method [4]. We have determined both S- and P-wave masses and determined upper bounds for the state's widths. We extend this work here to include three new analysis techniques to determine the bottomonium spectrum, including the widths of the states. These approaches are: a Maximum Likelihood approach using a Gaussian spectral function for the ground state (see sec. 3), the Backus Gilbert method (sec. 4), and the Kernel Ridge Regression machine learning procedure (sec. 5). This contribution contains a progress report on this

*e-mail: 996870@swansea.ac.uk

project, in which the long-term aim is to test and compare these and other methods, using the same lattice dataset in order to extract the best estimates of the bottomonium spectrum at finite temperature.

2 Lattice Method

Because the b-quark's mass is larger than any other mass scale, it can be approximated as a non-relativistic particle and the NRQCD effective theory can be used for its dynamics. Note that this effective theory also holds in the case of thermal QCD since the temperature reached in heavy-ion collisions is less than the b-quark mass. We use an $O(v^4)$ lattice implementation of NRQCD where $v = |\mathbf{p}|/M$ is the velocity of the b-quark in the bottomonium's rest frame. Details of the action and the tuning of the b-quark parameters can be found in [3].

NRQCD has simpler time evolution properties compared to relativistic theories, it is an "initial value" problem with the quark propagating forward in time only. The missing backward movers lead to a simpler spectral decomposition. The meson correlator, $G(\tau)$, can be expressed in terms of the spectral function, $\rho(\omega)$, via

$$G(\tau) = \int_{\omega_{min}}^{\omega_{max}} \frac{d\omega}{2\pi} K(\tau,\omega)\rho(\omega), \qquad K(\tau,\omega) = e^{-\omega\tau}. \tag{1}$$

In NRQCD the kernel $K(\tau, \omega)$ takes the simple exponential form, whereas in the relativistic case it is $\cosh[\omega(\tau - 1/2T)]/\sinh(\omega/2T)$.

The challenge is to reconstruct $\rho(\omega)$ from $G(\tau)$. Eq. (1) illustrates the "ill-posed" nature of this problem: $G(\tau)$ is typically known at $O(10-100)$ data points, whereas $\rho(\omega)$ is a continuous function which requires $O(1000)$ points to correctly represent it. This is a well-known problem and many techniques have been developed across several research fields to solve this problem. In the lattice context, it is important to compare and test these methods to ascertain which one(s) are most applicable.

FASTSUM's "Generation 2L" anisotropic lattice ensembles are used, with a physical size of $(32a_s)^3 \times (N_\tau a_\tau)$ where $a_\tau^{-1} = 5.997(34)$ GeV and $a_s/a_\tau = 3.453(6)$ [5–7]. The simulation is performed with 2+1 dynamical Wilson-clover quark flavours where the pion mass is $M_\pi = 236(2)$ MeV. There are $O(1000)$ configurations for all temperatures and the pseudocritical temperature is 162(1) MeV, as measured from the inflection point of the renormalised chiral condensate [8]. The temperatures and corresponding N_τ values are shows in Table 1.

N_τ	16	20	24	28	32	36	40	48	56	64	128
T [MeV]	375	300	250	214	187	167	150	125	107	94	47

Table 1: Temporal lattice sizes and corresponding temperatures for the FASTSUM Generation 2L ensembles [8]. The $N_\tau = 128$ ensemble was kindly provided by the Hadron Spectrum Collaboration [5–7].

To start the discussion we show in Fig.1 the effective mass in lattice units for the Υ meson, defined via

$$M_{\text{eff}}(\tau) = \log\left(\frac{G(\tau)}{G(\tau+1)}\right), \tag{2}$$

for all the temperatures considered. Some evidence of thermal effects can be observed.

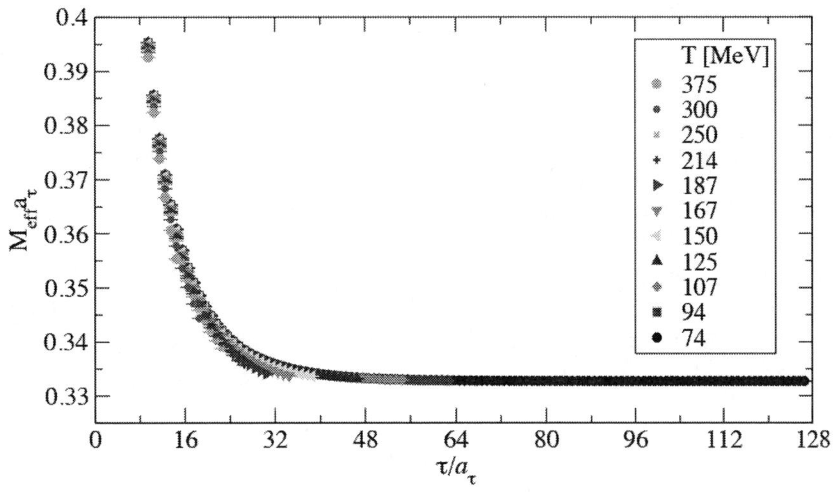

Figure 1: Effective mass plot in the ϒ channel, for all temperatures considered.

3 Gaussian Maximum Likelihood

In this approach the bottomonium spectral function is parametrised by a Gaussian ground state (see also [9]) and a single δ-function to account for all spectral weight beyond the ground state,

$$\rho_{\text{ansatz}}(\omega) = A_{\text{ground}} \exp\left(\frac{(\omega - M_{\text{ground}})^2}{2\sigma^2}\right) + A_{\text{excited}}\, \delta(\omega - M_{\text{excited}}). \quad (3)$$

The Gaussian ansatz for the ground state is chosen to allow for a finite width, which is expected in a thermal medium. The spectral weight beyond the ground state could be modelled with a function more sophisticated than a δ-function, but the extra parameters in any such function would reduce the method's predictability. Inserting Eq. (3) into Eq. (1) leads to a closed form expression for the correlation function $G_{\text{ansatz}}(\tau)$. We then apply the standard Maximum Likelihood method to determine the best fit parameters $\{M_{\text{ground/excited}}, A_{\text{ground/excited}}, \sigma\}$.

In the Maximum Likelihood method, we need to chose a time window $\tau \in [\tau_1, \tau_2]$ to perform the fits. This introduces a systematic effect which we study in Fig. 2 (left). Here the full width at half maximum (FWHM) is plotted against $1/\tau_2$ for the ϒ channel at $T = 47$ MeV with $\tau_1 = 8$ throughout. Note that we use temporal lattice units here, $a_\tau \equiv 1$, such that the τ's are integers. We expect the best estimate of the width to be given as $1/\tau_2 \to 0$, since this isolates the ground state. However, at non-zero temperature the temporal extent of the lattice is naturally restricted. As can be seen, the width heavily depends on the time window. We consider two extrapolations $1/\tau_2 \to 0$; the blue line is a linear fit to the eight leftmost

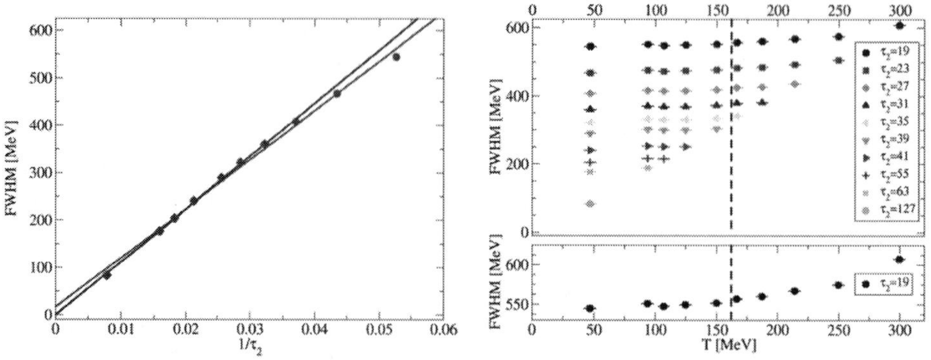

Figure 2: Left: Full width at half maximum (FWHM) for the Υ at $T = 47$ MeV as a function of the inverse time window parameter $1/\tau_2$, with $\tau_1 = 8$ fixed. The red (blue) line is a linear extrapolation on all (eight leftmost) data points. Right: FWHM of the Υ state for all temperatures for various τ_2 values. The pseudo-critical temperature, $T_{pc} = 162(1)$ MeV, is shown by vertical dashed line. The lower pane contains a close-up using $\tau_1 = 8$ and $\tau_2 = 19$ for all temperatures.

(blue) data points, whereas the red line is a linear fit to all (blue and red) points. Taking these extrapolations at face-value yields a width estimate of ~ 10 MeV.

The effect of varying the time window $\tau \in [8, \tau_2]$ is shown in Fig. 2 (right) for all temperatures. Note that results from the same time window are plotted with the same colour. The variation in the width as the fit parameter τ_2 is varied is substantial. Further details can be found elsewhere [10], including a more detailed analysis of the systematic effects. As a preliminary result we note from the right lower pane the indication that the width increases above T_{pc} when results from the same time window but at different temperatures are compared.

4 Backus Gilbert

The Backus-Gilbert method was introduced originally to solve an ill-posed problem relevant to geology [12]. It estimates a solution to Eq. (1), $\hat{\rho}(\omega)$, in a point-wise manner by sampling the target spectrum, $\rho(\omega)$, using a basis of resolution functions, $A(\omega, \omega_0)$, that are peaked around some $\omega_0 \in [\omega_{min}, \omega_{max}]$,

$$\hat{\rho}(\omega_0) = \int_{\omega_{min}}^{\omega_{max}} A(\omega, \omega_0)\rho(\omega)d\omega. \tag{4}$$

Ideally, these resolution functions are as close to the delta function $\delta(\omega - \omega_0)$ as possible. They are expressed as a linear combination of the kernel function,

$$A(\omega, \omega_0) = \sum_\tau c_\tau(\omega_0)K(\omega, \tau), \tag{5}$$

where the coefficients, $c_\tau(\omega_0)$ are to be determined. This means that, by combining Eqs. (4) and (5), the solution estimate $\hat{\rho}(\omega_0)$ is obtained from a linear combination of the original correlation functions, $G(\tau)$,

$$\hat{\rho}(\omega_0) = \sum_\tau c_\tau(\omega_0)G(\tau). \tag{6}$$

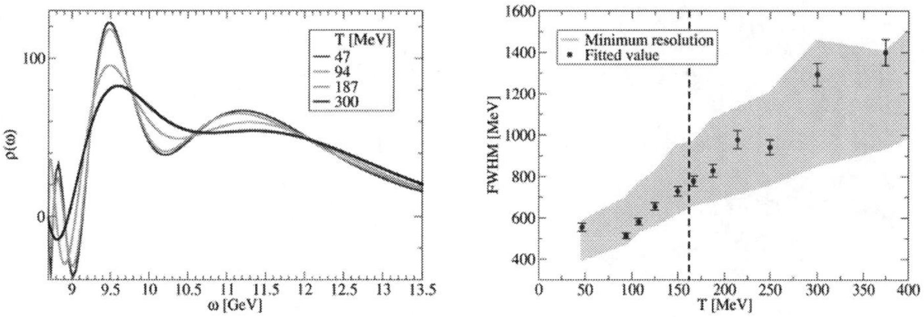

Figure 3: Left: Reconstructed spectral functions of Υ from the Backus-Gilbert method for a sample of temperatures. Right: The ground state's FWHM against temperature with the vertical dashed line indicating $T_{\rm pc}$. The grey band shows the minimum resolution, and its error band, that the resolution functions could achieve at a given temperature.

The coefficients $c_\tau(\omega_0)$ can be determined by the Dirichlet least-squares criterion of minimising the distance between the resolution functions and the delta function,

$$J(\omega_0) = \int_{\omega_{min}}^{\omega_{max}} [A(\omega, \omega_0) - \delta(\omega - \omega_0)]^2 d\omega, \tag{7}$$

leading to

$$\mathcal{K}_{\tau\tau'} \cdot c_{\tau'}(\omega_0) = K(\omega_0, \tau) \qquad \text{where } \mathcal{K}_{\tau\tau'} = \int_{\omega_{min}}^{\omega_{max}} K(\omega, \tau) K(\omega, \tau') d\omega. \tag{8}$$

The variance in the solution is also a function of $c_\tau(\omega_0)$ and is given by

$$\mathrm{Var}[\hat{\rho}(\omega_0)] = \sum_{\tau,\tau'} c_\tau \Sigma_{\tau,\tau'} c_{\tau'}, \tag{9}$$

where $\Sigma_{\tau,\tau'}$ is the covariance in $G(\tau)$.

Reconstructed spectral functions for a sample of temperatures are shown in Fig. 3 (left). From this we can see little change in the spectrum between 47 and 97 MeV, but then an increase in mass and a large apparent increase in width at larger temperatures. We note that the features below 9 GeV are artefacts of the method. In Fig. 3 (right) the FWHM is plotted against temperature where the error bars are statistical only. At first sight, there appears a clear increase in the FWHM with temperature. However, also plotted in the grey band is the FWHM of the resolution function $A(\omega, \omega_0)$ and its error bound calculated at $\omega_0 = M_{\rm ground}$. This represents the lower bound in the width that the method can resolve – the method is incapable of predicting widths below this bound, and only widths which lie above this bound can be taken as predictions. Since the bottomonium widths calculated using this method follow this lower bound, the Backus Gilbert method is currently incapable of resolving the bottomonium width. Further discussion can be found in [11, 13].

5 Kernel Ridge Regression

As is common in machine learning paradigms, kernel ridge regression (KRR) infers a prediction of a quantity based on previously observed training data. In the case of solving the

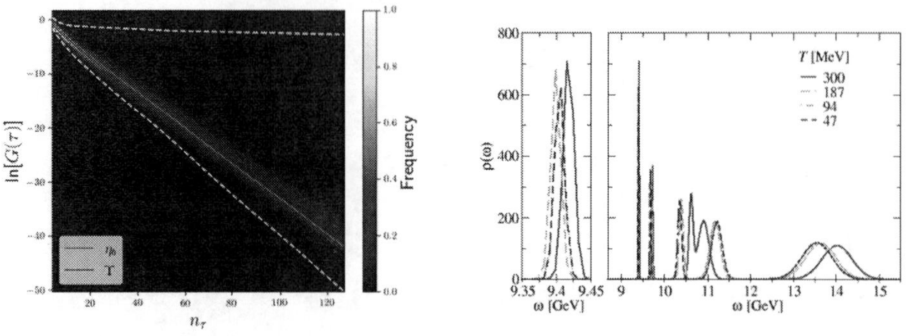

Figure 4: Left: Frequency of central values generated to form the set of mock correlation functions overlaid with the actual correlation functions, the data is constrained to lie within the dashed lines and be most frequent near the true data. Right: Reconstructed spectral functions at a sample of temperatures with a zoom in of the ground state peak highlighted to observe the temperature dependence.

inversion problem for NRQCD, mock data with known spectral functions are generated as training data.

Details of the KRR approach in the context of NRQCD can be found in [15]. Here we note that we modified our method how the input data, the set of correlators $G_i(\tau)$ (where $i = 1, 2, \ldots, N_{\text{train}}$), is combined into the matrix (or kernel) $\mathbf{C}$ before the regression procedure. In contrast to [15], we use here the following encoding for the matrix elements of $\mathbf{C}$,

$$C_{ij} = \exp\left(-\gamma \sum_n \left[\frac{G_i(\tau_n) - G_j(\tau_n)}{\overline{G}(\tau_n)}\right]^2\right), \tag{10}$$

where the sum n runs over all time slices (starting at $n = 4$), γ is a hyperparameter that sets the correlation length in the space of correlators, and $\overline{G}(\tau_n) = 1/N_{\text{train}} \sum_i G_i(\tau_n)$ is the average correlator of the training data at time τ_n.

We continue with some remarks on the generation of mock data, which has applications beyond KRR. KRR works by inferring a spectral function from a training set of mock spectral functions weighted by how much the concomitant correlation function resembles the true lattice correlation function. Clearly then, the closer the training correlation functions are to the actual lattice correlation functions obtained via numerical simulations, the more relevant training spectral functions the KRR has to draw upon to infer a spectral function. Building on [15], work has been done to increase the overlap of the training data and the lattice data. The overlap between the new training data and the actual lattice correlators is shown in Fig. 4 (left). Here, the solid lines are the lattice correlators obtained in the simulations at a given temperature ($N_\tau = 128$ in this case). The heatmap shows the distribution of the training data around these values and the dashed lines enclose the region inside which all training data lies. As can be seen, the mock data tracks the actual correlators well. More details can be found in [16].

Resulting spectral functions are shown in Fig. 4 (right). We observe ground state peaks at all four temperatures shown, with some temperature dependence visible in the left pane. It is noted that the mass of both the ground state and first excited state lie below the experimental values. Further discussion is presented in [11, 16].

6 Conclusion

This work presents a progress report of the FASTSUM collaboration's programme to apply a number of methods to reconstruct the bottomonium spectral functions from NRQCD lattice correlation functions. The three methods used are a Maximum Likelihood approach using a Gaussian spectral function for the ground state; the Backus Gilbert method; and the Kernel Ridge Regression machine learning procedure. The results presented are preliminary and indicate the substantial systematic uncertainties.

Further work to analyse these systematic effects is in progress [11].

Acknowledgements

This work is supported by STFC grant ST/T000813/1. RHD has been supported by a Maynooth University SPUR scholarship. SK is supported by the National Research Foundation of Korea under grant NRF-2021R1A2C1092701 funded by the Korean government (MEST). BP has been supported by a Swansea University Research Excellence Scholarship (SURES). This work used the DiRAC Extreme Scaling service at the University of Edinburgh, operated by the Edinburgh Parallel Computing Centre on behalf of the STFC DiRAC HPC Facility (www.dirac.ac.uk). This equipment was funded by BEIS capital funding via STFC capital grant ST/R00238X/1 and STFC DiRAC Operations grant ST/R001006/1. DiRAC is part of the National e-Infrastructure. This work was performed using PRACE resources at Cineca via grants 2015133079 and 2018194714. We acknowledge the support of the Supercomputing Wales project, which is part-funded by the European Regional Development Fund (ERDF) via Welsh Government, and the University of Southern Denmark for use of computing facilities. We are grateful to the Hadron Spectrum Collaboration for the use of their zero temperature ensemble.

References

[1] T. Matsui and H. Satz, Phys. Lett. B **178** (1986), 416-422.

[2] S. Chatrchyan *et al.* [CMS], Phys. Rev. Lett. **107** (2011), 052302 [arXiv:1105.4894 [nucl-ex]].

[3] G. Aarts, C. Allton, T. Harris, S. Kim, M. P. Lombardo, S. M. Ryan and J. I. Skullerud, JHEP **07** (2014), 097 [arXiv:1402.6210 [hep-lat]].

[4] M. Asakawa, T. Hatsuda and Y. Nakahara, Prog. Part. Nucl. Phys. **46** (2001), 459-508 [arXiv:hep-lat/0011040 [hep-lat]].

[5] R. G. Edwards, B. Joo and H. W. Lin, Phys. Rev. D **78** (2008), 054501 doi:10.1103/PhysRevD.78.054501 [arXiv:0803.3960 [hep-lat]].

[6] H. W. Lin *et al.* [Hadron Spectrum], Phys. Rev. D **79** (2009), 034502 doi:10.1103/PhysRevD.79.034502 [arXiv:0810.3588 [hep-lat]].

[7] G. K. C. Cheung *et al.* [Hadron Spectrum], JHEP **12** (2016), 089 doi:10.1007/JHEP12(2016)089 [arXiv:1610.01073 [hep-lat]].

[8] G. Aarts, C. Allton, J. Glesaaen, S. Hands, B. Jäger, S. Kim, M. P. Lombardo, A. A. Nikolaev, S. M. Ryan and J. I. Skullerud, *et al.* [arXiv:2007.04188 [hep-lat]].

[9] R. Larsen, S. Meinel, S. Mukherjee and P. Petreczky, Phys. Rev. D **100** (2019) no.7, 074506 [arXiv:1908.08437 [hep-lat]].

[10] T. Spriggs *et al.*, PoS **LATTICE2021** 077.

[11] FASTSUM, in preparation.

[12] G. Backus and F. Gilbert. Geophysical Journal International, 16(2) (1968) 169-205.

[13] B. Page *et al.*, PoS **LATTICE2021** 134.

[14] D. W. Oldenburg, "An introduction to linear inverse theory," IEEE Transactions on Geoscience and Remote Sensing, GE-22 (1984) 665-674.

[15] S. Offler, G. Aarts, C. Allton, J. Glesaaen, B. Jäger, S. Kim, M. P. Lombardo, S. M. Ryan and J. I. Skullerud, PoS **LATTICE2019** (2019), 076 [arXiv:1912.12900 [hep-lat]].

[16] S. Offler *et al.*, PoS **LATTICE2021** 509.

EPJ Web of Conferences **258**, 05012 (2022) https://doi.org/10.1051/epjconf/202225805012
A Virtual Tribute to Quark Confinement and the Hadron Spectrum (vConf21)

Finite temperature QCD phase transition and its scaling window from Wilson twisted mass fermions

A.Yu. Kotov[1,2,*], *M.P.* Lombardo[3,**], and *A.* Trunin[4,***]

[1]Jülich Supercomputing Centre, Forschungszentrum Jülich, D-52428 Jülich, Germany
[2]Bogoliubov Laboratory of Theoretical Physics, Joint Institute for Nuclear Research, Dubna, 141980 Russia
[3]INFN, Sezione di Firenze, 50019 Sesto Fiorentino (FI), Italy
[4]Samara National Research University, Samara, 443086 Russia

Abstract. We study the properties of finite temperature QCD using lattice simulations with $N_f = 2 + 1 + 1$ Wilson twisted mass fermions for pion masses from physical up to heavy quark regime. In particular, we investigate the scaling properties of the chiral phase transition close to the chiral limit. We found compatibility with $O(4)$ universality class for pion masses up to physical and in the temperature range [120 : 300] MeV. We also discuss other alternatives, including mean field behaviour or Z_2 scaling. We provide an estimation of the critical temperature in the chiral limit, $T_0 = 134^{+6}_{-4}$ MeV, which is stable against various scaling scenarios.

1 Introduction

The phase diagram of Quantum Chromodynamics (QCD) has been actively investigated for many years. Nevertheless, it still contains a lot of interesting open problems. One particular question is the interplay between confinement and chiral symmetry breaking. It has been suggested that in QCD they may be related, and this belief receives some support from the observed correlation among some chiral and gauge observables. However, as it is well known, chiral symmetry is well defined only for massless quarks, and confinement can be described as a phase transition with an associated order parameter only in pure gauge systems, or, equivalently, in the infinite quark mass limit, where quarks decouple. In the intermediate region, which comprises the physical point, both phenomena are crossovers. It is then conceivable that the two phenomena are not related, and indeed there are suggestions of a possible large splitting between deconfinement and chiral symmetry restoration phase transitions, including specific quarkyonic phase [1] at high density or a possible onset of deconfinement in QCD at temperatures higher than chiral phase transition [2, 3].

In the present study we investigate chiral properties of QCD. In particular we look at the onset of the scaling window, related to the behaviour of QCD in the chiral limit $m \to 0$. Our study involves ensembles generated with Wilson twisted mass fermions for several masses,

*e-mail: a.kotov@fz-juelich.de
**e-mail: lombardo@fi.infn.it
***e-mail: amtrnn@gmail.com

for physical to moderately high. We plan in the future to investigate the interrelation of chiral symmetry and confinement properties, further exploiting results in our set of pion masses.

The properties of QCD in the chiral limit attract a lot of attention. They are tightly related to the fate of chiral and axial symmetry and their breaking/restoration. If the $U_A(1)$ axial symmetry remains broken after the chiral phase transition, the thermal QCD phase transition is in the three dimensional $O(4)$-universality class. If it is effectively restored at the chiral phase transition, then in the chiral limit the transition is in another universality class [4] or is the first order. In the latter case, the transition remains of first order for small nonzero quark mass, ending in a second order endpoint in the Z_2 universality class, for details see review of lattice results [5, 6].

Table 1. Pion masses and lattice spacings of the ensembles used in this study.

	M140	D210	D370	B270
m_π [MeV]	139.3(7)	225(5)	383(11)	376(14)
a [fm]	0.0801(4)	0.0619(18)	0.0619(18)	0.0815(30)

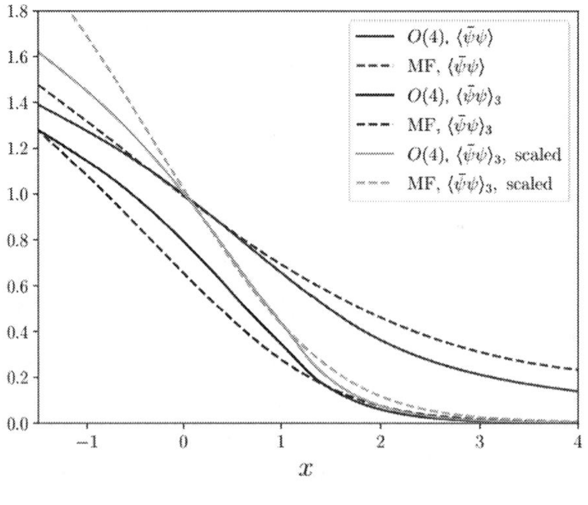

Figure 1. The universal scaling behaviour for the chiral condensate and the new order parameter $\langle\bar\psi\psi\rangle_3$ according to $O(4)$ universality class and mean field scaling. We also present rescaled dependence of $\langle\bar\psi\psi\rangle_3$, going via point (0, 1). Figure from [7].

We study the properties of QCD thermal phase transition for several pion masses starting from physical up to a heavy quark regime, $m_\pi \simeq 400$ MeV. Particular attention is drawn to the possible scaling behaviour close to the chiral limit. We discuss possible scenarios, including $O(4)$ scaling, Z_2 scaling and a possible crossover to a mean field behaviour. We also provide an estimation of critical temperature in the chiral limit, which is robust under various scenarios. In [8] it was argued that critical temperature in the chiral limit is an upper bound for the temperature of the critical end point at non-zero density[8].

First results of our study were reported in [9]. A more detailed discussion can be found in [7].

2 Lattice details and studied observables

We performed simulations with $N_f = 2 + 1 + 1$ Wilson twisted mass fermions at maximal twist. Simulations have been carried out in a fixed scale approach. A short summary of used ensembles is presented in Tab. 1.

To study the chiral phase transition we measured the chiral condensate $\langle\bar\psi\psi\rangle$ and the chiral susceptibility $\chi = \frac{\partial\langle\bar\psi\psi\rangle}{\partial m}$. In order to assess possible scaling window we introduce a new order parameter[7], combining the chiral condensate and the chiral susceptibility:

$$\langle\bar\psi\psi\rangle_3 = \langle\bar\psi\psi\rangle - m\chi, \quad (1)$$

where m is the light quark mass [10]. Note that any term linear in mass in the chiral condensate cancels out in this combination. In particular, the leading order divergence $\sim m/a^2$, as well as any regular linear terms are absent in $\langle\bar\psi\psi\rangle_3$. Taylor expansion of $\langle\bar\psi\psi\rangle_3$ in mass in the chirally symmetric phase starts from cubic terms $\sim m^3$.

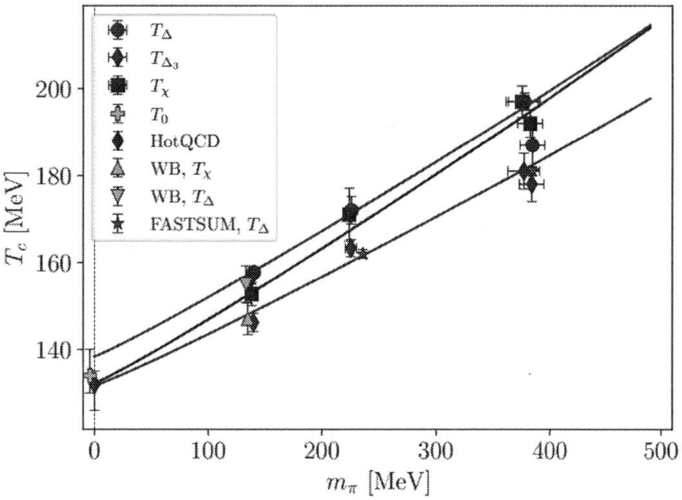

Figure 2. The dependence of the pseudocritical temperature on the pion mass extracted from chiral condendsate (red circles), chiral susceptibility (greed rombi) and new order parameter (blue squares) together with fits by $O(4)$ universality class. The results by Budapest-Wuppertal collaboration [11], HotQCD collaboration [12], FASTSUM collaboration [13, 14] are also presented. Green cross indicates the estimation of the critical temperature $T_0 = 134^{+6}_{-4}$ MeV in the chiral limit. Figure from [7].

From the universal Equation of State (EoS) for the chiral condensate

$$\frac{\langle\bar\psi\psi\rangle}{m^{1/\delta}} = f(x = t/m^{1/\beta\delta}) \quad (2)$$

one can easily obtain the EoS for the new order parameter

$$\frac{\langle\bar\psi\psi\rangle_3}{m^{1/\delta}} = f(x)(1 - 1/\delta) + \frac{x}{\beta\delta} f'(x). \quad (3)$$

Here δ and β are the critical exponents, $f(x)$ is the universal scaling function. In Fig. 1 we present the dependence of the chiral condensate and $\langle\bar\psi\psi\rangle_3$ on the scaling parameter x

Table 2. Critical temperature in the chiral limit T_0 extracted from three studied observables.

Observable	Chiral condensate	Chiral susceptibility	$\langle\bar{\psi}\psi\rangle_3$
T_0 [MeV]	138(2)	132(4)	132(3)

according to $O(4)$ universality class and mean field scaling. The pseudo-critical temperature extracted from the inflection point of the new order parameter corresponds to $x = 0.55(1)$, which is lower than the pseudocritical temperature determined by the inflection point of the chiral condensate, $x = 0.74(4)$. It implies that pseudo critical temperature determined from the new order parameter is closer to the critical temperature in the chiral limit. We also note that $\langle\bar{\psi}\psi\rangle_3$ falls off much faster as compared to the chiral condensate in the high-temperature region: $\langle\bar{\psi}\psi\rangle_3 \sim t^{-\gamma-2\beta\delta}$ and $\langle\bar{\psi}\psi\rangle \sim t^{-\gamma}$.

3 Critical temperature in the chiral limit

3.1 Scaling of pseudocritical temperatures

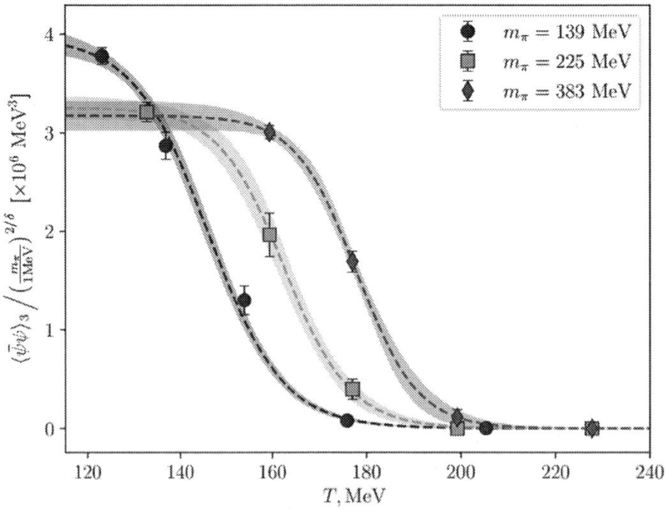

Figure 3. New order parameter divided by $m_\pi^{2/\delta}$. The crossing point for two lower pion masses provides the estimation of the critical temperature in the chiral limit $T_0 = 138(2)$ MeV.

We extract the pseudo-critical temperature from the inflection point of the chiral condensate and new order parameter $\langle\bar{\psi}\psi\rangle_3$ and from the peak of susceptibility. In Fig. 2 we present the determined pseudo-critical temperatures as functions of pion mass. Apart from our results we also show the results of other groups: Budapest-Wuppertal collaboration [11], HotQCD collaboration [12], FASTSUM collaboration [13, 14]. Lines in Fig. 2 correspond to the chiral extrapolation $m \to 0$ of our data $T_c = T_0 + Am_\pi^{2\beta\delta}$ with exponents β and δ predicted by the $O(4)$ universality class. The critical temperatures T_0 in the chiral limit obtained from the extrapolation are presented in Tab. 2. We note that critical temperatures, extracted from the chiral susceptibility and from the new observable $\langle\bar{\psi}\psi\rangle_3$, extrapolate to the same values

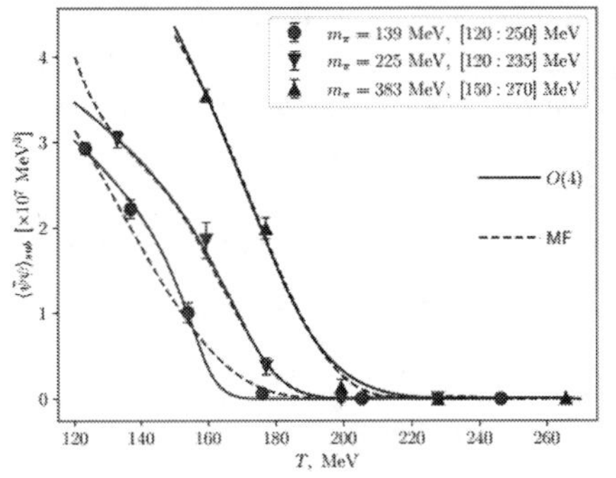

Figure 4. Fit of the new order parameter to the $O(4)$ and mean field Equation of State. Figure from [7].

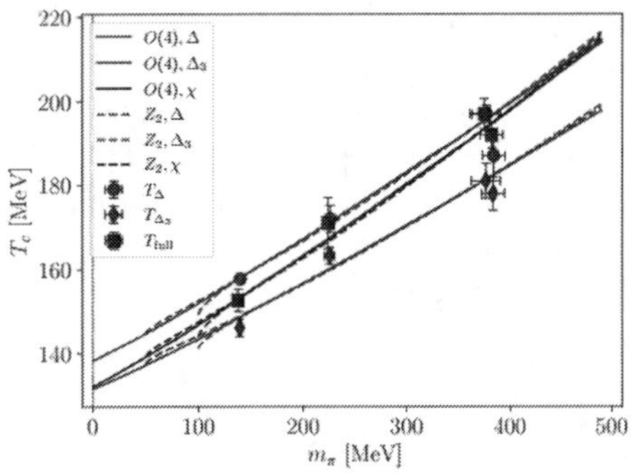

Figure 5. Pseudocritical temperatures extracted from three observables together with $O(4)$-based extrapolation and Z_2 scaling fits with fixed critical pion mass $m_\pi^{\rm cr}$ = 50 and 100 MeV. Figure from [7].

of T_0, while the critical temperature obtained by extrapolation of the data for the chiral condensate is slightly higher. It might indicate possible larger violations of scaling in the chiral condensate. As a final conservative estimation we take an average over all three extrapolated values:

$$T_0 = 134^{+6}_{-4}\ \text{MeV}. \tag{4}$$

3.2 Behaviour of $\langle\bar\psi\psi\rangle_3$

Alternative method for the estimation of the critical temperature in the chiral limit is based on the universal behaviour of $\langle\bar\psi\psi\rangle_3$. According to Eq. (3), for the critical temperature at the chiral limit $t = 0$: $\langle\bar\psi\psi\rangle_3/m^{1/\delta} \sim \langle\bar\psi\psi\rangle_3/m_\pi^{2/\delta}$ = const does not depend on the pion mass. In Fig. 3 we present the rescaled data for $\langle\bar\psi\psi\rangle_3/m^{1/\delta}$ for three studied values of the pion mass. We note that all three curves go through one point, although for the largest pion mass it might be a mere coincidence. From the crossing point of the curves for two light pion masses, we obtain another estimation of the critical temperature in the chiral limit T_0 = 138(2) MeV. Note that this number is in agreement with the estimation of T_0 obtained from the extrapolation of critical temperatures, Eq. (4).

4 Scaling behaviour

4.1 Fitting with the $O(4)$ EoS

It would be interesting to see, whether the behaviour of the new order parameter $\langle\bar\psi\psi\rangle_3$ can be described by universal scaling functions. In Fig. 4 we present $\langle\bar\psi\psi\rangle_3$ together with the fits given by the Equation of State. Following standard reasoning, one may expect that any critical behaviour would eventually cross-over to mean field. We consider two possible scenarios: $O(4)$ universality class and mean field scaling. We note that the fits describe our data. One of the parameters of the fitting function is the critical temperature in the chiral limit T_0, thus within this procedure one can determine the T_0 using the data only for one particular pion mass. The values of the T_0 determined from the fits to $O(4)$ universality class are: $T_0 = 142(2), 159(3), 174(2)$ MeV for the pion masses $m_\pi = 139$, 225 and 386 MeV, correspondingly. Only values for the lowest pion mass are compatible with values of T_0 obtained from other two procedures, possibly indicating large violations to the universal scaling behaviour for pion masses $m_\pi > m_\pi^{\rm phys}$ larger then physical. At the same time, the fits to the mean field behaviour work also pretty decently, only for the lowest pion mass there is some tension between fits and numerical results. We conclude that our data are compatible with $O(4)$ scaling for pion masses up to physical $m_\pi \lesssim m_\pi^{\rm phys}$, where for largish masses the agreement with O(4) is likely accidental and the data are simply following a mean field behaviour.

4.2 Z_2 scaling

One of the alternative possible scenarios include a first order transition terminating in a critical endpoint at a critical pion mass $m_{\rm cr}$ with Z_2 scaling. In Fig. 5 we present the pseudo-critical temperature versus pion mass extracted from all three observables together with the fits motivated by Z_2 behaviour. In the fits we constrain $m_{\rm cr}$ to two values 50 and 100 MeV. Closeness of the critical exponent for Z_2 and $O(4)$ universality classes leads to very similar, almost indistinguishable behaviour of these two fits. For this reason, we cannot exclude the possibility of Z_2 scaling with some critical pion mass up to a physical $m_{\rm cr} \lesssim m_\pi$. On the positive side, the extrapolated predictions for the (pseudo)-critical temperatures for $O(4)$ and Z_2 scenarios are very close to each other.

4.3 High temperature behaviour

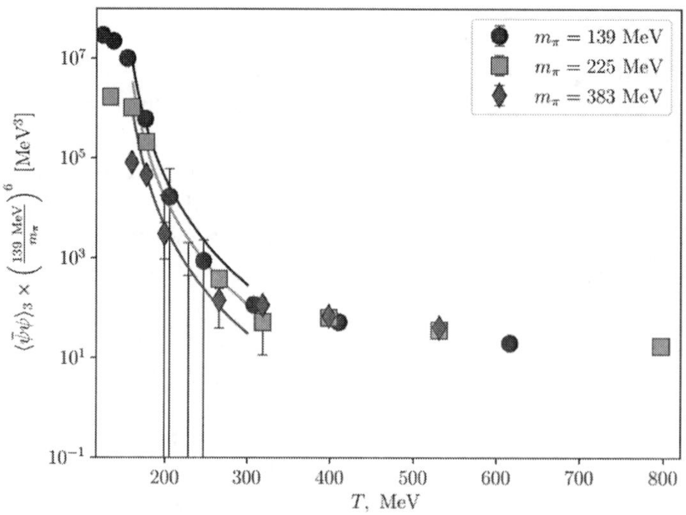

Figure 6. New order parameter $\langle\bar\psi\psi\rangle_3$ divided by m_π^6. Lines correspond to the $O(4)$ high-temperature fit in the range [160 : 300] MeV. At temperatures higher than 300 MeV points fall onto one curve, indicating behaviour described by leading order Griffith analyticity.

At large temperatures $\langle\bar{\psi}\psi\rangle_3$ falls off quite fast: $\langle\bar{\psi}\psi\rangle_3 \sim t^{-\gamma-2\beta\delta}$. In Fig. 6 we present the fits of the new observables to this function in the range [160 : 300] MeV, which describe data quite well. In the fit we have fixed critical temperature in the chiral limit $T_0 = 138$ MeV. At the temperature $T \sim 300$ MeV this behaviour changes, giving way to a leading order Griffith analyticity $\langle\bar{\psi}\psi\rangle_3 \sim m^3 \sim m_\pi^6$, thus rescaled $\langle\bar{\psi}\psi\rangle_3/m_\pi^6$ lie on one curve for temperatures $T > 300$ MeV. Remarkably, in the previous paper [15] temperature $T \sim 300$ MeV was found to be an onset of the dilute instanton gas behaviour. The existence of possible thresholds in quark-gluon plasma has been actively discussed in several works [3, 16, 17]. In particular, in [3] it was proposed that $T \sim 300$ MeV can be a pseudo-critical temperature for magnetic monopole condensation, which may be associated with the confinement-deconfinement phase transition. We plan to study the possible connection of the observed upper boundary of the scaling window with confinement properties in the future.

In Fig. 7 we sketch the $O(4)$ scaling window consistent with our results: it extends for temperatures from 120 MeV to 300 MeV and pion masses up to physical $m_\pi \lesssim m_\pi^{\text{phys}}$.

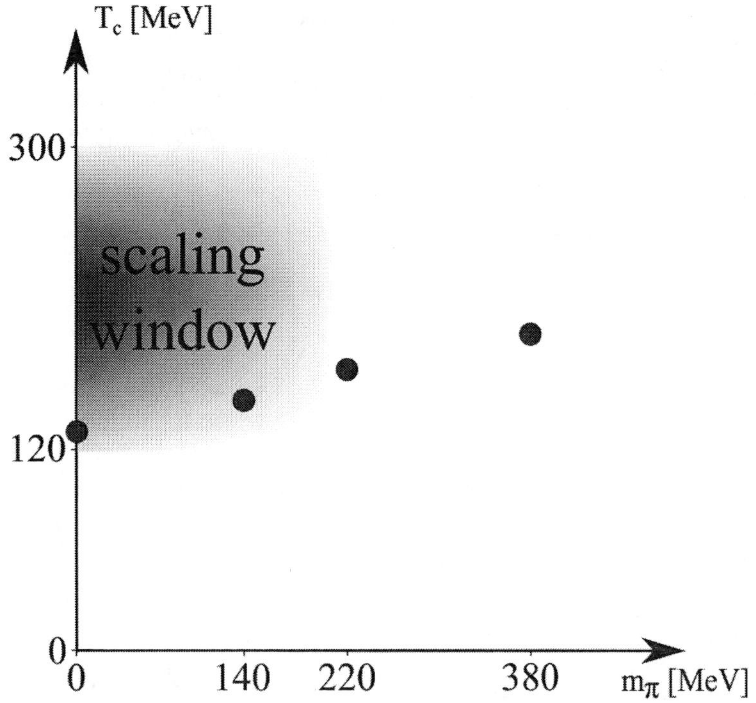

Figure 7. Sketch of the $O(4)$ scaling window consistent with the results of our simulations. Blue points indicate the (pseudo)critical temperatures extracted determined from the new observable for the studied pion masses and in the chiral limit. Figure from [7].

5 Conclusions

The properties of QCD and its thermal phase transition have been the subject of various studies. In this Proceeding we present the results of our lattice study of finite temperature

QCD with $N_f = 2 + 1 + 1$ Wilson twisted mass fermions for pion masses from physical up to heavy quark regime. We propose a novel order parameter (1), which is free from linear terms in mass and turns out to be quite useful for study of the scaling behaviour. We measured the pseudo-critical temperatures and made an extrapolation to the chiral limit, $T_0 = 134^{+6}_{-4}$ MeV. An alternative method, based on the scaling properties of the new observable, picks up slightly larger value $T_0 = 138(2)$ MeV, although compatible within errorbars. The dependence of the new order parameter on the temperature can be described by $O(4)$ universal functions, however only for the lowest pion mass $m_\pi = 139.3(7)$ MeV the extracted value of T_0 is compatible with previous estimation. At the same time, our data can be described by mean field motivated fits, only for the lowest pion mass there is a tension between the fits and our data. Based on this we conclude that our data are compatible with $O(4)$ universality class for temperatures in $[120, 300]$ MeV and pion masses up to physical, although an alternative scanario, based on Z_2 scaling, cannot be ruled out. We sketch our estimation of the $O(4)$ scaling window in Fig. 7.

Acknowledgements

The work is partially supported by STRONG-2020, a European Union's Horizon 2020 research and innovation programme under grant agreement No. 824093. A.Yu.K. and A.T. acknowledge support from RFBR grant 18-02-40126. A.T. was supported by the "BASIS" foundation. Numerical simulations were carried out using computing resources of CINECA (based on the agreement between INFN and CINECA, on the ISCRA project IsB20), the supercomputer of Joint Institute for Nuclear Research "Govorun", and the computing resources of the federal collective usage center Complex for Simulation and Data Processing for Mega-science Facilities at NRC "Kurchatov Institute", http://ckp.nrcki.ru/.

References

[1] L. McLerran, R.D. Pisarski, Nucl. Phys. A **796**, 83 (2007), 0706.2191
[2] L.Y. Glozman, Phys. Rev. D **102**, 014509 (2020), 2005.10538
[3] M. Cardinali, M. D'Elia, A. Pasqui (2021), 2107.02745
[4] A. Pelissetto, E. Vicari, Phys. Rev. D **88**, 105018 (2013), 1309.5446
[5] A.Y. Kotov, M.P. Lombardo, A. Trunin, Symmetry **13**, 1833 (2021), 2111.00569
[6] O. Philipsen, Symmetry **13**, 2079 (2021), 2111.03590
[7] A.Y. Kotov, M.P. Lombardo, A. Trunin, Phys. Lett. B **823**, 136749 (2021), 2105.09842
[8] F. Karsch, PoS **CORFU2018**, 163 (2019), 1905.03936
[9] A.Y. Kotov, M.P. Lombardo, A.M. Trunin, Eur. Phys. J. A **56**, 203 (2020), 2004.07122
[10] W. Unger, Ph.D. thesis, U. Bielefeld (main) (2010)
[11] S. Borsanyi, Z. Fodor, J.N. Guenther, R. Kara, S.D. Katz, P. Parotto, A. Pasztor, C. Ratti, K.K. Szabo, Phys. Rev. Lett. **125**, 052001 (2020), 2002.02821
[12] A. Bazavov et al. (HotQCD), Phys. Lett. **B795**, 15 (2019), 1812.08235
[13] G. Aarts et al., *Spectral quantities in thermal QCD: a progress report from the FAST-SUM collaboration*, in *37th International Symposium on Lattice Field Theory* (2019), 1912.09827
[14] G. Aarts et al., arXiv:2007.04188 (2020)
[15] F. Burger, E.M. Ilgenfritz, M.P. Lombardo, A. Trunin, Phys. Rev. D **98**, 094501 (2018), 1805.06001
[16] C. Rohrhofer, Y. Aoki, G. Cossu, H. Fukaya, C. Gattringer, L.Y. Glozman, S. Hashimoto, C.B. Lang, S. Prelovsek, Phys. Rev. D **100**, 014502 (2019), 1902.03191

[17] A. Alexandru, I. Horváth, Phys. Rev. D **100**, 094507 (2019), 1906.08047

EPJ Web of Conferences **258**, 06001 (2022) https://doi.org/10.1051/epjconf/202225806001
A Virtual Tribute to Quark Confinement and the Hadron Spectrum (vConf21)

Electromagnetic finite-size effects beyond the point-like approximation

Matteo Di Carlo[1,*], *Maxwell T.* Hansen[1,**], *Nils* Hermansson-Truedsson[2,***], and *Antonin* Portelli[1,****]

[1]Higgs Centre for Theoretical Physics, School of Physics and Astronomy, University of Edinburgh, Peter Guthrie Tait Road, Edinburgh, EH9 3FD, United Kingdom
[2]Albert Einstein Center for Fundamental Physics, Institute for Theoretical Physics, Universität Bern, Sidlerstrasse 5, 3012 Bern, Switzerland

Abstract. We present a model-independent and relativistic approach to analytically derive electromagnetic finite-size effects beyond the point-like approximation. The key element is the use of electromagnetic Ward identities to constrain vertex functions, and structure-dependence appears via physical form-factors and their derivatives. We apply our general method to study the leading finite-size structure-dependence in the pseudoscalar mass (at order $1/L^3$) as well as in the leptonic decay amplitudes of pions and kaons (at order $1/L^2$). Knowledge of the latter is essential for Standard Model precision tests in the flavour physics sector from lattice simulations.

*e-mail: matteo.dicarlo@ed.ac.uk
**e-mail: maxwell.hansen@ed.ac.uk
***Speaker, e-mail: nils@itp.unibe.ch
****e-mail: antonin.portelli@ed.ac.uk

1 Introduction

Lattice quantum chromodynamics (QCD) allows for systematically improvable Standard Model (SM) precision tests from numerical simulations performed in a finite-volume (FV), discretised Euclidean spacetime. In order to reach (sub-)percent precision in lattice predictions, also strong and electromagnetic isospin breaking corrections have to be included. The latter are encoded via quantum electrodynamics (QED), but the inclusion of QED in a FV spacetime is complicated because of Gauss' law [1]. This problem is related to zero-momentum modes of photons and the absence of a QED mass-gap. Several prescriptions of how to include QED in a finite volume have been formulated and the one used here is $\mathrm{QED_L}$ where the spatial zero-modes are removed on each time-slice. The long-range nature of QED in addition enhances the FV effects (FVEs), which typically leads to power-law FVEs that are larger than the exponentially suppressed ones for single-particle matrix elements in QCD alone.

The FVEs for a QCD+QED process depend on properties of the involved particles, including masses and charges, but also structure-dependent quantities such as electromagnetic form-factors and their derivatives. In order to analytically capture the finite-volume scaling fully, one cannot neglect hadron structure, and in the following we develop a relativistic and model-independent method to go beyond the point-like approximation at order e^2 in $\mathrm{QED_L}$.

We consider a space-time with periodic spatial extents L but with infinite time-extent. To exemplify the method, we first consider the pseudoscalar mass in Sec. 2, and then proceed to leptonic decays in Sec. 3. The discussion is based on the results in Ref. [2], and the reader is referred there for further technical details.

2 Pseudoscalar Mass

To study the finite-size scaling in the mass $m_P(L)$ of a charged hadronic spin-0 particle P, we first define the full QCD+QED infinite-volume two-point Euclidean correlation function

$$C_2^\infty(p) = \int d^4x \, \langle 0| \mathrm{T}[\phi(x)\phi^\dagger(0)] \,|0\rangle \, e^{-ipx} \,. \tag{1}$$

Here ϕ is an interpolating operator coupling to P, and $p = (p_0, \mathbf{p})$ is the momentum. We denote the finite-volume counterpart of this correlator $C_2^L(p)$, but for the moment only consider $C_2^\infty(p)$. This can be diagrammatically represented as

$C_2^\infty(p) =$ $= Z_P \cdot D(p) \cdot Z_P \,, \qquad D(p) = \dfrac{Z(p^2)}{p^2 + m_P^2} \,, \qquad Z_P = \langle 0| \phi(0) \,|P, \mathbf{p}\rangle \,, \quad (2)$

where the double-line represents the QCD+QED propagator $D(p)$, the ϕ-blob is the overlap between ϕ and P and $Z(p^2) = 1 + O(p^2 + m_P^2)$ is the residue of the propagator. Expanding $C_2^\infty(p)$ in (2) around $e = 0$ yields

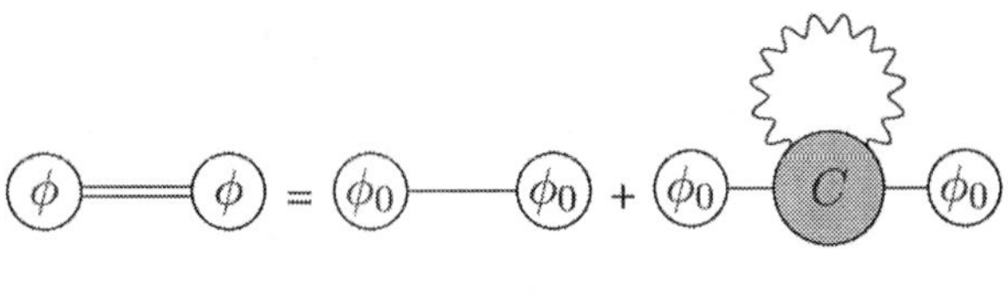 $+ O(e^4) \,, \quad (3)$

where quantities with subscript 0 are evaluated in QCD alone. The grey blob is the Compton scattering kernel defined via

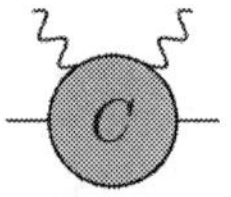

$$= C_{\mu\nu}(p,k,q) = \int d^4x\, d^4y\, d^4z\, e^{ipz+ikx+iqy} \frac{\langle 0|\, \mathrm{T}[\phi(0)J_\mu(x)J_\nu(y)\phi^\dagger(z)]\, |0\rangle}{Z_{P,0}^2\, D_0(p) D_0(p+k+q)} . \quad (4)$$

Here k and q are incoming photon momenta and $J_\mu(x)$ is the electromagnetic current. Note that the unphysical dependence on the arbitrary interpolating operator ϕ must cancel for any physical quantity, and when the external legs in $C_{\mu\nu}(p,k,q)$ go on-shell the kernel is nothing but the physical forward Compton scattering amplitude. Using (3) the electromagnetic mass-shift of the meson is readily obtained in terms of an integral over the photon loop-momentum k. One may follow an equivalent procedure for the finite-volume correlation function $C_2^L(p)$, where the integral over the spatial momentum $\mathbf{k}$ is replaced by a sum. The leading electromagnetic FVEs in the mass, $\Delta m_P^2(L)$, are thus given by the sum-integral difference

$$\Delta m_P^2(L) = -\frac{e^2}{2} \lim_{p_0^2 \to -m_P^2} \left(\frac{1}{L^3} \sum_{\mathbf{k}}{}' - \int \frac{d^3\mathbf{k}}{(2\pi)^3} \right) \int \frac{dk_0}{2\pi} \frac{C_{\mu\mu}(p,k,-k)}{k^2} \bigg|_{\mathbf{p}=0} , \quad (5)$$

where the rest-frame $\mathbf{p} = 0$ was chosen for convenience and the primed sum indicates the omission of the photon zero-mode $\mathbf{k} = 0$ in $\mathrm{QED_L}$. The analytical dependence on $1/L$ including structure-dependence can now be obtained from this formula through a soft-photon expansion of the integrand, i.e. an expansion order by order in $|\mathbf{k}|$ which is directly related to the expansion in $1/L$ via $|\mathbf{k}| = 2\pi|\mathbf{n}|/L$ where $\mathbf{n}$ is a vector of integers. The first step is to decompose $C_{\mu\nu}(p,k,q)$ into two irreducible electromagnetic vertex functions Γ_1 and Γ_2 according to

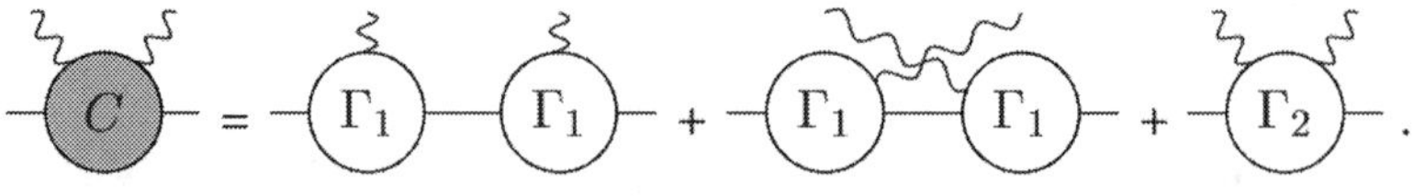

(6)

The vertex functions depend in general on the structure of the particle, as can be seen from e.g. the form-factor decomposition

$$\Gamma_1 = \Gamma_\mu(p,k) = (2p+k)_\mu\, F(k^2,(p+k)^2,p^2) + k_\mu\, G(k^2,(p+k)^2,p^2) , \quad (7)$$

where $F(k^2,(p+k)^2,p^2)$ and $G(k^2,(p+k)^2,p^2)$ are structure-dependent electromagnetic form-factors depending on three virtualities. This means that F and G contain off-shell effects, but we stress that these non-physical quantities always cancel in the FVEs. The cancellation occurs since the vertex functions $\Gamma_{1,2}$ are related to each other and the propagator $D_0(p)$ via Ward identities. An example of an off-shell relation is $F(0,p^2,-m_{P,0}^2) = Z_0(p^2)^{-1}$. The derivatives of $Z_0(p^2)$ are already known in the literature as $\delta D^{(n)}(0)$ [3] and z_n [4], but these could in principle be set to zero as they always cancel in the final results. The Ward identities further yield G as a function of F. The form-factor F also contains physical information, and for our purposes it suffices to know that $F^{(1,0,0)}(0,-m_{P,0}^2,-m_{P,0}^2) = F'(0) = -\langle r_P^2 \rangle/6$, where $\langle r_P^2 \rangle$ is the physical electromagnetic charge radius of P which is well-known experimentally [5].

Using our definitions of the vertex functions in $C_{\mu\nu}(p,k,q)$ in (5) we obtain the FVEs

$$\Delta m_P^2(L) = e^2 m_P^2 \left\{ \frac{c_2}{4\pi^2 m_P L} + \frac{c_1}{2\pi (m_P L)^2} + \frac{\langle r_P^2 \rangle}{3 m_P L^3} + \frac{C}{(m_P L)^3} + O\left[\frac{1}{(m_P L)^4} \right] \right\} , \quad (8)$$

where the c_j are finite-volume coefficients specific to $\mathrm{QED_L}$ arising from the sum-integral difference in (5). These are discussed in detail in Ref. [2]. Here we see the charge radius $\langle r_P^2 \rangle$ appearing at order $1/L^3$ and its coefficient agrees with that derived within non-relativistic scalar QED [6]. However, there is an additional structure-dependent term C related to the branch-cut of the forward, on-shell Compton amplitude. This contribution can be found, in other forms, also in Refs. [3, 7], and only arises because of the $\mathrm{QED_L}$ prescription with the subtracted zero-mode. Its value is currently unknown but one can show $C > 0$ [2], meaning that it cannot cancel the charge radius contribution. Note that all unphysical off-shell contributions from the form-factors F and G have vanished.

3 Leptonic Decays

Leptonic decay rates of light mesons are of the form $P^- \to \ell^-\bar{\nu}_\ell$, where P is a pion or kaon, ℓ a lepton and ν_ℓ the corresponding neutrino. These are important for the extraction of the Cabibbo-Kobayashi-Maskawa matrix elements $|V_{us}|$ and $|V_{ud}|$ [8, 9]. The leading virtual electromagnetic correction to this process yields an infrared (IR) divergent decay rate Γ_0. One must therefore add the real radiative decay rate $\Gamma_1(\Delta E)$ for $P^- \to \ell^-\bar{\nu}_\ell\gamma$, where the photon has energy below ΔE, to cancel the IR-divergence in Γ_0. The IR-finite inclusive decay rate is thus $\Gamma(P^- \to \ell^-\nu_\ell[\gamma])$, and following the lattice procedure first laid out in Ref. [8] we may write

$$\Gamma_0 + \Gamma_1(\Delta E_\gamma) = \lim_{L\to\infty}[\Gamma_0(L) - \Gamma_0^{\mathrm{uni}}(L)] + \lim_{L\to\infty}[\Gamma_0^{\mathrm{uni}}(L) + \Gamma_1(L, \Delta E_\gamma)]\,. \tag{9}$$

Here, Ref. [8] chose to add and subtract the universal finite-volume decay rate $\Gamma^{\mathrm{uni}}(L)$, calculated in point-like scalar QED in Ref. [4], to cancel separately the IR-divergences in Γ_0 and Γ_1. In the following we are interested in only the first term in brackets. The subtracted term $\Gamma_0^{\mathrm{uni}}(L)$ cancels the FVEs in $\Gamma_0(L)$ through order $1/L$, and hence $\Gamma_0(L) - \Gamma_0^{\mathrm{uni}}(L) \sim O(1/L^2)$. Structure-dependence enters at order $1/L^2$. With the goal of systematically improving the finite-volume scaling order by order including structure-dependence, we replace the universal contribution by

$$\Gamma_0^{\mathrm{uni}}(L) \longrightarrow \Gamma_0^{(n)}(L) = \Gamma_0^{\mathrm{uni}}(L) + \sum_{j=2}^{n} \Delta\Gamma_0^{(j)}(L)\,, \tag{10}$$

where $n \geq 2$ and $\Delta\Gamma_0^{(j)}(L)$ contains the FVEs at order $1/L^j$. This means that the finite-volume residual instead scales as $\Gamma_0(L) - \Gamma_0^{(n)}(L) \sim O(1/L^{n+1})$. We may parametrise $\Gamma_0^{(n)}(L)$ in terms of a finite-volume function $Y^{(n)}(L)$ according to

$$\Gamma_0^{(n)}(L) = \Gamma_0^{\mathrm{tree}}\left[1 + 2\frac{\alpha}{4\pi}\, Y^{(n)}(L)\right] + O\left(\frac{1}{L^{n+1}}\right), \tag{11}$$

where Γ_0^{tree} is the tree-level decay rate.

Since we are interested in the leading structure-dependent contribution we consider $Y^{(2)}(L)$. In order to derive it, we define the QCD+QED correlation function

$$C_W^{rs}(p, p_\ell) = \int d^4z\, e^{ipz}\, \langle \ell^-, \mathbf{p}_\ell, r; \nu_\ell, \mathbf{p}_{\nu_\ell}, s|\, \mathrm{T}[O_W(0)\phi^\dagger(z)]\,|0\rangle\,, \tag{12}$$

where $p_\ell = (p_\ell^0, \mathbf{p}_\ell)$ is the momentum of the on-shell lepton of mass m_ℓ, $p_{\nu_\ell} = (p_{\nu_\ell}^0, \mathbf{p}_{\nu_\ell})$ is the momentum of the massless neutrino and $O_W(0)$ is the four-fermion operator of the decay in question. We may diagrammatically represent this in a similar way as for the mass according to

$C^{rs}_W(p, p_\ell) =$ 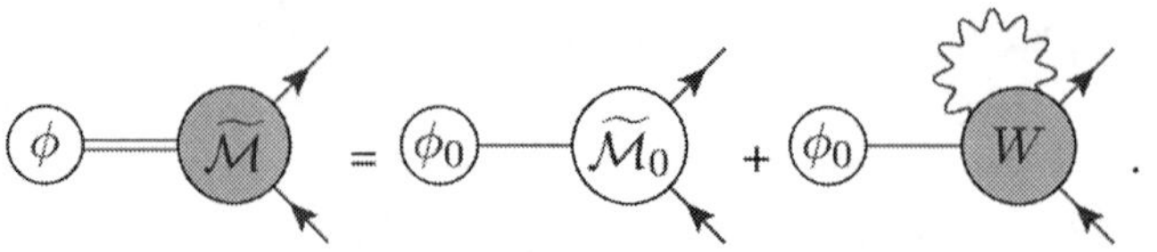 (13)

The grey blob containing W is of order e^2 and can be separated, just like the Compton amplitude, into several irreducible vertex functions. The exact definitions of these vertex functions are quite involved and can be found in Ref. [2], but several comments can be made. First of all, the vertex functions are related to various structure-dependent form-factors containing both on-shell and off-shell information. Again, the off-shellness must cancel. The vertex functions also contain physical structure-dependent information (similar to how Γ_1 depends on the charge radius) and for $Y^{(2)}(L)$ this is the axial-vector form-factor $F_A(-m_P^2) = F_A^P$ from the real radiative decay $P^- \to \ell^-\bar{\nu}_\ell\gamma$.

By performing the amputation on the external meson leg in (12) to obtain the matrix element needed for the decay rate in (11), one finds the finite-volume function $Y^{(2)}(L)$ to be

$$\begin{aligned} Y^{(2)}(L) = {} & \frac{3}{4} + 4\log\left(\frac{m_\ell}{m_W}\right) + 2\log\left(\frac{m_W L}{4\pi}\right) + \frac{c_3 - 2\,(c_3(\mathbf{v}_\ell) - B_1(\mathbf{v}_\ell))}{2\pi} - \\ & - 2A_1(\mathbf{v}_\ell)\left[\log\left(\frac{m_P L}{2\pi}\right) + \log\left(\frac{m_\ell L}{2\pi}\right) - 1\right] - \frac{1}{m_P L}\left[\frac{(1+r_\ell^2)^2\,c_2 - 4\,r_\ell^2\,c_2(\mathbf{v}_\ell)}{1-r_\ell^4}\right] + \\ & + \frac{1}{(m_P L)^2}\left[-\frac{F_A^P}{f_P}\,\frac{4\pi\,m_P\,[(1+r_\ell^2)^2\,c_1 - 4\,r_\ell^2\,c_1(\mathbf{v}_\ell)]}{1-r_\ell^4} + \frac{8\pi\,[(1+r_\ell^2)\,c_1 - 2\,c_1(\mathbf{v}_\ell)]}{(1-r_\ell^4)}\right]. \end{aligned} \tag{14}$$

Here, $r_\ell = m_\ell/m_P$, $\mathbf{v}_\ell = \mathbf{p}_\ell/E_\ell$ the lepton velocity in terms of the energy E_ℓ, and m_W the W-boson mass. Also, c_k, $A_1(\mathbf{v}_\ell)$, $B_1(\mathbf{v}_\ell)$ and $c_j(\mathbf{v}_\ell)$ are finite-volume coefficients defined in Ref. [2]. Note that no unphysical quantities appear. At order $1/L^2$, there is one structure-dependent contribution proportional to F_A^P and the other term is purely point-like. This result is in perfect agreement with Ref. [4] for the universal terms up to $O(1/L)$, which we derived in a completely different approach. The numerical impact of the $1/L^2$-corrections is studied in Ref. [2].

4 Conclusions

We have presented a relativistic and model-independent method to derive electromagnetic FVEs beyond the point-like approximation. We are currently working to obtain the leading FVEs for semi-leptonic kaon decays, relevant for future precision tests in the SM flavour physics sector.

Acknowledgements

M.D.C, M.T.H., and A.P. are supported in part by UK STFC grant ST/P000630/1. Additionally M.T.H. is supported by UKRI Future Leader Fellowship MR/T019956/1. A.P. additionally received funding from the European Research Council (ERC) under the European Union's Horizon 2020 research and innovation programme under grant agreements No 757646 & 813942. N. H.–T. is funded by the Albert Einstein Center for Fundamental Physics at the University of Bern.

References

[1] M. Hayakawa, S. Uno, Prog. Theor. Phys. **120**, 413 (2008), 0804.2044

[2] M. Di Carlo, M.T. Hansen, N. Hermansson-Truedsson, A. Portelli (2021), 2109.05002

[3] S. Borsanyi et al., Science **347**, 1452 (2015), 1406.4088

[4] V. Lubicz, G. Martinelli, C.T. Sachrajda, F. Sanfilippo, S. Simula, N. Tantalo, Phys. Rev. **D95**, 034504 (2017), 1611.08497

[5] P.D. Group, P.A. Zyla et al., Progress of Theoretical and Experimental Physics **2020** (2020), 083C01, https://academic.oup.com/ptep/article-pdf/2020/8/083C01/33653179/ptaa104.pdf

[6] Z. Davoudi, M.J. Savage, Phys. Rev. **D90**, 054503 (2014), 1402.6741

[7] N. Tantalo, V. Lubicz, G. Martinelli, C.T. Sachrajda, F. Sanfilippo, S. Simula, PoS **LATTICE2016** (2016), 1612.00199

[8] N. Carrasco, V. Lubicz, G. Martinelli, C. Sachrajda, N. Tantalo, C. Tarantino, M. Testa, Phys. Rev. D **91**, 074506 (2015), 1502.00257

[9] M. Di Carlo, D. Giusti, V. Lubicz, G. Martinelli, C. Sachrajda, F. Sanfilippo, S. Simula, N. Tantalo, Phys. Rev. D **100**, 034514 (2019), 1904.08731

$N\bar{N}$ interaction from chiral effective field theory and its application to neutron-antineutron oscillations

Johann Haidenbauer[1,*] and *Ulf-G.* Meißner[1,2,**]

[1]Institute for Advanced Simulation, Institut für Kernphysik and Jülich Center for Hadron Physics, Forschungszentrum Jülich, D-52425 Jülich, Germany
[2]Helmholtz-Institut für Strahlen- und Kernphysik and Bethe Center for Theoretical Physics, Universität Bonn, D-53115 Bonn, Germany

Abstract. An $\bar{N}N$ potential is introduced which is derived within chiral effective field theory and fitted to up-to-date $\bar{N}N$ phase shifts and inelasticities, provided by a proper phase-shift analysis of available $\bar{p}p$ scattering data. As an application of this interaction neutron-antineutron oscillations in the deuteron are considered. In particular, results for the deuteron lifetime are presented, evaluated in terms of the free-space $n-\bar{n}$ oscillation time, utilizing that $\bar{N}N$ potential together with an NN interaction likewise derived within chiral effective field theory.

1 Introduction

Recently the Jülich-Bonn group has established a high-precision $\bar{N}N$ potential within chiral effective field theory (EFT) [1]. Starting point for that work was Ref. [2] where a new generation of NN potentials derived in the framework of chiral EFT was presented. In that publication a novel local regularization scheme was introduced and applied to the pion-exchange contributions of the NN force. Furthermore, an alternative scheme for estimating the theoretical uncertainty was proposed that no longer depends on a variation of the cutoffs [3, 4]. Those concepts were adopted and implemented in the study of the $\bar{N}N$ interaction [1]. Specifically, an $\bar{N}N$ potential was derived up to next-to-next-to-next-to-leading order (N^{3}LO) in the perturbative expansion, thereby extending a previous work by the Jülich-Bonn group that had considered the $\bar{N}N$ force up to N^{2}LO [5]. In both cases the strength parameters of the contact terms that arise in the EFT framework [4, 6] (i.e. the low-energy constants or simply LECs) have been fixed by a fit to the phase shifts and inelasticities provided by a proper phase-shift analysis of $\bar{p}p$ scattering data [7].

As an application of this interaction neutron-antineutron oscillations in the deuteron have been considered [8]. Neutron-antineutron ($n-\bar{n}$) oscillations involve a change of the baryon number (B) by two units ($|\Delta B| = 2$). An experimental observation would allow a glimpse on physics beyond the standard model, see e.g. [9–11]. Since in such oscillations B is violated the process satisfies one of the Sakharov conditions [12] that have been formulated in order to explain the observation that there is more matter than anti-matter in the universe [13].

*e-mail: j.haidenbauer@fz-juelich.de
**e-mail: meissner@hiskp.uni-bonn.de

The key quanitity in this subject is the free $n-\bar{n}$ oscillation time, $\tau_{n-\bar{n}}$. The presently best experimental limit on it is $\tau_{n-\bar{n}} > 0.86 \times 10^8$ s ≈ 2.7 yr (with 90 % C.L.) [14]. Additional information can be deduced from studies of $n-\bar{n}$ oscillations in a nuclear environment. Corresponding experiments have been performed, e.g., for ^{56}Fe [15], ^{16}O [16], and for the deuteron (^{2}H) [17], while others are planned [18]. In such a case the oscillation process is suppressed as compared to the free situation. The pertinent lifetime $\tau_{\rm nuc}$ is commonly expressed in terms of the one in free space as [10]

$$\tau_{\rm nuc} = R\,\tau^2_{n-\bar{n}}\,, \tag{1}$$

where R is an intranuclear suppression factor, also called reduced lifetime, that depends on the specific nucleus. It can be calculated from nuclear theory and then can be used to relate the measured lifetimes of those nuclei with the free $n-\bar{n}$ oscillation time [10], see, e.g., Refs. [19–23].

For a long time the suppression factors published in 1983 [20] have been used as standard by experimentalists in the interpretation of their measurements [15, 17]. For example, in case of the deuteron the corresponding value is $R \sim (2.40-2.56)\times 10^{22}$ s^{-1}, a prediction based on the phenomenological antinucleon-nucleon ($\bar{N}N$) potentials by Dover and Richard [24, 25]. Recently, however, those values have been challenged in a work by Oosterhof et al. [26]. In that study an effective field theory for the $|\Delta B| = 2$ interaction is constructed and the quantity R is evaluated within the power counting scheme proposed by Kaplan, Savage, and Wise (KSW) [27, 28] for the nucleon-nucleon (NN) and $\bar{N}N$ interactions. The value of R for the deuteron obtained in that approach is $(1.1 \pm 0.3) \times 10^{22}$ s^{-1}, about a factor 2 smaller than the one by Dover et al. [20].

In the light of this controversal situation a new calculation of the suppression factor for the deuteron has been performed by us [8]. It was prompted by the aim to utilize the discussed modern chiral interactions for the involved NN [2] and $\bar{N}N$ [1] systems. Specifically, in case of the $\bar{N}N$ interaction most of the available precise $\bar{p}p$ scattering data (mostly from the LEAR facility at CERN [29]) have appeared only after the publication of the potentials used in Ref. [20]. Therefore, an update is long overdue. Of course, the main motivation was the aforementiond discrepancy reported in Ref. [26] and the prospect to find a plausible explanation for that difference.

The paper is structured in the following way: In Sect. 2 the employed $\bar{N}N$ potential is introduced and some results for $\bar{n}p$, relevant for the calculation of $n-\bar{n}$ oscillations in the deuteron, are presented. In Sect. 3 a basic description of the formalism for evaluating the $n-\bar{n}$ oscillations in the deuteron is provided. Results for the oscillations, specifically for the suppression factor R, are presented in Sect. 4. The paper closes with a brief summary.

2 The $\bar{N}N$ interaction in chiral EFT

A detailed description of the derivation of the chiral $\bar{N}N$ potential using the Weinberg power counting can be found in Ref. [1]. For information on the NN potential employed in the present calculation see Ref. [2]. As already indicated above, the chiral potential contains pion exchanges and a series of contact interactions with an increasing number of derivatives. Up to N^{3}LO there are contributions from one-, two- and three-pion exchanges. Those are identical to the ones that appear in the NN potential [2]. However, there is a sign change in case of an odd number of exchanged pions due to its negative G parity, i.e. $V^{\bar{N}N} = -V_\pi + V_{2\pi} - V_{3\pi}$, with V_π etc. being the corresponding contributions to the NN force. On the other hand, the contact interaction, $V^{\bar{N}N}_{cont}$, cannot be taken over simply from the NN case. Those contact terms represent effectively the short-range part of the interaction and, therefore, the G parity of the indivudal contributions remains unresolved. Thus, the strength parameters associated with the arising contact terms, the LECs, need to be determined in a fit to $\bar{N}N$ data. We fix them by fitting to the phase shifts and inelasticity parameters of the PWA of Zhou and Timmermans [7]. How this is done is described in detail in Ref. [1]. Note that there are more independent LECs in the $\bar{N}N$ case than in NN because in the former system there is no restriction from the Pauli principle. In addition, in our approach $\bar{N}N$ annihilation (into multi-meson channels) is likewise parameterized by contact terms, see Refs. [1, 5] for explicit expressions.

Once the potential is established, the reaction amplitude is obtained from the solution of a relativistic Lippmann-Schwinger (LS) equation. It reads in partial-wave projected form

$$\begin{aligned} T_{L''L'}(p'',p';E_k) &= V_{L''L'}(p'',p') \\ &+ \sum_L \int_0^\infty \frac{dpp^2}{(2\pi)^3} V_{L''L}(p'',p) \frac{1}{2E_k - 2E_p + i0^+} T_{LL'}(p,p';E_k) \,. \quad (2) \end{aligned}$$

Here p'' and p' are the (moduli of the) center-of-mass $\bar{N}N$ momenta in the final and initial states, respectively, and $E_k = \sqrt{M_p^2 + k^2}$, where k is the on-shell momentum. We adopt a relativistic scattering equation so that our amplitudes fulfill the relativistic unitarity condition at any order, as done also in the NN sector [3, 6]. On the other hand, relativistic corrections to the potential are calculated order by order. They appear first at N^{3}LO in the Weinberg scheme, see Appendix A in Ref. [1]. L, L', etc., specifies the orbital angular momentum, considering that the $\bar{N}N$ system can be in uncoupled (spin-singlet or triplet) states where $L'' = L' = L = J$ or in coupled partial waves with $L'', L', L = J-1, J+1$.

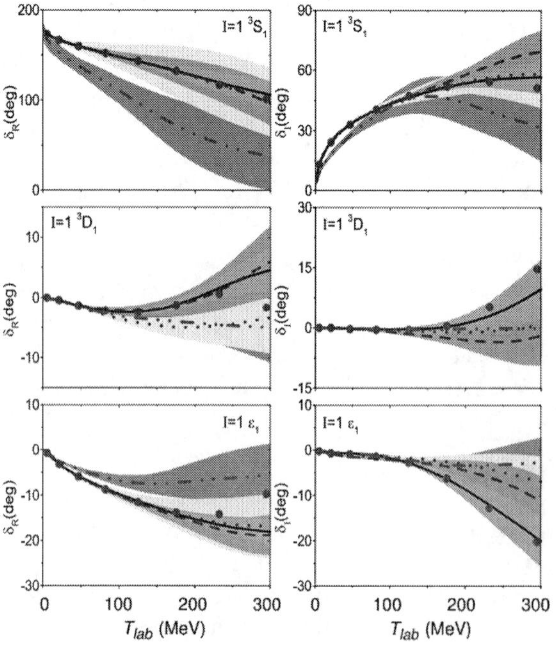

Figure 1. Real and imaginary parts of the phase shifts in the $^3S_1$-$^3D_1$ partial wave with isospin $I = 1$. Results at N^{3}LO (black solid line), N^{2}LO (blue dashed line), and NLO (red dotted line) are shown. The estimated uncertainty is indicated by bands: N^{3}LO (magenta), N^{2}LO (cyan), and NLO (yellow). The filled circles represent the solution of the $\bar{p}p$ PWA [7].

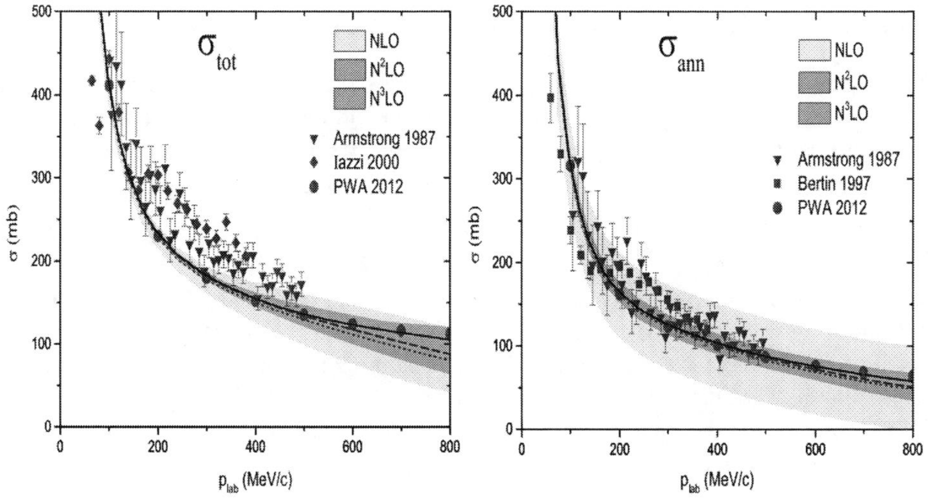

Figure 2. Total (σ_{tot}) and integrated annihilation (σ_{ann}) cross sections for $\bar{n}p$ scattering. Same description of curves and bands as in Fig. 1.

Since the integral in the LS equation (2) is divergent for the chiral potentials [3, 6] a regularization needs to be introduced. For that the regularization scheme of Ref. [2] is utilized, where a local regulator is used for the pion-exchange contributions and a nonlocal regulator for the contact terms:

$$V_{n\pi}(q) \rightarrow V_{n\pi}(r) \times f_R(r) \rightarrow V_{n\pi}^{reg}(q); \qquad (\vec{q} = \vec{p}\,' - \vec{p})$$
$$V_{cont} = V(p', p) \rightarrow V(p', p) \times f_\Lambda(p', p) = V_{cont}^{reg} \,. \tag{3}$$

The explict form of the regulation functions is $f_R(r) = \left[1 - exp(-r^2/R^2)\right]^6$ and $f_\Lambda(p, p') = exp(-(p'^2 + p^2)/\Lambda^2)$. The cutoff radius R is varied in the range $R = 0.7$-1.2 fm [2] where $\Lambda = 2/R$ is used for relating the momentum-space cutoff parameter with the cutoff radius.

A complete overview of our results for $\bar{N}N$ scattering up to N^{3}LO in chiral EFT can be found in Ref. [1] while possible $\bar{N}N$ bound states are discussed in Ref. [30]. Here we focus on quantities that are relevant for the study of $n - \bar{n}$ oscillations in the deuteron. Specifically, we look at the $^3S_1$-$^3D_1$ phase shifts in the $\bar{n}p$ (isospin $I = 1$) channel (Fig. 1) and the $\bar{n}p$ cross section, see Fig. 2. Note that data for the latter reaction were not included in the PWA [7] because they are less precise than the ones for $\bar{p}p$. The bands in Figs. 1 and 2 represent the estimated uncertainty of our results. Here we follow the suggestion of Ref. [2] and use as measure the expected size of the higher-order corrections together with the actual size of the higher-order corrections, see Ref. [1] for details.

In the study of $n - \bar{n}$ oscillations below we employ also an earlier (N^{2}LO) $\bar{N}N$ potential published by our group. A description of that interaction can be found in Ref. [5]. We consider these two $\bar{N}N$ interactions because they are based on rather different regularization schemes. In the earlier potential [5] a non-local exponential exponential regulator was employed for the whole potential while, as outlined above, in the N^{3}LO interaction [1] a local regulator was adopted for the evaluation of the one- and two-pion contributions. Comparing the pertinent results allows us to shed light on the question in how far the choice of the regulator influences the predictions. For exploring the sensitivity of the results to the deuteron wave function we employ also those from two meson-exchange potentials [31, 32], besides the ones calculated consistently within chiral EFT.

3 Formalism for $n - \bar{n}$ oscillations

The evaluation of $n - \bar{n}$ oscillations in the deuteron is done along the formalism presented in Refs. [19, 20]. However, since our study is performed in momentum space we describe the main steps below. The starting point is the eigenvalue (Schrödinger) equation [19]

$$\begin{pmatrix} H_0 + V_{np} & V_{n-\bar{n}} \\ V_{n-\bar{n}} & H_0 + V_{\bar{n}p} \end{pmatrix} \begin{pmatrix} |\psi_{np}\rangle \\ |\psi_{\bar{n}p}\rangle \end{pmatrix} = (E - i\Gamma/2) \begin{pmatrix} |\psi_{np}\rangle \\ |\psi_{\bar{n}p}\rangle \end{pmatrix} . \tag{4}$$

Here, V_{np} and $V_{\bar{n}p}$ are the potentials in the np and $\bar{n}p$ systems and $|\psi_{np}\rangle$ and $|\psi_{\bar{n}p}\rangle$ are the corresponding wave functions. The systems are coupled via $V_{n-\bar{n}}$ which is given by the $n - \bar{n}$ transition matrix element $\delta m_{n-\bar{n}}$ where the latter is proportional to the inverse of the $n - \bar{n}$ oscillation time, i.e. $V_{n-\bar{n}} = \delta m_{n-\bar{n}} = \hbar/\tau_{n-\bar{n}}$ [10].

To leading order the $\bar{n}p$ component $|\psi_{\bar{n}p}\rangle$ obeys the equation

$$(H_0 + V_{\bar{n}p} - E_d)|\psi_{\bar{n}p}\rangle = -V_{n-\bar{n}}|\psi_d\rangle , \tag{5}$$

where E_d is the unperturbed energy of the deuteron and $|\psi_d\rangle$ is the corresponding deuteron wave function. The decay width of the deuteron, Γ_d, is then [19]

$$\Gamma_d = -2\, V_{n-\bar{n}}\, \mathrm{Im}\langle\psi_d|\psi_{\bar{n}p}\rangle . \tag{6}$$

We solve Eq. (5) in momentum space. Performing a partial wave decomposition and taking into account the coupling of the $^3S_1$ and $^3D_1$ channels, the above integral equation reads

$$(2E_p - E_d)\psi^L_{\bar{n}p}(p) + \sum_{L'} \int \frac{dq\, q^2}{(2\pi)^3} V^{L,L'}_{\bar{n}p}(p,q)\psi^{L'}_{\bar{n}p}(q) = -V_{n-\bar{n}}\, \psi^L_d(p) , \tag{7}$$

with L, $L' = 0, 2$. Note that E_d is the total energy corresponding to the deuteron, i.e. $E_d - 2m_N = 2\sqrt{m_N^2 - \kappa^2} - 2m_N = -B_d$ where B_d is the standard binding energy of 2.225 MeV and $\kappa = \sqrt{m_N B_d} \simeq 45.7$ MeV is the binding momentum. The deuteron wave function is normalized by

$$\int dpp^2 \left[(\psi^0_d(p))^2 + (\psi^2_d(p))^2\right] = 1 , \tag{8}$$

and the width is provided by

$$\Gamma_d = -2\, V_{n-\bar{n}}\, \mathrm{Im} \sum_L \int dpp^2\, \psi^L_d(p)\, \psi^L_{\bar{n}p}(p) . \tag{9}$$

The deuteron lifetime τ_d is given by $\tau_d = \hbar/\Gamma_d$. The interesting quantity is the so-called reduced lifetime R [19, 20, 23] which relates the free $n - \bar{n}$ oscillation lifetime with that of the deuteron,

$$\tau_d = R\, \tau^2_{n-\bar{n}} \quad \text{i.e.} \quad R = \frac{\hbar}{\Gamma_d \tau^2_{n-\bar{n}}} . \tag{10}$$

4 Results and discussion

Our results for the reduced lifetime R for the deuteron are summarized in Table 1. They are based on our N^{3}LO interaction with cutoff $R_0 = 0.9$ fm [1] and the N^{2}LO interaction with cutoff $\{\Lambda, \tilde{\Lambda}\} = \{450, 500\}$ MeV [5]. For technical details see the corresponding publications. Besides the predictions for R based on the chiral $\bar{N}N$ interactions we list also the values given in Ref. [20] where the $\bar{N}N$ potentials DR$_1$ and DR$_2$ by Dover-Richard [24, 25] have been utilized. Furthermore we include results from the calculation of Oosterhof et al. performed directly within EFT on the basis of the KSW approach. In this case R can be represented in a compact analytical form which reads up to NLO [26]

$$R = -\frac{\kappa}{m_N} \frac{1}{\operatorname{Im} a_{\bar{n}p}} \frac{1}{1 + 0.4 + 2\kappa \operatorname{Re} a_{\bar{n}p} - 0.13 \pm 0.4} . \tag{11}$$

Obviously, the only parameter here is the $\bar{n}p$ $^3S_1$ scattering length. All other quantities that enter are well established NN observables, cf. Ref. [26] for details. Note that in that paper, the scattering length $\operatorname{Re} a_{\bar{n}p}$ was taken from Ref. [1].

Table 1. Reduced lifetime R calculated for the χEFT $\bar{N}N$ potentials from Refs. [1, 5], together with information on the pertinent $\bar{n}p$ $^3S_1$ scattering length. Results for the Dover-Richard potentials DR$_1$ and DR$_2$ are taken from Ref. [20]. The corresponding scattering lengths are from Ref. [33]. Predictions based on Eq. (11), i.e. on the KSW approach applied in Ref. [26], are indicated too.

	χEFT N^{2}LO [5]	χEFT N^{3}LO [1]	DR$_1$ [20]	DR$_2$ [20]
R [s^{-1}]	2.49×10^{22}	2.56×10^{22}	2.56×10^{22}	2.40×10^{22}
(Eq. (11))	$(1.1 \pm 0.3) \times 10^{22}$	$(1.2 \pm 0.3) \times 10^{22}$	$(1.4 \pm 0.4) \times 10^{22}$	$(1.3 \pm 0.3) \times 10^{22}$
$a_{^3S_1}$ [fm]	$0.44 - \mathrm{i}\,0.91$	$0.44 - \mathrm{i}\,0.96$	$0.87 - \mathrm{i}\,0.66$	$0.89 - \mathrm{i}\,0.71$

Table 1 reveals that the values for R predicted by the chiral $\bar{N}N$ interactions are fairly similar to those obtained for the DR potentials in the past. By contrast, the results based on the framework employed by Oosterhof et al. [26] are rather different. Since the scattering length from the N^{3}LO chiral $\bar{N}N$ interaction [1] is utilized in that work, the large discrepancy observed in Ref. [26] is certainly not due to differences in Im $a_{\bar{n}p}$ but must be primarily a consequence of the different approaches.

For investigating the sensitivity of our results to the used ingredients we performed various exploratory calculations. Specifically, we employed the NLO and N^{2}LO variants of the considered $\bar{N}N$ (and NN) interactions. The corresponding predictions for R were found to lie within a range of $(2.48 - 2.65) \times 10^{22}$ s^{-1}. If one takes this variation as measure for the uncertainty due to the nuclear structure, i.e. the NN and $\bar{N}N$ interactions (wave functions), a value of roughly $R = (2.6 \pm 0.1) \times 10^{22}$ s^{-1} can be deduced. Application of the method proposed in Ref. [2] for estimating the uncertainty to the calculation based on the $\bar{N}N$ interaction from 2017 [1], say, leads to a slightly smaller uncertainty. We have also varied the deuteron wave functions alone. As an extreme case we even took wave functions from phenomenological NN potentials derived in an entirely different framework, namely in the meson-exchange picture [31, 32]. Also here the obtained values for R remained within the range given above. Finally, omitting the D-wave component of the deuteron wave function in our calculation causes a 5 % variation. It leads to an increase of R and, thus, does not bring the result closer to the values presented by Oosterhof et al. Overall, we confirm the observation by Dover et al. that the predictions for R are fairly insensitive to the details of the employed $\bar{N}N$ potentials [20], provided that these potentials describe the $\bar{p}p$ data at low energies.

Admittedly, we do not have a ready explanation for the difference of our results (and those of Ref. [20]) to the ones of Oosterhof et al. [26]. However, we believe that it is due to the fact that in the latter work the width Γ_d is evaluated following the perturbative scheme developed by Kaplan, Savage, and Wise [28]. In that scheme there is no proper deuteron wave function. Rather one works with an effectively constructed wave function that is represented in terms of an irreducible two-point function [26, 28]. This seems to work well for some electromagnetic form factors of the deuteron, at least at low momentum transfer [28, 34]. On the other hand, the quadrupole moment of the deuteron is overestimated by 40 % [28], which hints that the properties of the wave function at large distances (small momenta) are not that well represented in this scheme. Clearly, this should have an impact on the quantity studied in the present work as well. Note that a comparable agreement (mismatch) with regard to the KSW scheme has been also observed in studies of the electric dipole moment (magnetic quadrupole moment) of the deuteron [35–37]. In any case, one should not forget that there is convergence problem of the KSW approach for NN partial waves where the tensor force from pion exchange is present [38]. It affects specifically the $^3S_1$-$^3D_1$ channel where difficulties appear already for momenta around 100 MeV/c, see [38] and also the discussions in Refs. [4, 39].

5 Summary

In this presentation we have introduced an $\bar{N}N$ potential [1], established within chiral effective field theory, which has been fitted to up-to-date $\bar{N}N$ phase shifts and inelasticities provided by a phase-shift analysis of available $\bar{p}p$ scattering data [7].

As an application of this interaction neutron-antineutron oscillations in the deuteron have been considered. In particular, results for the deuteron lifetime have been presented, evaluated in terms of the free-space $n-\bar{n}$ oscillation time, utilizing that $\bar{N}N$ potential in combination with an NN interaction likewise derived within chiral effective field theory. The value obtained for the so-called reduced lifetime R which relates the free-space $n-\bar{n}$ oscillation time $\tau_{n-\bar{n}}$ with the deuteron lifetime is found to be $R = (2.6 \pm 0.1) \times 10^{22}$ s^{-1}, where the quoted uncertainty is due to the NN and $\bar{N}N$ interactions (wave functions).

Our prediction for R agrees with the value obtained by Dover and collaborators almost four decades ago [20] but deviates from recent EFT calculations, based on the perturbative scheme proposed by Kaplan, Savage, and Wise [26], by about a factor of 2. A possible explanation for the difference could be that the KSW scheme does not involve a proper deuteron wave function. Rather this ingredient is represented effectively in terms of an irreducible two-point function. It is known from past studies that the KSW approach fails to describe quantities that depend more sensitively on the wave function like, for example, the quadrupole moment of the deuteron [28].

Acknowledgements

This work is supported in part by the DFG and the NSFC through funds provided to the Sino-German CRC 110 "Symmetries and the Emergence of Structure in QCD" (DFG grant. no. TRR 110) and the VolkswagenStiftung (grant no. 93562). The work of UGM was supported in part by The Chinese Academy of Sciences (CAS) President's International Fellowship Initiative (PIFI) (grant no. 2018DM0034).

References

[1] L. Y. Dai, J. Haidenbauer and U.-G. Meißner, JHEP **1707**, 078 (2017).
[2] E. Epelbaum, H. Krebs and U.-G. Meißner, Eur. Phys. J. A **51**, 53 (2015).
[3] E. Epelbaum, W. Glöckle and U.-G. Meißner, Nucl. Phys. A **747**, 362 (2005).

[4] E. Epelbaum, H.-W. Hammer and U.-G. Meißner, Rev. Mod. Phys. **81**, 1773 (2009).
[5] X. W. Kang, J. Haidenbauer and U.-G. Meißner, JHEP **1402**, 113 (2014).
[6] R. Machleidt and D. R. Entem, Phys. Rept. **503**, 1 (2011).
[7] D. Zhou and R. G. E. Timmermans, Phys. Rev. C **86**, 044003 (2012).
[8] J. Haidenbauer and U.-G. Meißner, Chin. Phys. C **44**, 033101 (2020).
[9] R. N. Mohapatra, J. Phys. G **36**, 104006 (2009).
[10] D. G. Phillips II *et al.*, Phys. Rept. **612**, 1 (2016).
[11] S. Gardner and X. Yan, Phys. Lett. B **790**, 421 (2019).
[12] A.D. Sakharov, JETP Lett. **5**, 24 (1967).
[13] C. Grojean, B. Shakya, J. D. Wells and Z. Zhang, Phys. Rev. Lett. **121**, 171801 (2018).
[14] M. Baldo-Ceolin *et al.*, Z. Phys. C **63**, 409 (1994).
[15] J. Chung *et al.*, Phys. Rev. D **66**, 032004 (2002).
[16] K. Abe *et al.* [Super-Kamiokande], Phys. Rev. D **103**, 012008 (2021).
[17] B. Aharmim *et al.* [SNO], Phys. Rev. D **96**, 092005 (2017).
[18] J. L. Barrow, E. S. Golubeva, E. Paryev and J.-M. Richard, Phys. Rev. D **101**, 036008 (2020).
[19] P. G. H. Sandars, J. Phys. G **6**, L161 (1980).
[20] C. B. Dover, A. Gal and J.-M. Richard, Phys. Rev. D **27**, 1090 (1983).
[21] W. M. Alberico, A. De Pace and M. Pignone, Nucl. Phys. A **523**, 488 (1991).
[22] J. Hufner and B. Z. Kopeliovich, Mod. Phys. Lett. A **13**, 2385 (1998).
[23] E. Friedman and A. Gal, Phys. Rev. D **78**, 016002 (2008).
[24] C. B. Dover and J.-M. Richard, Phys. Rev. C **21**, 1466 (1980).
[25] C. B. Dover and J.-M. Richard, Phys. Rev. C **25**, 1952 (1982).
[26] F. Oosterhof, B. Long, J. de Vries, R. G. E. Timmermans and U. van Kolck, Phys. Rev. Lett. **122**, 172501 (2019).
[27] D. B. Kaplan, M. J. Savage and M. B. Wise, Phys. Lett. B **424**, 390 (1998).
[28] D. B. Kaplan, M. J. Savage and M. B. Wise, Phys. Rev. C **59**, 617 (1999).
[29] E. Klempt, F. Bradamante, A. Martin and J.-M. Richard, Phys. Rept. **368**, 119 (2002).
[30] J. Haidenbauer, EPJ Web Conf. **181**, 01028 (2018).
[31] J. Haidenbauer, K. Holinde and M. B. Johnson, Phys. Rev. C **48**, 2190 (1993).
[32] R. Machleidt Phys. Rev. C **63**, 024001 (2001).
[33] J. Carbonell, J.-M. Richard and S. Wycech, Z. Phys. A **343**, 325 (1992).
[34] M. Walzl and U.-G. Meißner, Phys. Lett. B **513**, 37 (2001).
[35] J. de Vries, E. Mereghetti, R. G. E. Timmermans and U. van Kolck, Phys. Rev. Lett. **107**, 091804 (2011).
[36] C.-P. Liu, J. de Vries, E. Mereghetti, R. G. E. Timmermans and U. van Kolck, Phys. Lett. B **713**, 447 (2012).
[37] J. Bsaisou, J. de Vries, C. Hanhart, S. Liebig, U.-G. Meißner, D. Minossi, A. Nogga and A. Wirzba, JHEP **1503**, 104 (2015) Erratum: [JHEP **1505**, 083 (2015)].
[38] S. Fleming, T. Mehen and I. W. Stewart, Nucl. Phys. A **677**, 313 (2000).
[39] P. F. Bedaque and U. van Kolck, Ann. Rev. Nucl. Part. Sci. **52**, 339 (2002).

EPJ Web of Conferences **258**, 06003 (2022) https://doi.org/10.1051/epjconf/202225806003
A Virtual Tribute to Quark Confinement and the Hadron Spectrum (vConf21)

A road towards a beyond the Standard Model model

Giancarlo Rossi[1,2,3,*]

[1]Università di Roma *Tor Vergata*, Via della Ricerca Scientifica - 00133 Roma - Italy
[2]INFN - Sezione di Roma *Tor Vergata*, Via della Ricerca Scientifica - 00133 Roma - Italy
[3]Centro Fermi - Piazza del Viminale, 1 - 00184 Roma - Italy

Abstract. In this talk we describe examples of renormalizable strongly interacting field theories where chiral symmetry, broken at the UV cutoff by the presence of some irrelevant $d > 4$ operators in the fundamental Lagrangian, is recovered at low energy owing to the tuning of certain Lagrangian parameters. The interference of UV effects with IR features coming from the spontaneous breaking of the recovered chiral symmetry yields non perturbatively generated elementary fermion masses parametrically expressed by formulae of the kind $m_q \sim C_q(\alpha)\Lambda_{\rm RGI}$ with α the gauge coupling constant and $\Lambda_{\rm RGI}$ the RGI scale of the theory. Upon introducing EW interactions, this mechanism can be extended to give mass to EW bosons and leptons and can thus be used as an alternative to the Higgs scenario. In order to give the top quark and the weak gauge bosons a mass of the phenomenologically correct order of magnitude, the model must necessarily include (yet unobserved) super-strongly interacting massive fermions endowed, besides ordinary Standard Model interactions, with super-strong interactions with a RGI scale, $\Lambda_T \gg \Lambda_{QCD}$ in the few TeV range. Though limited in its scope (here we ignore hypercharge and leptons and discuss only the case of one family neglecting weak isospin splitting), the model opens the way to a solution of the naturalness problem and an understanding of the fermion mass hierarchy.

1 Introduction

We show in this talk that in strongly interacting field theories where chiral symmetry, broken at the UV cutoff by the presence of some irrelevant $d > 4$ operators of the Wilson type in the fundamental Lagrangian, is recovered at low energy owing to the tuning of certain Lagrangian parameters, mass terms for elementary fermions are dynamically generated. They are a consequence of a sort of "interference" between such UV chiral breaking terms and non-perturbative (NP) effects coming from the spontaneous breaking of the (recovered) chiral symmetry occurring in the Nambu–Goldstone (NG) phase of the theory. and are standardly induced by the strong interaction dynamics [1, 2]. A detailed analysis of this remarkable field-theoretical feature shows that such NP masses have the parametric form

$$m_q \sim C_q(\alpha)\Lambda_{\rm RGI}\,, \tag{1}$$

where $\Lambda_{\rm RGI}$ is the RGI scale of the theory and $C_q(\alpha)$ is a function of the gauge coupling constants $\{\alpha\}$ of the theory. If we take the irrelevant chiral breaking Wilson-like term to be

*e-mail: rossig@roma2.infn.it

a $d = 6$ operator, one finds to lowest loop order $C_q(\alpha) = O(\alpha_f^2)$, where α_f is the strongest among the gauge interactions the particle is subjected to.

An exact symmetry protects elementary masses against power divergencies, at variance with Wilson lattice QCD (WLQCD). Masses are "small", i.e. $O(\Lambda_{RGI})$, as the massless theory enjoys an enhanced (chiral) symmetry, consistently with the 't Hooft notion of naturalness [3].

A consequence of the formula (1), if applied to the top quark, is that since $m_{top} = O(\Lambda_{RGI})$, to match its experimental values a new sector of super-strongly interacting particles, gauge-invariantly coupled to standard matter, needs to exist so that the complete theory, including the new sector and Standard Model (SM) particles, will have an RGI scale $\Lambda_{RGI} \equiv \Lambda_T \gg \Lambda_{QCD}$ and of the order of a few TeV's. We immediately notice that the reason for assuming the existence of a super-strongly interacting sector here is very different from the reason why Technicolor was introduced [4, 5]. Technicolor was invoked to give mass to the EW bosons in the first place, while in the present approach super-strong interactions are introduced to give the right order of magnitude to the top quark and W masses. Anyway to avoid confusion and following Glashow [6], we will refer to this new set of particles as Tera-particles.

The issue of "families" and how eq. (1) can be extended to deal with lighter families (and to cope with weak isospin splitting) is outside the scope of this talk and, as we shall discuss in sect. 4, is related to the question of the universality of the structure of the NP mass formulae.

The model can be naturally extended to incorporate EW interactions and leptons. EW bosons as well as leptons will acquire NP masses $O(\Lambda_T)$ via the same mechanism that leads to eq. (1) [1], but with coefficient functions that scale like powers of the EW gauge coupling. For instance, to the leading loop order one gets for the W mass (see sect. 3 and refs. [7, 8])

$$M_W \sim \sqrt{\alpha_w} c_w(\alpha)\Lambda_{RGI}\,, \quad c_w(\alpha) = O(\alpha)\,, \tag{2}$$

where α in eq. (2) is a short-hand for the set of gauge couplings $[\alpha_s, \alpha_T]$ with α_s and α_T the strong and Tera-strong gauge couplings, respectively.

Eqs. (1) and (2) are somewhat similar to the expression of the Higgs-masses of fermions and W's with, however, two fundamental differences. The first is that the scale of the mass is not the vev of the Higgs field, but a dynamical scale related to a new interaction. The second is that the modulation of the Yukawa couplings that in the SM is introduced by hand to fit the values of fermion masses, here, as we said before, is controlled to leading order by the magnitude of the gauge coupling of the strongest among the interactions the particle feels.

We conclude from this analysis that the NP scenario for mass generation we advocate here can be considered as a valid alternative to the Higgs mechanism, with the extra advantage that, as we shall see, we will not have to deal with power divergencies in the Higgs mass. A further conceptual bonus is that, in view of the above mass formulae, we have a natural interpretation of the magnitude of the EW scale, as the dynamical physical parameter, Λ_T.

There is a number of further interesting features of the approach we are describing in this talk that are worth mentioning, but which we are not going to develop here for lack of space.

First of all, the dependence of the NP fermion masses upon the gauge couplings ($C(\alpha_f) = O(\alpha_f^2)$ to leading order for a $d = 6$ Wilson-like operator, see eq. (6)) may offer a hint to understand the fermion mass hierarchy $m_\tau \ll m_t \ll m_T$, as due to the ranking among weak, strong and Tera-strong gauge couplings, $\alpha_w \ll \alpha_s \ll \alpha_T$ [2].

Secondly, lacking the need for a Higgs boson, we propose to interpret the 125 GeV resonance, recently identified at LHC, as a W^+W^-/ZZ composite state, bound by exchanges of Tera-particles. Since this bound state is light on the Λ_T scale, it should be incorporated in the effective low energy theory that one gets by integrating out the Tera-degrees of freedom. As

[1]With the SM hypercharge assignment, in our model neutrinos are massless because ν_R is sterile.

[2]In this talk we limit ourselves to discuss only the heaviest fermion family.

a result, at (momenta)$^2 \ll \Lambda_T^2$ the Quantum Effective Lagrangian (QEL) [3] of the model can be argued to resemble very much to the SM.

Finally in ref. [9] it has been shown that with a reasonable choice of the elementary particle content, a theory extending the SM with the inclusion of the new super-strong sector leads to gauge coupling unification at a $\sim 10^{18}$ GeV scale, yielding a proton life-time comfortably larger than the present limit $\tau_{\rm prot} > 1.7 \times 10^{34}$ years.

2 A simple toy-model

The simplest model enjoying the NP mass generation mechanism we outlined before is described by a Lagrangian where an SU(2) fermion doublet, subjected to non-abelian gauge interactions (of the QCD type), is coupled to a complex scalar via $d = 4$ Yukawa and "irrelevant" $d = 6$ Wilson-like chiral breaking terms. The Lagrangian of this toy-model reads

$$\mathcal{L}_{\rm toy}(q, A, \Phi) = \mathcal{L}_{kin}(q, A, \Phi) + \mathcal{V}(\Phi) + \mathcal{L}_{Yuk}(q, \Phi) + \mathcal{L}_{Wil}(q, A, \Phi) \tag{3}$$

$$\bullet\ \mathcal{L}_{kin}(q, A, \Phi) = \frac{1}{4}(F^A \cdot F^A) + \bar{q}_L \mathcal{D}\!\!\!/^A q_L + \bar{q}_R \mathcal{D}\!\!\!/^A q_R + \frac{1}{2}\mathrm{Tr}\,[\partial_\mu \Phi^\dagger \partial_\mu \Phi] \tag{4}$$

$$\bullet\ \mathcal{V}(\Phi) = \frac{\mu_0^2}{2}\mathrm{Tr}\,[\Phi^\dagger \Phi] + \frac{\lambda_0}{4}(\mathrm{Tr}\,[\Phi^\dagger \Phi])^2 \tag{5}$$

$$\bullet\ \mathcal{L}_{Yuk}(q, \Phi) = \eta\,(\bar{q}_L \Phi q_R + \bar{q}_R \Phi^\dagger q_L) \tag{6}$$

$$\bullet\ \mathcal{L}_{Wil}(q, A, \Phi) = \frac{b^2}{2}\rho\,(\bar{q}_L \overleftarrow{\mathcal{D}}_\mu^A \Phi \mathcal{D}_\mu^A q_R + \bar{q}_R \overleftarrow{\mathcal{D}}_\mu^A \Phi^\dagger \mathcal{D}_\mu^A q_L)\,, \tag{7}$$

where $\Phi = \varphi_0 1\!\!1 + i\varphi_j \tau^j = [-i\tau_2 \varphi^\star | \varphi]$ with $\varphi = (\varphi_2 - i\varphi_1, \varphi_0 - i\varphi_3)^T$ is a complex scalar doublet, singlet under SU(3), $b^{-1} \sim \Lambda_{UV}$ is the UV cutoff, η is the Yukawa coupling, ρ is to keep track of $\mathcal{L}_{Wil}$ and $\mathcal{D}_\mu^A$ is the covariant derivative. The Lagrangian (3) is power counting renormalizable (like WLQCD is, despite the presence of the $d = 5$ Wilson term).

Among other obvious symmetries, $\mathcal{L}_{\rm toy}$ is invariant under the (global) transformations involving fermions and scalars ($\Omega_{L/R} \in$ SU(2))

$$\chi_L \times \chi_R = [\tilde{\chi}_L \times (\Phi \to \Omega_L \Phi)] \times [\tilde{\chi}_R \times (\Phi \to \Phi \Omega_R^\dagger)]\,, \quad \text{where} \tag{8}$$

$$\tilde{\chi}_{L/R} : q_{L/R} \to \Omega_{L/R} q_{L/R}\,, \quad \bar{q}_{L/R} \to \bar{q}_{L/R} \Omega_{L/R}^\dagger\,. \tag{9}$$

The $\chi_L \times \chi_R$ symmetry can be realized either *á la* Wigner or *á la* NG depending on the shape of the scalar potential. In any case no power divergent fermion mass can be generated by quantum corrections as the mass operator $(\bar{q}_L q_R + \bar{q}_R q_L)$ is not invariant under $\chi_L \times \chi_R$.

The operators $\mathcal{L}_{Wil}$ and $\mathcal{L}_{Yuk}$ break $\tilde{\chi}_L \times \tilde{\chi}_R$ and mix. Thus for generic values of η, $\mathcal{L}_{\rm toy}$ is not invariant under the fermionic chiral transformations $\tilde{\chi}_L \times \tilde{\chi}_R$. However, one can show [1, 10] that chiral invariance can be recovered (up to vanishingly small O(b^2) terms) at a critical value $\eta = \eta_{cr}$, where the Wilson-like term and the Yukawa terms "compensate", much like chiral symmetry is recovered (up to O(a) cutoff effects) in WLQCD by tuning the bare quark mass to a critical value, m_{cr}, where the Wilson and the mass term "compensate".

In the Wigner phase (where $\langle|\Phi|^2\rangle = 0$) the condition determining η_{cr} corresponds to the vanishing of the effective Yukawa term in the QEL of the theory. Diagrammatically we represent in fig. 1 this condition to lowest loop order.

In the NG phase (where $\langle|\Phi|^2\rangle = v^2 = \mu_0^2/\lambda_0$) the same condition implies that the Higgs-mass of the fermion is cancelled out. This is schematically shown in fig 2.

It is important to note that the quadratic divergency of the loop integrals both in fig. 1 and 2 is exactly compensated by the b^2 factor coming from the insertion of the Wilson-like vertex. In the following we will encounter other instances of this UV vs. IR compensation.

[3] By Quantum Effective Lagrangian we mean the generating functional of the 1PI vertices of the theory.

Figure 1. The condition determining η_{cr} in the Wigner phase. The box represents the insertion of the Wilson-like vertex.

Figure 2. The condition determining η_{cr} in the NG phase.

2.1 NP mass generation

Now the key observation is that at $\eta = \eta_{cr}$ where in the NG phase the fermion Higgs-mass is cancelled, the quark acquires a non-vanishing NP mass via a UV vs. IR compensation mechanism, reminiscent of the one that makes finite the 1-loop diagrams of figs. 1 and 2.

Because of this kind of subtle UV vs. IR compensations, it is crucial to analyze and determine the structure of the formally $O(b^2)$ terms of the regularized theory. This can be properly done by making recourse to the Symanzik expansion technique [11].

This analysis was carried out in ref. [1] where it was shown that because of the occurrence of the spontaneous breaking of the (restored) $\tilde{\chi}_L \times \tilde{\chi}_R$ symmetry, triggered by residual $O(b^2)$ chiral breaking terms, one must include in the Symanzik expansion also formally $O(b^2)$ NP-ly generated operators of the form

$$O_{6,\bar{q}q} \propto b^2 \Lambda_s \alpha_s |\Phi| \left[\bar{q}\, \mathcal{D}\!\!\!/^A q\right], \quad O_{6,FF} \propto b^2 \Lambda_s \alpha_s |\Phi| \left[F^A \cdot F^A\right]. \tag{10}$$

The form of these operators is completely fixed by symmetries (in particular $\chi_L \times \chi_R$) and dimensional considerations. The presence of the Λ_s factor signals their NP origin. The effect of the terms in eq. (10) can be elegantly taken into account by including in the calculations the new diagrams generated by the augmented Lagrangian

$$\mathcal{L}_{\rm toy} \to \mathcal{L}_{\rm toy} + \Delta\mathcal{L}_{NP}^{Sym}, \qquad \Delta\mathcal{L}_{NP}^{Sym} = b^2 \Lambda_s \alpha_s |\Phi| \left[c_{FF} F^A \cdot F^A + c_{\bar{q}q} \bar{q}\, \mathcal{D}\!\!\!/^A q\right] + \ldots \tag{11}$$

A remarkable fact is that new diagrams do exist that contribute, among other correlators, to the quark self-energy. A couple of these diagrams are drawn in fig. 3. A straightforward calculation gives for the amputated zero momentum diagram in the left panel of fig. 3

Figure 3. Two lowest loop order quark self-energy diagrams. The blobs represent insertions of $\Delta\mathcal{L}_{NP}^{Sym}$.

$$m_q^{NP} \propto \alpha_s^2 \int^{1/b} \frac{d^4k}{k^2} \frac{\gamma_\mu k_\mu}{k^2} \int^{1/b} \frac{d^4\ell}{\ell^2 + m_{\zeta_0}^2} \frac{\gamma_\nu (k+\ell)_\nu}{(k+\ell)^2} \cdot b^2 \gamma_\rho (k+\ell)_\rho\, b^2 \Lambda_s \gamma_\lambda (2k+\ell)_\lambda \sim \alpha_s^2 \Lambda_s \, . \tag{12}$$

Here again we get a finite result, owing to an exact compensation between the UV power divergencies of the 2-loop integrals and the IR behaviour determined by the expression of the Wilson-like vertex (7) and the physics of the spontaneous breaking of the chiral $\tilde{\chi}_L \times \tilde{\chi}_R$ symmetry encoded in the form of the Symanzik operators (10). A similar compensation occurs in the calculation of the diagram in the right panel of fig. 3.

2.2 The QEL of the critical theory in the NG phase

The resulting non-vanishing quark mass, the occurrence of which was checked in explicit lattice simulations of the model (3) in ref. [2], can be incorporated in the QEL, Γ^{NG}, of the theory, upon introducing the non-analytic field U in the polar decomposition

$$\Phi = (v + \zeta_0)U\,, \quad U = \exp[i\vec{\tau}\vec{\zeta}/v]\,, \tag{13}$$

As U transforms like Φ, new $\chi_L \times \chi_R$ invariant operators can indeed be constructed, yielding

$$\Gamma^{NG}_{d=4} = \frac{1}{4}(F^A \cdot F^A) + \bar{q}_L \mathcal{D}\!\!\!/^A q_L + \bar{q}_R \mathcal{D}\!\!\!/^A q_R + \tag{14}$$
$$+ c_q \Lambda_s [\bar{q}_L U q_R + \bar{q}_R U^\dagger q_L] + \frac{c^2 \Lambda_s^2}{2} \mathrm{Tr}\,[\partial_\mu U^\dagger \partial_\mu U]\,.$$

Naturally, of special interest is the first term in the second line of eq. (14) because, upon expanding $U = 1\!\!1 + \vec{\tau}\vec{\zeta}/v + \ldots$, it gives raise to a mass for the fermion plus a wealth of NG boson interactions [4].

3 Introducing weak and Tera-interactions

To proceed to the construction of a possible, realistic beyond-the-SM-model (bSMm) we clearly need to introduce EW interactions. At the same time, as mentioned in the Introduction, it is also necessary to extend the model by incorporating a super-strongly interacting sector, so that the whole theory will have an RGI scale, Λ_T, much larger than Λ_{QCD} and of the order of a few TeVs. Only in this way there is the chance that eq. (1) can yield the correct order of magnitude of the top quark mass. The extended Lagrangian is obtained by doubling the structure of quarks in order to encompass Tera-particles (Q=Tera-quarks and G=Tera-gluons) and gauging the exact χ_L symmetry to introduce weak interactions. The whole Lagrangian will read

$$\mathcal{L}(q, Q; \Phi; A, G, W) = \mathcal{L}_{kin}(q, Q; \Phi; A, G, W) + \mathcal{V}(\Phi) + \mathcal{L}_{Yuk}(q, Q; \Phi) + \mathcal{L}_{Wil}(q, Q; \Phi; A, G, W) \tag{15}$$

- $\mathcal{L}_{kin}(q, Q; \Phi; A, W) = \frac{1}{4}\Big(F^A \cdot F^A + F^G \cdot F^G + F^W \cdot F^W\Big) +$
$$+\Big[\bar{q}_L \mathcal{D}\!\!\!/^{AW} q_L + \bar{q}_R \mathcal{D}\!\!\!/^A q_R\Big] + \Big[\bar{Q}_L \mathcal{D}\!\!\!/^{AGW} Q_L + \bar{Q}_R \mathcal{D}\!\!\!/^{AG} Q_R\Big] + \frac{k_b}{2} \mathrm{Tr}\,[(\mathcal{D}^W_\mu \Phi)^\dagger \mathcal{D}^W_\mu \Phi] \tag{16}$$
- $\mathcal{V}(\Phi) = \frac{\mu_0^2}{2} k_b \mathrm{Tr}\,[\Phi^\dagger \Phi] + \frac{\lambda_0}{4}(k_b \mathrm{Tr}\,[\Phi^\dagger \Phi])^2 \qquad (17)$
- $\mathcal{L}_{Yuk}(q, Q; \Phi) = \eta_q\,(\bar{q}_L \Phi\, q_R + \bar{q}_R \Phi^\dagger q_L) + \eta_Q\,(\bar{Q}_L \Phi\, Q_R + \bar{Q}_R \Phi^\dagger Q_L) \qquad (18)$
- $\mathcal{L}_{Wil}(q, Q; \Phi; A, G, W) = \frac{b^2}{2} \rho_q\,(\bar{q}_L \overleftarrow{\mathcal{D}}^{AW}_\mu \Phi \mathcal{D}^A_\mu q_R + \bar{q}_R \overleftarrow{\mathcal{D}}^A_\mu \Phi^\dagger \mathcal{D}^{AW}_\mu q_L) +$
$$+ \frac{b^2}{2} \rho_Q\,(\bar{Q}_L \overleftarrow{\mathcal{D}}^{AGW}_\mu \Phi \mathcal{D}^{AG}_\mu Q_R + \bar{Q}_R \overleftarrow{\mathcal{D}}^{AG}_\mu \Phi^\dagger \mathcal{D}^{AGW}_\mu Q_L) \tag{19}$$

with obvious notations for the covariant derivatives. Besides the Yukawa (eq. (18)) and the Wilson-like (eq. (19)) operators, now also the kinetic term of the scalar (last term in eq. (16)) breaks $\tilde{\chi}_L \times \tilde{\chi}_R$ and mixes with $\mathcal{L}_{Yuk}$ and $\mathcal{L}_{Wil}$. Thus to get to the critical theory (invariant under $\tilde{\chi}_L \times \tilde{\chi}_R$), on top of η_q and η_Q, a further parameter, k_b, needs to be introduced and appropriately tuned. The conditions determining the critical theory correspond to have vanishing effective Yukawa interactions and vanishing scalar kinetic term in the QEL. Notice that the coefficient k_b appears also in the expression of the scalar potential (17). The reason for this is that in this

[4] We may notice that in the NP mass, U has a role similar to the one the non-analytic $r/|r|$ ratio has in WLQCD, as if we had promoted the Wilson r parameter to a field, Φ.

way the (bare) vev^2 (in the NG phase) and the quartic coupling of the canonically normalized scalar field will have the standard definitions, $v^2 = |\mu_0^2|/\lambda_0$ and λ_0.

The tuning conditions that determine $\eta_{q\,cr}$, $\eta_{Q\,cr}$ and $k_{b\,cr}$ in the Wigner phase are schematically reported in the figs. 4, 5 and 6. To the lowest loop order one finds

$$\eta_{q\,cr}^{(1-loop)} = \rho_q\,\eta_{1q}\alpha_s\,, \qquad \eta_{Q\,cr}^{(1-loop)} = \rho_Q\,\eta_{1Q}\alpha_T\,, \tag{20}$$

$$k_{b\,cr}^{(1-loop)} = [\rho_q^2 N_c + \rho_Q^2 N_c N_T]k_1\,, \tag{21}$$

where η_{1q}, η_{1Q} and k_1 are computable coefficients and SU(N_T) is the Tera-gauge group. As before, UV loop divergencies are exactly compensated by the IR behaviour of the inserted Wilson-like vertices.

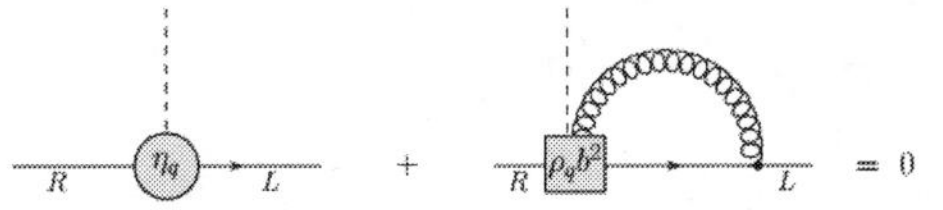

Figure 4. The tuning condition determining $\eta_{q\,cr}$ at lowest order. The box represents the insertion of the quark Wilson-like vertex.

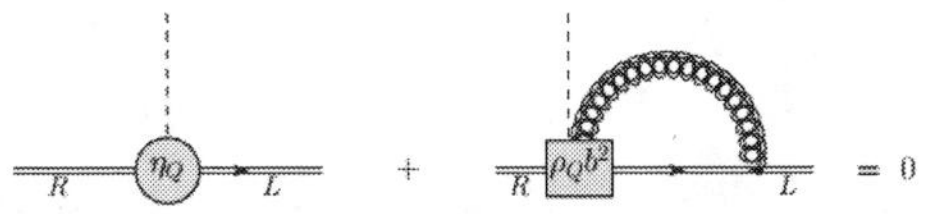

Figure 5. The tuning condition determining $\eta_{Q\,cr}$ at lowest order. Double lines represent Tera-particles. The box represents the insertion of the Tera-quark Wilson-like vertex.

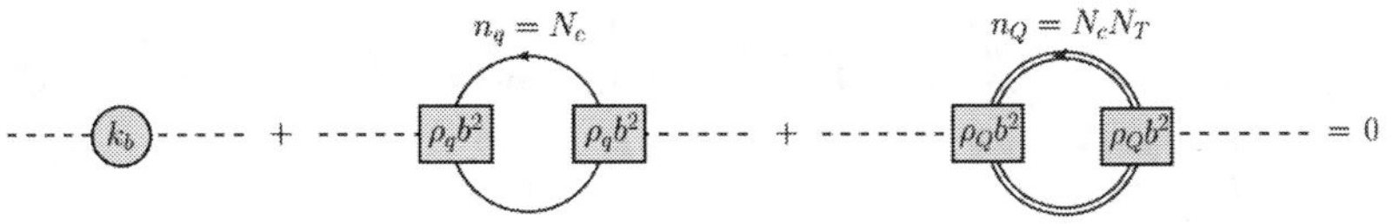

Figure 6. The tuning condition determining $k_{b\,cr}$ at lowest order. Notations are as in figs. 4 and 5.

In the NG phase the above tuning conditions imply the cancellation of the Higgs-like mass of quarks, Tera-quarks and W's, as schematically represented in the figs. 7, 8 and 9.

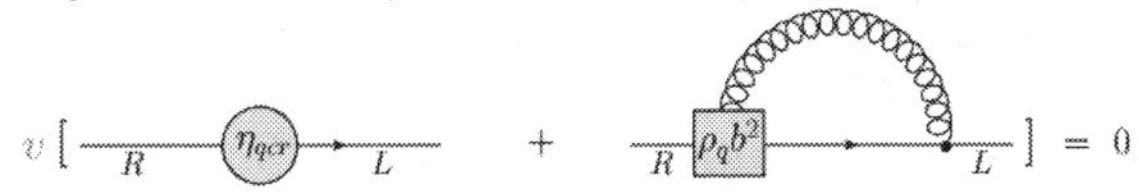

Figure 7. The mechanism behind the cancellation of the Higgs-like quark mass term, $v\,\bar{q}q$, operating in the NG phase of the critical theory.

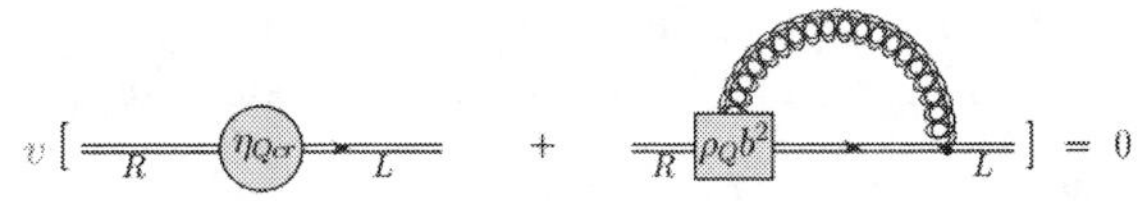

Figure 8. The mechanism behind the cancellation of the Higgs-like Tera-quark mass term, $v\,\bar{Q}Q$, operating in the NG phase of the critical theory.

Extending to the present case the Symanzik analysis of the formally O(b^2) operators necessary to describe the NP effects related to the spontaneous breaking of the (recovered) $\tilde{\chi}_L \times \tilde{\chi}_R$ symmetry in the regularized theory, one finds the following set of operators

$$\bullet\; O_{6,\bar{Q}Q}^{T} = b^2\alpha_T\rho_Q\Lambda_T|\Phi|\Big[\bar{Q}_L\mathcal{D}^{AGW}Q_L + \bar{Q}_R\mathcal{D}^{AG}Q_R\Big] \tag{22}$$

$$\bullet\; O_{6,\bar{Q}Q}^{s} = b^2\alpha_s\rho_Q\Lambda_T|\Phi|\Big[\bar{Q}_L\mathcal{D}^{AGW}Q_L + \bar{Q}_R\mathcal{D}^{AG}Q_R\Big] \tag{23}$$

$$\bullet\; O_{6,GG} = b^2\alpha_T\rho_Q\Lambda_T|\Phi|F^G\cdot F^G \qquad \bullet\; O_{6,AA} = b^2\alpha_s\rho_Q\Lambda_T|\Phi|F^A\cdot F^A \tag{24}$$

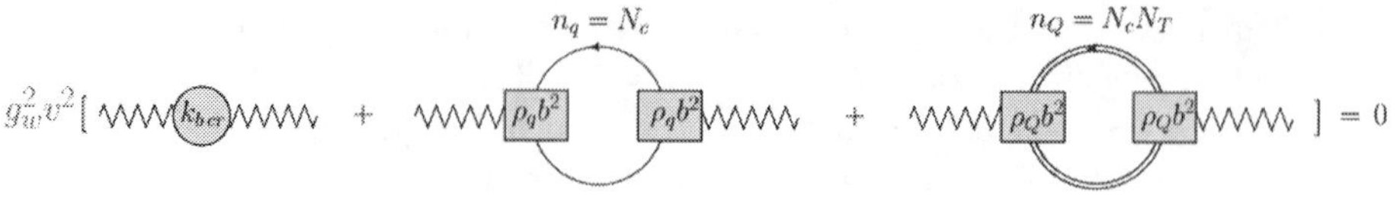

Figure 9. The mechanism behind the cancellation of the Higgs-like W mass term, $g_w^2 v^2 \mathrm{Tr}\,[W_\mu W_\mu]$, operating in the NG phase of the critical theory.

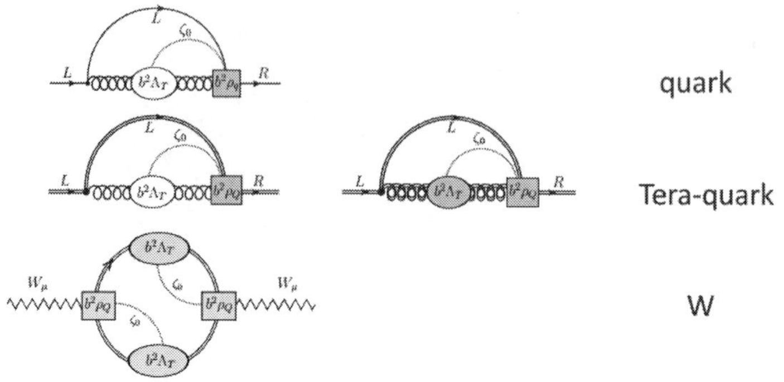

Figure 10. Examples of self-energy diagrams giving NP masses to quarks, Tera-quarks and W. The blobs represent the insertion of the NP Symanzik operators of eqs. (22), (23) and (24),

NP masses emerge from the kind of diagrams shown in fig. 10. They are all finite owing to the by now well-known UV-IR compensation and all of $\mathrm{O}(\Lambda_T)$ times gauge coupling dependent coefficient functions (see next section).

3.1 The critical QEL in the NG phase

With the same line of arguments we used before to derive eq. (14), we get for the $d = 4$ piece of the QEL of the critical theory in the NG phase the expression

$$\Gamma^{NG}_{4\,cr}(q,Q;\Phi;A,G,W) = \frac{1}{4}\Big(F^A\cdot F^A + F^G\cdot F^G + F^W\cdot F^W\Big) + \tag{25}$$
$$+\Big[\bar q_L \mathcal{D}\!\!\!/^{WA} q_L + \bar q_R \mathcal{D}\!\!\!/^{A} q_R\Big] + C_q\Lambda_T\left(\bar q_L U q_R + \bar q_R U^\dagger q_L\right) +$$
$$+\Big[\bar Q_L \mathcal{D}\!\!\!/^{WAG} Q_L + \bar Q_R \mathcal{D}\!\!\!/^{AG} Q_R\Big] + C_Q\Lambda_T\left(\bar Q_L U Q_R + \bar Q_R U^\dagger Q_L\right) + \frac{1}{2}c_w^2\Lambda_T^2\mathrm{Tr}\,[(\mathcal{D}^W_\mu U)^\dagger \mathcal{D}^W_\mu U]\,.$$

From a detailed analysis of the diagrams in fig. 10 one can read-off the parametric dependence of the mass of quarks, Tera-quarks and Ws to the leading loop order. One finds

$$m_q^{NP} = C_q\,\Lambda_T\,, \qquad C_q = \mathrm{O}(\alpha_s^2) \tag{26}$$
$$m_Q^{NP} = C_Q\,\Lambda_T\,, \qquad C_Q = \mathrm{O}(\alpha_T^2,\ldots) \tag{27}$$
$$M_W^{NP} = C_w\,\Lambda_T\,, \qquad C_w = \sqrt{\alpha_w}\,c_w\,,\ c_w = k_w\mathrm{O}(\alpha_T,\ldots)\,. \tag{28}$$

4 Universality

A key point that needs to be thoroughly discussed is to what extent the NP mass predictions we have derived above are "universal", or in other words to what extent their value depends on the exact form of the $d > 4$ Wilson-like terms that one decides to introduces in the fundamental Lagrangian. Actually it turns out that the power of the gauge coupling multiplying the RGI in the leading loop order in the formula expressing the NP masses may depend on the dimensions of such chiral breaking terms. For instance, one can see that generically the higher is the dimension of the Wilson-like terms the larger will be the power of the gauge couplings in front of the RGI scale in the mass formulae [5].

This may appear as a blunt violation of universality, however, one can imagine to exploit this feature to try to interpret family mass hierarchy (from heavy to light) as due to Wilson-like terms of increasing dimensions. Implementing this idea is not at all trivial, if we want to get flavour as an emerging quantum number that identifies different families, somehow related to the "geometry" of the Wilson-like terms.

5 Conclusions and Outlook

We have shown that, as an alternative to the Higgs mechanism, masses can be NP-ly generated in a strongly interacting theory where chiral symmetry, broken at the UV cutoff level by irrelevant Wilson-like terms, is recovered at low energy owing to the tuning of certain Lagrangian parameters. Since an exact symmetry prevents the existence of standard mass terms of the kind $m\bar{\psi}\psi$ and there is no Higgs field around, the mechanism we are advocating here allows a neat solution of the naturalness problem. In fact, no power divergencies can affect physical masses and masses are "small", i.e. $O(\Lambda_{\rm RGI})$, as the massless theory enjoys an enhanced (chiral) symmetry, consistently with the 't Hooft idea of naturalness [3].

We have seen that to cope with the magnitude of the W and top mass, a super-interacting sector, gauge invariantly coupled to standard matter, must exists leading to a theory with an RGI scale $\Lambda_{\rm RGI} = \Lambda_T$ of the order of a few TeVs. The simplest model discussed in ref. [1] enjoying the NP mass generation mechanism can be straightforwardly extended to incorporate weak interactions and the new Tera-degrees of freedom. In a similar way one could introduce leptons and the hypercharge interaction, an issue that we defer to a forthcoming investigation.

Acknowledgements

We thank R. Frezzotti for infinitely many discussions on the issues presented in this talk. We would also like to acknowledge some exchange of ideas with M. Garofalo.

References

[1] R. Frezzotti and G. C. Rossi, Phys. Rev. D **92** (2015) no.5, 054505.
[2] S. Capitani *et al.*, Phys. Rev. Lett. **123** (2019) no.6, 061802.
[3] G. 't Hooft, "*Naturalness, Chiral Symmetry and Spontaneous Chiral Symmetry Breaking*" in "Recent Developments in Gauge Theories" (Plenum Press, 1980).
[4] L. Susskind, Phys. Rev. D **20** (1979) 2619.
[5] S. Weinberg, Phys. Rev. D **19** (1979) 1277.
[6] S. L. Glashow, [arXiv:0504287 [hep-ph]].
[7] R. Frezzotti and G. C. Rossi, PoS LATTICE **2018** (2018) 190.
[8] R. Frezzotti and G. Rossi, in *Strong dynamics for physics within and beyond the Standard Model at LHC and Future Colliders* (LFC19) 9-13 September 2019, Trento, Italy.
[9] R. Frezzotti, M. Garofalo and G. C. Rossi Phys. Rev. D **93** (2016) no.10, 105030.
[10] R. Frezzotti and G. C. Rossi, Pos(Lattice2013) 354 (2013), [arXiv:1312.1069 [hep-lat]].
[11] K. Symanzik, Nucl. Phys. B **226** (1983) 187 and 205.
[12] M. Bochicchio, L. Maiani, G. Martinelli, G. Rossi, M. Testa, Nucl. Phys. **B262** (1985) 331.

[5] As for the ρ dependence of the NP masses, one can prove that in the case of a theory with only quarks and W's (and not Tera-quarks or leptons) their mass is actually ρ independent. If other fermions are present, be them Tera-particles or leptons, NP masses will be functions of the ratios of the ρ parameters of the various fermions.

Recent progress in hadronic light-by-light scattering

Martin Hoferichter[1,*]

[1]Albert Einstein Center for Fundamental Physics, Institute for Theoretical Physics, University of Bern, Sidlerstrasse 5, 3012 Bern, Switzerland

Abstract. In recent years, significant progress in the calculation of the HLbL contribution to the anomalous magnetic moment of the muon has been achieved both with data-driven methods and in lattice QCD. In these proceedings I will discuss current developments aimed at controlling HLbL scattering at the level of 10%, as required for the final precision of the Fermilab E989 experiment.

1 Introduction

The 4.2σ tension between the experimental world average [1–5]

$$a_\mu^{\text{exp}} = 116\,592\,061(41) \times 10^{-11} \tag{1}$$

and the Standard-Model prediction for the anomalous magnetic moment of the muon [6–33]

$$a_\mu^{\text{SM}} = 116\,591\,810(43) \times 10^{-11} \tag{2}$$

continues to motivate further improvements in controlling the hadronic contributions, both hadronic vacuum polarization (HVP) and hadronic light-by-light scattering (HLbL). While for the HVP contribution a key challenge concerns reconciling the data-driven determination as reviewed in Ref. [6] and the lattice-QCD calculation by BMWc [34], see Refs. [35–39], for HLbL scattering the phenomenological determination, based on Refs. [19–31, 40–45], agrees well with the lattice calculation by RBC/UKQCD [32] already included in Ref. [6] and the more recent calculation by the Mainz group [46]. Here we discuss mainly the prospects and challenges involved in improving the data-driven determination further towards 10% precision, as required for the final goal of the Fermilab E989 experiment [47] and beyond [48, 49].

2 White paper status

The current status of the phenomenological determination as described in the White Paper (WP) [6] is given in Table 1 in comparison with previous compilations [42, 50–52]. The final recommended value was obtained by combining errors for the well-controlled low-energy contributions—pseudoscalar poles [23, 26–28] and leading two-pion intermediate states [24, 25]—in quadrature, to the effect that the subleading contributions—scalar, tensor, and axial-vector states as well as short-distance contributions [19, 29–31, 40–45]—dominate

*e-mail: hoferichter@itp.unibe.ch

Contribution	PdRV(09)	N/JN(09)	J(17)	WP20
π^0, η, η'-poles	114(13)	99(16)	95.45(12.40)	93.8(4.0)
π, K-loops/boxes	−19(19)	−19(13)	−20(5)	−16.4(2)
S-wave $\pi\pi$ rescattering	−7(7)	−7(2)	−5.98(1.20)	−8(1)
subtotal	88(24)	73(21)	69.5(13.4)	69.4(4.1)
scalars	–	–	–	} − 1(3)
tensors	–	–	1.1(1)	}
axial vectors	15(10)	22(5)	7.55(2.71)	6(6)
u, d, s-loops/short-distance	–	21(3)	20(4)	15(10)
c-loop	2.3	–	2.3(2)	3(1)
total	105(26)	116(39)	100.4(28.2)	92(19)

Table 1. Estimates for the various contributions to HLbL scattering from WP20 [6], PdRV(09) [50], N/JN(09) [51, 52], J(17) [42]; from Ref. [6].

the error budget. In particular, the uncertainties for these contributions were added linearly, to account for the fact that, contrary to the low-energy part, the bookkeeping is not yet based on a dispersive reasoning [20–22, 53], which introduces ambiguities and thus sizable correlations among the errors. The total number a_μ^{HLbL} (phenomenology) $= 92(19) \times 10^{-11}$ agrees well with lattice QCD (for $q = u, d, s$)—a_μ^{HLbL} (RBC/UKQCD) $= 79(35) \times 10^{-11}$ [32] and a_μ^{HLbL} (Mainz) $= 106.8(14.7) \times 10^{-11}$ [46]. The combination of phenomenology and RBC/UKQCD (accounting for the role of the c-loop) then leads to the final recommendation a_μ^{HLbL} (WP) $= 90(17) \times 10^{-11}$ [6].

3 Pseudoscalar poles

For the π^0 pole evaluations from dispersion relations [26, 27], Canterbury approximants [23], and lattice QCD [28] are available

$$a_\mu^{\pi^0\text{-pole}}\Big|_{\text{dispersive}} = 63.0^{+2.7}_{-2.1} \times 10^{-11}, \qquad a_\mu^{\pi^0\text{-pole}}\Big|_{\text{Canterbury}} = 63.6(2.7) \times 10^{-11},$$
$$a_\mu^{\pi^0\text{-pole}}\Big|_{\text{lattice+PrimEx}} = 62.3(2.3) \times 10^{-11}, \qquad a_\mu^{\pi^0\text{-pole}}\Big|_{\text{lattice}} = 59.7(3.6) \times 10^{-11}, \tag{3}$$

which agree at a level well below the required precision goal. The agreement is further improved if the experimental normalization for $\pi^0 \to \gamma\gamma$ from the PrimEx experiment [54] is imposed in the lattice calculation. In addition, the singly-virtual limit of the pion transition form factor (TFF) agrees with the measurement from the BESIII experiment [55].

The situation differs for η, η', for which the WP number currently derives from Canterbury approximants alone [23], calling for further corroboration from dispersion relations and lattice QCD, both of which are in progress. On the lattice side, calculations for the η, η' are noisier than for the π^0, but are being addressed by (at least) two collaborations [56, 57]. On the dispersive side, the main challenge concerns so-called factorization breaking contributions, which can arise from a left-hand-cut structure involving the a_2 resonance. Such contributions cannot be addressed in the Canterbury approach, and thus it becomes critical to clarify their impact. A first step in this direction, relying on data for $e^+e^- \to \eta\pi\pi$, was recently performed in Ref. [58], with the main result that more differential data are required to resolve these contributions conclusively.

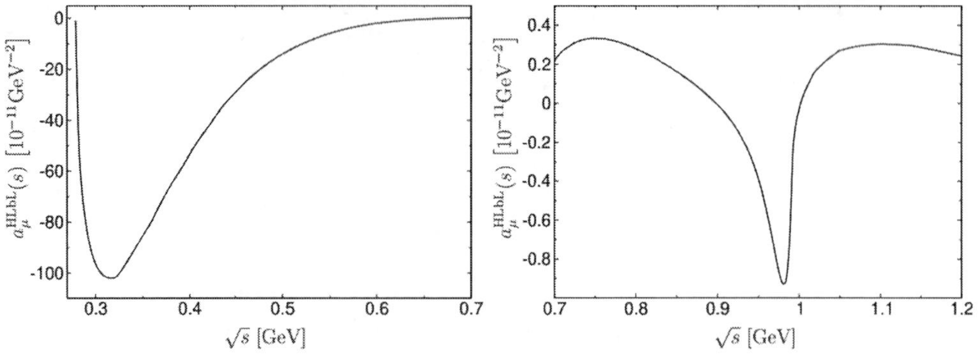

Figure 1. Spectral shape in the $f_0(500)$ (left) and the $f_0(980)$ (right) regions, as a function of center-of-mass energy $\sqrt{s}$; from Ref. [59]. The integration over s gives the contribution to a_μ^{HLbL}.

4 Scalar contributions and $\pi\pi$ rescattering

In general, the contribution from single-particle poles to HLbL scattering depends on the choice of the basis for the HLbL tensor [25], with only the entire HLbL tensor basis independent by virtue of sum rules that may receive contributions from several (narrow) resonances of different quantum numbers at a time. Only pseudoscalar poles constitute an exception, making scalar resonances the first non-trivial test case. In Refs. [24, 25] it was shown that the dominant such contribution—S-wave $\pi\pi$ rescattering, which for isospin $I = 0$ implements the $f_0(500)$ resonance in a model-independent way—is largely basis independent by itself, since the two helicity components cancel in the corresponding sum rule to a large extent. Recently, this approach was generalized to the partial-wave helicity amplitudes for $\gamma^*\gamma^* \to \pi\pi/\bar{K}K$ [60–65], which allows one to also describe the $f_0(980)$ and compare to a narrow-width approximation (NWA) [59]:

$$a_\mu^{\text{HLbL}}[f_0(980)]\big|_{\text{rescattering}} = -0.2(1)\times 10^{-11}, \qquad a_\mu^{\text{HLbL}}[f_0(980)]\big|_{\text{NWA}} = -0.37(6)\times 10^{-11}, \quad (4)$$

where the TFFs from the quark model of Ref. [66] have been used, in line with the expected asymptotic behavior [67]. This shows that the result from the NWA comes out reasonably close to the full dispersive implementation, suggesting that at least for sufficiently narrow states a similar approach should be viable as well for axial-vector and tensor resonances. In addition, it was found that the combined contribution from S-waves, including also the $a_0(980)$, amounts to $a_\mu^{\text{HLbL}}[\text{scalars}] = -9(1)\times 10^{-11}$ [59], with the effect of yet heavier states small and very uncertain due to the lack of reliable TFF input. Given that heavy scalar contributions are not expected to play a special role in the implementation of short-distance constraints (SDCs), contrary to axial-vector resonances, and first enter $\gtrsim 1.5\,\text{GeV}$, such effects are best considered part of the asymptotic matching [29, 68, 69].

5 Axial-vector contributions and short-distance constraints

SDCs are available in two kinematic regimes, (i) one in which all three non-vanishing virtualities are large, and (ii) another in which one is much smaller than the other two. The result for the former case has only recently been put onto solid footing by demonstrating that the perturbative QCD quark loop does arise as the first term in a well-defined operator product expansion (OPE) [29]. While non-perturbative corrections prove negligible [68], the

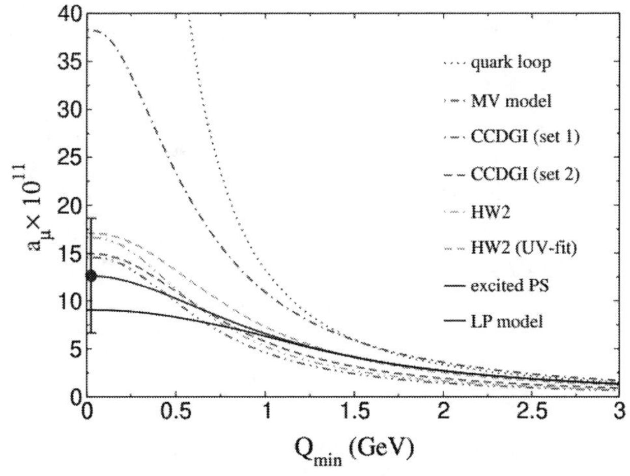

Figure 2. Comparison of the longitudinal short-distance contributions as a function of a common lower cutoff $Q_{\rm min}$ on the virtualities in the HLbL loop integral. The red dashed line refers to the perturbative QCD quark loop, the others to Ref. [19] (MV), Ref. [71] (CCDGI), Ref. [70] (HW2), Refs. [30, 31] (excited PS), and Ref. [74] (LP); from Ref. [75].

perturbative α_s corrections provide valuable insights into the onset of the asymptotic regime, typically scaling approximately as $1 - \alpha_s/\pi$ [69].

The SDCs in regime (ii) were derived in Ref. [19], with subsequent discussions how the respective constraints should be implemented. The simple model proposed in Ref. [19] remains valid in the chiral limit, but its simplicity comes at the price of neglecting 2π and 3π singularities that strongly affect the low-energy region of the HLbL integral. This was pointed out in Refs. [30, 31], in which a Regge model based on excited pseudoscalar mesons was proposed, exploiting the fact that these are well controlled theoretically and can provide a viable implementation at physical quark masses. Alternatively, a model for a tower of resonances in holographic QCD has been proposed [70, 71], which allows one to identify the model from Ref. [19] in a particular limit, while demonstrating at the same time that an additional contribution is necessary to fulfill the Landau–Yang theorem [72, 73], thus avoiding a sizable effect in the low-energy region. Third, an approach based on interpolants has been proposed in Ref. [74]. There is broad agreement among all three approaches as concerns the impact of the longitudinal SDCs, leading to contributions significantly smaller than the model from Ref. [19] would predict. A critical comparison of the different approaches can be found in Ref. [75], leading to the situation illustrated in Fig. 2.

In contrast, the role of the transverse SDCs, closely related to axial-vector degrees of freedom, is less well understood. There are a number of papers that analyze axial-vector contributions in a Lagrangian model [40, 42, 45, 76], but the combination with contributions evaluated in a dispersive framework remains to be understood. In particular, while it is possible to find a HLbL basis in which the axial-vector contributions coincide with the Lagrangian model [75], due to sum rules this choice of basis affects other contributions as well, and these consequences need to be carefully investigated.

In addition, information on the axial-vector TFFs is scarce, with the main source of information from the L3 measurements [77, 78] of $e^+e^- \to e^+e^-A$, $A = f_1, f_1'$, with additional input from $f_1 \to \rho\gamma$ [79, 80], $f_1 \to \phi\gamma$ [80, 81], and, most recently, $f_1 \to e^+e^-$ [82]. In particular, an improved measurement of the latter process would be extremely valuable to

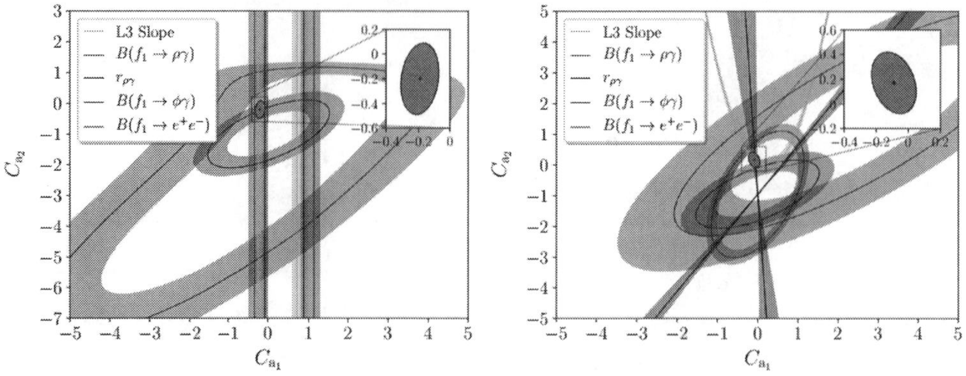

Figure 3. Constraints on the antisymmetric TFFs of the $f_1(1285)$ in terms of couplings C_{a_1}, C_{a_2}, for two variants of vector-meson-dominance parameterizations; from Ref. [83].

disentangle different TFFs, given that $e^+e^- \to e^+e^-A$ is primarily sensitive to the symmetric TFF only. A global analysis for the $f_1(1285)$, for which the experimental situation is best, was recently presented in Ref. [83], see Fig. 3. Estimates for the $f_1(1420)$ and the $a_1(1260)$ can then be obtained from $U(3)$ symmetry, with the mixing angle determined by the L3 measurements of the equivalent two-photon decay widths.

6 Conclusions

In these proceedings I reviewed the current status of HLbL scattering and discussed recent developments that aim at improving the precision towards the 10% level. While controlling subleading contributions in a data-driven way becomes increasingly challenging, recent work shows that it should be possible to achieve this goal with a combination of rigorous short-distance constraints and experimental input on the subleading intermediate states. Anticipating similar progress in lattice QCD, there should be two independent methods available to determine the HLbL contribution at the required level, with detailed comparisons potentially allowing for further improvements in precision.

I am grateful to Gilberto Colangelo, Bastian Kubis, and Peter Stoffer for longstanding collaboration on the subjects reviewed here and for carefully reading this manuscript. Support by the Swiss National Science Foundation, under Project No. PCEFP2_181117, is gratefully acknowledged.

References

[1] G.W. Bennett et al. (Muon $g-2$), Phys. Rev. D **73**, 072003 (2006), `hep-ex/0602035`
[2] B. Abi et al. (Muon $g-2$), Phys. Rev. Lett. **126**, 141801 (2021), `2104.03281`
[3] T. Albahri et al. (Muon $g-2$), Phys. Rev. D **103**, 072002 (2021), `2104.03247`
[4] T. Albahri et al. (Muon $g-2$), Phys. Rev. A **103**, 042208 (2021), `2104.03201`
[5] T. Albahri et al. (Muon $g-2$), Phys. Rev. Accel. Beams **24**, 044002 (2021), `2104.03240`
[6] T. Aoyama et al., Phys. Rept. **887**, 1 (2020), `2006.04822`
[7] T. Aoyama, M. Hayakawa, T. Kinoshita, M. Nio, Phys. Rev. Lett. **109**, 111808 (2012), `1205.5370`
[8] T. Aoyama, T. Kinoshita, M. Nio, Atoms **7**, 28 (2019)

[9] A. Czarnecki, W.J. Marciano, A. Vainshtein, Phys. Rev. D **67**, 073006 (2003), [Erratum: Phys. Rev. D **73**, 119901 (2006)], hep-ph/0212229
[10] C. Gnendiger, D. Stöckinger, H. Stöckinger-Kim, Phys. Rev. D **88**, 053005 (2013), 1306.5546
[11] M. Davier, A. Hoecker, B. Malaescu, Z. Zhang, Eur. Phys. J. C **77**, 827 (2017), 1706.09436
[12] A. Keshavarzi, D. Nomura, T. Teubner, Phys. Rev. D **97**, 114025 (2018), 1802.02995
[13] G. Colangelo, M. Hoferichter, P. Stoffer, JHEP **02**, 006 (2019), 1810.00007
[14] M. Hoferichter, B.L. Hoid, B. Kubis, JHEP **08**, 137 (2019), 1907.01556
[15] M. Davier, A. Hoecker, B. Malaescu, Z. Zhang, Eur. Phys. J. C **80**, 241 (2020), [Erratum: Eur. Phys. J. C **80**, 410 (2020)], 1908.00921
[16] A. Keshavarzi, D. Nomura, T. Teubner, Phys. Rev. D **101**, 014029 (2020), 1911.00367
[17] B.L. Hoid, M. Hoferichter, B. Kubis, Eur. Phys. J. C **80**, 988 (2020), 2007.12696
[18] A. Kurz, T. Liu, P. Marquard, M. Steinhauser, Phys. Lett. B **734**, 144 (2014), 1403.6400
[19] K. Melnikov, A. Vainshtein, Phys. Rev. D **70**, 113006 (2004), hep-ph/0312226
[20] G. Colangelo, M. Hoferichter, M. Procura, P. Stoffer, JHEP **09**, 091 (2014), 1402.7081
[21] G. Colangelo, M. Hoferichter, B. Kubis, M. Procura, P. Stoffer, Phys. Lett. B **738**, 6 (2014), 1408.2517
[22] G. Colangelo, M. Hoferichter, M. Procura, P. Stoffer, JHEP **09**, 074 (2015), 1506.01386
[23] P. Masjuan, P. Sánchez-Puertas, Phys. Rev. D **95**, 054026 (2017), 1701.05829
[24] G. Colangelo, M. Hoferichter, M. Procura, P. Stoffer, Phys. Rev. Lett. **118**, 232001 (2017), 1701.06554
[25] G. Colangelo, M. Hoferichter, M. Procura, P. Stoffer, JHEP **04**, 161 (2017), 1702.07347
[26] M. Hoferichter, B.L. Hoid, B. Kubis, S. Leupold, S.P. Schneider, Phys. Rev. Lett. **121**, 112002 (2018), 1805.01471
[27] M. Hoferichter, B.L. Hoid, B. Kubis, S. Leupold, S.P. Schneider, JHEP **10**, 141 (2018), 1808.04823
[28] A. Gérardin, H.B. Meyer, A. Nyffeler, Phys. Rev. D **100**, 034520 (2019), 1903.09471
[29] J. Bijnens, N. Hermansson-Truedsson, A. Rodríguez-Sánchez, Phys. Lett. B **798**, 134994 (2019), 1908.03331
[30] G. Colangelo, F. Hagelstein, M. Hoferichter, L. Laub, P. Stoffer, Phys. Rev. D **101**, 051501 (2020), 1910.11881
[31] G. Colangelo, F. Hagelstein, M. Hoferichter, L. Laub, P. Stoffer, JHEP **03**, 101 (2020), 1910.13432
[32] T. Blum, N. Christ, M. Hayakawa, T. Izubuchi, L. Jin, C. Jung, C. Lehner, Phys. Rev. Lett. **124**, 132002 (2020), 1911.08123
[33] G. Colangelo, M. Hoferichter, A. Nyffeler, M. Passera, P. Stoffer, Phys. Lett. B **735**, 90 (2014), 1403.7512
[34] S. Borsanyi et al., Nature **593**, 51 (2021), 2002.12347
[35] C. Lehner, A.S. Meyer, Phys. Rev. D **101**, 074515 (2020), 2003.04177
[36] A. Crivellin, M. Hoferichter, C.A. Manzari, M. Montull, Phys. Rev. Lett. **125**, 091801 (2020), 2003.04886
[37] A. Keshavarzi, W.J. Marciano, M. Passera, A. Sirlin, Phys. Rev. D **102**, 033002 (2020), 2006.12666

[38] B. Malaescu, M. Schott, Eur. Phys. J. C **81**, 46 (2021), 2008.08107
[39] G. Colangelo, M. Hoferichter, P. Stoffer, Phys. Lett. B **814**, 136073 (2021), 2010.07943
[40] V. Pauk, M. Vanderhaeghen, Eur. Phys. J. C **74**, 3008 (2014), 1401.0832
[41] I. Danilkin, M. Vanderhaeghen, Phys. Rev. D **95**, 014019 (2017), 1611.04646
[42] F. Jegerlehner, Springer Tracts Mod. Phys. **274**, 1 (2017)
[43] M. Knecht, S. Narison, A. Rabemananjara, D. Rabetiarivony, Phys. Lett. B **787**, 111 (2018), 1808.03848
[44] G. Eichmann, C.S. Fischer, R. Williams, Phys. Rev. D **101**, 054015 (2020), 1910.06795
[45] P. Roig, P. Sánchez-Puertas, Phys. Rev. D **101**, 074019 (2020), 1910.02881
[46] E.H. Chao, R.J. Hudspith, A. Gérardin, J.R. Green, H.B. Meyer, K. Ottnad, Eur. Phys. J. C **81**, 651 (2021), 2104.02632
[47] J. Grange et al. (Muon $g-2$) (2015), 1501.06858
[48] M. Abe et al., PTEP **2019**, 053C02 (2019), 1901.03047
[49] M. Aiba et al. (2021), 2111.05788
[50] J. Prades, E. de Rafael, A. Vainshtein, Adv. Ser. Direct. High Energy Phys. **20**, 303 (2009), 0901.0306
[51] A. Nyffeler, Phys. Rev. D **79**, 073012 (2009), 0901.1172
[52] F. Jegerlehner, A. Nyffeler, Phys. Rept. **477**, 1 (2009), 0902.3360
[53] M. Hoferichter, G. Colangelo, M. Procura, P. Stoffer, Int. J. Mod. Phys. Conf. Ser. **35**, 1460400 (2014), 1309.6877
[54] I. Larin et al. (PrimEx-II), Science **368**, 506 (2020)
[55] M. Ablikim et al. (BESIII), Chin. Phys. C **44**, 040001 (2020), 1912.05983
[56] S. Burri (2021), https://indico.cern.ch/event/1006302/contributions/4373340/attachments/2287903/3888993/burri_lattice21.pdf
[57] W. Verplanke (2021), https://indico.cern.ch/event/1006302/contributions/4375437/attachments/2288167/3889500/VERPLANKE_LATTICE2021.pdf
[58] S. Holz, J. Plenter, C.W. Xiao, T. Dato, C. Hanhart, B. Kubis, U.G. Meißner, A. Wirzba, Eur. Phys. J. C **81**, 1002 (2021), 1509.02194
[59] I. Danilkin, M. Hoferichter, P. Stoffer, Phys. Lett. B **820**, 136502 (2021), 2105.01666
[60] R. García-Martín, B. Moussallam, Eur. Phys. J. C **70**, 155 (2010), 1006.5373
[61] M. Hoferichter, D.R. Phillips, C. Schat, Eur. Phys. J. C **71**, 1743 (2011), 1106.4147
[62] B. Moussallam, Eur. Phys. J. C **73**, 2539 (2013), 1305.3143
[63] I. Danilkin, M. Vanderhaeghen, Phys. Lett. B **789**, 366 (2019), 1810.03669
[64] M. Hoferichter, P. Stoffer, JHEP **07**, 073 (2019), 1905.13198
[65] I. Danilkin, O. Deineka, M. Vanderhaeghen, Phys. Rev. D **101**, 054008 (2020), 1909.04158
[66] G.A. Schuler, F.A. Berends, R. van Gulik, Nucl. Phys. B **523**, 423 (1998), hep-ph/9710462
[67] M. Hoferichter, P. Stoffer, JHEP **05**, 159 (2020), 2004.06127
[68] J. Bijnens, N. Hermansson-Truedsson, L. Laub, A. Rodríguez-Sánchez, JHEP **10**, 203 (2020), 2008.13487
[69] J. Bijnens, N. Hermansson-Truedsson, L. Laub, A. Rodríguez-Sánchez, JHEP **04**, 240 (2021), 2101.09169
[70] J. Leutgeb, A. Rebhan, Phys. Rev. D **101**, 114015 (2020), 1912.01596

[71] L. Cappiello, O. Catà, G. D'Ambrosio, D. Greynat, A. Iyer, Phys. Rev. D **102**, 016009 (2020), 1912.02779
[72] L.D. Landau, Dokl. Akad. Nauk SSSR **60**, 207 (1948)
[73] C.N. Yang, Phys. Rev. **77**, 242 (1950)
[74] J. Lüdtke, M. Procura, Eur. Phys. J. C **80**, 1108 (2020), 2006.00007
[75] G. Colangelo, F. Hagelstein, M. Hoferichter, L. Laub, P. Stoffer, Eur. Phys. J. C **81**, 702 (2021), 2106.13222
[76] P. Masjuan, P. Roig, P. Sánchez-Puertas (2020), 2005.11761
[77] P. Achard et al. (L3), JHEP **03**, 018 (2007)
[78] P. Achard et al. (L3), Phys. Lett. B **526**, 269 (2002), hep-ex/0110073
[79] D.V. Amelin et al., Z. Phys. C **66**, 71 (1995)
[80] P.A. Zyla et al. (Particle Data Group), PTEP **2020**, 083C01 (2020)
[81] S.I. Bityukov et al., Phys. Lett. B **203**, 327 (1988)
[82] M.N. Achasov et al. (SND), Phys. Lett. B **800**, 135074 (2020), 1906.03838
[83] M. Zanke, M. Hoferichter, B. Kubis, JHEP **07**, 106 (2021), 2103.09829

Short-distance constraints for HLbL in muon g-2

Johan Bijnens[1,*], *Nils* Hermansson-Truedsson[2,**], and *Antonio* Rodríguez-Sánchez[3,***]

[1]Department of Astronomy and Theoretical Physics, Lund University, Box 43, SE221-00 Lund, Sweden
[2]Albert Einstein Center for Fundamental Physics, Institute for Theoretical Physics, Universität Bern, Sidlerstrasse 5, CH–3012 Bern, Switzerland
[3]Université Paris-Saclay, CNRS/IN2P3, IJCLab, 91405 Orsay, France

Abstract. Model-independent short-distance constraints allow for a reduction of theoretical uncertainties associated to the analytic evaluation of Hadronic Light-by-Light contributions to the muon g-2. In this talk we focus on the region where the three loop virtualities are large. Even when the fourth photon leg is soft, we show how a precise Operator Product Expansion can be applied in that region. The leading contribution is found to be given by the quark loop, while the evaluation of both gluonic and power corrections show how the expansion is well behaved at relatively low energies, where significant contributions to the muon g-2 remain. Numerical values for them are also presented.

1 Introduction

A new measurement of the anomalous magnetic moment of the muon, coming from the FNAL Muon g-2 Experiment, was released this year [1], updating the combined experimental average to

$$a_\mu^{\rm exp} = 11659\,2061(41)\cdot 10^{-11}\,, \tag{1}$$

and confirming the tension with the SM prediction of Ref. [2],

$$a_\mu^{\rm SM} = 11659\,1810(43)\cdot 10^{-11}\,. \tag{2}$$

Such a remarkable theoretical precision is in part a consequence of the strong dominance of pure QED contributions, whose theory uncertainties are completely negligible compared to the experimental result. Electroweak corrections are very small and well known too, so they cannot, within the SM, account for such a tension. A discrepancy of $\sim 2.5\cdot 10^{-8}$ with perfectly controlled corrections from strong interactions would immediately translate into evidence of BSM physics. In order to achieve such a control, strong efforts are being made by both the data-driven and lattice approaches in the two relevant contributions: Hadronic Vacuum Polarization (HVP) and Hadronic Light-by-Light contributions (HLbL). The focus of our work, [3–5], has been oriented to improve the data-driven evaluation of the latter.

2 Hadronic Light-by-Light contributions to the muon g-2

The Hadronic Light-by-Light contribution to g-2 is depicted in Fig. 1. The key piece for its calculation is the corresponding HLbL tensor, defined as

*e-mail: johan.bijnens@thep.lu.se
**e-mail: nils@itp.unibe.ch
***Speaker, e-mail: arodriguez@ijclab.in2p3.fr

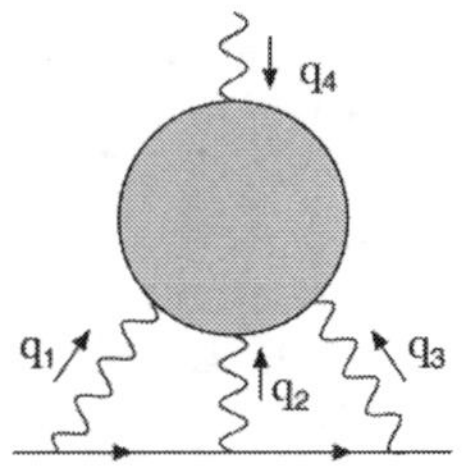

Figure 1. HLbL contribution to g-2.

$$\Pi^{\mu_1\mu_2\mu_3\mu_4} \equiv -i \int \frac{d^4 q_4}{(2\pi)^4} \left(\prod_{i=1}^{4} \int d^4 x_i \, e^{-iq_i x_i} \right) \langle 0 | T \left(\prod_{j=1}^{4} J^{\mu_j}(x_j) \right) | 0 \rangle \,, \tag{3}$$

where $J^\mu(x) = \bar{q}(x)\, Q_q \gamma^\mu q(x)$ is the electromagnetic quark current. The correponding g-2 number is obtained when convoluting

$$\lim_{q_4 \to 0} \frac{\partial \Pi^{\mu_1\mu_2\mu_3\nu_4}}{\partial q_4^{\mu_4}} \,, \tag{4}$$

with the other two loops [6, 7]

$$a_\mu^{\rm HLbL} = \frac{2\alpha^3}{3\pi^2} \int_0^\infty dQ_1 \int_0^\infty dQ_2 \int_{-1}^{1} d\tau \, \sqrt{1-\tau^2}\, Q_1^3 Q_2^3 \sum_{i=1}^{12} T_i(Q_1, Q_2, \tau) \overline{\Pi}_i(Q_1, Q_2, \tau) \,. \tag{5}$$

The $\overline{\Pi}_i$ functions only depend on the five independent scalar functions of the Lorentz decomposition of Eq. (4) [5]. The main difficulty in the evaluation of this contribution is finding those functions. The weights T_i enhance the low-energy contributions, since the scale of the problem is the muon mass. Perturbative QCD cannot be used for the tensor of Eq. (3) at that energy and non-perturbative methods have to be applied. QCD-inspired evaluations of it gave a first assessment of this contribution [8–14], but a more model-independent evaluation of it was desirable. That possibility arrived with the data-driven approach of Ref. [6], where a model-independent way of evaluating the different leading long-distance contributions was assessed.[1]

While systematic improvements on them are possible, the residual contributions from medium and short distances are more challenging. One of the expected contributions at short distances was the quark loop, but the low-energy scale of the muon mass and the static g-2 photon requires some lower cut-off, since the logarithmic mass divergences associated to the light-by-light contributions do not make much sense for light-quarks. There are no physical hadronic states with such small masses. A constituent quark mass was typically used as a regulator (see for example Ref. [9]). However, in the context of a data-driven evaluation, such a model-dependent solution was not satisfactory. Our work has been focused on the interplay of the short-distance contributions to g-2 within QCD, aiming to see whether and how the quark loop is the leading term of any asymptotic expansion in the g-2 kinematics.

[1] Very recently lattice methods have achieved a similar precision [15]. The very good agreement between two largely model-independent and completely different approaches reinforces the idea that a miss-evaluation of HLbL cannot be behind the current tension in muon g-2.

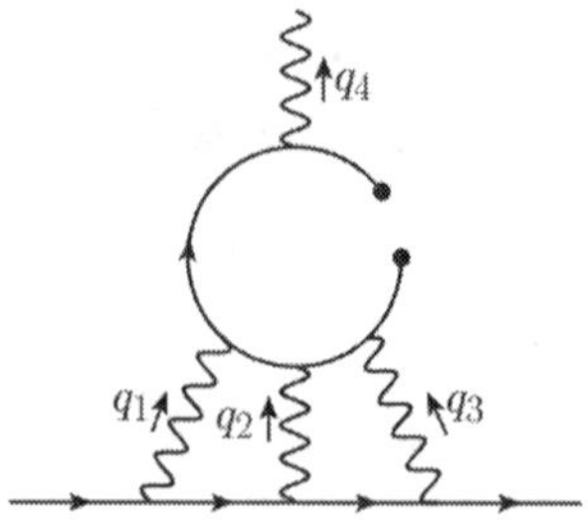

Figure 2. Quark condensate contribution to HLbL. In the static $q_4 \to 0$ limit the quark propagator on the left diverges.

3 The Operator Product Expansion in the QCD vacuum

Two-point correlation functions of quark currents J_1 and J_2,

$$\Pi(q) = \int d^4x\, e^{-iqx} \langle 0|T(J_1(x)J_2(0))|0\rangle \,, \tag{6}$$

are described at large Euclidean Momenta by its Operator Product Expansion (OPE) in the presence of the QCD vacuum [16],

$$\Pi \approx \sum_{i,D} \frac{c_{i,D} O_{i,D}}{Q^D} \,, \tag{7}$$

where D is the dimension of the associated operator $O_{i,D}$, $c_{i,D}$ are c-numbers, the Wilson coefficients, and $Q^2 = -q^2$. Any possible operator with the same quantum numbers as the QCD vacuum can give a nonzero contribution, starting by unity, which provides the perturbative series, and supplemented by operators with quarks and gluons such as the scalar current $\bar{q}q$, which gives rise to the well-known quark condensate, $\langle \bar{q}q \rangle$. A known limitation of this description for some of these expansions is in the separation of the tail of the perturbative expansion, of asymptotic nature, and the effect of vacuum condensates. Some possible prescriptions on how to separate those effects can be found in the literature (e.g. see [17]). The same OPE can be applied to correlation functions with a higher number of external legs. A detailed study of the OPE of three-point correlation functions can be found in Ref. [18].

We then have an asymptotic expansion describing the behaviour of the HLbL tensor of Eq. (3) at large Euclidean momenta, whose leading contribution is clearly given by the quark loop. Its applicability to the muon g-2 kinematics is, however, not obvious even when we explore the region with three large (Euclidean, $Q_{1,2,3}$) loop momenta. This is because the fourth (g-2) momentum is defined in the static $q_4 \to 0$ limit, where the use of such a vacuum OPE is not justified. Indeed, an explicit exploration shows how the first dimensional correction, the one associated to the quark condensate, presents divergences in that limit from the diagram shown in Fig. 2 [3]. This observation may cast some doubts on the validity of the quark loop as the leading term of the (subleading) short-distance contributions to HLbL g-2.

4 The Operator Product Expansion in the QCD vacuum with an external background photon field

Completely analogous problems for extrapolating the QCD vacuum OPE to the static-momentum regime were found when trying to describe the nucleon magnetic moments

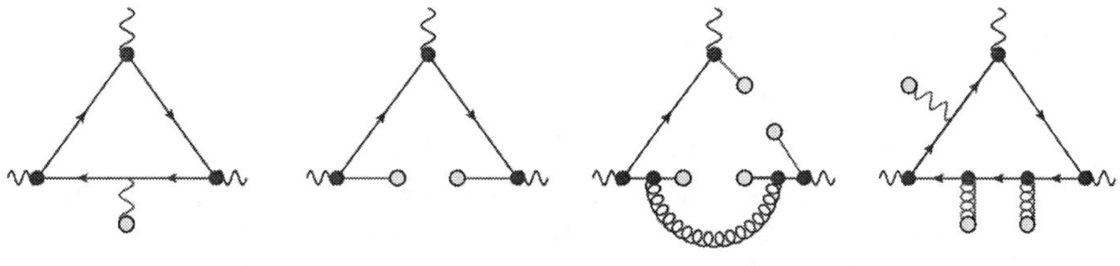

Figure 3. Topologies giving the different contributions to the studied OPE.

through sum rules. A very elegant solution was presented in Refs. [19, 20], and some applications to other muon contributions can actually be found in the literature, e.g. [21]. The essential point is that the OPE should be performed keeping the external g-2 photon as an explicit external (background) state. This is applying the OPE to the (equivalent) tensor

$$\Pi^{\mu_1\mu_2\mu_3}(q_1,q_2) \equiv -\frac{1}{e}\int \frac{d^4q_3}{(2\pi)^4}\left(\prod_{i=1}^{3}\int d^4x_i\, e^{-iq_ix_i}\right)\langle 0|T\left(\prod_{j=1}^{3} J^{\mu_j}(x_j)\right)|\gamma_E(q_4)\rangle\,, \tag{8}$$

where the $\gamma_E(q_4)$ is the external static photon. Essentially, from such an OPE one needs to keep not those local operators with the same quantum numbers as the QCD vacuum, but those operators with the same quantum numbers as the external g-2 photon, $F_{\mu\nu}$,

$$Q_{1,\mu\nu} \equiv e\, e_q F_{\mu\nu}, \quad Q_{2,\mu\nu} \equiv \bar{q}\sigma_{\mu\nu}q \quad , \ldots \tag{9}$$

The leading contribution is associated to the lowest dimensional operator, Q_1. This is represented in the first diagram from the left of Fig. 3. We showed in Ref. [3] how, up to gluonic corrections, it corresponds to the massless quark loop, which becomes the leading term in a well-defined expansion.

5 Power corrections

Some remarkable things occur when trying to assess the leading quark mass corrections. One would naively expect that the perturbative mass correction should be suppressed by a factor

$$\frac{\sum_q m_q^2 e_q^4}{Q^2 \sum_q e_q^4}\,, \tag{10}$$

which at $Q_1 \approx Q_2 \approx Q_3 = 1\,\text{GeV}$ gives, taking $m_u = m_d = 5\,\text{MeV}$ and $m_s = 100\,\text{MeV}$, $\sim 6\cdot 10^{-4}$. An explicit computation gives instead $\sim 5\cdot 10^{-3}$. A careful inspection of the quark loop mass corrections shows that they scale as [4]

$$\frac{\sum_q m_q^2 e_q^4 \ln \frac{Q^2}{m_q^2}}{Q^2 \sum_q e_q^4}\,, \tag{11}$$

which explains the enhancement. However, perturbative QCD series with such mass singularities diverge and then such a enhancement is, in principle, unphysical. The standard operator renormalization allows for a complete separation of the contributions in short-distance pieces $c_O(\mu)$ and long-distance matrix elements $\langle O(\mu)\rangle$ that must be evaluated nonperturbatively [22]. In the OPE in the vacuum there are no (relative to the leading order pieces) $D = 2$

operators to regularize $m_q^2 \ln \frac{Q^2}{m_q^2}$ divergences, but they do not appear when dealing with physical observables [22, 23]. They start at $m_q^4 \ln \frac{Q^2}{m_q^2}$ and are regularized by the quark and gluon condensates. In this case they are regularized by the QED mixing of the tensor current $\bar{q}\sigma^{\mu\nu}q$ with the leading $F_{\mu\nu}$ contribution [4]. The long-distance quark mass dependence gets absorbed into the mass corrections of the nonperturbative matrix element $\langle 0|\bar{q}\sigma^{\mu\nu}q(\mu)|\gamma_E(q_4)\rangle$.

However, in the neighbourhood of the $m_q \to 0$ limit, the leading quark mass correction is not quadratic but linear in the mass and is given by the contribution of the second diagram in Fig 3 to the tensor current, which has a non-zero expectation value in the presence of an external background photon even in the strict chiral limit. This effect of spontaneous chiral symmetry breaking is known as the magnetic susceptibility of the QCD vacuum and is trivially related to the zero momentum limit of $\Pi_{VT}(q)$. The explicit results associated to these contributions, which scale as

$$\frac{(4\pi)^2 m_q X_q}{Q^2} \sim \frac{\Lambda_\chi^2}{M_\rho^2}\frac{M_{\pi(K)}^2}{Q^2}, \tag{12}$$

can be found in Refs. [3, 4] and may be used as short-distance constraints for possible evaluations of mass corrections to chiral limit evaluations of HLbL.

Beyond the chiral limit, the leading power corrections are suppressed by four powers of the large loop momenta. The associated topologies are the ones of Fig. 3. Details in the computation and explicit expressions for them can be found in Ref. [4]. We have checked that, as far as one is above 1 GeV, the power corrections are typically suppressed by at least two orders of magnitude with respect to the leading massless quark loop and can be safely neglected.

6 The two-loop corrections

In principle the gluonic corrections to the HLbL tensor contain many different two-loop diagrams which depend on where to connect the gluon lines, also with respect to the soft static photon. However, a simplification occurs when realizing that the color structure carried by the gluon can be factored out and it can be regarded as an extra "photon". Then, before contracting the gluon propagator and setting the external g-2 momentum to zero, we simply have a fully symmetric sum of hexagons. It is then a natural step to take advantage of this strong symmetry to eventually break it with both the soft-photon and the second (gluon-propagator) loop. This very simple set-up leads to a very large number of two-loop integrals. But then we can use general projectors $P^{\tilde{\Pi}_i}_{\mu_1\mu_2\mu_3\mu_4\nu_4}$ to the five independent scalar functions $\tilde{\Pi}_i$ of the HLbL g-2 tensor. One finds [5]

$$\begin{aligned}\tilde{\Pi}_i &= P^{\tilde{\Pi}_i}_{\mu_1\mu_2\mu_3\mu_4\nu_4} \lim_{q_4\to 0} \frac{\partial \Pi^{\mu_1\mu_2\mu_3\nu_4}}{\partial q_4^{\mu_4}} \\ &= -\frac{(N_c^2-1)g_s^2 e_q^4}{4}\int \frac{d^4 q_5}{(2\pi)^4}\frac{g_{\mu_5\mu_6}}{q_5^2} \lim_{\substack{q_4\to 0\\ q_6\to -q_5}} P^{\tilde{\Pi}_i}_{\mu_1\mu_2\mu_3\mu_4\nu_4}\frac{\partial}{\partial q_4^{\nu_4}} H^{\mu_1\mu_2\mu_3\mu_4\mu_5\mu_6},\end{aligned} \tag{13}$$

where

$$\begin{aligned}H^{\mu_1\mu_2\mu_3\mu_4\mu_5\mu_6} \equiv \int \frac{d^4 p}{(2\pi)^4} \sum_{\sigma(1,2,4,5,6)} \mathrm{Tr}\Big(\gamma^{\mu_3} S(p+q_1+q_2+q_4+q_5+q_6)\gamma^{\mu_1} S(p+q_2+q_4+q_5+q_6) \\ \times \gamma^{\mu_2} S(p+q_4+q_5+q_6)\gamma^{\mu_4} S(p+q_5+q_6)\gamma^{\mu_5} S(p+q_6)\gamma^{\mu_6} S(p)\Big).\end{aligned} \tag{14}$$

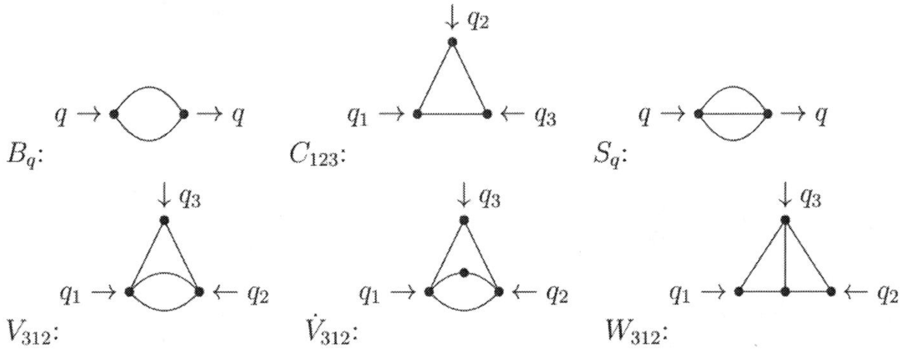

Figure 4. Master integrals entering into the two-loop calculation after applying scalar reduction.

	$\tilde{\Pi}_1$	$\tilde{\Pi}_7$	$\tilde{\Pi}_{10}$	$\tilde{\Pi}_{13}$	$\tilde{\Pi}_{19}$
Quark loop	−0.0816	0.123	0.0363	0.0274	0.0263
Gluon corrections ($\times\pi/\alpha_s$)	0.0781	−0.136	−0.0376	−0.0398	−0.0411

Table 1. Values for the quark loop and gluonic correction contributions to the $\tilde{\Pi}$ in GeV units for a benchmark tuple $(Q_1^2, Q_2^2, Q_3^2) = (1, 1.3, 1.7)$ GeV^2. Sum over the three flavours has been made. The last line is in units of α_s/π.

$S(p) = \frac{\not{p}}{p^2}$ is the massless quark propagator and $\sigma(1,2,4,5,6)$ the set of pairwise permutations of μ_i and q_i for $i = 1, 2, 3, 5, 6$.

Now we "simply" need to deal with $\sim 10^{3,4}$ different scalar two-loop integrals, depending on the set of used projectors (which are related by gauge invariance). In order to get the task done we take advantage of KIRA [24], which is a software that allows for a reduction of multi-loop scalar integrals into a set of master integrals. We find that there is a set of them, corresponding to the integrals of Fig 4, for which all the needed $\frac{1}{D-4}$ expansions have been worked out analytically in terms of classical poly-logs [25].

We then have fully analytic results for the five independent scalar functions of the HLbL tensor encoding the g-2 kinematics, which are valid for the regime of large momenta, including the leading term, power and gluonic corrections. We find that gluonic corrections are typically suppressed with respect to the quark loop by a factor of approximately $-\frac{\alpha_s}{\pi}$.[2] Numerical results for a benchmark point are shown in Table 1.

Now it is a matter of convoluting with the other two loops and numerically integrate gluonic corrections, using Eq. (5), for the kinematics regions where the expansion is valid. This is $Q_{1,2,3} > Q_{\min}$. Having computed the power and gluonic corrections, we can actually assess from which values of $Q_{\min}$ the expansion is safe. The result for the leading massless quark loop and the gluonic corrections is displayed in Fig. 5. We observe how, as far as the running of the strong coupling is well defined, the gluonic corrections are rather small and negative. As a consequence, the obtained expansion gives a very precise description of the (subleading, but not negligible) fully short-distance contributions of HLbL muon g-2.

[2]Before realizing of a typo in one of the analytic expressions for the master integrals presented in Ref. [25], the obtained prefactor in front of $\frac{\alpha_s}{\pi}$ was $O(10^{2,3})$, which would have completely killed the applicability of this expansion.

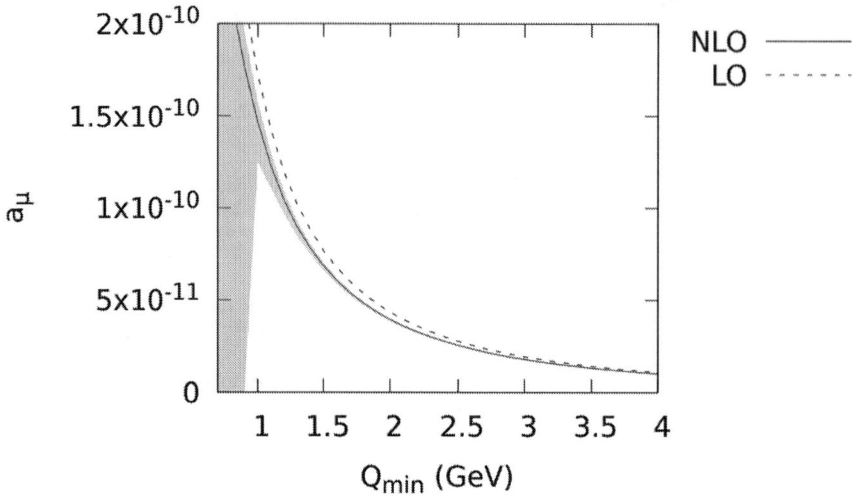

Figure 5. Fully short-distance contributions for HLbL muon g-2, as obtained by our expansion. The leading term (dashed-line) corresponds to the massless quark loop and the NLO one includes the gluonic corrections.

In order to improve the numerical precision, explicit analytic expansions have been performed in some kinematic regions to cancel spurious singularities. One of them corresponds to a sub-region of the Melnikov-Vainshtein short-distance constraints [26]. We have checked how we recover the correct limits for both the quark loop and the gluonic corrections [27, 28].

7 Conclusions

We have shown how, in spite of the soft-momentum associated to the external g-2 photon, an OPE can be applied for the HLbL muon g-2 contribution when the three loop momenta are large. The leading contribution is given by the massless quark loop [3].

The leading power correction is found to be suppressed by two powers of the energy, from which one corresponds to the quark mass. In the chiral limit, the first power corrections appear suppressed by four powers of the energy. We have worked out explicit expressions for them in Ref. [4]. They are found to be negligible as far as one is above ~ 1 GeV.

Finally, we have computed the gluonic corrections. The relatively small prefactors in the $\frac{\alpha_s}{\pi}$ expansion confirms that it is well behaved from relatively low energies and can be used to improve the precision of the data-driven evaluation of muon g-2.

Acknowledgements

We thank Laetitia Laub for a fruitful and enjoyable collaboration. This research is supported in part by the Albert Einstein Center for Fundamental Physics at Universität Bern (NHT), the Swedish Research Council grants contract numbers 2016-05996 and 2019-03779 (JB) and the Agence Nationale de la Recherche (ANR) under grant ANR-19-CE31-0012 (project MORA) (ARS).

References

[1] B. Abi et al. (Muon g-2), Phys. Rev. Lett. **126**, 141801 (2021), 2104.03281

[2] T. Aoyama et al., Phys. Rept. **887**, 1 (2020), 2006.04822

[3] J. Bijnens, N. Hermansson-Truedsson, A. Rodríguez-Sánchez, Phys. Lett. **B798**, 134994 (2019), 1908.03331

[4] J. Bijnens, N. Hermansson-Truedsson, L. Laub, A. Rodríguez-Sánchez, JHEP **10**, 203 (2020), 2008.13487

[5] J. Bijnens, N. Hermansson-Truedsson, L. Laub, A. Rodríguez-Sánchez, JHEP **04**, 240 (2021), 2101.09169

[6] G. Colangelo, M. Hoferichter, M. Procura, P. Stoffer, JHEP **09**, 074 (2015), 1506.01386

[7] G. Colangelo, M. Hoferichter, M. Procura, P. Stoffer, JHEP **04**, 161 (2017), 1702.07347

[8] J. Bijnens, E. Pallante, J. Prades, Phys. Rev. Lett. **75**, 1447 (1995), [Erratum: Phys.Rev.Lett. 75, 3781 (1995)], hep-ph/9505251

[9] J. Bijnens, E. Pallante, J. Prades, Nucl. Phys. B **474**, 379 (1996), hep-ph/9511388

[10] M. Hayakawa, T. Kinoshita, Phys. Rev. D **57**, 465 (1998), [Erratum: Phys.Rev.D 66, 019902 (2002)], hep-ph/9708227

[11] J. Bijnens, E. Pallante, J. Prades, Nucl. Phys. B **626**, 410 (2002), hep-ph/0112255

[12] M. Hayakawa, T. Kinoshita (2001), hep-ph/0112102

[13] J. Prades, E. de Rafael, A. Vainshtein, Adv. Ser. Direct. High Energy Phys. **20**, 303 (2009), 0901.0306

[14] F. Jegerlehner, *The Anomalous Magnetic Moment of the Muon*, Vol. 274 (Springer, Cham, 2017)

[15] E.H. Chao, R.J. Hudspith, A. Gérardin, J.R. Green, H.B. Meyer, K. Ottnad, Eur. Phys. J. C **81**, 651 (2021), 2104.02632

[16] M.A. Shifman, A.I. Vainshtein, V.I. Zakharov, Nucl. Phys. **B147**, 385 (1979)

[17] M. Jamin (2021), 2106.01614

[18] T. Kadavý, K. Kampf, J. Novotny, JHEP **10**, 142 (2020), 2006.13006

[19] B.L. Ioffe, A.V. Smilga, Nucl. Phys. **B232**, 109 (1984)

[20] I.I. Balitsky, A.V. Yung, Phys. Lett. **129B**, 328 (1983)

[21] A. Czarnecki, W.J. Marciano, A. Vainshtein, Phys. Rev. **D67**, 073006 (2003), [Erratum: Phys. Rev. **D73**, 119901 (2006)], hep-ph/0212229

[22] E. Braaten, S. Narison, A. Pich, Nucl. Phys. **B373**, 581 (1992)

[23] A. Pich, J. Prades, JHEP **06**, 013 (1998), hep-ph/9804462

[24] P. Maierhöfer, J. Usovitsch, P. Uwer, Comput. Phys. Commun. **230**, 99 (2018), 1705.05610

[25] F. Chavez, C. Duhr, JHEP **11**, 114 (2012), 1209.2722

[26] K. Melnikov, A. Vainshtein, Phys. Rev. **D70**, 113006 (2004), hep-ph/0312226

[27] J. Lüdtke, M. Procura, Eur. Phys. J. C **80**, 1108 (2020), 2006.00007

[28] G. Colangelo, F. Hagelstein, M. Hoferichter, L. Laub, P. Stoffer, Eur. Phys. J. C **81**, 702 (2021), 2106.13222

EPJ Web of Conferences **258**, 06006 (2022) https://doi.org/10.1051/epjconf/202225806006
A Virtual Tribute to Quark Confinement and the Hadron Spectrum (vConf21)

CP violation in charm

Jakub Ryżka on behalf of the LHCb Collaboration[1,*]

[1]AGH, University of Science and Technology, Cracow

Abstract. The existence of *CP* violation in the decays of strange and beauty mesons is very well established experimentally. On the contrary, *CP* violation in the decays of charmed particles has been elusive for a long time and has been observed for the first time in 2019 by the LHCb experiment. Since then several studies have been performed in the charm sector. During the LHC Run 1 and Run 2, the LHCb collaboration has collected large samples containing charm hadron decays, on a scale never seen before. Collected data enabled physicists to obtain several new results, most of which surpassed previous results and became new world's best measurements. Presently the LHCb spectrometer is being upgraded to enhance readout system, improve subdetector components and increase integrated luminosity to 50 fb^{-1} by the end of Run 4.

1 Introduction

CP symmetry (charge conjugation parity symmetry) states that laws of physics should be the same if a particle is interchanged with its antiparticle (charge conjugation symmetry) and its spatial coordinates are inversed (parity symmetry). However, in case of weak interactions, the matter-antimatter symmetry can be violated under *CP* transformation. The weak interactions of the quarks are described by Cabibbo-Kobayashi-Masakawa (CKM) matrix [1, 2]. *CP* violation in the Standard Model (SM) of particle physics is included through a complex phase in the CKM quark-mixing matrix. There are three ways to violate *CP* symmetry. The first one is induced by neutral mesons mixing and is related to the difference in the time-dependent transition probability of mesons and antimesons. Secondly, *CP* symmetry can be broken by the difference in the decay amplitude of a particle in comparison to that of its own antiparticle. Third way is related to the interference between decay and mixing [3] to the same final state. The existence of *CP* violation in the decays of strange and beauty mesons is very well established experimentally. Nonetheless, the level of *CP* violation in the standard model is not sufficient to fully explain the disparity of matter and antimatter in the observable universe, implying the necessity of the presence of additional sources of *CP* violation beyond those known in the SM.

2 LHCb detector

The LHCb detector [4, 5], is a forward spectrometer covering the pseudorapidity range $2 < \eta < 5$. It operates at Large Hadron Collider (LHC) at the European Organization for Nuclear Research (CERN). The LHCb detector was built to perform precise measurement of matter -

*e-mail: jakub.ryzka@cern.ch

anti-matter asymmetry in the heavy flavour sector (beauty and charm quarks) and search for New Physics beyond the SM.

3 Selected LHCb results

During last few years LHCb experiment has provided several new results in beauty and charm sectors, some of which represent new world's best measurements. The latest selected charm results are presented below.

3.1 First observation of *CPV* in charm

After decades of experimental searches, the *CP* violation in the charm sector has been observed in the $D^0 \to K^+K^-$ and $D^0 \to \pi^+\pi^-$ decays, using the large samples of charmed hadrons collected in Run 1 and Run 2.

The time-dependent *CP* asymmetry $A_{CP}(f,t)$ between states produced as D^0 or $\bar{D}^0$ mesons decaying to a *CP* eigenstate f ($f = K^+K^-$ or $\pi^+\pi^-$) at time t is defined as:

$$A_{CP} = \frac{\Gamma(D^0(t) \to f) - \Gamma(\bar{D}^0(t) \to f)}{\Gamma(D^0(t) \to f) + \Gamma(\bar{D}^0(t) \to f)}, \tag{1}$$

where Γ stands for the time-dependent rate of a given decay. The result of this analysis has been reported as the difference in *CP* asymmetries (ΔA_{CP}) of the decays, which is defined as:

$$\Delta A_{CP} \equiv A_{CP}(K^-K^+) - A_{CP}(\pi^-\pi^+) = A_{raw}(K^-K^+) - A_{raw}(\pi^-\pi^+). \tag{2}$$

By combining the Run 2 results with the earlier LHCb measurements [6, 7], value of ΔA_{CP} is obtained as:

$$\Delta A_{CP} = (-15.4 \pm 2.9) \times 10^{-4}. \tag{3}$$

Both statistical and systematic uncertainties are incorporated in one value. The significance of the deviation from zero corresponds to 5.3 standard deviations. This is the first observation of *CP* violation in the charm meson decays [8].

3.2 Time-dependent $D^0 - \bar{D}^0$ asymmetry

The Cabibbo-suppressed $D^0 \to f$ decays, where typical D^0 and $\bar{D}^0$ final states consist of either a pair of charged kaons or pions, provide one of the most sensitive tests of time-dependent *CP* violation between D^0 and $\bar{D}^0$ decay rates as defined in eq. (1). The D^0 meson is required to originate from the $D^*(2010)^+ \to D^0\pi^+_{tag}$ decay and its flavour is determined by the charge of the accompanying pion. The decay rate can be affected by meson oscillations. The mixing parameters are smaller than 1%, which allows the asymmetry formula to be expanded and approximated by its linear term only:

$$A_{CP}(f,t) \approx a^d_f + \Delta Y_f \frac{t}{\tau^0_D}, \tag{4}$$

where a^d_f is the *CP* asymmetry in the analysed decay, τ^0_D is the D^0 meson lifetime, and the ΔY_f is a parameter which approximately equals to the negative of the asymmetry of the effective decay widths of D^0 and $\bar{D}^0$ mesons into the final state (A^f_T). In the Standard Model, the value of ΔY_f is estimated to be of the order of 10^{-5} or less [9].

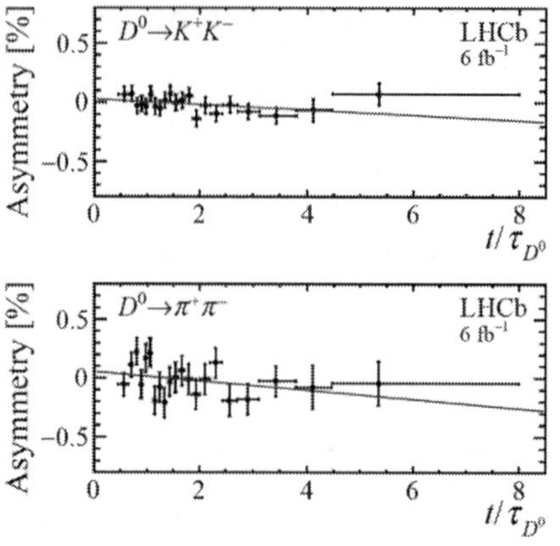

Figure 1. The measured asymmetry (A_{CP} defined in 4) as a function of decay time for $D^0 \to K^+K^-$ decays (top) and $D^0 \to \pi^+\pi^-$ decays (bottom). A linear fit to data is superimposed.

Nuisance asymmetries are removed by agreeing the kinematics of D^0 and $\bar{D}^0$ decays and π^+_{tag} and π^-_{tag} candidates. Secondary contributions originated from B^0 or $B^\pm$ mesons were subtracted with additional cut applied for D^0 impact parameter.

The measured time-dependent asymmetries of the $D^0 \to K^+K^-$ and $D^0 \to \pi^+\pi^-$ decays are presented in Fig. 1. The linear fits to data are overlapped and their slopes are [10]:

$$\begin{aligned} \Delta Y_{K^+K^-} &= (-2.3 \pm 1.5 \pm 0.3) \times 10^{-4}, \\ \Delta Y_{\pi^+\pi^-} &= (-4.0 \pm 2.8 \pm 0.4) \times 10^{-4}. \end{aligned} \tag{5}$$

These results are compatible with the conservation of the *CP* symmetry within two standard deviations. Moreover, systematic and statistical uncertainties are improved with respect to previous measurements by approximately a factor of two.

The difference of ΔY_f between the two final states, combined with previous LHCb measurements is calculated to be [10]:

$$\Delta Y_f = (-1.0 \pm 1.1 \pm 0.3) \times 10^{-4}. \tag{6}$$

This result is consistent with *CP* conservation hypothesis and improves by a factor 2 the uncertainty on the world average.

3.3 Measurement of *CP* asymmetry in $D^0 \to K^0_S K^0_S$ decays

Similarly to eq. (1) *CP* violation is also measured in the $D^0 \to K^0_s K^0_s$ decays. Its *CP* asymmetry is defined as:

$$A^{CP}(K^0_s K^0_s) \equiv \frac{\Gamma(D^0 \to K^0_s K^0_s) - \Gamma(\bar{D}^0 \to K^0_s K^0_s)}{\Gamma(D^0 \to K^0_s K^0_s) + \Gamma(\bar{D}^0 \to K^0_s K^0_s)}. \tag{7}$$

The analysed D^0 mesons come from $D^{*+} \to D^0\pi^+$ decays. The charge of the accompanying so-called tagging pion allows the flavour of the D^0 meson to be determined. Nuisance production and detection asymmetries are removed using the $D^0 \to K^+K^-$ calibration channel, for which *CP* asymmetry is known with high precision. The final result after taking into account previous results from whole Run 2 data sample is:

$$A^{CP}(K_s^0 K_s^0) = (-3.1 \pm 1.2 \pm 0.4 \pm 0.2). \quad (8)$$

The current value provides improvement over earlier LHCb result [12]. It is currently the most accurate measurement of this quantity and remains in agreement with *CP* conservation hypothesis at the level of 2.4 standard deviations.

3.4 *CP* violation in $D^+_{(s)} \to h^+\pi^0$ and $D^+_{(s)} \to h^+\eta$ decays

The two-body decays $D^+_{(s)} \to h^+\pi^0$ and $D^+_{(s)} \to h^+\eta$, where h^+ denotes either π^+ or K^+ are proceed through by Cabibbo favoured (CF), singly-Cabibbo suppressed (SCS) or doubly Cabibbo suppressed (DCS) processes, which are presented in Fig. 2. The SCS $D^+ \to \pi^+\pi^0$ is especially interesting since their *CP* asymmetry is expected to be zero in the SM. The definition of this asymmetry is analogous to eq. (7).

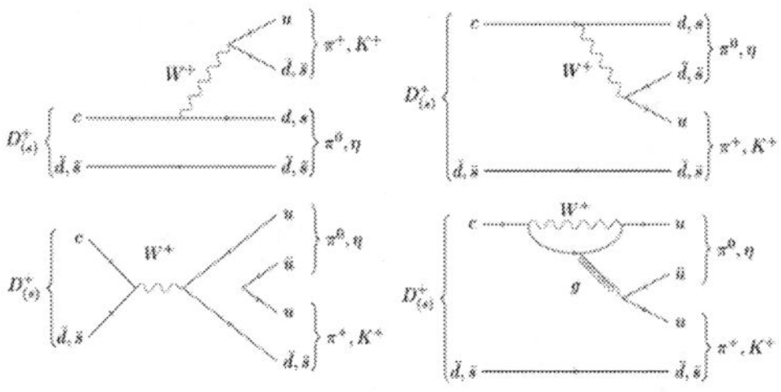

Figure 2. Feynmann diagrams of $D^+_{(s)} \to h^+\pi^0/\eta$ decays: tree-level: colour-favored (top left), colour suppressed (top right) , annihilation (bottom left) and penguin decay (bottom right).

The *CP* asymmetries of seven $D^+_{(s)} \to h^+\pi^0$ and $D^+_{(s)} \to h^+\eta$ modes have been measured using samples corresponding respectively to 9 fb^{-1} or 6 fb^{-1}. The raw asymmetries are measured using the following formula:

$$A_{raw}(D^+_{(s)} \to h^+h^0) \equiv \frac{N(D^+_{(s)} \to h^+h^0) - N(D^-_{(s)} \to h^-h^0)}{N(D^+_{(s)} \to h^+h^0) + N(D^-_{(s)} \to h^-h^0)}, \quad (9)$$

where N is the signal yield of the given decay mode. This asymmetry can by approximated by:

$$A_{raw}(D^+_{(s)} \to h^+h^0) \approx A_{CP}(D^+_{(s)} \to h^+h^0) + A_{prod}(D^+_{(s)}) + A_{det}(h^+), \quad (10)$$

where $A_{prod}(D^+_{(s)})$ and $A_{det}(h^+)$ stand for nuisance asymmetries (production and detection), which are eliminated by subtracting the raw asymmetry obtained in the control decay $D^+_s \to K^0_s h^+$ after kinematically weighted to match the signal modes.

The final *CP* asymmetries are resolved to be:

$$\begin{aligned}
A_{CP}(D^+ \to \pi^+\pi^0) &= (-1.3 \pm 0.9 \pm 0.6)\%,\\
A_{CP}(D^+ \to K^+\pi^0) &= (-3.2 \pm 4.7 \pm 2.1)\%,\\
A_{CP}(D^+ \to \pi^+\eta) &= (-0.2 \pm 0.8 \pm 0.4)\%,\\
A_{CP}(D^+ \to K^+\eta) &= (-6 \pm 10 \pm 4)\%, \quad (11)\\
A_{CP}(D^+_s \to K^+\pi^0) &= (-0.8 \pm 3.9 \pm 1.2)\%,\\
A_{CP}(D^+_s \to \pi^+\eta) &= (0.8 \pm 0.7 \pm 0.5)\%,\\
A_{CP}(D^+_s \to K^+\eta) &= (0.9 \pm 3.7 \pm 1.1)\%
\end{aligned}$$

All of the above results are consistent with no *CP* violation. The first five values are the world best measurements.

3.5 Mass difference between neutral charm meson eigenstates

A neutral meson can oscillate into their own antiparticle. Their mass eigenstates can be expressed as linear combinations of the flavour eigenstates:

$$|M_{1,2}\rangle = p|M^0\rangle \pm q|\bar{M}^0\rangle, \tag{12}$$

where p and q are complex parameters satisfying $|p|^2 + |q|^2 = 1$. In case of no *CP* violation in mixing $|p/q| = 1$. $|M_1\rangle$ and $|M_2\rangle$ are defined as *CP* even and odd eigenstates respectively. There are two mixing parameters: $x \equiv (m_1 - m_2)c^2/\Gamma$ and $y \equiv (\Gamma_1 - \Gamma_2)/2\Gamma$, where $m_{1,2}$ and $\Gamma_{1,2}$ are the mass and decay width of the $M_{1,2}$ states, respectively, and the average decay width is denoted as Γ. Mixing and *CP* violation can be described using z_{CP} and Δz observable, where: $z_{CP} \pm \Delta z \equiv -(q/p)^{\pm 1}(y + ix)$. The results are expressed with the *CP*-even mixing parameters $x_{CP} \equiv -Im(z_{CP})$ and $y_{CP} \equiv -Re(z_{CP})$, and of the *CP*-violating differences $\Delta x \equiv -Im(\Delta z)$ and $\Delta y \equiv -Re(\Delta z)$. Conservation of *CP* symmetry indicates that $x_{CP} = x$ and $y_{CP} = y$. In such scenario both Δx and Δy are equal to 0.

The last results on mixing parameters are based on 30.6 million $D^0 \to K_s^0\pi^+\pi^-$ decays collected in the LHCb experiment from 2016 to 2018 and analysed using the "bin-flip" method [14]. This technique is based on dividing data into disjoint regions (bins) in the Dalitz space. Within each bin strong-phase differences $\Delta\delta(m_-^2 m_+^2)$ between the D^0 and $\bar{D}^0$ amplitudes is almost constant [15]. Two groups of eight bins are assembled symmetrically about the $m_+^2 = m_-^2$ bisector, where the squared invariant mass $m^2(K_s^0\pi^\pm)$ is denoted as $m_\pm^2$ for D^0 decays and $m_\mp^2$ for $\bar{D}^0$ decays, and presented in Fig. 3. The region satisfying $m_+^2 > m_-^2$ is given a positive index $+b$, while the opposite region is given a negative index $-b$. For each decay-time interval, the ratio of the number of decays in each negative Dalitz plot bin $(-b)$ to its positive counterpart $(+b)$ is calculated. The benefit from using ratios is that they reduce the need for precise modeling of the efficiency variation across the phase space and the decay time.

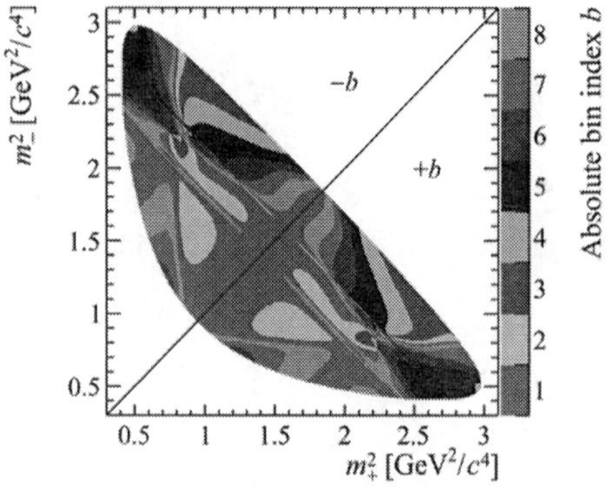

Figure 3. The Dalitz plot binning scheme for the $D^0 \to K_s^0\pi^+\pi^-$ decay.

The mixing and *CP*-violating parameters are measured to be [16]:

$$\begin{aligned} x_{CP} &= (3.97 \pm 0.46 \pm 0.29) \times 10^{-3}, \\ y_{CP} &= (4.59 \pm 1.20 \pm 0.85) \times 10^{-3}, \\ \Delta x &= (-0.27 \pm 0.18 \pm 0.01) \times 10^{-3}, \\ \Delta y &= (0.20 \pm 0.36 \pm 0.13) \times 10^{-3}. \end{aligned} \tag{13}$$

The first uncertainty is statistical and second is systematic. The nonzero mass difference $x = 3.98^{+0.56}_{-0.54} \times 10^{-3}$ in the D^0 decays is measured for the first time with a significance of this observation exceeding 7 standard deviations. The results are in agreement with *CP* symmetry and provide improvement for limits on the mixing-induced *CP* violation in the charm sector.

3.6 Searches for direct *CP* violation in Ξ_c decays

So far, the *CP* violation in baryon decays has not been observed. The particularly interesting decays wherein *CP* symmetry is suspected to be broken are $\Xi_c^+ \to pK^-\pi^+$ decays. The searches [17] are done in model-independent ways with the: binned S_{CP} [18] method and the unbinned k-nearest neighbour method (kNN) [19–21].

The S_{CP} method is based on dividing the Dalitz plot into n bins. For every i^{th} bin the significance of the difference between number of particles (N_i^+) and antiparticles (N_i^-) is calculated using the following formula:

$$S^i_{CP} = \frac{N_i^+ - \alpha N_i^-}{\sqrt{\alpha(N_i^+ + N_i^-)}}, \tag{14}$$

where $\alpha = N^+/N^-$ parameter accounts for the global nuisance asymmetries such as production asymmetries. The N^+ and N^- are total number of particles and antiparticles, respectively. The purpose of the S_{CP} method is to detect local asymmetries. If there are no local *CP* asymmetries the S_{CP} distribution follows a Gaussian distribution with $\mu = 0$ and $\sigma = 1$. Finally, a χ^2 test is performed in order to measure p-value. In case of *CP* violation p-value is smaller than 10^{-5} (corresponding to 5 standard deviations). The obtained results are presented in Fig. 4. The measured p-values are equal to 32% and 72%. The calculated mean and sigma are in agreement with values 0 and 1. The results are consistent with *CP* symmetry conservation.

The second method, the kNN, is based on the concept of accounting nearest neighbour events (n_k) in the mixed sample of baryons and antibaryons to determine whether they share the same parent distribution function. The test statistic T is defined as follows:

$$T = \frac{1}{n_k(n_+ + n_-)} \sum_{i=1}^{n_+ + n_-} \sum_{k=1}^{n_k} I(i,k), \tag{15}$$

where $I(i,k) = 1$ if the ith event and its kth nearest neighbour are from the same set of particles or antiparticles and $I(i,k) = 0$ if otherwise, n_+ and n_- are the numbers of particles and antiparticles. Under the hypothesis of no *CP* violation the distribution of T follows a normal distribution with a mean μ_T and a variance σ_T computed from known parameters of the distributions. There are two options in which the kNN method is employed to search for *CP* asymmetries in a given region of the Dalitz plot: first is looking for "normalization" asymmetry ($n_+ \neq n_-$) using pull $(\mu_T - \mu_{TR})/\Delta(\mu_T - \mu_{TR})$, where μ_{TR} is reference value and second is searching for "shape" asymmetry using another pull $(T - \mu_T)/\sigma_T$ from the measured pulls. Their corresponding p-values are shown in Fig. 5. The values of pulls describing the shape (pdf) asymmetry lay within the range -3 and $+3$ standard deviations and $(\mu_T - \mu_{TR})$ pulls are different from zero in all regions. The results are in agreement with

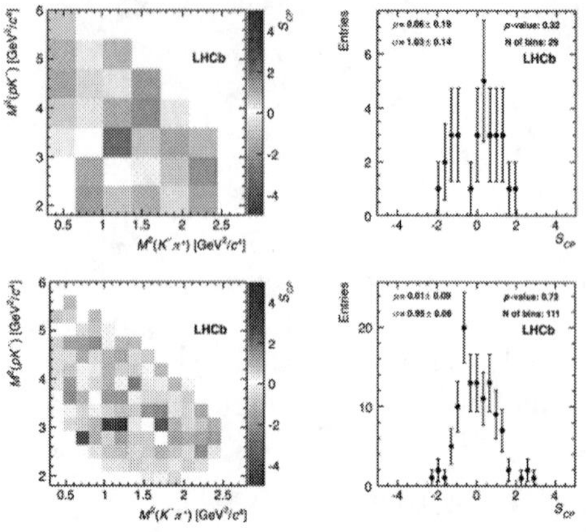

Figure 4. Distributions of S^i_{CP} and matching one-dimensional distributions for $\Xi_c^+ \to pK^-\pi^+$ decays for the united samples collected 2011 and 2012: (top row) 29 uniform bins and (bottom row) 111 uniform bins of the Dalitz plot.

no observation of *CP* violation. An update of the measurement by using data collected in Run 2 is currently ongoing.

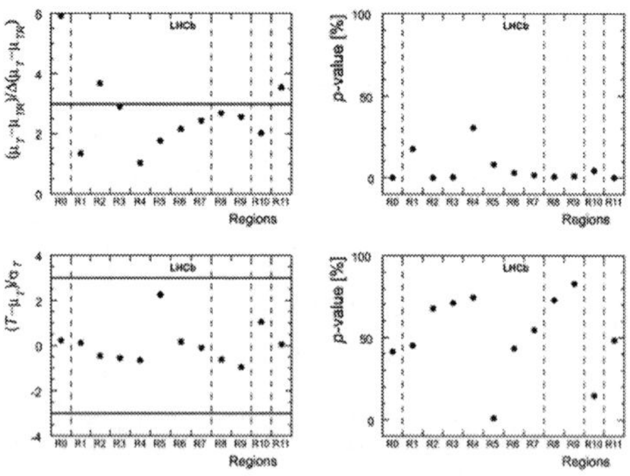

Figure 5. (Top row) $(\mu_T - \mu_{TR})/\Delta(\mu_T - \mu_{TR})$ pulls and their corresponding *p*-values and (bottom row) $(T - \mu_T)/\sigma_T$ pulls and matching *p*-values obtained using the kNN method with $n_k = 50$ for the combined data collected in 2011 and 2012. The horizontal lines in the left plots display −3 and +3 pull values.

3.7 Prospects of the LHCb upgrade

The LHCb proved itself to be a general purpose forward physics detector with a rich physics program extending significantly beyond its primary goals. Presently, the LHCb spectrometer is being upgraded and the main goal is focused to increase luminosity to at least 50 fb^{-1} by the end of Run 4. To achieve that plan many enhancements will be implemented such as VELO and tracking detectors will be improved, completely new readout system is meant to be installed and other spectrometer components would be updated as well.

4 Summary

Recently, the number of charm particle decay studies has increased significantly. The essential part in the conducted analyses was played in the LHCb experiment. The collected data allowed for the first observation of *CP* violation in charm and led to update the measurements: the time-dependent $D^0 - \bar{D}^0$ asymmetry, the *CP* asymmetry in $D^0 \rightarrow K_s^0 K_s^0$ decays, *CP* violation in $D^+_{(s)} \rightarrow h^+\pi^0$ and $D^+_{(s)} \rightarrow h^+\eta$ and mass difference between neutral charm-meson eigenstates. Most of them exceeded previous outcomes and became new world's best measurements. So far, *CP* asymmetry has not been observed in any baryon decays, but the $\Xi_c^+ \rightarrow pK^-\pi^+$ decays are being investigated using a new technique based on kernel density estimator.

Run 2 phase of data taking has ended and currently the LHCb spectrometer is undergoing deep modernisation. The most significant changes are the new flexible fully software trigger system, new readout electronics able to process the full detector information at the LHC clock (40 MHz) and brand new tracking detectors. The experiment aims at collecting the data sample corresponding to the integrated luminosity of at least 50 fb^{-1} by the end of LHC Run 4 and *CPV* observables in charm will be measured with improved precision with Run 3 – 4.

References

[1] N. Cabibbo, Phys. Rev. Lett. **10**, 531 (1963)
[2] M. Kobayashi and T. Maskawa, Prog. Theor. Phys. **49**, 652 (1973)
[3] M. Gersabeck, arXiv:1503.00032 (2015)
[4] LHCb Collaboration, JINST **3**, S08005 2008
[5] LHCb Collaboration, Int J. Mod. Phys. A **30**, 1530022 (2015)
[6] LHCb Collaboration, JHEP **07**, 041 (2014)
[7] LHCb Collaboration, Phys. Rev. Lett. **116**, 191601 (2016)
[8] LHCb Collaboration, Phys. Rev. Lett. **122**, 211803 (2019)
[9] A. L. Kagan and L. Silvestrini, Phys. Rev. D **103**, 053008 (2021)
[10] LHCb Collaboration, Phys. Rev. D **104**, 072010 (2021)
[11] LHCb Collaboration, Phys. Rev. D **104**, L031102 (2021)
[12] LHCb Collaboration, JHEP **11**, 048 (2018)
[13] LHCb Collaboration, JHEP **06**, 019 (2021)
[14] A. Di Canto et al., Phys. Rev. D **99**, 012007, (2019)
[15] CLEO Collaboration, Phys. Rev. D **82**, 112006, (2010)
[16] LHCb Collaboration, Phys. Rev. Lett. **127**, 111801 (2021)
[17] LHCb Collaboration, Eur. Phys. J. C **80**, 986 (2020)
[18] I. Bediaga et al., Phys. Rev. D **80**, 096006 (2009).
[19] M. Williams, JINST **5**, P09004 (2010)
[20] N. Henze, The Annals of Statistics **16**, 2 (1988)
[21] M. Schilling, J. Am. Stat. Assoc. **81**, 799 (1986)

Electric dipole moment of light nuclei in chiral effective field theory

Alex Gnech[1,*], *Jordy* de Vries[2,3], *Sachin* Shain[3], and *Michele* Viviani[4]

[1]Theory Center, Jefferson Lab, Newport News, Virginia 23606, USA
[2]Amherst Center for Fundamental Interactions, Department of Physics, University of Massachusetts Amherst, Amherst, MA 01003, USA
[3]RIKEN BNL Research Center, Brookhaven National Laboratory, Upton, New York 11973-5000, USA
[4]INFN-Pisa, 56127 Pisa, Italy

Abstract.
CP-violating interactions at quark level generate CP-violating nuclear interactions and currents, which could be revealed by looking at the presence of a permanent nuclear electric dipole moment. Within the framework of chiral effective field theory, we discuss the derivation of the CP-violating nuclear potential up to next-to-next-to leading order (N^2LO) and the preliminary results for the charge operator up to next-to leading order (NLO). Moreover, we introduce some renormalization argument which indicates that we need to promote the short-distance operator to the leading order (LO) in order to reabsorb the divergences generated by the one pion exchange. Finally, we present some selected numerical results for the electric dipole moments of ^{2}H, ^{3}He and ^{3}H discussing the systematic errors introduced by the truncation of the chiral expansion.

1 Introduction

The measurement of electric dipole moments (EDMs) of particles is the most promising observable for studying new sources of charge (C) and parity (P) violation [1, 2]. The standard model (SM) has a natural source of CP violation in the Cabibbo-Kobayashi-Maskawa quark mixing matrix; however, it only induces EDMs through multiple electroweak loops and lead to immeasurably small values for EDMs [3, 4]. Therefore, any non-vanishing EDM of a nuclear or an atomic system would highlight CP violating effects due to the so-far undiscovered quantum chromo dynamic (QCD) θ-term or from beyond-the-SM (BSM) sources of CP violation. Current experimental EDM limits [5–7] constrain large set of CP-odd dimension-six operators at the multi- TeV scale at the level of SM effective field theory (SMEFT), that translate in limits on BSM models.

The connection between the SMEFT and the EDMs of complex objects such as nuclei is not trivial. However, the combination of lattice QCD (LQCD) [8–10], chiral effective field theory (χEFT) [11–13] and *ab-initio* nuclear calculation [14–16] permits to exploit this connection, improving significantly model-independent first principle calculations. The starting point is the SMEFT which permits to derive CP-violating operator that involves only light

*e-mail: agnech@jlab.org

quarks, gluons and photons. Using then χEFT it is possible to construct CP-violating Lagrangians where the relevant degrees of freedom are pions, nucleons and photons. Associated to each Lagrangian there is a low-energy constant (LEC) that take care of all the higher energy degrees of freedom. The LECs can be ideally calculated by LQCD. From the χEFT Lagrangian is then possible to derive the CP-violating nuclear interactions and currents, which, in combination with modern few-body *ab-initio* techniques, can be used to compute the EDM of light nuclei. It is important to note that also the LECs have different chiral order based on the source of CP-violation [11, 12]. In this sense only several independent measurements of different nuclear EDMs can permit to fix all the LECs and then to distinguish between the various possible sources of CP-violation.

The χEFT approach is based on the observation that the chiral symmetry exhibited by QCD has a noticeable impact in the low-energy regime. Therefore, the CP-violating interactions among pions, nucleons and photons are constrained in order to reproduce the same behavior under chiral transformation that is given at the SMEFT level. Moreover, the χEFT framework provides an expansion of nuclear amplitudes in terms of p/Λ_χ where $p \sim m_\pi \sim O(100\,\text{MeV})$ and $\Lambda_\chi \sim 4\pi F_\pi \sim O(1\,\text{GeV})$ [17–19], where $F_\pi \simeq 92.4$ MeV is the pion decay constant.

EDMs of light nuclei were computed using CP-violating interactions that were derived up to N^{2}LO [16, 20]. However, a full consistent calculation of the EDMs would require the development of the corresponding electromagnetic current (in term of the charge operator) up to the same order. As the best of our knowledge, within χEFT, this work has been only partially carried out in Ref. [21], and only for the deuteron. Moreover, some recent results [22] showed that the Weinberg's power counting scheme, based on naive dimensional analysis (NDA) of the LECs, is not reliable in the nuclear sector for CP-violating interactions. Therefore, we need to reconsider the chiral order at which some of the interactions appear.

In this proceeding, after discussing some basic definitions for the computation of the EDM of nuclei in Section 2, we will briefly discuss the derivation of the CP-violating nuclear interactions and currents in Section 3. In this section we will also present the arguments contained in Ref. [22] in order to reconsider the NDA power counting scheme. Finally selected numerical results are provided in Section 4.

2 Basics on nuclear EDMs

The EDM d_A of a nucleus is given by the matrix element

$$d_A = \lim_{q^2\to 0} \frac{1}{iq} \langle \psi^{A,J,J} | \rho(q) | \psi^{A,J,J} \rangle \,, \tag{1}$$

where q is the momenta carried by the exchanged photon, $\psi^{A,J,J}$ the wave function of the nucleus of mass A with total angular momentum $J_z = J$ and $\rho(q)$ is the electromagnetic charge operator. The limit written in this way indicates that only the first multipole of the charge operator (C1) is considered. In absence of the CP-violating term this matrix element is zero due to the different parity generated by the C1 multipole.

In this matrix element the CP-violation can be introduced in two different ways. The first is through the wave function. The CP-violating interactions between pions and nucleons at the Hamiltonian level is expressed as a CP-violating term in the nuclear potential. Therefore, the solution of the Hamiltonian contains a CP-violating part, and the final wave function can be then written as

$$\psi^{A,J,J} = \psi^{A,J,J}_{PC} + \psi^{A,J,J}_{\not{C}\not{P}} \,, \tag{2}$$

where with PC we indicate the parity conserving and time-reversal conserving quantities and with $\not{C}\not{P}$ the CP-violating ones. In this case the charge operator that will appear in the matrix element will be the standard PC charge operator. At leading order (LO) the only relevant operator correspond to the electric dipole moment operator it-self, namely

$$\rho_{PC}^{\mathrm{LO}}(q) = ie \sum_{i=1}^{A} \frac{(1+\tau_z(i))}{2} \boldsymbol{r}_i \cdot \boldsymbol{q}\,, \tag{3}$$

where e is the electric charge, $\boldsymbol{r}_i$ the position of the nucleon i and $\tau_z(i)$ the isospin projection along the z-axis of the nucleon i. The complete expression for the charge operator up to N^{3}LO can be found in Ref. [23]. We will indicate this contribution to the nuclear EDM as d_A^V and it is explicitly given by

$$d_A^V = 2 \times \lim_{q^2 \to 0} \frac{1}{iq} \langle \psi_{PC}^{A,J,J} | \rho_{PC}(q) | \psi_{\not{C}\not{P}}^{A,J,J} \rangle\,. \tag{4}$$

The second way, it is to introduce the CP-violation through the charge operator it-self. In this case only the PC term of the wave function is needed since the CP-violating wave function would generate higher order contributions in the CP-violating LECs that can be discarded. The LO contribution of $\rho_{\not{C}\not{P}}(q)$ is given by

$$\rho_{\not{C}\not{P}}^{\mathrm{LO}}(q) = i \sum_{i=1}^{A} \left(d_n \left(\frac{1-\tau_z(i)}{2} \right) + d_p \left(\frac{1+\tau_z(i)}{2} \right) \right) \boldsymbol{\sigma}_i \cdot \boldsymbol{q}\,, \tag{5}$$

where $\boldsymbol{\sigma}$ indicates the Pauli matrices and d_n and d_p are the intrinsic EDM of the neutron and the proton respectively. The higher order terms of the charge operator up to NLO are discussed in Section 3. We can write this contribution to the EDM as

$$d_A^\rho = \lim_{q^2 \to 0} \frac{1}{iq} \langle \psi_{PC}^{A,J,J} | \rho_{\not{C}\not{P}}(q) | \psi_{PC}^{A,J,J} \rangle\,. \tag{6}$$

The final nuclear EDM is then given by the sum of Eqs. (4) and (6), namely

$$d_A = d_A^V + d_A^\rho\,. \tag{7}$$

3 CP-violating interaction and charge operator

Starting from the chiral Lagrangians derived in Refs. [11, 12] it is possible to obtain the nuclear interaction using the time ordered perturbation theory (TOPT) [16] or the heavy-barion (HB) expansion [20]. The TOPT diagrams which contributes to the nuclear potential order by order up to N^{2}LO are given in Figure 1. The results are identical to the ones obtained by using the HB expansion except that the relativistic corrections are typically counted one order lower. Note that we do not considered at this level the different power counting that comes from the LECs them-selves that depends explicitly on the source of CP-violation at the fundamental level.

The explicit formula for the nuclear potential order by order can be found in Refs. [16, 20]. Here we briefly comment some details. In Figure 1 we separate the various diagrams on the base of the LEC that is attached to them. The LECs g_0, g_1 and g_2 present the typical topology of the one-pion exchange (OPE) at LO and two-pions exchange at N^{2}LO. A characteristic of the CP-violating interaction is the presence of a three-pions vertex that generate at NLO a new topology. At the same order also three-body forces arises due to the same vertex.

By considering NDA the contact interactions appear at N^{2}LO. However, following the analysis of Ref. [22], the iteration of the OPE CP-violating term proportional to g_0 in the Lippmann-Schwinger equation (that is the analogous of solving the Hamiltonian) generate divergences that cannot be cured at LO, if not by promoting a counter-term (*i.e.* the contact interaction) at LO as well. In Figure 1 this is indicated by the red arrow. The same promotion is not necessary for the OPE proportional to g_1 [22]. There are no analysis about the OPE diagrams proportional to g_2 so far, even if the similar operator structure to the g_0 term suggest that the promotion of the relative counter-term at LO could be necessary. While for the numerical computation of the contribution of the single diagrams the change of the power counting has no impact, it has an enormous impact when the LECs are fitted to reproduce the experimental data.

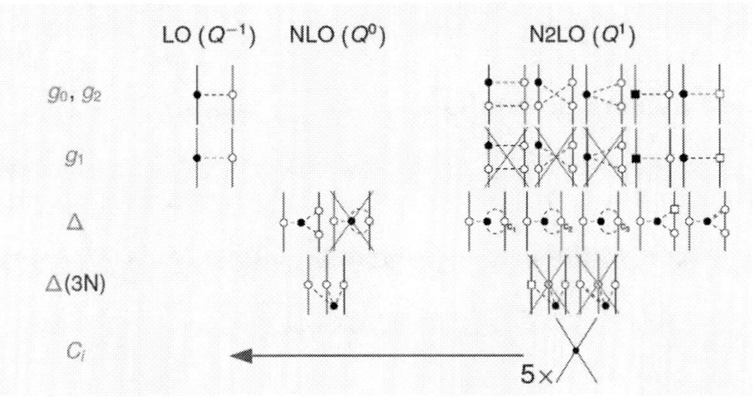

Figure 1. Diagrams contributing to the CP-violating nuclear potential for each different LEC following the NDA. The full (empty) circles indicates CP-violating (conserving) vertices at LO. The full (empty) squares indicates CP-violating (conserving) vertices at NLO. The double dashed lines indicate the second order in the pion propagator. Note that the red crossed diagrams cancel out when all the time ordered diagrams are summed. The red arrow indicate the necessity of promoting some of the contact interactions as counter-term at LO. See the text for more details.

In Figure 2 we present the preliminary results for the time ordered diagrams of the CP-violating charge operator up to NLO. As can be seen by inspecting the figure at LO we have the contribution of the single nucleon EDM (see Eq. (5)) and also of the OPE exchange with the photon attached to the pion on flight. However the time-order diagrams of this topology cancel out. Therefore, the first contributions beyond the single nucleon EDM come at NLO where we have again OPE with the photon in the nucleon-pion vertices or with the nucleon only. Note that there is a new LEC related to a photon-pion-nucleon CP-violating vertex. At the same order also different three-pion diagrams with a photon appear. The TPE and the contact diagrams that contribute to the charge current are instead at N^{2}LO. The explicit formula of the CP-violating charge operator will be object of a future publication [24].

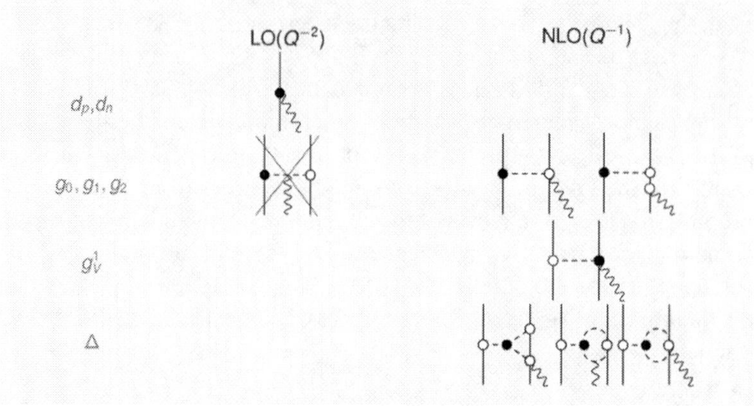

Figure 2. Diagrams contributing to the CP-violating charge current for each different CP-violating LEC. The full (empty) circles indicates CP-violating (conserving) vertices at LO. Note that the red crossed diagrams cancel out when all the time ordered diagrams are summed.

4 Selected numerical results

The EDM of nuclei can be then expressed as a product of the LECs with some numerical coefficients that contain the dynamic of the potential and the charge operator, namely

$$\begin{aligned} d_A &= g_0(a_0^V + a_0^{\rho}) + g_1(a_1^V + a_1^{\rho}) + g_2(a_2^V + a_2^{\rho}) + \Delta(a_\Delta^V + a_\Delta^{\rho}) \\ &+ \sum_{i=1}^{5} C_i(A_i^V + A_i^{\rho}) + d_p a_p + d_n a_n + g_V^1 a_{1V}^{\rho}, \end{aligned} \tag{8}$$

where the numerical coefficients a_i^V are the contributions that come from Eq. (4) while the coefficients a_i^{ρ} are the ones coming from Eq. (6). For now we only have results for the first ones, computed with only the LO charge operator (see Eq. (3)). While some tests showed that next orders corrections coming form the chiral expansion of the *PC* charge operator are numerically very small and almost negligible, this is not true for the CP-violating charge operator. Therefore, a complete calculation where also the a_i^{ρ} coefficients are computed is necessary to have a fully consistent description of nuclear EDMs.

In order to perform the calculations, we need to introduce a regularization cut-off. In our calculation as regularization function we used

$$C(\Lambda) = \exp\left(-(k/\Lambda)^4\right), \tag{9}$$

where k is the exchanged momentum between the two nucleons. As PC nuclear potential in the Hamiltonian we use the N4LO/N2LO- 500, the N4LO/N2LO-450, and the N4LO/N2LO-550 models from Refs. [25, 26], where the final number is the value of the cut-off Λ that we matched also for the CP-violating part.

We also evaluated the errors on the numerical coefficients due to the truncation of the chiral expansion as

$$(\delta a_i)^2 = \left(\delta a_i^{\rm PC}\right)^2 + \left(\delta a_i^{\not{CP}}\right)^2 + \left(\delta a_i^{\psi}\right)^2 , \tag{10}$$

where δa_i^{PC} is the error associated to the chiral expansion of the PC potential, $\delta a_i^{\not{CP}}$ the error associated with the chiral expansion of the CP-violating potential computed following

Table 1. Values of the numerical coefficients for the ^{2}H, the ^{3}H and the ^{3}He in units of e fm (a_n (a_p) in units of d_n (d_p)) for a choice of the cutoff $\Lambda = 500$ MeV.

	^{2}H	^{3}H	^{3}He
a_n	0.939 ± 0.001	-0.033 ± 0.001	0.908 ± 0.009
a_p	0.939 ± 0.001	0.909 ± 0.009	-0.033 ± 0.001
a_0^V	$---$	-0.055 ± 0.013	0.056 ± 0.013
a_1^V	0.192 ± 0.006	0.154 ± 0.004	0.155 ± 0.004
a_2^V	$---$	-0.120 ± 0.009	0.118 ± 0.009
a_Δ^V	-0.306 ± 0.174	-0.388 ± 0.101	-0.383 ± 0.102
A_1^V	$---$	0.006 ± 0.001	-0.006 ± 0.002
A_2^V	$---$	-0.010 ± 0.003	0.010 ± 0.003
A_3^V	0.013 ± 0.004	-0.008 ± 0.002	-0.008 ± 0.002
A_4^V	-0.013 ± 0.004	0.013 ± 0.004	0.013 ± 0.004
A_5^V	$---$	-0.022 ± 0.006	0.022 ± 0.006

Ref. [27] and δa_i^ψ is the error associated to the numerical accuracy of the 3-body wave function which we estimated to be of the order of $\sim 1\%$. The calculations of the wave functions were performed using the Hyperspherical Harmonics approach [28].

In Table 1 as an example we report the results obtained for the a_i^V coefficients and a_n and a_p computed for a cut-off of $\Lambda = 500$ MeV for the deuteron, the tritium and the ^{3}He. The detailed tables with all the coefficients for all the different cutoffs can be found in Ref. [16].

5 Conclusions

In this proceeding we presented the calculation of the EDMs of nuclei within χEFT. We discussed the derivation of the CP-violating nuclear interaction with particular attention on the necessity of overcoming the Weinberg's power counting by adding the CP-violating counterterm at LO in order to reabsorb the divergences generated by the OPE in the Schrödinger equation. We also presented a preliminary analysis of the diagrams that contribute to the CP-violating charge operator up to NLO.

Finally we reported some numerical results for which we computed the errors coming from the chiral expansion truncation. The calculation is for now limited on considering contributions to the EDM that come only from the CP-violating nuclear potential. However, the contribution coming from the CP-violating charge density is fundamental in order to have a full consistent description of the nuclear EDMs within χEFT. This work is already ongoing.

To conclude we want to remark that the calculations of this dynamical coefficients together with possible future experimental results on EDMs of light nuclei will permit to determine the LECs and then to individuate the sources of CP-violation at the fundamental level.

Acknowledgments

This work is supported by the U.S. Department of Energy, Office of Nuclear Science, under contracts DE-AC05-06OR23177 (A.G.).

References

[1] N. Yamanaka, B. Sahoo, N. Yoshinaga, T. Sato, K. Asahi, B. Das, Eur. Phys. J. A **53**, 54 (2017), 1703.01570

[2] T. Chupp, P. Fierlinger, M. Ramsey-Musolf, J. Singh, Rev. Mod. Phys. **91**, 015001 (2019), 1710.02504
[3] C.Y. Seng, Phys. Rev. C **91**, 025502 (2015), 1411.1476
[4] Y. Yamaguchi, N. Yamanaka, Phys. Rev. Lett. **125**, 241802 (2020), 2003.08195
[5] B. Graner, Y. Chen, E. Lindahl, B. Heckel, Phys. Rev. Lett. **116**, 161601 (2016), [Erratum: Phys.Rev.Lett. 119, 119901 (2017)], 1601.04339
[6] V. Andreev et al. (ACME), Nature **562**, 355 (2018)
[7] C. Abel et al. (nEDM), Phys. Rev. Lett. **124**, 081803 (2020), 2001.11966
[8] T. Bhattacharya, V. Cirigliano, R. Gupta, H.W. Lin, B. Yoon, Phys. Rev. Lett. **115**, 212002 (2015), 1506.04196
[9] M. Abramczyk, S. Aoki, T. Blum, T. Izubuchi, H. Ohki, S. Syritsyn, Phys. Rev. **D96**, 014501 (2017), 1701.07792
[10] J. Dragos, T. Luu, A. Shindler, J. de Vries, A. Yousif, Phys. Rev. C **103**, 015202 (2021), 1902.03254
[11] E. Mereghetti, W. Hockings, U. van Kolck, Annals Phys. **325**, 2363 (2010), 1002.2391
[12] J. de Vries, E. Mereghetti, R.G.E. Timmermans, U. van Kolck, *The Effective Chiral Lagrangian From Dimension-Six Parity and Time-Reversal Violation* (2013), 1212.0990
[13] J. Bsaisou, U.G. Meißner, A. Nogga, A. Wirzba, Annals Phys. **359**, 317 (2015), 1412.5471
[14] J. de Vries, R. Higa, C.P. Liu, E. Mereghetti, I. Stetcu, R.G.E. Timmermans, U. van Kolck, Phys. Rev. **C84**, 065501 (2011), 1109.3604
[15] J. Bsaisou, J. de Vries, C. Hanhart, S. Liebig, U.G. Meißner, D. Minossi, A. Nogga, A. Wirzba, JHEP **03**, 104 (2015), [Erratum: JHEP05,083(2015)], 1411.5804
[16] A. Gnech, M. Viviani, Phys. Rev. C **101**, 024004 (2020), 1906.09021
[17] S. Weinberg, Physica A **96**, 327 (1979)
[18] E. Epelbaum, H.W. Hammer, U.G. Meissner, Rev. Mod. Phys. **81**, 1773 (2009), 0811.1338
[19] H.W. Hammer, S. König, U. van Kolck, Rev. Mod. Phys. **92**, 025004 (2020), 1906.12122
[20] C. Maekawa, E. Mereghetti, J. de Vries, U. van Kolck, Nucl. Phys. A **872**, 117 (2011), 1106.6119
[21] J. Bsaisou, C. Hanhart, S. Liebig, U.G. Meissner, A. Nogga, A. Wirzba, Eur. Phys. J. A **49**, 31 (2013), 1209.6306
[22] J. de Vries, A. Gnech, S. Shain, Phys. Rev. C **103**, L012501 (2021), 2007.04927
[23] S. Pastore, L. Girlanda, R. Schiavilla, M. Viviani, Phys. Rev. C **84**, 024001 (2011), 1106.4539
[24] A. Gnech, R. Schiavilla (2021), to be published
[25] D.R. Entem, N. Kaiser, R. Machleidt, Y. Nosyk, Phys. Rev. C **91**, 014002 (2015), 1411.5335
[26] L.E. Marcucci, F. Sammarruca, M. Viviani, R. Machleidt, Phys. Rev. C **99**, 034003 (2019), 1809.01849
[27] E. Epelbaum, H. Krebs, U.G. Meißner, Phys. Rev. Lett. **115**, 122301 (2015), 1412.4623
[28] L.E. Marcucci, J. Dohet-Eraly, L. Girlanda, A. Gnech, A. Kievsky, M. Viviani, Frontiers in Physics **8**, 69 (2020)

Physics of warped dimensions and continuous spectra *

Eugenio Megías[1,**], *Manuel* Pérez-Victoria[2,***], and *Mariano* Quirós[3,****]

[1]Departamento de Física Atómica, Molecular y Nuclear and Instituto Carlos I de Física Teórica y Computacional, Universidad de Granada, Avenida de Fuente Nueva s/n, 18071 Granada, Spain
[2]Centro Andaluz de Partículas Elementales (CAFPE) and Departamento de Física Teórica y del Cosmos, Universidad de Granada, Avenida de Fuente Nueva s/n, 18071 Granada, Spain
[3]Institut de Física d'Altes Energies (IFAE), Barcelona Institute of Science and Technology (BIST), Campus UAB, 08193 Bellaterra, Barcelona, Spain

Abstract. We study some features of a warped five-dimensional model that solves the hierarchy problem and exhibits a continuum of Kaluza-Klein (KK) modes with a mass gap at the TeV scale. We compute the propagators and spectral functions for massless bulk gauge bosons, and study how the continuum can be reached as the limit of a set of models with discrete spectrum. Finally, we study the low energy effective theory and provide explicit results for the Wilson coefficients.

1 Introduction

No clear deviations from the predictions of the Standard Model (SM) of electroweak (EW) and strong interactions have been found so far at past and present particle physics experiments [1]. In spite of that, the SM fails to describe a number of observational facts, such as dark matter and dark energy or the baryon asymmetry of the universe, and does not incorporate quantum gravity. So, it should be considered an effective theory, valid for energies below a certain cutoff scale. If this cutoff is near the Planck scale $M_{\rm Pl}$, then the large hierarchy with respect to the electroweak scale $M_{\rm EW}$ requires fine tuning of the parameters in the ultraviolet (UV). This is known as the hierarchy problem, and is addressed by several extensions of the SM. Here, we are interested in models that generate hierarchical scales by a red-shift effect in a warped extra dimension [2]. In most of them, the extra dimension is compact and gives rise to a discrete tower of Kaluza-Klein (KK) states for each 5D field.

The lowest KK modes that do not correspond to SM fields are expected to have masses of the order of a few TeV, so they could in principle give rise to resonant signals at the LHC and future colliders. However, no such signals have been observed so far. These TeV KK modes would actually be more elusive to direct searches if their contribution to cross sections were not peaked, and hence difficult to distinguish from the background. Proposals in this sense include models with broad resonances [3], narrowly-spaced KK modes that cannot be resolved [4, 5] (see also [6] for a four-dimensional version) or, in the extreme case of

*Talk given by E. Megías at the vConf21: A Virtual Tribute to Quark Confinement and the Hadron Spectrum, 2-6 August 2021, online.
**e-mail: emegias@ugr.es
***e-mail: mpv@ugr.es
****e-mail: quiros@ifae.es

vanishing spacing, a continuum of KK modes. Unparticle models with a conformal sector that gives rise to continuous spectra were proposed in [7]. When the unparticle stuff is charged under the SM gauge group, then the phenomenological viability of the model requires a mass gap of the order of the TeV [8]. This can be realized in extra-dimensional theories with scalar-gravity backgrounds with an admissible singularity at a finite proper distance [9, 10] (see [11] for a detailed analysis of stability). The phenomenology of this scenario has been explored in [12, 13] and, more recently, in [14–16]. Here, we discuss some basic features of this setup in the particular model proposed in [17], which allows for an exact analytical treatment.

2 The extra-dimensional model

Let us consider a 5D theory with a domain-wall warped geometry characterized by a metric

$$ds^2 = g_{MN}dx^M dx^N \equiv \bar{g}_{\mu\nu}dx^\mu dx^\nu - dy^2, \qquad \text{with} \qquad \bar{g}_{\mu\nu} = e^{-2A(y)}\eta_{\mu\nu}\,, \tag{1}$$

where $y \in [y_0, y_s]$ parametrizes the extra dimension. The background we are interested in can be dynamically generated in a scalar-gravity theory with two three-branes at the boundary $y = y_0 = 0$ (UV brane) and at the point $y = y_1$ (IR brane), where $0 < y_1 < y_s \leq \infty$. The coordinate y is chosen such that $A(0) = 0$. The 5D action of the model reads [11]

$$S = \int d^5x \sqrt{|\det g_{MN}|}\left[-\frac{1}{2\kappa^2}R + \frac{1}{2}g^{MN}(\partial_M\phi)(\partial_N\phi) - V(\phi)\right]$$
$$- \sum_\alpha \int_{B_\alpha} d^4x \sqrt{|\det \bar{g}_{\mu\nu}|}\lambda_\alpha(\phi) + S_{\text{GHY}} + S_{\text{SM}}\,, \tag{2}$$

where $V(\phi)$ is the bulk scalar potential, $\lambda_\alpha(\phi)$ are the UV ($\alpha = 0$) and IR ($\alpha = 1$) 4D brane potentials and $\kappa^2 = 1/(2M_5^3)$ with M_5 the 5D Planck scale. S_{GHY} is the standard Gibbons-Hawking-York boundary term, while S_{SM} refers to terms involving the SM fields. The values of y_1 and y_s are eventually determined by the dynamics of this action. The equations of motion (EoM) for ϕ and A can be written in terms of a superpotential $W(\phi)$, as [18]

$$\phi'(y) = \frac{1}{2}W'(\phi)\,, \qquad A'(y) = \frac{\kappa^2}{6}W(\phi)\,, \tag{3}$$

while the brane potentials impose some boundary/jumping conditions on the branes.

A simple model that addresses the hierarchy problem was defined in Refs. [11, 15, 17] by the choice

$$W(\phi) = \frac{6k}{\kappa^2}\left(1 + e^{\nu\phi}\right). \tag{4}$$

A hierarchy of scales is generated dynamically: $k \lesssim M_5$ gives rise to $A(y_1) \approx \mathcal{O}(35)$, which implies $M_{\text{Pl}} \simeq 10^{15} M_{\text{EW}}$ if the Higgs is localized at or close to y_1. The spectrum of the KK modes of fields propagating in this background depends crucially of the value of the parameter ν. There is a critical value $\nu_c \equiv \kappa/\sqrt{3}$ such that: i) $\nu < \nu_c$ corresponds to ungapped continuum KK spectra; ii) $\nu = \nu_c$ corresponds to continuum KK spectra with a mass gap m_g; and iii) $\nu > \nu_c$ leads to discrete KK spectra. In the following we will focus on the critical case $\nu = \nu_c$, in which there is a metric singularity at $y_s < \infty$. Moreover, we consider an approximation to the superpotential (4) that gives rise to a warp factor $A(y)$ [17]

$$A(y) \simeq ky\,\Theta(y_1 - y) + [ky_1 - \log(k(y_s - y))]\,\Theta(y - y_1)\,, \tag{5}$$

where $\Theta(x)$ is the step function. This simplified metric allows to get exact analytical expressions for the propagators of 5D fields. The geometry between 0 and y_1 is AdS$_5$, as in the Randall-Sundrum model [2], while the behaviour between y_1 and y_s is as in the linear dilaton model [4, 5, 16]. The natural scales measured by an observer on the IR brane is $\rho \equiv k\,e^{-A(y_1)} \sim$ TeV, as required to explain the hierarchy, and continuity across y_1 imposes the relation $k(y_s - y_1) = 1$, with $y_s < \infty$ defined as the position of the singularity in (5). This turns out to imply $m_g \sim \rho$, and therefore opens possibility of observing the continuum at large colliders. Note that in conformal coordinates, $ds^2 = a(z)^2(dx^2 - dz^2)$, the singularity is located at infinite value of z. As a summary, the positions of the branes and the singularity in proper and conformal coordinates is given by

(UV brane)		(IR brane)		(IR singularity)
$y_0 = 0$	$<$	$y_1 = A(y_1)/k$	$<$	$y_s = y_1 + 1/k$
$z_0 = 1/k$	$<$	$z_1 = 1/\rho$	$<$	$+\infty$

3 Propagators

Next, we study the 5D propagators for fields propagating in this background. We work at the classical level (see [19] for a discussion of one-loop resummed propagators in a scenario with continuous spectrum). For definiteness we consider here the case of gauge bosons of an unbroken 5D symmetry. That is, we study the sector of $S_{\rm SM}$ with Lagrangian

$$\mathcal{L}_{\rm g} = \int_0^{y_s} dy \left[-\frac{1}{4} \mathrm{tr}\, F_{\mu\nu}F^{\mu\nu} - \frac{1}{2} e^{-2A}\, \mathrm{tr}\, \partial_y A_\mu \partial_y A_\mu \right]. \tag{6}$$

The boundary conditions for the A_μ field are Neumann at $y = 0$ and the requirement of regularity at y_s. We have already imposed the gauge condition $A_5 = 0$, which is compatible with that boundary condition. The type of spectrum for this field can be easily read from the Schrödinger form of the EoM, $-\partial_z^2 \tilde{A}_\mu(p,z) + V_A(z)\tilde{A}_\mu(p,z) = p^2 \tilde{A}_\mu(p,z)$, where we are using a mixed momentum-position representation in conformal coordinates and rescaled fields $\tilde{A}_\mu(p,z) = a(z)^{1/2} A_\mu(p,z)$ [10]. The effective Schrödinger potential for the metric in Eq. (5) is

$$V_A(z) = \frac{3}{4z^2}\Theta(z_1 - z) + \frac{\rho^2}{4}\Theta(z - z_1)\,. \tag{7}$$

The fact that it asymptotes to a constant $\frac{\rho^2}{4}$ at large z indicates a continuous spectrum with mass gap $m_g = \rho/2$.

3.1 Continuum of gauge boson modes

To see this explicitly, we compute the 5D gauge boson propagators in the mixed representation, which are Green's functions of the EoM. They are given by

$$G_A^{\mu\nu}(y, y'; p) = \left[\eta^{\mu\nu} - (1 - \xi)\frac{p^\mu p^\nu}{p^2} \right] G_A(y, y'; p)\,, \tag{8}$$

where ξ is a gauge parameter for the remaining gauge freedom and G_A obeys

$$p^2 G_A(y, y'; p) + \partial_y \left(e^{-2A(y)} \partial_y G_A(y, y'; p) \right) = \delta(y - y')\,. \tag{9}$$

The solution is required to satisfy the boundary conditions, with regularity enforced at Euclidean momenta. The explicit solution, given in [17], has an isolated simple pole at $s = 0$ and a branch cut along $[m_g^2, +\infty)$. Subtracting the pole part $G_A^0 \equiv \lim_{p\to 0} G_A(y, y'; p) = 1/(y_s p^2)$,

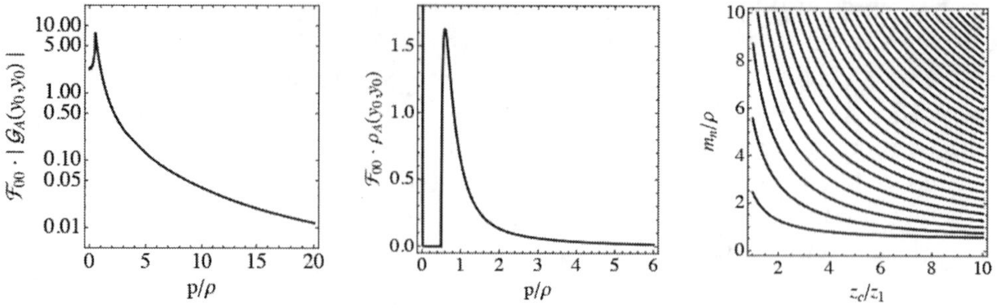

Figure 1. *Plots of the UV-UV Green's function* $\mathcal{F}_{00} \cdot |\mathcal{G}_A(y_0, y_0; p)|$ *(left panel), spectral function* $\mathcal{F}_{00} \cdot \rho_A(y_0, y_0; p)$ *(middle panel), and spectrum of the lightest KK modes as a function of* z_c/z_1 *in the discretized model of Sec. 3.2 (right panel). For comparison with the numerical result (blue lines) in the right panel, we display the result for* $n = 6$ *by using the approximate analytical formula of Eq. (18) (dashed red line). We have used* $A(y_1) = 35$ *in all panels.*

we define the continuum part of the propagator as $\mathcal{G}_A(y, y'; p) \equiv G_A(y, y'; p) - G_A^0$. We display in Fig. 1 (left) the behavior of $|\mathcal{G}_A(y_0, y_0; p)|$, normalized by the factor $\mathcal{F}_{00} \equiv \frac{\rho^2}{k}(ky_s)$ to make it almost invariant under changes of $A(y_1)$ [17]. A 5D spectral function can be defined by

$$\rho_A(y, y'; s) = -\frac{1}{\pi}\text{Im } G_A(y, y'; s + i0^+) . \tag{10}$$

In our model it has the form $\rho_A(y, y'; s) = \frac{1}{y_s}\, \delta(s) + \eta_A(y, y'; s)\Theta(s - m_g^2)$, for a certain function η_A [17]. The pole at $s = 0$ produces the Dirac delta (an isolated zero mode), while the branch cut gives rise to η_A which is a continuous distribution above m_g. We plot $\mathcal{F}_{00} \cdot \rho_A(y_0, y_0; p)$ in Fig. 1 (middle). The Kramers-Kronig relations allow to recover the whole Green's function from its imaginary part, leading to the spectral representation

$$G_A(y, y'; s) = \int_0^\infty dm^2 \frac{\rho_A(y, y'; m^2)}{s - m^2 + i0^+} . \tag{11}$$

The spectral function can be related to the profiles of the KK modes and a 4D spectral density σ [20]:

$$\rho_A(y, y'; s) = \sigma(s) f_s(y) f_s^*(y') . \tag{12}$$

The profiles are normalizable (non-normalizable) for s in the discrete (continuous) spectrum.

Rather than writing the complete expression for σ, we write here the $s \gg \rho^2$ limit of the brane-to-brane spectral functions:

$$\rho_A(y_0, y_0; p^2) \underset{\rho \ll p}{\simeq} \frac{1}{2[\log(p/k)]^2} \frac{k}{p^2} , \tag{13}$$

$$\rho_A(y_0, y_1; p^2) \underset{\rho \ll p}{\simeq} \frac{1}{2\sqrt{\pi}} \frac{\sin(p/\rho) - \cos(p/\rho)}{\log(p/k)} \frac{k}{\rho^{1/2} p^{3/2}} , \tag{14}$$

$$\rho_A(y_1, y_1; p^2) \underset{\rho \ll p}{\simeq} \frac{1}{2\pi} \left[1 - \sin(2p/\rho)\right] \frac{k}{\rho p} . \tag{15}$$

One might worry about the fact that $\rho_A(y, y'; s)$ with $y \neq y'$ is not positive for all s. In fact, there is no problem as the 5D spectral operator $\hat{\rho}_A$, with matrix elements $\langle y, p|\hat{\rho}_A|y', p\rangle = \rho_A(y, y'; p^2)$, is definite positive. Indeed, for any 5D function ϕ for which the following integrals are convergent, we have

$$\begin{aligned}\langle\phi|\hat{\rho}_A|\phi\rangle &= \int dp\, dy\, dy'\, \phi^*(y,p)\rho_A(y,y';p^2)\phi(y',p)\\ &= \int dp\sigma(p^2)\left(\int dy\,\phi^*(y,p)f_{p^2}(y)\right)\left(\int dy'\, f^*_{p^2}(y')\phi(y',p)\right)\\ &= \int dp\sigma(p^2)\left|\int dy\,\phi^*(y,p)f_{p^2}(y)\right|^2 \geq 0\,. \end{aligned} \tag{16}$$

The expression (12) also explains the fact, observed in [17, 21], that the non-vanishing formal eigenvalue $\lambda(s)$ of $\hat{\rho}_A(s)$ (note that $\hat{\rho}_A(s)$ has rank one due to its factorized form) is given by a divergent integral, when s belongs to the continuum spectrum:

$$\lambda(s) = \mathrm{tr}\,\hat{\rho}_A = \int_0^{y_s} dy\,\rho_A(y,y;s) = \sigma(s)\int_0^{y_s} dy\,|f_s(y)|^2 \sim \infty\,, \qquad s > m_g^2\,. \tag{17}$$

Actually, a finite integral would indicate that s belongs to the discrete spectrum! Three kinds of contributions to σ appear in the continuum model, i.e. $\sigma(s) = \frac{1}{\|f_0\|^2}\delta(s) + [\sigma_{\mathrm{res}}(s) + \sigma_{\mathrm{un}}(s)]\,\Theta(s - m_g^2)$, where $\sigma_{\mathrm{res}}(s)$ is a continuous positive function vanishing at $s = m_g^2$, while $\sigma_{\mathrm{un}}(s) \propto 1/(s - m_g^2)^{1/2}$ corresponds to a contribution from unparticles with a dimension $d_U = 3/2$ and a mass gap, see e.g. Ref. [19].

3.2 Gapped continuum spectrum as the limit of a discrete spectrum

The models with continuous spectra can be recovered as the zero-spacing limit of versions of them with discrete spectra [22]. This can be useful in understanding some properties of the continuum in terms of ordinary particles, or to build models with narrowly-spaced resonances with a similar phenomenology to the corresponding continuum model. For instance, for the model with superpotential (4) we could consider $\nu > \nu_c$ and then take the limit $\nu \to \nu_c$. To find a discrete version of the simplified model considered here, we simply end the extra dimension before the singularity y_s is reached. In conformal coordinates this amounts to restricting the coordinate z to a compact domain $[z_0, z_c]$, $z_c \geq z_1$ and imposing Neumann boundary conditions at the new boundary $z = z_c$ (Dirichlet conditions would work similarly). Adjusting z_c, we can smoothly evolve from a Randall-Sundrum model when $z_c = z_1$ to our continuum model when $z_c \to \infty$. The mass spectrum in the discretized version of the model of Eq. (5) has the approximate analytical behaviour

$$m_n^2 \simeq m_g^2 + \left(n - \frac{1}{4}\right)^2 \frac{\pi^2}{z_c^2}\,, \qquad n = 1, 2, 3, \cdots\,. \tag{18}$$

This formula has a relative error $\lesssim 1\%$ which gets smaller for higher values of m_n^2. We show in Fig. 1 (right) the spectrum of the lightest KK modes as a function of z_c/z_1. Notice that the mass spacing Δm_n is governed by z_c, so that the eigenvalues pile-up as z_c increases. By using this approach, it is possible to recover the propagator and spectral functions in the continuum model as a limit of its discretized version:

$$G_{A(\mathrm{d})}(y,y';s) \equiv \sum_n \frac{1}{\|f_n\|^2}\frac{f_n(y)f_n^*(y')}{s - m_n^2 + i0^+} \quad \underset{z_c\to\infty}{\longrightarrow} \quad G_A(y,y';s)\,, \tag{19}$$

$$\rho_{A(\mathrm{d})}(y,y';s) \equiv \sum_n \frac{f_n(y)f_n^*(y')}{\|f_n\|^2}\delta(s - m_n^2) \quad \underset{z_c\to\infty}{\longrightarrow} \quad \rho_A(y,y';s)\,, \tag{20}$$

where $G_A(y,y';s)$ and $\rho_A(y,y';s)$ are given by Eqs. (11) and (12), respectively. In these expressions $f_n(y)$ is the eigenfunction for the n-th KK mode. A more detailed analysis will be provided in [20].

4 Low energy operators generated by the continuum

By integrating out the modes heavier than a certain scale μ, we can obtain a low-energy effective theory that contains only the light degrees of freedom with mass $m < \mu$. Let us consider the quark operators $(\bar{q}\gamma^\mu \lambda^a q)(\bar{t}\gamma_\mu \lambda_a t)$ induced by the exchange of gluon modes heavier than μ, where q and t stand for quarks localized on the branes or zero modes of quark bulk fields. The Wilson coefficient is in general given by a convolution

$$C_{qt}(\mu) \equiv \frac{1}{4\mu^2} \int dy dy' C_{qt}(y, y'; \mu) \, |f_0^q(y)|^2 |f_0^t(y')|^2 , \tag{21}$$

with $f_0^{q,t}$ the profiles of the corresponding zero modes or the delta functions that localize the quarks.

The position-dependent coefficients $C_{qt}(y, y'; \mu)$ can be easily computed using the 5D propagators[1] :

$$C_{qt}(y, y'; \mu) = \lim_{p\to 0} g_5^2 \mu^2 \left(G_A(y, y'; p) - \int_0^{\mu^2} ds \frac{\rho_A(y, y'; s)}{p^2 - s + i0^+} \right) = -g_4^2 y_s \mu^2 \int_{\mu^2}^{\infty} \frac{ds}{s} \rho_A(y, y'; s) , \tag{23}$$

where g_4 is the 4D strong gauge coupling, related to the 5D one by $g_5 = \sqrt{y_s} g_4$. In the limit $\mu \gg m_g$ we can use Eqs. (13)-(15) to approximate

$$C_{qt}(y_0, y_0; \mu) \underset{m_g \ll \mu}{\simeq} -g_4^2 \frac{k y_s}{2} \frac{\log(\mu/k) - 1}{[\log(\mu/k)]^3} , \tag{24}$$

$$C_{qt}(y_0, y_1; \mu) \underset{m_g \ll \mu}{\simeq} -g_4^2 \frac{k y_s}{\sqrt{\pi}} \frac{\sin(\mu/\rho) + \cos(\mu/\rho)}{\log(\mu/k)} \left(\frac{\rho}{\mu}\right)^{1/2} , \tag{25}$$

$$C_{qt}(y_1, y_1; \mu) \underset{m_g \ll \mu}{\simeq} g_4^2 \frac{k y_s}{2\pi} \left[\cos(2\mu/\rho) - 2\mu/\rho\right] . \tag{26}$$

We display in Fig. 2 the exact and approximate results of these coefficients $C_{qt}(y_\alpha, y_\beta; \mu)$ as a function of μ/ρ. The approximate analytical results of Eqs. (24)-(26) reproduce quite well the numerical exact results for $\mu/\rho > 2$. More relevant for today's indirect searches is the effective theory for energies sufficiently smaller than m_g, which contains no continuous degrees of freedom. This effective theory, corresponding to $\mu = m_g$, is obtained by integrating out all the KK gluons except the zero mode. The position-dependent Wilson coefficient is given by (23) with $\mu = m_g$ or, equivalently, by

$$C_{qt}(y, y'; m_g) = \lim_{p\to 0} g_5^2 m_g^2 \mathcal{G}_A(y, y'; p) . \tag{27}$$

By using the low momentum behavior of the brane-to-brane Green's functions of the model of Eq. (5) (cf. Ref. [17]), the brane-to-brane Wilson coefficients are found out to be

$$C_{qt}(y_0, y_0; m_g) = -g_4^2 \frac{9}{16 k y_s} \underset{1 \ll k y_s}{\simeq} -g_4^2 \frac{9}{16 A(y_1)} , \tag{28}$$

$$C_{qt}(y_0, y_1; m_g) = g_4^2 \frac{3(-3 + k y_s)}{16 k y_s} \underset{1 \ll k y_s}{\simeq} \frac{3}{16} g_4^2 , \tag{29}$$

$$C_{qt}(y_1, y_1; m_g) = -g_4^2 \frac{9 + 2 k y_s(-3 + k y_s)}{16 k y_s} \underset{1 \ll k y_s}{\simeq} -\frac{1}{8} g_4^2 A(y_1) . \tag{30}$$

[1]Notice that the result of Eq. (23) implies that $\bar{C}_{qt}(y, y'; \mu) \equiv C_{qt}(y, y'; \mu)/\mu^2$ obeys the evolution equation

$$\mu^2 \frac{\partial}{\partial \mu^2} \bar{C}_{qt}(y, y'; \mu) = g_4^2 y_s \cdot \rho_A(y, y'; \mu^2) , \tag{22}$$

which is a first order differential equation that determines $\bar{C}_{qt}(y, y'; \mu)$ in terms of $\bar{C}_{qt}(y, y'; m_g)$ (see below).

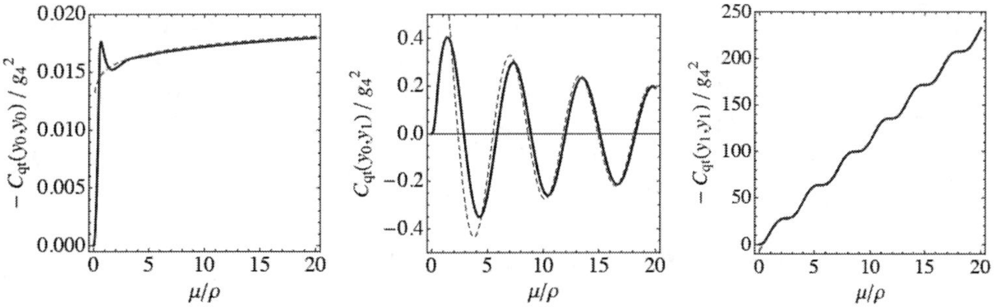

Figure 2. *Wilson coefficients of the effective theory below the scale* μ: $-C_{qt}(y_0, y_0; \mu)/g_4^2$ *(left panel),* $C_{qt}(y_0, y_1; \mu)/g_4^2$ *(middle panel) and* $-C_{qt}(y_1, y_1; \mu)/g_4^2$ *(right panel); as functions of* μ/ρ. *The results of the numerical integration of Eq. (23) (blue lines) are compared with the analytic results of Eqs. (24)-(26) (dashed red lines). We have used* $A(y_1) = 35$ *in all panels.*

The factors $A(y_1) = \log(k/\rho)$ produce a hierarchy of values of the Wilson coefficients, $|C_{qt}(y_0, y_0; m_g)| < |C_{qt}(y_0, y_1; m_g)| < |C_{qt}(y_1, y_1; m_g)|$. Finally, notice that $G_A(y, y'; p)$ has real values for $0 < p < m_g$, so that $C_{qt}(y, y'; m_g)$ is also real. An analysis similar to the one presented above has been provided in Ref. [23] within the linear dilaton model, leading to similar conclusions.

5 Conclusions

We have studied a warped extra-dimensional model that generates a hierarchy between the Planck scale and the electroweak scale by means of a warp factor. The geometry of the model is AdS_5 near the UV brane, and it behaves like a linear dilaton model towards the IR. The KK tower of massless gauge bosons (photon and gluon) has an isolated massless mode corresponding to a 4D SM gauge boson, and a continuum with mass gap $m_g \sim$ TeV.

We have computed in this model the propagators and spectral functions for massless gauge bosons, and have related them to a continuum version of the KK expansion. We have shown that the continuum can be obtained as the limit of models defined on a compact interval of the conformal coordinate, which have only discrete spectrum. Finally, we have found the Wilson coefficients of low-energy operators generated by the continuous modes. Other features of this model not discussed in the present work include the appearance of broad resonances already at tree level, as well as the increase of the cross sections $\sigma(pp \to Q\bar{Q})$ with respect to the SM prediction [15, 17]. A study of its phenomenology at the LHC is in progress. Another interesting application of this scenario is the idea of continuum dark matter [24, 25].

Acknowledgement

The work of EM is supported by the project PID2020-114767GB-I00 financed by MCIN/AEI/10.13039/501100011033, by the FEDER/Junta de Andalucía-Consejería de Economía y Conocimiento 2014-2020 Operational Program under Grant A-FQM-178-UGR18, by Junta de Andalucía under Grant FQM-225, and by the Consejería de Conocimiento, Investigación y Universidad of the Junta de Andalucía and European Regional Development Fund (ERDF) under Grant SOMM17/6105/UGR. The research of EM is also supported by the Ramón y Cajal Program of the Spanish MCIN under Grant RYC-2016-20678. The work of MPV is partially supported by the MCIN project PID2019-106087GB-C22, and by Junta de Andalucía projects FQM-101, A-FQM-211-UGR18, P18-FR-4314 and SOMM17/6104/UGR (including ERDF). The work of MQ is partly supported by the Spanish MCIN under Grant FPA2017-88915-P, by the Catalan Government under Grant 2017SGR1069, and by Severo Ochoa Excellence Program of MCIN under Grant SEV-2016-0588. IFAE is partially funded by the CERCA program of the Generalitat de Catalunya.

References

[1] P.A. Zyla et al. (Particle Data Group), PTEP **2020**, 083C01 (2020).
[2] L. Randall, R. Sundrum, Phys. Rev. Lett. **83**, 3370 (1999).
[3] R. Escribano, M. Mendizabal, M. Quirós, E. Royo, JHEP **05**, 121 (2021).
[4] I. Antoniadis, A. Arvanitaki, S. Dimopoulos, A. Giveon, Phys. Rev. Lett. **108**, 081602 (2012).
[5] P. Cox, T. Gherghetta, JHEP **05**, 149 (2012).
[6] G.F. Giudice, Y. Kats, M. McCullough, R. Torre, A. Urbano, JHEP **06**, 009 (2018).
[7] H. Georgi, Phys. Rev. Lett. **98**, 221601 (2007).
[8] G. Cacciapaglia, G. Marandella, J. Terning, JHEP **01**, 070 (2008).
[9] G. Cacciapaglia, G. Marandella and J. Terning, JHEP **02**, 049 (2009).
[10] A. Falkowski, M. Pérez-Victoria, JHEP **12**, 107 (2008).
[11] J.A. Cabrer, G. von Gersdorff, M. Quirós, New J. Phys. **12**, 075012 (2010).
[12] A. Falkowski, M. Pérez-Victoria, JHEP **12**, 061 (2009).
[13] A. Falkowski, M. Pérez-Victoria, Phys. Rev. **D79**, 035005 (2009).
[14] C. Csáki, G. Lee, S.J. Lee, S. Lombardo, O. Telem, JHEP **03**, 142 (2019).
[15] E. Megías, M. Quirós, JHEP **08**, 166 (2019).
[16] E. Megías, M. Quirós, Acta Phys. Polon. B **52**, 711 (2021).
[17] E. Megías, M. Quirós, JHEP **09**, 157 (2021).
[18] O. DeWolfe, D.Z. Freedman, S.S. Gubser, A. Karch, Phys. Rev. **D62**, 046008 (2000).
[19] A. Delgado, J.R. Espinosa, J.M. No, M. Quirós, Phys. Rev. **D79**, 055011 (2009).
[20] E. Megías, M. Pérez-Victoria, M. Quirós, `work in progress` (2021).
[21] E. Megías, M. Quirós (2021), `2107.09102`.
[22] M. A. Stephanov, Phys. Rev. **D76**, 035008 (2007).
[23] E. Megías, M. Quirós, Int. J. Mod. Phys. A **36**, 2141002 (2021).
[24] C. Csáki, S. Hong, G. Kurup, S.J. Lee, M. Perelstein, W. Xue (2021), `2105.07035`.
[25] I. Chaffey, S. Fichet, P. Tanedo, JHEP **06**, 008 (2021).

Present status of radiative and rare kaon decays

Oscar Catà[1,*]

[1]Center for Particle Physics Siegen (CPPS), Theoretische Physik 1, Universität Siegen, Walter-Flex-Straße 3, D-57068 Siegen, Germany

Abstract. I will review recent developments in rare and radiative kaon decays from the theory side, with emphasis on those modes that are actively analyzed by the experimental collaborations.

1 Introduction

Rare and radiative kaon decays have a long history in particle physics as extremely useful tests of the Standard Model (SM) at long distances but also as probes of short-distance physics, including physics beyond the SM.

Currently the main focus of attention is on the golden modes, $K^+ \to \pi^+ \bar{\nu}\nu$ and $K_L \to \pi^0 \bar{\nu}\nu$, which are the object of devoted experiments by the NA62 and the KOTO collaborations, respectively. Such short-distance dominated decays can be computed in the SM using perturbation theory and are very clean probes of new physics. Aside from this, NA62 is regularly bringing up new analyses, either in the form of more precise measurements of long-distance dominated modes or as exclusion bounds for processes forbidden in the SM, such as lepton-flavor violating kaon decays. LHCb has also a kaon physics program that will soon start producing new results.

In this review I will concentrate on a number of rare and radiative kaon decays which have experienced recent developments both on the experimental and theoretical side. In most of the cases, these developments have greatly benefited from a very fluent communication between experimental and theoretical collaborations.

I will not report here on exotic or forbidden channels, which are a very active part of the NA62 program. For a review on these activities, I direct the reader to two recent talks at the EPS-HEP conference this year [1, 2].

2 Current status of $K^+ \to \pi^+ \bar{\nu}\nu$ and $K_L \to \pi^0 \bar{\nu}\nu$

The measurement of the golden modes $K^+ \to \pi^+ \bar{\nu}\nu$ and $K_L \to \pi^0 \bar{\nu}\nu$ is presently the main experimental endeavour in experimental kaon physics. The NA62 experiment aims at a determination of $K^+ \to \pi^+ \bar{\nu}\nu$ within an accuracy of 5% and the KOTO experiment is devoted to the first measurement of $K_L \to \pi^0 \bar{\nu}\nu$.

Both modes are theoretically extremely clean. Their prediction was revisited some years ago [3]. $K^+ \to \pi^+ \bar{\nu}\nu$ is now computed including QCD and electroweak corrections at NLO

*e-mail: oscar.cata@uni-siegen.de

for the top contribution, while for the charm contribution the QCD corrections are computed up to NNLO. The theoretical prediction yields

$$\mathcal{B}(K^+ \to \pi^+\bar{\nu}\nu) = (8.4 \pm 1.0) \times 10^{-11} \tag{1}$$

This has to be compared with the most recent experimental determination from NA62 earlier this year, based on 20 signal events, mostly collected in 2018 (see fig. 1). Their final result is [4]

$$\mathcal{B}(K^+ \to \pi^+\bar{\nu}\nu) = (10.6^{+4.0}_{-3.4} \pm 0.9) \times 10^{-11}\,, \tag{2}$$

which is the most precise determination to date. Their precision is foreseen to improve to 10% with the data from Run 2, which started this year and will end in 2023. The final goal is to achieve a 5% precision. From the results available right now one can conclude that sizeable deviations from the SM are to be excluded.

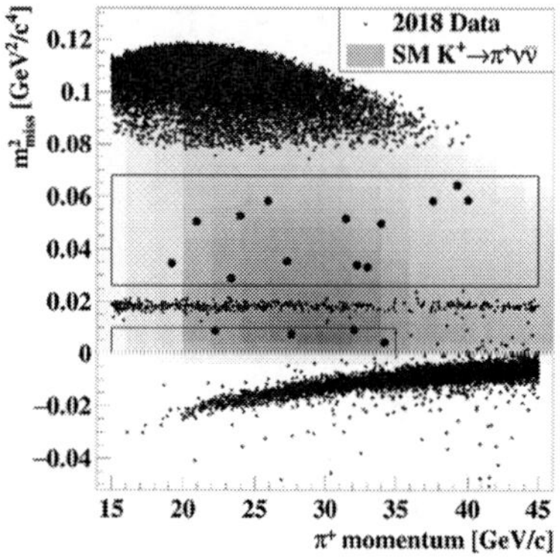

Figure 1. Kinematical details of the NA62 data for $K^+ \to \pi^+\bar{\nu}\nu$ corresponding to their 2018 dataset. Figure borrowed from [4].

The theoretical prediction for $K_L \to \pi^0\bar{\nu}\nu$ was also updated in [3]. The charm contribution is extremely suppressed and was neglected. The top contribution was computed including QCD and electroweak corrections at NLO. The final result reads

$$\mathcal{B}(K_L \to \pi^0\bar{\nu}\nu) = (3.4 \pm 0.6) \times 10^{-11} \tag{3}$$

This is an extremely challenging mode from the experimental side. The KOTO collaboration reported in 2018 preliminary results including 3 signal events from the data-taking period 2016-2018. After reexamination of the backgrounds these signals are now deemed consistent with backgrounds. Their result from the analysis of the 2016-2018 data set is [5]

$$\mathcal{B}(K_L \to \pi^0\bar{\nu}\nu) < 4.9 \times 10^{-9}\,, \tag{4}$$

which is slightly higher than the bound reported for the 2015 data, $\mathcal{B}(K_L \to \pi^0\bar{\nu}\nu) < 3.0 \times 10^{-9}$.

3 $K \to \pi\gamma^*$ and $K \to \pi\pi\gamma^*$

Radiative 3- and 4-body decays are long-distance dominated modes, with the main contribution coming from Bremsstrahlung. The subleading hadronic contributions are typically termed electric and magnetic contributions, depending on their transformation properties under parity. These modes are important as tests of the Standard Model at low energies, which is described by Chiral Perturbation Theory (ChPT) in the $\Delta S = 1$ sector:

$$\mathcal{L}_{\Delta S=1} = G_8 f_\pi^4 \,\mathrm{tr}[\lambda_6 D_\mu U^\dagger D^\mu U] + G_8 f_\pi^2 \sum_j N_j\, W_j(U, D_\mu U, \lambda_6) + O(p^6)\,, \tag{5}$$

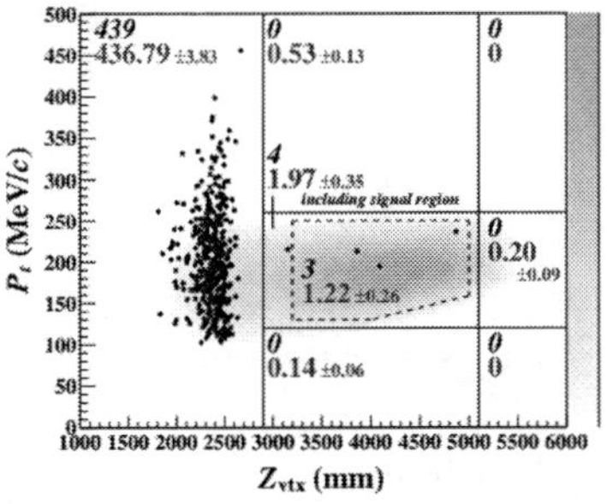

Figure 2. Data from the KOTO experiment for the period 2016-2018. All the data points are consistent with the background estimates. Figure borrowed from [5].

where $U = \exp\left[\frac{i}{f_\pi}\phi^a \tau^a\right]$ is the chiral matrix containing the Goldstone fields ϕ^a. At NLO there are 37 unknown coefficients N_j, which capture the effects of nonperturbative hadronic states at the kaon mass scale. The general form of the amplitudes for these decay modes is therefore

$$\mathcal{M}(K \to X\gamma^{(*)}) = \underbrace{\mathcal{M}_B(O(p^2))}_{\text{Brems.}} + \underbrace{\mathcal{M}_E(O(p^4))}_{\text{electric, CP-even}} + \underbrace{\mathcal{M}_M(O(p^4))}_{\text{magnetic, CP-odd}} \,, \tag{6}$$

where the electric and magnetic contributions are functions of the N_j low-energy couplings. These couplings are related to the slopes of the differential decay rates and can be accessed by measuring the interference term between Bremsstrahlung and electric/magnetic emission.

Experimentally this is a rather fertile field, with efforts that started roughly 15 years ago and are still ongoing. The present experimental situation can be summarized as follows:

$$\begin{array}{llllll}
K^\pm \to \pi^\pm \gamma^* & [10^{-7}]_{3\%}\,; & K_S \to \pi^0 \gamma^* & [10^{-9}]_{50\%}\,; & K_L \to \pi^0 \gamma^* & [< 10^{-10}] \\
K^\pm \to \pi^\pm \pi^0 \gamma & [10^{-6}]_{7\%}\,; & K_S \to \pi^+ \pi^- \gamma & [10^{-3}]_{3\%}\,; & K_L \to \pi^+ \pi^- \gamma & [10^{-5}]_{4\%} \\
K^\pm \to \pi^\pm \gamma\gamma & [10^{-6}]_{6\%}\,; & K_S \to \pi^0 \gamma\gamma & [10^{-8}]_{37\%}\,; & K_L \to \pi^0 \gamma\gamma & [10^{-6}]_{3\%} \\
K^\pm \to \pi^\pm \pi^0 \gamma^* & [10^{-6}]_{3\%}\,; & K_S \to \pi^+ \pi^- \gamma^* & [10^{-5}]_{3\%}\,; & K_L \to \pi^+ \pi^- \gamma^* & [10^{-7}]_{6\%}\,,
\end{array} \tag{7}$$

where the value in brackets indicates the order of magnitude of each branching ratio and the percentage the current experimental precision.

In the following I will discuss two recent lines of research related to these decay modes. The first one describes a strategy, mostly using experimental input, to determine in a closed form the set of NLO coefficients relevant for radiative kaon decays. The second line of research is more ambitious and aims at a theoretical prediction of the NLO coefficients.

However, before discussing these issues I would like to emphasize that radiative 3-body but especially 4-body decays can also be used as probes of short-distance physics, within and beyond the Standard Model. 4-body decays depend on 5 kinematic variables and therefore have very rich kinematical distributions. The differential amplitudes can be parametrized as

$$\begin{aligned}
\frac{d^5\Gamma}{dE_\gamma^* dT_c^* dq^2 d\cos\theta_\ell d\phi} &= \mathcal{A}_1 + \mathcal{A}_2 \sin^2\theta_\ell + \mathcal{A}_3 \sin^2\theta_\ell \cos^2\phi + \mathcal{A}_4 \sin 2\theta_\ell \cos\phi \\
&+ \mathcal{A}_5 \sin\theta_\ell \cos\phi + \mathcal{A}_6 \cos\theta_\ell + \mathcal{A}_7 \sin\theta_\ell \sin\phi \\
&+ \mathcal{A}_8 \sin 2\theta_\ell \sin\phi + \mathcal{A}_9 \sin^2\theta_\ell \sin 2\phi \,,
\end{aligned} \tag{8}$$

where $\mathcal{A}_i$ are form factors that depend on the dynamical variables E_γ^*, T_c^* and q^2. The first line contains the terms responsible for P-conserving interactions, with only the first three terms contributing to the total cross section. The second and third lines instead describe P-violating interactions induced by short-distance and long-distance dynamics, respectively. They can be isolated with appropriate angular asymmetries and most of them yield very clean tests of new physics.

3.1 Towards a determination of the radiative NLO chiral couplings

Out of the 37 N_j NLO operators of the $\Delta S = 1$ ChPT Lagrangian, radiative kaon decays are sensitive to a rather small subset of them, namely $W_{14} - W_{18}$ (CP-even sector) and $W_{28} - W_{31}$ (CP-odd sector).

The observation made in [6] was that an individual determination of them all was feasible after the measurement of the electric interference terms either of $K^\pm \to \pi^\pm\pi^0\gamma^*$ or of $K_S \to \pi^+\pi^-\gamma^*$ (in combination with $K_L \to \pi^+\pi^-\gamma^*$). These decay modes were already in the programs of NA48/2 and LHCb and the observation thus provided an extra motivation for their analysis.

Out of the measured radiative kaon modes listed in (7) there are only 4 non-redundant combinations of NLO parameters. In particular, for the modes $K^\pm \to \pi^\pm\gamma^*$, $K_S \to \pi^0\gamma^*$, $K^\pm \to \pi^\pm\pi^0\gamma$ and $K^\pm \to \pi^\pm\gamma\gamma$ one finds the combinations

$$
\begin{aligned}
\mathcal{N}_E^{(1)} &\equiv N_{14}^r - N_{15}^r = \frac{3}{64\pi^2}\left(\frac{1}{3} - \frac{G_F}{G_8}a_+ - \frac{1}{3}\log\frac{\mu^2}{m_K m_\pi}\right) - 3L_9^r = -0.0167(13)\\
\mathcal{N}_S &\equiv 2N_{14}^r + N_{15}^r = \frac{3}{32\pi^2}\left(\frac{1}{3} + \frac{G_F}{G_8}a_S - \frac{1}{3}\log\frac{\mu^2}{m_K^2}\right) = +0.016(4)\\
\mathcal{N}_E^{(0)} &\equiv N_{14}^r - N_{15}^r - N_{16}^r - N_{17} = -\frac{|\mathcal{M}_K|f_\pi}{2G_8}X_E = +0.0022(7)\\
\mathcal{N}_0 &\equiv N_{14}^r - N_{15}^r - 2N_{18}^r = \frac{3}{128\pi^2}\hat{c} - 3(L_9^r + L_{10}^r) = -0.0017(32)\,,
\end{aligned}
\tag{9}
$$

where in the second equality on each line one can see how the NLO counterterms are related to the slopes a_+, a_S, X_E and $\hat{c}$. The numbers on the right-hand side are obtained by using the experimental numbers for the slopes.

One can easily see that N_{14}, N_{15} and N_{18} can be individually determined, but only the combination $N_{16} + N_{17}$ is constrained. When $K^\pm \to \pi^\pm\pi^0 e^+e^-$ is considered, one is sensitive to $\mathcal{N}_E^{(0)}$, $\mathcal{N}_E^{(1)}$ and to a new independent combination of counterterms, namely

$$
\mathcal{N}_E^{(2)} = N_{14}^r + 2N_{15}^r - 3(N_{16}^r - N_{17}) \tag{10}
$$

This combination leads to $\mathcal{N}_E^{(2)} = +0.089(11) + 6N_{17}$ when only polynomial terms are considered. Since N_{17} is predicted to be small by different hadronic models, one can assume that indeed $\mathcal{N}_E^{(2)} \sim +\mathcal{O}(10^{-1})$ and it is sizeable. The interference term has a characteristic pattern than depends mostly on $\mathcal{N}_E^{(0)}$ and $\mathcal{N}_E^{(2)}$. This is illustrated in fig. 3, where the upper and lower dotted lines correspond to setting $\mathcal{N}_E^{(0)} = 0$ and $\mathcal{N}_E^{(2)} = 0$, respectively.

Even though $K^\pm \to \pi^\pm\pi^0 e^+e^-$ is a 4-body decay and has the corresponding phase-space suppression, the decay rate is comparable in size to $K^+ \to \pi^+\gamma^*$. The challenge is to extract information on the structure-dependent terms. The magnetic decay width is 70 times smaller than the Bremsstrahlung, and for the interference term one typically needs to have a sensitivity below the percent level. However, the Bremsstrahlung is mostly localized at very small values of the dilepton invariant mass. Using cuts on this variable one can then increase the required sensitivity above the percent level.

An analysis of this decay has been published recently [7], based on the complete data sets of the NA48/2 experiment. The Bremsstrahlung and magnetic contributions could be measured and agreement was found with the theoretical prediction. Unfortunately, in order to pin down the interference term more statistics would be needed. Progress on this side will have to come from a new experiment.

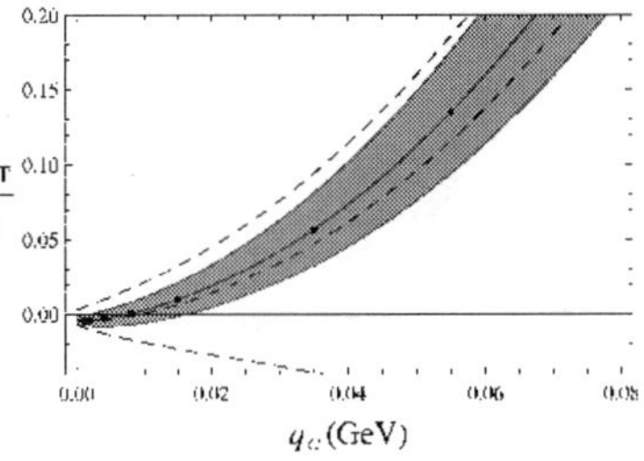

Figure 3. Ratio between the interference and the Bremsstrahlung contributions for the $K^\pm \to \pi^\pm\pi^0 e^+e^-$ decay. The solid line and the error bar correspond to the values displayed in the main text. The upper and lower dashed curves correspond to setting $N_E^{(0)} = 0$ and $N_E^{(0)} = 0$, respectively. Figure borrowed from [6].

An important alternative to determine the set of NLO counterterms is to combine data on $K_S \to \pi^+\pi^-e^+e^-$ and $K_L \to \pi^+\pi^-e^+e^-$. Information on the decay rates exists for both modes, but not for the interference term, which would require typically a percent precision. Right now a new measurement of $K_S \to \pi^+\pi^-e^+e^-$ is in the LHCb agenda but I am not aware of any new measurement planned for $K_L \to \pi^+\pi^-e^+e^-$. One should probably stress that the counterterm structure for $K_S \to \pi^+\pi^-e^+e^-$ can be predicted from other radiative kaon decays to be

$$BR(K_S \to \pi^+\pi^-e^+e^-) = \underbrace{4.74 \cdot 10^{-5}}_{\text{Brems.}} + \underbrace{4.39 \cdot 10^{-8}}_{\text{Int.}} + \underbrace{1.33 \cdot 10^{-10}}_{\text{DE}}, \tag{11}$$

but in order to determine N_{16} and N_{17} one needs to measure the slope of $K_L \to \pi^+\pi^-e^+e^-$.

3.2 $K^+ \to \pi^+\ell^+\ell^-$

In the previous section the NLO counterterms were extracted from experiment. However, they can in principle be estimated theoretically. In [8] a new method was proposed and applied to $K^+ \to \pi^+\ell^+\ell^-$.

It is customary to parametrize this differential decay rate as

$$\frac{d^2\Gamma}{dzdx} = \frac{\alpha^2 m_K}{(4\pi)^5)} f(x,z)|W(z)|^2, \qquad W(z) = G_F m_K^2(a_+ + b_+ z) + W^{\pi\pi}(z), \tag{12}$$

where a_+ and b_+ are slope parameters. A determination of such parameters requires a description of $W(z)$ for all values of z. As a first approximation, the authors of [8] suggested the split

$$W(z) = W^{\pi\pi}(z) + W^{\text{res}}(z;\nu) + W^{\text{SD}}(z;\nu), \tag{13}$$

where the first term is the pion rescattering contribution, which can be extracted from experimental data, and the last piece is the short-distance tail, which can be computed with the operator product expansion. The intermediate-energy region is modelled with an infinite tower of states, whose parameters are chosen to match the high- and low-energy ends. If this matching is done at 1-loop level, one is sensitive to the scale dependence of the slope parameters. This scale-dependence is illustrated in fig. 4.

The value found for a_+ with this method is clearly off, but one should take into account that the model was rather crude. In particular, no kaon rescattering effects were taken into account. Work in this direction is currently underway.

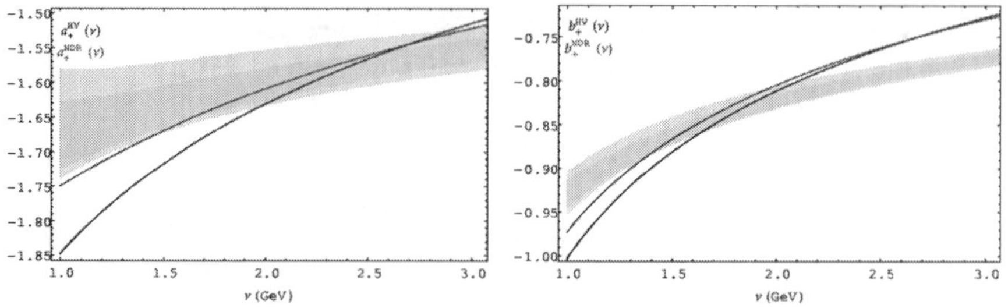

Figure 4. Scale dependence of the slope parameters a_+ and b_+. Figure borrowed from [8].

The $K^+ \to \pi^+\mu^+\mu^-$ mode has been recently reanalyzed by NA62 with the result [9]

$$a_+ = -0.592(13)(7)(1); \qquad b_+ = -0.699(46)(35)(3) \tag{14}$$

The results are still preliminary, but show a preference for negative values of the slopes. There exists a positive solution at $a_+ = 0.368$, $b_+ = 2.045$, but it is statistically disfavored. The existence of two sets of solutions with similar statistical significance is a feature of the data which was already pointed out in [8]. The new result of NA62 agrees with previous determinations and, once compared with the existing data for the electron channel, shows no indication of lepton-flavor universality violation.

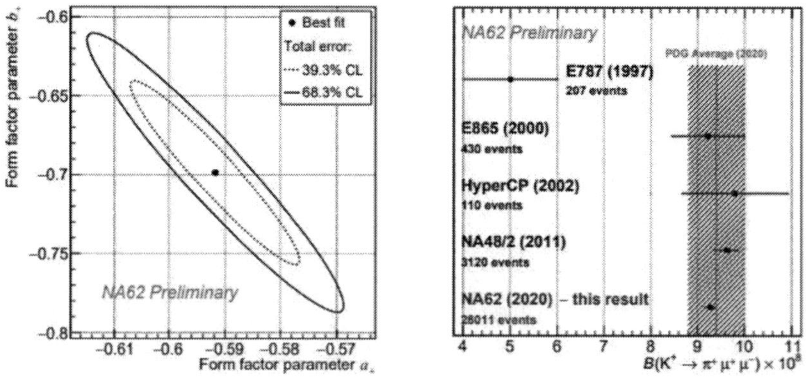

Figure 5. Left panel: NA62 new determination of the branching ratio for $K^+ \to \pi^+\mu^+\mu^-$ and comparison with previous results. Right panel: NA62 results for the slope parameters a_+ and b_+. Figures borrowed from [9].

4 $K^+ \to \pi^0 e^+ \nu \gamma$

I will finally review briefly the status of the $K_{\ell 3\gamma}$ decays. These modes are interesting for different reasons. Rather generically, they can be used as tests of the Standard Model at low energies, but the charged modes are also sensitive to the chiral anomaly and can probe T violation.

Similarly to $K \to \pi\gamma(*)$ and $K \to \pi\pi\gamma(*)$, these modes are long-distance dominated and can be analyzed in the context of ChPT. The Bremsstrahlung piece is the dominant contribution (of $O(p^2)$ in the chiral counting) and through Low's theorem it can be expressed in terms of the $K_{\ell 3}$ decay modes. The structure-dependent pieces start at $O(p^4)$ in the chiral expansion and are typically probed with the ratios

$$R_j = \frac{\mathcal{B}(K\ell 3\gamma_j)}{\mathcal{B}(K\ell 3)}, \tag{15}$$

where the subindex j denotes some specific cuts on the photon energy and the angle spanned between the lepton and the photon. It is common to use the benchmark ratios $R_1(E_\gamma > 10$ MeV; $\theta > 10^o)$, $R_2(E_\gamma > 30$ MeV; $\theta > 20^o)$ and $R_3((E_\gamma > 10$ MeV; $0.6 < \cos\theta < 0.9)$.

The decay mode $K_{e3\gamma}$ has been recently measured by the OKA collaboration [10] and NA62 [11]. The results of the two collaborations for the three reference ratios R_j can be found in Table 1. The values are compared with the theoretical estimate, which were carried out up to $O(p^6)$ in ChPT [12].

The results for R_2 and R_3 are largely compatible between NA62 and OKA, and show a notable deficit with respect to the theoretical result. The situation with R_1 is more puzzling: OKA shows a marked excess and NA62 a marked deficit with respect to the theoretical number.

Table 1. Experimental numbers for R_j for the $K_{e3\gamma}$ decay compared to the theoretical estimate.

	$O(p^6)$ ChPT [12]	OKA collaboration [10]	NA62 (preliminary) [11]
$R_1(\times 10^2)$	1.804(21)	1.990(17)(21)	1.684(5)(10)
$R_2(\times 10^2)$	0.640(8)	0.587(10)(15)	0.599(3)(5)
$R_3(\times 10^2)$	0.559(6)	0.532(10)(12)	0.523(3)(3)

It is however hard to assess how serious these discrepancies are. The NA62 results are preliminary and one should wait for the published version of their analysis. The values released so far corresponds to the data sets of 2017 and 2018, so there is room to enhance their statistics. If their numbers are confirmed, one could claim a sizeable deficit for all ratios with respect to the ChPT estimate. The $O(p^6)$ analysis was performed taking into account only the polynomial pieces and the NNLO low-energy constants were estimated with power-counting arguments. A discrepancy with experiment would motivate a revision of the theory estimate to include the non-polynomial pieces. A better estimate of the NNLO constants would be hard to achieve, given how poorly known they are.

Right now, the R_j ratios are the only realistic tests for this decay mode. Experiments are not sensitive to the T-odd parameter: the present upper bounds are two orders of magnitude larger than the theoretical prediction.

5 Summary and outlook

Rare and radiative kaon decays are an active field of research thanks to the ongoing scientific programs of the NA62 and KOTO experiments. Most of the energies are currently devoted to the $K^+ \to \pi^+\bar{\nu}\nu$ and $K_L \to \pi^0\bar{\nu}\nu$ modes. For the former, NA62 will soon achieve a 10% precision, but sizeable new-physics effects are already excluded. For the latter, there exist so far upper bounds. In parallel, there is an intense experimental activity on exotic and forbidden modes by NA62. Regarding radiative kaon decays, there is still room for improvement on the experimental side, especially with the $K \to \pi\pi\gamma^*$ modes, where there is very little information aside from the branching ratios. The recent analysis of $K^+ \to \pi^+\pi^0 e^+e^-$ has shown that the statistics collected by NA48/2 were not enough for a determination of the electric NLO chiral counterterms. LHCb will soon measure $K_S \to \pi^+\pi^-e^+e^-$, but it would be interesting to have also a more precise measurement of $K_L \to \pi^+\pi^-e^+e^-$ somewhere in the near future.

Acnowledgements

I would like to thank Evgueni Goudzovski and Rainer Wanke for kindly guiding me to the latest NA62 results.

References

[1] E. Minucci, "Searches for lepton flavour/number violation in K+ and pi0 decays at the NA62 experiment," contribution to the EPS-HEP2021 conference, https://indico.desy.de/event/28202/contributions/105897/

[2] C. J. Parkinson, "Search for heavy neutral lepton production at the NA62 experiment," contribution to the EPS-HEP2021 conference, https://indico.desy.de/event/28202/contributions/105801/

[3] A. J. Buras, D. Buttazzo, J. Girrbach-Noe and R. Knegjens, JHEP **11**, 033 (2015) [arXiv:1503.02693 [hep-ph]].

[4] E. Cortina Gil *et al.* [NA62], JHEP **06**, 093 (2021) [arXiv:2103.15389 [hep-ex]].

[5] J. K. Ahn *et al.* [KOTO], Phys. Rev. Lett. **126**, no.12, 121801 (2021) [arXiv:2012.07571 [hep-ex]].

[6] L. Cappiello, O. Catà and G. D'Ambrosio, Eur. Phys. J. C **78**, no.3, 265 (2018) [arXiv:1712.10270 [hep-ph]].

[7] J. R. Batley *et al.* [NA48/2], Phys. Lett. B **788**, 552-561 (2019) [arXiv:1809.02873 [hep-ex]].

[8] G. D'Ambrosio, D. Greynat and M. Knecht, JHEP **02**, 049 (2019) [arXiv:1812.00735 [hep-ph]]; G. D'Ambrosio, D. Greynat and M. Knecht, Phys. Lett. B **797**, 134891 (2019) [arXiv:1906.03046 [hep-ph]].

[9] L. Bician, PoS **ICHEP2020**, 364 (2021)

[10] A. Y. Polyarush *et al.* [OKA], Eur. Phys. J. C **81**, no.2, 161 (2021) [arXiv:2010.07983 [hep-ex]].

[11] F. Brizioli, "New measurement of radiative decays $K^+ \to e^+\nu\gamma$ and $K^+ \to \pi^0 e^+\nu\gamma$ at the NA62 experiment," contribution to the EPS-HEP2021 conference, https://indico.desy.de/event/28202/contributions/107307/

[12] B. Kubis, E. H. Muller, J. Gasser and M. Schmid, Eur. Phys. J. C **50**, 557-571 (2007) [arXiv:hep-ph/0611366 [hep-ph]].

Hybrid stars from a constrained equation of state

Márcio Ferreira[1,*], *Renan Câmara Pereira*[1,**], and *Constança Providência*[1,***]

[1]CFisUC, Department of Physics, University of Coimbra, P-3004 - 516 Coimbra, Portugal

Abstract. We determine, within a meta-model, the properties of the nuclear matter equation of state (EoS) that allow for a phase transition to deconfinement matter. It is shown that the properties that define the isoscalar channel are the ones that are affected, in particular, a phase transition implies much larger values of the skewness and kurtosis. The effect of multi-quark interaction channels in the description of the quark phase in hybrid stars is also studied. NS properties, such as the mass and radius of the quark core, show an interplay dependence between the 8-quark vector and the 4-quark isovector-vector interactions. We show that low mass NS, $M \sim 1.4M_\odot$, may already contain a quark core, and satisfy all existing NS observational constraints. We discuss the strangeness content of the quark core and its influence on the speed of sound.

1 Introduction

The physics of neutron stars (NS) can be studied by combining astrophysical observations of electromagnetic radiation, astro-particles and gravitational waves (GW) – multi-messenger astrophysics. The recent observation of the massive pulsars, PSR J0348+0432 with $M = 2.01\pm0.04\ M_\odot$ [1] and or MSP J0740+6620, a mass $2.08^{+0.07}_{-0.07}M_\odot$ [2, 3] impose strong constraints on the equation of state (EoS) of nuclear matter. The estimation of an upper bond on tidal deformability of a NS star by the LIGO/Virgo collaboration from the GW170817 event is important in further constraining the high density behavior of the NS EoS [4, 5]. While a not too stiff EoS is favored from the upper bond on the tidal deformability, massive pulsars require, on the other hand, a considerably stiff EoS for the NS. Such apparent tension might indicate the presence of exotic matter in NS, i.e., non-nucleonic degrees of freedom. The transition, via a first-order phase transition, from hadronic to quark matter at high densities could be an explanation for the existence of low tidal deformabilities for NS masses around $1.4M_\odot$. There is a great interest in finding observational signatures sensible to the presence of exotic matter inside NS [6–8].

Herein, hybrid NS are described by applying a two-model approach: we choose an hadronic model for the confined phase and a quark model for the quark phase, which are connected through a first-order phase transition (applying the Maxwell construction). The quark model chosen is the well known Nambu-Jona–Lasinio (NJL) model, an effective model of QCD, that considers chiral symmetry preserving interactions [9, 10]. The use of the NJL model in such two-model approaches has been widely explored both considering a first-order phase transition [11–16] and with a smooth (crossover) transition between both phases

*e-mail: marcio.ferreira@uc.pt
**e-mail: renan.pereira@student.uc.pt
***e-mail: cp@fis.uc.pt

[17, 18]. Starting from the three-flavor NJL model, we study the effect of several interaction channels in the existence and stability of hybrid stars, and specifically on quark core properties [19–21].

2 Equation of state of hybrid NS

The matter inside hybrid stars is modelled via a two-model approach: an hadronic phase is connected to a quark phase through a first-order phase transition, i.e., a Maxwell construction is implemented. Therefore, we need to choose both an hadronic model and a quark model.

2.1 Hadronic EoS

In the following, we consider two types of hadronic EoS: a) a set of metamodels based on a Taylor expansion around saturation density [22]; b) a relativistic mean-field model with density dependent couplings DDME2 [23], which satisfies well accepted nuclear matter properties at saturation (K_{sat} = 250 MeV, E_{sym} = 32.3 MeV, L_{sym} = 51.2 MeV, K_{sym} = −87 MeV), and besides describes two solar mass stars.

In the following, we summarize the metamodel considered. The energy per particle functional of homogeneous nuclear matter, $\mathcal{E}$, can be written as:

$$\mathcal{E}(n_n, n_p) = e_{sat}(n) + e_{sym}(n)\delta^2. \tag{1}$$

where n_n and n_p are neutron and proton densities, $n = n_n + n_p$ is the baryonic density and $\delta = (n_n - n_p)/n$ is the asymmetry. Following previous studies, e.g. [22], we parametrize the hadronic EoS as a Taylor expansion around the saturation density. We consider a fourth order expansion around saturation density, n_0:

$$e_{sat}(x) = \sum_{n=0}^{4} \frac{1}{n!} P_{IS}^{(n)} x^n, \qquad e_{sym}(x) = \sum_{n=0}^{4} \frac{1}{n!} P_{IV}^{(n)} x^n \tag{2}$$

where, $x = \frac{(n-n_0)}{3n_0}$. The coefficients of $e_{sat}(n)$ and $e_{sym}(n)$ can be identified, respectively, with the isoscalar and isovector empirical parameters,

$$P_{IS}^{(k)} = (3n_0)^k \left.\frac{\partial^k e_{sat}}{\partial n^k}\right|_{\{\delta=0,n=n_0\}}, \qquad P_{IV}^{(k)} = (3n_0)^k \left.\frac{\partial^k e_{sym}}{\partial n^k}\right|_{\{\delta=0,n=n_0\}}. \tag{3}$$

The correspondence between the coefficients and the empirical parameters can then be written as: $E_{sat}, K_{sat}, Q_{sat}, Z_{sat} \rightarrow P_{IS}^{(0)}, P_{IS}^{(2)}, P_{IS}^{(3)}, P_{IS}^{(4)}$, and $E_{sym}, L_{sym}, K_{sym}, Q_{sym}, Z_{sym} \rightarrow P_{IV}^{(0)}, P_{IV}^{(1)}, P_{IV}^{(2)}, P_{IV}^{(3)}, P_{IV}^{(4)}$.

The advantages of taking a meta-model for the hadron phase can be summarized in the following two points: (i) its flexibility to build an hybrid EoS with $1.97M_\odot$ maximum mass; (ii) the possibility of choosing the empirical nuclear parameters that satisfy uncertainties on the EoS parameters known from experiments and other theoretical studies. From all the meta-model EoSs built, 354 of them allow for a phase transition to quark matter. In Table 1, the mean and standard deviation of the properties of the both sets of EoS, nucleonic and hybrid, are given. We also include a third set of EoS that allows for a phase transition to quark matter at a quite low baryonic density, i.e. the transition density n_t satisfies $1.3 < n_t/n_0 < 2.5$. Comparing the three meta-models sets we conclude that a phase transition to quark matter requires larger values of isoscalar properties, in particular, Q_{sat} and Z_{sat}, and a wider spread, while the isovector channel is essentially not affected.

Table 1. The mean $\overline{P}_i$ and standard deviation $\sqrt{\sigma_{P_i}}$ of the multivariate Gaussian, where σ_{P_i} is the variance of the parameter P_i. Our EoSs are sampled using the initial distribution for P_i assuming that there are no correlations among the parameters. All the quantities are in units of MeV. The values of E_{sat} and n_0 are fixed to -15.8 MeV and 0.155 fm^{-3}, respectively. The parameters are given for three scenarios always considering that a two solar mass star is described by the models: (i) purely nucleonic EoS; (ii) hybrid EoS, all without furhter filters; (iii) hybrid stars with a phase transition in the range $1.3n_0 - 2.5n_0$.

(i) Nucleonic	K_{sat}	Q_{sat}	Z_{sat}	E_{sym}	L_{sym}	K_{sym}	Q_{sym}	Z_{sym}
mean	233.95	58.62	-181.97	33.32	51.56	-43.96	238.21	371.91
std	18.75	123.33	143.04	1.89	11.83	63.02	300.33	698.56
(ii) Hybrid (all)	K_{sat}	Q_{sat}	Z_{sat}	E_{sym}	L_{sym}	K_{sym}	Q_{sym}	Z_{sym}
mean	236.52	241.38	362.11	33.17	49.99	-37.48	191.54	503.39
std	17.73	242.03	593.75	1.84	12.55	67.23	310.48	734.69
(iii) Hybrid ($1.3 < n_t/n_0 < 2.5$)	K_{sat}	Q_{sat}	Z_{sat}	E_{sym}	L_{sym}	K_{sym}	Q_{sym}	Z_{sym}
mean	244.55	434.37	557.83	33.38	49.90	-18.07	195.39	400.46
std	16.91	174.6	743.94	2.00	11.55	57.47	300.13	672.24

To build an hadronic EoS, we use random sampling to choose a point in the 8-dimensional space of parameters from a multivariate Gaussian with zero covariance:

$$\mathrm{EoS}_i = (E_{sym}, L_{sym}, K_{sat}, K_{sym}, Q_{sat}, Q_{sym}, Z_{sat}, Z_{sym})_i \sim N(\boldsymbol{\mu}, \boldsymbol{\Sigma}) \quad (4)$$

with the mean defined by $\boldsymbol{\mu}^T = (\overline{E}_{sym}, \overline{L}_{sym}, \overline{K}_{sat}, \overline{K}_{sym}, \overline{Q}_{sat}, \overline{Q}_{sym}, \overline{Z}_{sat}, \overline{Z}_{sym})$, and the covariance matrix given by $\boldsymbol{\Sigma} = diag(\sigma_{E_{sym}}, \sigma_{L_{sym}}, \sigma_{K_{sat}}, \sigma_{K_{sym}}, \sigma_{Q_{sat}}, \sigma_{Q_{sym}}, \sigma_{Z_{sat}}, \sigma_{Z_{sym}})$.

2.2 Quark EoS

For the quark matter phase, we use a SU(3) NJL-type model, given by the following multi-quark interaction Lagrangian density:

$$\begin{aligned}
\mathcal{L} = {} & \bar{\psi}(i\partial - \hat{m} + \hat{\mu}\gamma^0)\psi + G_S \sum_{a=0}^{8} [(\bar{\psi}\lambda^a\psi)^2 + (\bar{\psi}i\gamma^5\lambda^a\psi)^2] - G_D[\det(\bar{\psi}(1+\gamma_5)\psi) + \det(\bar{\psi}(1-\gamma_5)\psi)] \\
& - G_\omega[(\bar{\psi}\gamma^\mu\lambda^0\psi)^2 + (\bar{\psi}\gamma^\mu\gamma_5\lambda^0\psi)^2] - G_\rho \sum_{a=1}^{8} [(\bar{\psi}\gamma^\mu\lambda^a\psi)^2 + (\bar{\psi}\gamma^\mu\gamma_5\lambda^a\psi)^2] \\
& - G_{\omega\omega}[(\bar{\psi}\gamma^\mu\lambda^0\psi)^2 + (\bar{\psi}\gamma^\mu\gamma_5\lambda^0\psi)^2]^2 \\
& - G_{\rho\rho} \sum_{a=1}^{8}\sum_{b=1}^{8} [(\bar{\psi}\gamma^\mu\lambda^a\psi)^2 + (\bar{\psi}\gamma^\mu\gamma_5\lambda^a\psi)^2][(\bar{\psi}\gamma^\mu\lambda^b\psi)^2 + (\bar{\psi}\gamma^\mu\gamma_5\lambda^b\psi)^2] \\
& - G_{\omega\rho}[(\bar{\psi}\gamma^\mu\lambda^0\psi)^2 + (\bar{\psi}\gamma^\mu\gamma_5\lambda^0\psi)^2] \sum_{a=1}^{8} [(\bar{\psi}\gamma^\mu\lambda^a\psi)^2 + (\bar{\psi}\gamma^\mu\gamma_5\lambda^a\psi)^2]. \quad (5)
\end{aligned}$$

The parameters of the quark model, $\hat{m} = diag(m_u, m_d, m_s)$, G_S, G_D, and Λ, are fit to reproduce the vacuum meson masses and decay constants (Table 2), while the remaining parameters are left as free.

We have analyzed the effect of some of these specific channels on hybrid stars in [19–21], where a detailed description can be found. In particular, in the following we discuss the effect of parameters G_ω, G_ρ and $G_{\omega\omega}$. The parameters $G_{\rho\rho}$ and $G_{\omega\rho}$ have only a residual or small

Table 2. Parameters of the quark model: Λ is the model cutoff, $m_{u,d,s}$ are the quark current masses, G_S and G_D are coupling constants. $M_{u,d}$ and M_s are the resulting constituent quark masses in the vacuum.

Λ [MeV]	$m_{u,d}$ [MeV]	m_s [MeV]	$G_S\Lambda^2$	$G_D\Lambda^5$	$M_{u,d}$ [MeV]	M_s [MeV]
623.58	5.70	136.60	1.67	13.67	332.2	510.7

effect on the EoS and NS properties. In the following, we will characterize each quark model by the coupling ratios: $\chi_\omega = G_\omega/G_S$, $\chi_{\omega\omega} = G_{\omega\omega}/G_S^4$, $\chi_\rho = G_\rho/G_S$, $\chi_{\rho\rho} = G_{\rho\rho}/G_S^4$, and $\chi_{\omega\rho} = G_{\omega\rho}/G_S^4$.

The thermodynamical potential is determined using the mean-field approximation, from which we calculate the pressure and energy density for cold quark matter, i.e., at zero temperature. The pressure P and energy density ϵ are defined up to an extra constant term B (the bag constant), i.e., $P \to P + B$ and $\epsilon \to \epsilon - B$. This extra parameter will allow us to control the baryonic density at the hadron-quark phase transition. As referred before, for the hadronic phase, we use the metamodel introduced above and the relativistic mean-field model with density dependent couplings model DDME2 [23]. The first-order phase transition from hadronic (H) to quark (Q) matter is accomplished by imposing the Maxwell construction, i.e., $\mu_B^H = \mu_B^Q$ and $P^H = P^Q$, where $\mu_B^{H,Q}$ is the baryon chemical potential of each phase.

3 Results and discussion

In the following, we study how the quark matter properties affect the existence of stable hybrid stars. First we analize the effect of the vector-isoscalar parameter $\chi_\omega = G_\omega/G_S$ on the values of the mass and radius, respectively, M_t and R_t, of the star where quarks start to nucleate (see Fig. 1 left panel) and the radius and mass of the quark core, respectively, M_{QC} and R_{QC}, as a function of the maximum NS mass (Fig. 1 center and right panels). In this figure each hybrid EoS is represented by one point whose color designates a specific value of χ_ω, shown in color scale. The radius and mass of the quark core, R_{QC} and M_{QC}, is determined from the quark content of the most massive stable NS that contains quark matter. As χ_ω increases, the appearance of quarks occurs for larger NS masses and radii. Quark cores in massive NS, with $M > 2.4M_\odot$, are attainable with $\chi_\omega > 0.6$. However, the quark core size (center and right panels), of the EoSs that sustain high values of M_{max}, is quite small. The distribution of M_{QC} has a mean value of $0.08M_\odot$ and a maximum value of $1.29M_\odot$, whereas the mean value of R_{QC} is 1.99 km and the maximum value of 8.78 km. Quark cores having radii as large as 7 km have been also obtained in [24] using a quite different approach.

We next analyze the impact of $\{\chi_{\omega\omega},\chi_\rho\}$ (with $\chi_\omega = 0$) in the $M(R)$ diagram (Fig. 2). Larger quark branches are realized for larger $\chi_{\omega\omega}$ values, where massive NS ($M > 2M_\odot$) are already attained for $\chi_{\omega\omega} = 15$ and 20. This is explained because the magnitude of the $\chi_{\omega\omega}$ term is still small at the hadron-quark transition, not affecting the transition density. However, at larger densities it contributes with a strong repulsive term, gives rise to a hard quark EoS, and larger star masses are reached.

The NS mass of the quark onset increases with χ_ρ (regardless of $\chi_{\omega\omega}$) because the coupling shifts the onset of strange quarks to lower densities, making the quark EoS softer. A second effect of the softening of the EoS is the decrease of the quark branch: the smoothing of the EoS caused by the appearance of strange quarks prevents too massive stars to occur.

We have determined the binary effective tidal deformability $\tilde{\Lambda}$ for $M_{\rm chirp} = 1.186M_\odot$ and $0.73 \leq q = M_1/M_2 \leq 1$ (GW170817 event [26]). Figure 3 shows the $q(\tilde{\Lambda})$ diagram

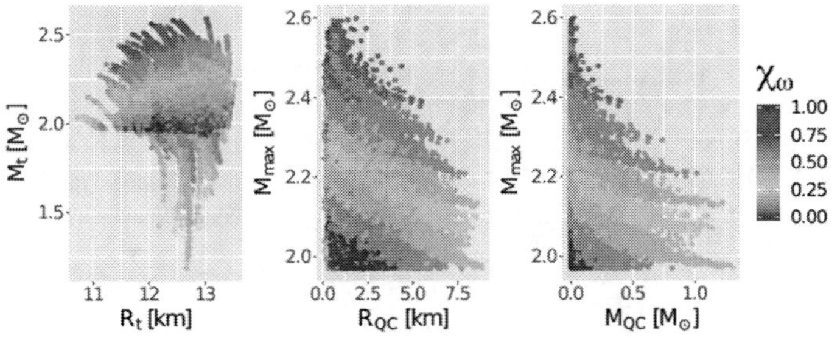

Figure 1. The M_t vs R_t (left), which corresponds to the (M, R) where the onset of quark matter occurs, and the diagrams of maximum NS mass vs. quark core radius (center) and mass (right) as a function of χ_ω (color scale) for each hybrid EoS.

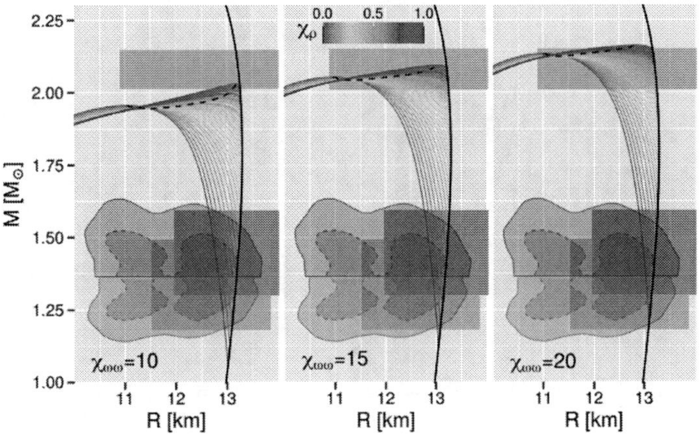

Figure 2. $M(R)$ diagrams as a function of χ_ρ (color scale) for three values of $\chi_{\omega\omega}$: 10 (left), 15 (center), and 20 (right). The solid black line represents the purely hadronic sequence, and the black dashed line indicates the maximum mass of each hybrid EoS. The top blue and bottom red regions indicate, respectively, the 90% (solid) and 50% (dashed) credible intervals of the LIGO/Virgo analysis for the GW170817 event [26]. The rectangular regions enclosed by dotted lines indicate the constraints from the millisecond pulsar PSR J0030+0451 NICER x-ray data [28, 29]. The top brown region is from [25].

as function of $\chi_{\omega\omega}$ (color scale) for three $\{\chi_\omega, \chi_\rho\}$ sets. The hybrid EoS are concentrated around $\tilde{\Lambda} = 650 - 750$ and fall inside the 90% credible region for $0.8 \leq q \leq 1$ when $(\chi_\omega = 0, \chi_\rho = 0)$. Highly asymmetric systems, $q < 0.8$, have a purely hadronic NS component with low M and large R, giving rise to larger $\tilde{\Lambda}$. For the other two sets, $(\chi_\omega = 0.1, \chi_\rho = 0)$ and $(\chi_\omega = 0, \chi_\rho = 0.1)$, (middle and right panels) $\tilde{\Lambda} \sim 800$, and, therefore, they are still compatible with the GW170817 event if moderate asymmetries are considered, i.e., $q > 0.8$.

Figure 4 shows the effect of $\{\chi_{\omega\omega}, \chi_\rho, \chi_\omega\}$ on various NS properties: the quark core mass $M_{\rm QC}$, the quark core radius $R_{\rm QC}$, the speed of sound at the central density $v_s^2(n_{\rm max})$, and the central density $n_{\rm max}$. Several conclusions can be drawn: i) while $M_{\rm QC}$ increases with an increase of $\chi_{\omega\omega}$, the opposite effect occurs for χ_ρ or/and χ_ω, i.e. for larger values of these last two parameters the mass of the quark core reduces a lot. As a result, large quark cores require

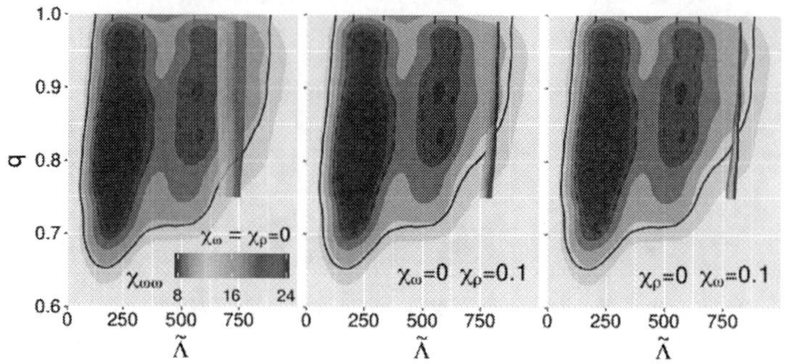

Figure 3. $q(\tilde{\Lambda})$ diagrams for binary systems, with $M_{\rm chirp} = 1.186M_\odot$, as a function of $\chi_{\omega\omega}$ (color scale) for three sets (χ_ω, χ_ρ): $(0,0)$ (left), $(0.1,0)$ (center), and $(0,0.1)$ (right). The probability distribution function $P(q,\tilde{\Lambda})$ from the LIGO/Virgo collaboration [27] is displayed in gray gradient, while the red dashed and solid lines indicate the 50% and 90% credible regions, respectively.

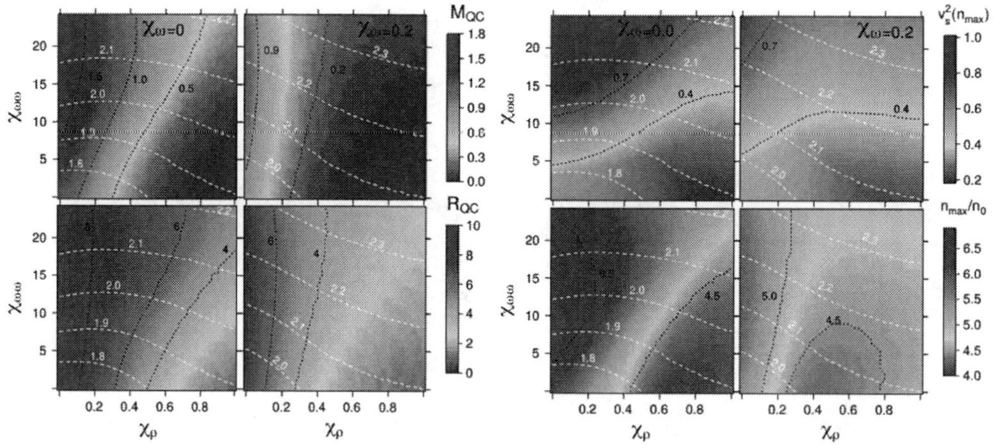

Figure 4. The quark core mass $M_{\rm QC}$ $[M_\odot]$ (top left), the quark core radius $R_{\rm QC}$ [km] (bottom left), the squared speed of sound at the central density $[c^2]$ (top right), and central densities $[n_0]$ of $M_{\rm max}$ (bottom right) as a function of $(\chi_{\omega\omega}, \chi_\rho)$. The white dashed and black dotted lines represent, respectively, the value of $M_{\rm max}$ $[M_\odot]$ and specific $\{M_{\rm QC}, R_{\rm QC}, v_s^2(n_{max}), n_{max}/n_0\}$ contour lines.

small values of χ_ρ and χ_ω and a large $\chi_{\omega\omega}$ coupling; ii) considering $\chi_\omega = \chi_\rho = 0$, massive NS are only realized for $\chi_{\omega\omega} \gtrsim 12$, with very large radii $R_{\rm QC}$ and speed of sound values $v_s^2(n_{\rm max})$ close to 1; iii) v_s^2 increases quite rapidly with $\chi_{\omega\omega}$ but decreases with χ_ρ; iv) the largest central densities occur for $\chi_\omega = \chi_\rho = 0$; v) it is possible to have a quark core having a mass of the order of 1/3 of the total star mass, and a speed of sound $v_s^2(n_{\rm max}) \sim 0.6$. The softening of the EoS with a finite χ_ρ due to the earlier onset of strange quarks explains this behavior.

The presence of strangeness inside the star can be understood considering Fig. 5, where the strangeness fraction at the center of the maximum mass star is plotted as a function of $\chi_{\omega\omega}$ and χ_ρ for two values of χ_ω, 0 and 0.1. The largest fraction of strangeness is attained for $\chi_\omega = 0$, $\chi_\rho = 0.5$ and $\chi_{\omega\omega} \approx 20$. It is interesting to notice that under these conditions the speed of sound squared is of the order of 0.7 and the strangeness fraction of the order of 0.25. Taking $\chi_\rho = 0$ the speed of sound square takes values close to 1 and the strangeness fractions

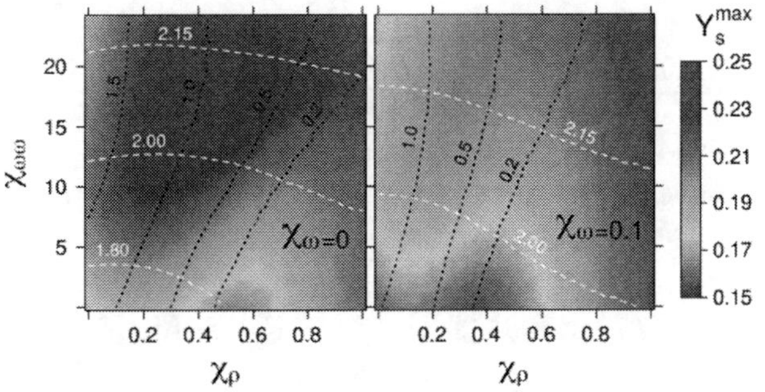

Figure 5. The fraction of *s*-quarks, Y_s^{max}, at the center of the maximum mass star, as a function of $\chi_{\omega\omega}$ and χ_ρ for $\chi_\omega = 0$ (left) and 0.1 (right). The white dashed and black dotted lines represent, respectively, the value of the star maximum mass M_{max} $[M_\odot]$ and specific M_{QC} $[M_\odot]$ contour lines.

drops to values of the order of 0.18. The presence of a larger amount of *s*-quarks prevents the speed of sound to take values close to one, but also reduces the mass of the quark core.

We have studied under which conditions two solar mass hybrid stars may exist. It was shown that a phase transition into a quark phase requires a harder isoscalar channel for the nucleonic EoS, than expected from just considering nucleonic degrees of freedom. We have shown that the present constraints on mass, radius, and tidal deformability from NS observations allow for the existence of hybrid stars with large quark cores. The $\chi_{\omega\omega}$ channel is important in generating massive quark cores, even at moderate quark masses, e.g., $1.4M_\odot$ stars, with $\tilde{\Lambda} < 800$. However, The $\chi_{\omega\omega}$ channel may originate very large values for the speed of sound. On the other hand, the χ_ρ channel controls the amount of strangeness inside the star, and reduces the speed of sound.

References

[1] J. Antoniadis et al., 10.1126/science.1233232 Science **340**, 6131 (2013), http://arxiv.org/abs/1304.6875 arXiv:1304.6875 [astro-ph.HE]

[2] H. T. Cromartie, E. Fonseca, S. M. Ransom, P. B. Demorest, Z. Arzoumanian, H. Blumer, P. R. Brook, M. E. DeCesar, T. Dolch, J. A. Ellis, R. D. Ferdman, E. C. Ferrara, N. Garver-Daniels, P. A. Gentile, M. L. Jones, M. T. Lam, D. R. Lorimer, R. S. Lynch, M. A. McLaughlin, C. Ng, D. J. Nice, T. T. Pennucci, R. Spiewak, I. H. Stairs, K. Stovall, J. K. Swiggum, and W. W. Zhu, 10.1038/s41550-019-0880-2 Nature Astronomy , 439 (2019), http://arxiv.org/abs/1904.06759 arXiv:1904.06759 [astro-ph.HE]

[3] E. Fonseca *et al.*, Astrophys. J.Lett. **915**, L12 (2021), arXiv:2104.00880 [astro-ph.HE]

[4] B. P. Abbott et al. (LIGO Scientific, Virgo), 10.1103/PhysRevLett.119.161101 Phys. Rev. Lett. **119**, 161101 (2017), http://arxiv.org/abs/1710.05832 arXiv:1710.05832 [gr-qc]

[5] B. A. et al. (The LIGO Scientific Collaboration and the Virgo Collaboration) (The LIGO Scientific Collaboration and the Virgo Collaboration), 10.1103/PhysRevLett.121.161101 Phys. Rev. Lett. **121**, 161101 (2018)

[6] M. G. Alford, S. Han, and K. Schwenzer, 10.1088/1361-6471/ab337a J. Phys. **G46**, 114001 (2019), http://arxiv.org/abs/1904.05471 arXiv:1904.05471 [nucl-th]

[7] E. R. Most, L. J. Papenfort, V. Dexheimer, M. Hanauske, S. Schramm, H. Stöcker, and L. Rezzolla, 10.1103/PhysRevLett.122.061101 Phys. Rev. Lett. **122**, 061101 (2019), http://arxiv.org/abs/1807.03684 arXiv:1807.03684 [astro-ph.HE]

[8] L. R. Weih, M. Hanauske, and L. Rezzolla, 10.1103/PhysRevLett.124.171103 Phys. Rev. Lett. **124**, 171103 (2020), http://arxiv.org/abs/1912.09340 arXiv:1912.09340 [gr-qc]

[9] T. Hatsuda and T. Kunihiro, Phys. Rept. **247** (1994), 221-367 doi:10.1016/0370-1573(94)90022-1 [arXiv:hep-ph/9401310 [hep-ph]].

[10] M. Buballa, Phys. Rept. **407** (2005), 205-376 doi:10.1016/j.physrep.2004.11.004 [arXiv:hep-ph/0402234 [hep-ph]].

[11] G. Pagliara and J. Schaffner-Bielich, Phys. Rev. D **77** (2008), 063004 doi:10.1103/PhysRevD.77.063004 [arXiv:0711.1119 [astro-ph]].

[12] S. Benic, Eur. Phys. J. A **50** (2014), 111 doi:10.1140/epja/i2014-14111-1 [arXiv:1401.5380 [nucl-th]].

[13] S. Benic, D. Blaschke, D. E. Alvarez-Castillo, T. Fischer and S. Typel, Astron. Astrophys. **577** (2015), A40 doi:10.1051/0004-6361/201425318 [arXiv:1411.2856 [astro-ph.HE]].

[14] A. Zacchi, M. Hanauske and J. Schaffner-Bielich, Phys. Rev. D **93** (2016) no.6, 065011 doi:10.1103/PhysRevD.93.065011 [arXiv:1510.00180 [nucl-th]].

[15] R. Câmara Pereira, P. Costa and C. Providência, Phys. Rev. D **94** (2016) no.9, 094001 doi:10.1103/PhysRevD.94.094001 [arXiv:1610.06435 [nucl-th]].

[16] X. Wu, A. Ohnishi and H. Shen, Phys. Rev. C **98** (2018) no.6, 065801 doi:10.1103/PhysRevC.98.065801 [arXiv:1806.03760 [nucl-th]].

[17] G. Baym, T. Hatsuda, T. Kojo, P. D. Powell, Y. Song and T. Takatsuka, Rept. Prog. Phys. **81** (2018) no.5, 056902 doi:10.1088/1361-6633/aaae14 [arXiv:1707.04966 [astro-ph.HE]].

[18] T. Kojo, Universe **4** (2018) no.2, 42 doi:10.3390/universe4020042

[19] M. Ferreira, R. Câmara Pereira and C. Providência, Phys. Rev. D **101** (2020) no.12, 123030 doi:10.1103/PhysRevD.101.123030 [arXiv:2005.10543 [nucl-th]].

[20] M. Ferreira, R. Câmara Pereira and C. Providência, Phys. Rev. D **102** (2020) no.8, 083030 doi:10.1103/PhysRevD.102.083030 [arXiv:2008.12563 [nucl-th]].

[21] M. Ferreira, R. Câmara Pereira and C. Providência, Phys. Rev. D **103** (2021) no.12, 123020 doi:10.1103/PhysRevD.103.123020 [arXiv:2105.06239 [nucl-th]].

[22] J. Margueron, R. H. Casali, and F. Gulminelli, Phys. Rev. C **97** (2018), 025805

[23] G. A. Lalazissis, T. Nikšić, D. Vretenar, and P. Ring, 10.1103/PhysRevC.71.024312 Phys. Rev. C **71**, 024312 (2005)

[24] E. Annala, T. Gorda, A. Kurkela, J. Nättilä, and A. Vuorinen, Nat. Phys. **16** (2020), 907.

[25] M. C. Miller, *et al.*, Astrophys. J. Lett. **918** (2021) no.2, L28.

[26] B. P. Abbott *et al.* [LIGO Scientific and Virgo], Phys. Rev. Lett. **121** (2018) no.16, 161101 doi:10.1103/PhysRevLett.121.161101 [arXiv:1805.11581 [gr-qc]].

[27] B. P. Abbott *et al.* [LIGO Scientific and Virgo], Phys. Rev. X **9** (2019) no.1, 011001 doi:10.1103/PhysRevX.9.011001 [arXiv:1805.11579 [gr-qc]].

[28] T. E. Riley, A. L. Watts, S. Bogdanov, P. S. Ray, R. M. Ludlam, S. Guillot, Z. Arzoumanian, C. L. Baker, A. V. Bilous and D. Chakrabarty, *et al.* Astrophys. J. Lett. **887** (2019) no.1, L21 doi:10.3847/2041-8213/ab481c [arXiv:1912.05702 [astro-ph.HE]].

[29] M. C. Miller, F. K. Lamb, A. J. Dittmann, S. Bogdanov, Z. Arzoumanian, K. C. Gendreau, S. Guillot, A. K. Harding, W. C. G. Ho and J. M. Lattimer, *et al.* Astrophys. J. Lett. **887** (2019) no.1, L24 doi:10.3847/2041-8213/ab50c5 [arXiv:1912.05705 [astro-ph.HE]].

Inferring the dense nuclear matter equation of state with neutron star tides

Pantelis Pnigouras[1,2,*], *Nils* Andersson[3], and *Andrea* Passamonti[4]

[1]Dipartimento di Fisica, "Sapienza" Università di Roma & Sezione INFN Roma1, Piazzale Aldo Moro 2, 00185 Roma, Italy
[2]Department of Physics, Aristotle University of Thessaloniki, 54124 Thessaloniki, Greece
[3]School of Mathematics and STAG Research Centre, University of Southampton, Southampton SO17 1BJ, UK
[4]Via Greve 10, 00146 Roma, Italy

Abstract. During the late stages of a neutron star binary inspiral finite-size effects come into play, with the tidal deformability of the supranuclear density matter leaving an imprint on the gravitational-wave signal. As demonstrated in the case of GW170817—the first direct detection of gravitational waves from a neutron star binary—this can lead to strong constraints on the neutron star equation of state. As detectors become more sensitive, effects which may have a smaller influence on the neutron star tidal deformability need to be taken into consideration. Dynamical effects, such as oscillation mode resonances triggered by the orbital motion, have been shown to contribute to the tidal deformability, especially close to the neutron star coalesence, where current detectors are most sensitive. We calculate the contribution of the various stellar oscillation modes to the tidal deformability and demonstrate the (anticipated) dominance of the fundamental mode. We show what the impact of the matter composition is on the tidal deformability, as well as the changes induced by more realistic additions to the problem, e.g. the presence of an elastic crust. Finally, based on this formulation, we develop a simple phenomenological model describing the effective tidal deformability of neutron stars and show that it provides a surprisingly accurate representation of the dynamical tide close to merger.

1 Introduction

The breakthrough detections of gravitational waves from binary neutron star inspirals [1, 2] have led to renewed focus on the elusive neutron star equation of state. The problem has a number of complicating aspects—both relating to the observational data and the theoretical underpinning—but the essential question is quite simple: to what extent can we use observations to constrain the state and composition of matter under the extreme conditions that neutron stars represent?

Much of the recent focus has been on the neutron star tidal deformability, essentially the extent to which the tidal interaction with a binary companion deforms the neutron star fluid. This is a useful measure as it can be extracted from (or, at least, constrained by) the gravitational-wave signal [3, 4]. Notably, the celebrated GW170817 event [1] has led to a

*e-mail: pantelis.pnigouras@uniroma1.it

constraint on a suitable weighted average tidal deformability, corresponding (roughly) to a neutron star radius in the range 10 – 13 km [5] (the result is somewhat model dependent). Moreover, as this radius range agrees well with the constraints obtained from the "Neutron Star Interior Composition Explorer" (NICER) observations of PSR J0030+0451 [6, 7] and PSR J0740+6620 [8, 9], a consistent picture is beginning to emerge.

The deformability (often expressed in terms of the dimensionless Love number, k_l) represents the static contribution to the neutron star's tidal response (equilibrium tide). In addition, there is a dynamical tide. This is traditionally represented by the excitation of the different oscillation modes of the star. The resonance problem was first considered some time ago [10–12], but the issue is back in focus following the suggestion that the (fundamental) f-mode of the star may be excited to a relevant level, even though it may not reach resonance during the inspiral [13–15]. The associated effect on the inspiral signal is weak, but its inclusion has been demonstrated to improve waveform models [16, 17].

An interesting question to pose is to what extent the composition of the neutron star matter enters the problem. As the star is deformed by the tidal interaction, matter is driven out of equilibrium and it is easy to argue that the relevant nuclear reactions are too slow to re-establish equilibrium on the time scale of inspiral. For neutron star cores dominated by a conglomerate of neutrons, protons, and electrons, the relevant equilibration time scales are

$$t_{mU} \sim \frac{2\,\text{months}}{T_9^6}, \quad t_{dU} \sim \frac{20\,\text{s}}{T_9^4}, \tag{1}$$

for the modified and direct Urca reactions, respectively. The temperature is scaled to hot systems, $T_9 \equiv T/10^9\,\text{K}$, but inspiraling neutron stars are old and cold, typically in the range $T_9 \leq 0.01$, which would make both t_{mU} and t_{dU} much longer than the time it takes a given system to move through the sensitivity band of a ground-based interferometer ($\sim$ minutes).

The upshot of the equilibration argument is that the equation of state is no longer barotropic, as has been assumed in virtually every previous analysis of the tidal problem. Hence, we want to establish to what extent a "frozen" matter composition leads to a noticeable effect on, for example, the Love number, and whether this in turn affects the extraction of neutron star parameters from an observed signal.

In the following, we outline the work presented in Refs. [15, 18, 19].

2 The effective Love number

The tide raised by a binary companion (here treated as a point particle) induces a linear response in the primary. In order to quantify this response, we solve the linearised fluid equations in Newtonian gravity. Assuming that the star is non-rotating, the tidal perturbation is described by the Euler equation

$$-\omega^2\xi_i + \frac{1}{\rho}\nabla_i\delta p - \frac{1}{\rho^2}\delta\rho\nabla_i p + \nabla_i\delta\Phi = -\nabla_i\chi, \tag{2}$$

where ξ_i denotes the tidal displacement vector, ρ is the density, p is the pressure, Φ is the gravitational potential, δ denotes Eulerian perturbations, and ω is the tidal frequency. In a coordinate system centred on the primary, which we will take to have mass $M_\star$, the tidal potential χ is given by

$$\chi = -GM' \sum_{l\geq 2}\sum_{m=-l}^{l} \frac{W_{lm}r^l}{D^{l+1}(t)} Y_{lm} e^{-im\psi(t)}, \tag{3}$$

where M' is the mass of the secondary. The orbit of the companion is taken to be in the plane $[D(t), \pi/2, \psi(t)]$ where D is the binary separation and ψ is the orbital phase. For $l =$

2 (which leads to the main contribution to the gravitational-wave signal), we have $W_{20} = -\sqrt{\pi/5}$, $W_{2\pm2} = \sqrt{3\pi/10}$, and $W_{2\pm1} = 0$.

We now aim to express the driven response of the fluid in terms of a set of normal modes, corresponding to solutions ξ_n (where n is a label that identifies the modes, say in terms of the number of nodes in the radial eigenfunction and the corresponding spherical harmonics). Letting the mode frequency be ω_n, we have

$$\xi^i = \sum_n a_n \xi^i_n. \tag{4}$$

The Love number is generally defined as

$$k_l = \frac{1}{2}\frac{\delta\Phi(R)}{\chi(R)}, \tag{5}$$

Then, after expressing the perturbed gravitational potential in terms of the displacement vector, we get

$$k_l = -\frac{1}{2} + \frac{2\pi}{2l+1}\sum_n \frac{\tilde{Q}_n^2}{\tilde{\omega}_n^2 - \tilde{\omega}^2}\left[1 - \tilde{\omega}^2\left(\frac{V_n}{W_n}\right)_R\right]\left[1 - \tilde{\omega}_n^2\left(\frac{V_n}{W_n}\right)_R\right]^{-1}, \tag{6}$$

where frequencies have been normalised as $\tilde{\omega} = \omega/(GM_\star/R^3)$, each mode is decomposed into its radial and horizontal components as

$$\xi^i_n = \left(W_n \frac{\nabla^i r}{r}\right) Y_{lm} + V_n \nabla^i Y_{lm}, \tag{7}$$

and we have introduced the "overlap integral"

$$\tilde{Q}_n = -\frac{1}{M_\star R^l}\int \delta\rho_n^* r^{l+2} \mathrm{d}r. \tag{8}$$

In the low-frequency limit (i.e., in the equilibrium tide approximation), we get

$$k_l \approx -\frac{1}{2} + \frac{2\pi}{2l+1}\sum_n \frac{\tilde{Q}_n^2}{\tilde{\omega}_n^2}\left[1 - \tilde{\omega}_n^2\left(\frac{V_n}{W_n}\right)_R\right]^{-1}. \tag{9}$$

As a quantitative test of Eq. (9), we compare results for three models corresponding to a background configuration with a polytropic equation of state of the form $p \propto \rho^\Gamma$, with $\Gamma = 2$. Since, as noted before, the nuclear reactions required to establish chemical equilibrium are too slow to act on the inspiral timescale, it would be reasonable to assume that the composition of a perturbed fluid element is held frozen, as the system sweeps through the sensitivity band of a ground based detector. This changes the response of the stellar fluid to the tidal driving which, in turn, allows us to estimate the impact the matter composition has on the problem. Composition gradients can be accounted for by adjusting the adiabatic index $\Gamma_1 = (\partial \ln p/\partial \ln \rho)_{\rm ad}$, which is evaluated at fixed composition. Our reference model is barotropic ($\Gamma_1 = \Gamma$) and we compare it to two stratified models, with $\Gamma_1 = 2.05$ and $7/3$, respectively.

The numerical results, listed in Table 1, demonstrate the relative importance of g-modes (buoyancy modes) for strongly stratified models. In each case, the mode sum converges to the expected value for the Love number which, for a barotropic model with $\Gamma = 2$, should be $k_l \approx 0.259909$ and the results from Table 1 do, indeed, converge towards this number. This tells us that the sum over the star's different oscillation modes provides an alternative representation

Table 1. The accumulated contribution to the Love number from the different modes, in order of relevance of the contribution, for the three different models we consider. We expect (for the barotropic $\Gamma = 2$ case) to have $k_l \approx 0.259909$. If we add up the contributions from the different modes in each case, the result converges to the expected answer. The mode sum is always dominated by the f-mode, with a modest correction from the other modes of the star, but the enhanced importance of the g-modes with increasing stratification is notable.

$\Gamma_1 = 2$		$\Gamma_1 = 2.05$		$\Gamma_1 = 7/3$	
mode	k_l	mode	k_l	mode	k_l
f	0.27528	f	0.27055	f	0.24685
$+p_1$	0.25887	$+p_1$	0.25526	$+g_1$	0.26115
$+p_2$	0.26021	$+p_2$	0.25653	$+p_1$	0.25052
$+p_3$	0.26015	$+g_1$	0.25878	$+g_2$	0.25556
		$+g_2$	0.25960	$+p_2$	0.25653
		$+g_3$	0.25993	$+g_3$	0.25856
		$+g_4$	0.26008	$+g_4$	0.25944
				$+g_5$	0.25983
	9×10^{-4}		7×10^{-4}		3×10^{-4}

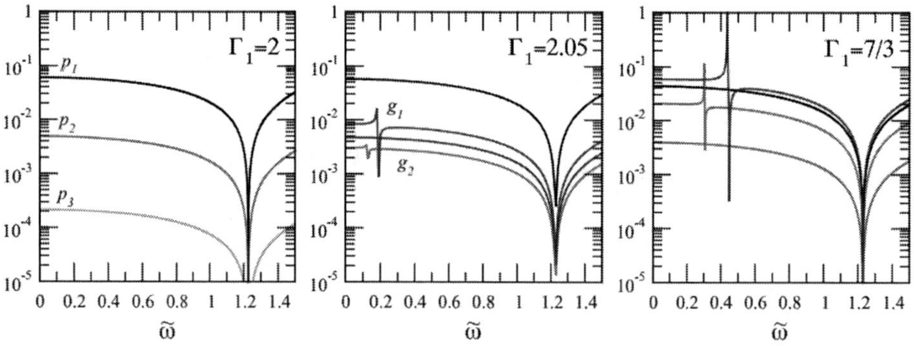

Figure 1. Relative contributions to the tidal deformability (compared to that of the f-mode alone). The three panels show, from left to right: the barotropic case $\Gamma_1 = \Gamma = 2$, $\Gamma_1 = 2.05$ and $\Gamma_1 = 7/3$. Individual modes are colour coded (as indicated in the panels), with the same colour representing the same mode in all panels.

for the Love number. It is important to note that this is true also for the stratified models. The f-mode provides the dominant contribution in all cases (as expected given that this mode most closely resembles the tidal driving force), but in order to have a precise representation we need to account for both p-modes (pressure modes) and g-modes.

Having demonstrated that the sum over the star's oscillation modes provides a precise description of the tidal response in the static limit, let us turn to the dynamical response associated with finite frequencies, given by Eq. (6). It provides a closed expression for the frequency dependent tidal response (encoded in k_l). This allows us to quantify the level at which each individual mode contributes to the overall result. Results for the three different values of Γ_1 we have considered are presented in Fig. 1. The different panels show the relative contributions to the tidal deformability (compared to that of the f-mode alone). The resonances associated with each mode, which occur when $\tilde{\omega} = \tilde{\omega}_n$, are easily distinguishable

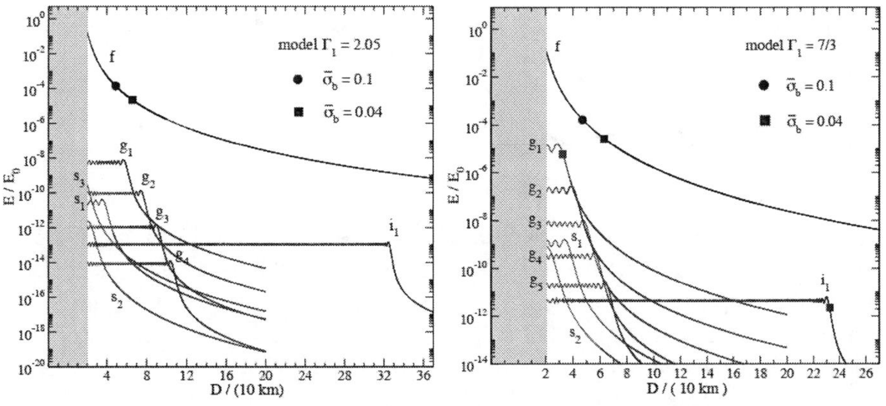

Figure 2. The evolution of the mode energies for varying binary separation. The energy is given in units of $E_0 = GM_\star^2/R$. The model with $\Gamma_1 = 2.05$ is shown in the left-hand panel, while the $\Gamma_1 = 7/3$ case is provided in the right-hand panel. The energy of a mode increases when it becomes resonant during the inspiral. The breaking energy limit E_b of the crust is represented by a circle or a square (see legend), which, respectively, denote the breaking energy for two choices of the crust breaking strain.

in each case and the resonance associated with the f-mode leads to a common feature in all panels. The results tell us that modes other than the f-mode contribute to the overall result at the few percent level. These results are important as they provide the first demonstration of the level at which frozen matter composition impacts on the tidal response across the range of frequencies relevant for a binary inspiral.

3 Adding the crust

When we turn to dynamical features of the tide, we need to be mindful of the fact that a neutron star interior is a little bit more complicated than a prescribed pressure-density relation. Nuclei in the lower density region are expected to freeze to form the so-called crust. This is important because the tidal response can be expressed as a sum over the stellar oscillation modes and the additional features of the interior physics may bring new modes into play and shift existing ones.

With this in mind, we studied the tidal response of a binary neutron star in models including an elastic crust and a fluid ocean. We showed that the presence of the crust does not significantly affect the fundamental, pressure, and buoyancy modes. As expected, the contribution to the Love number from the (crustal) shear modes is negligible, while the interface and surface g-modes have an impact similar to the first core g-modes. The influence of these modes, albeit small compared to the f-mode, increases for strongly stratified models.

Oscillations may be amplified by tidal resonances during the binary inspiral. This amplification is not only important for the gravitational-wave signal, but also for the impact that mode resonances can have on the crust. Thus, we studied the dynamical tidal evolution and determined the mode energy during the orbital shrinking. Our results confirm that the fundamental mode dominates the dynamical tides even when it is far from resonance. However, the interface modes are resonantly excited at an earlier stage and may accumulate enough energy to fracture the crust. This is mainly due to their peaked radial displacement at the crust-core or crust-ocean surface transition. The fundamental mode reaches the crust breaking limit in all our models, but not until the final part of the inspiral. Finally, we showed that the first

g-mode can fracture the crust only for strongly stratified stars and (again) in the very final phase of inspiral. These results are shown in Fig. 2.

4 A phenomenological model

Based on the formalism presented in Sec. 2, we also developed a "phenomenological" model describing the dynamical tides in neutron star binaries. Our results demonstrate that the dynamical contribution to the tide is dominated by the excitation of the f-mode of the star, something that was established a long time ago [10] in work aimed at quantifying the impact of mode resonances on the gravitational-wave signal. Based on this, the basic idea is to include only the f-mode contribution to the mode sum and accept the contribution from other modes as a "systematic error." Based on the stratified Newtonian models considered before, we expect this error to be below the 5% level.

After doing this, the relation for the effective Love number that we end up with is

$$k_l^{\text{eff}} \approx \frac{\bar{\omega}_f^2 k_l}{\bar{\omega}_f^2 - \delta(2\bar{\Omega})^2} + \frac{(2\bar{\Omega})^2}{\bar{\omega}_f^2 - \delta(2\bar{\Omega})^2}\left[\frac{\delta}{2} - \frac{\bar{\omega}_f^2}{C^3}\frac{\epsilon}{l}\left(k_l + \frac{1}{2}\right)\right]. \tag{10}$$

Here, the f-mode frequency is normalised as $\bar{\omega}_f \equiv M_\star \omega_f$ and can be obtained by a well-known "universal relation" in terms of the (static) tidal deformability k_l [20]. The orbital frequency is denoted by $\bar{\Omega}$ and the stellar compacness by C, which may also be expressed in terms of k_l via another universal relation [21]. The factor ϵ/l represents the ratio $(V_f/W_f)_R$, with $\epsilon = 1$ in a homogeneous star (and $\epsilon \approx 0.9$ for the polytropic models considered in Sec. 2). Finally, the parameter δ accounts for two relativistic effects: the gravitational redshift of the mode frequencies and the rotational frame-dragging induced by the orbital motion. An argument by Ref. [14] suggests that the two effects almost cancel, so—as a first approximation—we are motivated to simply remove the gravitational redshift altogether, implying that we may take $\delta = 1 - 2C$. Hence, we now have an explicit analytic formula for the effective Love number in terms of the result in the static limit, k_l, the orbital frequency $\bar{\Omega}$ and the (to some extent) free parameter ϵ.

As a suitable comparison (in fact, the only comparable results in the literature), we consider the results for the dynamical tide from Refs. [13, 14], which are similar in spirit as they introduce the notion of an effective tidal deformability. However, the main focus of Refs. [13, 14] was to extend the effective-one-body framework to account for the dynamical tide. In addition to this, Ref. [14] provides an approximate analytical formula from matched asymptotics, which has been tested against numerical relativity simulations. As it appears to perform well in such comparisons, it provides a natural benchmark against which to test our simple closed-form expression.

Focussing on the example illustrated by Ref. [14], i.e., an equal mass neutron star binary with $M_\star = 1.35\,M_\odot$ and $R = 13.5\,\text{km}$, it is easy to demonstrate that we obtain an accurate representation of the dynamical tide throughout the relevant frequency range (up to close to merger) by tuning the parameter ϵ. The effective Love number obtained from Eq. (10) is compared to the results from Ref. [14] in Fig. 3, for both $l = 2$ and $l = 3$. As our formula still leaves ϵ as a free parameter, we show results for the range $\epsilon = 0.85 - 0.9$ and, as is clear from the result in the figure, the corresponding curves for $\epsilon = 0.875$ provide an excellent match to the results from Ref. [14]. The essence of the comparison is that our formula (10) performs (perhaps surprisingly) well. It may be phenomenological in origin, but there can be little doubt that our simple expression provides an effective representation of the required behaviour, and hence reflects the underlying physics.

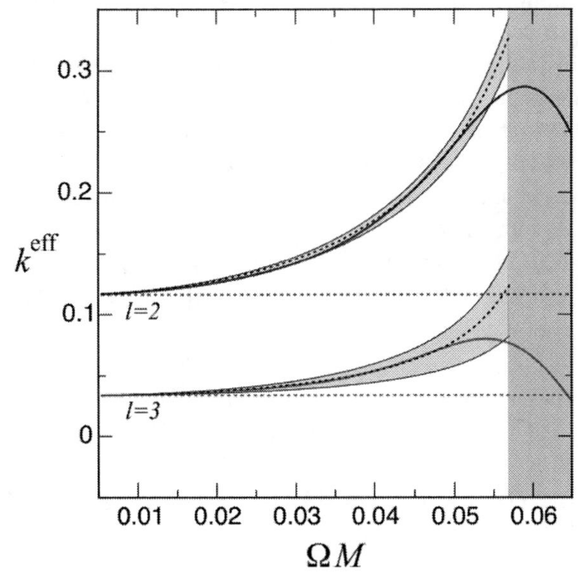

Figure 3. Comparing the effective Love number k_l^{eff} from Eq. (10) with $\delta = 1 - 2C$ to the results from Ref. [14], for both $l = 2$ and $l = 3$. The dashed horizontal lines represent the static Love number (k_l^{eff} in the $\Omega \to 0$ limit). The results of Ref. [14] are shown as solid curves (blue for $l = 2$ and red for $l = 3$). Estimates from Eq. (10) are shown for the range $\epsilon = 0.85 - 0.9$, with the latter representing the lower edge of the filled band in each case. The particular choice $\epsilon = 0.875$ (dashed black curves) provides an an excellent fit to the results from Ref. [14]. Finally, we indicate the region beyond the (approximate) merger frequency, $\Omega M \gtrsim 0.057$ in this case, by the shaded area in the figure.

It is worth noting that, while the two sets of results diverge for large values of ΩM in Fig. 3 (where $M = M_\star + M'$), the corresponding frequencies are close to (or indeed beyond) the merger frequency. As the picture of two separate, tidally deformed, bodies breaks down there is no reason to expect the model to make sense beyond this point. The post-merger region is indicated by the shaded area in Fig. 3.

5 Implications

We have discussed the tidal response of a neutron star during the late stages of neutron star binary inspiral. In particular, we have focussed on the role of the matter composition. This issue has previously been ignored as studies have almost exclusively focussed of barotropic fluid models. However, it is natural to argue (given the time scale involved) that the matter composition should remain "frozen" during the late stages of binary inspiral, leading to a stratified perturbation problem (where the adiabatic index of the perturbation is different from that of the equilibrium background). This connects with previous work on tidal resonances, which has quantified the role of the g-modes (which rely on stratification for their existence).

Our numerical results indicate that the difference is at (or below) the level of a few percent. However, it is nevertheless important to quantify this contribution. We need to do this in order to understand systematic "errors" associated with the assumed physics, which ultimately determines the accuracy with which we can hope to extract stellar parameters like the radius from observations. Today's gravitational-wave detectors are not at a level where a change of a percent in the tidal response makes much difference, but one might want to keep an eye on these issues for future reference.

The problem is also important from the physics point of view. Having quantified the level at which the matter composition enters the discussion, we can compare to the role of other elements of neutron star physics, like an elastic crust. Moreover, this formalism allowed us to develop a phenomenological, physically motivated, model for the effective tidal deformability of a neutron star binary, adding frequency dependence that comes into play during the late stages of inspiral. A comparison against alternative descriptions suggests that we have at hand a simple, yet accurate, description of the tidal imprint. This should make the model an attractive alternative for an implementation of the matter effects in gravitational-wave data analysis algorithms.

P.P. and N.A. acknowledge support from the Science and Technology Facilities Council (STFC) via grant ST/R00045X/1. P.P. acknowledges support from the "Ministero dell'istruzione, dell'università e della ricerca" (MIUR) PRIN 2017 programme (CUP: B88D19001440001) and from the Amaldi Research Center funded by the MIUR programme "Dipartimento di Eccellenza" (CUP: B81I18001170001).

References

[1] B.P. Abbott et al., Phys. Rev. Lett. **119**, 161101 (2017), 1710.05832
[2] B.P. Abbott et al., Astrophys. J. **892**, L3 (2020), 2001.01761
[3] É.É. Flanagan, T. Hinderer, Phys. Rev. D **77**, 021502 (2008), 0709.1915
[4] T. Hinderer, B.D. Lackey, R.N. Lang, J.S. Read, Phys. Rev. D **81**, 123016 (2010), 0911.3535
[5] B.P. Abbott et al., Classical Quantum Gravity **35**, 065009 (2018), 1711.06843
[6] M.C. Miller et al., Astrophys. J. **887**, L24 (2019), 1912.05705
[7] T.E. Riley et al., Astrophys. J. **887**, L21 (2019), 1912.05702
[8] M.C. Miller et al., Astrophys. J. **918**, L28 (2021), 2105.06979
[9] T.E. Riley et al., Astrophys. J. **918**, L27 (2021), 2105.06980
[10] D. Lai, Mon. Not. R. Astron. Soc. **270**, 611 (1994), astro-ph/9404062
[11] A. Reisenegger, P. Goldreich, Astrophys. J. **426**, 688 (1994)
[12] K.D. Kokkotas, G. Schäfer, Mon. Not. R. Astron. Soc. **275**, 301 (1995), gr-qc/9502034
[13] T. Hinderer et al., Phys. Rev. Lett. **116**, 181101 (2016), 1602.00599
[14] J. Steinhoff, T. Hinderer, A. Buonanno, A. Taracchini, Phys. Rev. D **94**, 104028 (2016), 1608.01907
[15] N. Andersson, P. Pnigouras, Mon. Not. R. Astron. Soc. **503**, 533 (2021), 1905.00012
[16] P. Schmidt, T. Hinderer, Phys. Rev. D **100**, 021501 (2019), 1905.00818
[17] G. Pratten, P. Schmidt, N. Williams (2021), 2109.07566
[18] N. Andersson, P. Pnigouras, Phys. Rev. D **101**, 083001 (2020), 1906.08982
[19] A. Passamonti, N. Andersson, P. Pnigouras, Mon. Not. R. Astron. Soc. **504**, 1273 (2021), 2012.09637
[20] T.K. Chan, Y.H. Sham, P.T. Leung, L.M. Lin, Phys. Rev. D **90**, 124023 (2014), 1408.3789
[21] A. Maselli, V. Cardoso, V. Ferrari, L. Gualtieri, P. Pani, Phys. Rev. D **88**, 023007 (2013), 1304.2052

Inhomogeneous pion condensed phase hosting topologically stable baryons

Massimo Mannarelli[1], *Fabrizio* Canfora[2], *Stefano* Carignano[3], *Marcela* Lagos[4], and *Aldo* Vera[4,*]

[1]INFN, Laboratori Nazionali del Gran Sasso, Via G. Acitelli, 22, I-67100 Assergi (AQ), Italy
[2]Centro de Estudios Científicos (CECS), Casilla 1469, Valdivia, Chile
[3]Departament de Física Quàntica i Astrofísica and Institut de Ciències del Cosmos, Universitat de Barcelona, Martí i Franquès 1, 08028 Barcelona, Catalonia, Spain
[4]Instituto de Ciencias Físicas y Matemáticas, Universidad Austral de Chile, Casilla 567, Valdivia, Chile

Abstract. We discuss the inhomogeneous pion condensed phase within the framework of chiral perturbation theory. We show how the general expression of the condensate can be obtained solving three coupled differential equations, expressing how the pion fields are modulated in space. Upon using some simplifying assumptions, we determine an analytic solution in (3+1)-dimensions. The obtained inhomogeneous condensate is characterized by a non-vanishing topological charge, which can be identified with the baryonic number. In this way, we obtain an inhomogeneous system of pions hosting an arbitrary number of baryons at fixed position in space.

1 Introduction

One of the main goals of the high-energy physics community is to determine the properties of hadronic matter at extremes of temperature and baryonic densities. The fundamental motivation behind these studies is that the high-energy environment is the ideal one for probing the constituents of hadronic matter. At the microscopic level, the strong interaction is described by Quantum Chromodynamics (QCD), which is an extremely challenging theory in the infrared, because at low energies the basic constituents, quarks and gluons, are confined. On the other hand, with increasing energy scale we expect that quarks and gluons are liberated [1], and that perturbative methods can be eventually used at asymptotic energy scales. Unfortunately, in terrestrial relativistic heavy-ion colliders, as well as in compact stellar objects, the energy scale is not sufficiently large to ensure that perturbative methods are under control. This means that different phenomenological methods should be used to infer the properties of hadronic matter.

The possible phases of hadronic matter can be depicted in a grand-canonical phase diagram, which we pictorially report in Fig. 1. The phases of matter change as a function of

*e-mail: massimo@lngs.infn.it
e-mail: canfora@cecs.cl
e-mail: carignano@ice.cat
e-mail: marcela.lagos@uach.cl
e-mail: aldo.vera@uach.cl

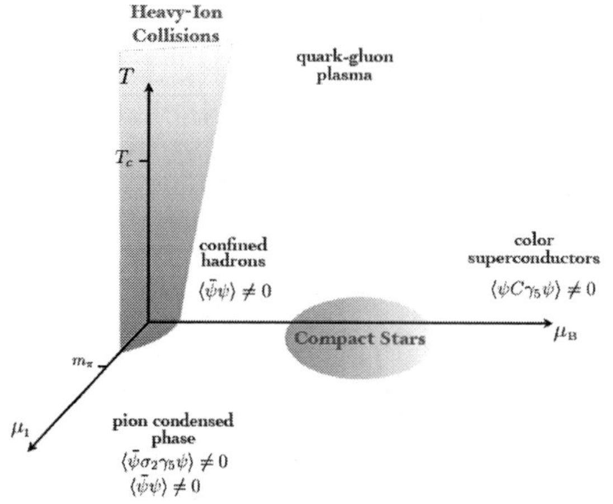

Figure 1. Cartoon of the QCD phase diagram. A grand-canonical description of the phases of matter obtained by changing the temperature, the baryonic chemical potential and the isospin chemical potential. The relevant di-quark condensates are shown.

temperature, T, baryonic chemical potential, μ_B, and isospin chemical potential, μ_I, and can be characterized by different values of the di-quark condensates.

In Fig. 1 are also shown the regions which roughly correspond to those accessible by relativistic heavy-ion collisions and compact stars. Although weak decays typically determine the isospin chemical potential at any temperature and baryonic chemical potential, it is useful to assume that μ_I is a free parameter. In this way one can explore the properties of matter at short time scales, that is shorter than the typical time-scale of weak interactions. Moreover, when the properties of hadronic matter are studied in lattice QCD simulation, it is typically assumed that the weak interaction is turned off.

Although lattice QCD simulations at non-vanishing μ_B are hampered by the so-called sign problem, they are feasible at non-vanishing isospin chemical potential [2]. This offers the opportunity to study deconfinement as a function of μ_I and to compare the results with the outcome of effective field theories, in particular with the results of chiral perturbation theory (χPT), see [3–11], which is a low-energy realization of hadronic matter having the same global symmetries of QCD. As an effective field theory, χPT perfectly describes the properties of pseudoscalar mesons, and it is therefore particularly useful for analyzing matter at vanishing μ_B and non-zero μ_I. In principle, χPT can be amended to describe baryons as well, see for instance [9–11], introducing the baryonic sources by hand. In [12] we have presented a different approach, developed with the aim of obtaining baryons as skyrmions embedded in an inhomogeneous gas of pions at non-vanishing μ_I. In this way, baryons emerge as solitonic solutions of the classical field equations. Since the obtained set of differential equations is quite difficult to solve, we have used a number of simplifying assumptions that make the problem almost analytically tractable.

2 Moving along the μ_I–axis

In χPT the three pions can be collected in a unimodular field

$$\Sigma = \mathbf{1}_2 \cos\alpha + i\,\mathbf{n}\cdot\boldsymbol{\sigma}\sin\alpha\,, \tag{1}$$

where the components of $\boldsymbol{\sigma}$ are the three Pauli matrices, and the three degrees of freedom are α and the two independent components of the unimodular field $\boldsymbol{n}$. This somehow unusual representation of the pion fields is particularly convenient for the analysis of different ground states. As we shall see, different vacua are indeed characterized by different values of the tilting angle α.

In χPT the interaction terms are grouped using an expansion in exchanged momenta. The leading order χPT Lagrangian including the electromagnetic field can be written as

$$\mathcal{L} = \frac{f_\pi^2}{4}\mathrm{Tr}\left[-\left(\Sigma^\mu \Sigma_\mu\right) + m_\pi^2\left(\Sigma + \Sigma^\dagger\right)\right] - \frac{1}{4}F_{\mu\nu}F^{\mu\nu}\,, \tag{2}$$

where the field strength is $F_{\mu\nu} = \partial_\mu A_\nu - \partial_\nu A_\mu$, with A_μ the electromagnetic field, and

$$\Sigma_\mu = \Sigma^{-1} D_\mu \Sigma\,, \tag{3}$$

with

$$D_\mu \Sigma = \partial_\mu \Sigma + i\widetilde{A}_\mu\left[\sigma_3, \Sigma\right]\,, \tag{4}$$

the covariant derivative. Here

$$\widetilde{A}^\mu = \left(\frac{\mu_I}{2} + A^0, \mathbf{A}\right)\,, \tag{5}$$

is a convenient definition of the shifted electromagnetic field that includes the effect of the isospin chemical potential. The low-energy phenomenological constants appearing in Eq. (2) are the pion decay constant, $f_\pi \simeq 93$ MeV, and the pion mass, $m_\pi \simeq 135$ MeV; we assume degenerate pion masses.

The homogeneous energetically favored phase can be determined assuming that α and $\boldsymbol{n}$ are variational parameters: the favored state corresponds to the values of these parameters that maximize the Lagrangian in (2), see [13] for a review. It results that for $|\mu_I| < m_\pi$ the standard vacuum with $\alpha = 0$ is energetically favored. In this case, the value of the vector $\boldsymbol{n}$ is not specified, and the only non-vanishing condensate is the so-called chiral condensate

$$\sigma \propto \langle\bar{\psi}\psi\rangle\,, \tag{6}$$

which in terms of the quark fields ψ corresponds to the locking of the chiral symmetry $SU(2)_L\times SU(2)_R$ to the vector SU(2)–isospin symmetry. Upon increasing μ_I the isospin symmetry is explicitly broken and the masses of the charged pions split, similarly to what happens with the Stark effect. However, the system has still a $U(1)$ global symmetry. When $|\mu_I| > m_\pi$ this symmetry is spontaneously broken, one the pion field becomes massless and the system becomes superfluid. In this case the vacuum is *tilted* in isospin space by an angle

$$\cos\alpha = \left(\frac{m_\pi}{\mu_I}\right)^2\,, \tag{7}$$

corresponding to the fact that in addition to the chiral condensate, the pion condensate

$$\pi \propto \langle\bar{\psi}\sigma\gamma_5\psi\rangle\,, \tag{8}$$

assumes a non-vanishing value. The energetically favored configuration is such that the third component of the field $\boldsymbol{n}$ vanishes. The remaining two components of $\boldsymbol{n}$ are not determined, and the only remaining independent degree of freedom corresponds to the Nambu-Goldstone field. When considering the effect of the electromagnetic interaction, the system becomes a superconductor and the electromagnetic interaction is screened, see [13, 14] for more details.

3 Inhomogeneous phase

The picture emerging so far is that of a homogeneous system that reacts to an isospin chemical potential exceeding the pion mass by the appearance of a pion condensate. This condensate coexists with the chiral condensate. In this situation, it is possible that the energetically favored configuration is such that the space distribution of the two condensates is not uniform. To study this possibility, we have to promote the variational parameters α and $\boldsymbol{n}$ to classical fields. For vanishing electromagnetic fields, the effective Lagrangian in Eq. (2) can be rewritten as

$$\begin{aligned}\mathcal{L}_{\rm m} =& \frac{f_\pi^2}{2}\Big[\partial_\mu\alpha\partial^\mu\alpha + \sin^2\alpha\partial_\mu\Theta\partial^\mu\Theta + 2m_\pi^2\cos\alpha \\ &+ \sin^2\alpha\sin^2\Theta(\partial_\mu\Phi - \mu_I\delta_{\mu 0})(\partial^\mu\Phi - \mu_I\delta^{\mu 0})\Big]\,, \end{aligned} \tag{9}$$

where we have parameterized the components of $\boldsymbol{n}$ as follows

$$n_1 = \sin\Theta\cos\Phi\,,\quad n_2 = \sin\Theta\sin\Phi\,,\quad n_3 = \cos\Theta\,. \tag{10}$$

The three classical fields α, Θ and Φ can be obtained as solutions of the Euler-Lagrange equations

$$\partial_\mu\partial^\mu\Phi = -(\partial_\mu\Phi - \mu_I\delta_{\mu_0})\partial^\mu(\log(\sin^2\alpha\sin^2\Theta))\,, \tag{11}$$

$$\partial_\mu\partial^\mu\Theta = -2\cot\alpha\,\partial^\mu\Theta\partial_\mu\alpha + \frac{\sin 2\Theta}{2}K\,, \tag{12}$$

$$\partial_\mu\partial^\mu\alpha = -m_\pi^2\sin\alpha + \frac{\sin(2\alpha)}{2}(\partial_\mu\Theta\partial^\mu\Theta + K\sin^2\Theta)\,, \tag{13}$$

which is a set of three coupled differential equations, with

$$K = (\partial_\mu\Phi - \mu_I\delta_{\mu_0})(\partial^\mu\Phi - \mu_I\delta^{\mu_0})\,. \tag{14}$$

Since we are interested in topologically protected phases, it is mandatory to take into account finite volume effects. For this reason, we define the line element as

$$ds^2 = dt^2 - \ell^2\left(dr^2 + d\theta^2 + d\phi^2\right)\,, \tag{15}$$

where the coordinates r, θ and ϕ vary in the ranges

$$0 \le r \le 2\pi\,,\quad 0 \le \theta \le \pi\,,\quad 0 \le \phi \le 2\pi\,, \tag{16}$$

while

$$\ell = \frac{b}{m_\pi}\,, \tag{17}$$

with b a real number, is the typical dimension of the system. The actual solution is expected to satisfy some appropriate Dirichlet boundary conditions. In particular, we demand that

$$n(t,r,0,\phi) = -n(t,r,\pi,\phi)\,,\quad n(t,r,\theta,0) = n(t,r,\theta,2\pi)\,, \tag{18}$$

and that

$$\Sigma(t,0,\theta,\phi) = \pm\Sigma(t,2\pi,\theta,\phi)\,, \tag{19}$$

where we have taken the freedom to have a flip of the Σ field at the boundary, see [12, 15–17] for more details. Since solving the general Cauchy problem is hard, we take the simplifying assumption that

$$\Phi = \mu_I\ell\phi + \Phi_0\,,\qquad \Theta = q\theta + \Theta_0\,, \tag{20}$$

with q odd and

$$\mu_I \ell = p\,, \tag{21}$$

with p an integer, to properly satisfy the above boundary conditions, see [12, 15–17]. The particular choice of the Φ field is such that K vanishes, see Eq. (14). In this way, the only nontrivial differential equation turns to be

$$\frac{\partial^2 \alpha}{\partial r^2} = m_\pi^2 \ell^2 \sin\alpha + \frac{q^2}{2}\sin(2\alpha)\,, \tag{22}$$

where we have assumed that α is stationary. The appropriate boundary conditions in the r–direction are then

$$\alpha(0) = 0 \qquad \text{and} \qquad \alpha(2\pi) = n\pi\,, \tag{23}$$

with n the integer winding number. Odd values of n correspond to the minus sign in Eq. (19), while even values of n correspond to a positive sign, that is anti-periodic and periodic boundary condition in the r–direction, respectively. Some solutions of this differential equation are shown in Fig. 2, for five different winding numbers.

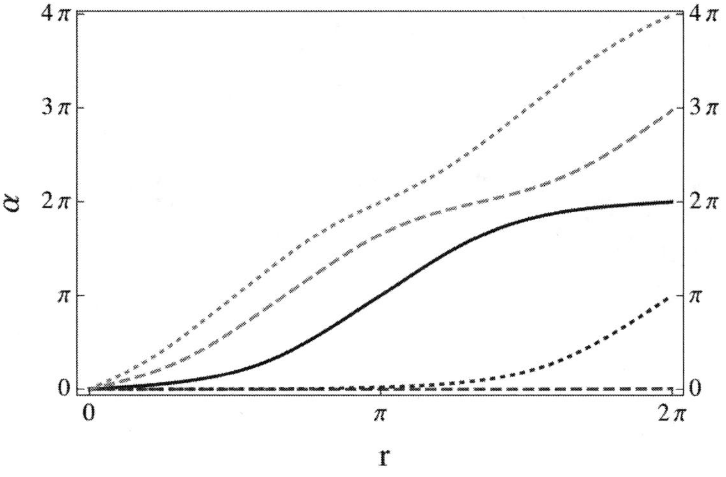

Figure 2. Profile of the radial mode, $\alpha(r)$, obtained by numerical integration of Eq. (22) for $q = 1$ and $\ell = 1/m_\pi$ and assuming the boundary conditions in Eq. (23). The reported results are obtained with five different values of the winding number n.

The obtained spatial modulations are characterized by a non-vanishing topological charge. This charge can indeed be written as

$$B_{\rm m} = \frac{\ell^3}{24\pi^2}\int_V dr d\theta d\phi\, \rho_{\rm m}\,, \tag{24}$$

where V is the considered volume and

$$\rho_{\rm m} = \epsilon^{ijk}{\rm Tr}\left\{\left(\Sigma^{-1}\partial_i\Sigma\right)\left(\Sigma^{-1}\partial_j\Sigma\right)\left(\Sigma^{-1}\partial_k\Sigma\right)\right\}\,, \tag{25}$$

is the topological density of the matter fields. The expression in Eq. (25) depends on the space derivatives of the fields, and it vanishes if the scalar field Σ depends on only one or two space coordinates. Upon substituting Eq. (1) in Eq. (25), the topological density can be rewritten as

$$\rho_{\rm m} = -12\frac{pq}{\ell^3}\sin(q\theta)\sin^2(\alpha)\frac{\partial\alpha}{\partial r}\,, \tag{26}$$

which can be readily integrated using the fact that

$$\int_0^{2\pi} dr \sin^2(\alpha)\frac{\partial\alpha}{\partial r} = \int_0^{n\pi} d\alpha \sin^2\alpha = n\frac{\pi}{2} . \tag{27}$$

According to the boundary condition in Eq. (23), the topological charge takes the values

$$B_{\rm m} = \begin{cases} np & \text{if} \quad q \text{ odd} , \\ 0 & \text{if} \quad q \text{ even} , \end{cases} \tag{28}$$

where p is defined in Eq. (21) and the q parameter has been introduced in Eq. (20). We report in Fig. 3 the topological charge densities for $q = 1$ and two different winding numbers: $n = 1$, in the left panel, and $n = 2$, in the right panel. In principle, we can obtain solutions with arbitrary large winding numbers, corresponding to a large number of solitons. Notice that having an odd value of q is mandatory to have a non-vanishing topological charge. An odd value of q implies that the unimodular field $\boldsymbol{n}$ flips at the θ boundary. One may actually enlarge the θ boundary, say from $[0, \pi]$ to $[0, 2\pi]$. In this way the unimodular field $\boldsymbol{n}$ does not flip at the θ boundary. The resulting solitonic configuration would consist of a crystal of solitons and anti-solitons with vanishing total topological charge.

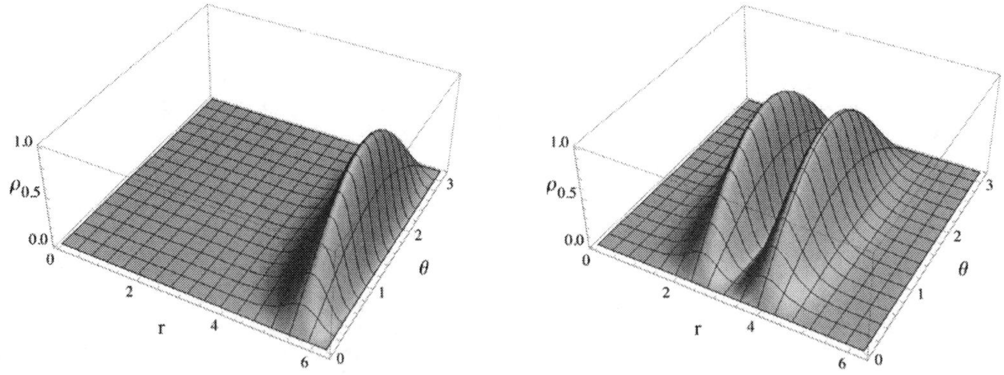

Figure 3. Topological density distribution, see Eq. (26), obtained with $q = 1$ and $n = 1$, left panel, and $n = 2$, right panel. The topological density has been normalized to the value at the maximum.

Different solutions arise for values of q different from 1. For instance, in Fig. 4 we report the solution obtained with $q = 3$ and $n = 4$. The system now consists of a grid of solitons and anti-solitons.

Since the topological charge can be understood as the baryonic charge of the system [18, 19], we have obtained a configuration in which baryons (and eventually anti-baryons) at fixed space points are embedded in a gas of pions. This means that we are actually probing the region of the QCD phase diagram in Fig. 1 at non-vanishing μ_B. The baryonic number density is given by

$$n_B = \frac{n}{V} , \tag{29}$$

where V is the volume. For $b = 1$, see Eq. (16) and (17), we have

$$V = 4\pi^3\ell^3 = \frac{4\pi^3}{m_\pi^3} \simeq 387\,\text{fm}^3 , \tag{30}$$

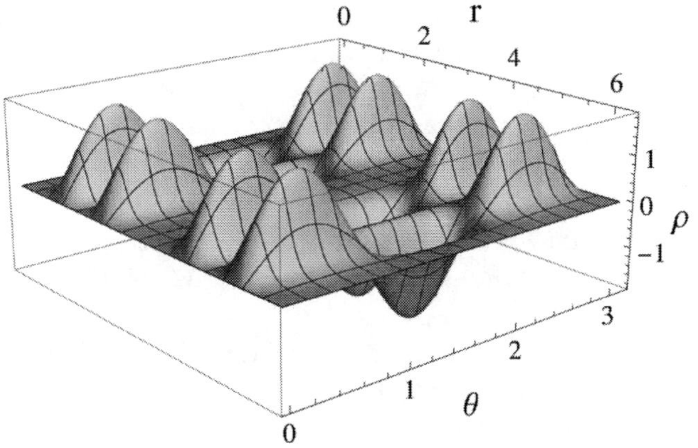

Figure 4. Topological density distribution, see Eq. (26), obtained with $q = 3$ and $n = 4$. The topological density assumes positive values, corresponding to solitons, as well as negative values, corresponding to anti-solitons.

which is a pretty large volume for baryonic matter. It follows that

$$n_B \simeq 2.6\,n \times 10^{-3}\,\mathrm{fm}^{-3}\,, \tag{31}$$

which for small n is rather low. For instance, to reach the nuclear saturation density it is necessary to have $n \simeq 60$ baryons within the volume V. Finally, we note that a non-vanishing baryonic density implies a non-vanishing baryonic chemical potential. The fact that the baryonic chemical potential is non-zero is not in contrast with the definition of the covariant derivative in the χPT Lagrangian, see Eqs. (4) and (5), because pions have zero baryonic charge, see also the discussion in [14].

4 Conclusions and outlook

We have determined an inhomogeneous solution of a system of pions at non-vanishing isospin chemical potential. Remarkably, the isospin chemical potential and the system size are related by Eq. (21), stating that the box length should be proportional to $1/\mu_I$. For appropriate boundary conditions, the solutions are characterized by a non-vanishing topological number, corresponding to the baryonic number of the system.

Regarding the stability of the obtained solution, the fact that a solitonic configuration has a non-vanishing topological charge does not in general prevent its decay to the standard vacuum. It is well known that in three or fewer space-time dimensions the thermal fluctuations can melt the condensates with one dimensional modulation [20]. This is indeed the reason why we have considered a (3+1)-dimensional modulation. On the other hand, the Derrick's scaling argument [21] implies that in flat, topologically trivial (3+1)-dimensional space-time, the (3+1)-dimensional modulations are unstable. We have circumvented this theorem considering a finite spatial volume with non-trivial boundary conditions, which makes the Derrick's scaling argument inapplicable.

Although the solitonic solution we have found is not trivially unstable, it has been determined by the particular ansatz in Eq. (20), therefore it should correspond to a stationary solution which may be metastable or even unstable to low-energy fluctuations. To analyze its stability we are currently working on the fluctuations around the background configuration,

$$\alpha \to \alpha + \hat{\alpha}\,, \qquad \Theta \to \Theta + \hat{\Theta}\,, \qquad \Phi \to \Phi + \hat{\Phi}\,, \tag{32}$$

where α, Θ and Φ indicate the background solutions and $\hat{\alpha}$, $\hat{\Theta}$ and $\hat{\Phi}$ are the fluctuations. The stability requirement is that upon inserting the above expressions in Eq. (9) all these fields are well behaved, in particular none of them should have a tachyonic mass. The main technical problem in conducting this analysis is that these fluctuations are non-linearly interacting. This analysis is presently underway.

References

[1] N. Cabibbo, G. Parisi, Phys. Lett. **59 B**, 67 (1975)
[2] M.G. Alford, A. Kapustin, F. Wilczek, Phys. Rev. **D 59**, 054502 (1999), hep-lat/9807039
[3] S. Weinberg, Physica A: Statistical Mechanics and its Applications **96**, 327 (1979)
[4] J. Gasser, H. Leutwyler, Ann. Phys. **158**, 142 (1984)
[5] H. Georgi, *Weak Interactions and Modern Particle Theory*, Dover Books on Physics Series (Dover Publications, 2009), ISBN 9780486469041, https://books.google.it/books?id=aN9PPQAACAAJ
[6] H. Leutwyler, Ann. Phys. **235**, 165 (1994), hep-ph/9311274
[7] G. Ecker, Prog. Part. Nucl. Phys. **35**, 1 (1995), hep-ph/9501357
[8] H. Leutwyler, Helv. Phys. Acta **70**, 275 (1997), hep-ph/9609466
[9] A. Pich, *Effective field theory: Course*, in *Probing the standard model of particle interactions. Proceedings, Summer School in Theoretical Physics, NATO Advanced Study Institute, 68th session, Les Houches, France, July 28-September 5, 1997. Pt. 1, 2* (1998), pp. 949–1049, hep-ph/9806303
[10] S. Scherer, Adv. Nucl. Phys. **27**, 277 (2003), hep-ph/0210398
[11] S. Scherer, M.R. Schindler (2005), hep-ph/0505265
[12] F. Canfora, S. Carignano, M. Lagos, M. Mannarelli, A. Vera, Phys. Rev. D **103**, 076003 (2021), 2012.05921
[13] M. Mannarelli, Particles **2**, 411 (2019), 1908.02042
[14] A. Mammarella, M. Mannarelli, Phys. Rev. **D 92**, 085025 (2015), 1507.02934
[15] F. Canfora, Eur. Phys. J. C **78**, 929 (2018), 1807.02090
[16] F. Canfora, S.H. Oh, A. Vera, Eur. Phys. J. C **79**, 485 (2019), 1905.12818
[17] F. Canfora, M. Lagos, A. Vera, Eur. Phys. J. C **80**, 697 (2020), 2007.11543
[18] C.G. Callan, Jr., E. Witten, Nucl. Phys. B **239**, 161 (1984)
[19] B.M.A.G. Piette, D.H. Tchrakian, Phys. Rev. D **62**, 025020 (2000), hep-th/9709189
[20] G. Baym, B.L. Friman, G. Grinstein, Nucl. Phys. B **210**, 193 (1982)
[21] G.H. Derrick, J. Math. Phys. **5**, 1252 (1964)

EPJ Web of Conferences **258**, 07004 (2022) https://doi.org/10.1051/epjconf/202225807004
A Virtual Tribute to Quark Confinement and the Hadron Spectrum (vConf21)

NICER view on holographic QCD

Niko Jokela[1,2,*]

[1]Department of Physics, P.O.Box 64, FIN-00014 University of Helsinki, Finland
[2]Helsinki Institute of Physics, P.O.Box 64, FIN-00014 University of Helsinki, Finland

Abstract. The holographic models for dense QCD matter work surprisingly well. A general implication seems that the deconfinement phase transition dictates the maximum mass of neutron stars. The nuclear matter phase turns out to be rather stiff which, if continuously merged with nuclear matter models based on effective field theories, leads to the conclusion that neutron stars do not have quark matter cores in the light of all current astrophysical data. We comment that as the perturbative QCD results are in stark contrast with strong coupling results, any future simulations of neutron star mergers incorporating corrections beyond ideal fluid should proceed cautiously. For this purpose, we provide a model which treats nuclear and quark matter phases in a unified framework at strong coupling.

1 Introduction

The QCD phase diagram is theoretically in most parts unknown. This is because the equations are too hard to solve using traditional perturbative methods and the lattice formulation is still insufficient to map out dense regimes. At the advent of multimessenger era we are unraveling, the astrophysical input has already given us hints on how the matter behaves under extreme conditions those where the matter is in the verge of collapsing into black holes. We direly need microscopic understanding of this behavior, which might be encoded in the equation of state (EoS).

We attempt to solve a QCD-like theory at strong coupling and in the Veneziano limit of large number of colors $N_c \to \infty$ and flavors $N_f \to \infty$, with $N_c/N_f = 1$ using holographic duality [1]. We will review recent progress both for the equilibrium quantities aka equation of state as well as transport quantities near equilibrium. We show that the holographic model, anchored with known QCD physics at low densities has predictive power when extrapolated at finite densities and low temperatures. The conclusions we draw are in stark contrast with perturbative QCD and, in particular, we underline the observation that even with all astrophysical constraints taken into account, the neutron stars do not have quark matter cores. Interestingly, this is at tension with recent claims on the contrary [2] and begs for further study.

This talk is mainly based on articles [7, 11] as well as on work in progress [24] dealing with holographic applications in neutron stars in sections 2 and 3 and on articles [32, 33] describing transport properties of strongly interacting quark matter phase discussed in section 4.

*e-mail: niko.jokela@helsinki.fi

2 Equilibrium

We begin with a brief discussion of the holographic framework which we use to model QCD-like matter. This framework is based on a holographic model called V-QCD [3]. Here the letter V stands for the Veneziano limit, $N_c \to \infty, N_f \to \infty$ while keeping their ratio fixed. In order to make contact with neutron stars we will in the end extrapolate down to $N_c \to 3, N_f \to 3$, since charm quarks are too heavy to be expected in neutron star densities. This model has since then been extended in many directions, notably to finite density and temperature [4, 5].

This is a bottom-up approach in holography, but our model is loosely based on string theory construction, so we aim to follow the rules of string theory as closely as possible. The idea is that we have introduced many new parameters, essentially parametrizing the ignorance by potentials for various bulk fields, the metric, the dilaton, and the (temporal part of) the gauge potential in the search for finite density solutions. The potentials are not completely arbitrary as they have been chosen by various criteria, *e.g.*, by demanding that the coupling constant (dual to the dilaton field) in the UV part of the theory is faithfully captured at two loops. Such constraints are robust enough that there is very little freedom in the parameter space left. The remaining freedom in the potentials are fixed by matching the thermodynamic quantities both for the gluon sector [6] and in the flavor sector to the 2+1 lattice QCD data [7]. In addition to, *e.g.*, reproducing the non-trivial bump in the interaction measure which comes out with no fine-tuning, in view towards finite density physics, the ability of V-QCD to match onto lattice data of the baryon number susceptibility is crucial.

The comparison with QCD data is essential and the V-QCD works surprisingly well at low densities in the QGP phase. Motivated by this success the model is then extrapolated to finite densities and very low temperatures, the environments in the interior of quiescent neutron stars. In addition to mimicking quark matter phase, there is also a baryonic nuclear matter phase. This phase emerges where expected, though, for simplicity we treat baryons in the homogeneous approximation [8]. At low temperatures, which can essentially be thought to mean $T \approx 0$, the model then describes (unpaired) quark matter (deconfined and chirally symmetric as we assume massless quarks) phase at high density with a first order transition to holographic nuclear matter phase (chirally broken confined) and then a further transition to a confined chirally broken zero density phase [9]. One remarkable feature of this framework is that it allows us to consider the same (holographic) model both for the nuclear and the quark matter phase; this is a clear advantage when we later discuss applications to neutron stars.

When one approaches low densities, the V-QCD nuclear matter description is not reliable anymore. We therefore adopt a hybrid construction such that we demand that the holographic nuclear matter continuously transitions to "traditional" nuclear matter models based on effective field theories at low densities. The matching density $n \sim 1.2 - 2n_s$ is left as a free parameter. Variations in this matching density, the selection of various nuclear matter models, as well as the margin in matching onto lattice QCD data of the V-QCD model at vanishing density give rise to a band in the obtained equations of state. In figure 1 we have described the construction [10, 11]. The blue band in figure 1 corresponds to all those EoSs that are physically sensible while the red band follows upon taking holography into account. It is important to notice that the red band is significantly smaller (log-scale) and hence the holographic model has predictive power. This is quite remarkable and a non-trivial result as the V-QCD model is *only* constrained at vanishing baryon densities.

3 Applications to neutron stars

The hybrid equations of state we have constructed suggest that indeed holography has predictive power and hence we can ask if one additionally also takes into account astrophysical

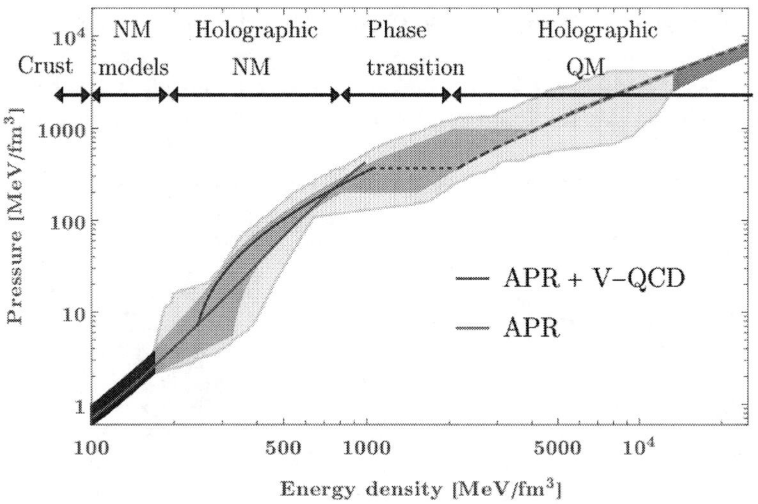

Figure 1. The hybrid equations of state and constructed such that at low densities we adopt the nuclear matter models from literature and merge them into holographic V-QCD at densities $n \sim 1.2 - 2n_s$ after of which it is just holography. It is noteworthy that the EoS from holography correctly matches onto pQCD constraints on the right, plotted in orange. As an example, V-QCD matched with APR is shown in curves.

constraints from neutron star this persists. The answer is affirmative and will lead to interesting results. We therefore insert the hybrid EoSs in the Tolman-Oppenheimer-Volkov (TOV) equations are compute the mass-radius relationships for self-gravitating droplets in the standard, asymptotically Minkowski spacetime. The result following from this procedure is depicted in figure 2. Similar to figure 1 the red band is significantly smaller than the blue, indicating that strongly interacting matter as described by holography is useful.

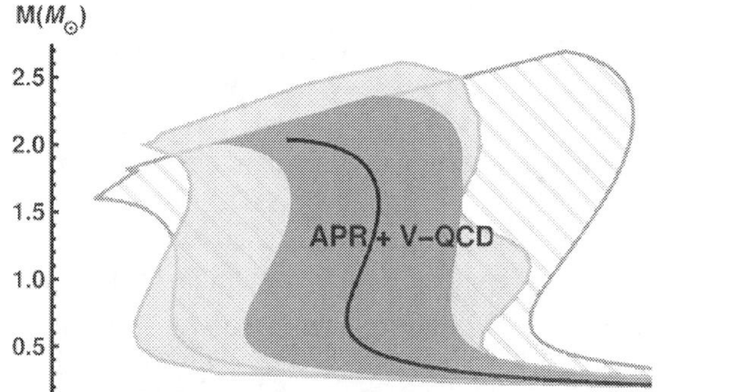

Figure 2. The mass-radius curves that follow from solving the TOV equations. The color coding is the same as in figure 1. The V-QCD model is predictive as the red band is smaller than the blue. We have also shown the representative V-QCD(APR) curve.

Let us then implement the astrophysical constraints, in particular in the light of recent NICER observations [12, 13]. We constrain the allowed regions by first of all demanding that the equations of state be stiff enough to accommodate two-solar mass stars as solutions, as those are detected [14–16]. Second, the LIGO/Virgo collaboration also has provided constraints on the tidal deformability, especially constraining EoSs to be soft enough [17]. These ($M \geq 2M_\odot, 580 \geq \Lambda(1.4M_\odot) \geq 70$) together with accommodating NICER observations lead to very narrow bands for the true mass-radius to reside. We learn that indeed V-QCD,

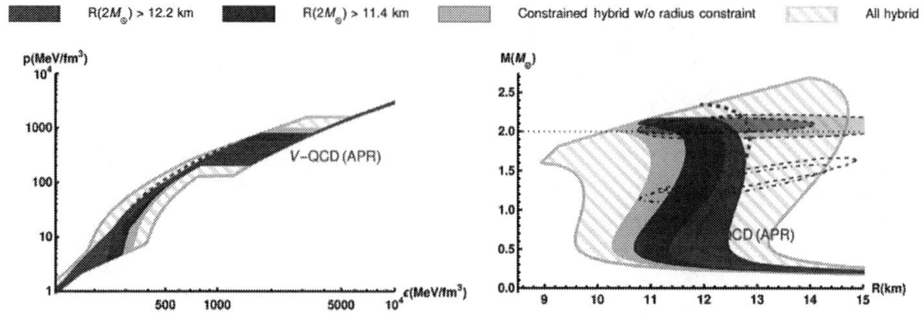

Figure 3. The constrained EoS and mass-radius bands using various constraints from astrophysics. The V-QCD(APR) curves are included as representatives of the hybrid EoSs and the corresponding MR-curves.

matched onto traditional nuclear matter models, after taking into account all current astrophysical constraints, leads to quite narrow allowed ranges. We thereby find that the model not only conforms with known physics but remains predictive. The maximum mass of the neutron stars is set by the underlying deconfinement phase transition, a phenomenon suggested in [18] to occur over a wide range of holographic models, see, *e.g.*, [7, 19, 20]. This is due to the phase transition being strongly first order from the stiff nuclear matter phase to the quite soft quark matter phase.

One of our main predictions is that the hybrid EoSs constrained with the available radius, mass, and tidal deformability measurements favor larger radii, giving tightly constrained limits 12.8km$\geq R(1.4M_\odot) \geq$ 12.0km. Moreover, by taking the most stringent NICER bound $R(2M_\odot) >$ 12.2km at face value, we find that the V-QCD matched onto APR EoS at $n_{\rm tr} = 1.6n_s$ emerges as being an extremely nice representative of one of the most realistic equations of state to date; see figure 3. We have marked this curve in the figures in solid, the other dashed and dotted curves correspond to the remaining span in the high density regime, the allowed parameter range after the fit to lattice QCD.[1] First numerical relativity simulations of neutron star mergers with V-QCD EoSs (matched onto SLy EoS at low density) [10] pave the road to betterment on interpreting future gravitational wave data.

In addition to static properties, we can also compute predictions for the peak frequencies of the gravitational wave signal using semi-universal relations [22, 23]. The characteristic frequencies of the merger and postmerger gravitational wave signal are those that correspond to the frequency at the instant of the merger and the peaks of the power spectral density of the postmerger signal, respectively. Similar to the mass-radius and EoS bands, the frequencies are tightly constrained and the clear outcome is that constraining the hybrid EoSs by radius in all of the cases is to favor lower frequencies as generic polytropic interpolations. This phenomenon is also understood by the fact that our hybrid EoSs are quite stiff. See figure 4 for examples of the prominent peak frequency f_2 and the frequency at the time of the merger $f_{\rm mrg}$; we have chosen to present the case when there is no prompt collapse to a black hole. The

[1]The equations of state V-QCD(APR) can be found in the online repository CompOSE https://compose.obspm.fr/ [21].

frequencies are in the detectable range by LIGO/Virgo in the afterglow of the future neutron star merger events [24].

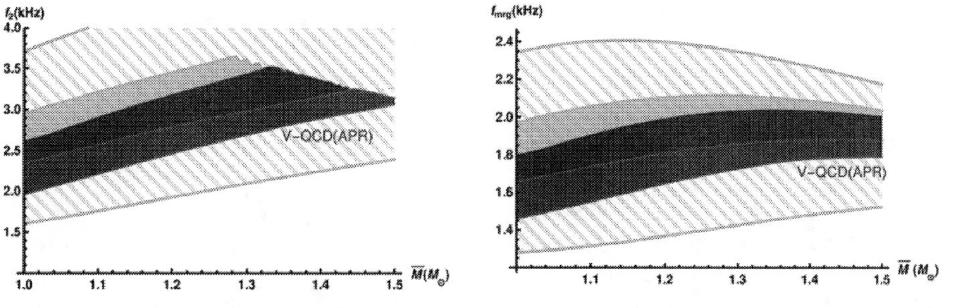

Figure 4. Characteristic frequencies f_2 and $f_{\rm mrg}$ of the gravitational wave signal for equal mass binaries. Also shown the V-QCD(APR) as representative curves. Notice that in the left plot the prompt collapse limit cuts the bands diagonally.

4 Out of equilibrium

The neutron star merger is a violent event where equilibrium is not maintained and the matter heats up significantly. The simulations suggest that the temperatures can reach several tens of MeVs. This means that there is dire need also for the equation of state at finite temperature [25]. This may not be enough, however, since the transport of heat and currents may become relevant also for the strongly interacting sector. Luckily, the computation of transport properties are also pretty straightforward using AdS/CFT duality, at least when the matter is in the deconfining phase.

So what are the most relevant transport properties of unpaired quark matter relevant for neutron stars? Due to the lack of systematic simulations of neutron star mergers incorporating microscopic physics beyond the equation of state this is largely unknown. See, however, [26] for one example, where viscous corrections are discussed. The review [27] discusses the boundary of our understanding drawn by the effects from corrections beyond ideal hydrodynamics. The expectation is that bulk viscosity is likely the most important transport coefficient that may have a noticeable effect on the neutron star merger dynamics. The shear viscosity of the quark matter has been largely associated with the mere existence of rapidly spinning neutron stars due to the pivotal role in suppressing the r-mode instabilities. Heat and electric conductivities are relevant for the neutron star cooling and equilibration after the merger. If the deconfinement transition is very sharp it can lead to observable signal in the gravitational wave spectrum and perhaps could be taken as a smoking gun evidence for the emergence of quark matter phase in the stellar ram [25, 28, 29].

While the EoS of dense and cold QCD matter has large uncertainties, even less is known about transport. Only available first-principles results for quark matter is the leading order pQCD analysis [30] in the unpaired phase for the viscosities as well as the heat and thermal conductivities.

The computation using the holographic approach proceeds quite straightforwardly. However, when computing the conductivities there is one stumbling block. This has to do with the fact that for a homogeneous ground state, the linear response would yield infinite results as charges accelerate indefinitely lacking any momentum dissipation. The key to finding finite results is to realize that the gradients of chemical potential and temperature can balance each other in a steady state configuration [31], *i.e.*, the transport does not occur via convection. Adapting these ideas in the context of holographic quark matter [32, 33], in particular the V-QCD model at hand, allows us to compare the results with those of pQCD. The results are qualitatively strikingly different and call for caution in the use of the perturbative results in neutron-star applications.

5 Conclusions

In this talk we showed that gauge/gravity duality, combined with other approaches, is useful to study dense QCD. Using the holograpic V-QCD model with simple approximations, many details work really well. For example, we obtained a precise fit to lattice thermodynamics at vanishing baryon chemical potential $\mu \approx 0$. The model was then extrapolated to finite density and small temperatures. The resulting equation of state is still reasonable and the model has predictive power. This model is one rare example where both nuclear and quark matter phases can be treated in a unified manner.

We constructed a hybrid equation of state, where we assumed traditional nuclear matter models based of effective field theories at low densities and then we swiftly transformed to the holographic description at densities where nucleons have significant overlap. The resulting equation of state thereafter for the holographic nuclear matter phase was quite stiff while that for the quark matter phase for soft. This lead to the observation that, as a whole, all astrophysical data is consistent with our approach. One interesting conclusion is that neutron stars do not have quark matter cores, contrary to the implications in [2]. The reason for this is that the polytropic index for the holographic nuclear matter reaches very small values before the deconfinement phase transition – such small polytropic index values are classified as quark matter in [2].

We also showed that the holographic framework is able to provide predictions that could be tested in future measurements. To this end we outlaid results for peak frequencies of gravitational wave spectrum. We also showed that transport properties of quark matter are quite distinct from those that one would naively obtain by extrapolating the perturbative results.

Acknowledgements We would like to thank the organizers of the (virtual) Quark Confinement and the Hadron Spectrum 2021 conference in Stavanger, Norway, for the invitation to present this talk. The author has been supported in part by the Academy of Finland grant no. 1322307.

References

[1] A.V. Ramallo, Springer Proc. Phys. **161**, 411 (2015), 1310.4319

[2] E. Annala, T. Gorda, A. Kurkela, J. Nättilä, A. Vuorinen, Nature Phys. **16**, 907 (2020), 1903.09121

[3] M. Järvinen, E. Kiritsis, JHEP **03**, 002 (2012), 1112.1261

[4] T. Alho, M. Järvinen, K. Kajantie, E. Kiritsis, K. Tuominen, JHEP **01**, 093 (2013), 1210.4516

[5] T. Alho, M. Järvinen, K. Kajantie, E. Kiritsis, C. Rosen, K. Tuominen, JHEP **04**, 124 (2014), [Erratum: JHEP02,033(2015)], 1312.5199

[6] U. Gürsoy, E. Kiritsis, L. Mazzanti, F. Nitti, Nucl. Phys. **B820**, 148 (2009), 0903.2859

[7] N. Jokela, M. Järvinen, J. Remes, JHEP **03**, 041 (2019), 1809.07770

[8] O. Bergman, G. Lifschytz, M. Lippert, JHEP **11**, 056 (2007), 0708.0326

[9] T. Ishii, M. Järvinen, G. Nijs, JHEP **07**, 003 (2019), 1903.06169

[10] C. Ecker, M. Järvinen, G. Nijs, W. van der Schee, Phys. Rev. D **101**, 103006 (2020), 1908.03213

[11] N. Jokela, M. Järvinen, G. Nijs, J. Remes, Phys. Rev. D **103**, 086004 (2021), 2006.01141

[12] M.C. Miller et al. (2021), 2105.06979

[13] T.E. Riley et al. (2021), 2105.06980

[14] J. Antoniadis et al., Science **340**, 6131 (2013), 1304.6875

[15] H.T. Cromartie et al., Nat. Astron. **4**, 72 (2019), 1904.06759

[16] E. Fonseca et al. (2021), 2104.00880

[17] B. Abbott et al. (LIGO Scientific, Virgo), Phys. Rev. Lett. **121**, 161101 (2018), 1805.11581

[18] C. Hoyos, D. Rodríguez Fernández, N. Jokela, A. Vuorinen, Phys. Rev. Lett. **117**, 032501 (2016), 1603.02943

[19] K. Bitaghsir Fadafan, J. Cruz Rojas, N. Evans, Phys. Rev. D **101**, 126005 (2020), 1911.12705

[20] L.A.H. Mamani, C.V. Flores, V.T. Zanchin, Phys. Rev. D **102**, 066006 (2020), 2006.09401

[21] S. Typel, M. Oertel, T. Klähn, Phys. Part. Nucl. **46**, 633 (2015), 1307.5715

[22] F. Zappa, S. Bernuzzi, D. Radice, A. Perego, T. Dietrich, Phys. Rev. Lett. **120**, 111101 (2018), 1712.04267

[23] M. Breschi, S. Bernuzzi, F. Zappa, M. Agathos, A. Perego, D. Radice, A. Nagar, Phys. Rev. D **100**, 104029 (2019), 1908.11418

[24] N. Jokela, M. Järvinen, J. Remes, **to appear** (2021)

[25] P.M. Chesler, N. Jokela, A. Loeb, A. Vuorinen, Phys. Rev. **D100**, 066027 (2019), 1906.08440

[26] M.G. Alford, L. Bovard, M. Hanauske, L. Rezzolla, K. Schwenzer, Phys. Rev. Lett. **120**, 041101 (2018), 1707.09475

[27] A. Schmitt, P. Shternin, Astrophys. Space Sci. Libr. **457**, 455 (2018), 1711.06520

[28] E.R. Most, L.J. Papenfort, V. Dexheimer, M. Hanauske, S. Schramm, H. Stöcker, L. Rezzolla, Phys. Rev. Lett. **122**, 061101 (2019), 1807.03684

[29] A. Bauswein, N.U.F. Bastian, D.B. Blaschke, K. Chatziioannou, J.A. Clark, T. Fischer, M. Oertel, Phys. Rev. Lett. **122**, 061102 (2019), 1809.01116

[30] H. Heiselberg, C.J. Pethick, Phys. Rev. **D48**, 2916 (1993)

[31] B. Goutéraux, N. Jokela, A. Pönni, JHEP **07**, 004 (2018), 1803.03089

[32] C. Hoyos, N. Jokela, M. Järvinen, J.G. Subils, J. Tarrio, A. Vuorinen, Phys. Rev. Lett. **125**, 241601 (2020), 2005.14205

[33] C. Hoyos, N. Jokela, M. Järvinen, J.G. Subils, J. Tarrio, A. Vuorinen (2021), 2109.12122

Equation of state for holographic nuclear matter as instanton gas *

Kazuo Ghoroku[1,**], *Kouji* Kashiwa[1,***], *Yoshimasa* Nakano[****], *Motoi* Tachibana[2,†], and *Fumihiko* Toyoda[3,‡]

[1]Fukuoka Institute of Technology, Wajiro, Fukuoka 811-0295, Japan
[2]Department of Physics, Saga University, Saga 840-8502, Japan
[3]Faculty of Humanity-Oriented Science and Engineering, Kinki University, Iizuka 820-8555, Japan

Abstract. In a holographic model, which was used to investigate the color superconducting phase of QCD, a dilute gas of instantons is introduced to study the nuclear matter. The free energy of the nuclear matter is computed as a function of the baryon chemical potential in the probe approximation. Then the equation of state is obtained at low temperature. Using the equation of state for the nuclear matter, the Tolman-Oppenheimer-Volkov equations for a cold compact star are solved. We find the mass-radius relation of the star, which is similar to the one for quark star. This similarity implies that the instanton gas given here is a kind of self-bound matter.

1 The Set-up

We start from the bottom-up holographic model which is proposed as Yang-Mills (YM) theory with color superconducting phase. It is given by the following gravitational theory [2–5].

$$S = \int d^{d+1}\xi \sqrt{-g}\,\mathcal{L}, \tag{1}$$

$$\mathcal{L} = \mathcal{L}_{\rm Gravity} + \mathcal{L}_{\rm CSC}, \tag{2}$$

$$\mathcal{L}_{\rm Gravity} = \mathcal{R} + \frac{d(d-1)}{L^2}, \tag{3}$$

$$\mathcal{L}_{\rm CSC} = -\frac{1}{4}F^2 - |D_\mu\psi|^2 - m^2|\psi|^2, \tag{4}$$

$$F_{\mu\nu} = \partial_\mu A_\nu - \partial_\nu A_\mu, \quad D_\mu\psi = (\partial_\mu - iqA_\mu)\psi. \tag{5}$$

It describes $d+1$ dimensional gravity coupled to a $U(1)$ gauge field, A_μ, and a charged scalar field, ψ. Here we consider the case of $d = 5$. The charge q denotes the baryon number of the scalar ψ, and it is chosen as $q = 2/N_c$ to represent the quark Cooper pair formation. The mass

*This is based on [1].
**e-mail: gouroku@fit.ac.jp
***e-mail: kashiwa@fit.ac.jp
****e-mail: ynakano@kyudai.jp
†e-mail: motoi@cc.saga-u.ac.jp
‡e-mail: f1toyoda@jcom.home.ne.jp

m is given to reproduce the corresponding conformal dimension of the diquark operator dual to the scalar field ψ with the charge $q = 2/N_c$. Here we put $1/2\kappa_6^2 = 1$, and the dimensionful $U(1)$ gauge coupling constant is set as unit. So the metric is dimension less and A_μ has dimension one as in the 3+1 dimensional case.

Here we add the sector of the $SU(N_f)$ gauge fields, which is used to build the nuclear matter through the instanton configuration, which generates the baryon number in QCD. The instanton configuration is here deformed since it is embedded in a higher dimensional curved space.

$$\mathcal{L} = \mathcal{L}_{\text{Gravity}} + \mathcal{L}_{\text{CSC}} + \mathcal{L}_{\text{V}}, \tag{6}$$

$$\mathcal{L}_{\text{V}} = -\frac{1}{4}\text{tr}F^2_{SU(N_f)} \tag{7}$$

where $F^2_{SU(N_f)}$ denotes the 2-form $SU(N_f)$ gauge field. The coupling constant g^2_{YM} is set as one. The mass dimension of the gauge field is taken to be one. Then, in order to pick up the coupling of the baryon number current and the instanton configuration, we add the Chern-Simons (CS) term,

$$S = \int d^{d+1}x\sqrt{-g}\,\mathcal{L} + S_{\text{CS}}. \tag{8}$$

Hereafter we proceed the analysis according to the probe approximation with $N_f = 2$.

1.1 Instanton in R^4

Before solving the equation of motion of our model, the ansatz for the $U(1)$ gauge field and the instanton configurations for $SU(2)$ gauge fields are given. The coordinates are denoted as $(\xi^0, \xi^1, \xi^2, \xi^3, \xi^4, \xi^5) = (x^0, x^1, x^2, x^3, w, z)$. Then, we make the ansatz for $U(1)$ gauge field as

$$A_b = A_b(z)\delta^0_b. \tag{9}$$

As for the $SU(2)$ vector fields, which are set as

$$\vec{f_1} = F^i_{ab}\tau_i, \tag{10}$$

and the following ansats are imposed,

$$(\vec{f_1})_{ij} = Q(x^m - a^m, \rho)\epsilon_{ijk}\tau^k, \qquad (\vec{f_1})_{iz} = Q(x^m - a^m, \rho)\tau^i, \tag{11}$$

where ρ (a^m) denotes the instanton size (position), $\epsilon_{123z} = 1$, $i, j = 1, 2, 3$ and $m = 1, \ldots, 4$, where $x^4 = z$. Then the $SU(N_f)$ vector part (7) is given for $N_f = 2$ as

$$\mathcal{L}_{\text{V}} = -\frac{3}{2}Q^2\left[\left(g^{11}\right)^2 + g^{11}g^{zz}\right]. \tag{12}$$

In order to see the baryon, Q is given as an instanton solution in flat 4D space $\{x^m\}$ [6, 7],

$$Q = \frac{2\rho^2}{((x^m - a^m)^2 + \rho^2)^2}. \tag{13}$$

Now let us make an ansatz for Q as follows:

$$Q^2 = \sum_i^{N_I} \frac{4\rho^4}{\left((x^m - a^m_i)^2 + \rho^2\right)^4}. \tag{14}$$

This corresponds to the dilute gas approximation of instantons. Then the size of instanton is determined by the variational principle so as to minimize the energy density.

Lastly, the CS term is given by

$$S_{\rm CS} = \kappa_{\rm CS}\epsilon^{m_1\cdots m_4}\int d^4x\, dz\, A_0{\rm tr}(F_{m_1m_2}F_{m_3m_4}). \tag{15}$$

2 Effective action and EoS for nuclear matter

Let us now evaluate the effective action of the instantons. To this end, the gravity background dual to the low temperature confinement phase is given as the following:

$$ds^2 = k^{2/5}\eta_{\mu\nu}dx^\mu dx^\nu + \frac{4}{25}k^{-3/5}(k^{-2/5}dz^2 + z^2d\theta^2), \tag{16}$$

where $k = 1 + z^2$. Then the matter action with embedded instantons is obtained as

$$\begin{aligned} S_{\rm matter} &= \int d^6\xi\,\sqrt{-g}\left(-\frac{1}{4}F^2 - \frac{1}{4}{\rm tr}F^2_{SU(2)}\right) + S_{\rm CS} \\ &= \int d^4x dz\left(\frac{1}{2}zk^{3/5}A_0'^{\,2} - nz\frac{12}{25}\bar{q}^2\left[k^{-4/5} + \frac{25}{4}k^{3/5}\right] + nn_0A_0\bar{q}^2\right), \end{aligned} \tag{17}$$

where $A_0' = \partial_zA_0(z)$, and $n_0 = 48\kappa_{\rm CS}$. In order to estimate this action, we solve the equation of motion of $A_0(z)$. It is obtained as

$$-\partial_z\left(zk^{3/5}A_0'\right) + nn_0\bar{q}^2 = 0. \tag{18}$$

Then we have

$$zk^{3/5}A_0' = \bar{d} = \frac{\pi^2}{6}nn_0\frac{2z^3 + 3z\rho^2}{(z^2+\rho^2)^{3/2}} + c. \tag{19}$$

Here, c denotes an integration constant of Eq. (18) over z. We take $c = 0$ to set as $\bar{d}(z = 0) = 0$. Due to this condition, the matter action $S_{\rm matter}$ of (17) is written as

$$S_{\rm matter} = \int d^4x\left\{\mu\bar{Q} - \int dz\left(\frac{\bar{d}^{\,2}}{2zk^{3/5}} + nz\frac{12}{25}\bar{q}^2\left[k^{-4/5} + \frac{25}{4}k^{3/5}\right]\right)\right\}, \tag{20}$$

where the chemical potential μ and the charge density $\bar{Q} \equiv \bar{d}(\infty)$ are given as

$$\mu = A_0(\infty) = \int_{z_1}^{\infty} dz\frac{\bar{d}}{zk^{3/5}} + A_0(z_1) = \int_{z_1}^{\infty} dz\frac{\bar{d}}{zk^{3/5}}, \tag{21}$$

$$\bar{Q} = \bar{d}(\infty) = \frac{\pi^2}{3}nn_0, \tag{22}$$

where z_1 is an arbitrary positive value and $A_0(z_1)$ is set as zero. which is the boundary condition in solving the differential equation of $A_0(z)$, Eq. (19).

Then the free energy density $\mathcal{E}$ of the instanton system is given as

$$S_{\rm matter} = -\int d^5\xi\ \mathcal{E}(\rho,\mu) \tag{23}$$

$$\mathcal{E}(\rho,\mu) = \int dz\left(\frac{\bar{d}^{\,2}}{2zk^{3/5}} + nz\frac{12}{25}\bar{q}^2\left[k^{-4/5} + \frac{25}{4}k^{3/5}\right]\right) - \mu\bar{Q} \tag{24}$$

2.1 EoS for nuclear matter

By giving n, n_0, z_1, which are the parameters of the present bottom-up model, $\mu(\rho)$ and $\mathcal{E}(\rho)$ are calculated according to the Eqs. (21) and (24) as the functions of ρ. Then a minimum point of $\mathcal{E}(\rho)$ is found at $\rho = \rho_{\rm min}$. Thus we can obtain $\mathcal{E}(\rho_{\rm min})$ and $\mu(\rho_{\rm min})$ for the given parameters n, n_0, z_1.

Table 1. EoS of nuclear system for $n_0 = 1.5, z_1 = 0.1$

n	$\mu(\rho_{\rm min})$	$\mathcal{E}(\rho_{\rm min})$	$\rho_{\rm min}$	$v_{\rm min}$
0.005	0.0688	0.00031	0.01	2.09×10^{-8}
0.01	0.138	0.0000393	0.02	3.35×10^{-7}
0.015	0.209	-0.00022	0.03	1.70×10^{-6}
0.02	0.275	-0.00178	0.04	5.36×10^{-6}
0.03	0.411	-0.00829	0.06	2.75×10^{-5}
0.05	0.68	-0.0368	0.09	1.52×10^{-4}
0.1	1.335	-0.2075	0.13	9.2×10^{-4}

Table 1 shows a resultant example of such calculations for $n_0 = 1.5, z_1 = 0.1$. The regions of the calculations are restricted to the region for $\mu < 1.7$ [3]. On the other hand, the pressure p, which is given by $p = -\mathcal{E}$, of the instanton gas is negative for $\mu < \mu_c \sim 0.17$. So in this region, the gas is in an undesirable phase as a nuclear matter considered here.

Thus we find that the stable nuclear matter exists in the region, $1.7 > \mu > \mu_c \sim 0.17$. In this region of the nuclear matter, its dilute gas picture is reasonable since $v_{\rm min} \sim O(10^{-4}) \ll 1$, where $v_{\rm min} = \frac{4}{3}\pi\rho_{\rm min}^3 n$ indicates the volume which is occupied by instantons in a unit 3D volume.

Here we proceed the analysis, and we can arrive at the following approximate formula

$$p = a\mu(\mu - \mu_c)\,, \tag{25}$$

where $a = 0.13$ and $\mu_c = 0.17$. Then using (25), the energy density is given at $T = 0$ as [8]

$$\epsilon = \mu q - p = \mu\frac{\partial p}{\partial\mu} - p = a\mu^2\,, \tag{26}$$

and the speed of sound is obtained as

$$C_s^2 = \frac{\partial p}{\partial\epsilon} = \frac{\partial p/\partial\mu}{\partial\epsilon/\partial\mu} = 1 - \frac{\mu_c}{2\mu}\,. \tag{27}$$

This leads to the following bound of C_s^2 as $\frac{1}{2} < C_s^2 < 1$. We notice that the lower bound $1/2$ is fairly large comparing to that of the ordinary nuclear matter. We should however notice that the constraint for C_s^2 comes from the formula (25), which is an approximate formula available at small μ. It is possible to improve (25) by adding correction terms of higher powers of μ. For example, by adding a term like μ^4, we can find a new formula, $p = a'\mu(\mu - \mu_c') + b'\mu^4$ with appropriate values for the parameters, a', μ_c', b'. In this case, we can see that the maximum value of C_s is realized at μ smaller than 1 and its maximum value is suppressed from 1, then it decreases to $C_s = 1/\sqrt{3}$ in the large μ region. However, as we can see, the correction term of μ^4 becomes small in the small μ region. Then it is difficult to suppress largely the lower bound 1/2. We find the small value of sound speed in region of

$0 < C_s^2 < 1/2$ for $\mu < \mu_c$, where $\mu_c(\neq \mu_c')$ denotes the critical point and it satisfies $p(\mu_c) = 0$. In this region of μ, however, the pressure p is negative, then the instanton gas cannot make a nucleon matter like a star. This implies that our model can not cover the low density part of the normal nuclear matter, in which the sound velocity decreases to zero from its upper bound $(1/\sqrt{3})$. How to overcome this point is remained here as a future problem.

In this sense, the nuclear matter given here might be a special one. Our model is considered in a restricted region of density or μ, $\mu_c < \mu < 1.7$, where 1.7 denotes the transition point to the deconfined RN phase. Here, therefore, we proceed analysis by using a simple model with (25) and (26), then we arrive at the EoS of the nuclear matter given as the instanton gas. It is written as,

$$p = \epsilon - \sqrt{a\epsilon}\,\mu_c\,. \tag{28}$$

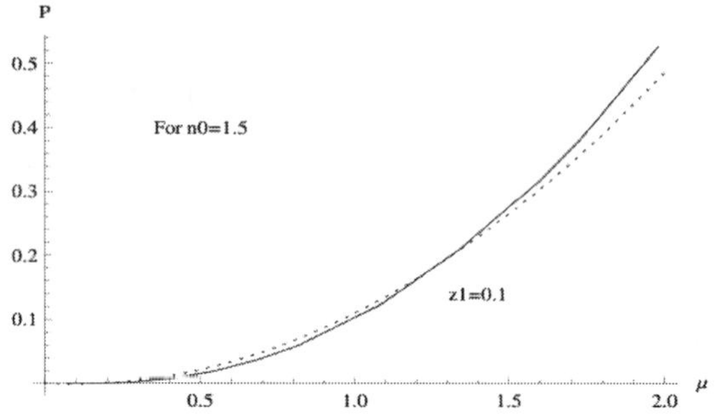

Figure 1. $p = -\mathcal{E}(\mu)$ versus μ at stable instanton size $\rho_{\min}$ near phase transition point $\mu_c = 0.17$. The solid curve denotes the numerical calculations, and the dotted curve represents $p = 0.13\mu(\mu - \mu_c)$.

3 TOV equation and M-R relation of neutron star

The TOV equations for a star with the mass m and p at the radius r in the star are given as,

$$\frac{dp}{dr} = -G(\epsilon + p)\frac{m + 4\pi r^3 p}{r(r - 2Gm)}\,, \tag{29}$$

$$\frac{dm}{dr} = 4\pi r^2 \epsilon\,. \tag{30}$$

Here, G denotes the gravitational constant. They can be solved by using the EoS given in (28) with the boundary condition, $p(r = 0) = p_c$ and $m(r = 0) = 0$. Solving these equations, we obtain the mass of the star, $M = m(R)$, and its radius R, where R is defined by $p(R) = 0$.

In order to obtain the numerical results with dimensionful quantities, we must adjust the scale parameters for each dimensionful quantities $(\tilde{x})$. To do so, the above TOV equations are rewritten by using the dimensionless quantities as [9]

$$\frac{d\tilde{p}}{d\tilde{r}} = -B(\tilde{\epsilon} + \tilde{p})\frac{\tilde{m} + 4\pi A\tilde{r}^3 \tilde{p}}{\tilde{r}(\tilde{r} - 2B\tilde{m})}\,, \tag{31}$$

$$\frac{d\tilde{m}}{d\tilde{r}} = 4\pi A\tilde{r}^2 \tilde{\epsilon}\,, \tag{32}$$

where

$$A = \frac{r_0^3 \epsilon_0}{m_0}\,, \quad B = \frac{G m_0 \epsilon_0}{p_0 r_0}\,, \tag{33}$$

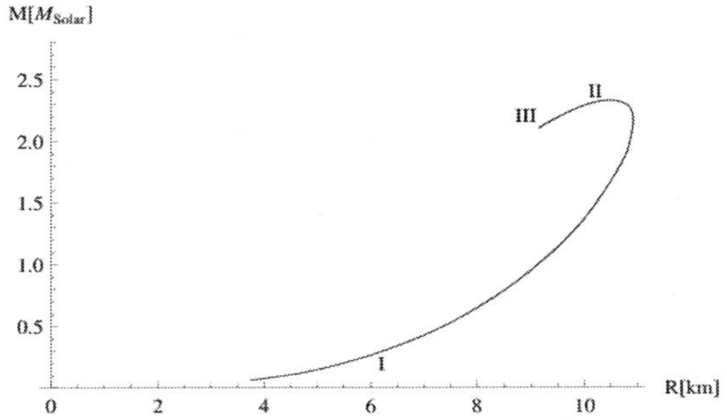

Figure 2. Plot of the neutron mass $M(R)$ (in unit of solar mass $M_\odot$) versus the radius R for $p = \epsilon - \sqrt{a\epsilon}\mu_c$ (or $p = a\mu(\mu - \mu_c)$), $\mu_c = 0.17$ and $a = 0.13$.

and the various variables (x) are replaced by the dimensionless quantities ($\tilde{x}$) by using its typical dimensionful value x_0 as $x = x_0\tilde{x}$. For example, we rewrite as $p = p_0\tilde{p}$.

Then the solutions of (31) and (32) with $A = B = 1$ are equal to the one of (29) and (30) with $G = 1$. Thus the solution of the latter equations are translated to the one of the former ones as shown in Fig. 2. It is given for $r_0 = 3.0$ km in natural unit. In this case, we find $m_0 = 2.03M_\odot$ [1] and $\epsilon_0^{1/4} = 0.896$ GeV, which provides physical unit to the quantities given in the Table 1. For example, the energy density of the center in the star, whose mass is about two solar mass, is given by 0.00242 GeV^4. This is about twice as large as the one of the normal nuclear matter.

4 Comparison with the observational data

There are some important observational data which can constrain the theoretical EoS. We now have the following at least three important constraints for the M-R relation;

1. Two solar mass ($2M_\odot$) neutron star observation from the Shapiro delay measurement [10, 11].

2. Radius constraint from GW170817 via the gravitational wave observation. The actual constraint is $9.0 < R < 13.6$ km for $M = 1.4M_\odot$ [12–14].

3. Upper bound of M for cold spherical neutron stars which is estimated from no detection of relativistic optical counterpart in the analysis of GW170817. The actual limit is estimated as the range $2.15 - 2.26M_\odot$ [15].

The maximum mass $M_{\rm max}$ of our result overshoots the $2.15 - 2.26M_\odot$ constraint, but we can expect that there are some phase transitions at high density such as the chiral, deconfined, and color superconducting phase transitions which we do not consider in this study. Then, EoS can be softer than the present one at high density and thus $M_{\rm max}$ should be smaller than the present value.

With our EoS, the M-R curve flows down to the left bottom, Such behavior is similar to the M-R curves obtained with EoS proposed in Ref. [16]; the actual behavior of the curves can be seen in Refs. [17, 18] denoted as "SQM1-3"; these EoS which contain the quark matter make self-bounded neutron stars which do not have minimum masses. The tendency of the

[1]The symbol $M_\odot$ denotes the solar mass.

M-R curves in the small *M* region in our model may be modified when we suitably introduce "interactions" to the present model because we employ the dilute instanton gas approximation; the baryonic contribution can be expected to have relatively strong effects on EoS at low density. In addition, we here use the dilute gas of the instanton of the $SU(2)$ gauge field and thus the number of the flavor is considered as two. But since the core of the neutron star is very dense, the hyperon degree of freedom can join the game as the baryonic mode in addition to nucleons. In the present study, we consider the simple symmetric nuclear matter because the difference between the flavors and effects of the electrons are difficult to include in the model at present. Therefore, the inclusion of the effects such as the beta equilibrium and charge neutrality are our future work.

5 Summary and discussions

In this paper, we have studied cold nuclear matter and its equation of state (EoS) based on the six dimensional holographic model. The nuclear matter was introduced as a dilute gas of deformed instantons in the AdS soliton background. The instantons are electric charge neutral and they are made of the $SU(2)$ gauge fields and the action includes the Chern-Simons term as well as the kinetic term. Owing to the energy balance between these two terms in the action, the size of the instanton can be determined for fixed parameters of the model. As the result, EoS for nuclear matter, *i.e.*, pressure as a function of energy density, is obtained.

Then, we applied the EoS to solve the Tolman-Oppenheimer-Volkov equations for a compact star numerically and found the mass-radius (*M-R*) relationship. The curve is somewhat similar to the one for strange quark matter and we provided a certain interpretation for that. In our approach, it might be possible that the mass of an instanton is very heavy compared with an ordinary nucleon, which is a specific property in holographic QCD since the nucleon mass is proportional to the number of colors N_c. So far we have no definite answer and this is a future issue.

In order to improve our current study, several ingredients are taken into account. One is to go beyond the dilute gas approximation of instantons. The other is to study whether the system enjoys the phase transition from nuclear matter to (perhaps color superconducting) quark matter at higher baryon density. If such a phase transition is present, the EoS gets soften and the resultant *M-R* curve could be modified. Furthermore it will be interesting to extend our current study into the case with hyperon degrees of freedom. To this end, the holographic treatment of heavy-light meson system (for instance, see [19]) must be instructive. These will be the future issues.

Acknowledgments

We are grateful D. Blaschke and A. Schmitt for their useful comments during the online conference "A Virtual Tribute to Quark Confinement and the Hadron Spectrum" (vConf2021). It is also a pleasure to thank Masayuki Matsuzaki for useful discussions on the neutron stars and TOV equations.

References

[1] Kazuo Ghoroku, Kouji Kashiwa, Yoshimasa Nakano, Motoi Tachibana, Fumihiko Toyoda "Stiff equation of state for a holographic nuclear matter as instanton gas", arXiv:2107.14450[hep-th]:

[2] Pallab Basu, Fernando Nogueira, Moshe Rozali, Jared B. Stang, Mark Van Raamsdonk, "Towards A Holographic Model of Color Superconductivity", New J.Phys.13:055001,2011, [arXiv:1101.4042 [hep-th]].

[3] Kazuo Ghoroku, Kouji Kashiwa, Yoshimasa Nakano, Motoi Tachibana, Fumihiko Toyoda "Color Superconductivity in Holographic SYM Theory", Phys. Rev. D 99, 106011 (2019), arXiv:1902.01093[hep-th]:

[4] Steven S. Gubser, "Breaking an Abelian gauge symmetry near a black hole horizon", Phys.Rev.D78:065034,2008, arXiv:0801.2977 [hep-th].

[5] Sean A. Hartnoll, Christopher P. Herzog, Gary T. Horowitz, "Building an AdS/CFT superconductor", Phys. Rev. Lett. 101.031601, [Xiv:0803.3295 [hep-th]];

[6] Kazuo Ghoroku, Kouki Kubo, Motoi Tachibana, Tomoki Taminato, Fumihiko Toyoda, "Holographic cold nuclear matter as dilute instanton gas", PhysRevD.87.066006, arXiv:1211.2499 [hep-th].

[7] H. Hata, T. Sakai, S. Sugimoto and S. Yamato, "Baryons from instantons in holographic QCD," Prog. Theor. Phys. **117**, 1157 (2007) [hep-th/0701280 [HEP-TH]].

[8] Kazem Bitaghsir Fadafan, Jesus Cruz Rojas, Nick Evans "Deconfined, Massive Quark Phase at High Density and Compact Stars: A Holographic Study", Phys. Rev. D 101, 126005 (2020), arXiv:1911.12705 [hep-ph]

[9] Kazem Bitaghsir Fadafan, Jesus Cruz Rojas, Nick Evans, "Holographic quark matter with colour superconductivity and a stiff equation of state for compact stars", Phys. Rev. D 103, 026012 (2021) [arXiv:2009.14079 [hep-ph]].

[10] P. Demorest, T. Pennucci, S. Ransom, M. Roberts and J. Hessels, Nature **467**, 1081-1083 (2010) doi:10.1038/nature09466 [arXiv:1010.5788 [astro-ph.HE]].

[11] J. Antoniadis, P. C. C. Freire, N. Wex, T. M. Tauris, R. S. Lynch, M. H. van Kerkwijk, M. Kramer, C. Bassa, V. S. Dhillon and T. Driebe, *et al.* Science **340**, 6131 (2013) doi:10.1126/science.1233232 [arXiv:1304.6875 [astro-ph.HE]].

[12] E. Annala, T. Gorda, A. Kurkela and A. Vuorinen, Phys. Rev. Lett. **120**, no.17, 172703 (2018) doi:10.1103/PhysRevLett.120.172703 [arXiv:1711.02644 [astro-ph.HE]].

[13] S. De, D. Finstad, J. M. Lattimer, D. A. Brown, E. Berger and C. M. Biwer, Phys. Rev. Lett. **121**, no.9, 091102 (2018) [erratum: Phys. Rev. Lett. **121**, no.25, 259902 (2018)] doi:10.1103/PhysRevLett.121.091102 [arXiv:1804.08583 [astro-ph.HE]].

[14] I. Tews, J. Margueron and S. Reddy, Phys. Rev. C **98**, no.4, 045804 (2018) doi:10.1103/PhysRevC.98.045804 [arXiv:1804.02783 [nucl-th]].

[15] M. Shibata, S. Fujibayashi, K. Hotokezaka, K. Kiuchi, K. Kyutoku, Y. Sekiguchi and M. Tanaka, Phys. Rev. D **96**, no.12, 123012 (2017) doi:10.1103/PhysRevD.96.123012 [arXiv:1710.07579 [astro-ph.HE]].

[16] M. Prakash, J. R. Cooke and J. M. Lattimer, Phys. Rev. D **52**, 661-665 (1995) doi:10.1103/PhysRevD.52.661

[17] J. M. Lattimer and M. Prakash, Astrophys. J. **550**, 426 (2001) doi:10.1086/319702 [arXiv:astro-ph/0002232 [astro-ph]].

[18] F. Ozel, D. Psaltis, T. Guver, G. Baym, C. Heinke and S. Guillot, Astrophys. J. **820**, no.1, 28 (2016) doi:10.3847/0004-637X/820/1/28 [arXiv:1505.05155 [astro-ph.HE]].

[19] J. Erdmenger, K. Ghoroku and I. Kirsch, "Holographic heavy-light mesons from non-Abelian DBI," J. High Energ. Phys. 09 (2007) 111, doi:10.1088/1126-6708/2007/09/111 [arXiv:0706.3978 [hep-th]].

EPJ Web of Conferences **258**, 07006 (2022) https://doi.org/10.1051/epjconf/202225807006
A Virtual Tribute to Quark Confinement and the Hadron Spectrum (vConf21)

Kaonic atoms measurements at the DAΦNE collider: the SIDDHARTA-2 experiment

Catalina Curceanu[1], *Marco* Miliucci[1,*], *Massimiliano* Bazzi[1], *Damir* Bosnar[2], *Mario* Bragadireanu[3,1], *Marco* Carminati[4], *Michael* Cargnelli[5], *Alberto* Clozza[1], *Griseld* Deda[4], *Luca* De Paolis[1], *Raffaele* Del Grande[6,1], *Carlo* Fiorini[4], *Carlo* Guaraldo[1], *Mihail* Iliescu[1], *Masahiko* Iwasaki[7], *Pietro* King[4], *Paolo* Levi Sandri[1], *Johann* Marton[5], *Paweł* Moskal[8], *Fabrizio* Napolitano[1], *Szymon* Niedźwiecki[8], *Kristian* Piscicchia[9], *Alessandro* Scordo[1], *Francesco* Sgaramella[1], *Hexi* Shi[5,1], *Michał* Silarski[8], *Diana* Sirghi[1,3], *Florin* Sirghi[1,3], *Magdalena* Skurzok[8,1], *Antonio* Spallone[1], *Marlene* Tüchler[5], *Oton* Vazquez Doce[1,6], and *Johann* Zmeskal[5]

[1]INFN, Laboratori Nazionali di Frascati, Frascati, 00044 Roma, Italy
[2]Department of Physics, Faculty of Science, University of Zagreb, 10000 Zagreb, Croatia
[3]Horia Hulubei National Institute of Physics and Nuclear Engineering, IFIN-HH, Magurele 077125, Romania
[4]Politecnico di Milano, Dipartimento di Elettronica, Informazione e Bioingegneria and INFN Sezione di Milano, 20133 Milano, Italy
[5]Stefan-Meyer-Institut für Subatomare Physik, 1090 Vienna, Austria
[6]Excellence Cluster Universe, Technische Universität München - Garching, Germany
[7]RIKEN - Tokyo, Japan
[8]The M. Smoluchowski Institute of Physics, Jagiellonian University, 30-348 Kraków, Poland
[9]Centro Ricerche Enrico Fermi - Museo Storico della fisica e Centro Studi e Ricerche "Enrico Fermi", 00184 Roma, Italy

Abstract. The X-ray spectroscopy measurements of light kaonic atoms' de-excitation towards the fundamental level provide unique information on the low-energy Quantum ChromoDynamics (QCD) in the strangeness sector, being a direct probe of the kaon/nucleon interaction at threshold, unobtainable through the scattering experiments. In this framework, the SIDDHARTA-2 collaboration is going to perform the first kaonic deuterium 2p → 1s transition measurement at the DAΦNE collider of Istituto Nazionale di Fisica Nucleare – Laboratori Nazionali di Frascati. Combining this measurement with the kaonic hydrogen one performed by SIDDHARTA in 2009 it will be possible to obtain, in a model-independent way, the isospin-dependent antikaon-nucleon scattering lengths. The paper introduces the SIDDHARTA-2 setup, an upgraded version with respect to the one used for the kaonic hydrogen measurement, dedicated to the ambitious kaonic deuterium measurement, together with the preliminary results obtained during the kaonic helium run, preparatory for the SIDDHARTA-2 data taking campaign.

*e-mail: μmarco.miliucci@lnf.infn.it

1 Light kaonic atoms X-ray spectroscopy

Kaonic atoms are systems in which a negatively charged kaon (K^-) is bound to the nucleus by electromagnetic interaction. When K^- enters a target, it is slowed down by losing its kinetic energy through the interaction with the medium and is captured by an atom replacing an electron. The system is in a highly excited state given by the about thousand times higher mass of the negative meson with respect to the replaced electron. The subsequent kaonic atom cascades down to a low n-state where the strong interaction between the kaon and the nucleus adds up to the electromagnetic one. The X-ray emissions generated by the de-excitation processes provide direct information on the KN strong interaction at threshold. Therefore, kaonic atoms X-ray spectroscopy is a perfect tool to directly investigate the strong interaction in the low-energy limit, the so-called non-perturbative regime, without the need of extrapolation to zero energy as for the scattering experiments.
In this context, the X-ray spectroscopy measurements of the strong interaction induced shift and width for the 1s level of light kaonic atoms, in particular for hydrogen and deuterium, is a key tool to experimentally probe the low-energy KN interaction at vanishing energies, providing information on the relationship between spontaneous and explicit chiral symmetry breaking in low-energy Quantum ChromoDynamics (QCD). In detail, combining the most precise measurement for the kaonic hydrogen obtained by SIDDHARTA in 2009 [1] with the analogous kaonic deuterium one, it is possible to extract the K^-p and K^-d scattering lengths using the Deser–Truemann-type formulae with isospin-breaking corrections [2, 3]. The SIDDHARTA-2 collaboration has built an upgraded experimental apparatus with respect to the SIDDHARTA one, making use of new veto and Silicon Drift Detectors (SDDs) systems, with the aim to perform the unprecedented kaonic deuterium precision X-ray spectroscopy measurement. The setup is presently installed at the DAΦNE collider of Istituto Nazionale di Fisica Nucleare – Laboratori Nazionali di Frascati (INFN-LNF), ready to measure the kaonic deuterium 1s level shift (ϵ) and width (Γ), important for a deeper understanding of the non-perturbative QCD in the strangeness sector, with implications from particle and nuclear physics to astrophysics [4]-[7].

2 The SIDDHARTA-2 experiment at the DAΦNE collider

The DAΦNE electron–positron collider at INFN-LNF [8, 9] generates K^--K^+ beams through the decay of the Φ-meson, with a branching ratio of about 50%. The INFN-LNF DAΦNE Φ-factory is unique in the world, since it provides an excellent quality low-energy and monochromatic K^- beam (16 MeV of kinetic energy, 0.1% energy spread) suitable to perform high precision kaonic atoms spectroscopy measurements. This beam has been used during the successful SIDDHARTA run in 2009 [10, 11]. SIDDHARTA-2, a major upgrade of SIDDHARTA setup, is presently installed at the DAΦNE collider, ready to perform the kaonic deuterium measurement with a precision comparable to the kaonic hydrogen measurement.
Figure 1 shows a schematic view of the SIDDHARTA-2 experimental apparatus (top) and the cross section layout of the target (bottom).
The major upgrades of the SIDDHARTA-2 setup are briefly, the following ones:
Luminometer: plastic scintillators read by Photo-Multipliers (PMs) placed on the lateral sides of the IR (Interaction Region), used to evaluate the beams' quality in terms of luminosity and background [12].

Target cell: the monochromatic, low-energy charged kaons are degraded and then stopped in a cryogenic gaseous target, producing kaonic atoms. The target is a critical item, since both its geometrical structure and the density gas density (temperature and pressure) must be optimized to maximize the kaonic atom X-ray signal. The cylindrically shaped target cell, with 75 μm thick Mylar wall reinforced by high purity aluminium bars, operates at around 30 K and a working pressure for the deuterium gas of 0.4 MPa (around 3% of Liquid Hydrogen Density, LHD).
SDDs and Front End Electronic (FEE): 48 large area fast SDDs detector array (2 x 4 matrix) are arranged in a head-to-head configuration all around the target cell, covering a total active surface of 245.76 cm^2. The 450 μm thick Silicon wafer is glued on a ceramic carrier which provides the polarization to the units, each one connected to the proper preamplifier (CUBE), then the signals are processed by a dedicated FEE (SFERA, [13, 14]). The SDDs system's spectroscopic response in terms of linearity, stability, energy and timing resolutions, has been proved to be suitable for high precision kaonic atoms X-ray spectroscopy [15]-[17].
Veto-1 system: scintillators read by PMs which externally surround the SIDDHARTA-2 vacuum chamber [18]. The functioning is based on the kaon moderation time in the gas, which is longer than the corresponding one in solid elements of the setup. Additionally, it allows the rejection of the background, both synchronous and asynchronous to the beams.
Veto-2 system: scintillators read by Silicon Photo-Multipliers (SiPMs) placed behind the SDDs [19], are used to reject the hadronic background coming from border hits of Minimum Ionizing Particles (MIPs), depositing energy in the X-ray range.
Kaon trigger: scintillators used as a trigger on K^--K^+ pairs from DAΦNE, to suppress the asynchronous (i.e. electromagnetic) background applying a timing window selection in coincidence with the SDDs' X-ray signals.

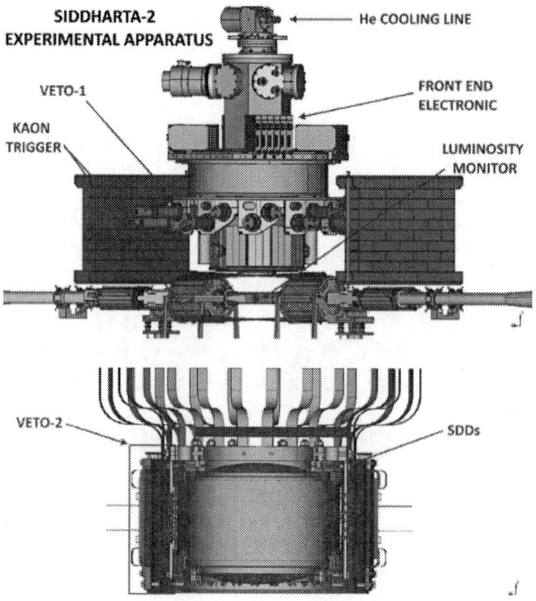

Figure 1. Top: Schematic view of the SIDDHARTA-2 experimental apparatus installed at the Interaction Region of the DAΦNE collider. Bottom: Cross section layout of the SIDDHARTA-2 target cell, red box highlights the SDDs arrays and the scintillators of the veto-2 behind them.

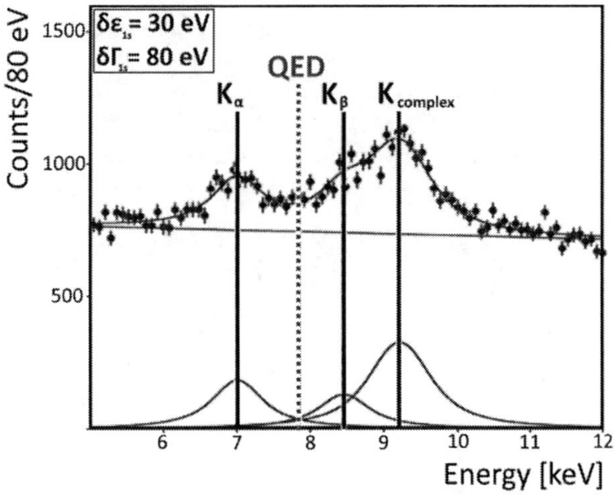

Figure 2. Simulated K⁻d Monte Carlo spectrum corresponding to an integrated luminosity of 800 pb^{-1}, assuming $\epsilon 1s$ = -800 eV, and $\Gamma 1s$ = 750 eV and a yield of 0.1%. The red-dotted line at 7834 eV corresponds to the pure QED value. The precision (δ) on the shift and the width of the $K\alpha$ peak resulting from the fit are reported on the top left inset

All the mentioned items have been included in the SIDDHARTA-2 GEANT4 simulation for 800 pb^{-1} acquired integrated luminosity, taking as input parameters: a yield of 0.1% of the 2p to 1s transition and a shift and width of -800 eV and 750 eV, respectively, from theoretical calculations [20, 21]. The fit on the Monte Carlo simulated signal shown in Figure 2 indicates that the K⁻d 1s level, modified by the strong interaction, can be determined with a precision of 30 eV for the shift ($\delta\epsilon$) and 80 eV for the width ($\delta\Gamma$). These precisions are comparable to those obtained for the kaonic hydrogen by SIDDHARTA.

3 Preliminary to SIDDHARTA-2: the SIDDHARTINO run

During the DAΦNE commissioning phase, preparatory for the kaonic deuterium data taking campaign, the SIDDHARTA-2 experimental apparatus has been installed housing 8 out of 48 SDDs arrays, in a reduced configuration called SIDDHARTINO. During this phase, concluded with the kaonic helium run in July 2021, the main SIDDHARTA-2 functionalities have been successfully tested.
The luminometer continuously monitored the beam quality for all the period, measuring the number of kaons generated in e^--e^+ collisions. Figure 3 (top) shows, as an example, the 2D-plot of the Time-to-Digital Converter (TDC) coincidence signals detected on the scintillators placed respectively on the internal ([TDC1+TDC4]/2) and external ([TDC2+TDC3]/2) side of the Interaction Region (IR). On the diagonal, the signals related to the MIPs are well separated from the kaons: the violet circles highlight the coincidence signals generated by all the particles lost from the e^- and e^+ bunches, while the red circles point out the hits associated to the kaons. The hits of the particles belonging to the bunches out from collision are, instead, distributed outside the diagonal. The projection in time of the diagonal elements allows to determine the number of kaons produced and to obtain a direct indication of the machine background.

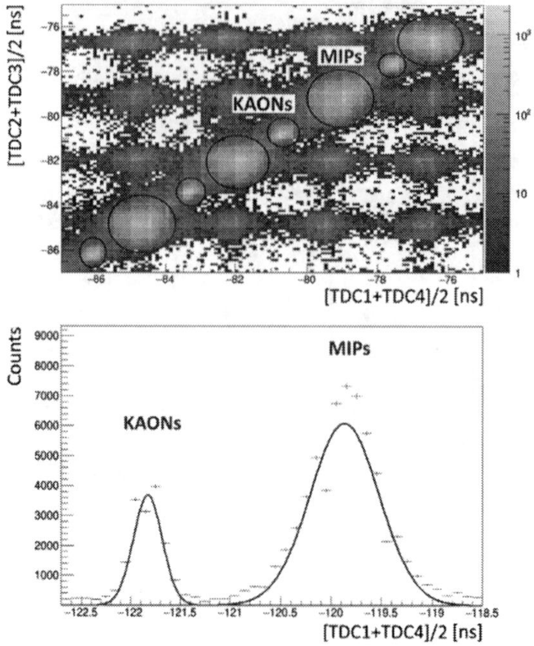

Figure 3. Top: 2D-plot of the TDCs coincidence detected on the internal ([TDC1+TDC4]/2) and external ([TDC2+TDC3]/2) side of the luminometer: red and violet circles indicate, respectively, the kaons and MIPs distributions. Bottom: Gaussian fit of the kaons and MIPs peaks on a timing selection of the 2D-plot diagonal projection.

Figure 3 (bottom) reports an example of the Gaussian fits of the kaons and MIPs signals referring to a selection of the diagonal elements of figure 3 top, used to obtain the number of kaons generated in the collision, compared to the background.
The continuous on-line feedback on the quality of the beams in terms of Kaons-over-MIPs ratio provided by the luminosity monitor, played a key role during all the DAΦNE beams optimization phase.
After having reached a satisfactory condition in terms of kaon signal over background ratio, the SIDDHARTINO data taking campaign has been concluded with the K^-He^4 measurement. Figure 4 reports, the preliminary calibrated spectrum for a subset of 6 pb^{-1} integrated luminosity, applying a $2\mu s$ timing cut on the SDDs' X-ray events in coincidence with the Kaon trigger signals to reject the asynchronous background. The calibrated spectrum clearly shows the K^-He^4 peak at around 6.4 KeV and satellite peaks given by the K^- interaction with the elements of the setup, namely the Carbon from the Kapton target walls and the Titanium placed at the top of the target cell.

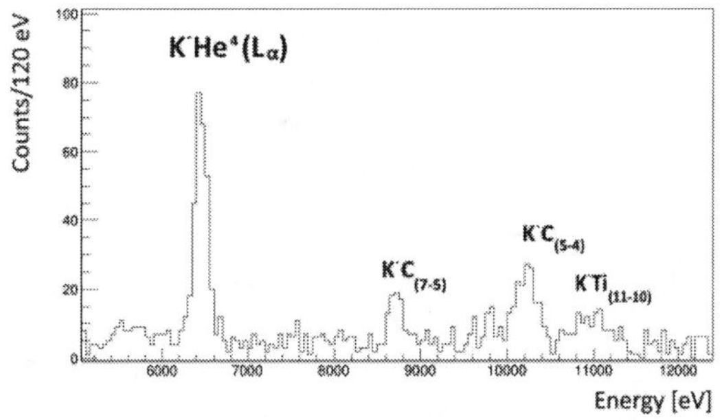

Figure 4. Preliminary calibrated SIDDHARTINO spectrum for 6 pb^{-1} integrated luminosity applying $2\mu s$ timing cut on the triggered events.

4 Conclusion

The SIDDHARTA-2 collaboration is prepared to perform the first ambitious kaonic deuterium precision X-ray spectroscopy measurement at the DAΦNE collider of LNF-INFN. In order to achieve this result, a dedicated experimental apparatus has been built and all the integrated systems have been successfully tested during the collider beam optimization phase, concluded with the measurement of the K^-He^4. The SIDDHARTA-2 kaonic deuterium data taking campaign is planned for 2021-2022. In the same time, the collaboration is developing new detector technologies and future proposals dedicated to kaonic atom measurements along the periodic table, for a deeper understanding of the QCD theory, with implications in astrophysics, particle and nuclear physics.

Funding

Part of this work was supported by the Austrian Science Fund (FWF): P24756-N20 and P33037-N; the Croatian Science Foundation under the project IP-2018-01-8570; EU STRONG-2020 project (grant agreement No. 824093), the Polish Ministry of Science and Higher Education grant No. 7150/E-338/M/2018 and the Foundational Questions Institute and Fetzer Franklin Fund, a donor advised fund of Silicon Valley Community Foundation (Grant No. FQXi-RFP-CPW-2008).

Acknowledgments

The authors acknowledge C. Capoccia from INFN-LNF and H. Schneider, L. Stohwasser, and D. Pristauz Telsnigg from Stefan-Meyer-Institut für Subatomare Physik for their fundamental contribution in designing and building the SIDDHARTA-2 setup. We thank as well the DANE staff for the excellent working conditions and permanent support.

References

[1] M. Bazzi et al., Nucl. Phys. A **881**, 88 (2012).
[2] U.G. Meißner, U. Raha, A. Rusetsky, Eur. Phys. J. C **35**, 349 (2004).
[3] U.G. Meißner, U. Raha, A. Rusetsky, Eur. Phys. J. C **47**, 473 (2006).
[4] Merafina M. et al., Phys. Rev. D **102**, 083015(1-12) (2020).

[5] Curceanu C. et al., Symmetry **12**, 1-12 (2020)
[6] Curceanu C. et al., Rev. Mod. Phys. **91**, 025006(1-36) (2019).
[7] De Pietri R. et al., Astrophys. J. **881**, 122(1-21) (2019).
[8] Milardi C. et al., IPAC-2018 **334-7** (2018).
[9] Zobov M. et al., Phys. Rev. Lett. **104**, 174801(1-5) (2010).
[10] Bazzi M. et al., Phys. Lett. B**704**, 113-117 (2011).
[11] Bazzi M. et al. , Nucl Phys. A. **907**, 69-77 (2013).
[12] Skurzok M. et al. , Journal of Instrumentation **15**, P10010(1-13) (2020).
[13] Schembari F. et al. , IEEE Trans. Nucl. Sci. **63**, 1797-1807 (2016).
[14] Quaglia R. et al. , Nucl. Instrum. Methods Phys. Res. A **824**, 449-451 (2016).
[15] Miliucci M. et al., Condens. Matter **4**, 31(1-8) (2019).
[16] Iliescu M. et al., IEEE Trans. Instrum. Meas. **70**, 9507807 (2021).
[17] Miliucci M. et al. , Meas. Sci. Techn. **32**, 095501(1-7) (2021).
[18] Bazzi M. et al., Journal of Instrumentation **8**, T11003(1-12) (2013).
[19] Tüchler M. et al. , J. Phys. Conf. Ser. **1138**, 012012(1-8) (2018).
[20] Doring M. and U. G. Meißner , Phys. Lett. B **704**, 663-666 (2011).
[21] Shevchenko N.V. , Nucl. Phys. A **890-891**, 50-62 (2017).

Maximum latent heat of neutron star matter without GR

Eva Lope-Oter[1,*]

[1]Dept. Física Teórica and IPARCOS, Univ. Complutense de Madrid

Abstract.
We show how the specific latent heat is relevant to characterize the first-order phase transitions in neutron stars. Our current knowledge of this dynamical quantity strongly depends on the uncertainty bands of Chiral Perturbation Theory and of pQCD calculations and can be used to diagnose progress on the equation of state. We state what is known to be hadron-model independent and without feedback from neutron star observations and, therefore, they can be used to test General Relativity as well as theories beyond GR, such as modified gravity.

1 Introduction

One essential point in the present discussion on strongly interacting matter at finite density is whether there may be a first order phase transition. At the present time, we do not know whether neutron-star matter undergoes such first-order phase transition to an exotic, perhaps nonhadronic phase, of great interest to nuclear and particle physics [1]. Many have been proposed, such as color-superconducting phases [2], inhomogeneous (crystalline-like) phases [3] or flavored [4] or mixed phases [5] among others; but which or any presents itself in neutron stars remains under investigation [6].

Observables including mass, radius, moment of inertia and tidal deformability are largely determined by the equation of state that abstracts microscopic properties of the phase of the material. However, if there is a first-order transition between phases of very different energy density ε then this could verifiably affect the mass-radius relation, for example. Possible first-order phase transitions would leave distinct observable traces, such as a kink in the mass-radius diagram (accessible when neutron star radii become more routinely measured.

A way to characterise the intensity of a phase transition is the latent heat that we here address.

2 Equations of state for Neutron Stars

Our approach consists in generating EoS [7] to be used to test GR as well as beyond GR, such as modified gravity. General Relativity requires further testing because Einstein's equations $G^{\mu\nu} = -8\pi G T^{\mu\nu}$ have not been exhaustively constrained at the large ε of neutron stars; prior knowledge of the EoS $P(\varepsilon)$ from nuclear and hadron physics is needed, to have $T^{\mu\nu}$ inside NS purely from hadron theory without using astrophysics observables nor GR. Our EoS sets [1] are

*e-mail: μ mariaevl@ucm.es

[1]`teorica.fis.ucm.es/nEoS`

constrained only by input from hadron physics and fundamental principles, without feedback from neutron star observations. They are obtained starting from Chiral Perturbative theories (ChPT) at low density and ending at perturbative Quantum Chromodynamics (pQCD), using first principles (thermodynamic stability and causality) alone. These EoS sets are not as constrained as others in the recent literature [8], but more reliable for testing gravity.

Basic properties of neutron stars such as the maximum allowed mass are most sensitive to the intermediate density range. Yet there, QCD is not easily tractable: we need to resort to basic theoretical properties, causality ($c_s < c$) and monotony ($cs \geq 0$). Therefore, we have developed a set of EoS reported in [7] to sample EoS at intermediate densities. We sample the uncertainty bands of Chiral Perturbation Theory (at low density), the band of perturbative QCD (at high density) and interpolate between both of them at intermediate density. First, we fix the low-density and high-density limits of the intermediate band as determined by monotony and causality, as shown in the left plot of figure 1. The low-density limits are determined by the different computations of ChPT (P_1, P_2 on that plot), while the choice of pQCD starting points (Q1 and Q2) is shown in the right plot of figure 2).

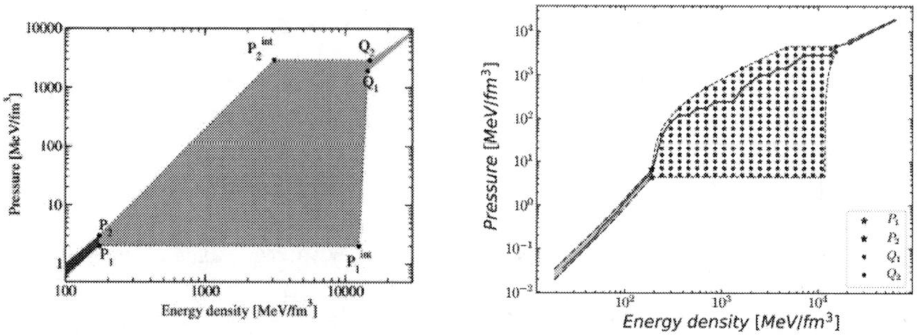

Figure 1. Left: simple sketch schematizing the construction of the boundary of the region at intermediate densities (as determined by causality and monotony) between chiral perturbation theory and high-density physics treated with pQCD (narrow bands at the left–bottom and right–top corner, respectively). **Right:** Additionally to the boundary (red dashed lines) we show the nodes (red circles) of the grid interpolating at intermediate densities between the chiral computation of [9] and the high–density physics treated with pQCD at $\mu_{\rm match} \sim 2.6$ (black dashed-dotted lines at the left–bottom and right top corner). A sample EoS is shown (solid line, blue online)).

Then, a grid is laid out in that band and for each energy density we choose a random value of the pressure from the values on the grid (right plot of figure 1). Multiple transitions are possible as in complex condensed matter systems.

3 Latent heat of first-order phase transitions

The Gibbs thermodynamic-equilibrium condition determines at what critical chemical potential μ_c will the pressure of the two phases be equal

$$T_H = T_E = 0\,, \quad \mu = \mu_c\,, \quad P_H = P_E := P_c\,. \tag{1}$$

First-order phase transitions (PT) at zero temperature in neutron stars are characterized by a jump in the energy density $\varepsilon_E - \varepsilon_H$ between the hadronic (H) and exotic (E) phases. This jump is usually quantified by the specific latent heat of the transition L; table 1 shows some specific values for some cases, from condensed matter to nuclear physics [14].

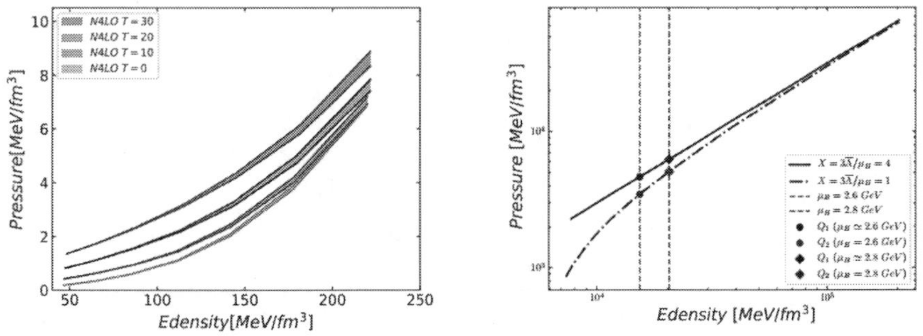

Figure 2. Asymptotic limits for the EoS of neutron stars. **Left:** neutron matter at low density, taken from the ChPT computations of the nuclear potential [10–12], with T increasing from bottom to top in 10 MeV steps. The effect of a typical temperature seems to be less than the uncertainty band. **Right:** High-density band from pQCD [13] showing Q_1 and Q_2 (for the two different sample values of μ_B). They are picked in a high density range where the P uncertainty is still narrow. Since we present $P(\varepsilon)$ and not $P(\mu_B)$, we first fix Q_2 from the higher ($X = 4$, (in the notation of [13]) line with either $\mu_B = 2.6$ or 2.8 GeV. We then use the resulting ε to obtain a slightly modified μ_B and from the pQCD pressure we obtain the low Q_1 band end with $X = 1$. Lowering the chosen values of Q_1, Q_2 would better constrain the EoS at intermediate density at a formal level, but would make the perturbativity of QCD less-reliable.

Table 1. Characteristic examples of latent heat in known phase transitions (in natural units, L being dimensionless).

Substance/transition	L
He-3 superfluid	1.5 μJ/mol = 5.5×10^{-24}
$NdCu_4Fe_4O_{12}$ perovskite	25.5 kJ kg^{-1} = 2.8×10^{-13}
Ice-water	79.7 cal/g = 3.71×10^{-12}
Nuclear evaporation	30 MeV/A = 3×10^{-2}
Neutron star matter?	> $O(0.1)$?

4 First order phase transition

The aim is to obtain L from the Equation of State (EoS) $P(\varepsilon)$ and our starting point is the latent heat per nucleon normalized to the vacuum neutron mass:

$$L := \frac{\Delta E}{N M_N}\ , \tag{2}$$

Integrating the first law of thermodynamics in terms of pressure we obtain the energy per nucleon difference $\Delta E = E_E - E_H$

$$\Delta E = P_H \frac{(n_E - n_H)}{n_E n_H} \tag{3}$$

Our approach is to use independent-model EoS by interpolating between low- and high-density regimes, but the difficulty is not knowing the energy per nucleon B/A and the density number in the area of intermediate energy from a matter Lagrangian. Therefore it has to be interpolated too, so as to quantify the latent heat. Introducing in Eq. (3) the energy density through

$$\varepsilon = n(M_N c^2 + B/A)\ , \tag{4}$$

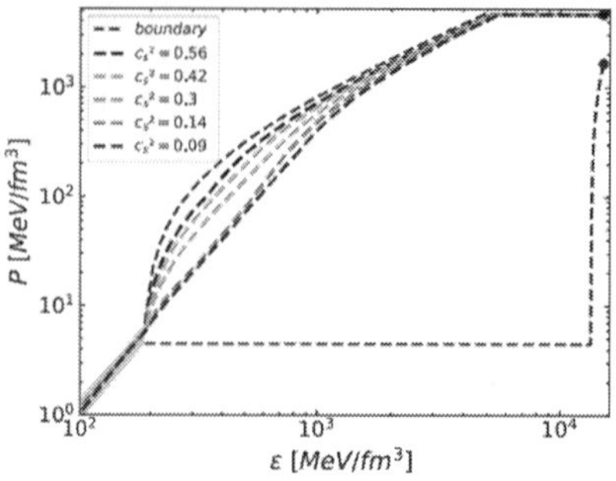

Figure 3. nEoS [7] theoretically allowed Eq. of state of neutron matter from hadron physics alone (pQCD and ChPT [9, 15]) with no astrophysics nor General Relativity input, labelled in the same vertical order by c_s^2 at the ChPT matching point. The low-density ChPT approximation is used up to $1.2n_s$, just above the nuclear saturation density ($n_s \simeq 0.16\text{fm}^{-3}$).

the latent heat ($L|_\varepsilon$) can be obtained from the EoS

$$L|_\varepsilon = P_H \frac{(\varepsilon_E - \varepsilon_H)}{\varepsilon_E \varepsilon_H} \; . \tag{5}$$

At low ε, Eq. (5) is equivalent to Eq. (2) save for the binding energy $0 \simeq (B/A) << M_N$, introducing a few percent error quantified below. Near the density allowing maximum L, Eq. (5) is a very practical definition of L.

Initially, we generated a set of EoS in order to look for the maximum latent heat in a PT, which are shown in Fig. 3.These are EoS of zero charge, β-stable neutron-star matter (NSM). At lowest number densities $n \leq 0.05n_{\text{sat}}$ (with saturation density $n_{\text{sat}} \simeq 0.16/\text{fm}^3$) nuclear data directly constrains the crustal EoS [16, 17]. (Depending on the order of perturbation theory, this corresponds to $B/A \simeq 16 \pm 1$ MeV added to the nucleon mass, and corresponds to a pressure around 153 MeV/fm^3).Shown here are the N^3LO bands of [9, 18] up to $n_m = 1.2n_{\text{sat}}$. The highest densities, at baryon chemical potential $\mu \geq 2.6\ GeV$ and above (this corresponds to an energy density of about 15 GeV/fm^3), can be studied with pQCD [15]. It is likely that such high densities cannot be achieved in neutron stars within GR, but these limits provide a powerful constraint on the shape to be adopted by the equation of state.

The intermediate ε region is enclosed in the figure by a soft-gray dotted line (occupying most of it in the log-scale employed). It is obtained by enforcing that the derivative of the EoS curves satisfy $0 \leq P'(\varepsilon) := c_s^2 \leq 1 = c^2$ (that is, respecting monotony and causality). Thus, the speed of sound along the dotted boundary is either 0 or 1.

We sample the contained region with a 500^2 grid. Any $P(\varepsilon)$ in this region is matched to ChPT at a number density n_m, with random slope $P' = c_{s,m}^2 \in (0, 1)$ (given in the figure legends). At each successive grid point, another random-slope step within those limits is taken.

For each one of this EoS, we computed the values for the specific latent heat L, starting by the simpler $L|_\varepsilon$. We proceeded by following each EoS that is compatible with all theoretical requirements (monotony and causality at every point, and satisfaction of both ChPT and pQCD constraints in their domains of validity), one at a time, from lower to higher ε.

At each grid point we computed the maximum stretch of $P' = 0$ (first order phase transition) that could take place without violating any of those requirements, that is, we momentarily assume that very point to be the lower end of a phase transition, (ε_H, P_H).

The largest maximum latent heat found for these conditions and with the current QCD understanding, $L|_\varepsilon \simeq 0.79$, would be reached for $\varepsilon \simeq 11\varepsilon_{\rm sat}$, (for an EoS matched to nuclear matter with maximum slope $c_s^2 \simeq 1$ at 1.2 $n_{\rm sat}$.

Assuming a small exotic matter core in the NS, this energy density discontinuity $\Delta\varepsilon$ has been quantified in terms of Seidov's limit [19]

$$\Delta\varepsilon := \varepsilon_E - \varepsilon_H = \varepsilon_H \left(\frac{1}{2} + \frac{3}{2}\frac{P_H}{\varepsilon_H}\right) \tag{6}$$

which is based on how much energy density a neutron star core of the exotic phase transition can take before collapsing to a black hole (in the perturbative, small-core approximation). Thus, it is a property that depends on General Relativity. A comparison with our purely hadronic latent heat is adequate, but Seidov's should not be used with theories other than General Relativity.

To compare with the latent heat that we are discussing, we define an analogous quantity that we dub "Seidov's Latent Heat" ($L|_\varepsilon$) in the same way as in Eq (3) taking as $\Delta\varepsilon$ the value obtained from Seidov´s limit. Figure 4 shows results obtained for both latent heats L and (L_S).

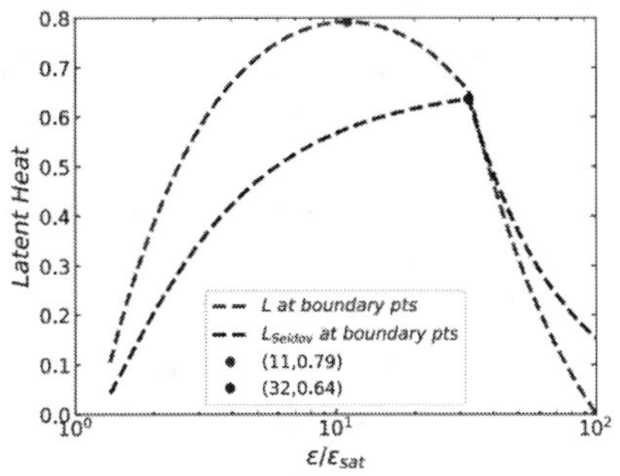

Figure 4. We compare the bound on the latent heat (upper line, red online) from microscopic hadron physics alone with the maximum latent heat that a static body in equilibrium can tolerate in General Relativity, the Seidov limit [19] (bottom line, blue online).

In our previous paper on this topic, the set of EoS was constructed to find the maximum latent heat in a PT, so that the stiffest equations possible have been generated by increasing smoothly slope up to $c_s^2 = 1$ and then flattening, up to the maximum energy density allowed by the causality band limit in figure 3, as if a PT happened. Subsequently, a new work [20] has shown that new constraints must be taken into account, regarding the density number n and the chemical potential μ. These constraints derive from causality in the (n, μ) plane instead of the (ε, P) that contains the EoS, and impose a condition on the first derivative of the number density

$$c_s^{-2} \geq \frac{\mu}{n}\frac{\partial n}{\partial \mu} \geq 1\,. \tag{7}$$

Consequently, we have to construct the allowed EoS by considering all possible functions $n(\mu)$, allowed by the above assumptions, connecting the low- and high-density limits related to the number density n (or μ). Taking into account these new constraints, the smaller boundary region obtained is shown in figure 5 and is an original contribution of this proceedings paper. Here, the new EoS are constructed following the causality constraints in both the (ε, P)

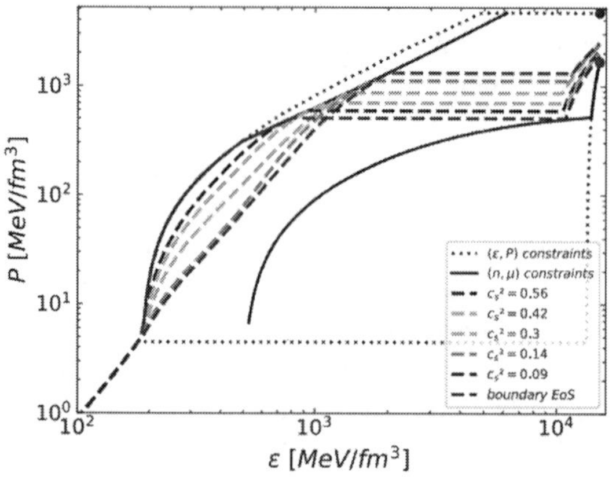

Figure 5. nEoS [7] theoretically allowed Eq. of state band of neutron matter from hadron physics alone (pQCD and ChPT [9, 15]) with no astrophysics nor General Relativity input, labelled in the same vertical order by c_s^2 at the ChPT matching point, taking into account new constraints from [20].

and (μ, n) planes simultaneously, as well as the limits of P and μ at the starting pQCD regime. We have used the same criteria to construct all these set of EoS, selecting for each density energy point of the grid the maximum pressure value allowed by the constraints to get the point where a PT happens, then flattening to the maximum energy density allowed. From this ending PT point to the starting pQCD regime, we construct the EoS applying the conform limit $c_s^2 \simeq 1/3$.

The maximum L resulting for the longest phase transition corresponding to the stiffest EoS in the entire construction (soft-grey dashed line) is $L = 0.88$. We compare in Figure 6 these results with the values obtained applying the Seidov's latent heat for the same EoS. There is an appreciable difference between both values, indicating either that both pQCD and ChPT calculations have to be improved, or that this relativistic limit should be extended to not so small nuclei of neutron stars.

Finally, in Figure 7 we compare the two different definitions L and $L|_\varepsilon$ in Eq. (2) and Eq. (3) for some of these EoS. Around $\varepsilon = 4.2\varepsilon_s$ our computations for the softest EoS differ between 4% ($c_s^2 = 0.09$) and 12% ($c_s^2 = 0.3$). For the stiffest EoS between ($c_s^2 = 1$ and $c_s^2 = 0.42$), the same variation of values happens at $\varepsilon = 2.6\varepsilon_s$. In the nonrelativistic limit they of course coincide.

5 Conclusion

Specific latent heat can be a useful quantity to characterize a phase transition in neutron stars. The results in this work show the sensitivity of this parameter to the construction of each EoS, which strongly depends on the choice of matching points at low and high density, due to the causality bounds at each point, as well as to the limitation of the maximum and minimum values (ε, P, μ, n) of the entry points to the pQCD regime. The maximum latent heat obtained with this set of EoS is around 0.9 (in natural units). In a near future, as ChPT and pQCD calculations improve, these results will be better constrained, by quantifying the maximum possible phase transition allowed by the uncertainty bands at low and high density. In fact, we have already updated our earlier results with some of the latest information from Kurkela

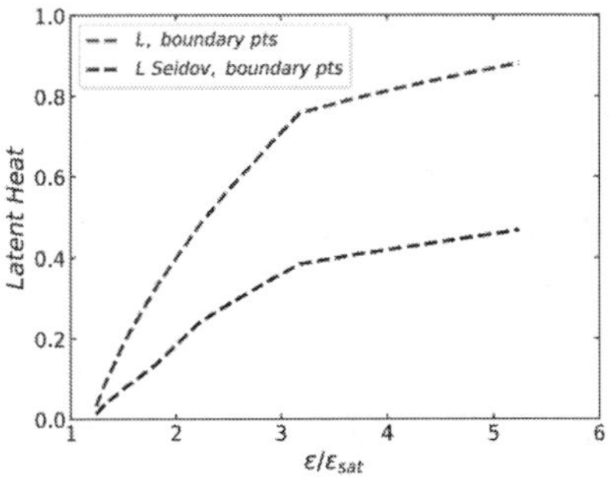

Figure 6. We compare the bound on the latent heat (upper line, red online) from microscopic hadron physics alone (see Figure 5) with the maximum latent heat that a static body in equilibrium can tolerate in General Relativity, the Seidov limit [19] (bottom line, blue online) taking into account new constraints from the (n, μ) plane [20]. While Seidov's bound is better, it is a small-core approximation valid in General Relativity only; our less tight bound is of wider generality.

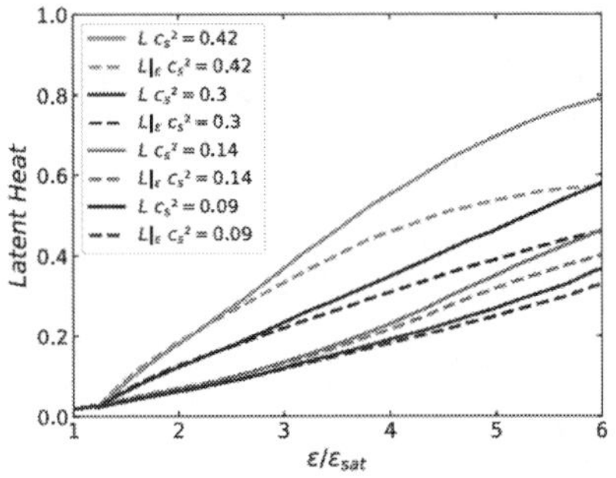

Figure 7. We compare the two definitions L and $L|_\varepsilon$ of Eq. (2) and Eq. (3) respectively for the EoS with sound speeds $(0.42 - 0.09)$ at the matching point. Below three times the energy density corresponding to nuclear saturation, one is in a nonrelativistic regime where the precise choice does not make a large difference. Above that, they start to diverge. If information from the (n, μ) plane is not at hand, that is, from $T^{\mu\nu}$ alone, our new definition in Eq. (3) is more adequate.

and Kolmotsev. Hadron physics predicts a maximum latent heat to any conceivable phase transition, even in theories that modify General Relativity.

Acknowledgment

I am thankful to Felipe J.Llanes-Estrada for assistance with the manuscript. We have received support from spanish grants MICINN: PID2019-108655GB-I00, PID2019-106080GB-C21 (Spain); the COST action CA16214 (Multimessenger Physics and Astrophysics of Neutron Stars); Univ. Complutense de Madrid under research group 910309 and IPARCOS.

References

[1] P. M. Chesler, N. Jokela, A. Loeb and A. Vuorinen, Phys. Rev. D **100**, 066027 (2019) doi:10.1103/PhysRevD.100.066027

[2] M. G. Alford, A. Schmitt, K. Rajagopal and T. Schäfer, Rev. Mod. Phys. **80**, 1455-1515 (2008) doi:10.1103/RevModPhys.80.1455.

[3] P. Fulde and R. A. Ferrell, Phys. Rev. **135**, A550-A563 (1964) doi:10.1103/PhysRev.135.A550; F. J. Llanes-Estrada and G. M. Navarro, Mod. Phys. Lett. A **27**, 1250033 (2012) doi:10.1142/S0217732312500332 M. Hoffberg, A. E. Glassgold, R. W. Richardson and M. Ruderman, Phys. Rev. Lett. **24**, 775 (1970) doi:10.1103/PhysRevLett.24.775

[4] M. Oertel, F. Gulminelli, C. Providência and A. R. Raduta, Eur. Phys. J. A **52**, 50 (2016) doi:10.1140/epja/i2016-16050-1

[5] H. Heiselberg, Contribution to the International Symposium on Strangeness and Quark Matter, 298-307 [arXiv:hep-ph/9501359 [hep-ph]].

[6] F. J. Llanes-Estrada and E. Lope-Oter, Prog. Part. Nucl. Phys. **109**, 103715 (2019) doi:10.1016/j.ppnp.2019.103715

[7] E. L. Oter, A. Windisch, F. J. Llanes-Estrada and M. Alford, J. Phys. G **46**, 084001 (2019) doi:10.1088/1361-6471/ab2567

[8] D. A. Godzieba, D. Radice and S. Bernuzzi, Astrophys. J. **908**, 122 (2021) doi:10.3847/1538-4357/abd4dd

[9] C. Drischler, A. Carbone, K. Hebeler and A. Schwenk, Phys. Rev. C **94**, 054307 (2016) doi:10.1103/PhysRevC.94.054307

[10] F. Sammarruca, R. Machleidt and R. Millerson, Mod. Phys. Lett. A **35**, 2050156 (2020) doi:10.1142/S0217732320501564

[11] F. Sammarruca and R. Millerson, Phys. Rev. C **104**, 034308 (2021) doi:10.1103/PhysRevC.104.034308

[12] F. Sammarruca, L. Coraggio, J. W. Holt, N. Itaco, R. Machleidt and L. Marcucci, PoS **CD15**, 026 (2016) doi:10.22323/1.253.0026

[13] A. Kurkela, E. S. Fraga, J. Schaffner-Bielich and A. Vuorinen, Astrophys. J. **789**, 127 (2014) doi:10.1088/0004-637X/789/2/127

[14] A. Carbone, A. Polls, A. Rios and I. Vidana, Phys. Rev. C **83**, 024308 (2011) doi:10.1103/PhysRevC.83.024308

[15] A. Kurkela, P. Romatschke and A. Vuorinen, Phys. Rev. D **81**, 105021 (2010) doi:10.1103/PhysRevD.81.105021.

[16] J. W. Negele and D. Vautherin, Nucl. Phys. A **207**, 298-320 (1973) doi:10.1016/0375-9474(73)90349-7

[17] G. Baym, C. Pethick and P. Sutherland, Astrophys. J. **170**, 299-317 (1971) doi:10.1086/151216

[18] C. Drischler, J. A. Melendez, R. J. Furnstahl and D. R. Phillips, Phys. Rev. C **102** 054315,(2020) doi :10.1103/PhysRevC.102.054315

[19] Z. Seidov, Sov. Astron, 15 (347) (1971).

[20] O. Komoltsev and A. Kurkela, [arXiv:2111.05350 [nucl-th]].

Confining density functional approach for color superconducting quark matter and mesonic correlations

Oleksii Ivanytskyi[1,*], *David* Blaschke[1,2,3,**], and *Konstantin* Maslov[2,***]

[1]Institute of Theoretical Physics, University of Wroclaw, Max Born Pl. 9, 50-204 Wroclaw, Poland
[2]Bogoliubov Laboratory of Theoretical Physics, JINR Dubna, Joliot-Curie Str. 6, 141980 Dubna, Russia
[3]National Research Nuclear University (MEPhI), Kashirskoe Shosse 31, 115409 Moscow, Russia

Abstract. We present a novel relativistic density-functional approach to modeling quark matter with a mechanism to mimic confinement. The quasiparticle treatment of quarks provides their suppression due to large quark selfenergy already at the mean-field level. We demonstrate that our approach is equivalent to a chiral quark model with medium-dependent couplings. The dynamical restoration of the chiral symmetry is ensured by construction of the density functional. Beyond the mean field, quark correlations in the pseudoscalar channel are described within the Gaussian approximation. This explicitly introduces pionic states into the model. Their contribution to the thermodynamic potential is analyzed within the Beth–Uhlenbeck framework. The modification of the meson mass spectrum in the vicinity of thee (de)confinement transition is interpreted as the Mott transition. Supplemented with the vector repulsion and diquark pairing the model is applied to construct a hybrid quark-hadron EoS of cold compact-star matter. We study the connection of such a hybrid EoS with the stellar mass-radius relation and tidal deformability. The model results are compared to various observational constraints including the NICER radius measurement of PSR J0740+6620 and the tidal deformability constraint from GW170817. The model is shown to be consistent with the constraints, still allowing for further improvement by adjusting the vector repulsion and diquark pairing couplings.

1 Introduction

A principal element of unified description of strongly interacting matter within effective theories is the hadronization of chiral quark models and the incorporation of a confinement mechanism into them, manifesting the switching between hadronic and quark degrees of freedom. The confining aspect of quark matter can be modeled by Cornell-type potentials. Within this picture deconfinement of colored degrees of freedom is associated with a decrease of the tension of the string connecting colored constituents and causing a correlation between them. Within the string-flip model (SFM), the energy of such a correlation is related to the string length distribution [1, 2]. The average string length is related to the average separation

*e-mail: oleksii.ivanytskyi@uwr.edu.pl
**e-mail: david.blaschke@uwr.edu.pl
***e-mail: maslov@theor.jinr.ru

between quarks. This separation is connected to the quark number density which, in turn, controls the quark mean-field self-energy. This idea was utilized in Ref. [3] in order to develop a density functional approach incorporating the confining aspect of quark matter via the divergence of the quark self-energy at vanishing number density of quarks.

In this work we do the next step and formulate the SFM-inspired density functional approach in a chirally symmetric way. This leads to the appearance of both scalar *and* pseudoscalar modes in the model. The pseudoscalar mode is most important for the phenomenology of strongly interacting matter in the confined phase since it represents the Goldstone bosons of spontaneously broken chiral symmetry. We treat this pseudoscalar mode beyond the mean-field approach within the Gaussian approximation. In the two-flavor case this introduces the pion degrees of freedom and in combination with the proposed phenomenological confinement mechanism of quarks in the chirally broken phase it provides a quark-hadron duality within our approach.

Another important and new aspect of this study is the introduction of the diquark pairing channels into the density functional approach. At sufficiently high densities and low temperatures the interaction in these channels causes the formation of a two-flavour color-superconducting phase of quark matter (2SC). We study the thermodynamics this 2SC phase of cold quark matter and derive its equation of state (EoS) within the mean-field approximation.

The recent analysis of observational data on the pulsar PSR J0740+6620 [4, 5] and the binary neutron star (NS) merger GW170817 [6] established challenging constraints on the mass-radius diagram of NS. Fulfilling these constraints requires the EoS of stellar matter to exhibit a significant softening in the density range typical for the intermediate-mass NS with a further stiffening at larger densities typical for NS with masses about two times the solar mass. Such a behavior is very unlikely for the scenario of purely hadronic stellar matter since the corresponding EoS consistent with the two solar mass constraint on the NS maximum mass are stiff at all densities. On the other hand, a first order phase transition from hadronic to quark matter can soften the stellar matter EoS in the intermediate density region. Therefore, we apply our EoS of cold color superconducting quark matter to model NS with quark cores within the framework of a hybrid quark-hadron EoS.

In the next section, we briefly describe the density functional approach, while sections 3 and 4 are devoted to the cases of zero chemical potential and zero temperature, respectively. The conclusions are given in section 5.

2 Density functional approach

We consider the case of two quark flavors labeled by the index $f = u, d$ and represented by the color-flavor-Dirac spinor $q = (u, d)^T$. Their masses and chemical potentials enter the matrices $\hat{m} = \text{diag}(m_u, m_d)$ and $\hat{\mu} = \text{diag}(\mu_u, \mu_d)$, respectively. The latter includes $\mu_f = \frac{\mu_B}{3} + Q_f \mu_Q$ given in terms of the baryon (μ_B) and electric (μ_Q) charge chemical potentials, and the electric charge of quarks Q_f. Our model accounts for the chirally symmetric interaction in scalar and pseudoscalar channels represented by the density functional $\mathcal{U}$ discussed below, as well as the vector and diquark pairing channels important for the astrophysical applications [7]. The Lagrangian is

$$\mathcal{L} = \overline{q}(i\partial\!\!\!/ - \hat{m})q - \mathcal{U} - G_V(\overline{q}\gamma_\mu q)^2 + G_D(\overline{q}i\gamma_5\tau_2\lambda_A q^c)(\overline{q}^c i\gamma_5\tau_2\lambda_A q). \quad (1)$$

Here G_V and G_D are couplings in vector and diquark channels, respectively, $q^c = i\gamma_2\gamma_0\overline{q}^T$ is the charge conjugate of the quark fields and the color index in the last term is implicitly summed over $A = 2, 5, 7$. Chiral symmetry of $\mathcal{U}$ is provided by its dependence on the argument $(\overline{q}q)^2 + (\overline{q}i\gamma_5\vec{\tau}q)^2$. Following Ref. [3] we expand this potential around the mean field expectation values $\langle \bar{q}q \rangle$ and $\langle \bar{q}i\gamma_5\vec{\tau}q \rangle = 0$. Below all the quantities evaluated at these expectation values are denoted by the subscript index "MF". In this study we use the expansion up to the second order terms, which accounts for the one-loop mesonic correlations of quarks, while the expansion up to the first order terms represents the mean-field approximation. Thus,

$$\mathcal{U}^{(2)} = \mathcal{U}_{MF} + (\bar{q}q - \langle \bar{q}q \rangle)\,\Sigma_{MF} - G_S(\bar{q}q - \langle \bar{q}q \rangle)^2 - G_{PS}(\bar{q}i\gamma_5\vec{\tau}q)^2. \tag{2}$$

Here the mean-field part of the quark self-energy Σ_{MF} and the medium-dependent effective couplings in the scalar G_S and pseudoscalar G_{PS} channels are defined as

$$\Sigma_{MF} = \frac{\partial \mathcal{U}_{MF}}{\partial \langle \bar{q}q \rangle}, \quad G_S = -\frac{1}{2}\frac{\partial^2 \mathcal{U}_{MF}}{\partial \langle \bar{q}q \rangle^2}, \quad G_{PS} = -\frac{1}{6}\frac{\partial^2 \mathcal{U}_{MF}}{\partial \langle \bar{q}i\gamma_5\vec{\tau}q \rangle^2}. \tag{3}$$

The Lagrangian with the second-order-expanded density functional $\mathcal{L}^{(2)} = \mathcal{L}_{\mathcal{U}=0} - \mathcal{U}^{(2)}$ is quadratic in all quark bilinears. This is equivalent to current-current quark interactions representing a chiral model of the NJL type [7–14] but with medium dependent couplings.

The next step is to bosonize the present model by means of the Hubbard-Stratonovich transformation (see, e.g., Ref. [8] for details). This introduces collective fields σ, $\vec{\pi}$, ω_μ and Δ_A, representing scalar, pseudoscalar, vector and diquark modes coupled to $\bar{q}q - \langle \bar{q}q \rangle$, $\bar{q}i\gamma_5\vec{\tau}q$, $\bar{q}\gamma_\mu q$ and $\bar{q}i\gamma_5\tau_2\lambda_A q$, respectively. They enter the bosonized Lagrangian as

$$\mathcal{L}^{bos} + q^+\hat{\mu}q = \bar{Q}\mathcal{S}^{-1}Q - \mathcal{U}_{MF} + \langle \bar{q}q \rangle\Sigma_{MF} + \sigma\langle \bar{q}q \rangle - \frac{\sigma^2}{4G_S} - \frac{\vec{\pi}^2}{4G_{PS}} + \frac{\omega_\mu\omega^\mu}{4G_V} - \frac{\Delta_A^*\Delta_A}{4G_D}. \tag{4}$$

It is written in terms of the Nambu-Gorkov bispinor $Q^T = \frac{1}{\sqrt{2}}(q\ q^c)$ and the propagator

$$\mathcal{S}^{-1} = \begin{pmatrix} S_+^{-1} - \sigma - i\gamma_5\vec{\tau}\cdot\vec{\pi} & i\Delta_A\gamma_5\tau_2\lambda_A \\ i\Delta_A^*\gamma_5\tau_2\lambda_A & S_-^{-1} - \sigma - i\gamma_5\vec{\tau}^T\cdot\vec{\pi} \end{pmatrix} \tag{5}$$

with $S_\pm^{-1} = i\partial\!\!\!/ \pm \omega\!\!\!/ - m^* \pm \gamma_0\hat{\mu}$ and the effective quark mass $m_f^* = m_f + \Sigma_{MF}$. The quark fields enter Eq. (5) quadratically and therefore can be integrated out leading to the effective mesonic Lagrangian

$$\mathcal{L}^{eff} = \frac{\mathrm{Tr}\ln(\beta\mathcal{S}^{-1})}{2\beta V} - \mathcal{U}_{MF} + \langle \bar{q}q \rangle\Sigma_{MF} + \sigma\langle \bar{q}q \rangle + \frac{\sigma^2}{4G_S} + \frac{\vec{\pi}^2}{4G_{PS}} - \frac{\omega_\mu\omega^\mu}{4G_V} + \frac{\Delta_A^*\Delta_A}{4G_D}, \tag{6}$$

where $\beta = \frac{1}{T}$ is the inverse temperature and V is the system volume.

Within our approach the quark confinement is mimicked by a rapid growth of quark self-energy Σ in the confining region. In agreement with the SFM argument, we assume Σ to be inversely proportional to the mean separation between quarks [1, 2], i.e. $\Sigma \sim (q^+q)^{-\frac{1}{3}}$ or, equivalently, $\mathcal{U} \sim (q^+q)^{\frac{2}{3}}$. The proportionality coefficients in these relations are related to the tension of the confining string. Using $\langle \bar{q}i\gamma_5\vec{\tau}q \rangle = 0$ and the approximate relation $\langle \bar{q}q \rangle \simeq \langle \bar{q}q \rangle_0 + \langle q^+q \rangle$ valid at low T and μ, we parametrize the chirally symmetric interaction potential as

$$\mathcal{U} = D_0\left[(1+\alpha)\langle \bar{q}q \rangle_0^2 - (\bar{q}q)^2 - (\bar{q}i\gamma_5\vec{\tau}q)^2\right]^{\frac{1}{3}}, \tag{7}$$

where $\langle \bar{q}q \rangle_0$ is the vacuum value of the chiral condensate and the coupling constant D_0 has the meaning of the confining string tension. The constant parameter α controls the behavior of the effective quark mass. The mean-field self-energy Σ_{MF} vanishes for $\alpha \to \infty$, thus making $\hat{m}^* = \hat{m}$. For $\alpha = 0$ this self-energy diverges for $\langle \bar{q}q \rangle = \langle \bar{q}q \rangle_0$ leading to an absolute "confinement" of quarks at low T and μ. Small positive values of this parameter provide sufficient suppression of quarks in the confining region. The adopted parameterization of

$\mathcal{U}$ provides a NJL-like expression for the effective mass $\hat{m}^* = \hat{m} - 2G_{PS}\langle\overline{q}q\rangle$. Moreover, replacing the SFM power "$\frac{1}{3}$" in Eq. (7) by unity we bring $\mathcal{L}^{(2)}$ to the NJL model form [9] up to an insignificant constant term. Thus, the present model can be considered as a generalization of the NJL model to the case of density dependent couplings. At the same time $G_S \neq G_{PS}$ in the general case due to the expansion of $\mathcal{U}$ around the mean-field solution, which is known to break the chiral symmetry. However, these couplings saturate to the same value $G_\infty = \frac{D_0}{3}(1+\alpha)^{-\frac{2}{3}}\langle\overline{q}q\rangle_0^{-\frac{4}{3}}$, when chiral symmetry gets dynamically restored for vanishing $\langle\overline{q}q\rangle$.

We fit the vacuum values of the chiral condensate $\langle\overline{q}q\rangle_0 = -2(251\text{ MeV})^3$, pion mass $M_\pi = 140$ MeV, pion decay constant $F_\pi = 90$ MeV, and the mean-field value of the pseudocritical temperature $T_c = 170$ MeV by choosing the current quark mass $m_u = m_d = m = 4.9$ MeV, momentum cutoff $\Lambda = 563$ MeV, $\alpha = 0.844$, and $D_0\Lambda^{-2} = 1.058$. Vector and diquark couplings, parameterized by $\eta_V \equiv \frac{G_V}{G_\infty}$ and $\eta_D \equiv \frac{G_D}{G_\infty}$, are treated as free parameters. Below the pairs of numbers corresponding to these scaled couplings are used in order to label the model parameterizations considered at finite chemical potentials. For example, (0.2, 1.5) stands for a model with $\eta_V = 0.2$ and $\eta_D = 1.5$.

3 Quark-hadron matter at zero chemical potential

In this regime vector and diquark fields have zero expectation values at the mean-field level. On the other hand, quark correlations in the pseudoscalar channel are the most important beyond mean-field. We limit the present consideration to these correlations only. For this we set $\sigma = \omega_\mu = \Delta_A = 0$. Note, that $\sigma = 0$ within the mean-field approximation since the expansion (2) is already performed around the corresponding solution. Thus, the effective Lagrangian becomes

$$\mathcal{L}^{eff} = \frac{\mathrm{Tr}\ln\left(\beta(S_0^{-1}+\Sigma_\pi)\right)}{\beta V} - \mathcal{U}_{MF} + \langle\overline{q}q\rangle\Sigma_{MF} + \frac{\vec{\pi}^2}{4G_{PS}} \tag{8}$$

with $S_0^{-1} = i\partial\!\!\!/ - m^*$ and $\Sigma_\pi = -i\gamma_5\vec{\tau}\cdot\vec{\pi}$ being the quark propagator in the quasiparticle approximation and self-energy caused by the pseudoscalar correlations. We treat these correlations within the Gaussian approximation (see e.g. Ref. [8] for details). For this we expand the trace of the logarithm in Eq. (8) up to the second order in Σ_π. The zeroth order term contributes to the mean-field thermodynamic potential

$$\Omega_{MF} = -12\int\frac{d\mathbf{k}}{(2\pi)^3}\left[\epsilon - 2T\ln(1-f)\right] + \mathcal{U}_{MF} - \langle\overline{q}q\rangle\Sigma_{MF}, \tag{9}$$

where $\epsilon = \sqrt{k^2+m^{*2}}$ and $f = [e^{\beta\epsilon}+1]^{-1}$ are the single particle energy and distribution function of quarks. The first order term vanishes due to the tracelessness of the Pauli matrices, while the quadratic term yields the one-loop polarization operator of pions with the four momentum p,

$$\Pi_\pi = -\frac{\mathrm{Tr}(i\gamma_5 S_0)^2}{\beta V} = -\frac{\langle\overline{q}q\rangle}{m^*} + 12p^2\int\frac{d\mathbf{k}}{(2\pi)^3}\frac{2f-1}{\epsilon(p^2-4\epsilon^2)}. \tag{10}$$

The poles of the propagator $D_\pi^{-1} = \frac{1}{2G_{PS}} - \Pi_\pi$ define the dispersion relation of the pseudoscalar mode, which at zero momentum gives the pion mass M_π. The present model being a generalization of the NJL one provides validity of the Gell-Mann-Oakes-Renner relation $M_\pi^2F_\pi^2 = -m\langle\overline{q}q\rangle$ with the pion decay constant

$$F_\pi^2 = 12m^*(m^*-m)\int\frac{d\mathbf{k}}{(2\pi)^3}\frac{2f-1}{\epsilon(M_\pi^2-4\epsilon^2)}. \tag{11}$$

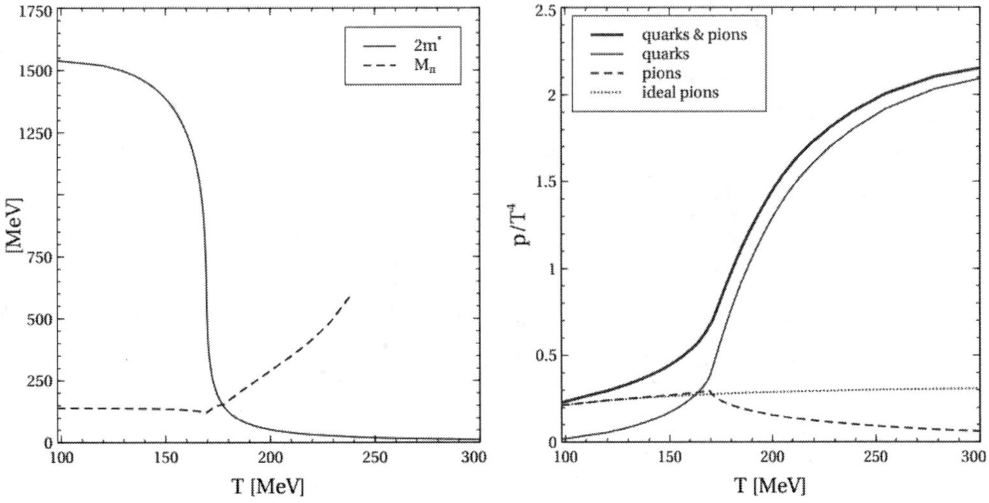

Figure 1. *Left panel:* temperature dependence of the pion mass M_π (blue dashed curve) compared to the two quark decay threshold $2m^*$ (red solid curve). *Right panel:* temperature dependence of the scaled pressure of quark-pion system (thick solid magenta curve) compared to partial contributions of quarks (thin solid red curve), pions (blue dashed curve) and noninteracting pions with mass $M_\pi = 140$ MeV (green dotted curve).

The temperature dependence of M_π is shown in the left panel of Fig. 1. At low T the pion mass remains almost constant and is significantly smaller than the two quark decay threshold $2m^*$. The increase of temperature leads to a rapid decrease of the effective quark mass around the pseudocritical temperature. Above this temperature $M_\pi > 2m^*$ and the pion is not a bound state but a resonance. Above a certain temperature the real part of D_π does not have zeros, which terminates the pion mass curve.

The polar representation of $D_\pi = |D_\pi| e^{i\delta_\pi}$ defines the phase shift of pions δ_π. It gives a direct access to the spectral function of pions $\rho_\pi = \frac{1}{\pi}\frac{\partial \delta_\pi}{\partial p_0}$. On the other hand, the field $\vec{\pi}$ enters $\mathcal{L}_{eff}$ quadratically through the term $\frac{\vec{\pi} D_\pi \vec{\pi}}{2}$. By integrating over the fields $\vec{\pi}$ we arrive to the thermodynamic potential

$$\Omega = \Omega_{MF} - \frac{3\mathrm{Tr}\ln(\beta^2 D_\pi)}{2\beta V} = \Omega_{MF} + 3T \int \frac{d\mathbf{p}}{(2\pi)^3} \int \frac{dp_0}{\pi} \ln(1 - e^{\beta p_0}) \frac{\partial \delta_\pi}{\partial p_0}, \tag{12}$$

where in the second step the pion contribution is given by the Beth-Uhlenbeck formula. The extremum of Ω with respect to $\langle \bar{q}q \rangle$ defines the chiral condensate. The system pressure can be found as $p = -\Omega + \Omega_0$, where the constant shift of energy Ω_0 is introduced in order to ensure $p = 0$ in the vacuum. The right panel of Fig. 1 shows the total pressure as a function of temperature, compared to partial contributions of quarks and pions. At low T the pressure of pions dominates the total one and is very close to the pressure of an ideal gas of particles with the mass $M_\pi = 140$ MeV. At $T \simeq T_c$ the quark pressure exhibits a fast growth while the pion pressure decreases due to the Mott dissociation. At high T quarks dominate the thermodynamics of the system.

4 Color superconducting quark matter at zero temperature

At zero temperature the beyond-mean-field bosonic correlations of quarks can be neglected. Therefore, in this section we treat the model within the mean-field approximation. In this case

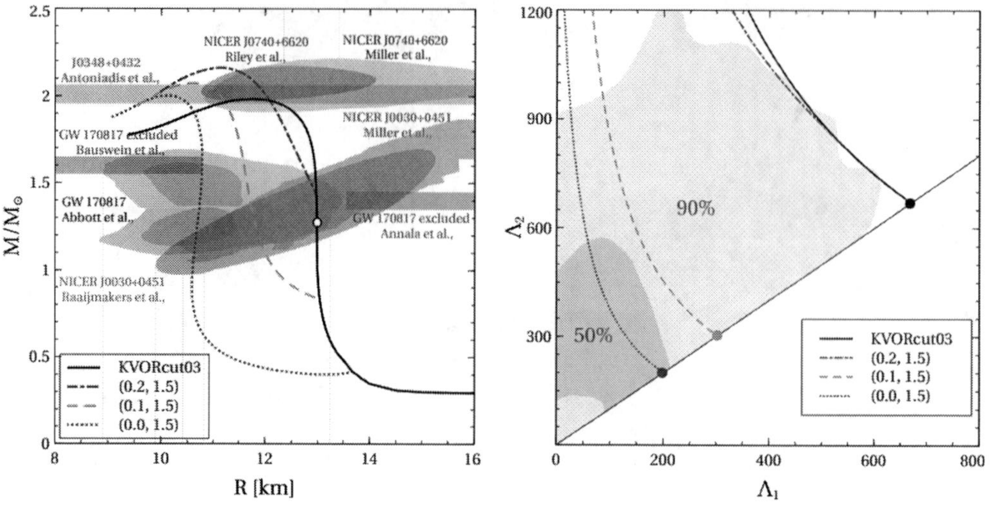

Figure 2. *Left panel:* mass-radius diagram calculated for the hybrid quark-hadron EoS discussed in the text (colored curves) and purely hadronic DD2npY - T EoS. Empty circle on the hadronic curve indicates the hyperon onset. The astrophysical constraints depicted by the colored bends and shaded areas are discussed in the text. *Right panel:* Tidal deformabilities of two components of the NS-NS merger with the chirp mass $\mathcal{M} = 1.188$ M$_\odot$. Colored filled circles represent the configurations with equal masses of two components $M_1 = M_2 = 1.3646$ M$_\odot$. Dark and light green shaded ares represent the regions falling into the 50 % and 90 % confidence levels.

all the fields introduced in Eq. (5) are set to some constant values. For scalar and pseudoscalar fields $\sigma = \vec{\pi} = 0$. By a proper Lorentz transformation only the zeroth component of the vector field attains a finite value ω, which is absorbed into the effective chemical potential of quarks $\mu_f^* = \mu_f + \omega$. Furthermore, there exists a global gauge transformation, which makes Δ_2 the only non-vanishing diquark field. Only its absolute value $\Delta = |\langle\Delta_2\rangle| = |\langle\Delta_2^*\rangle|$ enters the thermodynamic potential

$$\Omega = -\frac{\mathrm{Tr}\ln(\beta S^{-1})}{2\beta V} + \mathcal{U}_{MF} - \langle\bar{q}q\rangle\Sigma_{MF} - \frac{\omega^2}{4G_V} + \frac{\Delta^2}{4G_D}. \tag{13}$$

Solving the trace of the logarithm of the inverse Nambu-Gorkov propagator is a technically demanding but straightforward procedure exhaustively described in the literature (see e.g. Refs. [7, 8]). We omit it for brevity the description of our approach. Since working within the mean-field approximation, we require chiral condensate, vector and diquark fields to minimize Ω. This gives us a set of three coupled equations to be solved with respect to $\langle\bar{q}q\rangle$, ω and Δ. Having the solutions found, we obtain pressure, baryonic charge and energy density using standard thermodynamic identities as $p = -\Omega + \Omega_0$, $n_B = \frac{\partial p}{\partial \mu_B}$ and $\varepsilon = \mu_B n_B - p$, respectively.

The equation for the diquark pairing gap has two solutions. The trivial one $\Delta = 0$ exists at any value of the chemical potential, while $\Delta \neq 0$ appears at the 2SC phase onset. It occurs when two solutions coincide. The condition $\frac{\partial^2\Omega}{\partial\Delta^2}\Big|_{\Delta=0} = 0$ should be solved with respect to the baryonic chemical potential in order to find its critical value μ_B^c corresponding to the 2SC phase onset. Below μ_B^c quark matter exists in a normal phase, while above it color superconductivity is energetically favorable. This critical value decreases with growing η_D since a stronger diquark pairing leads to earlier onset of the 2SC phase. The requirement of the absence of color superconductivity in the vacuum sets an upper limit constraint on the diquark coupling $\eta_D < \eta_D^*$. This η_D^* can be found from the condition $\frac{\partial^2 \Omega}{\partial \Delta^2}\Big|_{\Delta=\mu_B=0} = 0$ combined with the vacuum mas gap equation. Thus

$$\eta_D^* = \frac{3}{2} \frac{m^*}{m^* - m} \frac{G_{PS}}{G_\infty} = 2.54, \tag{14}$$

which is about 70 % larger than the corresponding NJL value [14]. This difference is caused by the fact that η_D^* is obtained from the analysis of the vacuum state where $G_{PS} \simeq 2G_\infty$. Therefore, $G_D = \eta_D^* G_\infty$ is about twice larger compared to the maximum value allowed by the NJL model. This makes the present approach applicable to describing the regime of strong diquark couplings.

We construct the hybrid EoS of NS matter by imposing the conditions of β-equilibrium and electric neutrality [7] as well as joining our EoS for the quark matter phase to the one for the hadronic phase by means of the Maxwell construction. The hadronic phase is described by the KVORcut03 model [15], which is a relativistic density functional model that includes hyperons and has a medium dependent ratio of masses and couplings being a function of the scalar density. Solving the TOV equations we construct the mass-radius diagram of NS with quark cores shown on the left panel of Fig. 2. For the chosen sets of the vector and diquark couplings the model is able to fit the constraints coming from the observation of PSR J0348+0432 [16], PSR J0740+6620 [4, 5] and PSR J0030+0451 [17, 18] pulsars as well as the gravitational wave signal from the merger GW170817 [19, 20]. The model also fits the region of the 2σ confidence level coming from the analysis of the same gravitational wave signal [6]. Getting to the 1σ region requires an adjustment of η_D, which is beyond the scope of the present contribution.

We also compare the results of our model for the tidal deformability Λ of NS with the constraints for it extracted from the gravitational wave signal of the inspiral phase of the binary neutron star merger event GW170817 [6]. For this we calculate Λ_1 and Λ_2 of the two components of the binary system with masses $M_1 > M_2$ providing the chirp mass [21] $\mathcal{M} = 1.188\ M_\odot$. The sets of η_V and η_D consistent with the discussed mass-radius constraints get only to the 90 % confidence level interval of the constraint on Λ_1 and Λ_2. The agreement, however, can be improved by simultaneous adjustment of the vector and diquark couplings.

5 Conclusions

We have studied a relativistic density functional approach to quark matter, which i) mimics the quark confinement by a rapid growth of the quark self-energy in the confining region, ii) respects chiral symmetry of strong interaction and iii) can be interpreted as a chiral quark model with the density dependent coupling constants. In addition to the vector repulsion channel, we have introduced the diquark pairing not studied before within the density functional approach.

At zero baryon chemical potential we introduce mesonic correlations beyond the mean-field. For this the most important pseudoscalar mode of quark correlations identified with the pion excitations is considered within the Gaussian approximation. The corresponding pion contribution to the thermodynamic potential is analyzed within the Beth-Uhlenbeck formalism. Since the Mott dissociation of pions occurs at roughly the same temperature where the effective quark mass experiences a rapid decrease, the present approach exhibits a switching between the hadron and quark degrees of freedom with increase of the temperature.

In order to model NS with quark cores we apply the present density functional approach to construct a hybrid EoS of hadronic and color superconducting quark matter with the vector repulsion. The results of this modeling are confronted to various constraints of the NS mass-radius relation and tidal deformability. The approach provides a reasonable agreement with these constraints, however, having a potential for further improvement by adjustment of the values of the vector and diquark couplings.

Acknowledgements

This work has been supported by the Polish National Science Centre (NCN) under grant No. 2019/33/B/ST9/03059. It was performed within a project that has received funding from the European Union's Horizon 2020 research and innovation program under grant agreement STRONG – 2020 - No 824093. K. M. acknowledges support by the Foundation for the Advancement of Theoretical Physics and Mathematics "BASIS", project No. 17-15-568-1. D.B. was supported by the Russian Foundation for Basic Research under grant No. 18-02-40137 and by the Federal Program "Priority-2030".

References

[1] C. J. Horowitz, E. J. Moniz, and J. W. Negele, Phys. Rev. D **31**, 1689 (1985).
[2] G. Röpke, D. Blaschke, and H. Schulz, Phys. Rev. D **34**, 3499 (1986).
[3] M. A. R. Kaltenborn, N.-U. F. Bastian, and D. B. Blaschke, Phys. Rev. D **96**, 056024 (2017).
[4] T. E. Riley et al., Astrophys. J. Lett. **918**, L27 (2021).
[5] M. C. Miller et al., Astrophys. J. Lett. **918**, L28 (2021).
[6] P. B. Abbott et al., Phys. Rev. Lett. **121**, 161101 (2018).
[7] G. Baym, T. Hatsuda, T. Kojo, P. D. Powell, Y. Song and T. Takatsuka, Rept. Prog. Phys. **81**, 056902 (2018).
[8] D. Blaschke, M. Buballa, A. Dubinin, G. Röpke, and D. Zablocki, Annals Phys. **348**, 228 (2014).
[9] S. P. Klevansky, Rev. Mod. Phys. **64**, 649 (1992).
[10] J. Hüfner, S.P. Klevansky, P. Zhuang, and H. Voss, Annals Phys. **234**, 225 (1994)
[11] P. Zhuang, J. Hüfner, and S.P. Klevansky, Nucl. Phys. A **576**, 525 (1994).
[12] M. Buballa, Phys. Rep. **407**, 205 (2005).
[13] C. Ratti, M. A. Thaler and W. Weise, Phys. Rev. D **73**, 014019 (2006).
[14] D. S. Zablocki, D. B. Blaschke, R. Anglani, and Yu. L. Kalinovsky, Acta Phys. Polon. Supp. **3**, 771 (2010).
[15] K. A. Maslov, E. E. Kolomeitsev and D. N. Voskresensky, Nucl. Phys. A **950**, 64 (2016).
[16] J. Antoniadis et al., Science, **340**, 6131 (2013).
[17] G. Raaijmakers et al., Astrophys. J. Lett. **887**, L22 (2019).
[18] M. C. Miller et al., Astrophys. J. Lett. **887**, L24 (2019).
[19] A. Bauswein, O. Just, H. T. Janka, and N. Stergioulas, Astrophys. J. Lett. **850**, L34 (2017).
[20] E. Annala, T. Gorda, A. Kurkela, and A. Vuorinen, Phys. Rev. Lett. **120**, 172703 (2018).
[21] P. C. Peters and J. Mathews, Phys. Rev. **131**, 435 (1963).

EPJ Web of Conferences **258**, 07009 (2022) https://doi.org/10.1051/epjconf/202225807009
A Virtual Tribute to Quark Confinement and the Hadron Spectrum (vConf21)

Locating the special point of hybrid neutron stars

Mateusz Cierniak[1,*] and *David* Blaschke[1,2,3,**]

[1]University of Wroclaw, 50-137 Wroclaw, Poland
[2]Bogoliubov Laboratory of Theoretical Physics, JINR Dubna, 141980 Dubna, Russia
[3]National Research Nuclear University (MEPhI), 115409 Moscow, Russia

Abstract. The special point is a feature unique to models of hybrid neutron stars. It represents a location on their mass–radius sequences that is insensitive to the phase transition density. We consider hybrid neutron stars with a core of deconfined quark matter that obeys a constant–sound–speed (CSS) equation of state model and provide a fit formula for the coordinates of the special point as functions of the squared sound speed (c_s^2) and pressure scale (A) parameters. Using the special point mass as a proxy for the maximum mass of the hybrid stars we derive limits for the CSS model parameters based on the recent NICER constraint on mass and radius of pulsar PSR J0740+6620, $0.36 < c^2_{s\,\mathrm{min}} < 0.43$ and $80 < A[\mathrm{MeV/fm^3}] < 160$. The upper limit for the maximum mass of hybrid stars depends on the upper limit for c_s^2 so that choosing $c^2_{s,max} = 0.6$ results in $M_{\mathrm{max}} < 2.7\ M_\odot$, within the mass range of GW190814.

1 Introduction

The recent observation of gravitational waves originating from the inspiral of a binary system of neutron stars has resulted in a significant step forward in the field of nuclear astrophysics. The event, labeled GW170817 [1], along with a subsequent observation of the resulting kilonova [2] signaled the beginning of the era of multi-messenger astronomy. As a direct result, for the first time, it became possible to determine limits on the neutron star tidal deformability, and due to its connection with the equation of state (EoS) of dense neutron star matter [3] also to obtain limits on the masses and radii of the merging stars [4, 5]. In addition, this event has reignited the discussion on signals of quark- matter in the cores of neutron stars [6, 7].

The possibility of finding this rare state of matter inside neutron stars provides a unique opportunity of accessing the cold and dense sector of the Quantum Chromodynamics (QCD) phase diagram, which remains out of reach for first–principle lattice QCD calculations. A large body of work exists on the possible states of cold, strongly interacting matter and their relation to neutron star properties (for recent reviews, see [8, 9] and references therein), which were triggered by the observation of pulsars with masses around 2 $M_\odot$ [10–13]. But without a radius measurement at this high mass, there are still too many EoS scenarios possible, with and without a deconfinement phase transition.

Efforts are being made to solve the problem of a reliable radius measurement by employing the X-ray timing method (cf. [14] and references therein). For that purpose, the

*e-mail: mateusz.cierniak@uwr.edu.pl
**e-mail: david.blaschke@uwr.edu.pl

NASA NICER mission was deployed to the International Space Station in June 2017 and has since then gathered luminosity data from a few millisecond pulsars. Results have been published for two objects,: the heaviest known pulsar with a precisely determined mass, PSR J0740+6620 [15, 16], and an intermediate mass pulsar, PSR J0030+0451 [17, 18]. The latter only marginally agrees with the analysis of GW170817 [19] in the same mass range, resulting in a narrow region of possible radii for intermediate-mass neutron stars [20], or perhaps hint at a signal of the third family of compact stars [21] which would give rise to the phenomenon of mass twin stars [22] that could occur in different mass ranges [23], also in the mass range of GW170817, see [9] and references therein.

The high-mass pulsar PSR J0740+6620 is observed to have a rather large radius, see figure 1. This creates tension with the idea of purely hadronic neutron stars. Hadronic matter tends to soften at higher densities due to the onset of hyperons [24] which creates the "hyperon puzzle" problem for nonrelativistic EoS models that are based on realistic nucleon-nucleon interactions like the APR EoS [25]. Adding a repulsive few-nucleon interaction in the form of a multi-pomeron potential solves the hyperon puzzle for such models [26, 27] but this cannot explain the large radius for PSR J0740+6620.

While not completely excluded, the scenario of purely hadronic neutron star interiors is being increasingly disfavored by observation. A possible alternative are multi–phase models, predicting a phase transition from purely hadronic neutron stars at low mass to intermediate and high mass hybrid stars with a core presenting exotic states of matter, for instance a quark–gluon plasma (QGP). Such constructions are not completely free of the shortcomings of hadronic models, as the common convention is to construct a Maxwell–type first order phase transition from a stiffer hadronic equation of state (EoS) to a softer one, represented by the QGP. Recently, a number of alternative approaches were proposed [8, 28, 29] which join a soft low-density hadronic EoS with stiff QGP models at high densities, thus easily fulfilling all observational constraints.

In this manuscript, we will restrict ourselves to the standard Maxwell construction of a phase transition and make use of the novel insights due to the special point property of hybrid neutron star sequences [30–33] in order to determine the impact of the recent observations on the available parameter space of a class of constant–speed–of–sound (CSS) QGP models [34], which were shown to provide an accurate description of color–superconducting quark matter [35, 36].

The manuscript is organized as follows. After introducing the two-phase hybrid EoS model for dense matter in the next section, we present results of an extensive scan of the possible locations of the special point in the mass-radius plane in section 3. In section 4, we introduce a fit formula that relates the coordinates of the special point to the CSS model parameters A and c_s^2 and obtain as the main result of this contribution the mapping of the new NICER mass-radius constraint for PSR J0740+6620 onto the $A - c_s^2$ plane. In section 5 we draw our Conclusions.

2 The two phase dense matter model

In the absence of a unified description of cold hadronic and quark matter with a phase transition at high densities, the hybrid star matter EoS must be constructed by matching separate EoS, comparing pressures derived using hadrons, $P_h(\mu)$, or deconfined quarks, $P_q(\mu)$, as degrees of freedom. The EoS with the higher pressure is the one that is preferred by nature and describes the bulk thermodynamics. A crossing of both alternative EoS at the critical chemical potential μ_c marks a first-order phase transition fulfilling the condition $P_h(\mu_c) = P_q(\mu_c)$ for the Maxwell construction.

The hadronic EoS discussed in this study is chosen from the class of relativistic density-functional (RDF) models based on the DD2 parametrization [37] with excluded volume effects modelled according to [38]. Its inputs are fitted to the properties of finite nuclei and it predicts the EoS for nuclear matter at low densities up to and at nuclear saturation in accordance with the phenomenological data. Therefore, it should provide a reliable basis for extrapolations to supersaturation densities.

For the deconfined quark matter phase, the class of CSS quark matter models will be used. The EoS in the form of pressure as a function of the baryochemical potential reads (see also [31] or the Appendix of [34]),

$$P(\mu) = A(\mu/\mu_0)^{1+c_s^{-2}} - B, \tag{1}$$

where the model parameters A, B and c_s^2 are constant and $\mu_0 = 1$ GeV. The baryon density n follows from the canonical relation

$$\frac{dP(\mu)}{d\mu} = n(\mu) = (1 + c_s^{-2})\frac{A}{\mu_0}\left(\frac{\mu}{\mu_0}\right)^{c_s^{-2}}. \tag{2}$$

Using the above, we arrive at the energy density

$$\varepsilon = \mu n - P = B + c_s^{-2}A(\mu/\mu_0)^{1+c_s^{-2}}. \tag{3}$$

The relation between pressure and energy density takes the form

$$P = c_s^2\varepsilon - (1 + c_s^2)B. \tag{4}$$

It has been shown in [35] that the above CSS model fits well the EoS of color superconducting quark matter in both, the 2SC and the CFL phases which were obtained from a self-consistent solution of the three-flavor NJL model in the mean field approximation [39, 40].

3 The special point

The special point of hybrid neutron stars is a property first described in [30] and recently studied in [31–33]. It is a characteristic feature of a broad range of two–phase models (see, e.g., [41–43]), presenting itself as a stationary point in the mass–radius diagram for hybrid star sequences. Any parametrization of the hybrid model, which varies only by the onset density of the second phase, will cross this point. This is illustrated in figure 1.

As was shown in [31–33], this feature can be used to characterize a broad family of EoS, without the need to fully scan the parameter space of the model, and to easily compare model predictions with observational data. Most recently, this has been demonstrated in [33], where a prediction on the radius range of a hybrid star fulfilling PSR J0740+6620 mass–radius measurement was derived.

A detailed map of the special point locations and their dependence on the CSS model parameters has been made in [33], shown here as figure 2, and will be used in the following chapter. It has been shown in [31, 33] that this map is independent of the underlying hadronic EoS, the phase transition density and the type of the phase transition (first order, crossover etc.). Therefore, it can be used to isolate the hybrid neutron star branch and analyze observational data in the context of constraints on the parameters of the microscopic model for deconfined quark matter.

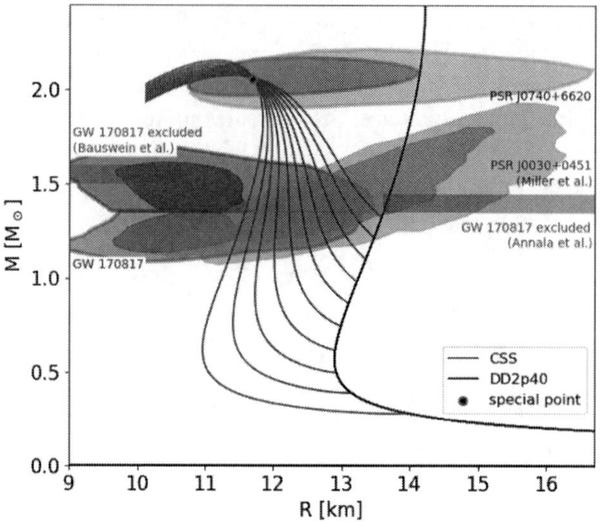

Figure 1. Special point as the intersection of hybrid star sequences (red solid lines) on the neutron star mass radius diagram. The black line represents the hadronic EoS with excluded volume DD2p40, naming convention analogous to [38]. The intersection of the red lines with the black one mark the onset of CSS quark matter in the core, see also [33]. Observational data taken from [4, 5, 15–18, 44]

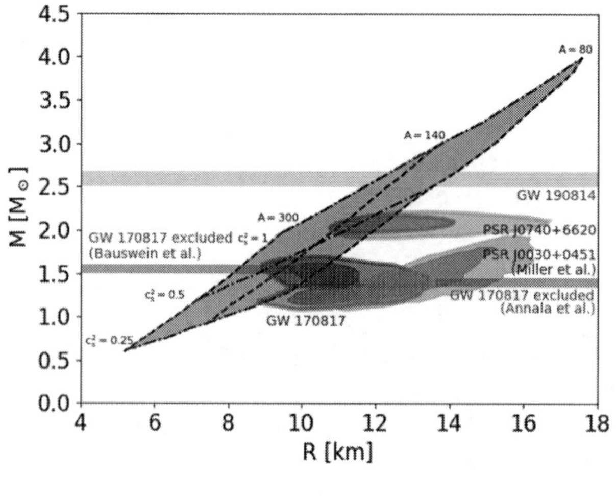

Figure 2. Location of the special point (green area) for $0.25 < c_s^2 < 1.0$ and $80 < A[\mathrm{MeV/fm^3}] < 300$, lines of constant c_s^2 and A are shown for comparison. Figure adopted from [33]. Observational data taken from [4, 5, 15–18, 44]

4 Parameter fit

The coordinates of the special point are to a good approximation described by the following relationship

$$\begin{cases} R_{SP}[\mathrm{km}] = a_R \cdot \left(c_s^2\right)^{b_R} \cdot \left(A[\mathrm{MeV/fm^3}]\right)^{c_R} + d_R, \\ M_{SP}[M_\odot] = a_M \cdot \left(c_s^2\right)^{b_M} \cdot \left(A[\mathrm{MeV/fm^3}]\right)^{c_M} + d_M. \end{cases} \tag{5}$$

Table 1. Parameters of the fit formulas Eq. (5) for the special point location in the $M - R$ diagram.

X	a_X	b_X	c_X	d_X
R	145.938	0.420	-0.47	-0.4
M	35.332	0.648	-0.47	-0.4

The accuracy of this approximation can be judged from figures 3 and 4, where the data points were taken from [33] (see also figure 2) and the black solid lines are obtained using the fit formulas of Eq. (5) with the parameters listed in table 1.

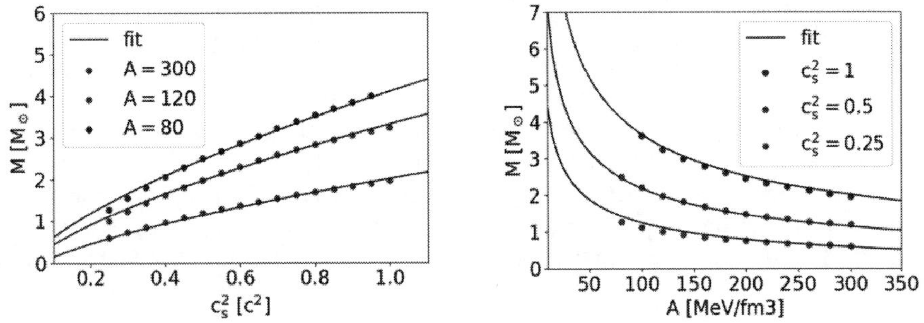

Figure 3. Masses of the special point location (colored points) as function of the CSS parameters c_s^2 (left panel) and A (right panel) compared to the fit formula (5) (black line). See text for details.

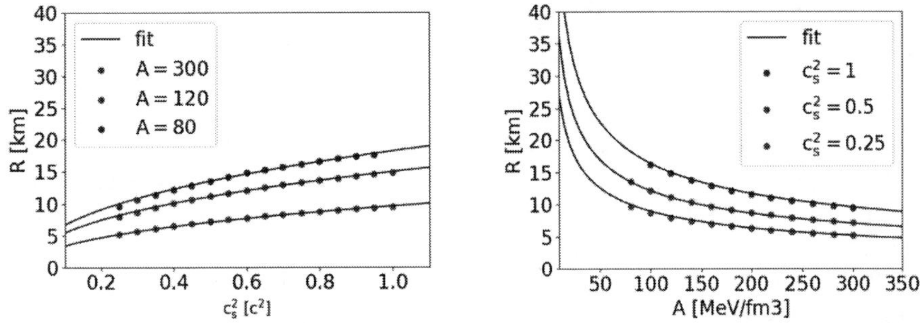

Figure 4. Radii of the special point location (colored points) as function of the CSS parameters c_s^2 (left panel) and A (right panel) compared to the fit formula (5) (black line). See text for details.

By inverting the fit (5) we were able to translate existing observational constraints on neutron star masses and radii to constraints on the CSS model parameter space. This is the main result of this work, shown in Fig. 5.

5 Conclusions

The special point typically corresponds to a mass slightly below $M_{\rm max}$, no less than $M_{\rm max} - M_{\rm SP} = 0.2\ M_\odot$ [31]. If we assume $M_{\rm SP} \approx M_{\rm max}$ and that PSR J0740+6620 represents the heaviest possible neutron star, we arrive at a narrow band of CSS model parameters that

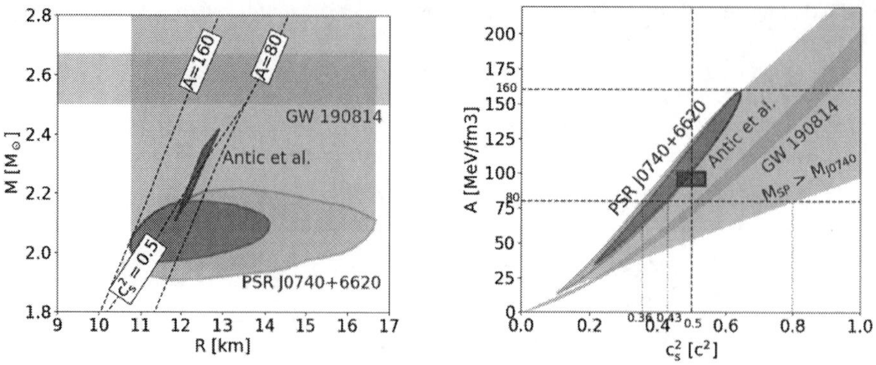

Figure 5. Left panel: Neutron star mass–radius diagram with the 1σ results of the NICER measurement of PSR J0740+6620 [15, 16]. The light green band is the mass range for the lighter object in the binary merger GW190814. The grey area approximates the region of special point locations for which the hybrid star sequence would fulfill the radius constraint from PSR J0740+6620. For orientation, we show limiting values of constant A as well as $c_s^2 = 0.5$. In the small red area lie all physical parametrizations of the color superconducting nonlocal NJL model of quark matter [36] Right panel: The same areas and lines as in the left panel, translated to the coordinate system spanned by the CSS model parameters c_s^2 and A.

remain in agreement with this observation. Further demanding the A parameter not to fall below a value of 80 MeV/fm^3, a value which ensures a phase transition not below the saturation density (cf. [33]), we can determine a narrow band of possible values for the speed of sound parameter c_s^2. For $A = 80$ MeV/fm^3, the possible range is $0.36 \leq c_s^2 \leq 0.43$. Higher values of A would subsequently drive the speed of sound to greater values with a limit in the vicinity of $c_s^2 \approx 0.6$.

Translating back the requirement of a transition at supersaturation density, we arrive at a limit represented by the black dashed line in the left panel of Fig. 5. Only the left side of this limit can be achieved with the CSS model assuming $M_{\max} \approx M_{J0740}$. A potential revision of the measurement, that would suggest a higher radius for the star would signal, that this is not the highest possible mass for a neutron star, or that the CSS model does not accurately approximate the cold and dense matter EoS.

Finally, the grey band of the right panel of figure 5 shows the available parameter space of the CSS model, assuming $M_{\max} > M_{J0740}$, for which the model would remain in agreement with the NICER observation. For $A = 80$ MeV/fm^3, this results in an upper limit on $c_{s,max}^2 = 0.8$, along with the previous lower limit of $c_{s,min}^2 = 0.36$.

We have demonstrated how the special point feature of hybrid neutron stars can be applied to provide direct limits on the chosen quark matter model parameters by translating the novel NICER measurement of PSR J0740+6620 to the parameter space of the CSS model. Further constraints can be derived from existing measurements, a task which we leave for subsequent studies. Due to the link between the phenomenological CSS model and the color–superconducting nlNJL model illustrated in [36], such limits can provide valuable insight into the phenomenon of color–superconductivity in quark matter at high densities.

Acknowledgements

M.C. and D.B. acknowledge support from the Polish National Science Centre (NCN) under grant number 2019/33/B/ST9/03059. The work of D.B. was supported by the Russian

Foundation for Basic Research under grant No. 18-02-40137 and by the Federal Program "Priority-2030".

References

[1] B.P. Abbott et al., Phys. Rev. Lett. **119**, 161101 (2017), 1710.05832
[2] B.P. Abbott et al., Astrophys. J. Lett. **848**, L12 (2017), 1710.05833
[3] T. Hinderer, B.D. Lackey, R.N. Lang, J.S. Read, Phys. Rev. D **81**, 123016 (2010), 0911.3535
[4] E. Annala, T. Gorda, A. Kurkela, A. Vuorinen, Phys. Rev. Lett. **120**, 172703 (2018), 1711.02644
[5] A. Bauswein, O. Just, H.T. Janka, N. Stergioulas, Astrophys. J. Lett. **850**, L34 (2017), 1710.06843
[6] A. Bauswein, N.U.F. Bastian, D.B. Blaschke, K. Chatziioannou, J.A. Clark, T. Fischer, M. Oertel, Phys. Rev. Lett. **122**, 061102 (2019), 1809.01116
[7] A. Bauswein, S. Blacker, V. Vijayan, N. Stergioulas, K. Chatziioannou, J.A. Clark, N.U.F. Bastian, D.B. Blaschke, M. Cierniak, T. Fischer, Phys. Rev. Lett. **125**, 141103 (2020), 2004.00846
[8] G. Baym, T. Hatsuda, T. Kojo, P.D. Powell, Y. Song, T. Takatsuka, Rept. Prog. Phys. **81**, 056902 (2018), 1707.04966
[9] D. Blaschke, N. Chamel, Astrophys. Space Sci. Libr. **457**, 337 (2018), 1803.01836
[10] P. Demorest, T. Pennucci, S. Ransom, M. Roberts, J. Hessels, Nature **467**, 1081 (2010), 1010.5788
[11] J. Antoniadis et al., Science **340**, 6131 (2013), 1304.6875
[12] H.T. Cromartie et al. (NANOGrav), Nature Astron. **4**, 72 (2019), 1904.06759
[13] E. Fonseca et al., Astrophys. J. Lett. **915**, L12 (2021), 2104.00880
[14] S. Bogdanov, G.B. Rybicki, J.E. Grindlay, Astrophys. J. **670**, 668 (2007), astro-ph/0612791
[15] M.C. Miller et al., Astrophys. J. Lett. **918**, L28 (2021), 2105.06979
[16] T.E. Riley et al., Astrophys. J. Lett. **918**, L27 (2021), 2105.06980
[17] M.C. Miller et al., Astrophys. J. Lett. **887**, L24 (2019), 1912.05705
[18] T.E. Riley et al., Astrophys. J. Lett. **887**, L21 (2019), 1912.05702
[19] B.P. Abbott et al. (LIGO Scientific, Virgo), Phys. Rev. Lett. **121**, 161101 (2018), 1805.11581
[20] C.D. Capano, I. Tews, S.M. Brown, B. Margalit, S. De, S. Kumar, D.A. Brown, B. Krishnan, S. Reddy, Nature Astron. **4**, 625 (2020), 1908.10352
[21] U.H. Gerlach, Phys. Rev. **172**, 1325 (1968)
[22] N.K. Glendenning, C. Kettner, Astron. Astrophys. **353**, L9 (2000), astro-ph/9807155
[23] J.E. Christian, A. Zacchi, J. Schaffner-Bielich, Eur. Phys. J. A **54**, 28 (2018), 1707.07524
[24] D. Chatterjee, I. Vidaña, Eur. Phys. J. A **52**, 29 (2016), 1510.06306
[25] A. Akmal, V.R. Pandharipande, D.G. Ravenhall, Phys. Rev. C **58**, 1804 (1998), nucl-th/9804027
[26] Y. Yamamoto, T. Furumoto, N. Yasutake, T.A. Rijken, Eur. Phys. J. A **52**, 19 (2016), 1510.06099
[27] Y. Yamamoto, H. Togashi, T. Tamagawa, T. Furumoto, N. Yasutake, T.A. Rijken, Phys. Rev. C **96**, 065804 (2017), 1708.06163

[28] K. Masuda, T. Hatsuda, T. Takatsuka, PTEP **2013**, 073D01 (2013), 1212.6803
[29] A. Ayriyan, D. Blaschke, A.G. Grunfeld, D. Alvarez-Castillo, H. Grigorian, V. Abgaryan, Eur. Phys. J. A **57**, 318 (2021), 2102.13485
[30] A.V. Yudin, T.L. Razinkova, D.K. Nadyozhin, A.D. Dolgov, Astron. Lett. **40**, 201 (2014), 1404.0865
[31] M. Cierniak, D. Blaschke, Eur. Phys. J. ST **229**, 3663 (2020), 2009.12353
[32] D. Blaschke, M. Cierniak, Astron. Nachr. **342**, 227 (2021), 2012.15785
[33] M. Cierniak, D. Blaschke, Astron. Nachr. **342**, 819 (2021), 2106.06986
[34] M.G. Alford, S. Han, M. Prakash, Phys. Rev. D **88**, 083013 (2013), 1302.4732
[35] J.L. Zdunik, P. Haensel, Astron. Astrophys. **551**, A61 (2013), 1211.1231
[36] S. Antić, M. Shahrbaf, D. Blaschke, A.G. Grunfeld (2021), 2105.00029
[37] S. Typel, G. Ropke, T. Klahn, D. Blaschke, H.H. Wolter, Phys. Rev. C **81**, 015803 (2010), 0908.2344
[38] S. Typel, Eur. Phys. J. A **52**, 16 (2016)
[39] D. Blaschke, S. Fredriksson, H. Grigorian, A. Öztas, F. Sandin, Phys. Rev. D **72**, 065020 (2005)
[40] T. Klähn, R. Lastowiecki, D. Blaschke, Phys. Rev. D **88 (8)**, 085001 (2013)
[41] D. Blaschke, T. Klahn, R. Lastowiecki, F. Sandin, J. Phys. G **37**, 094063 (2010), 1002.1299
[42] M.A.R. Kaltenborn, N.U.F. Bastian, D.B. Blaschke, Phys. Rev. D **96**, 056024 (2017), 1701.04400
[43] D. Blaschke, A. Ayriyan, D.E. Alvarez-Castillo, H. Grigorian, Universe **6**, 81 (2020), 2005.02759
[44] R. Abbott et al. (LIGO Scientific, Virgo), Astrophys. J. Lett. **896**, L44 (2020), 2006.12611

Holography of Strongly Coupled Gauge Theories

Nick Evans[1,*]

[1]Department of Physics and Astronomy, University of Southampton, Southampton, SO17 1BJ, UK

Abstract. The "periodic table" of strongly coupled gauge theories remains only sketchily understood. Holography has developed to the point where bottom up constructions can describe the spectrum of individual gauge theories (based on assumptions of their running) including quarks in different representations and higher dimension operators. I highlight the method with a "perfected" version of an AdS dual of QCD and results for composite higgs models with two representations of quarks. The method raises questions about the degree to which energy scales can be split in generic gauge theories including whether confinement and chiral symmetry breaking are linked.

1 Introduction

Asymptotically free gauge theories are a key element of the Standard Model of particle physics - SU(3) gauge theory describes the strong force and the SU(2) theory the weak force. They likely also play a role beyond the Standard Model and possible examples include technicolour [1], composite Higgs models [2], strongly coupled dark matter [3] or dynamical supersymmetry breaking [4]. Despite this our knowledge of the spectrum of the set of asymptotically free theories is rather poor. Lattice studies only exist for a smattering of such theories and there remains debate over issues such as where the edge of the conformal window lies for SU(3) theories with fundamental quarks [5, 6]. Even for theories that are known to break chiral symmetries the spectrum has frequently not yet been computed. It seems like we should do a better job here of understanding the "periodic table" of these theories.

Here I want to show that holographic modelling of these gauge theories [7, 8] is a useful tool that can quickly provide an estimate of the spectrum. Although these models are not first principle, they can provide a helpful exchange of ideas with lattice studies that can highlight important questions to address.

2 Holographic Magic

Holography [9, 10] is a weakly coupled dual description of strongly coupled gauge dynamics that has emerged from string theory. Formally the dual should be a string theory but for ground states of glueballs and meson a point particle, gravitational theory works well for the dual description [11, 12] (it is formally justified in the large N_c limit where the QCD string tension is very high). Full descriptions only exist for a small number of gauge theories close to supersymmetric cases but some generic lessons have emerged that can be used for model building. Let's briefly review three of these pieces of "magic".

*e-mail: evans@southampton.ac.uk

- **Renormalization Group Flow:** RG scale emerges as an extra holographic direction in these descriptions. In a five dimensional Anti-de-Sitter (AdS) space

$$ds^2 = \frac{dr^2}{r^2} + r^2 dx_{3+1}^2 \tag{1}$$

  The radial direction r is the RG scale and slices at some fixed r describe the gauge theory on the 3+1d sub-space. Fields in the AdS bulk have solutions that describe gauge invariant obervables ie operators O and sources $\mathcal{J}$.

- **Chiral Symmetry Breaking:** In QCD the chiral flavour symmetry is broken $SU(2)_L \times SU(2)_R \rightarrow SU(2)_V$ by the formation of a vev for the operator $\bar{u}_L u_R + \bar{d}_L d_R + h.c.$ Holographically [13] we describe the operator as a dimension one scalar field, L, in AdS_5 satisfying the Klein-Gordon equation [14]

$$\partial_r[r^3\partial_r L] - r\Delta m^2 L = 0, \qquad L = \frac{m}{r^\gamma} + \frac{\langle \bar{q}q\rangle}{r^{2-\gamma}}, \qquad \gamma(\gamma-2) = \Delta m^2 \tag{2}$$

  When $\Delta m^2 = 0$ the solutions describe a dimension one mass, m, and the dimension three quark condensate. For non-zero Δm^2 these operators develop an anomalous dimension γ as shown.

  In AdS_5 there is an instability bound for a scalar called the Breitenlohner Freedman bound [15] which here corresponds to $\Delta m^2 = -1$. If Δm^2 passes through this bound then the massless $L = 0$ solution is unstable. Interestingly here this corresponds to $\gamma = 1$ - if the dimension of $\bar{q}q$ falls to two then there is an instability to the generation of a quark condensate. This criteria precisely matches that from gap equation analysis [16] and which has been used to motivate the first guesses for the edge of the conformal window [17].

  We can thus make holographic models of chiral symmetry breaking which are essentially a higgs theory, the scalar field is L, but whose potential changes with RG scale (radial direction r). The potential (Δm^2) is determined by the running of the anomalous dimension which in turn is determined by the gauge dynamics. At the scale where $\gamma = 1$ the BF bound is violated and chiral symmetry breaking occurs.

- **Transitioning to the IR Meson Theory:** finally holography allows us to determine the low energy mesonic theory. For a given background vacuum, for example, a solution for the field L, we can allow infinitesimal fluctuations of the form $\delta(r)e^{ik.x}$, $k^2 = -M^2$ which describe fluctuations of the $\bar{q}q$ operator (the σ or f_0 state). Generically one obtains a Sturm-Louville system which only has normalizable solutions for particular M^2. Other states such as the ρ meson can be included via a gauge field dual to $\bar{q}\gamma^\mu q$ and so forth. This method determines the meson spectrum. Substituting these solutions back into the action and integrating over the radial direction returns a 3+1d theory of the mesons and their interaction couplings. This is the standard AdS/QCD methodology [11, 12].

2.1 Dynamic AdS/QCD

The simplest model that incorporates these points is the Dynamic AdS/QCD theory [18]

$$S = \int d^4x\, d\rho\, \mathrm{Tr}\, \rho^3 \left[\frac{1}{\rho^2 + |X|^2}|DX|^2 + \frac{\Delta m^2}{\rho^2}|X|^2 + \frac{1}{2\kappa^2}(F_V^2 + F_A^2) + \bar{\Psi}(\not{D}_{\rm AAdS} - m)\Psi\right] \tag{3}$$

where $X = Le^{2i\pi^a T^a}$ describes the quark condensate/σ and pion fields, F_V the ρ meson and F_A the a meson. Ψ is a Dirac field that can describe the nucleon. The factors of X in the metric (here $r^2 = \rho^2 + |X|^2$) are deduced from top down (probe D7 brane) models and are the

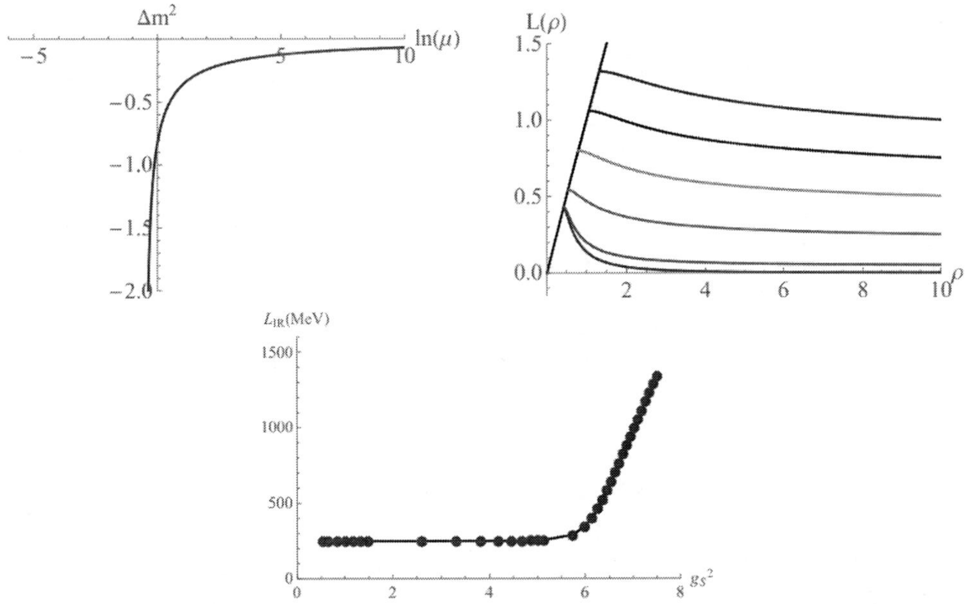

Figure 1. Top left: the running of the mass in the Dynamic AdS/QCD theory for 2 flavour SU(3) QCD. Top right: the resulting vacuum $L(\rho)$ profiles which asymptote at large ρ to the quark mass. Bottom: the IR quark mass against NJL coupling in two flavour QCD using the solutions above and (5).

simplest method to communicate the background quark condensate to the fluctuation fields. Δm^2 can be input using the perturbative results for the running of γ in a theory

$$\Delta m^2 = -2\gamma = -\frac{3(N_c^2 - 1)}{2N_c\pi}\alpha \tag{4}$$

where here α runs according to, for example, the two loop perturbative results with the holographic radial direction ρ directly set to be the RG scale μ (which we extend into the non-perturbative regime as an ansatz for the running that includes conformal window behaviour [17]). A key benefit of the model is that the only parameters are N_c, N_f and Λ_{QCD} as in the true QCD theory. More complex models exist – especially note the VQCD theory of Kiritsis and Jarvinen [19] - that attempt to decouple the quarks, include confinement below the chiral symmetry breaking scale, and explain stringy Regge trajectories. These are heroic pieces of work but necessarily contain more assumptions than this simple model.

The top figures in Figure 1 show the running of Δm^2 for 2 flavour SU(3) QCD and the resulting profiles for L, asymptoting at large ρ to the quark mass m.

2.2 Higher Dimension Operators (HDOs)

Holography is a strong weak duality and so the gravitational dual should only describe the IR region of an asymptotically free gauge theory where the coupling is large. Models such as AdS/QCD have some remnant of the N=4 super Yang-Mills theory left in the description that enforces conformality in the UV (large r). That matches the weak coupling physics of these theories - so people bluff and let the dual extend to the far UV. In reality one should cut off the description, eg around 3 GeV or so in QCD, where the coupling becomes weak [20, 21]. One can simply impose a cut off in the bulk. There is though then a matching problem - one

Table 1. The predictions for masses and decay constants (in MeV) for $N_f = 2$ SU(3) gauge theory. The columns show the QCD value, the holographic prediction for the theory with massless quarks and the perfected holographic QCD theory with HDOs. The ρ-meson mass has been used to set the scale and other quantities that are used to fix the HDO couplings are indicated by the *s.

	QCD	AdS/SU(3) 2 F 2 $\bar{F}$		AdS/SU(3) 2 F 2 $\bar{F}$	with HDO
M_ρ	775	775*		775*	
M_A	1230	1183	- 4%	1230*	$g_A^2 = 5.76149$
M_S	500/990	973	+64%/-2%	597	+20%/ − 40%
M_B	938	1451	+43%	938*	$g_B^2 = 25.1558$
f_π	93	55.6	-50%	93*	$g_S^2 = 4.58981$
f_ρ	345	321	- 7%	345*	$g_V^2 = 4.64807$
f_A	433	368	-16%	444	+2.5%
$M_{\rho,n=1}$	1465	1678	+14%	1532	+4.5%
$M_{A,n=1}$	1655	1922	+19%	1789	+8%
$M_{S,n=1}$	990 /1200-1500	2009	+64%/+35%	1449	+46%/0%
$M_{B,n=1}$	1440	2406	+50%	1529	+6%

should align the dual to meet QCD in the intermediate coupling period which would mean including some higher dimension operators at the UV boundary. We do now know how to include such operators using Witten's multi-trace prescription [22].

Consider adding a Nambu Jona Lasinio operator [23] to our description of QCD (Figure 1) [24–26]

$$\Delta\mathcal{L} = \frac{g^2}{\Lambda_{UV}^2}\bar{q}_L q_R \bar{q}_R q_L \quad \rightarrow \frac{g^2}{\Lambda_{UV}^2}\langle\bar{q}_L q_R\rangle\bar{q}_R q_L \quad \rightarrow m = \frac{g^2}{\Lambda_{UV}^2}\langle\bar{q}_L q_R\rangle \tag{5}$$

The logic shown is that in the presence of an NJL term, if the quark condensate forms then there is an effective UV mass. Witten's prescription is to reinterpret the vacuum solutions of Figure 1. We now consider solutions that asymptote to a non-zero mass to represent the theory with no bare mass but an NJL term. We read off m and $\langle\bar{q}_L q_R\rangle$ from the solution at the UV cut off and then compute g^2 from the relation in (5).We show the output in the model of two flavour QCD in the lower figure in Figure 1 - there is the usual NJL behaviour - at low couplings only the QCD dynamics generates an IR quark mass but as g goes through a critical value it sharply enhances the dynamical symmetry breaking scale.

This prescription can be used for all the fields in the Dynamic AdS/QCD action allowing one to include operators $|\bar{q}_L q_R|^2$, $|\bar{q}_L\gamma^\mu q_R|^2$, $|\bar{q}_L\gamma^\mu\gamma_5 q_R|^2$ and $|qqq|^2$. We will not try to match their couplings to perturbative QCD but instead use them as parameters that can be fitted to the low energy meson data - this is like the idea of a perfect action on the lattice [27].

3 Holographic Description of QCD

The Dynamic AdS/QCD, holographic model based on the QCD running shown in Figure 1 leads to the predictions shown in Table 1 [8]. The massless theory compares well to QCD data in that it gets the broad pattern of the masses and couplings correct. However, the pion decay constant is typically low, the baryon mass high and the excited state masses high. The quark and glue content of scalar states is very hard to identify and mixing with glueball states is neglected here, so these identifications are difficult. Including HDOs clearly reduces the

Table 2. Ground state masses and couplings for the vector, axial-vector, and scalar mesons for the two representations, F and A_2, for an Sp(4) and an SU(4) gauge theory. Results are shown from the lattice (Sp(4) from [30] and SU(4) from [31]) and from holography [7, 8].

	AdS Sp(4) $4F, 6A_2$	AdS Sp(4) quench	Lattice Sp(4) quench	Lattice SU(4) $4A_2, 2F, 2\bar{F}$	AdS SU(4) $4A_2, 2F, 2\bar{F}$	AdS SU(4) $5A_2, 3F, 3\bar{F}$
$f_{\pi A_2}$	0.120	0.103	0.1453(12)	0.15(4)	0.0997	0.111
$f_{\pi F}$	0.0701	0.0756	0.1079(52)	0.11(2)	0.0953	0.109
M_{VA_2}	1*	1*	1.000(32)	1.00(4)	1*	1*
f_{VA_2}	0.517	0.518	0.508(18)	0.68(5)	0.489	0.516
M_{VF}	0.814	0.962	0.83(19)	0.93(7)	0.939	0.904
f_{VF}	0.364	0.428	0.411(58)	0.49(7)	0.461	0.491
M_{AA_2}	1.35	1.28	1.75 (13)		1.37	1.32
f_{AA_2}	0.520	0.524	0.794(70)		0.505	0.521
M_{AF}	1.19	1.36	1.32(18)		1.37	1.23
f_{AF}	0.399	0.462	0.54(11)		0.504	0.509
M_{SA_2}	0.375	1.14			0.873	0.684
M_{SF}	0.902	1.25			1.02	0.798

predictivity of the theory but also greatly improves the fit including to the radially excited state masses. Including HDOs allows the description to be systematically improved in the spirit of low energy effective field theory.

4 Beyond QCD - Multi-Flavour Theories

Recently my collaborators and I have moved to modelling theories with quarks in two different representations of the gauge symmetry [7, 8] . These have been of interest for the construction of composite higgs models with baryons that play the role of top partners [28, 29], but here I will just consider them as field theory problems. There has also been some lattice work in this area [30, 31] and I will concentrate on those theories where we have something to compare to.

To model two different representations we simply repeat the Dynamic AdS/QCD model twice with the appropriate runnings of $\gamma/\Delta m^2$ for each of the two condensates we wish to study (and fluctuations of those operator with the various γ-matrix structures inserted). The only linkage between the sectors we include is that both representations enter the running of the gauge coupling. There is then an interesting question of how the quarks decouple when they go on mass-shell at strong coupling which future lattice work could investigate. We simply remove quarks from the running when they hit their IR mass-shell condition as one would at weak coupling.

4.1 Sp(4) 4F 6A_2

A first example model is Sp(4) gauge theory with 4 Weyl fermions in the fundamental representation and 6 Weyl fermions in the two-index anti-symmetric representation (of dimension 6). Here there is a lattice study with quenched gauge configurations. We compare to this by using the pure glue running of the Sp(4) theory. The results for the spectrum are shown in Table 2 and require a little digestion.

The first broad point to make about the results is that the predictions for the spectrum of the quenched theory on the lattice and using the Dynamic AdS/QCD model agree well on the pattern of masses and couplings (which are similar to that of QCD). In fact the numerical values here are better than our first attempt at QCD above. Note that the fundamental fermion sector is lighter than the A_2 sector, which in the case of the holographic model is because γ reaches one first for the A_2 representation so the condensation is at a higher scale.

Given that the holographic model is not a first principles computation, we do not claim to be directly competing with the lattice results - as the lattice groups improve their errors one should look there for the true answer. However, the holographic approach allows us to compute the σ meson masses in this model that was beyond the first lattice computation (states with the quantum numbers of the vacuum are hard). Further we can rather simply unquench the holographic model by just including the fermions in the running of Δm^2 whereas on the lattice this is computationally very expensive - the holographic results are again in Table 2. The holographic model suggest the changes in the spectrum will mostly in this case be at the 10% level or so except in the σ case where as the running slows the σ masses fall sharply. The holographic model has therefore contributed additional information that complements the lattice results.

4.2 SU(4) 3F 3F̄ 5A$_2$

SU(4) gauge theory with three Dirac fermions in the fundamental representation (or equivalently 3 Weyl fermions in the fundamental and 3 Weyls in the anti-fundamental) and five Weyl fermions in the two-index antisymmetric representation (again a colour sextet) is also a theory of interest for composite higgs models with top partners. Here though for technical reasons the lattice community has had to instead work on the theory with two Dirac fundamentals and four A_2s. This lattice simulation is unquenched though. In Table 2 we show both the lattice and Dynamic AdS/QCD results for the spectrum - there's a basic agreement that the A_2 sector is heavier and on the spectrum in each of the A_2 and fundamental sectors. The holographic model can easily add new information for the axial vector and scalar sectors.

Again though the holographic approach allows us to go further - here we can move to the truly desired fermion content (3 Dirac fermions and 5 A_2). The results are shown in Table 2. For the majority of the spectrum the shifts are at below the 10% level. This is good news because it means the simulations on the lattice are useful, as they would have hoped, for telling us about the true theory even though they don't quite match its fermion content. Again the holographic model shows a larger shift in the σ masses which fall as the running slows.

5 Discussion

We've looked at holographically modelling QCD and two theories with two different representations of fermions. I've tried to argue that the holographic models provide additional insight into the dynamics and spectrum beyond currently achieved lattice simulations. The main benefit is that one is simply solving some classical differential equations so it is very swift to compute in any given model once the code is set up (you just input the appropriate running for γ of the operator you are interested in). As an example of this, in [8] we in addition provided spectrum computations for twenty six of the proposed composite higgs models with top partner baryons in [32] (the remaining cases lie in the conformal window at the level of our approximations). A wider survey of theories yet is being constructed on the SCGT Forum web site which the reader is invited to use and contribute to:

http://www.southampton.ac.uk/~evans/SCGT/

A further product of modelling these gauge theories is that it throws up interesting dynamical questions that could be answered with future lattice work. One crucial assumption

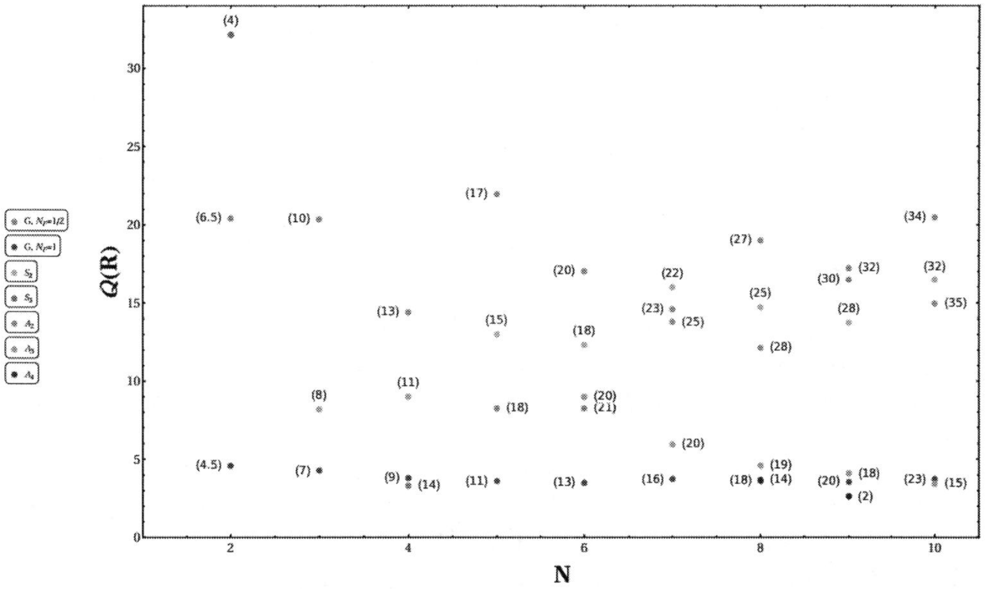

Figure 2. Plot of $Q(R) = \Lambda_{\chi SB\ R}/\Lambda_{\chi SB\ F}$ against the number of colours in theories with the minimal number of fermions in the higher dimensional representation (either 1 or 1/2 for real representations) and N_f^F in the fundamental representation. N_f^F has been tuned to maximize $Q(R)$ and its value is next to the point. We have used red for the fundamental, green for the adjoint, cyan for the rank-2 symmetric, gray for three-rank symmetric, gold for the rank-2 antisymmetric, pink for the A_3, maroon for the A_4, and blue, orange, black for the $R_{1,2,3}$ respectively.

we had to make was that quarks decouple from the dynamics when they go on mass shell at strong coupling in a similar fashion to at weak coupling - if this isn't true the running of these theories will be weaker to lower scales, changing the spectrum.

Another interesting aspect of multi-representation theories is that there is a gap between chiral symmetry breaking scales. If one assumes that the confinement scale lies below the lowest chiral symmetry breaking scale then one could imagine a gap between the highest chiral symmetry breaking scale and confinement. That separation in a full theory would be very interesting. This led Rigatos and I to perform a simple analysis [33], based on perturbative two loop running results, across all asymptotically free gauge theories to try to maximise the gap. In particular we included the minimum number of Weyl fermions allowed by anomaly constraints of some higher dimension representation R and then added N_f fundamental quarks. We estimated the chiral symmetry breaking scale for R as when $\gamma_R = 1$, then integrated out the R representation and ran the coupling with just fundamentals to where $\gamma_F = 1$. We then varied N_f to maximize the ratio

$$Q(R) = \frac{\Lambda_{\chi SB\ R}}{\Lambda_{\chi SB\ F}} \tag{6}$$

ie the ratio of the scales at which the two fermion species condense. We show the results in Figure 2 for a range of SU(N) gauge theories. We can immediately see that there are many theories where $5 < Q < 20$ which are theories that would have very different phase structure from QCD (for example potentially breaking the chiral symmetry of the R fermions whilst not confining at some finite temperatures). It's worth noting though that these theories are necessarily quite walking since we have dialled the number of fermions to widen the running time gaps between phenomena. They could be quite hard to study on the lattice therefore. There are less walking cases though where one might hope to see different phase structure than QCD although with not such wide gaps.

In conclusion then, I hope I have successfully argued that holography is a useful tool as we explore the periodic table of gauge theories and their application to beyond the Standard Model physics.

References

[1] C. T. Hill and E. H. Simmons, Phys. Rept. **381** (2003), 235-402.

[2] G. Cacciapaglia, C. Pica and F. Sannino, Phys. Rept. **877** (2020), 1-70.

[3] G. D. Kribs and E. T. Neil, Int. J. Mod. Phys. A **31** (2016) no.22, 1643004.

[4] I. Affleck, M. Dine and N. Seiberg, Nucl. Phys. B **241** (1984), 493-534.

[5] A. Hasenfratz, C. Rebbi and O. Witzel, Phys. Rev. D **100** (2019) no.11, 114508

[6] Z. Fodor, K. Holland, J. Kuti, S. Mondal, D. Nogradi and C. H. Wong, Phys. Rev. D **94** (2016) no.9, 091501.

[7] J. Erdmenger, N. Evans, W. Porod and K. S. Rigatos, Phys. Rev. Lett. **126** (2021) no.7, 071602.

[8] J. Erdmenger, N. Evans, W. Porod and K. S. Rigatos, JHEP **02** (2021), 058.

[9] J. M. Maldacena, Adv. Theor. Math. Phys. **2** (1998), 231-252.

[10] E. Witten, Adv. Theor. Math. Phys. **2** (1998), 253-291.

[11] J. Erlich, E. Katz, D. T. Son and M. A. Stephanov, Phys. Rev. Lett. **95** (2005), 261602.

[12] L. Da Rold and A. Pomarol, Nucl. Phys. B **721** (2005), 79-97.

[13] J. Erdmenger, N. Evans, I. Kirsch and Threlfall, Eur. Phys. J. A **35** (2008), 81-133.

[14] R. Alvares, N. Evans and K. Y. Kim, Phys. Rev. D **86** (2012), 026008.

[15] P. Breitenlohner and D. Z. Freedman, Annals Phys. **144** (1982), 249.

[16] A. G. Cohen and H. Georgi, Nucl. Phys. B **314** (1989), 7-24.

[17] T. Appelquist, J. Terning and L. C. R. Wijewardhana, Phys. Rev. Lett. **77** (1996), 1214.

[18] T. Alho, N. Evans and K. Tuominen, Phys. Rev. D **88** (2013), 105016.

[19] M. Jarvinen and E. Kiritsis, JHEP **03** (2012), 002.

[20] N. Evans, J. P. Shock and T. Waterson, Phys. Lett. B **622** (2005), 165-171.

[21] N. Evans and A. Tedder, Phys. Lett. B **642** (2006), 546-550.

[22] E. Witten, [arXiv:hep-th/0112258 [hep-th]].

[23] Y. Nambu and G. Jona-Lasinio, Phys. Rev. **122** (1961), 345-358.

[24] M. Jarvinen, JHEP **07** (2015), 033

[25] N. Evans and K. Y. Kim, Phys. Rev. D **93** (2016) no.6, 066002.

[26] W. Clemens and N. Evans, Phys. Lett. B **771** (2017), 1-4.

[27] P. Hasenfratz and F. Niedermayer, Nucl. Phys. B **414** (1994), 785-814.

[28] J. Barnard, T. Gherghetta and T. S. Ray, JHEP **02** (2014), 002.

[29] G. Ferretti, JHEP **06** (2014), 142.

[30] Bennett, D. K. Hong, J. W. Lee, C. J. D. Lin, B. Lucini, M. Mesiti, M. Piai, J. Rantaharju and D. Vadacchino, Phys. Rev. D **101** (2020) no.7, 074516.

[31] V. Ayyar, T. Degrand, D. C. Hackett, W. I. Jay, E. T. Neil, Y. Shamir and B. Svetitsky, Phys. Rev. D **97** (2018) no.11, 114505.

[32] G. Ferretti and D. Karateev, JHEP **03** (2014), 077.

[33] N. Evans and K. S. Rigatos, Phys. Rev. D **103** (2021), 094022.

Compositeness above the electroweak scale and a proposed test at LHCb

Gabriele Ferretti[1,*]

[1]Department of Physics, Chalmers University of Technology, Fysikgården, 41296 Göteborg, Sweden

Abstract. I review attempts to construct models of partial compositeness from strongly coupled gauge theories. A few minimal assumptions allow one to isolate a small number of representative models. After presenting the main idea, I discuss a recent proposal to detect a light pseudo-scalar, predicted in all these models, at the LHCb detector.

1 Introduction

The Higgs boson is nine years old but shows none of the restless behavior typical of that age, obeying all the rules of the Standard Model (SM). Still, many of us hope that it will eventually help us breaking the shackles of the SM and lead us into a new world. The reason for this hope is, of course, that its mass is not "Natural" and I still unapologetically embrace this argument.

There are basically two symmetry-based ways of attacking this problem. The first one is supersymmetry, pairing the Higgs field h with a fermionic field ψ whose mass is protected by chiral symmetry. In this case the pairing occurs via the fermionic transformation $\delta h \approx \epsilon\psi$ and the Higgs mass is protected indirectly via the fermion mass. The second mechanism employs a regular bosonic symmetry, may that be a scale $\delta h \approx \lambda h$ or a shift symmetry $\delta h \approx h + a$ protecting directly the Higgs mass. The last case arises in models where the Higgs is a composite (pseudo-)Nambu-Goldstone boson (pNGB) of a spontaneously broken approximate global symmetry [1].

Supersymmetry is a weakly coupled theory with natural elementary scalars and its problem is that it predicts too much in the sense that there are many unwanted dim < 4 operators that need to be suppressed somehow. On the contrary, compositeness is a strongly coupled theory without elementary scalars that predicts too little, in the sense that we lack the dim = 4 Yukawa operators needed to give fermion masses.

In this talk I concentrate on some model of compositeness, with the Higgs boson arising as a pNGB. While one should remain aware of the above central difficulty for all these constructions, one can at least argue that Nature already does make use of this mechanism in the description of pion and kaon dynamics with chiral lagrangians. It seems a shame not to use this idea more often!

*e-mail: ferretti@chalmers.se

$4\,(\psi,\tilde{\psi}) \in$ Complex irrep	$SU(4)\times SU(4)'/SU(4)_D$
$4\,\psi \in$ Pseudoreal irrep	$SU(4)/Sp(4)$
$5\,\psi \in$ Real irrep	$SU(5)/SO(5)$

Table 1. Minimal fermionic content for realistic EW symmetry breaking. All fermions are thought of as left-handed Weyl spinors.

2 Basic idea

The basic idea is to start with the Higgsless (and thus massless) part of the SM lagrangian

$$\mathcal{L}_{\rm SM0} = -\frac{1}{4}\sum_{F=GWB} F^2_{\mu\nu} + i\sum_{\Psi=QudLe} \bar{\Psi}\,\not{D}\Psi \tag{1}$$

and couple it to a theory $\mathcal{L}_{\rm comp.}$ with hyper-color gauge group $G_{\rm HC}$ and global symmetry structure $G_{\rm F}$ at some "flavor" scale $\Lambda_{\rm UV} \geq 10^3$ TeV spontaneously broken to $G_{\rm F} \to H_{\rm F}$ at energies Λ of a few TeV such that the Higgs doublet $H \in G_{\rm F}/H_{\rm F}$ and

$$\mathcal{L}_{\rm comp.} + \mathcal{L}_{\rm SM0} + \mathcal{L}_{\rm int.} \longrightarrow \mathcal{L}_{\rm SM} + \cdots \tag{2}$$

(where $\mathcal{L}_{\rm SM} + \cdots$ is the full SM plus possibly extra light matter from bound states of $\mathcal{L}_{\rm comp.}$ at energies below Λ.)

In order to realize the symmetry breaking mechanism above without reincurring in problems with naturalness one is led to use a fermionic bilinear condensate. The main difference with technicolor models is that the vacuum condensate is only mildly misaligned w.r.t. the electroweak (EW) group leading to acceptable corrections to the S-parameter. As far the EW sector is concerned, the minimal cosets of this type are those in table 1. The required misalignment can be induced by coupling the SM fermions to the hyper-color theory [2]. In order to accomplish this, one introduces in $\mathcal{L}_{\rm int.}$ four-fermion couplings between one SM fermion and three hyper-fermions, effectively giving rise, at low energies, to a linear coupling between SM fermions and composite bound-states from the strongly-coupled sector carrying the same SM quantum numbers as the SM fermion. This also gives rise to a novel mechanism to generate fermion masses known as partial compositeness [3]. A convenient way to realize this [4, 5] is to introduce two different types of hyper-fermions ψ and χ transforming under different irreps of $G_{\rm HC}$. The ψs are the same fermions that were employed to construct the EW coset of table 1, while the χs carry the necessary QCD color quantum numbers to allow coupling to the quarks, most notably the top quark.

In previous work [6, 7], we narrowed down the possibilities to a set of twelve models, summarized in table 2. The models in table 2 were selected (using the multi-loop beta-function) as those likely to lay outside of the conformal window, and thus in the confining region. The rationale for this choice is that one can always add more degrees of freedom to bring the theory inside the conformal window at a higher scale. If the scaling dimension of the fermionic composite operator is sufficiently large this allows the SM fermion mass (most crucially that of the top quark) to remain sizable until the confinement scale Λ is reached. Below this scale, these additional degrees of freedom decouple and the theory reduces to that described in table 2 giving rise to composite resonances at the TeV scale, among which the top partners.

$G_{\rm HC}$	ψ	χ	$G_{\rm F}/H_{\rm F}$
$SO(7)$	$5\times\mathbf{F}$	$6\times\mathbf{Spin}$	$\frac{SU(5)}{SO(5)}\frac{SU(6)}{SO(6)}U(1)$
$SO(9)$	$5\times\mathbf{F}$	$6\times\mathbf{Spin}$	
$SO(7)$	$5\times\mathbf{Spin}$	$6\times\mathbf{F}$	
$SO(9)$	$5\times\mathbf{Spin}$	$6\times\mathbf{F}$	
$Sp(4)$	$5\times\mathbf{A}_2$	$6\times\mathbf{F}$	$\frac{SU(5)}{SO(5)}\frac{SU(6)}{Sp(6)}U(1)$
$SU(4)$	$5\times\mathbf{A}_2$	$3\times(\mathbf{F},\overline{\mathbf{F}})$	$\frac{SU(5)}{SO(5)}\frac{SU(3)\times SU(3)'}{SU(3)_D}U(1)$
$SO(10)$	$5\times\mathbf{F}$	$3\times(\mathbf{Spin},\overline{\mathbf{Spin}})$	
$Sp(4)$	$4\times\mathbf{F}$	$6\times\mathbf{A}_2$	$\frac{SU(4)}{Sp(4)}\frac{SU(6)}{SO(6)}U(1)$
$SO(11)$	$4\times\mathbf{Spin}$	$6\times\mathbf{F}$	
$SO(10)$	$4\times(\mathbf{Spin},\overline{\mathbf{Spin}})$	$6\times\mathbf{F}$	$\frac{SU(4)\times SU(4)'}{SU(4)_D}\frac{SU(6)}{SO(6)}U(1)$
$SU(4)$	$4\times(\mathbf{F},\overline{\mathbf{F}})$	$6\times\mathbf{A}_2$	
$SU(5)$	$4\times(\mathbf{F},\overline{\mathbf{F}})$	$3\times(\mathbf{A}_2,\overline{\mathbf{A}}_2)$	$\frac{SU(4)\times SU(4)'}{SU(4)_D}\frac{SU(3)\times SU(3)'}{SU(3)_D}U(1)$

Table 2. Fermionic content of the twelve models of strongly coupled gauge theories passing the basic requirements of giving rise to a composite Higgs sector respecting custodial symmetry and a top quark partner. The models are referred to as M1 through M12. These models are conjectured to be outside of the conformal window.

3 Questions for the Lattice

In trying to estimate the viability of these models the first questions to be addressed concern the composite sector in isolation, before coupling it to the SM. For this task, the lattice can provide invaluable information. Looking at table 2 from this point of view one sees that the models can be grouped into three main classes:

- $SU(4)$ with N_F Fundamentals and N_A Antisymmetric (possibly also $SU(5)$)
- $Sp(4)$ with N_F Fundamentals and N_A Antisymmetric
- $SO(N)$ with N_F Fundamentals and N_S Spin (with $N = 7, 9, 10, 11$)

Lattice simulations of models in the first two classes have been performed (e.g. [8–11]) and more are underway. In figure 1 I give a schematic overview of the positioning of the four models of table 2 involving the hyper-color groups $Sp(4)$ and $SU(4)$. The positioning of the lower edge of the conformal window is an educated guess at best, but one can see that the models of interest fall very close to it, raising the hope that strong coupling effects give rise to the large anomalous dimensions needed for a realistic top quark mass.

Among the questions that one would like to answer are: Where does the boundary of the conformal window start? For models inside the window, can one find a fermionic composite operator of scaling dimension $\approx 5/2$? What are the masses of the composite fermionic resonances for the models in table 2? Can one estimate the low energy coefficients in the pNGB potential? Can one estimate the top Yukawa coupling? Some of these questions are starting to be addressed for $SU(4)$ [8, 9] and $Sp(4)$ [10, 11].

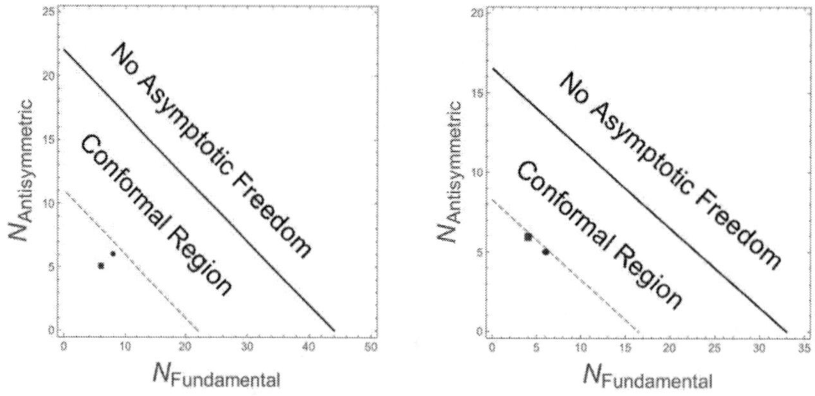

Figure 1. Positioning of some of the models in figure 2. Left: $G_{\text{HC}} = SU(4)$ (red square=M6, blue dot=M11). Right: $G_{\text{HC}} = Sp(4)$ (red square=M8, blue dot=M5). All fermions are counted as Weyl fermions.

4 An additional light ALP

In this section I describe a common feature [12, 13] of all of these models showing some promising potential for exclusion/discovery in hadronic collisions, most notably at LHCb [14]. At the classical level, all models in table 2 posses two axial $U(1)$ symmetries that are spontaneously broken by the hyper-quark condensates and weakly explicitly broken by the coupling to the SM fields. We can denote these two symmetries by $U(1)_\psi$ and $U(1)_\chi$ acting uniformly on all the flavor and gauge components of the fields, e.g. $\psi \to e^{i\alpha}\psi$. In the case of ψ transforming under a complex representation of G_{HC} this symmetry acts in the same way on the conjugate Weyl spinor as well: $(\psi, \tilde{\psi}) \to (e^{i\alpha}\psi, e^{i\alpha}\tilde{\psi})$. In this case there is an additional vector like symmetry $(\psi, \tilde{\psi}) \to (e^{i\beta}\psi, e^{-i\beta}\tilde{\psi})$, unbroken by the condensate $\langle\tilde{\psi}\psi\rangle$, that does not lead to additional pNGBs and shall not be discussed further.

Both axial $U(1)_{\psi,\chi}$s are subjected to a $U(1)_{\psi,\chi}G_{\text{HC}}G_{\text{HC}}$ ABJ anomaly, but one linear combination remains anomaly free. Denoting by $q_{\psi,\chi}$ the $U(1)$ charges of this combination, $N_{\psi,\chi}$ the number of flavor components (always counting Weyl degrees of freedom) and $T_{\psi,\chi}$ the index of their G_{HC} irrep ($T_{\psi,\chi} = T_{\tilde{\psi},\tilde{\chi}}$ for complex irreps) we require $q_\psi N_\psi T_\psi + q_\chi N_\chi T_\chi = 0$ for the cancellation of the anomaly. The pNGB a associated to the breaking of this combination $U(1)_a$ is light, since its mass comes from the explicit terms coupling the hyper-color theory to the SM. Let us denote the mass and "decay constant" of this pNGB by m_a and f respectively.

After canceling the $U(1)_a G_{\text{HC}} G_{\text{HC}}$ anomaly, there remains non-zero $U(1)_a SU(3)SU(3)$ and $U(1)_a U(1)_{\text{e.m.}} U(1)_{\text{e.m.}}$ anomalies with the QCD and QED gauge group, thus warranting calling this pNGB an "axion-like" particle (ALP). However, I hasten to point out that, because of the presence of the explicit breaking terms, this particle cannot solve the strong CP problem of QCD. For the same reason, its mass m_a is freed from obeying the relation $f m_a \approx f_\pi m_\pi$ typical of the proper QCD axion. This allows it to evade the usual constraints that exclude axions with decay constants $f < 10^{11}$ GeV. Needless to say, this particle decays promptly and has no relevance to Dark Matter. For an ALP decay constant $f = 1$ TeV, for the models in table 2, the widths are in the MeV range for masses $m_a \approx 60$ GeV, decreasing to few keV for masses $m_a \approx$ few GeV.

Because of its coupling to the gluon, this ALP could be the target of searches at hadron colliders. While the mass and decay constant depend on the strong coupling physics and on the specific details of the explicit symmetry breaking, the remaining couplings to the SM fermions and gauge bosons are more under control and can be estimated form the data in table 2. This leads to the following phenomenological lagrangian whose only undetermined parameters are m_a and f.

$$\mathcal{L}_{\text{eff}} = \frac{1}{2}(\partial_\mu a)^2 - \frac{1}{2}m_a^2 a^2 - i\sum \frac{C_\Psi m_\Psi}{f} a\bar{\Psi}\gamma^5\Psi + \frac{a}{16\pi^2 f}\left(g_s^2 K_g G^a_{\mu\nu}\tilde{G}^{a\mu\nu} + e^2 K_\gamma F_{\mu\nu}\tilde{F}^{\mu\nu}\right). \tag{3}$$

In (3) the sum is over the SM fermions Ψ of mass m_Ψ, $G^a_{\mu\nu}$ and $F_{\mu\nu}$ are the QCD and QED field strength and g_s and e their coupling constants. The couplings in (3) for the various models can be found in [14].

The strongest current bounds at low masses $m_a < 60$ GeV come from the $a \to \mu^+\mu^-$ channel and are given in [15]. They are presented as bounds on the mixing angle θ_H of the type IV two Higgs doublet model (2HDM) of [16, 17] at $\tan\beta = 0.5$. They can be easily recast to the models presented here by the following observation. In the case of the 2HDM models, both the production cross-sections and the decay rates are proportional to $\sin^2\theta_H$, while in the present models they scale as $(v/f)^2$, with $v = 246$ GeV the Higgs vacuum expectation value. This means that the recast can be easily obtained by the formula

$$\frac{v}{f} = \sqrt{\frac{\sigma(p\,p \to a \to \mu^+\mu^-)_{\sin\theta_H=1}}{\sigma(p\,p \to a \to \mu^+\mu^-)_{v=f}}}\,\sin\theta_H, \tag{4}$$

where the cross-sections are those of the 2HDM and (3) in the numerator and denominator respectively. The results of this recast are shown in figure 2, taken from [14].

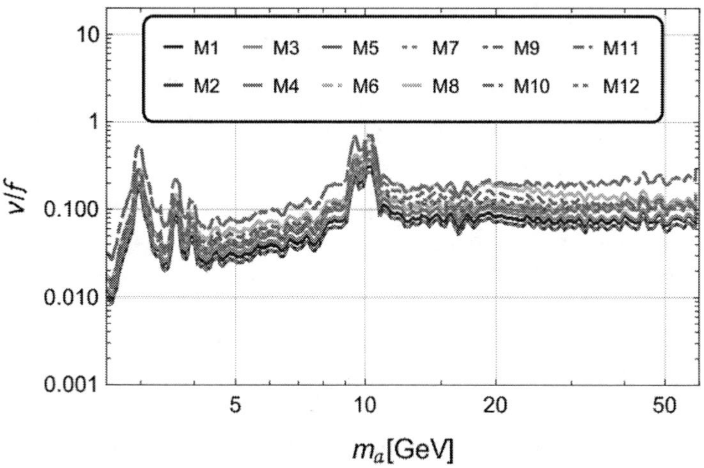

Figure 2. Recast of the LHCb bounds [15] for the present models. Taken from [14].

LHCb has also great potential to investigate other channels that have so far not been studied. One such channel is the $a \to \tau^+\tau^-$ decay mode, which is enhanced by a factor $(m_\tau/m_\mu)^2 \approx 286$ compared to the di-muon channel, under the assumption of $C_\tau \approx C_\mu$ in (3). The gain from this enhancement is however mitigated by the well known difficulties of working with taus, namely the unavoidable presence of missing energy (due to neutrinos) and the presence of hadronic tau decays that suffer from a large background. While all tau decay modes deserve attention, in [14] we showed that the most sensitive final state is the one in which the two taus decay to an electron and a muon respectively (the so-called opposite sign opposite flavor mode, (OSOF)). This comprises only 6.2% of the available final states from tau pairs but has the advantage of allowing for a significant reduction of the background at LHCb.

The full details of the analysis of the OSOF channel are given in [14]. Here I limit myself to point out that the two main sources of background for this mode are Drell-Yan production ($p\,p \to \tau^+\tau^-$) and bottom pairs production ($p\,p \to b\,\bar{b}$). The light (including charm) QCD jet background, which is the leading source of background for the semi-leptonic and hadronic tau decays is subdominant in this case. As an example, with the cut-flow discussed in [14], the signal efficiency for $m_a = 20$ GeV is 1.08×10^{-3}, while the Drell-Yan production is suppressed by 1.20×10^{-4} and $b\,\bar{b}$ production by 2.65×10^{-9}.

The projected sensitivity in v/f at 15 fb^{-1} is shown in figure 3. Comparing the two anal-

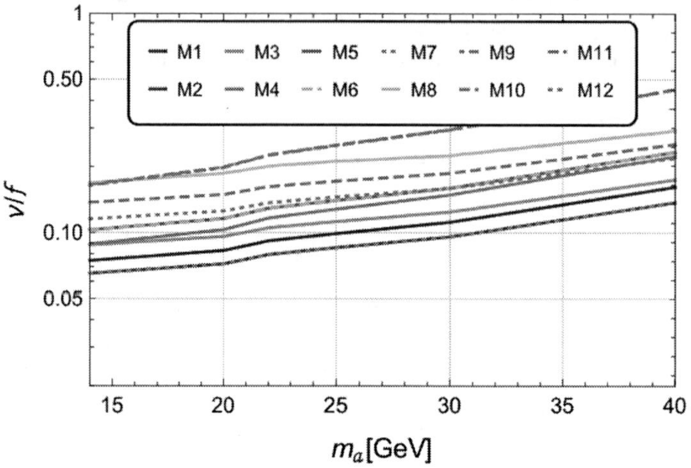

Figure 3. Projected sensitivity for integrated luminosity $L = 15$ fb^{-1}. Taken from [14].

ysis we see that muons still perform slightly better, but it should be remembered that the di-muon results come from a highly optimized experimental analysis. Also, di-tau resonances are interesting in their own right and would allow to probe models where $C_\mu \ll C_\tau$ accidentally.

5 Conclusions

In this short contribution I argued that realizing partial compositeness via ordinary 4D gauge theories provides a self contained concrete class of models to address the hierarchy problem. There are lots of open questions that go to the heart of strongly coupled theories, such as the range of the conformal window, anomalous dimensions and low energy coefficients, for which the lattice can provide invaluable information. These models are quite predictive, since many couplings can be computed from the underlying gauge theory. In particular, the presence of a ubiquitous light neutral pseudo-scalar a can be tested in various decay channels such as di-muon or di-tau.

6 Acknowledgments

I wish to thank the organizers of this workshop for their kind invitation and their big effort in the face of the current difficult situation. I thank my colleagues Diogo Buarque Franzosi,

Giacomo Cacciapaglia, Xabier Cid Vidal, Thomas Flacke and Carlos Vázquez Sierra for collaboration on many of the results presented in this talk. This work is supported by the Knut and Alice Wallenberg foundation under the grant KAW 2017.0100 (SHIFT project).

References

[1] D.B. Kaplan, H. Georgi, Phys. Lett. **136B**, 183 (1984)
[2] N. Arkani-Hamed, A.G. Cohen, E. Katz, A.E. Nelson, JHEP **07**, 034 (2002), hep-ph/0206021
[3] D.B. Kaplan, Nucl. Phys. **B365**, 259 (1991)
[4] J. Barnard, T. Gherghetta, T.S. Ray, JHEP **02**, 002 (2014), 1311.6562
[5] G. Ferretti, D. Karateev, JHEP **03**, 077 (2014), 1312.5330
[6] G. Ferretti, JHEP **06**, 107 (2016), 1604.06467
[7] A. Belyaev, G. Cacciapaglia, H. Cai, G. Ferretti, T. Flacke, A. Parolini, H. Serodio, JHEP **01**, 094 (2017), 1610.06591
[8] V. Ayyar, T. Degrand, D.C. Hackett, W.I. Jay, E.T. Neil, Y. Shamir, B. Svetitsky, Phys. Rev. D **97**, 114505 (2018), 1801.05809
[9] V. Ayyar, T. DeGrand, D.C. Hackett, W.I. Jay, E.T. Neil, Y. Shamir, B. Svetitsky, Phys. Rev. D **99**, 094502 (2019), 1812.02727
[10] J.W. Lee, E. Bennett, D.K. Hong, C.J.D. Lin, B. Lucini, M. Piai, D. Vadacchino, PoS **LATTICE2018**, 192 (2018), 1811.00276
[11] E. Bennett, D.K. Hong, J.W. Lee, C.J.D. Lin, B. Lucini, M. Piai, D. Vadacchino, JHEP **12**, 053 (2019), 1909.12662
[12] G. Cacciapaglia, G. Ferretti, T. Flacke, H. Serodio, Eur. Phys. J. C **78**, 724 (2018), 1710.11142
[13] G. Cacciapaglia, G. Ferretti, T. Flacke, H. Serôdio, Front. in Phys. **7**, 22 (2019), 1902.06890
[14] D. Buarque Franzosi, G. Cacciapaglia, X. Cid Vidal, G. Ferretti, T. Flacke, C. Vázquez Sierra (2021), 2106.12615
[15] R. Aaij et al. (LHCb), JHEP **10**, 156 (2020), 2007.03923
[16] U. Haisch, J.F. Kamenik, Phys. Rev. D **93**, 055047 (2016), 1601.05110
[17] U. Haisch, J.F. Kamenik, A. Malinauskas, M. Spira, JHEP **03**, 178 (2018), 1802.02156

EPJ Web of Conferences **258**, 08003 (2022) https://doi.org/10.1051/epjconf/202225808003
A Virtual Tribute to Quark Confinement and the Hadron Spectrum (vConf21)

Sp(4) gauge theories and beyond the standard model physics

Biagio Lucini[1,2,*], *Ed* Bennett[2], *Jack* Holligan[3,4], *Deog Ki* Hong[5], *Ho* Hsiao[6], *Jong-Wan* Lee[5,7], *C.-J. David* Lin[6,8,9], *Michele* Mesiti[2], *Maurizio* Piai[3], and *Davide* Vadacchino[10]

[1]Department of Mathematics, Swansea University, Fabian Way, SA1 8EN Swansea, Wales, UK
[2]Swansea Academy of Advanced Computing, Swansea University, Fabian Way, SA1 8EN Swansea, Wales, UK
[3]Department of Physics, Swansea University, Singleton Park, SA2 8PP Swansea, Wales, UK
[4]Department of Physics, University of Maryland, College Park, Maryland, USA, 20742
[5]Department of Physics, Pusan National University, Busan 46241, Korea
[6]Institute of Physics, National Yang Ming Chiao Tung University, 1001 Ta-Hsueh Road, Hsinchu 30010, Taiwan
[7]Extreme Physics Institute, Pusan National University, Busan 46241, Korea
[8]Center for High Energy Physics, Chung-Yuan Christian University, Chung-Li 32023, Taiwan
[9]Centre for Theoretical and Computational Physics, National Yang Ming Chiao Tung University, 1001 Ta-Hsueh Road, Hsinchu 30010, Taiwan
[10]School of Mathematics and Hamilton Mathematics Institute, Trinity College, D02 PN40 Dublin 2, Ireland

Abstract. We review numerical results for models with gauge group $Sp(2N)$, discussing the glueball spectrum in the large-N limit, the quenched meson spectrum of $Sp(4)$ with Dirac fermions in the fundamental and in the antisymmetric representation and the $Sp(4)$ gauge model with two dynamical Dirac flavours. We also present preliminary results for the meson spectrum in the $Sp(4)$ gauge theory with two fundamental and three antisymmetric Dirac flavours. The main motivation of our programme is to test whether this latter model is viable as a realisation of Higgs compositeness via the pseudo Nambu Goldstone mechanism and at the same time can provide partial top compositeness. In this respect, we report and briefly discuss preliminary results for the mass of the composite baryon made with two fundamental and one antisymmetric fermion (*chimera baryon*), whose physical properties are highly constrained if partial top compositeness is at work. Our investigation shows that a fully non-perturbative study of Higgs compositeness and partial top compositeness in $Sp(4)$ is within reach with our current lattice methodology.

1 Introduction

$Sp(4)$ gauge theory with two fundamental and three antisymmetric Dirac fermion flavours has been suggested as a possible template [1–3] for beyond the standard model strong interactions that can give rise to a composite Higgs boson through the breaking of the global fundamental flavour symmetry [4] and at the same time to a partial composite top quark state

*Speaker; e-mail: b.lucini@swansea.ac.uk

that explains why the observed mass of the latter particle is at the electroweak scale [5]. Partial top compositeness results from the mixing of the standard model top quark with a chimera baryon, i.e. a baryonic state formed with two fermions in the fundamental and one fermion in the antisymmetric representation. A necessary ingredient for partial top compositeness is the generation of a large anomalous dimension for the chimera baryon. Lattice gauge theory can be used as a framework to study non-perturbatively these phenomena, in order to assess their viability beyond semi-quantitative arguments.

In this work, we shall discuss our previous and current lattice investigations of the composite Higgs mechanism and of partial top compositeness using $Sp(4)$ gauge theory as a concrete realisation (for studies involving $SU(4)$, see, e.g., [6–8]), starting from special cases of the target action and approaching the latter in a *crescendo* of complexity. The work is organised as follows. In Sect. 2 we introduce the lattice model. Numerical results are presented in Sect. 3. Finally, we draw our conclusions and we give a brief outlook of forthcoming numerical investigations in Sect. 4.

2 Lattice formulation and observables

The lattice action we have used in our calculations in $Sp(2N)$ gauge theories with fermions in multiple representations R can be written as

$$S = \beta \sum_{x} \sum_{\mu<\nu} \left(1 - \frac{1}{2N_c} \Re \mathrm{Tr} \mathcal{P}_{\mu\nu}\right) + \sum_{j=1}^{N_f^R} \sum_{x,y} \overline{\Psi}_j^R(x) D^R(x,y) \Psi_j^R(y) \,. \tag{1}$$

The first term is the Wilson plaquette action, with $\beta = 2N_c/g^2$, where $N_c = 2N$ is the number of colours and g is the gauge coupling, $\mathcal{P}_{\mu\nu}(x)$ is the path exponential of the link variables $U_\mu(x) \in Sp(2N)$ around the elementary plaquette stemming from point x in directions μ, ν and $\Re$Tr indicates the real part of the trace. The second term is the fermionic part of the action. This term includes a sum over the representations R and over the flavours N_f^R at fixed representation. For the Dirac operator $D(x, y)$ we use the Wilson discretisation, given by

$$D^R(x,y) = (4 + am_0^R)\delta_{x,y} - \frac{1}{2}\sum_{\mu}\left\{ (1-\gamma_\mu)U_\mu^R(x)\delta_{x+\hat{\mu},y} + (1+\gamma_\mu)U_\mu^{R\dagger}(y)\delta_{x-\hat{\mu},y} \right\}, \tag{2}$$

where $U_\mu^R(X)$ is the gauge link variable in the representation R, a is the lattice spacing and m_0^R is the bare mass of the N_f^R degenerate flavours in representation R.

The path integral of the theory can be expressed as

$$Z = \int (\mathcal{D}U)\left(\mathcal{D}\Psi^R\right)\left(\mathcal{D}\overline{\Psi}^R\right) e^{-S} \,, \tag{3}$$

where the integral is performed over all fields and the action, S, is given by Eq. (1). In this work, we consider two flavours of fermions in the fundamental representation $R \equiv F$ and three flavours of fermions in the antisymmetric representation $R \equiv A$.

Given an operator $O(x_1, \ldots, x_n)$, which is a function of fields at points $x_1, \ldots, x_n$, its vacuum expectation value is computed as

$$\langle O(x_1,\ldots,x_n)\rangle = \frac{1}{Z}\int (\mathcal{D}U)\left(\mathcal{D}\Psi^R\right)\left(\mathcal{D}\overline{\Psi}^R\right) O(x_1,\ldots,x_n) e^{-S} \,, \tag{4}$$

where the path integral is calculated using Monte Carlo importance sampling.

We shall consider the following observables:

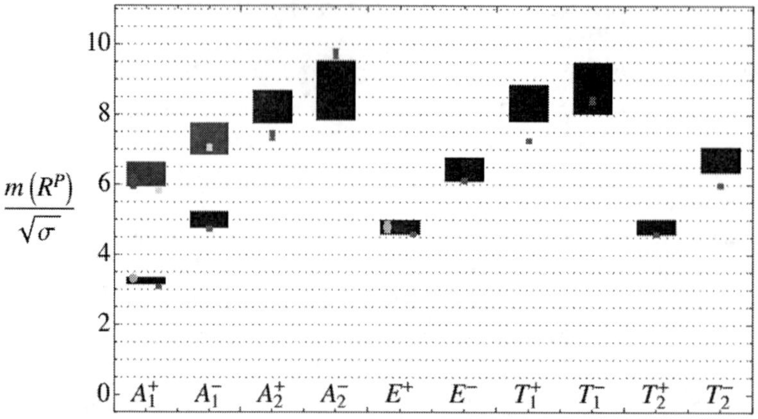

Figure 1. The $Sp(2N)$ glueball spectrum in the large-N limit (boxes, with ground states in each channel depicted in black and excitations in red) compared with $SU(N)$ data extrapolated to the same limit (circles with error bars). See text for further details.

- Glueball correlators, which take the form

$$C_{ij}(x_1, x_2) = \langle O_i(x_1) O_j^\dagger(x_2)\rangle - \langle O_i(x_1)\rangle\langle O_j^\dagger(x_2)\rangle\,, \tag{5}$$

where $O_i(x_1)$ and $O_j(x_2)$ are combinations of Wilson loops transforming in an irreducible representation of the octahedral group, at zero momentum and with defined parity;

- Meson correlators, which are expressed as

$$C^R_{\Gamma_1\Gamma_2}(x_1, x_2) = \left\langle \overline{\Psi}^R(x_1)\Gamma_1\Psi^R(x_1)\left(\overline{\Psi}^R(x_2)\Gamma_2\Psi^R(x_2)\right)^\dagger\right\rangle\,, \tag{6}$$

with Γ_1 and Γ_2 combinations of Euclidean Dirac matrices, whose specific form determines the J^{PC} quantum numbers of the states that saturate the propagator;

- Chimera baryon correlator, written as

$$C_T^{FFA}(x_1, x_2) = \left\langle \left(T\Psi^F(x_1)\Psi^F(x_1)\Psi^A(x_1)\right)\left(T\Psi^F(x_2)\Psi^F(x_2)\Psi^A(x_2)\right)^\dagger\right\rangle\,, \tag{7}$$

where T is a tensor contracting flavour and colour indices, which, for the sake of conciseness, will be left implicit. More details on this construction will be provided in a forthcoming publication.

3 Numerical Results

Before tackling the full theory, relevant operators have been studied in notable limits of the latter. Since the properties of the model in these limits are better known, the simplified calculations have enabled us to develop the methodology and at the same time to provide a set of safe reference measurements against which to check the complete calculations. Here, we report a selection of our numerical results. For additional measurements and their discussion, we refer to the quoted original papers.

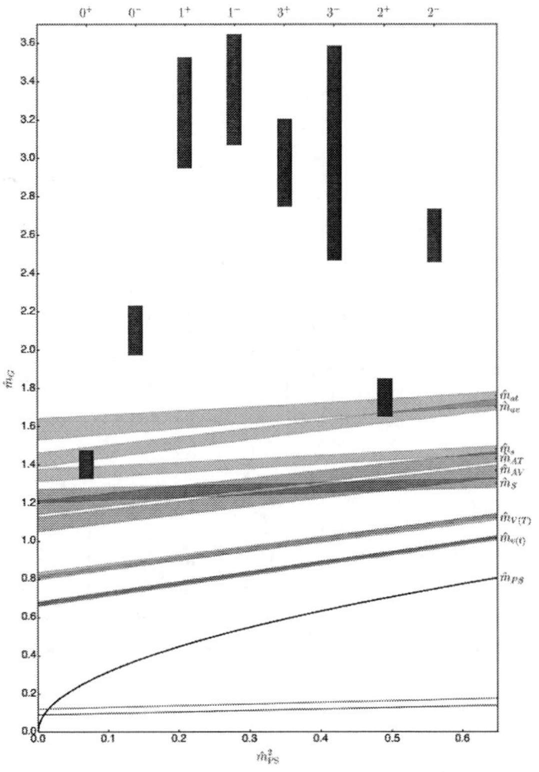

Figure 2. Quenched spectrum for $Sp(4)$ gauge theory with fermions in the fundamental and in the antisymmetric representation extrapolated to the continuum limit as a function of the pseudoscalar mass squared (in black). The bands account for the statistical errors. The fundamental spectrum is drown in blue, while the antisymmetric spectrum is represented in red. Observables are labelled in the right vertical axis. Glueball masses are also reported (boxes), with quantum numbers labelled on the top horizontal axes. Note that, while there is a dependency of mesonic quantities from the pion mass, since the theory is quenched, the glueball spectrum is constant as a function of the constituent fermion mass.

The first model we have considered is the pure gauge system (first investigated in [9]). In this case, relevant observables are glueball masses and the string tension. These are extracted from correlators of the type given in Eq. (5), looking at the asymptotic behavior

$$C_{ij}(x_1, x_2) \underset{|x_1 - x_2| \to \infty}{\simeq} C e^{-M|x_1 - x_2|} , \tag{8}$$

where M is the mass of the lowest-lying state carrying the quantum numbers of the source operators. In practice, the calculation of single correlators is very noisy. The bad signal-to-noise ratio can be overcome with a thorough variational calculation. The methodology we have used is explained in [10]. The string tension can be extracted similarly, considering correlators of Polyakov loops.

We have computed in [10] the lowest-lying glueball spectrum in units of the string tension in the continuum limit for $Sp(2N)$ models with $N = 2, 3, 4$ and we have extrapolated the latter to $N \to \infty$. For the large-N extrapolation, we have found that the leading expected $1/N$ correction provides a good description of the data. Our results are displayed by the boxes of Fig. 1. The black boxes indicate the masses of the groundstate in each channel, with the red

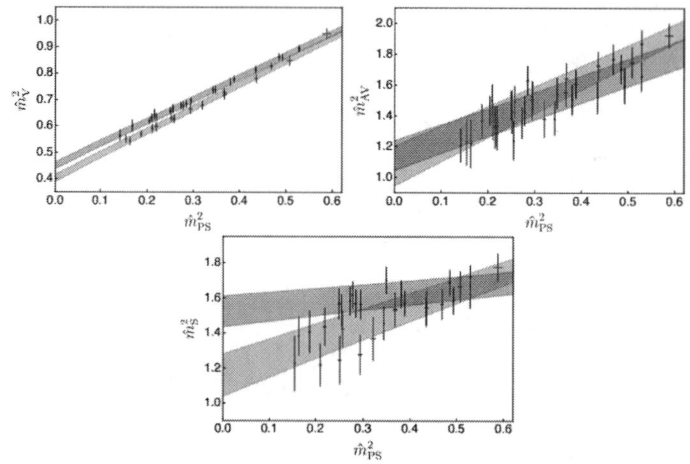

Figure 3. Squared masses of the vector (top left), the axial vector (top right) and the isotriplet scalar meson (bottom) as a function of the squared mass of the pseudoscalar for the fundamental representation in the continuum limit. We plot in red dynamical data and in blue quenched data, with the corresponding bands representing the chiral extrapolation.

boxes representing masses of the excited states (for the channels for which the latter turned out to be measurable on our samples). The quantum numbers refer to the irreducible representations of the octahedral group. The continuum J^P quantum numbers[1] can be reconstructed following the method described, e.g., in [11].

It is worth remarking that, since the large-N limit of $SU(N)$ and $Sp(2N)$ coincide, for the majority of the glueball states, our work is the first extrapolation of the glueball spectrum to $N = \infty$. Indeed, the only states whose $SU(N)$ large-N mass had been determined before our calculations are the groundstate and the first excitation in the spin zero channel (in our notation, the A_1 channel) and the spin two (which appears in both the E and the T_2 channel) [12, 13] (see also [14]). We report in Fig. 1 the more constraining measurements of [13] (in cyan for the groundstates and in green for the excitation in the A_1^+ channel), which are in full agreement with ours. After our work was published, a comprehensive study of the large-N $SU(N)$ glueball spectrum was published [15]. This enables us to perform a broader comparison of the two extrapolations. The results of [15] are also reported in Fig. 1 (in magenta for the groundstates, and in yellow for the excitations). While we notice that the authors of [15] quote significantly smaller errors, the agreement with our measurements is within the error bars (as measured on our own data) in most of the cases, with the largest discrepancies being at most around two standard deviations. This comparison provides evidence of the universality of the large-N glueball spectrum across the $SU(N)$ and the $Sp(2N)$ series.

We now discuss the quenched spectrum of the $Sp(4)$ gauge theory, comparing masses of glueballs and mesons, the latter studied as the mass of the constituent fermions is decreased. We consider a mesonic correlators of the form (6). Its asymptotic behaviour is given by

$$C^F_{\Gamma_1\Gamma_2}(x_1, x_2) \simeq A\left(e^{-Mt} + e^{-M(T-t)}\right) , \tag{9}$$

[1]We remind the reader that, since $Sp(2N)$ groups are pseudoreal, the charge conjugation quantum number of glueballs in $Sp(2N)$ is always $+1$.

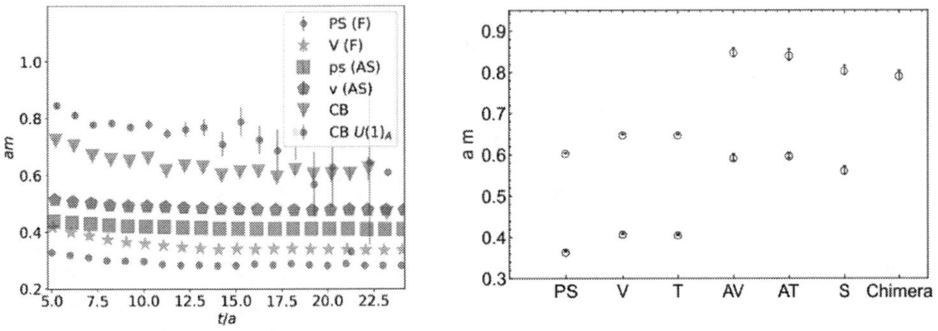

Figure 4. Left: The lowest-lying meson spectrum in $Sp(4)$ with three dynamical antisymmetric fermions. Taking quenched fundamental sources in this model, we also show the pseudoscalar and the vector meson masses in the fundamental representation, the chimera baryon (CB) and its $U(1)_A$ partner. The lattice parameters are discussed in the text. Right: The meson spectrum for both representations and the chimera baryon in the $Sp(4)$ gauge theory with two fundamental dynamical flavours and three antisymmetric dynamical flavours, at the lattice parameters discussed in the text.

where the second term (due to waves travelling across the antiperiodic boundary of our lattice) has now been inserted explicitly, as, unlike in the glueball case, neglecting it would give a measurable systematic error. In the previous formula, T is the temporal extension of the lattice. Points x_1 and x_2 have spacial coordinates summed over $|\vec{x}_1 - \vec{x}_2|$ at fixed $\vec{x}_2$ (i.e., the correlator has zero net momentum). The temporal coordinates are respectively τ and $t + \tau$ at fixed $\tau < T$. The decay constant of the considered states can be extracted from A, either directly or (in the case of the pseudoscalar meson) considering two appropriate combinations of correlators. We refer to the quoted original works for the details, including renormalisation of decay constants and how the choice of the Γ matrices identifies the J^{PC} quantum numbers.

The continuum quenched meson spectrum (in both the fundamental and antisymmetric representation), originally discussed in [16], is reported in Fig. 2, together with a selection of glueball states taken from [17]. The scale is set using the gradient flow derived quantity w_0, as described in [18]. The meson spectrum in both representations is plotted as a function of the pseudoscalar mass in that representation. Glueball states are also reported for the same model. The data show some clear indications: (1) in both representations, the general features of the quenched spectrum are qualitatively similar to those of QCD; (2) while the meson spectrum has the same behaviour in both representations, at fixed pseudoscalar mass antisymmetric states are around 10-20% heavier than the corresponding fundamental states, with the percent difference increasing towards the light fermion regime; (3) glueballs are generally significantly heavier than mesons, but the lightest glueball states have masses that are comparable with masses of states in the heaviest mesonic channels.

We have then studied the theory with fermionic matter, comparing the two-flavour dynamical model [19] to the quenched case [18] for fundamental representation fermions in the continuum limit. The masses of the vector, axial vector and isotriplet scalar meson are plotted in Fig. 3. With the exception of the latter, typically unquenching effects are very small, and hardly visible in the range of the obtained pion masses.

In the subsequent stage of our investigation programme of $Sp(4)$ gauge models, we have studied the system in a partially quenched setup, considering three dynamical antisymmetric fermions and two static fundamental flavours. This gives us an opportunity to perform a first study of the chimera baryon in a model that is less complicated to simulate than the target mixed-representation system. Figure 4 (left) displays our preliminary results for the two lowest-lying mesonic states (the pseudoscalar and the scalar) in both representations, together with the chimera baryon and its parity partner. Results have been obtained on a 48×24^3 lattice at gauge coupling $\beta = 6.65$, antisymmetric fermion bare mass $am_0^A = -1.07$ and quenched fundamental mass $am_0^F = -0.734$. Statistical noise has been reduced using Wuppertal smearing [20]. The data show that in our theory and for the used lattice parameters, chimera baryons are heavier than the antisymmetric vector meson. Beyond the (not unimportant) numerical details, likely to be dependent on our particular setup, the main conclusion of our investigation is that the technology we have developed enables us to determine masses of chimera baryons. Note in particular the visible splitting between the chimera baryon and its $U(1)_A$ partner.

Finally, we are now performing calculations in $Sp(4)$ gauge theory with two dynamical fundamental fermions and three dynamical antisymmetric fermions. A preliminary plot the meson spectrum and of the chimera baryon is reported in Fig. 4, right. The lattice parameters are $\beta = 6.5$, $am_0^A = -1.01$ and $am_0^F = -0.71$, and the lattice size is 48×24^3. The results show that for our choice of parameters the chimera baryon has a mass comparable to that of the axial vector and of the isotriplet scalar mesons in the antisymmetric representation.

4 Conclusions and outlook

Following the programme we started in [18], we have performed a set of extensive simulations aimed to assess the viability of $Sp(4)$ gauge theory with two fundamental and three antisymmetric fermions as a realisation of the composite Higgs and of the partial top compositeness mechanism. So far we have developed the needed technology by studying the glueball spectrum [17], the quenched meson spectrum [16] and the dynamical theories respectively with two fundamental [19] and three antisymmetric fermion flavours. A selection of the results is reported in this work. A more systematic calculation in the target phenomenological model, which includes dynamical fermions in both representations, is now within reach. We have presented some preliminary results for the latter model. A more detailed account of its features will be reported in a forthcoming publication. Another interesting direction, pursued by the authors of [21], is to explore an $Sp(4)$ gauge models with two non-degenerate fundamental fermions in the context of SIMP dark matter.

Acknowledgements - The work of E. B. has bee n funded by the Supercomputing Wales project, which is part-funded by the European Regional Development Fund (ERDF) via Welsh Government. J. H. is supported by the STFC Consolidated Grant No. ST/P00055X/1, by the College of Science, Swansea University, and by the Grant No. STFC-DTG ST/R505158/1. The work of D. K. H. was supported by Basic Science Research Program through the National Research Foundation of Korea (NRF) funded by the Ministry of Education (NRF-2017R1D1A1B06033701). The work of J. W. L. is supported in part by the National Research Foundation of Korea funded by the Ministry of Science and ICT (NRF-2018R1C1B3001379) and in part by Korea Research Fellowship program funded by the Ministry of Science, ICT and Future Planning through the National Research Foundation of Korea (2016H1D3A1909283). The work of C. J. D. L. is supported by the Taiwanese MoST Grant No. 105-2628-M-009-003- MY4. The work of B. L. and M. P. has been supported in part by the STFC Consolidated Grant No. ST/T000813/1. B. L. and M. P. received funding from the European Research Council (ERC) under the European Union's Horizon 2020 research and innovation program under Grant Agreement No. 813942. The work of B. L. is further supported in part by the Royal Society Wolfson Research Merit Award No. WM170010 and by the Leverhulme Trust Research Fellowship

No. RF-2020-4619. The work of D. V. is supported in part by the INFN HPCHTC project and in part by the Simons Foundation under the program "Targeted Grants to Institutes" awarded to the Hamilton Mathematics Institute. Numerical simulations have been performed on the Swansea SUNBIRD cluster (part of the Supercomputing Wales project) and AccelerateAI A100 GPU system, on the local HPC clusters in Pusan National University (PNU) and in National Chiao-Tung University (NCTU), and on the Cambridge Service for Data Driven Discovery (CSD3). The Supercomputing Wales project and AccelerateAI are part funded by the European Regional Development Fund (ERDF) via Welsh Government. CSD3 is operated in part by the University of Cambridge Research Computing on behalf of the STFC DiRAC HPC Facility (www.dirac.ac.uk). The DiRAC component of CSD3 was funded by BEIS capital funding via STFC capital Grants No. ST/P002307/1 and No. ST/R002452/1 and STFC operations Grant No. ST/R00689X/1. DiRAC is part of the National e-Infrastructure.

References

[1] J. Barnard, T. Gherghetta, T.S. Ray, JHEP **02**, 002 (2014), 1311.6562
[2] G. Ferretti, D. Karateev, JHEP **03**, 077 (2014), 1312.5330
[3] G. Cacciapaglia, G. Ferretti, T. Flacke, H. Serôdio, Front. in Phys. **7**, 22 (2019), 1902.06890
[4] D.B. Kaplan, H. Georgi, Phys. Lett. B **136**, 183 (1984)
[5] D.B. Kaplan, Nucl. Phys. B **365**, 259 (1991)
[6] V. Ayyar, T. DeGrand, M. Golterman, D.C. Hackett, W.I. Jay, E.T. Neil, Y. Shamir, B. Svetitsky, Phys. Rev. D **97**, 074505 (2018), 1710.00806
[7] V. Ayyar, T. Degrand, D.C. Hackett, W.I. Jay, E.T. Neil, Y. Shamir, B. Svetitsky, Phys. Rev. D **97**, 114505 (2018), 1801.05809
[8] G. Cossu, L. Del Debbio, M. Panero, D. Preti, Eur. Phys. J. C **79**, 638 (2019), 1904.08885
[9] K. Holland, M. Pepe, U.J. Wiese, Nucl. Phys. B **694**, 35 (2004), hep-lat/0312022
[10] E. Bennett, J. Holligan, D.K. Hong, J.W. Lee, C.J.D. Lin, B. Lucini, M. Piai, D. Vadacchino, Phys. Rev. D **102**, 011501 (2020), 2004.11063
[11] B. Lucini, A. Rago, E. Rinaldi, JHEP **08**, 119 (2010), 1007.3879
[12] B. Lucini, M. Teper, JHEP **06**, 050 (2001), hep-lat/0103027
[13] B. Lucini, M. Teper, U. Wenger, JHEP **06**, 012 (2004), hep-lat/0404008
[14] B. Lucini, M. Panero, Phys. Rept. **526**, 93 (2013), 1210.4997
[15] A. Athenodorou, M. Teper (2021), 2106.00364
[16] E. Bennett, D.K. Hong, J.W. Lee, C.J.D. Lin, B. Lucini, M. Mesiti, M. Piai, J. Rantaharju, D. Vadacchino, Phys. Rev. D **101**, 074516 (2020), 1912.06505
[17] E. Bennett, J. Holligan, D.K. Hong, J.W. Lee, C.J.D. Lin, B. Lucini, M. Piai, D. Vadacchino, Phys. Rev. D **103**, 054509 (2021), 2010.15781
[18] E. Bennett, D.K. Hong, J.W. Lee, C.J.D. Lin, B. Lucini, M. Piai, D. Vadacchino, JHEP **03**, 185 (2018), 1712.04220
[19] E. Bennett, D.K. Hong, J.W. Lee, C.J.D. Lin, B. Lucini, M. Piai, D. Vadacchino, JHEP **12**, 053 (2019), 1909.12662
[20] S. Gusken, Nucl. Phys. B Proc. Suppl. **17**, 361 (1990)
[21] A. Maas, F. Zierler, PoS(LATTICE2021) to appear (2021), 2109.14377

Dilaton chiral perturbation theory and applications

Maarten Golterman[1,*] and *Yigal* Shamir[2,**]

[1]Department of Physics and Astronomy, San Francisco State University, San Francisco, CA 94132, USA
[2]Raymond and Beverly Sackler School of Physics and Astronomy, Tel Aviv University, 69978, Tel Aviv, Israel

Abstract. We review dilaton chiral perturbation theory (dChPT), the effective low-energy theory for the light sector of near-conformal, confining theories. dChPT provides a systematic expansion in both the fermion mass and the distance to the conformal window. It accounts for the pions and the light scalar, the approximate Nambu–Goldstone bosons for chiral and scale symmetry, respectively. A unique feature of dChPT is the existence of a large-mass regime in which the theory exhibits approximate hyperscaling, while the expansion nevertheless remains systematic. We discuss applications to lattice data, presenting successes as well as directions for future work.

1 Introduction

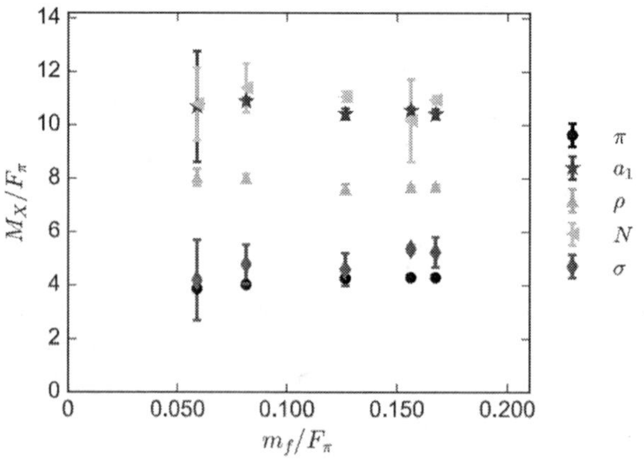

Figure 1. *Hadron masses as a function of fermion mass, all in units of F_π, in the SU(3) gauge theory with 8 fundamental flavors, from Ref. [2].*

Let us start with a look at spectral data for the SU(3) gauge theory with $N_f = 8$ fundamental fermions obtained by the LSD collaboration [1, 2], shown in Fig. 1. In comparison

*e-mail: maarten@sfsu.edu
**e-mail: shamir@tauex.tau.ac.il

with QCD, several salient differences stand out: First, we note the presence of a stable 0^{++} state, denoted by the points labeled σ, with a mass virtually degenerate with the pions, over this mass range. Second, within errors, the spectrum shows hyperscaling: hadronic mass ratios are essentially independent of the fermion mass. In addition, pion "taste splittings" associated with the use of staggered fermions in all these simulations, *i.e.*, non-degeneracies in the pion spectrum due to scaling violations, behave very differently as a function of the fermion mass in comparison with QCD (not shown in the figure). Similar results have been found in Ref. [3] by the LatKMI collaboration for the same theory, and in the SU(3) theory with two sextet fermions by the LatHC collaboration [4, 5], or with four light and six [6] or eight [7] heavy fundamental fermions. In this talk, we will discuss to what extent these data are decribed by tree-level dilaton Chiral Perturbation Theory (dChPT).

After a brief review of dChPT and its lowest-order lagrangian, we will explain the existence of a "large-mass" regime in which dChPT predicts approximate hyperscaling, as seen in Fig. 1. Going beyond hyperscaling, we will then apply dChPT to fit the LSD and LatKMI data, including the measured staggered taste splittings. We conclude with a brief discussion of what might be next.

2 Lowest-order dChPT

Dilaton ChPT is based on the following set of assumptions [8, 9]:

- Every gauge theory below the conformal sill contains Nambu–Goldstone bosons, or "pions," associated with the spontaneous breaking of chiral symmetry. The pions become massless in the limit $m \to 0$, where m is the (degenerate) fermion mass.
- Scale invariance gets restored in the infrared as we approach the conformal window, with the trace anomaly being proportional to the distance to the conformal window. In the Veneziano limit $N_f \to \infty$, $N_c \to \infty$, with $n_f = N_f/N_c$ fixed, this happens when n_f approaches a critical value n_f^* from below. The difference $n_f - n_f^*$ is a new small parameter.
- The theory contains a *dilaton, i.e.*, a Nambu–Goldstone boson associated with scale symmetry breaking, which becomes massless in the double limit $n_f - n_f^* \to 0$ and $m \to 0$.
- In addition, some technical assumptions on the dilaton potential are needed, see Ref. [10].

With these assumptions, one can prove that a systematic power counting in the small parameters

$$p^2 \sim m \sim n_f - n_f^* \sim 1/N_c \tag{1}$$

exists, and construct the lowest-order ($O(p^2)$) lagrangian,

$$\begin{aligned}\mathcal{L} \;=\; & \frac{1}{4} f_\pi^2 e^{2\tau}\, \mathrm{tr}(\partial_\mu \Sigma^\dagger \partial_\mu \Sigma) + \frac{1}{2} f_\tau^2 e^{2\tau} \partial_\mu \tau \partial_\mu \tau \\ & - \frac{1}{2} f_\pi^2 B_\pi m e^{(3-\gamma_*)\tau} \mathrm{tr}(\Sigma + \Sigma^\dagger) + f_\tau^2 B_\tau e^{4\tau} c_1 \left(\tau - \frac{1}{4}\right) .\end{aligned} \tag{2}$$

Here $\Sigma = \exp(2i\pi/f_\pi)$ is the non-linear pion field and τ is the dilaton field. c_1 is a parameter proportional to $n_f - n_f^*$, of order p^2 in our power counting. At leading order, five low-energy constants (LECs) appear: $f_{\pi,\tau}$, $B_{\pi,\tau}$, and γ_*, the mass anomalous dimension at the infra-red fixed point at the conformal sill. For a detailed discussion of the lagrangian and the power counting, see Ref. [8]. We have used the freedom to shift the field τ such that $v(m) = \langle\tau\rangle$ vanishes for $m = 0$, at this order.

The saddle point equation for the classical solution $v(m)$ is

$$\frac{m}{c_1 \mathcal{M}} = v(m) e^{(1+\gamma_*)v(m)} , \qquad \mathcal{M} \equiv \frac{4 f_\tau^2 B_\tau}{f_\pi^2 B_\pi N_f (3-\gamma_*)} . \tag{3}$$

The tree-level pion and dilaton masses, and their decay constants, are, using Eq. (3),

$$\begin{aligned} M_\pi^2 &= 2B_\pi m e^{(1-\gamma_*)v(m)} = 2B_\pi c_1 \mathcal{M} v(m) e^{2v(m)} , \\ M_\tau^2 &= 4B_\tau c_1 e^{2v(m)} (1 + (1+\gamma_*)v(m)) , \\ F_{\pi,\tau} &= f_{\pi,\tau} e^{v(m)} . \end{aligned} \tag{4}$$

Unlike in ordinary ChPT, here the tree-level decay constant(s) depend on the fermion mass via the factor $e^{v(m)}$.

3 The small- and large-mass regimes

An important observation is that in our power counting the ratio $\frac{m}{c_1\mathcal{M}}$ is parametrically $O(1)$. But this ratio can still be small, corresponding to the *small-mass regime*, or large, corresponding to the *large-mass regime*.

In the small-mass regime, $\frac{m}{c_1\mathcal{M}} \ll 1$, and $v \propto m$ while $e^v = 1 + O(m)$. One has $M_\pi^2 = 2B_\pi m \ll M_\tau^2 = 4B_\tau c_1 \propto |n_f - n_f^*|$. Even though, relative to the pions, the dilaton decouples, its mass still remains parameterically smaller than that of all other hadrons. For energies below the scale set by the dilaton mass the physics of pions only is described by standard ChPT.

The large-mass regime, $\frac{m}{c_1\mathcal{M}} \gg 1$, is more interesting [10]. In Eq. (3) the exponential dominates, and this equation has the approximate solution

$$e^{v(m)} \sim \left(\frac{m}{c_1\mathcal{M}}\right)^{1/1+\gamma_*} . \tag{5}$$

In turn, this implies that Eq. (4) predicts approximate hyperscaling:

$$M_\pi \sim M_\tau \sim F_\pi \sim F_\tau \sim M_\mathrm{h} \sim m^{1/1+\gamma_*} , \tag{6}$$

which can also be shown to apply to other hadron masses generically denoted as M_h. dChPT thus predicts that, in the large-mass regime, the theory behaves approximately as a mass-deformed conformal theory. Given the spectral results shown in Fig. 1, this is the first success of dChPT. The intuitive reason for hyperscaling is that, with $m \gg c_1\mathcal{M}$, the fermion mass is the dominant source of the breaking of scale invariance.

While all particle masses exhibit hyperscaling, the pion and the dilaton are still parametrically the lightest particles in the theory: $M_\pi \sim M_\tau \sim c_1$, which, by assumption, is a small parameter. In fact,

$$\frac{M_\pi^2}{(4\pi F_\pi)^2} \sim c_1 v(m) \sim c_1 \log\frac{m}{c_1\mathcal{M}} , \tag{7}$$

which implies that the expansion underlying dChPT is systematic as long as $c_1 \log\frac{m}{c_1\mathcal{M}} \ll 1$, even though $m/\mathcal{M}$ may be large. By contrast, ordinary ChPT for QCD is only valid when m is small relative to the infrared scale of the (massless) theory.

4 Fits to LSD data [1, 2]

One would like to test dChPT beyond the semi-quantitative observation of hyperscaling. Here we review some of the fits performed in Ref. [11] of the tree-level predictions to data for the pion mass and decay constant, as well as the dilaton mass, as a function of the fermion mass.[1] The ensembles of Ref. [2] all have the same bare coupling, and fermion masses

[1] The dilaton decay constant was not measured in Refs. [1–3].

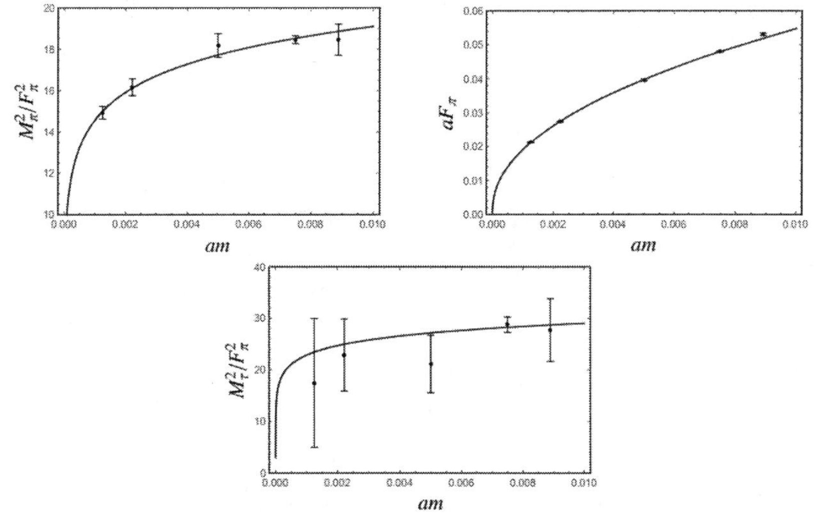

Figure 2. *M_π^2/F_π^2, aF_π and M_τ^2/F_π^2 as a function of the bare fermion mass in lattice units, am. Taken from Ref. [11].*

$10^3 am = (1.25,\ 2.22,\ 5.00,\ 7.50,\ 8.89)$. In Fig. 2 we show results of one of the fits of Ref. [11], using data for M_π^2/F_π^2, aF_π and M_τ^2/F_π^2 at the lowest four am values. The fit has a p-value of 0.89. Including also the largest mass leads to a less good fit, but the p-value is still 0.29. We conclude that tree-level dChPT provides an excellent description of the LSD data.[2]

The parameters controlling the mass dependence are well determined by these fits: we find $\gamma_* = 0.94(2)$ and $aB_\pi = 2.1(1)$. In contrast, finding af_π requires a long extrapolation to the chiral limit, and the result of the fits is $af_\pi = 0.0006(3)$.

We emphasize that the data all have $0.02 \lesssim aF_\pi(m) \lesssim 0.06$, with $F_\pi L \gtrsim 1$ and certainly $M_\pi L \gg 1$, so that the data are all in the p-regime. However, as already follows from the approximate hyperscaling, the LSD fermion masses are solidly in the large-mass regime, hence the long extrapolation needed to reach the chiral limit. This long extrapolation may make the value of af_π very sensitive to higher orders in dChPT, even if next-to-leading order (NLO) corrections at the LSD masses are relatively small. The (tentative) conclusion is that reaching the chiral limit while keeping $f_\pi L \gtrsim 1$ would require unrealistically large volumes, if the same bare coupling is kept.

The LSD collaboration also measured two of the staggered pion taste splittings (specifically, the "axial" and "tensor" ones). The taste splittings look very different from those found in QCD, which are explained well by the staggered extension of ordinary ChPT. Once we extend dChPT to include the discretization effects of staggered fermions (SdChPT for short), we find that these taste splittings can be quantitatively understood using SdChPT. For details, see Ref. [11]; we will show an example of taste splittings in the next section.

5 Fits to LatKMI data [3]

Next, we consider fitting dChPT to the data of Ref. [3]. The LatKMI collaboration considered the same theory, with fermion masses $10^2 am = (1.2,\ 1.5,\ 2.0,\ 3.0,\ 4.0,\ 5.0,\ 6.0,\ 7.0,\ 8.0,\ 10.0)$, significantly larger than those of

[2]For other applications of tree-level dChPT, see Refs. [5, 6].

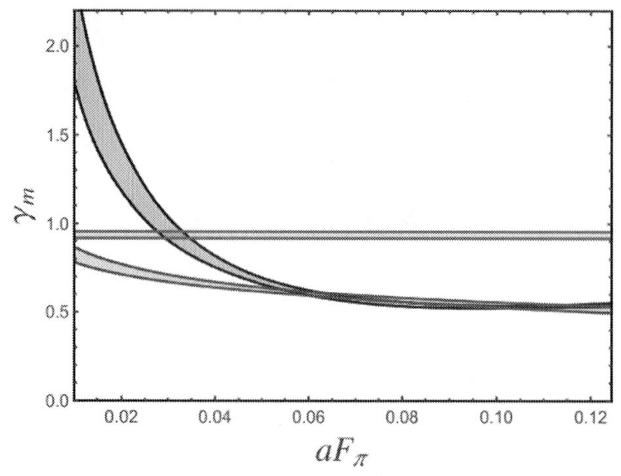

Figure 3. *The running mass anomalous dimension $\gamma(m)$, obtained from various fits of the LatKMI data, as a function of $aF_\pi(m)$, see text. Taken from Ref. [12].*

the LSD collaboration. Since the LatKMI and LSD collaborations used different lattice actions, and both collaborations did simulations at a single value of the bare coupling, we cannot compare the two data sets in physical units. In other words, we do not know the relative size of the LatKMI and LSD lattice spacings.

In Ref. [12] we found that we cannot fit the LatKMI data over their full mass range with tree-level dChPT. Fits of subsets suggest that NLO corrections will have to be taken into account. However, at NLO the effective theory has a large number of new LECs [8], and there are insufficient data to resolve these, even if the actual fits contain a smaller number of linear combinations of the new LECs.

Instead, we found that the data are well described with an m-dependent mass anomalous dimension[3]

$$\gamma(m) = \gamma_0 - bv(m) + cv^2(m)\,. \tag{8}$$

For a detailed description of the fits we refer to Ref. [12]. Here we just show the results we obtain for $\gamma(m)$ of Eq. (8), in Fig. 3. The blue band shows a fit (p-value = 0.48) to data at all but the largest mass of the LatKMI range, with all three parameters, γ_0, b and c. The magenta band shows a fit (p-value = 0.19) with the two largest masses omitted, and $c = 0$. For comparison, the gray band represents $\gamma_* = 0.94(2)$ obtained from the LSD data. The actual data all lie in the range $0.045 \lesssim aF_\pi \lesssim 0.12$, where the two bands overlap, as expected. Again, the LatKMI masses are in the large-mass regime, requiring a long extrapolation to the chiral limit. These results corroborate the sensitivity of the chiral limit to higher orders, even if those higher orders can be small in the range of the data. Indeed, small corrections to $v(m)$ over a limited m-interval make a larger effect on the "hyperscaling factor" $e^{v(m)}$ in the tree-level expressions (4), which magnifies if m is taken outside the range of the actual data.

To conclude this section, we show in Fig. 4 a fit of the taste splittings to tree-level SdChPT augmented by the varying $\gamma(m)$ of Eq. (8).[4] These taste splittings behave very different from

[3]In Ref. [12] we proved that this choice is still consistent with the Ward–Takahashi identities for scale invariance. One can think of Eq. (8) as arising from a partial resummation of higher orders.

[4]Ref. [3] measured taste splittings only for a subset of seven fermion masses with a maximum value 0.08. Setting $c = 0$ yields a good fit (p-value = 0.44) to tree-level SdChPT with $\gamma(m)$ of Eq. (8).

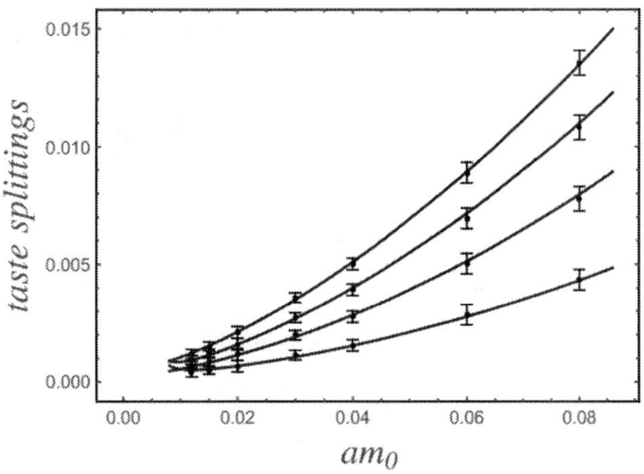

Figure 4. *Fit of tree-level SdChPT to the taste splittings of Ref. [3]. Taken from Ref. [12].*

those in QCD with staggered fermions, where the taste splittings are essentially independent of *am*. The goodness of the fit is a successful test of the dChPT framework.

6 Conclusion

We identified a "large-mass regime" in dChPT that has no equivalent in ordinary ChPT. This is the regime with $c_1\mathcal{M} \ll m \ll c_1\mathcal{M}e^{1/c_1}$. Much like a mass-deformed infrared-conformal theory, the large-mass regime of a "walking" theory exhibits approximate hyperscaling, while the expansion underlying dChPT remains systematic, thanks to the smallness of $c_1 \propto |n_f - n_f^*|$, the distance to the conformal sill. We found that current simulations of the $N_f = 8$, SU(3) theory are deep in the large-mass regime.

Consequently, reaching the chiral limit from the data of Refs. [2, 3] requires a very long extrapolation. This extrapolation appears to be very sensitive to the mass range used in the fits, as well as to higher orders in dChPT. For example, because of this sensitivity, the chiral-limit value $af_\pi = 0.0006(3)$ found from the LSD data may be afflicted with a larger systematic error than reflected by the statistical fit error qouted above. It is thus very difficult to determine whether af_π is actually different from zero! Likewise, it is difficult to obtain a precise value for $c_1\mathcal{M}$. The difficulty of controlling the long extrapolation raises the intriguing question whether, effectively, dChPT may also apply *inside* the conformal window, as long as the theory is mass-deformed.

Both Refs. [2] and [3] measured the dilaton mass, finding it to be nearly degenerate with the pions. But the errors are so large that the dilaton mass has little influence on the fit, as can be seen in Fig 2. Given the questions raised above it would be helpful if more precise data become available, especially at the lower fermion-mass range. In particular, this might allow for the probing of NLO corrections. Also, the size of the taste splittings indicates that scaling violations are large, and it would be nice to add another lattice spacing to study the approach to the continuum limit, about which no information is available at present.

We end with a separate comment. In Refs. [13, 14], a more general class of dilaton potentials

$$V_\Delta(\tau) \propto \frac{e^{4\tau}}{4-\Delta}\left(1 - \frac{4}{\Delta}e^{(\Delta-4)\tau}\right) \tag{9}$$

was considered. For $\Delta \to 4$, one recovers the dilaton potential in Eq. (2), while for $\Delta = 2$ one obtains the potential for the σ-model. However, as we pointed out in Ref. [12], for values of Δ not close to 4, there is no consistent power counting, and thus the lagrangian with this generalized potential does not constitute an effective field theory, contrary to what is claimed in Ref. [14].

Acknowledgments

MG's work is supported by the U.S. Department of Energy, Office of Science, Office of High Energy Physics, under Award Number DE-SC0013682. YS is supported by the Israel Science Foundation under grant no. 491/17.

References

[1] T. Appelquist *et al.* [LSD Collaboration], *Lattice simulations with eight flavors of domain wall fermions in SU(3) gauge theory,* Phys. Rev. D **90** 114502 (2014) [arXiv:1405.4752 [hep-lat]]; *Strongly interacting dynamics and the search for new physics at the LHC,* Phys. Rev. D **93** 114514 (2016) [arXiv:1601.04027 [hep-lat]].

[2] T. Appelquist *et al.* [Lattice Strong Dynamics], *Nonperturbative investigations of SU(3) gauge theory with eight dynamical flavors,* Phys. Rev. D **99** 014509 (2019) [arXiv:1807.08411 [hep-lat]].

[3] Y. Aoki *et al.* [LatKMI Collaboration], *Light flavor-singlet scalars and walking signals in $N_f = 8$ QCD on the lattice,* Phys. Rev. D **96** 014508 (2017) [arXiv:1610.07011 [hep-lat]].

[4] Z. Fodor, K. Holland, J. Kuti, D. Nogradi, C. Schroeder and C. H. Wong, *Can the nearly conformal sextet gauge model hide the Higgs impostor?,* Phys. Lett. B **718**, 657 (2012) [arXiv:1209.0391 [hep-lat]]; Z. Fodor, K. Holland, J. Kuti, D. Nogradi and C. H. Wong, *Can a light Higgs impostor hide in composite gauge models?,* PoS LATTICE **2013**, 062 (2014) [arXiv:1401.2176 [hep-lat]]; *The twelve-flavor β-function and dilaton tests of the sextet scalar,* EPJ Web Conf. **175**, 08015 (2018) [arXiv:1712.08594 [hep-lat]].

[5] Z. Fodor, K. Holland, J. Kuti and C. H. Wong, *Tantalizing dilaton tests from a near-conformal EFT,* PoS LATTICE **2018**, 196 (2019) [arXiv:1901.06324 [hep-lat]]; *Dilaton EFT from p-regime to RMT in the ϵ-regime,* PoS **LATTICE2019**, 246 (2020) arXiv:2002.05163 [hep-lat].

[6] T. Appelquist *et al.* [Lattice Strong Dynamics], *Near-conformal dynamics in a chirally broken system,* Phys. Rev. D **103** 014504 (2021) [arXiv:2007.01810 [hep-ph]].

[7] R. C. Brower, A. Hasenfratz, C. Rebbi, E. Weinberg and O. Witzel, *Composite Higgs model at a conformal fixed point,* Phys. Rev. D **93** 075028 (2016) [arXiv:1512.02576 [hep-ph]].

[8] M. Golterman and Y. Shamir, *Low-energy effective action for pions and a dilatonic meson,* Phys. Rev. D **94** 054502 (2016) [arXiv:1603.04575 [hep-ph]].

[9] M. Golterman and Y. Shamir, *Effective field theory for pions and a dilatonic meson,* PoS LATTICE **2016**, 205 (2016) [arXiv:1610.01752 [hep-ph]]

[10] M. Golterman and Y. Shamir, *Large-mass regime of the dilaton-pion low-energy effective theory,* Phys. Rev. D **98** 056025 (2018) [arXiv:1805.00198 [hep-ph]].

[11] M. Golterman, E. T. Neil and Y. Shamir, *Application of dilaton chiral perturbation theory to $N_f = 8$, SU(3) spectral data,* Phys. Rev. D **102** 034515 (2020) [arXiv:2003.00114 [hep-ph]].

[12] M. Golterman and Y. Shamir, *Explorations beyond dilaton chiral perturbation theory in the eight-flavor SU(3) gauge theory,* Phys. Rev. D **102**, 114507 (2020) [arXiv:2009.13846 [hep-lat]].

[13] T. Appelquist, J. Ingoldby and M. Piai, *Dilaton EFT Framework For Lattice Data,* JHEP **1707**, 035 (2017) [arXiv:1702.04410 [hep-ph]]; *Analysis of a Dilaton EFT for Lattice Data,* JHEP **1803**, 039 (2018) [arXiv:1711.00067 [hep-ph]];

[14] T. Appelquist, J. Ingoldby and M. Piai, *Dilaton potential and lattice data,* Phys. Rev. D **101** 075025 (2020) [arXiv:1908.00895 [hep-ph]].

EPJ Web of Conferences **258**, 08005 (2022) https://doi.org/10.1051/epjconf/202225808005
A Virtual Tribute to Quark Confinement and the Hadron Spectrum (vConf21)

The Fragmented Glueball: A Personal View

Eberhard Klempt[1,*]

[1]Helmholtz–Institut für Strahlen– und Kernphysik der Universität Bonn, Nussallee 14-16, 53115 Bonn, Germany

Abstract. A coupled-channel analysis has been performed to identify the spectrum of scalar mesons. The data include BESIII data on radiative J/ψ decays into $\pi^0\pi^0$, K_SK_S, $\eta\eta$, and $\omega\phi$, 15 Dalitz plots from $\bar{p}N$ annihilation at rest at LEAR, the CERN-Munich multipoles for $\pi\pi$ elastic scattering, the S-wave from BNL data on $\pi\pi$ scattering into K_SK_S, from GAMS data on $\pi\pi \to \pi^0\pi^0, \eta\eta$, and $\eta\eta'$, and NA48/2 data on low-mass $\pi\pi$ interactions from $K^\pm \to \pi\pi e^\pm\nu$ decays. The analysis reveals the existence of ten scalar isoscalar resonances. The resonances can be grouped into two classes: resonances with a large SU(3) singlet component and those with a large octet component. The production of isoscalar resonances with a large octet component should be suppressed in radiative J/ψ decays. However, in a limited mass range centered at 1900 MeV, these mesons are produced abundantly. Mainly-singlet scalar resonances are produced over the full mass range but with larger intensity at 1900 MeV. The total scalar isoscalar yield in radiative decays into scalar mesons shows a clear peak which is interpreted as the scalar glueball of lowest mass.

1 Introduction

Glueballs, bound states of gluons with any constituent quarks, are a firm prediction of QCD. Lattice gauge theories predict the lowest-mass glueball to have scalar quantum numbers and to have a mass of 1710±50±80 MeV [1]. Tensor and pseudoscalar glueballs are expected well above 2000 MeV. Analytic approximations to QCD find the scalar glueball in the range from 1850 to 1980 MeV [2–4]. Several observed scalar mesons have been proposed to contain a large glueball fraction but no firm conclusions have been reached so far [5–11]. In a recent paper, a partial-wave analysis of data carrying information on scalar mesons was reported [12]. Here, I present these results and give my personal interpretation.

2 Data and PWA

Decisive for the interpretation of the glueball as part of the spectrum of scalar mesons are the BESIII data on radiative J/ψ decays. In this process, the primary $c\bar{c}$ converts into one photon and two gluons (see Fig. 1). The gluons interact thus forming a glueball that decays into the observed final-state particles. Figure 2 shows the invariant mass spectra of $\pi^0\pi^0$, K_sK_s, $\eta\eta$, and $\phi\omega$ in a relative S wave produced in radiative J/ψ decays. In parti-

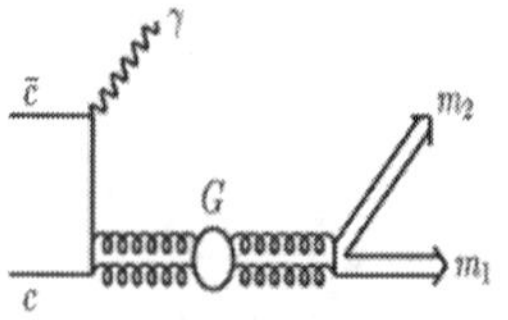

Figure 1. Radiative J/ψ decay

*e-mail: klempt@hiskp.uni-bonn.de

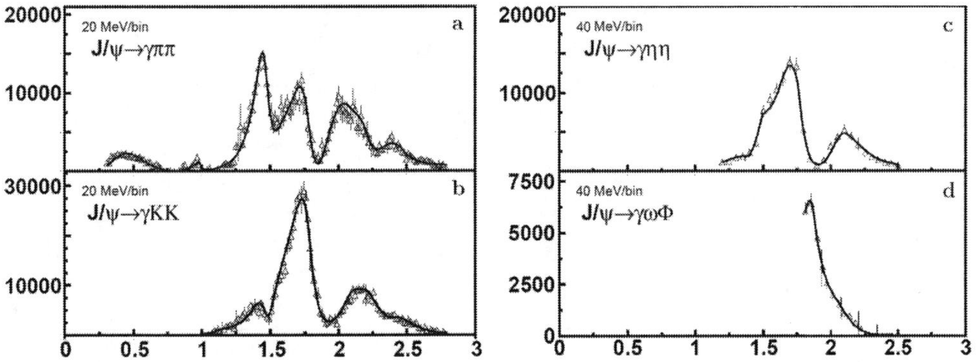

Figure 2. BESIII data on radiative J/ψ data into (a) $\pi^0\pi^0$, (b) K_sK_s (b), $\eta\eta$ (c), and $\phi\omega$ (d). Shown is the meson-meson S-wave contribution.

cular the $\pi^0\pi^0$ and K_sK_s mass distributions show a series of peaks and valleys. These are often seen at different mass positions emphasizing the importance of interference effects and the need of coupled-channel analyses.

These data are important for the interpretation. They are, however, not sufficient to constrain the number and the properties of the contributing resonances. Therefore we added several further data: 15 different Dalitz plots from $\bar{p}N$ annihilation at rest into three pseudoscalar mesons, the multipoles from the CERN-Munich analysis of elastic $\pi\pi$ scattering data – these data allow us to determine the *missing intensity* of scalar mesons – the GAMS data on $\pi^+\pi^-$ scattering into $\pi^0\pi^0$, $\eta\eta$, $\eta\eta'$, BNL data on $\pi^+\pi^-$ scattering into K_sK_s, and NA48/2 data on low-mass $\pi\pi$ interactions from $K^\pm \to \pi\pi e^\pm\nu$ decays.

The analysis finds ten scalar isoscalar resonances. Table 1 shows their masses and widths and compares them to PDG values. The five mesons below 1750 MeV are considered to be established by the Particle Data Group, $f_0(1770)$ is "new" (even though evidence had been reported earlier), the four higher-mass mesons were not regarded as being established. The agreement between our new values and those reported earlier is remarkable. Table 2 presents the yields of scalar mesons in the various final states. The missing intensity is derived from the CERN-Munich data and compared to the intensities reported for $J/\psi \to \gamma 4\pi$ and $J/\psi \to \gamma\omega\omega$. The last column gives the sum of all intensity assigned to a specific resonance.

Table 1. Pole masses and widths (in MeV) of scalar mesons. The RPP values are listed as small numbers for comparison.

Name	$f_0(500)$	$f_0(1370)$	$f_0(1710)$	$f_0(2020)$	$f_0(2200)$
M	410±20	1370±40	1700±18	1925±25	2200±25
	400→550	1200→1500	1704±12	1992±16	2187±14
Γ	480±30	390±40	255±25	320±35	150±30
	400→700	100→500	123±18	442±60	~ 200
Name	$f_0(980)$	$f_0(1500)$	$f_0(1770)$	$f_0(2100)$	$f_0(2330)$
M	1014±8	1483 ± 15	1765±15	2075±20	2340±20
	990±20	1506 ± 6		2086^{+20}_{-24}	~2330
Γ	71±10	116±12	180±20	260±25	165±25
	10→100	112±9		284^{+60}_{-32}	250±20

Table 2. Yields of scalar mesons in radiative J/ψ decays (in units of 10^{-5}). The RPP values are listed as small numbers for comparison.

$BR_{J/\psi\to\gamma f_0\to}$	$\gamma\pi\pi$	$\gamma K\bar{K}$	$\gamma\eta\eta$	$\gamma\eta\eta'$	$\gamma\omega\phi$	missing $\gamma 4\pi$	$\gamma\omega\omega$	total
$f_0(500)$	105±20	5±5	4±3	~0	~0	~0		114±21
$f_0(980)$	1.3±0.2	0.8±0.3	~0	~0	~0	~0		2.1±0.4
$f_0(1370)$	38±10	13±4	3.5±1	0.9±0.3	~0	14±5		69±12
		42±15				27±9		
$f_0(1500)$	9.0±1.7	3±1	1.1±0.4	1.2±0.5	~0	33±8		47±9
	10.9±2.4	2.9±1.2	$1.7^{+0.6}_{-1.4}$	$6.4^{+1.0}_{-2.2}$		36±9		
$f_0(1710)$	6±2	23±8	12±4	6.5±2.5	1±1	7±3		56±10
$f_0(1770)$	24±8	60±20	7±1	2.5±1.1	22±4	65±15		181±26
$f_0(1750)$	38±5	99^{+10}_{-6}	24^{+12}_{-7}		25±6	97±18	31±10	
$f_0(2020)$	42±10	55±25	10±10			(38±13)		145±32
$f_0(2100)$	20±8	32±20	18±15			(38±13)		108±25
$f_0(2200)$	5±2	5±5	0.7±0.4			(38±13)		49±17
$f_0(2100)/f_0(2200)$	62±10	109^{+8}_{-19}	$11.0^{+6.5}_{-3.0}$			115±41		
$f_0(2330)$	4±2	2.5±0.5	1.5±0.4					8±3
		20±3						

The sum of the yields of $f_0(1710)$ and $f_0(1770)$ should be compared to the RPP entry for $f_0(1750)$. RPP reports yields for $f_0(2100)$ and $f_0(2200)$. The yields can be compared to the sum of three resonances found by us. Above 1900 MeV, no data on $\pi\pi$ elastic scattering are known. The total $\gamma 4\pi$ intensity is therefore distributed between the three states observed in the 2000 to 2200 MeV range.

3 Interpretation

The mesons in Table 1 are grouped pairwise. There are two mesons close in mass followed by a mass gap. Figure 3 shows the squared mass values of the "higher-mass" and the "lower-mass" mesons as a function of a consequtive number. The squared masses depend about linearly on the consecutive number. The numbering starts with (-1) since we do not want to imply that $f_0(500)$ is the ground state of the scalar-isoscalar-meson spectrum. The "lower" and "higher"-mass mesons are separated by the mass-square difference between η and η' mesons but the spectrum is inverted. This is expected for instanton-induced interactions [17]. For scalar mesons, $f_0(500)$ is mainly singlet, $f_0(980)$ is mainly octet [18, 19]. The decay pattern identifies $f_0(1370)$ as mainly singlet, $f_0(1500)$ as mainly octet state. Thus we conjecture that the $q\bar{q}$ component of the "higher-mass" mesons is mainly in the octet, the "lower-mass" mesons in the singlet configuration.

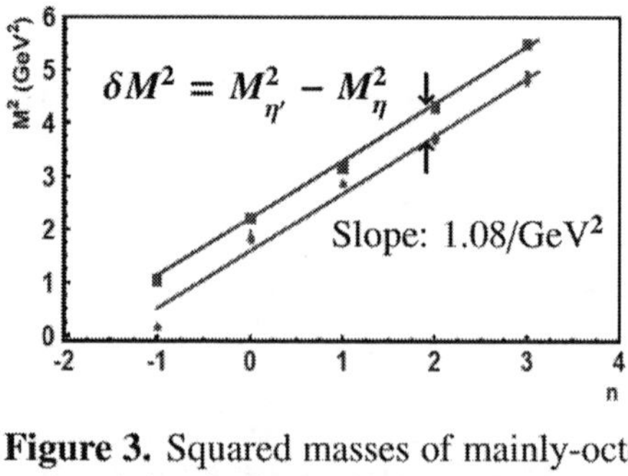

Figure 3. Squared masses of mainly-octet and mainly-singlet scalar isoscalar mesons as functions of a consecutive number.

Mainly-octet mesons should not be produced in radiative J/ψ decays but they are. Figure 4 shows the total yield of scalar isoscalar mesons in radiative J/ψ decays (from Table 2) as a function of the invariant mass. Open circles represent the yield of "higher-mass", full squares of "lower-mass" scalar mesons. Even though the quark content of the "higher-mass" scalar mesons is supposed to be mainly in the octet configuration, their is significant production in the mass range from 1500 to 2300 MeV. This high yield requires a significant singlet contribution in their wave functions. No selection rule forbids SU(3) singlet scalar mesons to be produced in radiative J/ψ decays, they are produced in the full mass range but their production is enhanced in the same mass range. We interpret this enhancement as scalar glueball.

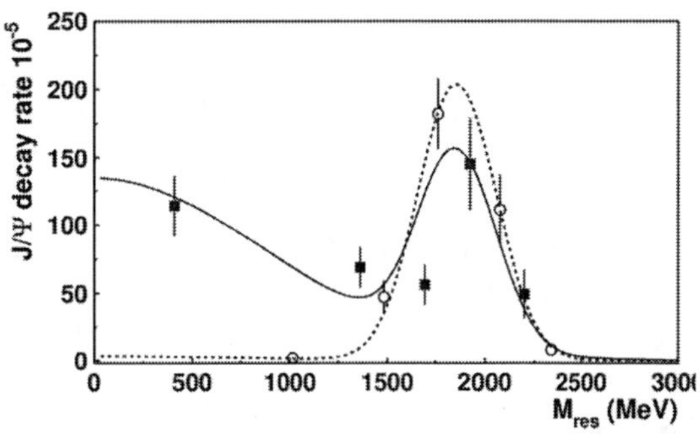

Figure 4. Squared masses of mainly-octet and mainly-singlet scalar isoscalar mesons as functions of a consecutive number.

This interpretation is confirmed when data on $J/\psi \rightarrow \gamma f_0$ and the LHCb data $\bar{B}^0_s \rightarrow J/\psi f_0$ [20, 21] are compared. In J/ψ decays, two gluons convert into the scalar meson that decays into the final-state particles. A montainous landscape is generated with high peaks and deep valleys. For $\bar{B}^0_s$ decays, the form factors from [22] are shown in Fig. 5. Their square is proportional to the cross section. Very little intensity is observed. This is particulary striking when two kaons are seen in the final state (the kaon form factor). In $\bar{B}^0_s$ decays an $s\bar{s}$ pair is generated in the initial state but no resonance is produced even though there is a strong $K\bar{K}$ peak when two gluons are in the initial state. The two montains in the $K\bar{K}$ invariant mass at 1700 and 2200 MeV are not generated – or at most weakly – by $s\bar{s}$ pairs but abundantly in gluon-gluon interactions.

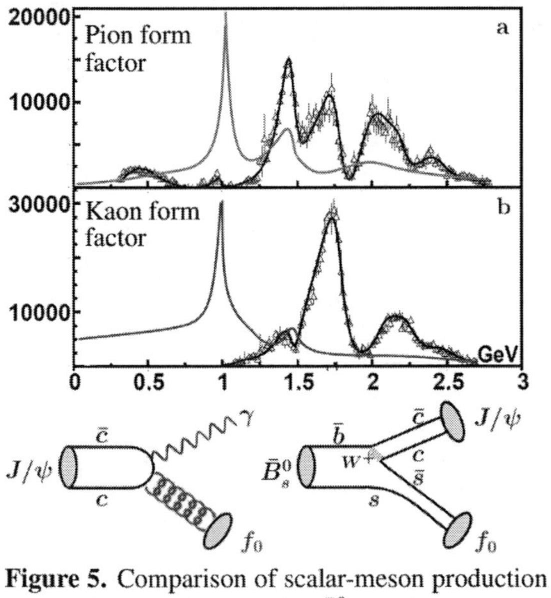

Figure 5. Comparison of scalar-meson production in radiative J/ψ decay and in $\bar{B}^0_s \rightarrow J/\psi f_0$.

Radiative J/ψ decays produce scalar isoscalar mesons abundantly. Those mesons that are supposed to have a mostly-octet $q\bar{q}$ structure (and that should not be produced in radiative J/ψ decays) are produced but only in a limited mass range, with a clear peak at about 1900 MeV. Mostly-singlet isoscalar mesons are produced in the full mass range but the yield is enhanced at about 1900 MeV. This peak is not seen in $\bar{B}^0_s$ decays, it is seen only with two gluons in the initial state. The enhancement can be fit with a Breit-Wigner resonance. The fit returns

$$\mathrm{M}_{\mathrm{glueball}} = (1865 \pm 25)\,\mathrm{MeV},\ \Gamma_{\mathrm{glueball}} = (370 \pm 50^{+30}_{-20})\,\mathrm{MeV}$$
$$\mathrm{Y}_{J/\psi\rightarrow\gamma\mathrm{G}_0} = (5.8 \pm 1.0)\cdot 10^{-3}$$

This is the largest yield in radiative J/ψ decays into a meson. It is the expected scalar glueball.

4 Discussion

Is this really the expected scalar glueball? At least one aspect is completely unexpected. There are ten scalar isoscalar resonances all falling on a linear (n, M^2) trajectory. All find a spectroscopic identification within the quark model – except perhaps $f_0(500)$ and $f_0(980)$. There is no extra state, no intruder that enters the spectrum, mixes with other states and increases the number of observed states by one. In Fig. 3, there is no additional resonance. How can that be? The scalar glueball seems to make up a fraction of the wave function of scalar isoscalar mesons without increasing the number of states.

We can write down the quark decomposition of the wave function of a mainly-octet scalar meson:

$$
\begin{aligned}
f_0 &= \alpha_1 \frac{1}{\sqrt{6}}(u\bar{u} + d\bar{d} - 2s\bar{s}) + \alpha_2 \frac{1}{\sqrt{6}}(u\bar{u}s\bar{s} + d\bar{d}s\bar{s} - 2u\bar{u}d\bar{d}) \\
&+ \alpha_3 \cdot (\text{meson} - \text{meson cloud}) + \alpha_4(\text{gg}) + \alpha_5(\text{q}\bar{\text{q}}\text{g}) \\
&+ \text{and some singlet contribution}: \ \{\alpha' \frac{1}{\sqrt{3}}(\text{u}\bar{\text{u}} + \text{d}\bar{\text{d}} + \text{s}\bar{\text{s}}) + \beta' \frac{1}{\sqrt{3}}(\text{u}\bar{\text{u}}\text{s}\bar{\text{s}} + \text{d}\bar{\text{d}}\text{s}\bar{\text{s}} + \text{u}\bar{\text{u}}\text{d}\bar{\text{d}})\}
\end{aligned}
$$

These are five different octet contributions, five Fock components. These could all be realized independently. They could mix, but the number of ground states – and the number of states in every excitation level – could be five. This seems not to be the case. The number of states is rather one per excitation level and not five. The large number of different Fock components does obviously not lead to a large number of states but only to components in the wave functions.

5 Summary

The BESIII collaboration reported data on radiative J/ψ decays with unprecedented statistics. The data reveal high intensities in the yield of scalar mesons. The data can be fit with ten scalar isoscalar resonances. The scalar resonances can be grouped into a class of mainly-singlet and mainly-octet states. The two groups fall onto linear (n, M^2)-trajectories. Octet scalar isoscalar resonances are produced mainly in the 1700 - 2100 MeV mass range. Singlet scalar resonances are produced over the full mass range but their intensity peaks in the 1700 - 2100 MeV mass range. The enhanced production of scalar mesons in the 1700 - 2100 MeV mass range is due to gluon-gluon in the initial state. The peak is the scalar glueball of lowest mass. The glueball does not intrude the spectrum of scalar states as additional resonance. It contributes to the wave function of several scalar isoscalar resonances.

References

[1] Y. Chen *et al.*, Phys. Rev. D **73**, 014516 (2006).
[2] A. P. Szczepaniak and E. S. Swanson, Phys. Lett. B **577**, 61-66 (2003).
[3] M. Q. Huber, C. S. Fischer and H. Sanchis-Alepuz, Eur. Phys. J. C **80** no.11, 1077 (2020).
[4] M. Rinaldi and V. Vento, Phys. Rev. D **104**, no.3, 034016 (2021).
[5] F. E. Close and A. Kirk, Eur. Phys. J. C **21**, 531-543 (2001).
[6] E. Klempt and A. Zaitsev, Phys. Rept. **454**, 1 (2007).
[7] V. Mathieu, N. Kochelev and V. Vento, Int. J. Mod. Phys. E **18**, 1 (2009).
[8] V. Crede and C. A. Meyer, Prog. Part. Nucl. Phys. **63**, 74 (2009).
[9] C. Amsler *et al.*, "Scalar Mesons below 2 GeV", in: P. A. Zyla *et al.* [Particle Data Group], PTEP **2020**, no.8, 083C01 (2020).

[10] W. Ochs, J. Phys. G **40**, 043001 (2013).
[11] F. J. Llanes-Estrada, Eur. Phys. J. ST **230**, no.6, 1575-1592 (2021).
[12] A. V. Sarantsev *et al.*, Phys. Lett. B **816**, 136227 (2021).
[13] M. Ablikim *et al.* [BESIII Collaboration], Phys. Rev. D **92** no.5, 052003 (2015).
[14] M. Ablikim *et al.* [BESIII Collaboration], Phys. Rev. D **98** no.7, 072003 (2018).
[15] M. Ablikim *et al.* [BESIII Collaboration], Phys. Rev. D **87**, no. 9, 092009 (2013).
[16] M. Ablikim *et al.* [[BESIII Collaboration], Phys. Rev. D **87** no.3, 032008 (2013).
[17] E. Klempt, B.C. Metsch, C.R. Münz and H.R. Petry, Phys. Lett. B **361**, 160-166 (1995).
[18] J. A. Oller, Nucl. Phys. A **727**, 353-369 (2003).
[19] E. Klempt, Phys. Lett. B **820**, 136512 (2021).
[20] R. Aaij *et al.* [LHCb], Phys. Rev. D **89**, no.9, 092006 (2014).
[21] R. Aaij *et al.* [LHCb], JHEP **08**, 037 (2017).
[22] S. Ropertz, C. Hanhart and B. Kubis, Eur. Phys. J. C **78**, no.12, 1000 (2018).

Equivariance and generalization in neural networks

Srinath Bulusu[1,*], *Matteo* Favoni[1,2,**], *Andreas* Ipp[1,***], *David I.* Müller[1,****], and *Daniel* Schuh[1,†]

[1]Institute for Theoretical Physics, TU Wien,
Wiedner Hauptstr. 8-10, 1040 Vienna, Austria
[2]Speaker and corresponding author

Abstract. The crucial role played by the underlying symmetries of high energy physics and lattice field theories calls for the implementation of such symmetries in the neural network architectures that are applied to the physical system under consideration. In these proceedings, we focus on the consequences of incorporating translational equivariance among the network properties, particularly in terms of performance and generalization. The benefits of equivariant networks are exemplified by studying a complex scalar field theory, on which various regression and classification tasks are examined. For a meaningful comparison, promising equivariant and non-equivariant architectures are identified by means of a systematic search. The results indicate that in most of the tasks our best equivariant architectures can perform and generalize significantly better than their non-equivariant counterparts, which applies not only to physical parameters beyond those represented in the training set, but also to different lattice sizes.

1 Introduction

The past century has firmly established symmetries as a cornerstone of physics and especially of field theories, tracing back to Noether's theorem [1]. Later on, the importance of local symmetries was understood, which then culminated in the development of the Standard Model of particle physics [2, 3], where the fundamental interactions are described as gauge theories.

In the past decade, machine learning techniques have been sharpened a great deal in many areas, especially in computer vision, and their success has attracted a lot of interest from the physics community, which progressively assimilated them into its arsenal of methods. In particular, neural networks (NNs) proved to be a very effective tool in phase detection [4] and regression tasks on observables and physical parameters [5, 6], but in many cases no structural adaptation from image processing to physical problems was put in place. Bearing in mind the central role of symmetries, one should take them into account ab initio when building a fitting

*e-mail: sbulusu@hep.itp.tuwien.ac.at
**e-mail: favoni@hep.itp.tuwien.ac.at
***e-mail: ipp@hep.itp.tuwien.ac.at
****e-mail: dmueller@hep.itp.tuwien.ac.at
†e-mail: schuh@hep.itp.tuwien.ac.at

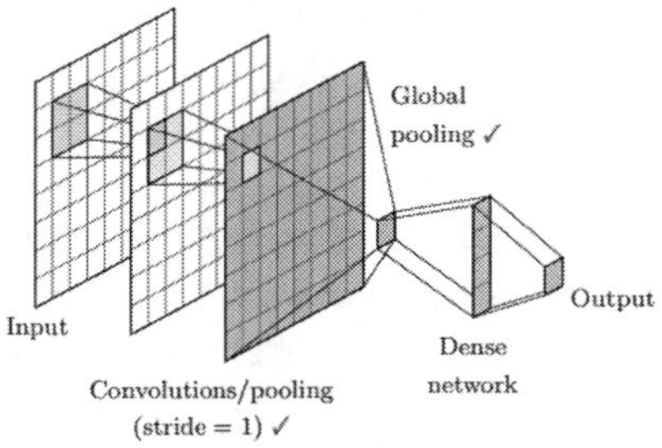

(a) Equivariant architecture (EQ)

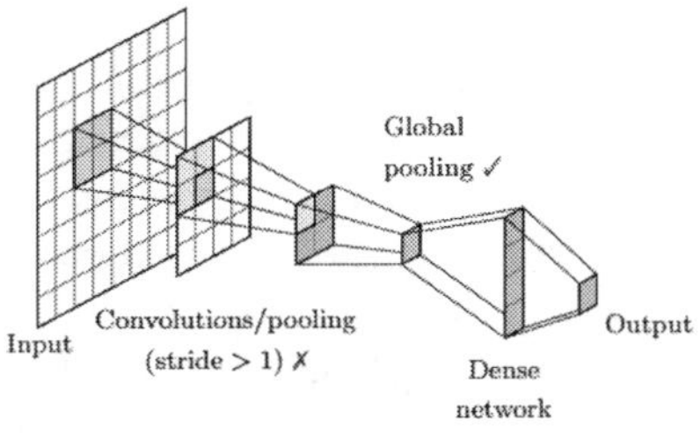

(b) Strided architecture (ST)

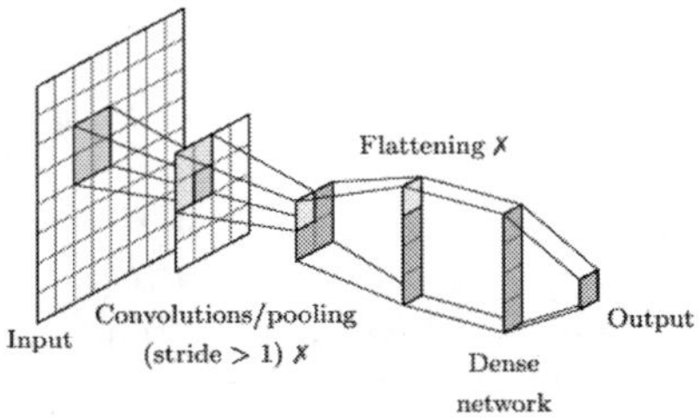

(c) Flattening architecture (FL)

Figure 1: The architecture types employed in this study. Translational equivariance is preserved by layers denoted by ✓ and broken by layers marked with ✗. Convolutions and spatial pooling layers with a stride of one respect equivariance (a), whereas a stride larger than one or a flattening layer ruin it, as in (b) and (c). Flattening (c) also forces the architecture to be used on only one specific lattice size. Figures from [12].

architecture for the theory in question. This means that, from a theoretical standpoint, a natural approach is to design the network architectures so that they inherently respect the desired symmetries. Whether a symmetry-preserving construction is beneficial in terms of actual performance can be problem dependent, but there have been recent indications that this is a promising strategy. In [7], group equivariant convolutional neural networks (G-CNNs) are proposed, which take care of the incorporation of translational, rotational and reflection symmetry in the architecture. The extension of this approach to local symmetries is formalized in [8]. Other proposals have been made in the context of lattice field theory, for example by designing a gauge equivariant normalizing flow [9, 10] and with the creation of specific layers that preserve gauge equivariance [11]. In these proceedings, we focus on invariance under spacetime translations and study the impact of different choices of architectures that do or do not break translational symmetry, investigating in particular the generalization to different physical parameters and lattice sizes, as was discussed in our paper [12].

2 Architecture choices

As mentioned, G-CNNs deal with all the spacetime isometries, while here our interest is restricted to translations only. A more natural choice is represented by convolutional neural networks (CNNs), which were precisely conceived to impose translational symmetry in NNs.

In order to study the influence of translational equivariance, we set up three different kinds of architectures, as depicted in fig. 1. A stride of one in the convolutional and spatial pooling layers guarantees that translational symmetry is not broken in the first architecture (EQ). The symmetry is exact if one uses circular padding, which is the case if the theory features periodic boundary conditions. It is also not broken by the particular type of global pooling chosen, nor by the optional linear layers that are appended afterwards. The second architecture (ST) breaks equivariance because of a stride larger than one in the convolutional part. A residual symmetry based on translations that are a multiple of the stride holds, but general equivariance is not satisfied. In the third architecture (FL), the global pooling layer is replaced by a flattening step, which breaks the symmetry completely and restricts the use of the architecture to a specific lattice size.

These three types of networks are typical examples of architectures used in literature.

3 Physical model

Given the previous discussion, a good candidate to test our considerations is a relatively simple model characterized by translational symmetry and periodic boundary conditions. We opt for a complex scalar field in 1+1 dimensions with nonzero chemical potential μ and quartic interaction, also because of a pre-existing study to make comparisons with [13]. The Euclidean lattice action has the following expression:

$$S_{lat} = \sum_x \left(\eta|\phi_x|^2 + \lambda|\phi_x|^4 - \sum_{\nu=1}^{2} \left(e^{\mu\delta_{\nu,2}} \phi_x^* \phi_{x+\hat{\nu}} + e^{-\mu\delta_{\nu,2}} \phi_x^* \phi_{x-\hat{\nu}} \right) \right), \tag{1}$$

with $\eta = 4 + m^2$. A sign problem arises as a consequence of the complex terms originating from the products of fields at different lattice sites. It can be removed, though, thanks to a dual formulation, called flux representation [14], which maps the field ϕ_x into the integer fields $k_{x,\nu}$ and $l_{x,\nu}$, with ν indicating the temporal or the spatial direction. Configurations of these fields are generated as in [14] and are the input of the NNs.

4 Prediction of observables

We tackle now a regression task, where the networks have to learn the relationship between a lattice configuration and two observables, which was already carried out in [13]. The observables are the particle number density and the average of the modulo squared field, respectively given by

$$n = \frac{1}{N} \sum_x k_{x,2}, \quad |\phi|^2 = \frac{1}{N} \sum_x \frac{W(f_x + 2)}{W(f_x)}, \tag{2}$$

with $f_x = \sum_\nu [|k_{x,\nu}| + |k_{x-\hat{\nu},\nu}| + 2(l_{x,\nu} + l_{x-\hat{\nu},\nu})]$ and $W(f_x) = \int_0^\infty \mathrm{d}x\, x^{f_x+1} \mathrm{e}^{-\eta x^2 - \lambda x^4}$. Given the intensive nature of these observables, a global average pooling at the end of the convolutional part is the most appropriate choice.

In order to drastically reduce human biases in the construction of specific architectures to compare, we employ an automatic hyperparameter optimizer, called Optuna [15]. We design a large search space in the three cases, allowing for networks ranging from very small to pretty large in terms of number of trainable parameters. We require Optuna to try the same hyperparameters three times in order to reduce the influence of the initialization. At the end of the search, we select the combination with the smallest average validation loss (which in this case is the mean squared error). Ten instances with such a combination are trained, and the procedure is repeated for each architecture and for various training set sizes. Tests are run on configurations with the same size used for training (60×4), while the chemical potential

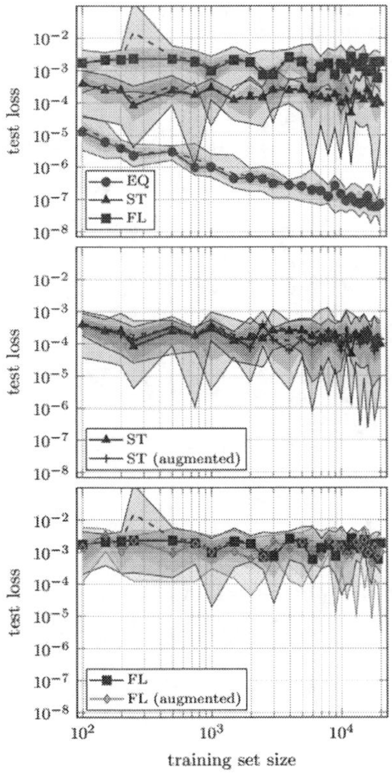

Figure 2: Performance comparison of best models suggested by Optuna. Ten instances with the most promising hyperparameter combination are trained at several training set sizes for each architecture type. The bands are delimited by the highest and lowest test loss, the dashed lines denote the average test losses and the continuous lines pass through the median test losses. The top plot shows the comparison of the three architecture types with no data augmentation, the middle and the bottom ones feature the test loss values with data augmentation for ST and FL, respectively. Image from [12].

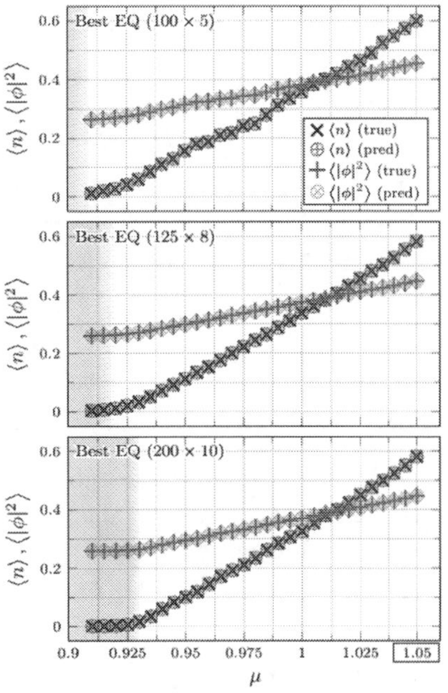

Figure 3: Silver blaze phase transition. The best EQ architecture is tested on chemical potentials and lattice sizes different from the ones it has been trained on (1.05 and 60×4, respectively). The silver blaze phase transition, which is better visible on larger lattices, is correctly predicted, even though training happened at only one phase. Image from [12].

ranges from 0.91 to 1.05 with steps of 0.005. Training takes place only at the uppermost value mentioned. The results are shown in fig. 2. The top plot reports the behavior of the test loss as a function of the training set size for all three architectures. The EQ architecture clearly outperforms the non-equivariant types, and it improves its accuracy when more samples are provided during training, while the other two do not benefit significantly with more samples. Remarkably, this is also the case when using data augmentation while training, as shown in the middle and bottom plots.

Figure 3 shows the generalization capabilities of the EQ models. As said, training occurs on lattices with dimension 60×4 with chemical potential $\mu = 1.05$, but the predictions remain

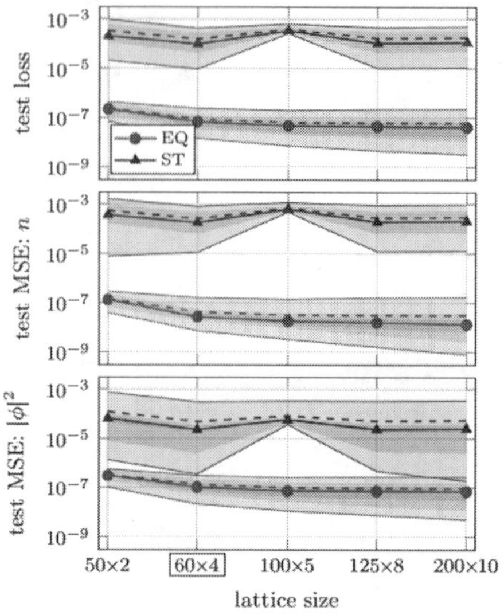

Figure 4: Comparison of the test loss as a function of the lattice size. The best EQ and ST architectures are tested on several lattice sizes, while having been trained on only one specific value (60×4). Image from [12].

very close to the true values also for different values of μ and of the lattice size. This even allows to detect the silver blaze phase transition [14] on the largest lattice even though training happened only in the disordered phase.

To further convince the reader, fig. 4 reports another comparison between architectures. It shows a test on various lattice sizes, which excludes FL architectures from this comparison. The EQ architectures are about three orders of magnitude more accurate than ST at every lattice size, which suggests that the incorporation of translational equivariance is advantageous. To support this even more, in [13] an FL architecture with about 10^7 parameters trained on the 200×10 lattice and on two values of μ reached a test loss of 10^{-6}, which is two orders of magnitude worse than the best model presented here.

5 Open worm detection

In the flux representation, the field k satisfies the conservation law $\sum_\nu (k_{x,\nu} - k_{x-\hat{\nu},\nu}) = 0$. The Prokof'ev-Svistunov algorithm [16] is employed to generate configurations that respect such an equation. It proposes local field updates in successive steps on adjacent lattice sites. One can keep track of where the field modifications occur, as is done in fig. 5a. The resulting path is referred to as a worm. Its endpoints exhibit flux violations, unless they coincide, in which case the flux conservation is restored at every site. Open worm configurations are characterized by two flux violations, and are therefore unphysical, while in closed worm configurations the flux is conserved everywhere, meaning they are physical. In the previous task, we have used only physical configurations, but here we deliberately create unphysical configurations and demand that the networks distinguish between the two situations.

In order to investigate if the networks are able to generalize, we create a dataset with several combinations of physical parameters and use a small subset in the training process. We choose squared lattices with $N_t = N_x \in \{8, 16, 32, 64\}$; the possible values of the chemical potential are $\mu \in \{1, 1.25, 1.5\}$; the parameter involving the mass term lies in the set $\eta \in$

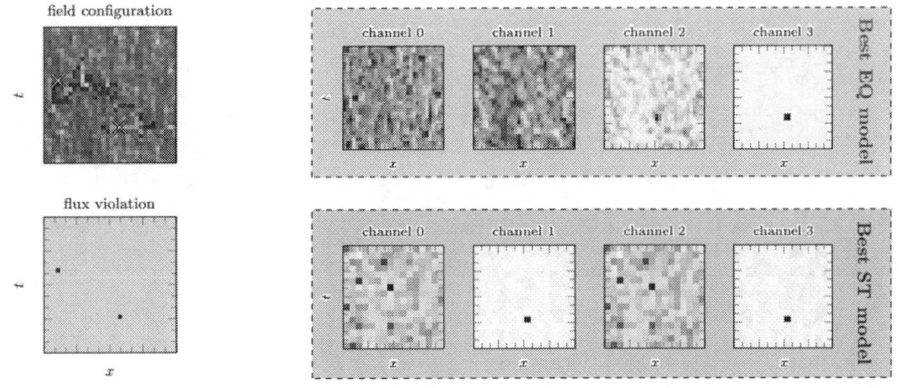

(a) Example field configuration (b) Feature maps of convolutional part in best EQ and ST models

Figure 5: Open worm and flux violation detection. The top plot in (a) is a configuration with an open worm, depicted in brown, on top of a physical configuration. The flux violations at the worm endpoints are visualized in the bottom plot of (a). The open worm detection in the networks is triggered by only one of the two flux violations. Image from [12].

$\{4.01, 4.04, 4.25\}$; the coupling constant is kept fixed at $\lambda = 1$. The training subset is given by the two combinations $(\eta, \mu) \in \{(4.01, 1.5), (4.25, 1)\}$ on 8×8 lattices. Open and closed worm configurations are equally represented in the dataset for each parameter combination.

We adopt the same approach used in the previous task with Optuna, in the attempt to compare the architecture types in the fairest possible way. The hyperparameter search space is slightly modified, for example it is not clear what global pooling type is the most favorable in a classification task, so we include it in the optimization procedure. Monitoring the binary cross entropy loss, Optuna indicates the most promising hyperparameter combination, which is used to train 50 new models using random parameter initialization. The test loss and test accuracy are shown in fig. 6 as functions of the chemical potential and of the lattice size.

While both metrics are not particularly sensitive to changes in the physical parameter μ, larger lattice sizes tend to reduce the model precision. In this case, FL models remain worse than EQ, but ST are more competitive and achieve an accuracy very close to the one of EQ. We also note that, interestingly, successful models learn to detect just one of the two flux violations, as fig. 5b suggests.

6 Multiple open worms counting

We revisit the previous task by adding more than one open worm on top of a physical configuration and requiring the networks to assess the right number of open worms. For this, we use the same physical parameters of the last section combined with a number of open worms $N_{\text{worms}} \in \{0, 1, \ldots, 10\}$. The training subset is characterized by the same physical parameters of the single open worm detection, except the number of open worms is 0 and 5. Using the same strategy with Optuna, we come to the results reported in fig. 7. In fig. 7a the test loss and test accuracy on 8×8 lattices are plotted against the number of open worms. Unlike EQ, which consistently maintains a good performance, ST and FL struggle to correctly predict small numbers of worms. Figure 7b shows metrics as functions of the lattice size, where EQ clearly outperforms ST, with all of its 20 instances scoring 100% test accuracy.

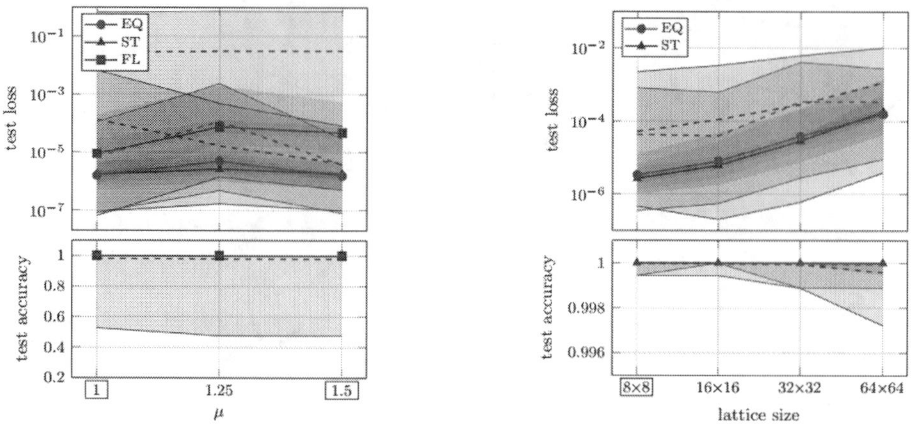

(a) Test loss and test accuracy vs. chemical potential on 8×8 lattices for each architecture.

(b) Test loss and test accuracy vs. lattice size for EQ and ST.

Figure 6: Results for detection of flux violations. The bands delimit the highest and lowest test metrics of 50 models of the architectures suggested by Optuna. In (a), the test loss and the test accuracy are given as a function of the chemical potential on 8×8 configurations, whereas in (b) they are reported as functions of the lattice size. Images from [12].

7 Conclusions

An extensive study on the implications of translational equivariance in neural networks was conducted. Three architecture types were designed and tested on three separate tasks, where the reliability of the equivariant type was evident, in particular when it came to the generalization capabilities, both in terms of different physical parameters and of different lattice sizes. The architectures that break translational symmetry led to a much worse performance in almost all cases and are also characterized by inherent drawbacks, such as the impossibility of using an architecture with a flattening layer on different lattice sizes. It is also worth pointing out that Optuna had a preference for small or medium-sized architectures, with a number of parameters between 10^2 and 10^5. To conclude, our work indicates that employing translationally equivariant neural networks in translationally symmetric problems can be extremely beneficial.

This work has been supported by the Austrian Science Fund FWF No. P32446-N27, No. P28352 and Doctoral program No. W1252-N27. The Titan V GPU used for this research was donated by the NVIDIA Corporation.

References

[1] E. Noether, Nachrichten von der Gesellschaft der Wissenschaften zu Göttingen, Mathematisch-Physikalische Klasse **1918**, 235 (1918)

[2] S. Weinberg, Phys. Rev. Lett. **19**, 1264 (1967)

[3] A. Salam, *Elementary Particle Theory: Relativistic Groups and Analyticity* (1968), p. 367, eighth Nobel Symposium. Stockholm: Almquvist and Wiksell.

[4] J. Carrasquilla, R.G. Melko, Nature Physics **13**, 431–434 (2017)

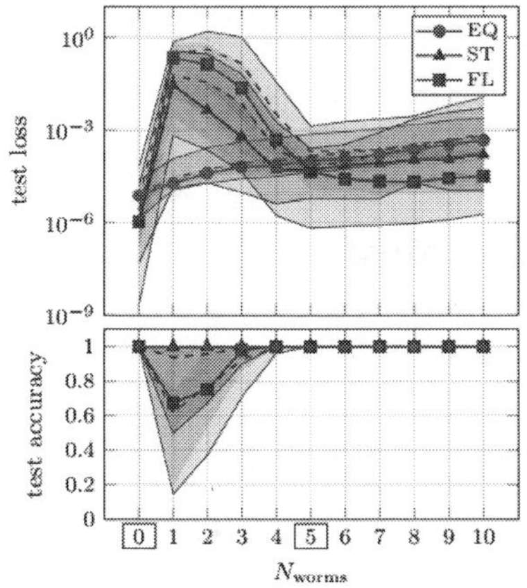

(a) Comparison of test loss and test accuracy vs. number of open worms on 8×8 lattices.

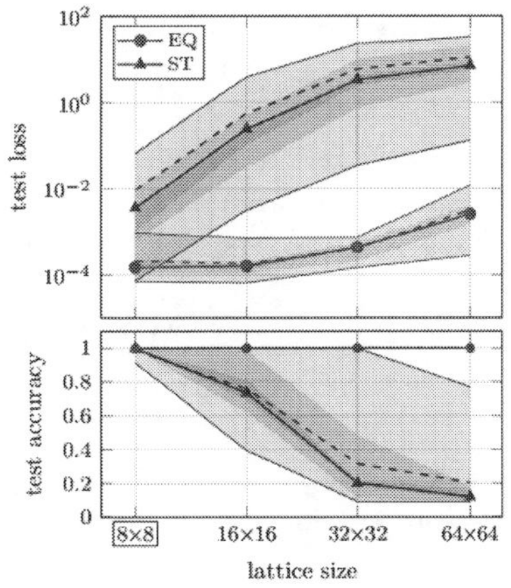

(b) Comparison of test loss and test accuracy vs. lattice size.

Figure 7: Results for worm counting. The bands delimit the highest and lowest test metrics of 20 instances of the architectures suggested by Optuna. In (a), the test loss and the test accuracy are given as a function of the chemical potential on 8×8 configurations, whereas in (b) they are reported as functions of the lattice size. Images from [12].

[5] B. Yoon, T. Bhattacharya, R. Gupta, Phys. Rev. D **100**, 014504 (2019)

[6] P.E. Shanahan, D. Trewartha, W. Detmold, Phys. Rev. D **97**, 094506 (2018)

[7] T.S. Cohen, M. Welling, *Group Equivariant Convolutional Networks*, in *Proceedings of The 33rd International Conference on Machine Learning* (JMLR, 2016), Vol. 48, pp. 2990–2999, `1602.07576`

[8] T.S. Cohen, M. Weiler, B. Kicanaoglu, M. Welling, *Gauge Equivariant Convolutional Networks and the Icosahedral CNN*, in *Proceedings of the 36th International Conference on Machine Learning* (JMLR, 2019), Vol. 97, pp. 1321–1330, `1902.04615`

[9] G. Kanwar, M.S. Albergo, D. Boyda, K. Cranmer, D.C. Hackett, S. Racanière, D.J. Rezende, P.E. Shanahan, Physical Review Letters **125** (2020), `2003.06413`

[10] D. Boyda, G. Kanwar, S. Racanière, D.J. Rezende, M.S. Albergo, K. Cranmer, D.C. Hackett, P.E. Shanahan, Phys. Rev. D **103**, 074504 (2021), `2008.05456`

[11] M. Favoni, A. Ipp, D.I. Müller, D. Schuh (2020), `2012.12901`

[12] S. Bulusu, M. Favoni, A. Ipp, D.I. Müller, D. Schuh, Phys. Rev. D **104**, 074504 (2021), `2103.14686`

[13] K. Zhou, G. Endrődi, L.G. Pang, H. Stöcker, Physical Review D **100** (2019), `1810.12879`

[14] C. Gattringer, T. Kloiber, Nuclear Physics B **869**, 56–73 (2013), `1206.2954`

[15] T. Akiba, S. Sano, T. Yanase, T. Ohta, M. Koyama, *Optuna: A Next-generation Hyperparameter Optimization Framework*, in *Proceedings of the 25th ACM SIGKDD International Conference on Knowledge Discovery and Data Mining* (Association for Computing Machinery, 2019), p. 2623–2631, ISBN 978-1-4503-6201-6, `1907.10902`

[16] N. Prokof'ev, B. Svistunov, Physical Review Letters **87** (2001), `cond-mat/0103146`

Effect of Systematic Uncertainty Estimation on the Muon $g-2$ Anomaly

Glen Cowan[1,*]

[1]Physics Department, Royal Holloway, University of London, Egham, TW20 0EX, U.K.

Abstract. The statistical significance that characterizes a discrepancy between a measurement and theoretical prediction is usually calculated assuming that the statistical and systematic uncertainties are known. Many types of systematic uncertainties are, however, estimated on the basis of approximate procedures and thus the values of the assigned errors are themselves uncertain. Here the impact of the uncertainty *on the assigned uncertainty* is investigated in the context of the muon $g-2$ anomaly. The significance of the observed discrepancy between the Standard Model prediction of the muon's anomalous magnetic moment and measured values are shown to decrease substantially if the relative uncertainty in the uncertainty assigned to the Standard Model prediction exceeds around 30%. The reduction in sensitivity increases for higher significance, so that establishing a 5σ effect will require not only small uncertainties but the uncertainties themselves must be estimated accurately to correspond to one standard deviation.

1 Introduction

The recent measurement of the muon's anomalous magnetic moment at Fermilab's Muon $g-2$ Experiment [1], when averaged with the 2006 value from Brookhaven [2] was found to be in disagreement with the Standard Model (SM) prediction [3] with a significance of 4.2σ.

The significance of the discrepancy treats both the measurements and the theoretical prediction as Gaussian distributed quantities whose standard deviations are exactly known. Although the statistical errors are no doubt estimated with negligible uncertainty, this is not necessarily the case for systematic errors, particularly those of the SM prediction.

Here the method for incorporating uncertainties on reported systematic errors of Ref. [5] is applied to the muon $g-2$ significance. Section 2 describes the input values for the analysis and Sec. 3 gives a brief description of the statistical model. Results are shown in Sec. 4 and finally some conclusions are drawn in Sec. 5.

The recent result from Borsanyi et al. [6] using lattice QCD gives an SM prediction for $g-2$ that differs less from the measured value. As the purpose of this note is to illustrate the importance of the assigned systematic uncertainty for the case of a significant discrepancy, we focus on the 4.2σ difference highlighted in Ref. [1] and leave the lattice result for future consideration.

*e-mail: g.cowan@rhul.ac.uk

2 Input values

To simplify the numerical treatment and presentation of the results, the values $a_\mu = (g-2)/2$ are transformed according to

$$y = a_\mu \times 10^9 - 1165900\,. \tag{1}$$

The recent FNAL measurement [1] when combined with the 2006 value from BNL [2] results in an averaged experimental value of

$$y_{\rm exp} = 20.61 \pm 0.41\,.$$

The uncertainty reflects both statistical (0.37) and systematic (0.17) errors. The Standard Model prediction is given in Ref. [3] as

$$y_{\rm SM} = 18.10 \pm 0.43\,.$$

The uncertainty is dominated by the hadronic vacuum polarization (0.40) and to a lesser extent the hadronic light-by-light contribution (0.18).

3 Including uncertainties in estimates of systematic errors

In this note, the gamma variance model of Ref. [5] is used to include an uncertainty in assigned systematic errors into the significance of the observed anomaly. This model treats values v of systematic variances (i.e., $v = s^2$ where s is the estimated standard deviation) not as fixed constants but rather as estimates that follow a gamma distribution. The expectation values $E[v]$ become adjustable parameters of the model, and the standard deviations σ_v are fitted to reflect the accuracy with which the systematic error is estimated. To do this, the analyst supplies parameters $r = \sigma_v/2E[v] \approx \sigma_s/E[s]$, which to first approximation represent the relative uncertainty in the assigned systematic errors.

Here we apply this method only to the uncertainty in the Standard Model prediction $y_{\rm SM}$. This is by far the largest systematic uncertainty and owing to its theoretical origin it is inherently difficult to estimate precisely. That is, we treat the value of $v_{\rm SM} = (0.43)^2$ as a gamma-distributed estimate of the true variance $\sigma^2_{\rm SM}$ of $y_{\rm SM}$, and we assign to its distribution a relative "error on the error" $r_{\rm SM}$. In principle, the model can be applied to any of the assigned uncertainties, including the experimental systematic error. As this is substantially smaller than the SM uncertainty we leave this for future investigation.

The uncertainties on the assigned uncertainties of the SM prediction were considered in Sec. 2 (in particular footnote 2) of [4] and also Sec. 2.3.6 of Ref. [3]. The authors conclude that the issue is relevant and in Ref. [3] approaches to address it are discussed that remain within the basic framework of the method of least squares.

It is not intended to imply here that the SM uncertainty of 0.43 assigned in Ref. [3] is necessarily incorrect, but rather that it could be uncertain and so one may ask what impact this will have on the significance of the observed discrepancy. Therefore in this note we investigate this significance as a function of an assumed $r_{\rm SM}$. Discussions are ongoing with authors of the SM prediction on application of the method presented here.

The likelihood function of the gamma variance model contains as free parameters the mean μ and variance $\sigma^2_{\rm SM}$ of $y_{\rm SM}$. The profile log-likelihood function is obtained by evaluating $\sigma^2_{\rm SM}$ with the value that maximizes the likelihood for a given μ. This is up to an additive constant a function $Q(\mu)$ that plays the role of the χ^2 in a least-squares average, but with the usual quadratic term for $y_{\rm SM}$ replaced by a logarithmic one:

$$Q(\mu) = \frac{(y_{\rm exp} - \mu)^2}{\sigma^2_{\rm exp}} + \left(1 + \frac{1}{2r^2_{\rm SM}}\right) \ln\left[1 + 2r^2_{\rm SM}\frac{(y_{\rm SM} - \mu)^2}{v_{\rm SM}}\right] . \tag{2}$$

If the variance of $y_{\rm SM}$ is estimated very accurately then $r_{\rm SM} \ll 1$ and by expanding the logarithm one recovers the usual quadratic constraint.

As discussed in Ref. [5], a usual least-squares average with known uncertainties is equivalent to having Gaussian distributed inputs. As the tails of a Gaussian decrease very rapidly, a Gaussian-distributed value is extremely unlikely to depart from its mean by, say, five standard deviations. The gamma variance model is equivalent to replacement of the Gaussian by a Student's t distribution (see, e.g., Ref. [7]), where the number of degrees of freedom is $\nu = 1/2r^2$. Thus a greater relative uncertainty on an assigned systematic error corresponds to a lower number of degrees of freedom and therefore to tails that are longer than those of a Gaussian.

By minimizing the function $Q(\mu)$ of Eq. (2) with respect to μ one obtains the maximum-likelihood estimator $\hat{\mu}$. This is a weighted average of the SM prediction and the experimental measurement, and in the current problem it is not of direct interest. The important result is rather the goodness of fit, which can be quantified by the statistic $q = Q(\hat{\mu})$.

A higher value of q represents increasing incompatibility between the measured and predicted values $y_{\rm exp}$ and $y_{\rm SM}$. This can be quantified with the p-value

$$p = \int_{q_{\rm obs}}^{\infty} f(q)\, dq , \tag{3}$$

where $f(q)$ is the probability density function (pdf) of q and $q_{\rm obs}$ is its observed value. For the usual least-squares case, this would be a chi-square distribution for one degree of freedom. Here, however, because of the logarithmic term in Eq. (2), the pdf of q is found to depart from chi-square when the error-on-error parameter $r_{\rm SM}$ increases.

By construction the pdf of q is independent of μ, but as $r_{\rm SM}$ increases, $f(q)$ acquires a dependence on $\sigma^2_{\rm SM}$. To find the desired p-value of the composite hypothesis, one should take the maximum p found for any $\sigma^2_{\rm SM}$. Here this is approximated by computing the pdf of q with the parameters set to their maximum likelihood estimators (MLEs) $\hat{\mu}$ and $\widehat{\sigma^2}_{\rm SM}$ (see, e.g., Refs. [8], [9] Sec. 40.3.2.1). After $\hat{\mu}$ is found by minimizing $Q(\mu)$ from Eq. (2), the corresponding estimator for $\sigma^2_{\rm SM}$ is given by

$$\widehat{\sigma^2}_{\rm SM} = \frac{v_{\rm SM} + 2r^2_{\rm SM}(y_{\rm SM} - \hat{\mu})^2}{1 + 2r^2_{\rm SM}} . \tag{4}$$

To find the p-value as a function of $r_{\rm SM}$, one can thus generate values of q with Monte Carlo using the MLEs from the real data for μ and $\sigma^2_{\rm SM}$. To determine a very small p-value corresponding to a significance of 5σ or higher, however, the required amount of simulated data becomes very large. In Ref. [5] it is shown that by using a simple correction due to Bartlett [10], one may obtain p-values with far less computation. This correction was used in the present analysis and found to agree well with the full Monte Carlo method in regions where the latter is computationally feasible.

As is the usual practice in Particle Physics, the p-value is converted into an equivalent significance Z according to

$$Z = \Phi^{-1}(1 - p/2) , \tag{5}$$

where Φ^{-1} is the standard Gaussian quantile (inverse of the standard Gaussian cumulative distribution). The formula used here is appropriate for a two-sided hypothesis test, i.e., a positive or negative difference between measurement and prediction is regarded as equally discrepant. In the limit where $r_{\rm SM} \ll 1$ and the logarithmic term in Eq. (2) becomes quadratic, the significance is given by $Z = \sqrt{q}$, i.e., the square root of the minimized chi-squared.

4 Significance of the muon $g-2$ anomaly

In this section the procedure outlined above is applied to the significance of the muon $g-2$ anomaly. Figure 1 shows the significance in σ of the discrepancy between the experimental and predicted values as a function of $r_{\rm SM}$ using the gamma-variance model described above. The figure also shows the significance that one would obtain from a naive model, in which one simply inflates the SM uncertainty $\sigma_{\rm SM}$ by

$$\sigma_{y_{\rm SM}} \rightarrow \sigma_{\rm SM}(1 + r_{\rm SM}) \,. \tag{6}$$

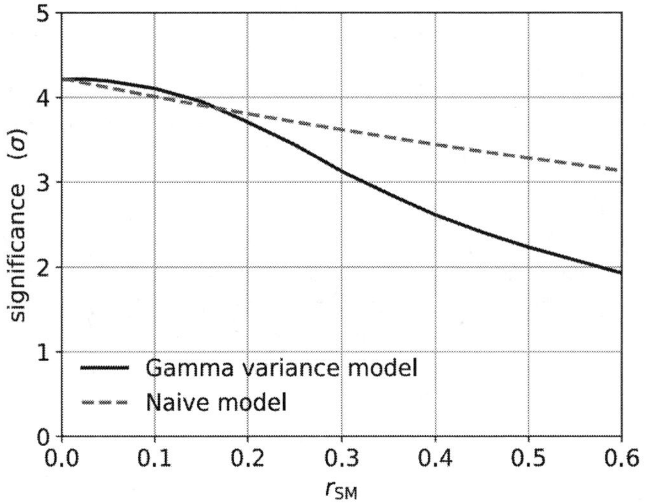

Figure 1. The significance of the muon $g-2$ discrepancy as a function of the relative uncertainty $r_{\rm SM}$ in the quoted systematic uncertainty of the SM prediction (the relative "error on the error").

As can be seen in the figure, the naive approach is quite close to the gamma variance model for relative uncertainties below around 20% (i.e., $r_{\rm SM} \leq 0.2$). But already for $r_{\rm SM} = 0.3$ the predictions differ substantially, with a significance of 3.12 from the gamma variance model and 3.63 from the naive model. The difference is increases at $r_{\rm SM} = 0.6$, with significances of 1.94 and 3.13.

Additional data taking from the Muon $g-2$ Experiment is expected reduce the experimental uncertainty. Suppose that the measured value remains the same but with half of its current uncertainty, i.e., $y_{\rm future} = 20.61 \pm 0.205$, and that the SM prediction remains $y_{\rm SM} = 20.40 \pm 0.43$. Figure 2 shows the significance of the discrepancy as a function of the relative uncertainty on the error assigned to the SM prediction $r_{\rm SM}$.

From Fig. 2 one sees that improved experimental accuracy is of almost no benefit unless the accuracy with which one assigns the SM uncertainty is kept below around 20% Also shown on the plot is the significance using an experimental uncertainty reduced by a factor

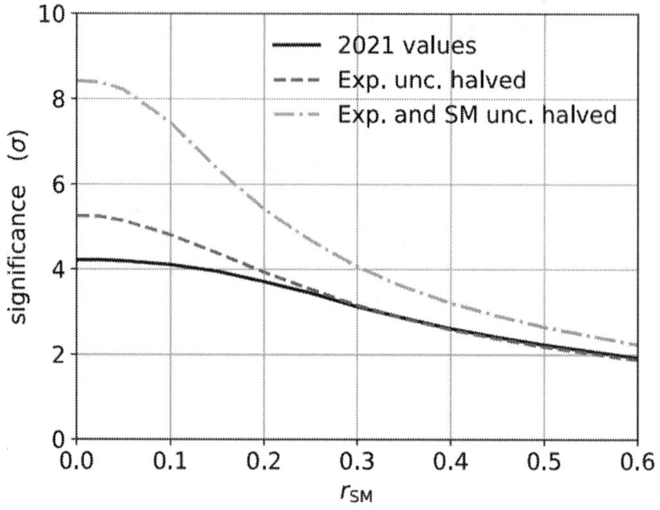

Figure 2. The significance of the muon $g-2$ discrepancy as a function of r_{SM} assuming the 2021 uncertainties (with SM uncertainty from [3]), with the experimental uncertainty reduced by a factor of two, and with the SM uncertainty also halved.

of two and an SM uncertainty σ_{SM} halved from 0.43 to 0.215. The situation is improved somewhat by a reduction in the SM uncertainty, but still only if this uncertainty is itself assigned to correspond accurately to one standard deviation. If the naive recipe $\sigma_{SM} \to (1 + r_{SM})\sigma_{SM}$ is applied to the case of halved uncertainties one finds a significance of 6.6σ at $r_{SM} = 0.5$, compared to only 2.6σ from the gamma variance model.

5 Discussion and conclusions

It should not be a surprise that the significance of the discrepancy between SM prediction and experiment decreases when one supposes an additional source of uncertainty, namely, the "error on the error" represented by r_{SM}. What is important to note, however, is that simply inflating the corresponding uncertainty by a factor of $(1 + r_{SM})$ does not adequately reflect the decrease in significance, as can seen by the curves in Fig. 1. Rather, the gamma variance model, which treats the assigned systematics as gamma-distributed estimates, results in a more rapid degradation of the significance for a relative uncertainty greater than a certain level, starting around 20% for this problem. If the relative uncertainty is 30%, then the significance drops to 3.1σ, and it goes below 2σ for a relative uncertainty of 60%.

If the all of the nominal uncertainties are reduced by a factor of two relative to their current values, the discovery significance under assumption of $r_{SM} = 0$ is 8.4σ. But this reduces to 5σ assuming a relative uncertainty in the theory error itself of 22%, and to 4σ for 30%. These results reflect the intrinsic difficulty in establishing a discovery at a high significance level if the uncertainties themselves are not accurately determined, and as a consequence the corresponding distributions acquire non-Gaussian tails. This underlines the importance of establishing appropriate procedures for quantifying systematic uncertainties both for the muon $g-2$ anomaly as well as for any investigations seeking to discover new phenomena.

Acknowledgements

Many thanks to the QCHS conference organizers for an interesting and productive meeting. Thanks for useful comments are due to Olaf Behnke, Kyle Cranmer, Louis Lyons, Veronique Boisvert and Bogdan Malaescu. This work was supported in part by the U.K. Science and Technology Facilities Council.

References

[1] B. Abi et al. (Muon $g-2$ Collaboration), *Measurement of the Positive Muon Anomalous Magnetic Moment to 0.46 ppm*, Phys. Rev. Lett. 126, 141801 (2021).

[2] G. W. Bennett et al. (Muon $g-2$ Collaboration), *Final report of the E821 muon anomalous magnetic moment measurement at BNL*, Phys. Rev. D 73, 072003 (2006).

[3] T. Aoyama, N. Asmussen, M. Benayoun, J. Bijnens, and T. Blum et al., *The anomalous magnetic moment of the muon in the standard model*, Phys. Rep. 887, 1 (2020).

[4] M. Davier, A. Hoecker, B. Malaescu and Z. Zhang, Eur. Phys. J. C 80 (2020) 241.

[5] G. Cowan, *Statistical Models with Uncertain Error Parameters*, Eur. Phys. J. C (2019) 79:133.

[6] Sz. Borsanyi et al., *Leading hadronic contribution to the muon magnetic moment from lattice QCD*, Nature vol. 593, (2021) 51-55.

[7] G. Cowan, *Statistical Data Analysis*, Oxford University Press, 1998.

[8] K. Cranmer, *Statistical challenges for searches for new physics at the LHC*, in Proceedings of PHYSTAT05, L. Lyons and M.K. Unel (eds.), Imperial College Press, pp. 112-123 (2005).

[9] P.A. Zyla et al. (Particle Data Group), Prog. Theor. Exp. Phys. 2020, 083C01 (2020).

[10] M.S. Bartlett, *Properties of sufficiency and statistical tests*, Royal Society of London Proceedings Series A 160, (1937) 268-282.

A machine learning pipeline for autonomous numerical analytic continuation of Dyson-Schwinger equations

Andreas Windisch[1,2,4,*], *Thomas* Gallien[3,4,**], and *Christopher* Schwarzlmüller[3,4,***]

[1]Department of Physics, Washington University in St. Louis, MO 63130, USA
[2]Know-Center GmbH, Inffeldgasse 13, 8010 Graz, Austria
[3]Silicon Austria Labs GmbH, Inffeldgasse 33, 8010 Graz, Austria
[4]RL Community, AI AUSTRIA, Wollzeile 24/12, 1010 Vienna, Austria

Abstract. Dyson-Schwinger equations (DSEs) are a non-perturbative way to express n-point functions in quantum field theory. Working in Euclidean space and in Landau gauge, for example, one can study the quark propagator Dyson-Schwinger equation in the real and complex domain, given that a suitable and tractable truncation has been found. When aiming for solving these equations in the complex domain, that is, for complex external momenta, one has to deform the integration contour of the radial component in the complex plane of the loop momentum expressed in hyper-spherical coordinates. This has to be done in order to avoid poles and branch cuts in the integrand of the self-energy loop. Since the nature of Dyson-Schwinger equations is such, that they have to be solved in a self-consistent way, one cannot analyze the analytic properties of the integrand after every iteration step, as this would not be feasible. In these proceedings, we suggest a machine learning pipeline based on deep learning (DL) approaches to computer vision (CV), as well as deep reinforcement learning (DRL), that could solve this problem autonomously by detecting poles and branch cuts in the numerical integrand after every iteration step and by suggesting suitable integration contour deformations that avoid these obstructions. We sketch out a proof of principle for both of these tasks, that is, the pole and branch cut detection, as well as the contour deformation.

1 Introduction

Computing Green's functions in the complex domain numerically turns out to be a tedious task, even when working in Euclidean space and when considering perturbative calculations at 1-loop level. The main complications one has to deal with are accounting for the poles and branch cuts that arise in the complex plane of the radial component of the square of the loop momentum, when the latter is expressed in hyper-spherical coordinates. Since the loop momentum is a 4-momentum, and because there is an external momentum flowing into the diagram, as well as the internal (loop-) momentum, two of the hyper-spherical angles can be integrated trivially, and one has to consider a two-fold, non-trivial integration over the square of the loop momentum, and over (the cosine of) the angle between the external and

*e-mail: windisch@physics.wustl.edu
**e-mail: thomas.gallien@silicon-austria.com
***e-mail: christopher.schwarzlmueller@silicon-austria.com

the internal momentum. As discussed e.g. in [1], branch cuts are induced by the angular integration and in conjunction with the current value of the external and internal momentum square, and they appear in the complex plane of the radial integration variable. Because these branch cuts often intersect the positive real axis, the original integration contour of the radial part, that starts at the origin (or a small infrared (IR) cut-off) and runs along the positive real half-axis to some ultraviolet (UV) cut-off, has to be deformed around the obstructive branch cut to connect the origin and the UV cut-off, and without picking up the residue of any pole in the plane that could appear between the deformed contour and the line between the origin and the UV cut-off. Put into other words, the integration contour has to be continuously deformable from the original contour to the new, deformed contour that avoids the branch cuts. While for perturbative loop diagrams usually a full, explicit form of the integrand is known, this normally does not hold true for DSEs (see e.g. [2] for a thorough treatment of the DSE formalism), as these are of the second Fredholm type and require e.g. iterative strategies to find a numerical solution. The dressing functions that result from a successful numerical solution of a Dyson-Schwinger equation appear on both sides of the equation, that is, explicitly on the left hand side, and in the integrand of the self-energy loop on the right hand side of the equation. The dressing functions are usually initialized to some real value, and given this as an initial condition, one could in principle analyze the integrand to find the poles and cuts for this given situation. One could then, in principle, find suitable contour deformations and avoid the cuts and poles, perform one iteration step, analyze the resulting, updated dressing functions, and try to deduce the newly arisen analytic obstructions for the next iteration step. Then, new contour deformations could be sought, and another iteration step could be performed. However, this is not feasible because of mainly two reasons. One, once we allow the dressing functions to take complex values for the external momentum square, the complex momenta will produce different branch cuts for each of its values, which results in a plethora of differently parametrized integration contours that have to be found. Usually one can cover all necessary contours by introducing a few parametrizations for different regions of the complex plane, but it is still a tedious task. The second reason why this is not feasible in practice is simply the fact that the system's solution is sought by an iterative approach, that is, after every iteration step such a full analysis would have to be conducted.

In these proceedings we propose a possible way to overcome these obstacles by introducing a machine learning pipeline that analyzes the current analytical properties of the integrand, and finds suitably deformed integration contours autonomously. In more concrete terms, here and in the following, we provide evidence for the feasibility of such a pipeline by providing proof (or proof of concept) for each and every of the following steps:

1. **PROOF (A):** Suitability of numerical contour deformation to compute the analytic properties of Green's functions

2. **PROOF OF CONCEPT (B):** Feasibility of determining the analytic properties of numerically computed functions by means of deep learning and computer vision

3. **PROOF OF CONCEPT (C):** Feasibility of autonomous contour deformation by means of a deep reinforcement learning agent

Here, the difference between **PROOF** and **PROOF OF CONCEPT** is, that the former has been successfully applied in a broad variety of contexts, which can thus be regarded as a successful strategy (and not as a proof in the mathematical sense). The latter means, that these steps have been introduced with the very application discussed here in mind, and we have published (and yet unpublished) knowledge that strongly suggests that the respective strategies will be successful. These proceedings are organized as follows. In the following three sections we discuss the points **(A)**, **(B)**, and **(C)** in detail, and then conclude with a summary and outlook.

2 Contour deformation for numerical computation of Green's functions in the complex domain (A)

In order to establish numerical proof that such a deformation can be obtained, we look at this study [3], in which a correlator of so-called i-particles was considered. These i-particles are discussed in [4], where also an exact solution of the analytic properties of the correlator, that has been treated numerically in [3], is presented. The exact solutions of [4], and the numerical results found in [3] by means of contour deformation, are in perfect agreement. While [3] is concerned with showing agreement between the analytic and numerical approaches, there is an abundance of studies in which this strategy has been deployed, see e.g. [5–8] for an exemplary but incomplete list of such studies.

3 Pole and branch cut detection with deep learning (B)

In this section we will discuss the pole and branch cut detection in numerical data by means of deep learning. In principle, and at least as far as pole detection is concerned, one could also do this by more conventional means, such as computing contours around possible pole locations in order to check for a non-vanishing residue, see e.g. [9]. This doesn't resolve the detection of branch cuts, though, which is why we chose a different approach here. Using a so-called U-Net, that has been originally developed for biomedical image segmentation [10], we created a similar pipeline for detecting poles and branch cuts in snippets of numerically computed functions. The U-Net architecture is comprised of a convolutional neural network (CNN) [11] with a contracting and an expanding path. In between, feature maps are being cropped and copied from the contracting to the expanding path. While images (in our case numerically computed snippets of functions in the complex plane) are being fed into the network, at its other end a segmentation mask is being produced, which, again in our case, marks the position of the poles and branch cuts. The network is trained by comparing the segmentation mask produced by the model with the ground truth that is provided as a 'label' for this learning setup.

3.1 U-Net based segmentation

In order to generate a data set used for training the U-Net for pole and branch-cut segmentation, we again considered the problem of the i-particle correlator mentioned above [4]. The numerical data has been produced by solving the angular integral for complex internal momenta (y) and complex external momenta (x) in equation (22) of [3] for $x, y \in [-10, 10] \times i[-10, 10]$. Once the cut has been produced, we added a random number of poles (between 0 and 9), with random positions and random residues additively to the function with the cut. The segmentation mask is a matrix with the same size as the data with the cut and poles (128×128 pixels), with its entries being ternary numbers. Using base-3 covers the three possible cases of every positional value, namely the 'pixel' being a pole location, belonging to a branch cut, or being a point that is neither of these two cases. The pole locations can be directly entered in the segmentation mask once the random location has been produced for a given training sample. The branch cut position can be marked by using the cut parametrization equation (25) in [3].

In Figure 1 we show an example of a training sample used to train the U-Net.

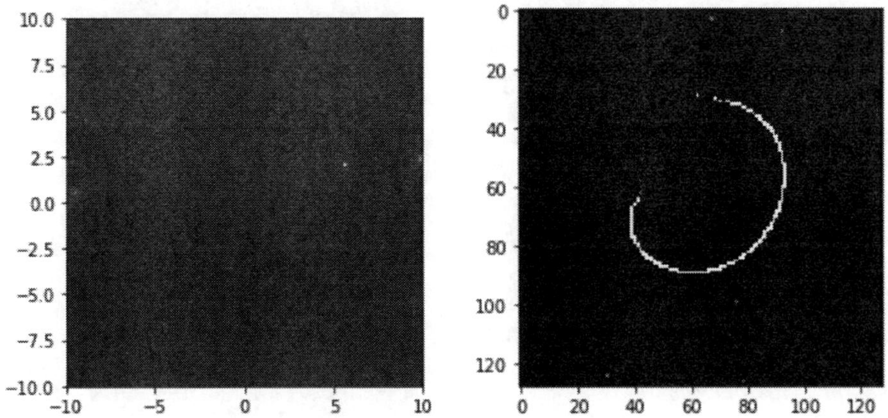

Figure 1. Left: A training sample with a branch cut and poles, produced as described in the text. The horizontal and vertical axis correspond to the real and imaginary part of the loop momentum square, y, expressed in GeV^2. Right: The corresponding segmentation mask. Note that the segmentation mask appears rotated here, because it has been expressed in pixel space rather than in the physical coordinate system.

In Figure 2 we show preliminary results of the first, trained U-Net using a training set generated as described above. We are currently investigating various extensions to our approach and will thus not provide any hyper parameters of the model at this point, as we are preparing a separate publication for the pole and cut detection with deep learning at the moment. The results of this first attempt look very promising, and suggest that this approach can be used in a Dyson-Schwinger setup to detect the cuts and poles in the integrand automatically. There is, of course, the problem of generalizability that is still to be addressed. In order to employ this approach to a Dyson-Schwinger setup, the variability encountered in the course of the iterative procedure must be captured by the distribution of the data set the U-Net has been trained with, otherwise it will most likely fail to recognize the respective patterns. This problem mostly affects the branch cuts, as poles are point-like and do not possess any relevant sub-structure, while cuts are essentially 1-dimensional objects. One possible way to generate cuts of arbitrary shape could be to produce a finite-length cut using a logarithm, and then apply an arbitrary distortion to the resulting cut-line to allow for arbitrary shapes. Another issue that is relevant in the context of the approach as described here is the sparsity of the ternary masks. Since there are only a few poles, and since also the branch cuts do not occupy a large amount of pixels, there is a class imbalance between pixels labeled as poles, cuts, or neither of the two. This skewness can become an issue when computing the loss function when training the system. To this end, one could compress the labels by deploying a compressive sensing approach, as discussed in [12, 13] in the context of biomedical imaging. We are currently also investigating this option, but do not have results yet.

4 Automated contour deformation through deep reinforcement learning (C)

In this section we discuss the next step in the pipeline, where we assume that the previous step has been successfully conducted and a segmentation mask that contains the information about the poles and cuts for the situation under consideration has been produced. Given this

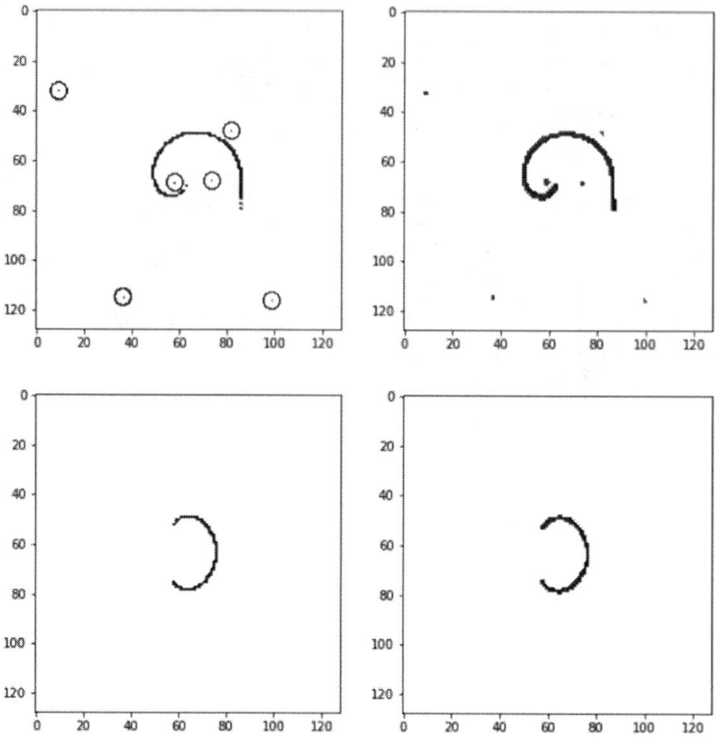

Figure 2. Preliminary results with the first U-Net attempt. Left column: Ground truth of segmentation masks. Since each pole only occupies a pixel in the mask, they have been marked by small circles for readability. Right column: Predictions produced by the trained U-Net.

information, we must now find a suitable integration contour that avoids the poles (i.e. is continuously deformable from the line along $\mathbb{R}_+ \cup \{0\}$), as well as the branch cuts. To this end we trained a deep reinforcement learning agent to perform this task (see [14]). Deep reinforcement learning is a learning approach inspired by behavioral science, in which a learning entity (the agent) observes and lives in an environment, conducts actions in this environment, and receives a reward or punishment by means of a scalar signal (see e.g. [15, 16] for an overview and a text book on the subject). The desired behavior of the agent can then be learned by maximizing the expected, cumulative reward, and the trained agent is then in possession of a policy, that takes in the state of the environment, and produces the most suitable action for a given situation.

4.1 Deep Reinforcement Learning agent for contour deformation

In [14], we considered a continuous scenario, in which we again used the *i*-particle correlator of [4]. This correlator possesses a branch cut, as well as two complex conjugate poles. Using a proximal policy optimization based approach [17], we successfully trained an agent that could produce meaningful contours in this environment, see Figure 3. The environment in this case consists of the complex plane of the radial integration variable, and the agent perceives this environment by means of a vector that holds relevant information, such as where the cut is, where the poles are located, in which direction the cut opens up, etc. In principle, one could also train a deep reinforcement learning agent solely on the numerical data as input, however, we decided against this approach, as we would lose a substantial amount of control over the whole setup, which, in addition, would be much more involved as well. In order to deploy this approach for DSEs, the DRL agent will be re-implemented in a discrete setup, as the continuous environment is harder to solve and not necessary for the relatively small sizes of discrete points the correlator will be solved for.

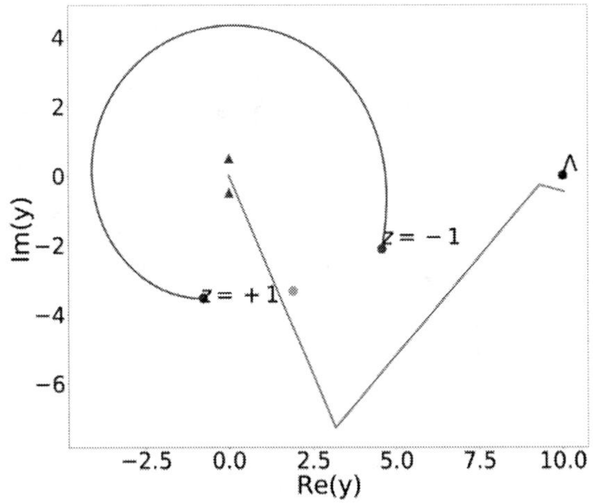

Figure 3. A successful contour found by the agent published in [14]. The red line represents the branch cut, the green triangles the complex conjugate poles located at $y = \pm\frac{i}{2}$. Note that this contour is continuously deformable from the line along the positive real half-axis to the cutoff, here represented by the Λ. The yellow point that is passed closely by the contour represents the direction of the opening of the cut, an information made available to the agent.

5 Conclusions and outlook

In the previous sections we provided proof, as well as proof of concept, for the main ingredients needed to set up an autonomous machine learning pipeline to perceive the analytic properties of the numerically computed, radial integration plane of the self energy loop in a DSE, as well as to produce the corresponding deformed integration contours. The first step, identifying the non-analyticities, will be implemented using either a U-Net architecture, or a regular CNN with dense labels produced by a compressive sensing approach. Once this information is available, it will be prepared for the DRL agent to observe, which will then produce a suitable contour. This could be done in every iteration step of the DSE, thereby solving the equation, while respecting the constraints as imposed by the non-analyticities. In order to study the quark propagator Dyson-Schwinger equation in Landau gauge in Euclidean space in the complex domain, we plan to deploy a suitable truncation scheme that establishes minimal constraints through the analytic properties of the self-energy loop.

Acknowledgements

AW acknowledges support through the U.S. Department of Energy, Office of Science, Office of Nuclear Physics under Award Number #DE-FG-02-05ER41375.

References

[1] A. Windisch, M.Q. Huber, R. Alkofer, Acta Phys. Polon. Supp. **6**, 887 (2013), 1304.3642
[2] R. Alkofer, L. von Smekal, Phys. Rept. **353**, 281 (2001), hep-ph/0007355
[3] A. Windisch, R. Alkofer, G. Haase, M. Liebmann, Comput. Phys. Commun. **184**, 109 (2013), 1205.0752
[4] L. Baulieu, D. Dudal, M.S. Guimaraes, M.Q. Huber, S.P. Sorella, N. Vandersickel, D. Zwanziger, Phys. Rev. **D82**, 025021 (2010), 0912.5153
[5] C.S. Fischer, M.Q. Huber, Phys. Rev. D **102**, 094005 (2020), 2007.11505
[6] G. Eichmann, P. Duarte, M.T. Peña, A. Stadler, Phys. Rev. **D100**, 094001 (2019), 1907.05402
[7] A.S. Miramontes, H. Sanchis-Alepuz, Eur. Phys. J. **A55**, 170 (2019), 1906.06227
[8] A. Windisch, M.Q. Huber, R. Alkofer, Phys. Rev. **D87**, 065005 (2013), 1212.2175
[9] A. Windisch, Phys. Rev. **C95**, 045204 (2017), 1612.06002
[10] O. Ronneberger, P. Fischer, T. Brox, *U-net: Convolutional networks for biomedical image segmentation* (2015), 1505.04597
[11] K. Fukushima, Biological Cybernetics **36**, 193 (1980)
[12] Y. Xue, N. Ray, CoRR **abs/1708.03307** (2017), 1708.03307
[13] Y. Xue, G. Bigras, J. Hugh, N. Ray, IEEE Transactions on Medical Imaging **38**, 2632 (2019)
[14] A. Windisch, T. Gallien, C. Schwarzlmüller, Phys. Rev. E **101**, 033305 (2020), 1912.12322
[15] V. François-Lavet, P. Henderson, R. Islam, M.G. Bellemare, J. Pineau, CoRR **abs/1811.12560** (2018), 1811.12560
[16] R.S. Sutton, A.G. Barto, *Reinforcement Learning: An Introduction* (MIT Press, 2018)
[17] J. Schulman, F. Wolski, P. Dhariwal, A. Radford, O. Klimov, CoRR **abs/1707.06347** (2017), 1707.06347

EPJ Web of Conferences **258**, 09004 (2022) https://doi.org/10.1051/epjconf/202225809004
A Virtual Tribute to Quark Confinement and the Hadron Spectrum (vConf21)

Preserving gauge invariance in neural networks

Matteo Favoni[1,*] *Andreas* Ipp[1,**] *David I.* Müller[1,2,***] and *Daniel* Schuh[1,****]

[1]Institute for Theoretical Physics, TU Wien,
Wiedner Hauptstr. 8-10, 1040 Vienna, Austria
[2]Speaker and corresponding author

Abstract. In these proceedings we present lattice gauge equivariant convolutional neural networks (L-CNNs) which are able to process data from lattice gauge theory simulations while exactly preserving gauge symmetry. We review aspects of the architecture and show how L-CNNs can represent a large class of gauge invariant and equivariant functions on the lattice. We compare the performance of L-CNNs and non-equivariant networks using a non-linear regression problem and demonstrate how gauge invariance is broken for non-equivariant models.

1 Introduction

The success of convolutional neural networks (CNNs) in image recognition has demonstrated that adapting machine learning methods to the specifics of the problem at hand can lead to better performing models with fewer parameters. For example, the problem of image classification (e.g. determining in a series of images if a particular image shows a certain animal or determining a number shown in an image of hand-written digits) often does not require knowledge about where in the image a certain feature is detected, but just that the image contains it somewhere. In that sense, these problems are invariant under spatial translations. The main idea behind CNNs is to make use of this symmetry by restricting neural network layers to be translationally equivariant. In this context, translational equivariance refers to the fact that applying a spatial translation to an input image yields an appropriately shifted output of the CNN. The concept of equivariance in neural networks has been extended to more general global symmetries (e.g. rotations, reflections) in the framework of group equivariant CNNs [1] or G-CNNs, where G refers to a symmetry group. Neural networks have also been generalized to local symmetry, specifically in the case of data defined on curved manifolds [2]. More generally, geometric deep learning [3, 4] is used as an umbrella term for the concept of incorporating the geometry of a machine learning problem in the choice of network architecture.

Neural network architectures that exhibit symmetry properties are well suited for applications in physics (see e.g. [5] for a pedagogical introduction), in particular methods that are tailored to preserve the symmetries of physical theories. For example, CNNs (or more

*e-mail: favoni@hep.itp.tuwien.ac.at
**e-mail: ipp@hep.itp.tuwien.ac.at
***e-mail: dmueller@hep.itp.tuwien.ac.at
****e-mail: schuh@hep.itp.tuwien.ac.at

generally G-CNNs) can be applied to problems in lattice field theory [6–12], which typically exhibit translation, rotation and reflection symmetry. Recently, machine learning models with built-in local symmetry have also been applied in the context of Abelian and non-Abelian lattice gauge theories, e.g. as generative models for Monte Carlo simulations [13–15]. In these proceedings we review some results of our recent work [16] on lattice gauge equivariant convolutional neural networks (L-CNNs), which is a general framework for networks that by construction preserve lattice gauge symmetry in pure $\mathrm{SU}(N_c)$ gauge theory. We first review some notation for lattice gauge theory in Sec. 2 and then introduce the L-CNN in Sec. 3. Finally, we demonstrate the performance of L-CNNs compared to non-equivariant CNNs in Sec. 4.

2 Lattice gauge theory

The link formalism of lattice gauge theory due to Wilson [17] allows us to construct lattice discretizations of non-Abelian Yang-Mills theory with exact lattice gauge symmetry. In the lattice formulation, gauge fields A_μ are replaced by (gauge) link variables $U_{x,\mu} \in \mathrm{SU}(N_c)$, defined on the edges of a hypercubic lattice Λ with lattice spacing a, which connect the starting lattice site x to the end at $x+\mu$.[1] For explicitness, we assume the lattice to be finite with periodic boundary conditions. We denote the dimension of the lattice by $D+1$, where $D \geq 1$ refers to the spatial dimensions and $\mu = 0$ is the (imaginary) time direction. The size of the lattice is given by $N_t \cdot N_s^D$. In terms of the gauge field $A_\mu(x) \in \mathfrak{su}(N_c)$, a gauge link $U_{x,\mu}$ is given by the path-ordered exponential[2]

$$U_{x,\mu} = \mathcal{P} \exp\left(i \int_0^1 \mathrm{d}s \frac{\mathrm{d}x^\nu(s)}{\mathrm{d}s} A_\nu(x(s)) \right), \tag{1}$$

where $x(s) : [0,1] \to \mathbb{R}^4$ defines the straight-line path connecting x to $x+\mu$, and $\mathbb{R}^4$ is the four-dimensional Euclidean spacetime. The geometric interpretation of gauge links is that they describe parallel transport along the edges of the lattice. Under general gauge transformations $\Omega : \mathbb{R}^4 \to \mathrm{SU}(N_c)$ of the gauge field,

$$A_\mu(x) \to \Omega(x) \left(A_\mu(x) - i\partial_\mu \right) \Omega^\dagger(x), \tag{2}$$

the gauge links transform according to

$$U_{x,\mu} \to \Omega_x U_{x,\mu} \Omega^\dagger_{x+\mu}. \tag{3}$$

For concreteness, we use the fundamental representation of $\mathfrak{su}(N_c)$ and $\mathrm{SU}(N_c)$ to represent A_μ and $U_{x,\mu}$ as complex matrices. The inverse link is denoted by $U_{x+\mu,-\mu} = U^\dagger_{x,\mu}$. In the limit of small lattice spacing $a \to 0$, gauge links can be approximated with the matrix exponential $U_{x,\mu} \simeq \exp\left(iaA_\mu(x + \frac{1}{2}\mu)\right)$ at the mid-point $x + \frac{1}{2}\mu$.

Multiple links connecting consecutive points can be multiplied to form Wilson lines along arbitrary paths on the lattice, and, in particular, closed paths or Wilson loops, where the start and end point coincide. The simplest loop is the 1×1 loop called plaquette

$$U_{x,\mu\nu} = U_{x,\mu} U_{x+\mu,\nu} U_{x+\mu+\nu,-\mu} U_{x+\mu,-\mu}, \tag{4}$$

[1] We use $x+\mu$ to denote the point $x + a\hat{e}_\mu$, where $\hat{e}_\mu$ is a Euclidean basis vector on the lattice.

[2] Here we use the convention that the path ordering operator $\mathcal{P}$ shifts fields earlier in the path to the left, and fields later to the right of the product, i.e. $\mathcal{P}(O(a)O(b)) = O(a)O(b)$ if $a < b$ and $\mathcal{P}(O(a)O(b)) = O(b)O(a)$ if $b < a$.

which under gauge transformations transforms according to

$$U_{x,\mu\nu} \to \Omega_x U_{x,\mu\nu} \Omega_x^\dagger. \tag{5}$$

In the continuum limit, the plaquette approximates the non-Abelian field strength tensor $F_{\mu\nu} = \partial_\mu A_\nu - \partial_\nu A_\mu - i\left[A_\mu, A_\nu\right]$ via $U_{x,\mu\nu} \simeq \exp\left(ia^2 F_{\mu\nu}\right)$. The Yang-Mills action can then be approximated by the Wilson action [17]

$$S_W[U] = \frac{2}{g^2} \sum_{x\in\Lambda} \sum_{\mu<\nu} \mathrm{Re}\,\mathrm{Tr}\left[\mathbb{1} - U_{x,\mu\nu}\right], \tag{6}$$

where $g > 0$ is the Yang-Mills coupling constant. Because of the trace over locally transforming plaquettes, the Wilson action is invariant under general lattice gauge transformations. This gauge invariant action can be used to perform Monte Carlo sampling of the path integral and compute expectation values of observables in lattice QCD at finite temperature [18].

3 Lattice gauge equivariant convolutional neural networks

This section reviews some aspects of the L-CNN architecture [16]. Similar to conventional CNNs, we construct our architecture from elementary layers, which explicitly respect lattice gauge symmetry. First, we define the data points for L-CNNs as tuples $(\mathcal{U}, \mathcal{W})$, where $\mathcal{U} = \{U_{x,\mu}\}$ is the set of gauge links of a particular lattice configuration, and $\mathcal{W} = \{W_{x,i}\}$ with $W_{x,i} \in \mathbb{C}^{N_c \times N_c}$, $\forall x \in \Lambda$, $1 \leq i \leq N_{\mathrm{ch}}$ is a set of locally transforming complex matrices with $N_{\mathrm{ch}} \in \mathbb{N}$ channels. Under lattice gauge transformations, we require that the matrices $\mathcal{W}$ transform locally

$$W_{x,i} \to \Omega_x W_{x,i} \Omega_x^\dagger. \tag{7}$$

Comparing L-CNNs to CNNs, the $\mathcal{W}$ matrices can be thought of as feature maps, while the links $\mathcal{U}$ provide the geometrical information to compare data at different lattice sites via parallel transport. We consider data points that are related by gauge transformations to be gauge equivalent.

Next, we need to specify what data the $\mathcal{W}$ matrices represent. In practice (see Sec. 4.1) we use plaquettes $U_{x,\mu\nu}$ at each point x on the lattice as input $\mathcal{W}$'s such that each face defined by the pair (μ, ν) is assigned to a particular channel. More generally, it is also possible to include all Polyakov loops (which are non-contractible loops wrapping around the periodic boundary of the lattice) at each lattice site in the input. We show why this is a particularly useful choice for input data in Sec. 3.2, where we prove that arbitrary contractible and non-contractible Wilson loops can be generated with L-CNNs using just two elementary layers.

3.1 Equivariant convolutions and bilinear layers

We introduce two gauge equivariant operations acting on data points $(\mathcal{U}, \mathcal{W})$: lattice gauge equivariant convolutions (L-Convs) and lattice gauge equivariant bilinear layers (L-Bilin). We formulate the layers in such a way that the output of these operations transforms consistently under Eqs. (3) and (7).

The L-Conv operation maps a data point $(\mathcal{U}, \mathcal{W})$ to a new data point $(\bar{\mathcal{U}}, \bar{\mathcal{W}})$ given by

$$\begin{aligned} \bar{U}_{x,\mu} &= U_{x,\mu}, \\ \bar{W}_{x,i} &= \sum_{j,\mu,k} \omega_{i,j,\mu,k} U_{x,k\cdot\mu} W_{x+k\cdot\mu,j} U^\dagger_{x,k\cdot\mu}, \end{aligned} \tag{8}$$

where $\omega_{i,j,\mu,k} \in \mathbb{C}$ are the trainable weight parameters (or kernel weights) with indices for output channels $1 \leq i \leq N_{\mathrm{ch,out}}$, input channels $1 \leq j \leq N_{\mathrm{ch,in}}$, lattice directions $0 \leq \mu \leq D$, and distances $-K \leq k \leq K$. Here, $K \in \mathbb{N}$ is the size of the kernel, which determines the receptive field, i.e. how many lattice points are considered when computing the convolution. The $\mathcal{W}$ matrices at different lattice sites are parallel transported to the common point x along straight paths using the Wilson lines constructed from links:

$$U_{x,k\cdot\mu} = \prod_{i=0}^{k-1} U_{x+i\cdot\mu,\mu} = U_{x,\mu} U_{x+\mu,\mu} U_{x+2\cdot\mu,\mu} \ldots U_{x+(k-1)\cdot\mu,\mu}. \tag{9}$$

Under gauge transformations, Eqs. (3) and (7), we find

$$\bar{W}_{x,i} \to \sum_{j,\mu,k} \omega_{i,j,\mu,k} \Omega_x U_{x,k\cdot\mu} \Omega^\dagger_{x+k\cdot\mu} \Omega_{x+k,\mu} W_{x+k\cdot\mu,j} \Omega^\dagger_{x+k\cdot\mu} \Omega_{x+k\cdot\mu} U^\dagger_{x,k\cdot\mu} \Omega^\dagger_x = \Omega_x \bar{W}_{x,i} \Omega^\dagger_x, \tag{10}$$

which shows that gauge equivariance is satisfied. Analogous to convolutional layers in standard CNNs, the output of L-Conv is also equivariant under lattice translations.

The second layer we introduce is a local bilinear operation. The L-Bilin layer maps two input data points $(\mathcal{U}, \mathcal{W})$ and $(\mathcal{U}', \mathcal{W}')$ with $\mathcal{U}' = \mathcal{U}$ to a new data point $(\bar{\mathcal{U}}, \bar{\mathcal{W}})$

$$\begin{aligned} \bar{U}_{x,\mu} &= U_{x,\mu}, \\ \bar{W}_{x,i} &= \sum_{j,k} \alpha_{i,j,k} W_{x,j} W'_{x,k}, \end{aligned} \tag{11}$$

where $\alpha_{i,j,k} \in \mathbb{C}$ are trainable weights with indices $1 \leq i \leq N_{\mathrm{ch,out}}$, $1 \leq j \leq N_{\mathrm{ch,in}}$, and $1 \leq k \leq N'_{\mathrm{ch,in}}$. The bilinear operation multiplies two locally transforming matrices at the same lattice site, which guarantees gauge equivariance:

$$\bar{W}_{x,i} \to \sum_{j,k} \alpha_{i,j,k} \Omega_x W_{x,j} \Omega^\dagger_x \Omega_x W'_{x,k} \Omega^\dagger_x = \Omega_x \bar{W}_{x,i} \Omega^\dagger_x. \tag{12}$$

As in the case of L-Conv, L-Bilin is equivariant under translations as well. The bilinear operation reduces to a quadratic layer when we use the same data point for both arguments, i.e. $(\mathcal{U}, \mathcal{W}) = (\mathcal{U}', \mathcal{W}')$.

Both operations only modify the $\mathcal{W}$ part of the tuple $(\mathcal{U}, \mathcal{W})$. It is also possible to formulate equivariant layers that modify the links $\mathcal{U}$ (see e.g. L-Exp in Ref. [16]), but for the purposes of these proceedings we only focus on L-Conv and L-Bilin. We also note that the expressivity of these layers can be further increased by adding additional channels to $\mathcal{W}$ prior to the operation. Specifically, before applying L-Conv or L-Bilin one may add unit elements and hermitian conjugates:

$$(W_{x,1}, W_{x,2}, \ldots, W_{x,N_{\mathrm{ch,in}}}) \to (\mathbb{1}, W_{x,1}, W_{x,2}, \ldots, W_{x,N_{\mathrm{ch,in}}}, W^\dagger_{x,1}, W^\dagger_{x,2}, \ldots, W^\dagger_{x,N_{\mathrm{ch,in}}}). \tag{13}$$

This increases the number of channels from N_{ch} to $2N_{\mathrm{ch}} + 1$. The unit elements generate bias terms for L-Conv, and bias and linear terms for L-Bilin. Hermitian conjugates correspond to changing the orientation of input plaquettes or Polyakov loops.

3.2 Generating arbitrary loops

It is important to address which class of functions can be represented by neural network architectures. Universality theorems exist for fully connected neural networks [19] and deep

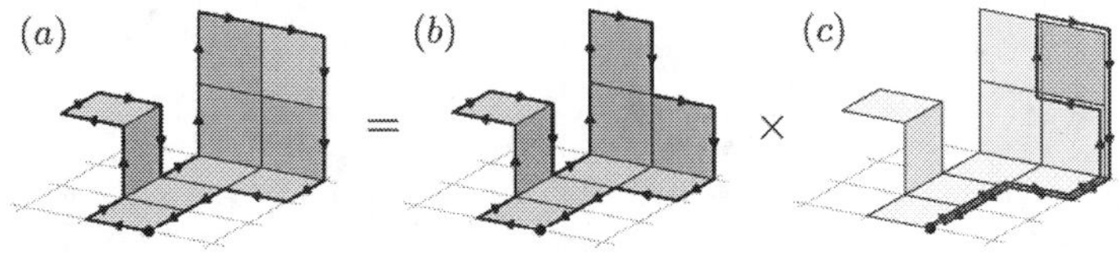

Figure 1. An arbitrary contractible Wilson loop on the lattice consisting of n squares (a) can be decomposed into a product of a loop with $n-1$ squares missing one plaquette (b) and a parallel transported single plaquette (c). Figure from [16].

CNNs [20], which demonstrate that neural networks are universal function approximators. In order to establish that L-CNNs are able to represent a large class of gauge equivariant functions, we prove by induction that arbitrary contractible and non-contractible Wilson loops can be constructed from repeated applications of L-Conv and L-Bilin.

Consider a data point tuple $(\mathcal{U}, \mathcal{W})$, where the set of $\mathcal{W}$ matrices are populated with all possible plaquettes at each lattice site. Initializing $\mathcal{W}$ in this manner trivially generates all possible 1×1 loops or loops of size 1. Next, consider an arbitrary contractible loop of size n (i.e. it consists of n squares) on the lattice, see Fig. 1 (a). Such a loop can be factorized into a loop of size $n-1$ (Fig. 1 (b)) with one missing plaquette and a transported single plaquette where the path traces along parts of the boundary of the original loop (Fig. 1 (c)). The factorization here refers to the composition of paths, which in the case of Wilson loops is realized by multiplying their matrix representations. L-CNNs can realize this factorization explicitly: Stacks of multiple L-Conv layers can generate arbitrary parallel transported plaquettes. For example, the parallel transported plaquette in Fig. 1 (c) can be generated with four L-Conv layers. On the other hand, the multiplication of the loop of size $n-1$ and the single transported plaquette can be realized by a single L-Bilin layer. By induction, every contractible loop can be constructed with L-Conv and L-Bilin operations starting from elementary plaquettes. This construction can be generalized to non-contractible loops (which wrap around the boundary of the lattice) by including, in addition to plaquettes, all possible straight line Polyakov loops at every lattice site in the initial set of $\mathcal{W}$ matrices.

Generally, L-CNNs consisting of stacks of L-Conv and L-Bilin layers can represent arbitrary linear combinations of Wilson loops of various shapes and sizes. The largest possible size and shape is determined by the kernel sizes and the number of layers. By including gauge equivariant non-linear activation functions in the L-CNN architecture (as introduced in our paper [16]), it is possible to represent non-linear functions of loops as well. Finally, if the desired output of an L-CNN is supposed to be gauge invariant (e.g. in the case of a regression problem for gauge invariant observables), it is possible to simply compute the trace of all matrices in $\mathcal{W}$. This renders the output of an L-CNN invariant because of the cyclic property of the trace.

4 Results and discussion

In order to test our new architecture, we designed a non-linear regression problem and tried out various L-CNN and non-equivariant CNN (baseline) architectures to solve it. The performance of L-CNNs and baseline networks are compared, and we investigate the breaking of gauge symmetry in baseline CNNs using adversarial attacks.

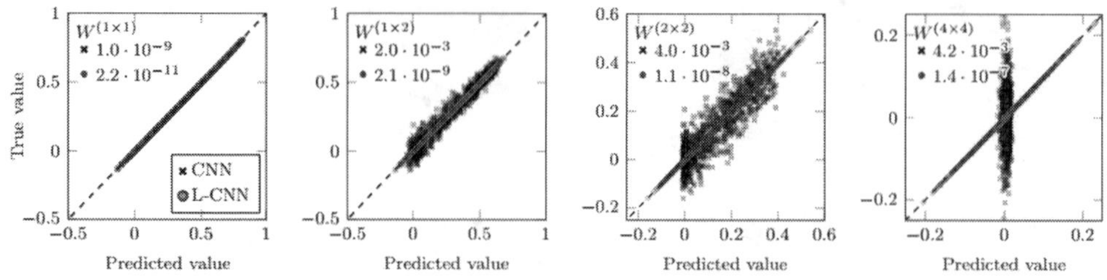

Figure 2. Scatter plots of our best L-CNN and baseline CNN models according to validation MSE on $8 \cdot 8$ lattices. True values of the Wilson loop are plotted against the predictions of the L-CNN model (green dots) and the baseline CNN (black crosses). High accuracy is achieved when all points are distributed close to the 45° line. The MSE is stated in the upper left corner of each panel. We observe that our best L-CNN models achieve low MSE for all studied loop sizes, while baseline CNNs perform worse at larger loop sizes. Figure adapted from [16].

4.1 Wilson loop regression

The proposed regression problem consists of two-dimensional ($D = 1$) lattice configurations sampled from an SU(2) Monte Carlo simulation, from which networks should compute the real value of the trace of $n \times m$ Wilson loops, i.e.

$$W^{(m\times n)}_{x,\mu\nu} = \frac{1}{N_c} \mathrm{Re}\,\mathrm{Tr}\left[U^{(m\times n)}_{x,\mu\nu}\right], \tag{14}$$

which is a gauge invariant observable. In our computational experiments we use 1×1 (as a trivial example), 1×2, 2×2 and 4×4 loops. We sample configurations for various values of the coupling constant on an $N_t \cdot N_s = 8 \cdot 8$ lattice to create the training and validation set. Test sets are generated for lattice sizes up to $64 \cdot 64$. For each architecture, trainable weights are randomly initialized multiple times and independently trained. We use mean squared error (MSE) as the objective (or loss) function to optimize during training. Both L-CNNs and baseline CNNs are provided with links and plaquettes in the input layer. For more details regarding architectures and the training process, we refer to our original paper and its forthcoming supplementary materials [16].

Summarizing the main results of our baseline study, we find that L-CNNs are indeed able to solve the regression problem for all sizes of the loop to a high degree of accuracy and also exhibit generalization to larger lattices. Baseline CNNs typically are not able to find adequate solutions, and the quality of their fits deteriorates with increased loop size. In Fig. 2 we show scatter plots for the best models of each type, where true values (i.e. labels in the dataset) are plotted against the predictions made by the networks.

4.2 Adversarial attacks for non-equivariant networks

While our L-CNN architectures are gauge invariant by construction (up to numerical precision), the baseline CNNs must learn this symmetry during the training process. Therefore, it is expected that even well-trained baseline CNNs at best only exhibit approximate gauge symmetry. The extent to which gauge symmetry is broken can be studied by applying gauge transformations to the input layer

$$U_{x,\mu} \to \Omega_x U_{x,\mu} \Omega^\dagger_{x+\mu}, \qquad W_{x,i} \to \Omega_x W_{x,i} \Omega^\dagger_x, \tag{15}$$

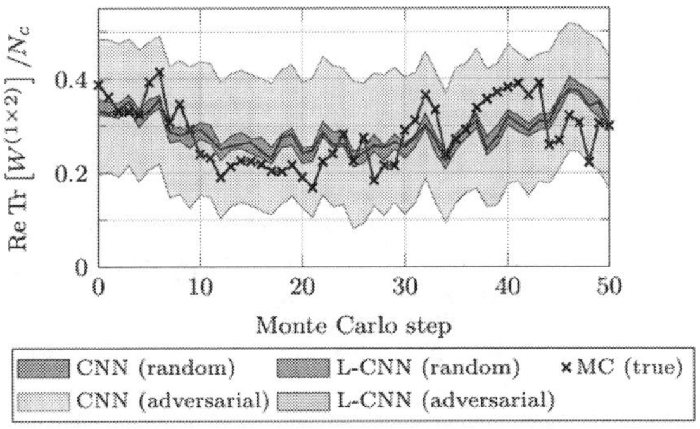

Figure 3. Uncertainty in predictions due to broken gauge symmetry in L-CNNs (green bands) and baseline CNNs (red bands) on 8·8 test data for the 1×2 loop regression problem. The true values (labels) are shown as black crosses. We use the same models as in Fig. 2. L-CNNs are invariant by construction and therefore only show deviations on the order of numerical precision. Baseline CNNs show much larger deviations, where the dark red band corresponds to errors due to random transformations, and the light red band shows the result of adversarial attacks. Figure from [16].

with $\Omega_x \in SU(N_c)$ and determining how much the predictions of a non-equivariant model change as a result. For a given model and a given lattice configuration, we apply two types of transformations: multiple random gauge transformations, where each Ω_x is sampled randomly, and adversarial attacks, where Ω_x is optimized to produce the largest deviation between the original prediction and the gauge transformed prediction. The adversarial attack is unique to each configuration and model, while random transformations do not depend on either. We show the results of such an attack for our best L-CNN and baseline CNN models for the 1×2 loop in Fig. 3. The L-CNN model is entirely unaffected by transformations (up to numerical precision), but the predictions of the baseline network show large errors in the case of adversarial attacks.

5 Conclusions

In these proceedings we have reviewed some aspects of the L-CNN architecture, particularly the L-Conv and L-Bilin layer, which in combination can be used to generate arbitrarily shaped Wilson loops on the lattice. We have also presented comparisons between L-CNNs and non-equivariant CNNs using a non-linear regression task for Wilson loops and demonstrated the breaking of gauge invariance in non-equivariant networks.

In this work we have mainly focused on the formulation and some properties of L-CNNs, but from a practical perspective, it would be interesting to apply the L-CNN architecture to normalizing [13–15] and continuous [12] flow models. On the other hand, in order to provide a more solid mathematical foundation, it would be worthwhile to understand L-CNNs as a special case of CNNs on principal bundles in the vein of gauge equivariant CNNs for curved manifolds [2] or gauge equivariant convolutions defined in [4].

DM thanks Jimmy Aronsson for many helpful discussions regarding equivariance, neural networks and geometric deep learning. This work has been supported by the Austrian Science Fund FWF No. P32446-N27, No. P28352 and Doctoral program No. W1252-N27. The Titan V GPU used for this research was donated by the NVIDIA Corporation.

References

[1] T.S. Cohen, M. Welling, *Group Equivariant Convolutional Networks*, in *Proceedings of The 33rd International Conference on Machine Learning* (JMLR, 2016), Vol. 48, pp. 2990–2999, 1602.07576

[2] T.S. Cohen, M. Weiler, B. Kicanaoglu, M. Welling, *Gauge Equivariant Convolutional Networks and the Icosahedral CNN*, in *Proceedings of the 36th International Conference on Machine Learning* (JMLR, 2019), Vol. 97, pp. 1321–1330, 1902.04615

[3] M.M. Bronstein, J. Bruna, T. Cohen, P. Veličković (2021), 2104.13478

[4] J.E. Gerken, J. Aronsson, O. Carlsson, H. Linander, F. Ohlsson, C. Petersson, D. Persson (2021), 2105.13926

[5] P. Mehta, M. Bukov, C.H. Wang, A.G. Day, C. Richardson, C.K. Fisher, D.J. Schwab, Physics Reports **810**, 1 (2019), a high-bias, low-variance introduction to Machine Learning for physicists

[6] K. Zhou, G. Endrődi, L.G. Pang, H. Stöcker, Phys. Rev. D **100**, 011501 (2019), 1810.12879

[7] D.L. Boyda, M.N. Chernodub, N.V. Gerasimeniuk, V.A. Goy, S.D. Liubimov, A.V. Molochkov, Phys. Rev. D **103**, 014509 (2021), 2009.10971

[8] S. Blücher, L. Kades, J.M. Pawlowski, N. Strodthoff, J.M. Urban, Phys. Rev. D **101**, 094507 (2020), 2003.01504

[9] D. Bachtis, G. Aarts, B. Lucini, Phys. Rev. E **102**, 053306 (2020), 2007.00355

[10] S. Bulusu, M. Favoni, A. Ipp, D.I. Müller, D. Schuh, Phys. Rev. D **104**, 074504 (2021), 2103.14686

[11] D. Bachtis, G. Aarts, B. Lucini, Phys. Rev. D **103**, 074510 (2021), 2102.09449

[12] P. de Haan, C. Rainone, M. Cheng, R. Bondesan (2021), 2110.02673

[13] G. Kanwar, M.S. Albergo, D. Boyda, K. Cranmer, D.C. Hackett, S. Racanière, D.J. Rezende, P.E. Shanahan, Phys. Rev. Lett. **125**, 121601 (2020), 2003.06413

[14] D. Boyda, G. Kanwar, S. Racanière, D.J. Rezende, M.S. Albergo, K. Cranmer, D.C. Hackett, P.E. Shanahan, Phys. Rev. D **103**, 074504 (2021), 2008.05456

[15] M.S. Albergo, D. Boyda, D.C. Hackett, G. Kanwar, K. Cranmer, S. Racanière, D.J. Rezende, P.E. Shanahan (2021), 2101.08176

[16] M. Favoni, A. Ipp, D.I. Müller, D. Schuh (2020), 2012.12901

[17] K.G. Wilson, Phys. Rev. D **10**, 2445 (1974)

[18] K.G. Wilson, *Monte-Carlo calculations for the lattice gauge theory*, in *Recent Developments in Gauge Theories. Proceedings, Nato Advanced Study Institute, Cargese, France, August 26 - September 8, 1979*, edited by G. 't Hooft, C. Itzykson, A. Jaffe, H. Lehmann, P. Mitter, I. Singer, R. Stora (1980), Vol. 59, pp. 363–402

[19] Z. Lu, H. Pu, F. Wang, Z. Hu, L. Wang (2017), 1709.02540

[20] D.X. Zhou, Applied and Computational Harmonic Analysis **48**, 787 (2020), 1805.10769

EPJ Web of Conferences **258**, 10001 (2022) https://doi.org/10.1051/epjconf/202225810001
A Virtual Tribute to Quark Confinement and the Hadron Spectrum (vConf21)

Highly excited pure gauge SU(3) flux tubes

Pedro Bicudo[1,*], *Nuno* Cardoso[1,**], and *Alireza* Sharifian[1,***]

[1]CeFEMA, Departamento de Física, Instituto Superior Técnico,
Av. Rovisco Pais, 1049-001 Lisboa, Portugal

Abstract. Flux tube spectra are expected to have full towers of levels due to the quantization of the string vibrations. We study a spectrum of flux tubes with static quark and antiquark sources with pure gauge $SU(3)$ lattice QCD in 3+1 dimensions up to a significant number of excitations. To go high in the spectrum, we specialize in the most symmetric case Σ_g^+, use a large set of operators, solve the generalized eigenvalue and compare different lattice QCD gauge actions and anisotropies.

[*]e-mail: bicudo@tecnico.ulisboa.pt
[**]e-mail: nuno.cardoso@tecnico.ulisboa.pt
[***]e-mail: alireza.sharifian@tecnico.ulisboa.pt

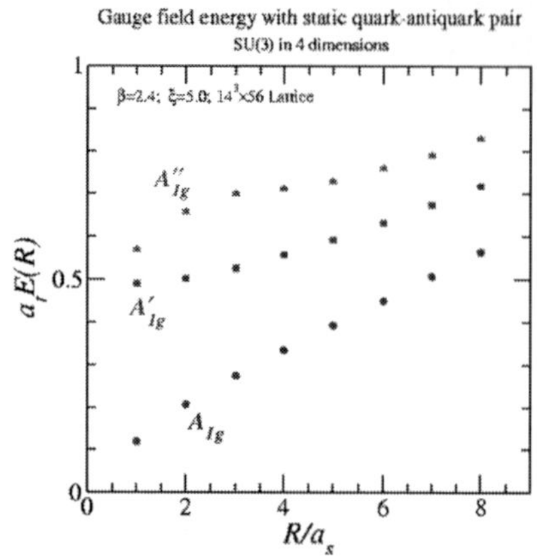

Figure 1. Previous excited Σ_g^+ flux tube spectrum [1].

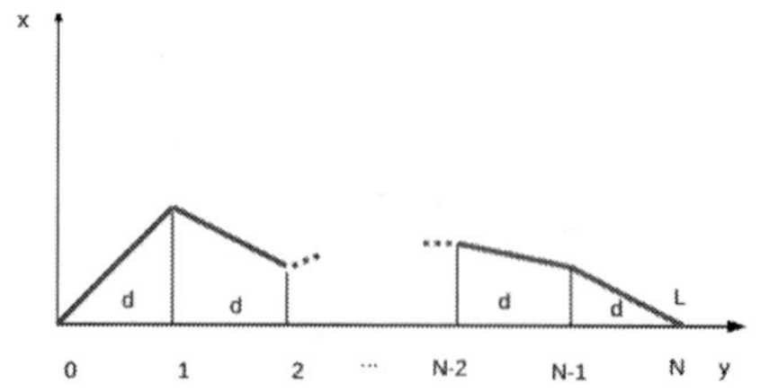

Figure 2. The discretization of the string transverse quantum fluctuations is sufficient to compute the Coulomb term of the string spectrum.

1 Introduction - Motivation

So far only energy levels up to $N = 2$ have been published for the pure gauge QCD flux tube [1] as in Fig. 1. Can we go higher in the spectrum?

To understand why there is a spectrum, we notice that if we neglect the flux tube intrinsic width, it is equivalent to a quantum string. Flux tubes and strings have been studied for a long time.

- In 1911 Onnes discovered superconductivity, the Meissner effect was discovered in 1933.
- In 1935, Rjabinin and Shubnikov experimentally discovered the Type-II superconductors. In 1950, Landau and Ginzburg, then continued by Abrikosov arrived at superconductor vortices, or flux tubes.
- When confinement was proposed for quarks inside hadrons in 1964 by Gell-Mann and Zweig, the analogy with flux tubes also led to a literature explosion in the quantum excitations of strings.
- The proposal of QCD in 1973 By Gross, Wilczek and Politzer shifted the interest back to particles.
- Nevertheless Lattice QCD by Wilson in 1974 was inspired in strings.
- The interest in strings returned in 1997 with Maldacena's and others AdS/CFT correspondence.
- The AdS/CFT and Holography has also been used as a model to compute spectra in hadronic physics.

An approximation to the flux tube spectrum is given by Effective String Theories (EST), say the Nambu-Goto model, which action is the area of the transverse bosonic string surface in time and space,

$$S = -T \int d^2\sigma \sqrt{-\det\gamma} \, , \gamma_{\alpha\beta} = \frac{\partial X^\mu}{\partial\sigma^\alpha} \frac{\partial X^\nu}{\partial\sigma^\beta} \eta_{\mu\nu} \tag{1}$$

it is classically equivalent to the Polyakov action which introduces an auxiliary *einbein* field, $g_{\alpha\beta}$ to remove the square root,

$$S = -\frac{T}{2} \int d^2\sigma \sqrt{-\det g} \, g^{\alpha\beta} \frac{\partial X^\mu}{\partial\alpha} \frac{\partial X^\nu}{\partial\beta} \eta_{\mu\nu} \, . \tag{2}$$

Its spectrum for an open string with ends fixed at distance R with Dirichlet boundary conditions is given by the Arvis potential [2],

$$\begin{aligned} V_n(R) &= \sigma \sqrt{R^2 + \frac{2\pi}{\sigma} \left(\sum_{a_i=1}^{\infty} n_{a_i} - \frac{D-2}{24} \right)} , \\ &= \sigma R + \frac{\pi}{R} \left(\sum_{a_i=1}^{\infty} n_{a_i} - \frac{D-2}{24} \right) + o\left(\frac{1}{R^3}\right) , \end{aligned} \tag{3}$$

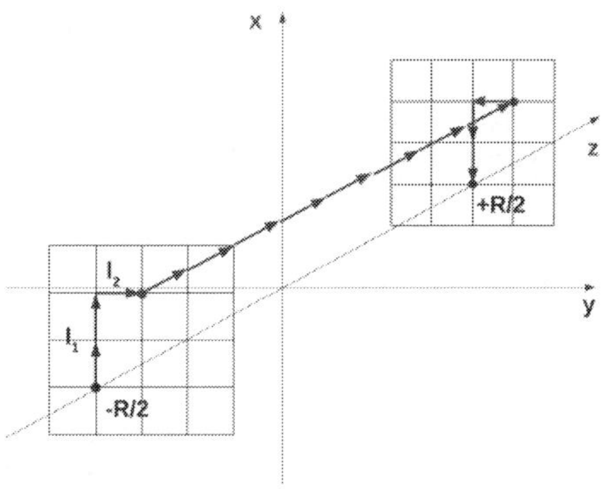

Figure 3. Example of a lattice QCD spacial operator, in this case $O(2,1)$.

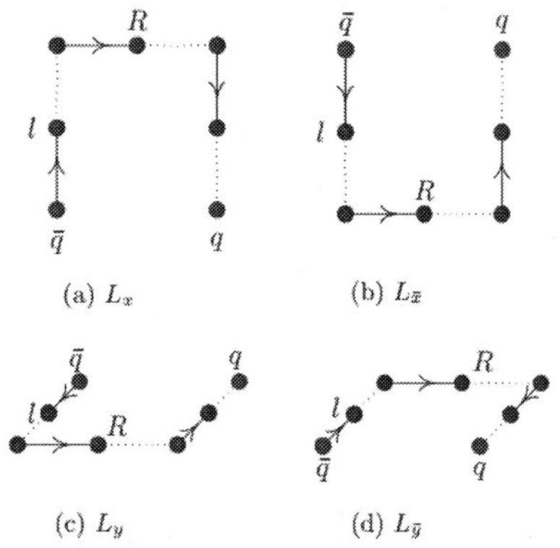

Figure 4. The four sub-operators, spatial Wilson line paths from the antiquark to the quark, used to construct the gauge field operators $O(l,0) = \frac{1}{\sqrt{4}}\left(L_x + L_{\bar{x}} + L_y + L_{\bar{y}}\right)$ at Euclidean time t_0. The inverse Wilson lines are used for the operators at time t.

where a_i are the principal transverse modes, and the continuum field theory computation of the zero mode energy is obtained using the Riemann Zeta regularization [3, 4],

$$1 + 2 + 3 + \cdots = \sum_{n=1}^{\infty} n \qquad \to \qquad \zeta(-1) = -\frac{1}{12}\,, \tag{4}$$

which in lattice QCD is provided by the lattice regularization. The Coulomb approximation in Eq. (3), Lüscher term [5], can also be computed with a discretization of the String as in Fig.2.

However the Nambu-Goto model is certainly not the EST of QCD flux tubes.

- There is a wider class of EST, the Nambu-Goto is just one of the possible EST [6–8].
- Contrary to Nambu-Goto, the zero mode of the QCD flux tubes has no tachyon *with negative square masses* at small distances R.
- There is lattice QCD evidence for an intrinsic width of the QCD flux tube [9].
- The QCD flux tube has a rich structure in chromoelectric and chromomagnetic field densities [10, 11].

Moreover, a quarkonium with flux tube excitations corresponds to an exotic, hybrid excitation of a meson. Thus we study the excited spectrum to learn more about the QCD flux tubes [12]. We restrict to Σ_g^+ flux tubes, the most symmetric ones, to go as high as possible in the spectrum.

2 Our lattice QCD framework for Σ_g^+ flux tubes

To compute the very excited spectrum, we use a large basis of spacial operators composed of generalized Wilson lines, and example of their spacial part is illustrated in Fig. 3 .

It turns out that using operators embedded only in axis planes, shown in Fig. 4, we decrease the degeneracy of states in our spectrum. This suppresses the Λ_g states. Although we do not want to study them, due to the cubic symmetry of the lattice they may be as well generated by our operators.

The first step to compute the energy levels, is to diagonalise the Generalized Eigenvalue Problem for the correlation matrix of our operators $O(l,0)$

$$C(t)v_n(t,t_0) = \lambda_n(t,t_0)C(t_0)v_n(t_0)\,, \tag{5}$$

ens.	action	operators	β	Volume	u_s	u_t	ξ	ξ_R	$a_s\sqrt{\sigma}$	$a_t\sqrt{\sigma}$	# conf.
O_1	Wilson	11 $O(l_1,l_2)$	6.2	$24^3\times48$	-	-	1	1	0.1610	0.1610	1180
O_2	Wilson	11 $O(l_1,l_2)$	5.9	$24^3\times48$	-	-	2	2.1737(4)	0.3088(4)	0.1421(2)	2630
W_1	Wilson	13 $O(l,0)$	6.2	$24^3\times48$	-	-	1	1	0.1610	0.1610	2500
W_2	Wilson	13 $O(l,0)$	5.9	$24^3\times48$	-	-	2	2.1737(4)	0.3093(2)	0.1423(1)	2170
W_4	Wilson	13 $O(l,0)$	5.6	$24^3\times96$	-	-	4	4.5459(9)	0.4986(4)	0.1097(1)	3475
S_4	S_{II}	13 $O(l,0)$	4.0	$24^3\times96$	0.82006	1.0	4	3.6266(32)	0.3043(3)	0.0839(1)	3575

Table 1. Our ensembles, for the isotropic Wilson action and the improved anisotropic S_{II} action. ξ is the bare anisotropy in the Lagrangian and ξ_R is the renormalized anisotropy.

for each time extent t of the Wilson loop, and get a set of time dependent eigenvalues $\lambda_i(t)$. With the time dependence, we study the effective mass plot

$$E_i \simeq \log\frac{\lambda_i(t)}{\lambda_i(t+1)}\,, \tag{6}$$

and search for clear plateaux consistent with a constant energy E_i in intervals $t\in[t_{i\text{ini}},t_{i\text{fin}}]$ between the initial and final time of the plateau.

We use the anisotropic Wilson action [13] computed with plaquettes,

$$S_{\text{Wilson}} = \beta\left(\frac{1}{\xi}\sum_{x,s>s'} W_{s,s'} + \xi\sum_{x,s} W_{s,t}\right) \tag{7}$$

where $W_c = \sum_c \frac{1}{3}\text{Re Tr}(1-P_c)$, $P_{s,s'}$ denotes the spatial plaquette, $P_{s,t}$ the spatial-temporal plaquettes and ξ is the (unrenormalized) anisotropy. Moreover, to improve our signal we also resort to the improved anisotropic action S_{II} developed in Ref. [14],

$$\begin{aligned} S_{\text{II}} &= \beta\left(\frac{1}{\xi}\sum_{x,s>s'}\left[\frac{5W_{s,s'}}{3u_s^4} - \frac{W_{ss,s'}+W_{s's',s}}{12u_s^6}\right] + \right.\\ &\left. +\xi\sum_{x,s}\left[\frac{4W_{s,t}}{3u_s^2u_t^2} - \frac{W_{ss,t}}{12u_s^4u_t^2}\right]\right), \end{aligned} \tag{8}$$

with $u_s = \langle\frac{1}{3}\text{Re Tr}P_{ss}\rangle^{1/4}$, $u_t = 1$. $W_{ss,s'}$ and $W_{ss,t}$, instead of plaquettes, include 2×1 rectangles. The results with more excited states shown in the literature [1], have been obtained with this action.

The anisotropy is used in order to have a smaller temporal lattice spacing a_t, to obtain more precise plateaux for excited energies since we have more time slices for the same time intervals.

For the S_4 ensemble and for the Wilson ensembles with anisotropy, we use MultiHit with 100 iterations in time followed by Stout smearing in space with $\alpha = 0.15$ and 20 iterations.

We use GPUS, and we find it is more economical to perform all our computations on the fly, rather than saving configurations. Our ensembles are summarized in Table 1.

3 Improved results for the Σ_g^+ spectrum

We obtain several plateaux with clear energy levels for each distance R. We first check for degeneracies. From Nambu-Goto, we expect degeneracies:
$n=0$, 1 state : $n_r=0$; $n=2$, 1 state : $n_r=1$; $n=4$, 3 states: $n_r=2$ or l=4, ...

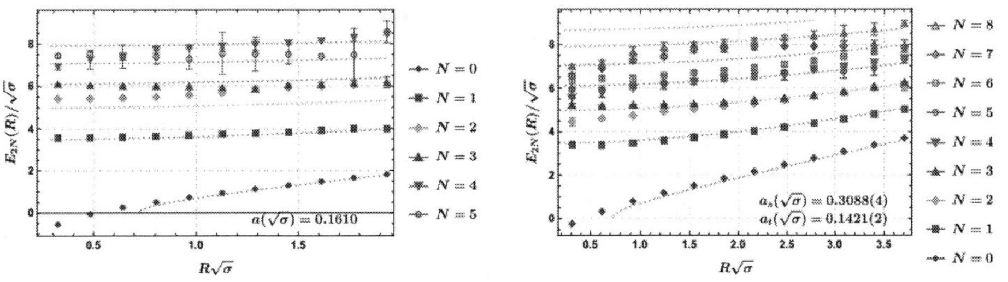

Figure 5. Results with the enlarged basis of 11 $O(l_1, l_2)$ operators with ensembles O_1 and O_2. The Nambu-Goto model spectrum is in dashed lines.

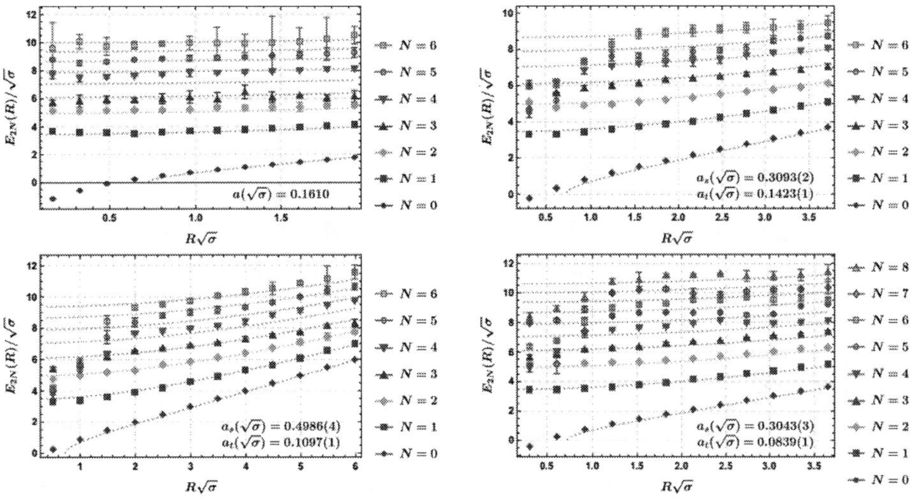

Figure 6. Results for the full spectrum, showing only the levels obtained with good Effective Mass Plots, from top to bottom respectively with ensembles W_1, W_2, W_4 and S_4. The Nambu-Goto model spectrum is in dashed lines.

Indeed, using off-axis operators, we find in Fig. 5 nearly degenerate states $N = 2$ and $N = 3$, compatible with $n_r = 2$, $l = 0$ and $n_r = 0$, $l = 4$ states.

Using only on axis operators, we are then able to suppress the $l = 4$ degeneracy. At least for larger distances, we find up to $N = 8$ levels for the S_4 ensemble, shown in Fig. 6.

At smaller distances, we are so far unable to avoid some degeneracy; possibly it is due to higher harmonics since $a_i = 1, 3, 5 \cdots$ are as well Σ_g^+ states.

4 Analysis of our Σ_g^+ spectrum

We cannot fit accurately the very high spectrum with the Nambu-Goto spectrum which energy levels are more compressed than ours. We thus generalize the Nambu-Goto model using two different string tensions, the σ_2 replaces the σ inside the square root. With a global non-linear fit, shown in Fig. 7, we extract as well the renormalized anisotropy ξ_R and the string tensions σ, σ_2.

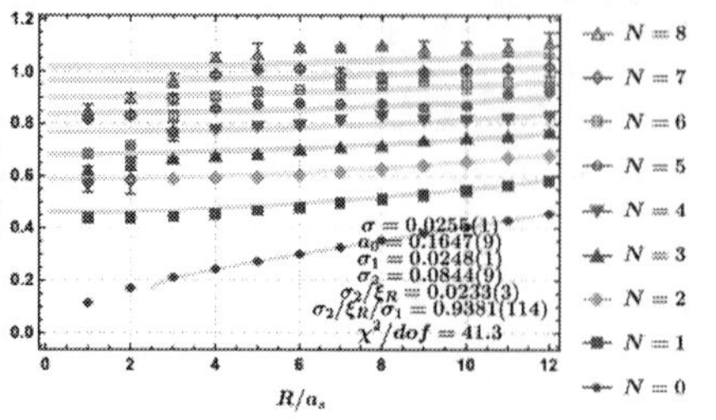

Figure 7. Very excited spectrum of the QCD flux tube fitted by a generalized Nambu-Goto.

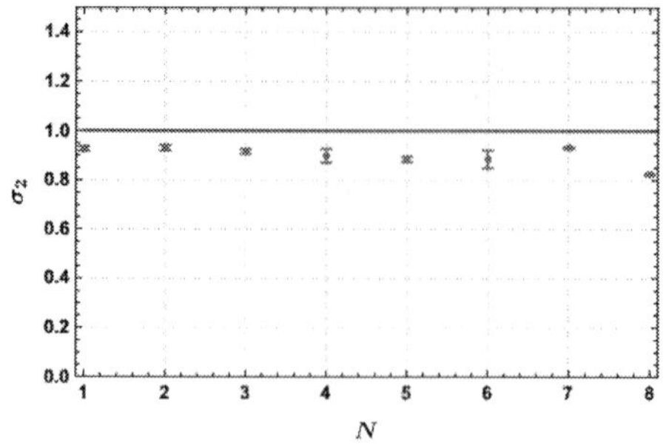

Figure 8. Small deviation from the Nambu-Goto spectrum, S_{II} action.

Parametrizing the deviation to the Nambu-Goto spectrum, the second σ_2 is apparently slightly smaller than σ. We find a deviation, in Fig. 8, of up to 10% for the S_4 ensemble. A deviation is also present for the other ensembles albeit smaller In the W_2 ensemble.

5 Conclusion and discussion

We compute the potentials for several new excitations of the pure $SU(3)$ flux tubes produced by two static 3 and $\bar{3}$ sources, specializing in the radial excitations of the groundstate Σ_g^+.

Using a large basis of operators, employing the computational techniques with GPUs of Ref. [10] and utilizing different actions with smearing and anisotropy we go up to $N = 8$ excitations.

In general the excited states of the the Σ_g^+ flux tubes are comparable to the Nambu-Goto EST with transverse modes, only depending on the string tension σ and the radial quantum number n_r.

A detailed analysis shows a deviation of up to 10% to the excited spectrum of the Nambu-Goto Model. We leave the confirmation of this deviation for futures studies.

A subsequent feasible study is the computation of the widening in the different wavefunctions in the spectrum, and comparing them to from the zero mode widening.

An important outlook would be the study of hybrid quarkonium resonances.

Acknowledgements

We acknowledge discussions on flux tubes and constituent gluons with our colleagues Sergey Afonin, Gunnar Bali, Daniele Binosi, Bastian Brandt, Richard Brower, Fabien Buisseret, Leonid Glozman, Alexei Kaidalov, Felipe Llanes-Estrada, Vincent Mathieu, Colin Morningstar, Lasse Müller, Fumiko Okiharu, Orlando Oliveira, Emilio Ribeiro and Marc Wagner. We thank the support of CeFEMA under the contract for R&D Units, strategic project No. UID/CTM/04540/2019, and the FCT project Grant No. CERN/FIS-COM/0029/2017. NC is supported by FCT under the contract No. SFRH/BPD/109443/2015 and AS is supported by the contract No. SFRH/PD/BD/135189/2017.

References

[1] C. Morningstar, *Excitations of the static-quark potential, su(3) in 4 dimensions*, https://www.andrew.cmu.edu/user/cmorning/static_potentials/SU3_4D/B30_AR3/plots/Rplots.html
[2] J.F. Arvis, Phys. Lett. **127B**, 106 (1983)
[3] C. Lovelace, Phys. Lett. B **34**, 500 (1971)
[4] R.C. Brower, Phys. Rev. D **6**, 1655 (1972)
[5] M. Luscher, Nucl. Phys. B **180**, 317 (1981)
[6] O. Aharony, E. Karzbrun, JHEP **06**, 012 (2009), 0903.1927
[7] B.B. Brandt, *Revisiting the flux tube spectrum of 3d SU(2) lattice gauge theory* (2021), 2102.06413
[8] C. Bonati, M. Caselle, S. Morlacchi (2021), 2106.08784
[9] N. Cardoso, M. Cardoso, P. Bicudo, Phys. Rev. **D88**, 054504 (2013), 1302.3633
[10] P. Bicudo, N. Cardoso, M. Cardoso, Phys. Rev. D **98**, 114507 (2018), 1808.08815
[11] L. Müller, O. Philipsen, C. Reisinger, M. Wagner, Phys. Rev. D **100**, 054503 (2019), 1907.01482
[12] P. Bicudo, N. Cardoso, A. Sharifian (2021), 2105.12159
[13] K.G. Wilson, Phys. Rev. D **10**, 2445 (1974)
[14] C. Morningstar, Nucl. Phys. B Proc. Suppl. **53**, 914 (1997), hep-lat/9608019

EPJ Web of Conferences **258**, 10002 (2022) https://doi.org/10.1051/epjconf/202225810002
A Virtual Tribute to Quark Confinement and the Hadron Spectrum (vConf21)

Hadrons as QCD Bound States

Paul Hoyer[1,*]

[1]Department of Physics, POB 64, FIN-00014 University of Helsinki, Finland

Abstract. Bound state perturbation theory is well established for QED atoms. Today the hyperfine splitting of Positronium is known to $O\left(\alpha^7 \log \alpha\right)$. Whereas standard expansions of scattering amplitudes start from free states, bound states are expanded around eigenstates of the Hamiltonian including a binding potential. The eigenstate wave functions have all powers of α, requiring a choice in the ordering of the perturbative expansion. Temporal ($A^0 = 0$) gauge permits an expansion starting from valence Fock states, bound by their instantaneous gauge field. This formulation is applicable in any frame and seems promising even for hadrons in QCD. The $O\left(\alpha_s^0\right)$ confining potential is determined (up to a universal scale) by a homogeneous solution of Gauss' law.

1 Motivations

Bound states – even QED atoms [1] – are inherently non-perturbative. Atomic wave functions have all powers of α and describe exponentially suppressed processes such as tunneling. No Feynman diagram has a bound state pole. The propagators contribute inverse powers of α at the Bohr momentum scale αm, thus reordering the perturbative expansion. The sum of Feynman "ladder" diagrams diverge at the bound state energies for arbitrarily small values of α. Atoms are at the borderline between hard perturbative and soft classical physics, featuring elementary quanta in a classical potential. The good news is that the truly hard as well as the soft, classical regime is familiar and under theoretical control.

The omission of basic bound state methods from modern field theory (QFT) textbooks seems unwarranted [2]. Quantum Mechanics is traditionally introduced via the Schrödinger equation, which in a field theory context should be derived from the QED action. Bound states give insights into QFT that are complementary to those of scattering amplitudes.

The non-perturbative aspects of atoms and hadrons may be related, as suggested by the successful quark model classification of the hadron spectrum. The QCD coupling $\alpha_s(Q)$ is Q-independent in a classical gluon field. There are indications that the coupling "freezes" at hadronic scales, $\alpha_s(Q \lesssim 1\,\text{GeV}) \simeq 0.5$ [3–5]. Then the strong binding of hadrons is not due to a large α_s, and a perturbative expansion can be relevant.

2 Bound state basics

The accuracy of QED applied to atoms is demonstrated by the hyperfine splitting of Positronium, the mass difference ΔE_b between the ground states of Orthopositronium ($J^{PC} = 1^{--}$)

*e-mail: paul.hoyer@helsinki.fi

and Parapositronium (0^{-+}) [6–8]

$$\frac{\Delta E_b}{m_e} = \frac{7}{12}\alpha^4 - \Big(\frac{8}{9} + \frac{\ln 2}{2}\Big)\frac{\alpha^5}{\pi} - \frac{5}{24}\alpha^6 \ln\alpha + \Big[\frac{1367}{648} - \frac{5197}{3456}\pi^2 + \Big(\frac{221}{144}\pi^2 + \frac{1}{2}\Big)\ln 2 - \frac{53}{32}\zeta(3)\Big]\frac{\alpha^6}{\pi^2}$$
$$- \frac{7\alpha^7}{8\pi}\ln^2\alpha + \Big(\frac{17}{3}\ln 2 - \frac{217}{90}\Big)\frac{\alpha^7}{\pi}\ln\alpha + O\big(\alpha^7\big) \tag{1}$$

The $\log\alpha$ dependence arises from the quenching of infrared singularities: Soft photons decouple from the neutral atom when their wavelength is larger than its Bohr radius $r = 2/\alpha m_e$.

Despite impressive achievements such as (1) the principles of bound state perturbation theory are not formulated at a level similar to the perturbative S-matrix in the Interaction Picture. The Poincaré invariance of interacting, spatially extended states is non-trivial. To illustrate, atoms in motion are often depicted as ellipses due to Lorentz contraction, with no discussion of how the contraction is realized in QED. The effective binding energy $\Delta E(P)$ of a Positronium atom of mass $M = 2m_e + E_b$, where $E_b = -\frac{1}{4}\alpha^2 m_e + O(\alpha^4)$, is inversely proportional to its momentum P in the limit of relativistic motion, $P \gg 2m_e$:

$$\Delta E(P) \equiv \sqrt{P^2 + (2m_e + E_b)^2} - \sqrt{P^2 + 4m_e^2} = \frac{2m_e E_b}{P} + O\big(\alpha^4\big) \tag{2}$$

The Coulomb energy $V(\boldsymbol{r}) = -\alpha/|\boldsymbol{r}|$ is independent of P for the uncontracted orthogonal separations, $\boldsymbol{r}\cdot\boldsymbol{P} = 0$. This seems to contradict the P-dependent energy difference (2). An explanation was given in [9] using the Bethe-Salpeter (B-S) equation [10, 11]. For $P > 0$ transverse photon exchange contributes at leading $O\big(\alpha^2\big)$ to the binding energy. This cancels the Coulomb energy (see Eq. (33) of [9]), leaving an interaction that ensures the correct dependence of the Positronium energy on its momentum, $E(P) = \sqrt{M^2 + P^2}$. The valence Fock state wave function $|e^+e^-\rangle$ contracts as expected, whereas $|e^+e^-\gamma\rangle$ behaves differently.

Feynman diagrams arise in an expansion around free states, which lack overlap with bound states. Consequently, no Feynman diagram for $e^+e^- \to e^+e^-$ has a pole at the Positronium energies. Bound state binding may be recovered by expanding the electron propagator in a photon field A^μ in a geometric series with free propagators,

$$\frac{i}{i\partial\!\!\!/ - m - eA\!\!\!/} = \frac{i}{i\partial\!\!\!/ - m} - \frac{i}{i\partial\!\!\!/ - m}ieA\!\!\!/\frac{i}{i\partial\!\!\!/ - m} + \ldots \tag{3}$$

In the non-relativistic limit Positronium dynamics is equivalent to that of the e^- in a fixed A^0 field. This allows to obtain the Schrödinger equation through a sum of Feynman ladder diagrams analogous to (3) (see, *e.g.*, Section III of [12]). The Positronium poles of $e^+e^- \to e^+e^-$ arise through the divergence of the sum of ladder diagrams.

Relativistic bound state dynamics cannot be reduced to scattering in an external field. Higher order corrections to Positronia were first evaluated [13] using the Bethe-Salpeter equation. It is based on the Dyson-Schwinger (D-S) identity for a $2 \to 2$ Green function G_T, whose external propagators are truncated,

$$G_T = K + G_T\, S\, K \tag{4}$$

Here S is a two-particle propagator and K a two-particle irreducible kernel. Any Feynman diagram contributes equally to the two-sides of (4). The D-S equation is thus formally exact when S and K are expanded to all orders in α. The B-S equation for bound state wave functions Φ_T is given by the residues of the poles in G_T,

$$\Phi_T = \Phi_T\, S\, K \tag{5}$$

Being based on Feynman diagrams the B-S equation is Poincaré covariant, but notoriously difficult to solve. There are retardation effects even at lowest order, with free Dirac propagators in S and a single photon exchange kernel K: The A^μ field at $e^-(t,\boldsymbol{x})$ depends on the distribution of $e^+(t-t',\boldsymbol{x}')$ at earlier times, due to the propagation time t' of transverse photons. No analytic solution of the B-S equation is known.

The rhs. of the D-S equation (4) involves a double perturbative expansion in S and K, whereas G_T has a unique expansion in α [14, 15]. This allows to fix S and let the expansion of K be determined by that of G_T. The B-S equation can thus at lowest order in the rest frame be reduced to the Schrödinger equation, and the higher order corrections systematically determined. More intuitively: There are many equivalent expansions in α when the first approximation of the wave function already has all powers of α. All reorderings must give the same result for physically measurable quantities such as the binding energies in (1).

The realization that the B-S equation is not unique led to Non-Relativistic QED [16, 17]. NRQED expands in powers of $|\boldsymbol{p}|/m_e$, the velocity of the $e^\pm$ in the rest frame. This reordering of the original B-S expansion is motivated by the nearly non-relativistic motion of atomic electrons. The expansion is uniquely defined when the Schrödinger equation is chosen as starting point. The higher order terms in (1) were derived using NRQED.

3 The instantaneity of gauge interactions

In the Schrödinger approximation Positronium is described by its valence $|e^+e^-\rangle$ Fock state, which is an eigenstate of the QED Hamiltonian with the instantaneous Coulomb potential. The tranverse photon interaction terms in $\mathcal{H}_{QED}$ create Fock states such as $|e^+e^-\gamma\rangle$, which contribute to the binding energy at higher orders in α. This suggests [12] a "Bound Fock expansion", where the constituents of each Fock state propagate in their instantaneous gauge field. Transitions between Fock states are given by the transverse photon interactions of $\mathcal{H}_{QED}$, and are suppressed by the coupling e. With all Fock states and the full Hamiltonian taken into account the method is formally exact and valid in any frame.

It is sometimes held that the concept of an instantaneous potential is valid only in a non-relativistic limit. However, gauge theories have instantaneous interactions even with relativistic dynamics. This seems surprising since the action is local and no constituent moves faster than light. The exception arises because the gauge can be fixed non-locally. Thus all instantaneous interactions are gauge dependent.

The QED action

$$\mathcal{S}_{QED} = \int d^4x\big[-\tfrac{1}{4}F_{\mu\nu}F^{\mu\nu} + \bar\psi(i\partial\!\!\!/ - m - eA\!\!\!/)\psi\big] \qquad F_{\mu\nu} = \partial_\mu A_\nu - \partial_\nu A_\mu \tag{6}$$

has no term involving $\partial_0 A^0$ since $F_{00} = 0$. Similarly, F_{ij} gives no space derivative of the longitudinal gauge field, which is defined by a scalar potential, $A_L^j = \partial_j\phi$. Hence A^0 and $\boldsymbol{A}_L$ do not propagate: Their values are fixed by the gauge.

Feynman gauge is convenient for scattering amplitudes, as the gauge fixing term $\mathcal{S}_{GF} = -\frac{1}{2}\int d^4x\,(\partial_\mu A^\mu)^2$ preserves full Poincaré invariance. $\mathcal{S}_{GF}$ adds the "missing" derivatives $\partial_0 A^0$ and $\boldsymbol{\nabla}\cdot\boldsymbol{A}_L$ to the action, allowing the gauge dependent fields to propagate similarly to the gauge invariant ones. For bound states, on the other hand, instantaneous interactions are welcome due to their lack of retardation. It is then advantageous to use Coulomb gauge ($\boldsymbol{\nabla}\cdot\boldsymbol{A}_L = 0$) or temporal gauge ($A^0 = 0$), which non-locally fix $\boldsymbol{A}_L$ and A^0, respectively.

The Hamiltonian generates evolution in time, leaving only space rotations and translations as explicit ("kinetic") symmetries. Time translations and boosts involve interactions and are thus "dynamic" symmetries. Both the Coulomb and temporal gauge conditions preserve the kinetic symmetries of the Hamiltonian.

The Hamiltonian approach defines a conjugate field π_α for each field φ_α in the action, and imposes canonical commutation relations between them,

$$\pi_\alpha(t,\boldsymbol{x}) \equiv \frac{\delta S(\varphi,\partial\varphi)}{\delta[\partial_0\varphi_\alpha(t,\boldsymbol{x})]} \qquad \left[\varphi_\alpha(t,\boldsymbol{x}),\pi_\beta(t,\boldsymbol{y})\right]_\pm = i\delta_{\alpha\beta}\delta^3(\boldsymbol{x}-\boldsymbol{y}) \tag{7}$$

In QED the conjugate field of A^0 vanishes. Quantization in Coulomb gauge can be realized by means of constraints [18], which are cumbersome especially in QCD [19]. Temporal gauge eliminates A^0 and thus avoids the problem with its lack of conjugate field. The canonical commutation relations can then be imposed for all space components $\boldsymbol{A}$, with the longitudinal electric field $\boldsymbol{E}_L = -\partial_0\boldsymbol{A}_L$ being conjugate to $\boldsymbol{A}_L$.

The classical nature of the instantaneous $\boldsymbol{E}_L$ field is apparent in temporal gauge. The gauge condition $A^0 = 0$ needs to be supplemented with a condition on the remaining gauge degrees of freedom [20–22]. Time independent gauge transformations which preserve $A^0 = 0$ are generated by Gauss' operator,

$$G(x) \equiv \frac{\delta S_{QED}}{\delta A^0(x)} = \partial_i E_L^i(x) - e\psi^\dagger\psi(x) \tag{8}$$

Since A^0 is fixed $G(x) = 0$ is not an operator equation of motion, but a condition which is imposed on physical states, $G(x)\,|phys\rangle = 0$ (similarly as $\boldsymbol{A}_L\,|phys\rangle = 0$ in the Gupta-Bleuler formalism). This ensures that the $|phys\rangle$ states are invariant under time independent gauge transformations and determines the value of $\boldsymbol{E}_L$,

$$\boldsymbol{E}_L(t,\boldsymbol{x})\,|phys\rangle = -\boldsymbol{\nabla}_x \int d\boldsymbol{y}\,\frac{e}{4\pi|\boldsymbol{x}-\boldsymbol{y}|}\,\psi^\dagger\psi(t,\boldsymbol{y})\,|phys\rangle \tag{9}$$

This is not an operator constraint, *e.g.*, it should be understood that $\psi^\dagger\psi(t,\boldsymbol{y})\,|0\rangle = 0$. The electric field $\boldsymbol{E}_L$ is classical, generated by the charged constituents and contributes an instantaneous ("V") potential to the QED Hamiltonian in temporal gauge,

$$\begin{aligned}
\mathcal{H}_{QED} &= \int d\boldsymbol{x}\,[E^i\partial_0 A_i + i\psi^\dagger\partial_0\psi - \mathcal{L}_{QED}] = \mathcal{H}_{QED}^{(V)} + \mathcal{H}_{QED}^{(KI)} \\
\mathcal{H}_{QED}^{(V)} &= \int d\boldsymbol{x}\,\tfrac{1}{2}\boldsymbol{E}_L^2 = \tfrac{1}{2}\int d\boldsymbol{x}d\boldsymbol{y}\,\frac{\alpha}{|\boldsymbol{x}-\boldsymbol{y}|}[\psi^\dagger\psi(t,\boldsymbol{x})][\psi^\dagger\psi(t,\boldsymbol{y})] \\
\mathcal{H}_{QED}^{(KI)} &= \int d\boldsymbol{x}\,[\tfrac{1}{2}\boldsymbol{E}_T^2 + \tfrac{1}{4}F^{ij}F^{ij} + \psi^\dagger[(-\boldsymbol{\alpha}\cdot(i\boldsymbol{\nabla}+e\boldsymbol{A}) + m\gamma^0)]\,\psi]
\end{aligned} \tag{10}$$

where $\boldsymbol{\alpha} \equiv \gamma^0\boldsymbol{\gamma}$. The kinetic energies of the electron and photon fields, as well as their $-e\int d\boldsymbol{x}\,\psi^\dagger\boldsymbol{\alpha}\cdot\boldsymbol{A}\,\psi$ interaction term, are in the "KI" part of the Hamiltonian.

4 Application to Positronium

The valence $|e^+e^-\rangle$ Fock state of Positronium is defined in terms of the electron field $\psi(x)$ as

$$\left|e^-e^+;\boldsymbol{P}\right\rangle \equiv \int d\boldsymbol{x}_1 d\boldsymbol{x}_2\,\bar{\psi}(\boldsymbol{x}_1)e^{i\boldsymbol{P}\cdot(\boldsymbol{x}_1+\boldsymbol{x}_2)/2}\,\Phi^{(\boldsymbol{P})}(\boldsymbol{x}_1-\boldsymbol{x}_2)\psi(\boldsymbol{x}_2)\,|0\rangle \tag{11}$$

where $t_1 = t_2 = 0$. The perturbative vacuum $|0\rangle$ projects on the creation operators $b^\dagger, d^\dagger$ in $\bar{\psi}$ and ψ, respectively. The (c-number) wave function $\Phi^{(\boldsymbol{P})}$ has 4×4 Dirac components and depends on the bound state momentum $\boldsymbol{P}$ and on the quantum numbers of the Positronium.

The action of $\mathcal{H}_{QED}^{(V)}$ (10) on the state (11) follows from $[\psi^\dagger\psi(\boldsymbol{x}), \bar{\psi}_\alpha(\boldsymbol{x}_1)] = \bar{\psi}_\alpha(\boldsymbol{x}_1)\delta(\boldsymbol{x}-\boldsymbol{x}_1)$ and $[\psi^\dagger\psi(\boldsymbol{y}), \psi_\beta(\boldsymbol{x}_2)] = -\psi_\beta(\boldsymbol{x}_2)\delta(\boldsymbol{y}-\boldsymbol{x}_2)$. Together with the $\boldsymbol{x} \leftrightarrow \boldsymbol{y}$ contribution this gives the Coulomb energy (the infinite "self-energies" with $\boldsymbol{x} = \boldsymbol{y} = \boldsymbol{x}_1$ and $\boldsymbol{x} = \boldsymbol{y} = \boldsymbol{x}_2$ are subtracted)

$$\mathcal{H}_{QED}^{(V)}\,\bar{\psi}_\alpha(\boldsymbol{x}_1)\psi_\beta(\boldsymbol{x}_2)\,|0\rangle = -\frac{\alpha}{|\boldsymbol{x}_1-\boldsymbol{x}_2|}\,\bar{\psi}_\alpha(\boldsymbol{x}_1)\psi_\beta(\boldsymbol{x}_2)\,|0\rangle \tag{12}$$

In the rest frame ($\boldsymbol{P} = 0$) only the kinetic energy of the electrons in $\mathcal{H}_{QED}^{(KI)}$ contributes at leading order, since the radiation of a transverse photon is suppressed by the $e^\pm$ Bohr momentum $\alpha m/2$. The eigenstate condition gives the bound state equation (BSE),

$$\mathcal{H}_{QED}\,\big|e^-e^+;\boldsymbol{P}=0\big\rangle = M\,\big|e^-e^+;\boldsymbol{P}=0\big\rangle$$

$$(i\boldsymbol{\alpha}\cdot\overrightarrow{\boldsymbol{\nabla}} + m\gamma^0)\Phi^{(0)}(\boldsymbol{x}) + \Phi^{(0)}(\boldsymbol{x})(i\boldsymbol{\alpha}\cdot\overleftarrow{\boldsymbol{\nabla}} - m\gamma^0) = [M - V(\boldsymbol{x})]\Phi^{(0)}(\boldsymbol{x}) \tag{13}$$

where $V(\boldsymbol{x}) = -\alpha/|\boldsymbol{x}|$. This BSE reduces to the Schrödinger equation in the non-relativistic limit, which may be taken at $O\big(\alpha^2\big)$ in the binding energy $E_b = M - 2m$.

For Positronium in motion ($P > 0$) the $e^\pm$ have momenta $\frac{1}{2}\boldsymbol{P} + O(\alpha m)$, so the $ee\gamma$ vertices are no longer suppressed by α. Including the $|ee\gamma\rangle$ Fock state [12] gives agreement with the B-S analysis [9]: The $|e^+e^-\rangle$ wave function Lorentz contracts and the bound state energy $E = \sqrt{M^2 + P^2}$ as required by Poincaré invariance.

Transverse photon exchange as well as annihilation $e^+e^- \to \gamma^* \to e^+e^-$ give the Positronium hyperfine splitting (1) at $O\big(\alpha^4\big)$. First checks of bound state scattering were also made for form factors and deep inelastic scattering [12]. The bound Fock expansion outlined here may seem cumbersome, but should remove the "art" (noted on p. 493 of [11]) from bound state calculations. Calculations can be streamlined when the foundation is solid.

5 Application to hadrons

The properties of charmonium and bottomonium states are successfully described using the Schrödinger equation with the "Cornell potential" [23, 24],

$$V(r) = V'r - \frac{4}{3}\frac{\alpha_s}{r} \quad \text{with } V' \simeq 0.18\ \text{GeV}^2,\ \ \alpha_s \simeq 0.39 \tag{14}$$

The confining potential $V'r$ in (14) involves a scale V' which is not present in the QCD action, and thus also not in Gauss' operator (8). The only way this scale can appear in a classical solution of Gauss' law is through a boundary condition. Analogous scenarios have been discussed before, notably in the context of the MIT Bag Model [25]. Can a non-trivial boundary condition arise in the bound state approach discussed above, applied to QCD in temporal gauge?

Gauss' operator in QCD is

$$G_a(x) \equiv \frac{\delta S_{QCD}}{\delta A_a^0(x)} = \partial_i E_a^i(x) + g f_{abc} A_b^i E_c^i - g\psi^\dagger T^a \psi(x) \tag{15}$$

As in QED, the color electric field $\boldsymbol{E}_L^a$ is constrained by $G_a(x)\,|phys\rangle = 0$ [20–22],

$$\boldsymbol{\nabla}\cdot\boldsymbol{E}_L^a(x)\,|phys\rangle = g\big[-f_{abc}A_b^i E_c^i + \psi^\dagger T^a\psi(x)\big]\,|phys\rangle \equiv g\,\mathcal{E}_a(x)\,|phys\rangle \tag{16}$$

In QED the $\bar{\psi}(\boldsymbol{x}_1)\psi(\boldsymbol{x}_2)\,|0\rangle$ component of the state (11) generates a dipole electric field, giving rise to the Coulomb potential $-\alpha/|\boldsymbol{x}_1 - \boldsymbol{x}_2|$ in (12). The analogous color potential $-\frac{4}{3}\alpha_s/r$ in

(14) arises from the color A quark components $|q_A(\boldsymbol{x}_1)\bar{q}_A(\boldsymbol{x}_2)\rangle$ of a meson state. However, a color singlet meson cannot generate a classical color octet electric field $E^a_L(\boldsymbol{x})$ *at any point* $\boldsymbol{x}$. Color symmetry ensures that the electric field cancels in the sum over quark colors.

The QED solution (9) for $\boldsymbol{E}_L$ was obtained with the obvious boundary condition $\lim_{|x|\to\infty} \boldsymbol{E}_L(\boldsymbol{x}) = 0$. The situation is less obvious in QCD, where the color electric field in any case vanishes at all $\boldsymbol{x}$ for color singlets. Confinement motivates to consider alternative boundary conditions, *i.e.*, homogeneous solutions of (16).

A homogenous solution should maintain the kinetic symmetries of space rotations and translations. The only possibility appears to be

$$\boldsymbol{E}^a_H(\boldsymbol{x})\,|phys\rangle = -\kappa\,\boldsymbol{\nabla}_x \int d\boldsymbol{y}\; \boldsymbol{x}\cdot\boldsymbol{y}\,\mathcal{E}_a(\boldsymbol{y})\,|phys\rangle \tag{17}$$

where $\mathcal{E}_a(\boldsymbol{y})$ is defined in (16). $\boldsymbol{E}^a_H$ trivially satisfies $\boldsymbol{\nabla}_x\cdot\boldsymbol{E}^a_H = 0$ when the normalization κ is independent of $\boldsymbol{x}$. Adding the homogeneous $\boldsymbol{E}^a_H$ to a particular solution of (16) (analogous to (9) in QED) defines the instantaneous interaction. After partial integration,

$$\mathcal{H}^{(V)}_{QCD} \equiv \tfrac{1}{2}\int d\boldsymbol{x}\,(\boldsymbol{E}_{a,L})^2 = \int d\boldsymbol{y}d\boldsymbol{z}\Big\{\boldsymbol{y}\cdot\boldsymbol{z}\Big[\tfrac{1}{2}\kappa^2\textstyle\int d\boldsymbol{x} + g\kappa\Big] + \tfrac{1}{2}\dfrac{\alpha_s}{|\boldsymbol{y}-\boldsymbol{z}|}\Big\}\mathcal{E}_a(\boldsymbol{y})\mathcal{E}_a(\boldsymbol{z}) \tag{18}$$

The term of $O(\kappa^2)$ is $\propto \int d\boldsymbol{x}$ (the volume of space) because the homogeneous solution generates an $\boldsymbol{x}$-independent field energy density. This (infinite) contribution is irrelevant provided it is universal, *i.e.*, the same for all components of all bound states. This determines the normalization κ in (17) for each state $|phys\rangle$, up to a common scale Λ.

Applying $\mathcal{H}^{(V)}_{QCD}$ to a meson Fock component $|q\bar{q}\rangle \equiv \sum_A \bar{\psi}^\alpha_A(\boldsymbol{x}_1)\psi^\beta_A(\boldsymbol{x}_2)\,|0\rangle$ gives

$$\mathcal{H}^{(V)}_{QCD}\,|q\bar{q}\rangle = \Big[\big(\tfrac{1}{2}\kappa^2\textstyle\int d\boldsymbol{x} + g\kappa\big)C_F\,(\boldsymbol{x}_1-\boldsymbol{x}_2)^2 - C_F\dfrac{\alpha_s}{|\boldsymbol{x}_1-\boldsymbol{x}_2|}\Big]\,|q\bar{q}\rangle \tag{19}$$

where $C_F = 4/3$ for QCD. For the coefficent of κ^2 to be independent of $\boldsymbol{x}_1$, $\boldsymbol{x}_2$ we must have $\kappa \propto 1/|\boldsymbol{x}_1-\boldsymbol{x}_2|$. A universal scale Λ may be defined by choosing

$$\kappa_{q\bar{q}} = \frac{\Lambda^2}{gC_F}\frac{1}{|\boldsymbol{x}_1-\boldsymbol{x}_2|} \tag{20}$$

The potential energy defined by $\mathcal{H}^{(V)}_{QCD}\,|q\bar{q}\rangle = V_{q\bar{q}}(\boldsymbol{x}_1-\boldsymbol{x}_2)\,|q\bar{q}\rangle$ is then, apart from the universal contribution of $O(\kappa^2)$,

$$V_{q\bar{q}}(\boldsymbol{x}_1-\boldsymbol{x}_2) = \Lambda^2|\boldsymbol{x}_1-\boldsymbol{x}_2| - C_F\frac{\alpha_s}{|\boldsymbol{x}_1-\boldsymbol{x}_2|} \tag{21}$$

Encouragingly, this agrees with the phenomenological Cornell potential (14). Note that the linear confinement was not assumed, it is a consequence of the required form (17) of the homogeneous solution of Gauss' constraint. The instantaneous potential (21) holds for quarks of any mass and momentum. Applying $\mathcal{H}_{QCD}$ to a meson valence state analogous to (11) and neglecting higher Fock states gives the bound state equation (13) in the rest frame, with the potential (21).

The Hamiltonian $\mathcal{H}^{(V)}_{QCD}$ (19) determines the instantaneous potential of any state with quarks and gluons. For a baryon component $|qqq\rangle = \sum_{A,B,C}\epsilon_{ABC}\,\psi^\dagger_A(\boldsymbol{x}_1)\psi^\dagger_B(\boldsymbol{x}_2)\psi^\dagger_C(\boldsymbol{x}_3)\,|0\rangle$,

$$V_{qqq}(\boldsymbol{x}_1,\boldsymbol{x}_2,\boldsymbol{x}_3) = \Lambda^2\,d_{qqq}(\boldsymbol{x}_1,\boldsymbol{x}_2,\boldsymbol{x}_3) - \frac{2}{3}\,\alpha_s\Big(\frac{1}{|\boldsymbol{x}_1-\boldsymbol{x}_2|} + \frac{1}{|\boldsymbol{x}_2-\boldsymbol{x}_3|} + \frac{1}{|\boldsymbol{x}_3-\boldsymbol{x}_1|}\Big)$$

$$d_{qqq}(\boldsymbol{x}_1,\boldsymbol{x}_2,\boldsymbol{x}_3) = \frac{1}{\sqrt{2}}\,\sqrt{(\boldsymbol{x}_1-\boldsymbol{x}_2)^2 + (\boldsymbol{x}_2-\boldsymbol{x}_3)^2 + (\boldsymbol{x}_3-\boldsymbol{x}_1)^2} \tag{22}$$

When two of the quarks in the baryon are close to each other the potential should (for $N_c = 3$) reduce to that of the color $3 \otimes \bar{3}$ meson potential $V_{q\bar{q}}$ (21). Setting $\boldsymbol{x}_2 = \boldsymbol{x}_3$ in V_{qqq} (and subtracting the infinite Coulomb energy) indeed gives

$$V_{qqq}(\boldsymbol{x}_1, \boldsymbol{x}_2, \boldsymbol{x}_2) = \Lambda^2|\boldsymbol{x}_1 - \boldsymbol{x}_2| - \frac{4}{3}\frac{\alpha_s}{|\boldsymbol{x}_1 - \boldsymbol{x}_2|} = V_{q\bar{q}}(\boldsymbol{x}_1, \boldsymbol{x}_2) \tag{23}$$

Corresponding potentials are obtained for color singlet $|q\bar{q}g\rangle$ and $|gg\rangle$ Fock states [12].

The above potentials depend only on the instantaneous positions $\boldsymbol{x}_i$ of the charges. They are independent of the bound state wave function and the CM momentum $\boldsymbol{P}$. In a general frame, for a $q\bar{q}$ state of the form (11), the bound state equation is

$$i\boldsymbol{\nabla} \cdot \{\alpha, \Phi^{(P)}(\boldsymbol{x})\} - \tfrac{1}{2}\boldsymbol{P} \cdot [\alpha, \Phi^{(P)}(\boldsymbol{x})] + m[\gamma^0, \Phi^{(P)}(\boldsymbol{x})] = [E - V(\boldsymbol{x})]\Phi^{(P)}(\boldsymbol{x}) \tag{24}$$

where $E = \sqrt{M^2 + \boldsymbol{P}^2}$ and the potential $V(\boldsymbol{x})$ is as in (21). Like in QED, Lorentz covariance requires for $P > 0$ to take into account also transverse gluon exchange, *i.e.*, the $O(g)$ $|q\bar{q}g\rangle$ Fock state created by $\mathcal{H}_{QCD}^{(KI)}$ (*cf.* (10) for QED). However, at $O\left(\alpha_s^0\right)$ the $|q\bar{q}\rangle$ state must satisfy covariance on its own, as any contribution from higher Fock states is of higher order in α_s. Remarkably, a linear $q\bar{q}$ potential is indeed boost covariant by itself. The $\boldsymbol{P}$-dependence of the wave function is not given by standard Lorentz contraction, however.

More generally, the $O(\alpha_s^0)$ meson states are fully determined by the $|q\bar{q}\rangle$ Fock states and the BSE (24) with $V(\boldsymbol{x}) = \Lambda^2|\boldsymbol{x}|$. Like for the Dirac equation the wave function has negative energy components, which reflect $q\bar{q}$ pairs due to time ordering ("Z"-diagrams). These pairs have features similar to sea quarks, contributing to deep inelastic scattering at small x_{Bj}. The $O(\alpha_s^0)$ states have no (transverse) gluon constituents – they arise at $O(g)$ via radiative effects. However, we may consider states with valence gluons at $O\left(\alpha_s^0\right)$, such as hybrid $|q\bar{q}g\rangle$ or glueball $|gg\rangle$ states.

For small quark masses m the BSE (24) at $O\left(\alpha_s^0\right)$ gives $|q\bar{q}\rangle$ states which lie on approximately linear Regge trajectories, with parallel daughter trajectories. There are no states with quantum numbers that would be exotic in the quark model. Chiral symmetry is exact for $m = 0$. There are $P^\mu = 0$ solutions whose mixing with the vacuum does not violate Poincaré invariance. This may allow to realize a spontaneous breaking of chiral symmetry.

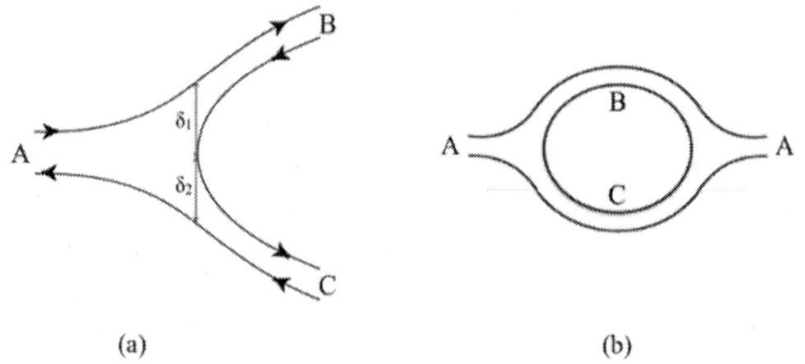

Figure 1. (a) The overlap $\langle B, C|A\rangle$, "string breaking", is determined by the wave functions of hadrons A, B and C. The linear potential is nearly the same before and after the split, $V(\delta_1 + \delta_2) \simeq V(\delta_1) + V(\delta_2)$. (b) Hadron loop correction through the creation and annihilation of a $q\bar{q}$ pair.

For quark separations $\boldsymbol{x}$ such that $V(\boldsymbol{x}) = \Lambda^2|\boldsymbol{x}| \ll E$ the BSE (24) describes free quarks. This gives rise to features of the parton model and duality. "String breaking" at large potentials $V(\boldsymbol{x})$ occur through an overlap of the hadron state with two or more hadrons (Fig. 1), as required for hadronization. These and other aspects require further study.

Acknowledgements. I thank the organizers of the (virtual) *Quark Confinement and the Hadron Spectrum 2021* conference in Stavanger, Norway for the opportunity to present this talk. I enjoy the privileges of Professor Emeritus at the Physics Department of Helsinki University. Travel grants from the Magnus Ehrnrooth Foundation have allowed me to maintain contacts and discuss my research with colleagues.

References

[1] G.T. Bodwin, D.R. Yennie, M.A. Gregorio, Rev. Mod. Phys. **57**, 723 (1985)
[2] A.S. Blum, Stud. Hist. Phil. Sci. B **60**, 46 (2017), 2011.05908
[3] Y.L. Dokshitzer, *Perturbative QCD theory (includes our knowledge of alpha(s))*, in *High-energy physics. Proceedings, 29th International Conference, ICHEP'98, Vancouver, Canada, July 23-29, 1998. Vol. 1*, 2 (1998), pp. 305–324, hep-ph/9812252
[4] Y.L. Dokshitzer, G. Marchesini, G.P. Salam, Eur. Phys. J. direct **C3**, 1 (1999), hep-ph/9812487
[5] A. Deur, S.J. Brodsky, G.F. de Teramond, Nucl. Phys. **90**, 1 (2016), 1604.08082
[6] A.A. Penin, PoS **LL2014**, 074 (2014)
[7] G.S. Adkins, Hyperfine Interact. **233**, 59 (2015)
[8] G. Adkins, J. Phys. Conf. Ser. **1138**, 012005 (2018)
[9] M. Järvinen, Phys. Rev. **D71**, 085006 (2005), hep-ph/0411208
[10] E.E. Salpeter, H.A. Bethe, Phys. Rev. **84**, 1232 (1951)
[11] C. Itzykson, J. Zuber, *Quantum Field Theory*, International Series In Pure and Applied Physics (McGraw-Hill, New York, 1980), ISBN 978-0-486-44568-7
[12] P. Hoyer, *Journey to the Bound States*, SpringerBriefs in Physics (Springer, 2021), ISBN 978-3-030-79488-0, 978-3-030-79489-7, 2101.06721
[13] T. Murota, Prog. Theor. Phys. Suppl. **95**, 46 (1988)
[14] W.E. Caswell, G.P. Lepage, Phys. Rev. **A18**, 810 (1978)
[15] G.P. Lepage, Ph.D. thesis, SLAC-R-0212 (1978), http://www-public.slac.stanford.edu/scidoc/docMeta.aspx?slacPubNumber=slac-R-212
[16] W.E. Caswell, G.P. Lepage, Phys. Lett. **167B**, 437 (1986)
[17] T. Kinoshita, *NRQED approach to the hyperfine structure of the muonium ground state*, in *International Workshop on Hadronic Atoms and Positronium in the Standard Model* (1998), hep-ph/9808351
[18] S. Weinberg, *The Quantum theory of fields. Vol. 1: Foundations* (Cambridge University Press, 2005), ISBN 9780521670531, 9780511252044
[19] N.H. Christ, T.D. Lee, Phys. Rev. **D22**, 939 (1980)
[20] J.F. Willemsen, Phys. Rev. **D17**, 574 (1978)
[21] J.D. Bjorken, *Elements of Quantum Chromodynamics*, in *Quantum chromodynamics: Proceedings, 7th SLAC Summer Institute on Particle Physics (SSI 79), Stanford, Calif., 9-20 Jul 1979* (1979), p. 219, http://www-public.slac.stanford.edu/sciDoc/docMeta.aspx?slacPubNumber=SLAC-PUB-2372
[22] G. Leibbrandt, Rev. Mod. Phys. **59**, 1067 (1987)
[23] E. Eichten, K. Gottfried, T. Kinoshita, K.D. Lane, T.M. Yan, Phys. Rev. **D21**, 203 (1980)
[24] E. Eichten, S. Godfrey, H. Mahlke, J.L. Rosner, Rev. Mod. Phys. **80**, 1161 (2008), hep-ph/0701208
[25] A. Chodos, R. Jaffe, K. Johnson, C.B. Thorn, V. Weisskopf, Phys. Rev. D **9**, 3471 (1974)

Quantum Anomalies (in) Matter

Karl Landsteiner[1,*]

[1]Instituto de Física Teórica UAM-CSIC
C/ Nicolás Cabrera 13-15
28049 Campus Cantoblanco
SPAIN

Abstract. Over the last decade it has bee realized that triangle anomalies give rise to dissipationless transport phenomena in hot and dense relativistic matter. I will review anomalous transport theory and then discuss its applications to the quark gluon plasma and the electronics of Weyl semimetals. Finally I briefly discuss the absence of genuine chiral torsional transport.

1 Introduction

Anomalies are one of the most emblematic properties of quantum field theories. An anomaly appears when not all the symmetries of the classical theory can be realized in the quantum theory [1, 2]. In four dimensional quantum field theory anomalies can be calculated in perturbation theory from Feynman diagrams of triangular shape. There one often speaks of triangle anomalies. For a general theory containing chiral fermions The general form of triangle diagrams can be summarized in the formulas

$$\langle D_\mu J^\mu_a\rangle = \epsilon^{\mu\nu\rho\lambda}\left(\frac{d_{abc}}{32\pi^2}F^b_{\mu\nu}F^c_{\rho\lambda} + \frac{b_a}{768\pi^2}R^\alpha_{\beta\mu\nu}R^\beta_{\alpha\rho\lambda}\right), \tag{1}$$

$$\langle D_\mu T^{\mu\alpha}\rangle = F^{\alpha\mu}_a J^a_\mu + \epsilon^{\mu\nu\rho\lambda}\frac{b_a}{384\pi^2}D_\beta\left(F^a_{\mu\nu}R^{\alpha\beta}{}_{\rho\lambda}\right). \tag{2}$$

Here we assume that the chiral fermions transform under a symmetry group generated by T_a. The theory is coupled covariantly to the sources for the currents J^μ_a and the energy-momentum tensor $T^{\mu\nu}$. These classical sources are the gauge fields A^a_μ and the metric $g_{\mu\nu}$. An anomaly is present if it is not possible to define currents and energy-momentum tensor such that the operators obey the Ward identities corresponding to the classical symmetries. Anomalies are properties of the theory itself. Their presence or absence does not depend on applying external electromagnetic gauge fields or putting the theory in a non-trivial geometry. Indeed the presence of the anomaly can be calculated by a triangle diagram in flat space in the absence of external gauge fields. It is uniquely determined by the non-vanishing of the anomaly coefficients

$$d_{abc} = \mathrm{str}(T_aT_bT_c)_{\mathrm{right}} - \mathrm{str}(T_aT_bT_c)_{\mathrm{left}}, \tag{3}$$

$$b_a = \mathrm{tr}(T_a)_{\mathrm{right}} - \mathrm{tr}(T_a)_{\mathrm{left}}, \tag{4}$$

*e-mail: karl.landsteiner@csic.es

where str denotes the symmetrized trace and the subscripts denotes right- and left-handed fermion representations. In the case where only $U(1)$ symmetries are present the traces becomes just the sums over the (products) of charges, $d_{abc} = \sum_{\text{right}} q_a q_b q_c - \sum_{\text{left}} q_a q_b q_c$ and $b_a = \sum_{\text{right}} q_a - \sum_{\text{left}} q_a$. The anomaly equations (1),(2) are operator equations thus independent of the state. They hold at finite temperature and chemical potential in the same form.

2 Anomaly induced currents

Since anomalies are understood as properties of the operators one could have naively expected that there is no additional effect due to anomalies depending on the state of the quantum system. This however is not true. It is very well established by now that anomalies give rise to new contributions to the currents and the energy-momentum tensor in a formal equilibrium state described by non-vanishing chemical potentials and at finite temperature (see [3] for a review and references therein for a guide to the literature)

$$\langle \vec{J}_a \rangle = \frac{d_{abc}}{4\pi^2}\mu_b \vec{B}_c + \left(\frac{d_{abc}}{4\pi^2}\mu_b\mu_c + \frac{b_a}{12}T^2 \right)\vec{\omega} \tag{5}$$

$$\langle \vec{J}_\epsilon \rangle = \left(\frac{d_{abc}}{8\pi^2}\mu_b\mu_c + \frac{b_a}{24}T^2 \right)\vec{B}_a + \left(\frac{d_{abc}}{6\pi^2}\mu_a\mu_b\mu_c + \frac{b_a}{6}\mu_a T^2 \right)\vec{\omega}\,. \tag{6}$$

The indexes number the symmetries, μ_a are chemical potentials corresponding to the symmetries numbered by a [1]. The magnetic fields $\vec{B}_a$ are the given by the curl of the source $\vec{\nabla} \times \vec{A}_a$ for the current $\vec{J}_a$ and $\omega = \vec{\nabla} \times \vec{v}$ is the curl of the velocity. We note that the latter can be understood as the chemical potentials for the conserved momenta. The state of the system is thus describe the the state variables T, μ_a, $\vec{v}$. Non of these appear in the fundamental Hamiltoniandescribing the quantum dynamics of the theory. In contrast the magnetic fields $\vec{B}_a$ are external couplings that do enter the Hamiltonian. Sometimes it is possible to substitute (gradients) of state variables by external fields. For example it is well known that a (the negative) gradient of a chemical potential has the same effect as an electric fields or the gradient of temperature has the same effect as a gradient in gravitational potential [4]. In the same vain the vorticity in eqs. (5), (6) can be substituted by a gravito-magnetic field [5]. This allows to calculate the anomalous currents in linear response theory via Kubo formulas and to pin down the appearance of the anomaly coefficients. Its a strong hint that the anomalies Eqs. (1) and (2) are the underlying deeper reason for these currents. Indeed this is by now well established. The dependence on the chemical potentials is fixed by the anomaly through by demanding a generalized second law to hold [6, 7] in the presence of anomalies. The temperature dependence can be fixed by geometrical arguments [8, 9]. We note that in the latter case it involves integration of the anomaly equations in a black hole geometry and ties the anomaly induced currents tightly to the phenomenon of Hawking radiation [10]. An argument based on global anomalies has also been presented in [11].

These formulas describe the spontaneous generation of currents at finite temperature T and chemical. These are only well defined in thermal equilibrium. We speak of a global symmetry if the current J^μ_a is not coupled to dynamical gauge fields whose dynamics is governed by gauge invariant Maxwell or Yang-Mills actions. For such a global current a spatially homogeneous expectation value is actually no an observable quantity. Such a current is only observable due to its divergence $\vec{\nabla}\cdot\vec{J}_a$ which describes accumulation of depletion of the global charge in a local manner.

[1]For non-Abelian symmetries they necessarily lie in the Cartan subalgabra.

If the current is a true gauge current (as happens for the electric current) then a theorem originally due to Bloch [2] states that the expectation value has to vanish exactly. At first sight this is puzzling since the anomaly coefficient d_{AVV} which describes the axial anomaly $\partial J^\mu_A = \frac{1}{16\pi^2}\epsilon^{\mu\nu\rho\lambda}F_{\mu\nu}F_{\rho\lambda}$ suggest the chiral magnetic effect $\vec{J} = \frac{\mu_5}{2\pi^2}\vec{B}$, with μ_5 being the chemical potential for the axial charge. Even such a homogeneous current would be observable because it would generate a magnetic field according to amperese law. By now it is however well known that the proper strict equilibrium expression contains a contribution arising from a counterterm needed to insure gauge invariance. Including this counterterm the chiral magnetic effect is described by

$$\vec{J} = \frac{\mu_5 - A^5_0}{2\pi^2}\vec{B}, \tag{7}$$

where A^μ_5 is the temporal component of the axial gauge field acting as source for the axial current. Defining strict equilibrium by the condition $A^5_0 = \mu_5$ we see that indeed the chiral magnetic effect vanishes with this strict and narrow definition of equilibrium [13].

This makes it clear that the anomaly induced currents can only lead to observable effects if a certain amount of non-equilibrium physics is present which nevertheless can still be described by (possibly space and time dependent) chemical potentials and temperature. Such situations are commonly referred to a as local thermal equilibrium. We expect then the effects of the anomaly induced currents are important in situations that can by described in hydrodynamic theory, i.e. an effective theory of dynamics of (approximately) conserved charged densities at in the limit in which gradients of these densities are small compared to the mean free path.

One aspect of the anomaly induced currents is especially worth emphasizing. Contrary to the actual anomalies eqs. (1) and (2) the expression for these currents do not depend on the presence of external electric and magnetic fields or of spacetime curvature! This means that if one can drive a system of quantum matter containing chiral fermions into a state of local thermal equilibrium in which the currents eqs. (5) and (6) apply one can probe for anomalies (even gravitational ones) by means of transport measurments! They truly represent quantum anomalies in matter.

Now we briefly will discuss two examples of such quantum matter systems: heavy ion collisions and Weyl semi-metals

3 Heavy Ion Collisions

In heavy ion collisions as performed at the Relativistic Heavy ion Collider RHIC and the Large Hadro Collider LHC ions are collided at high energies in order to create a new state of matter: the quark gluon plasma. In the initial fire ball of the collision very high energy densities are achieve and the subsequent time evolution of the quark gluon plasma can be described by relativistic hydrodynamics. In non-central collisions a strong magnetic field is generated by the spectators not participating in the collision. The third ingredient is the vacuum structure of QCD and the gluonic part of the axial anomaly

$$\partial_\mu J^\mu_5 = \frac{N_f}{16\pi^2}\epsilon^{\mu\nu\rho\lambda}\mathrm{tr}(G_{\mu\nu}G_{\rho\lambda})\,. \tag{8}$$

Topologically non-trivial gluon field configurations can be excited at high energies (so called strong sphalerons) and by the axial anomaly they will induce a non-vanishing axial charge. In such a chirally imbalanced medium the chiral magnetic effect is expected to lead to a signal

[2]See [12] for a modern formulation based on gauge symmetry.

of charge separation along the direction of the magnetic field [14, 15]. We note that this is a true non-equilibrium realization of the chiral magnetic effect. The chiral charge is generated by the strong sphalerons at an early stage [16]. However it also raises the question if the chiral magnetic current is immediately effective even far from equilibrium.

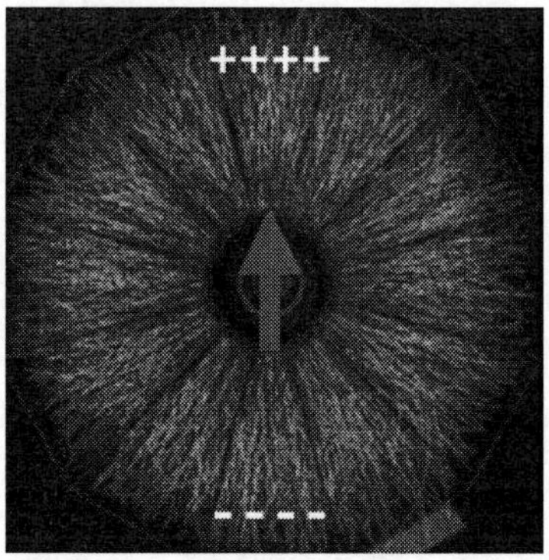

Figure 1. Particles are created in an heavy ion collision. The strong magnetic field at the early stages can lead to charge separation along the magnetic field (red arrow) via the Chiral Magnetic Effect.

The gauge/gravity duality (holography) has played a major role in the establishment and deeper understanding of anomaly induced transport phenomena. Furthermore it has become a standard tool to model hydrodynamics and far from equilibrium evolution of strongly coupled quantum systems. Therefore it is natural to investigate the question how fast the chiral magnetic current builds up in a holographic setup. This has been recently investigated in [17].

The results from this study indicate that the chiral magnetic current indeed shows a delayed build up. Adjusting the parameters of the holographic model to reproduce the entropy and the anomaly of three flavor QCD the build up times for typical RHIC energies can be estimated. For typical RHIC energies densities it is $\tau \approx 0.38 fm/c$ whereas for typical LHC energies it turns out to be $\tau \approx 0.114 fm/c$. This is high significance since the estimated lifetimes of the magnetic field at RHIC is $\tau_B \approx 0.6 fm/c$ and at LHC energies $\tau_B = 0.02 fm/c$. This is a strong hint that the CME will not be observable at the higher LHC energies simply because the magnetic field decays before the chiral magnetic current exists. On the other hand at the lower RHIC energies there seems to be no obstacle for the CME to be effective.

In view of the recent discussion about the ambigious results of the RHIC isobar run [18] it is clear that a much better understanding of the chiral magnetic effect in the dynamical setup of a heavy ion collision is needed. Holographic toy models of the strongly coupled quark gluon plasma can be a very useful tool in this respect.

4 Weyl Semimetals

Weyl semimetals are a new class of materials whose low energy electronic quasiparticle excitations can be described by a Dirac equations [19]. In Figure 2 we show a schematic view of the band structure of a Weyl semi-metal. In this simple description, the band structure has two band crossing points near the Fermi level. By the Nielsen-Ninomiya theorem [20] the two nodal points have opposite chirality. To lowest order in a derivative expansion near the Fermi points, the electronics can be effectively described by two Weyl fermions of opposite chirality. A useful equivalent view is to combine both Weyl fermions into one Dirac fermion.

Then the Dirac like wave equation for this four-component spinor is

$$(i\gamma^\mu \partial_\mu + \gamma^\mu \lambda_\mu \gamma_5)\psi = 0\,. \tag{9}$$

The four vector $\lambda_\mu = (\lambda_0, \vec{\lambda})$ describes the separation of the Weyl nodes of opposite chirality in momentum space. A useful view is to consider λ_μ to be the background value of an axial gauge field $A^5_\mu = \lambda_\mu$. Now we can immediately see that in equilibrium eq. (7) applies and with $\mu_5 = A^5_0 = \lambda_0$ results in a vanishing chiral magnetic effect in equilibrium. Here it is important to note that the relativistic convention for defining the chemical potential applies. This means that the chemical potentials have to be measured with respect to the band-crossing points and not with respect to the bottom of band in the Brillouin zone.

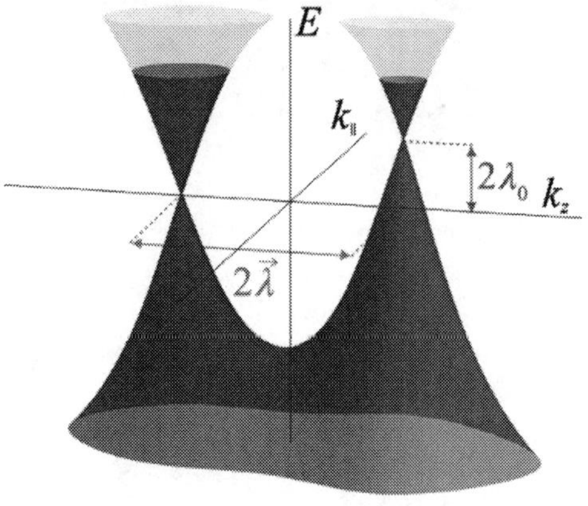

Figure 2. A schematic view of a Weyl semimetal's band structure.

Since the Weyl nodes in the Brillouin zone are separated by the distance $\vec{\lambda}$ and the interaction between the electronic quasiparticles is weak one can effectively treat the Weyl cones independently. The dominant relaxation processes are elastic scattering processes in which momentum of the quasiparticles is randomized and somewhat rarer intervalley scattering processes in which a quasiparticles is scattered from one Weyl cone to another. In this latter case the chirality changes. The relevant relaxation time scale for this process is the intervalley scattering time τ_5. This leads to the idea to write down an effective transport theory in which one only keeps track of the particle number and energy of the quasiparticles in each Weyl cone. In a magnetic field and under the application of an electric field and gradients in the chemical potential and temperature one can solve the anomalous transport equations and finds the induced conductivity matrix [21]

$$\begin{pmatrix} \delta J^\epsilon_i \\ \delta J_i \end{pmatrix} = \Sigma_{ij} \cdot \begin{pmatrix} -\partial_j T \\ E_j - \partial_j \mu \end{pmatrix}. \tag{10}$$

Here (J_i, J^ϵ_i) are the components of the charge current and of the energy current. This result does not depend on the chirality. Both chiralities contribute in the same way. For a gas of free chiral fermions in a weak magnetic field the conductivity matrix is

$$\Sigma_{ij} = \begin{pmatrix} \kappa_{ij} & \beta_{ij} \\ \alpha_{ij} & \sigma_{ij} \end{pmatrix}, \tag{11}$$

where the different entries are the anomaly induced contributions to the different conductivities

$$\text{thermal conductivity} : \kappa_{ij} = \tau_5 B_i B_j \frac{15\mu^2 T - 5\pi^2 T^3}{24\Delta} \tag{12}$$

$$\text{electro} - \text{thermal conductivity} : \beta_{ij} = \tau_5 B_i B_j \frac{11\pi^2 T^2 \mu + 15\mu^3}{8\pi^2 \Delta} \tag{13}$$

$$\text{thermo} - \text{electric conductivity} : \alpha_{ij} = -\tau_5 B_i B_j \frac{5\mu T}{4\Delta} \tag{14}$$

$$\text{electric conductivity} : \sigma_{ij} = \tau_5 B_i B_j \frac{21\pi^2 T^2 + 15\mu^2}{8\pi^2 \Delta}\,, \tag{15}$$

with $\Delta = 7\pi^4 T^4 + 6\pi^2 T^2 \mu^2 + 15\mu^4$. Let us define the heat current

$$\vec{J}_Q = \vec{J}^\epsilon - \mu \vec{J}. \tag{16}$$

The thermal conductivity is define throught the heat current by

$$J_{Q,i} = -\hat{\kappa}_{ij} \partial_j T\,. \tag{17}$$

It is partcilularly interesting to compute the Lorentz ratio

$$\frac{\hat{\kappa}_{ii}}{\sigma_{ii}} = \frac{5k_B^2 \pi^2 (\pi^2 k_B^2 T^3 + 3\mu^2 T)}{e^2 9(7\pi^2 k_B^2 T^2 + 5\mu^2)} = \frac{\pi^2 k_B^2}{3e^2} T + O(T^2/\mu^2)\,. \tag{18}$$

Here we have restored the unit of electric charge and the Boltzmann constant. In the limit of large chemical potential we see that we recover the Wiedemann-Franz law. This is quite interesting since the anomaly induced currents depend on the anomaly coefficients in (5), (6). For the special case of a single chiral fermion they are just right to reproduce the Wiedemann-Franz law indicative of the quasiparticle nature. In a similar vein the Mott formula for the Seebeck coefficient can be recovered from anomaly induced transport. Due to the quasiparticle nature these transport formulas can also be calculated from a kinec theory approach [22–24].

For large magnetic field in which the fermions reside completely in the lowest Landau level the conductivity matrix is

$$\Sigma_{ij} = \frac{\tau_5}{4\pi^2 |B|} \begin{pmatrix} \frac{\pi^2 T}{3} & \mu \\ 0 & 1 \end{pmatrix} B_i B_j \tag{19}$$

Again it can be checked that the Wiedemann-Franz law holds. For intermediate field strengths the conductivities can be calculated by summing over Landau levels in kinetic theory [25].

Generically the formulas predict a strong enhancement of the conductivities at small magnetic field with quadratic increase as the magnetic field is increase. This is indeed seen in experiments such as [26–28].

5 Torsion

Spacetime torsion is the geometric property that a parallelogram spanned by parallel vectors does not form a a closed loop [29]. In the Cartan formalism torsion and curvature are expressed via vielbein $e^a = e^z_\mu dx^\mu$ and spin connection $\omega^a_b = \omega^a_{\mu,b} dx^\mu$ by

$$de^a + \omega^a_b e^b = \theta^a\,, \tag{20}$$

$$d\omega^{ab} + \omega^a_c \wedge \omega^c_b = R^{ab}\,. \tag{21}$$

It is sometimes stated that there is a torsional contribution to the axial anomaly given by the Nieh-Yan tensor [30]

$$d(e^a \wedge \theta_a) = \theta^a \wedge \theta_a + R_{ab} \wedge e^a \wedge e^b \,. \tag{22}$$

However, this term has the wrong dimension for an anomaly, it is only of dimension two whereas anomalies have dimension four. This is related to the fact that the vielbein is not a connection but a well defined tensor under local Lorentz transformations and also a well defined one form. Nevertheless it is tempting to conjecture that torsion might induce similar chiral transport phenomena as the chiral magnetic and chiral vortical effects. For a general current J there are two candidate terms

$$J = c_T^{\|} u_a u_b \theta^a \wedge e^b + c_T^{\perp} (\eta_{ab} - u_a u_b) \theta^a e^b + c_V u \wedge du \,, \tag{23}$$

with a priori the right quantum numbers. While a torsional anomaly would be proportional to the square of some UV scale Λ^2 in order to compensate for the wrong dimension, the transport terms might depend on IR scales such as temperature T and chemical potential μ just as it happens for the CME and CVE currents. We emphasize that at finite temperature and chemical potential one must implicitly specify a frame u_a in which the system is at rest. This frame field is essential in writing down all possible torsion dependent response terms in (23). This means that one also has to include in the response terms the chiral vortical effect proportional to $u \wedge du$ with the frame one-form $u = u_a e^a$. This is important since in any attempt to compute the transport coefficients $c_T^{\|}$, $c_T^{\perp}$ and c_V via Kubo formulas there is a certain amount of mixing between them. In a recent calculation [31] this mixing was considered and it was found that for a minimally coupled chiral fermion

$$c_T^{\|} = c_T^{\perp} = 0 \,. \tag{24}$$

This does not mean that torsion has no effect, but it does state that the effect is perfectly accounted for the the chiral vortical effect with transport coefficient c_V. Indeed in a background with $u_a = (1, 0, 0, 0)$, vanishing spin connection but non-trivial vielbein $e^a = dx^a + \delta e^a$ one has to first order in the variation of the vielbein

$$J = c_V d(\delta e^a) = c_V \delta\theta^a \,. \tag{25}$$

One might be tempted to attribute this to a genuine torsional transport, but it is really just a manifestation of the chiral vortical effect due to frame dragging which happens even in spacetimes without torsion [5].

References

[1] R.A. Bertlmann, *Anomalies in quantum field theory* (Oxford, UK: Clarendon, 1996)
[2] K. Fujikawa, H. Suzuki, *Path integrals and quantum anomalies* (2004)
[3] K. Landsteiner, Acta Phys. Polon. **47**, 2617 (2016)
[4] J.M. Luttinger, Phys. Rev. **135**, A1505 (1964)
[5] K. Landsteiner, E. Megias, F. Pena-Benitez, *Anomalies and Transport Coefficients: The Chiral Gravito-Magnetic Effect*, in *11th Workshop on Non-Perturbative Quantum Chromodynamics* (2011), 1110.3615
[6] D.T. Son, P. Surowka, Phys. Rev. Lett. **103**, 191601 (2009), 0906.5044
[7] Y. Neiman, Y. Oz, JHEP **03**, 023 (2011), 1011.5107
[8] K. Jensen, R. Loganayagam, A. Yarom, JHEP **02**, 088 (2013), 1207.5824
[9] M. Stone, J. Kim, Phys. Rev. **D98**, 025012 (2018), 1804.08668

[10] S.P. Robinson, F. Wilczek, Phys. Rev. Lett. **95**, 011303 (2005), gr-qc/0502074
[11] S. Golkar, S. Sethi, JHEP **05**, 105 (2016), 1512.02607
[12] N. Yamamoto, Phys. Rev. **D92**, 085011 (2015), 1502.01547
[13] A. Gynther, K. Landsteiner, F. Pena-Benitez, A. Rebhan, JHEP **02**, 110 (2011), 1005.2587
[14] D.E. Kharzeev, L.D. McLerran, H.J. Warringa, Nucl. Phys. A **803**, 227 (2008), 0711.0950
[15] K. Fukushima, D.E. Kharzeev, H.J. Warringa, Phys. Rev. D **78**, 074033 (2008), 0808.3382
[16] D.E. Kharzeev, J. Liao, Nature Rev. Phys. **3**, 55 (2021), 2102.06623
[17] J.K. Ghosh, S. Grieninger, K. Landsteiner, S. Morales-Tejera, Phys. Rev. D **104**, 046009 (2021), 2105.05855
[18] M. Abdallah et al. (STAR) (2021), 2109.00131
[19] N.P. Armitage, E.J. Mele, A. Vishwanath, Rev. Mod. Phys. **90**, 015001 (2018)
[20] H.B. Nielsen, M. Ninomiya, Phys. Lett. B **130**, 389 (1983)
[21] M.N. Chernodub, Y. Ferreiros, A.G. Grushin, K. Landsteiner, M.A.H. Vozmediano (2021), 2110.05471
[22] D.T. Son, B.Z. Spivak, Phys. Rev. B **88**, 104412 (2013), 1206.1627
[23] B.Z. Spivak, A.V. Andreev, Phys. Rev. B **93**, 085107 (2016)
[24] R. Lundgren, P. Laurell, G.A. Fiete, Phys. Rev. B **90**, 165115 (2014)
[25] K. Das, A. Agarwal, Phys. Rev. Research **2**, 013088 (2020)
[26] Q. Li, D.E. Kharzeev, C. Zhang, Y. Huang, I. Pletikosic, A.V. Fedorov, R.D. Zhong, J.A. Schneeloch, G.D. Gu, T. Valla, Nature Phys. **12**, 550 (2016), 1412.6543
[27] J. Gooth, A.C. Niemann, T. Meng, A.G. Grushin, K. Landsteiner, B. Gotsmann, F. Menges, M. Schmidt, C. Shekhar, V. Süß et al., Nature **547** (2017)
[28] D. Vu, W. Zhang, C. Şahin, M.E. Flatté, N. Trivedi, J.P. Heremans, Nature Materials (2021)
[29] F.W. Hehl, P. von der Heyde, G.D. Kerlick, J.M. Nester, Rev. Mod. Phys. **48**, 393 (1976)
[30] H. Nieh, C. Yang, Int. J. Mod. Phys. A **22**, 5237 (2007)
[31] Y. Ferreiros, K. Landsteiner, arXiv e-prints arXiv:2011.10535 (2020), 2011.10535

Confinement and Pseudoscalar Glueball Spectrum in the $2+1$D QCD-Like Theory from the Non-Susy D2 Brane

Adrita Chakraborty[1,*]

[1]Department of Physics, Indian Institute of Technology Kharagpur, Kharagpur, India-721302

Abstract. We study two important properties of 2+1D QCD, namely confinement and Pseudoscalar glueball spectrum, using holographic approach. The confined state of the bounded quark-antiquark pair occurs in the self-coupling dominated nonperturbative regime, where the free gluons form the bound states, known as glueballs. The gauge theory corresponding to low energy decoupled geometry of isotropic non-supersymmetric D2 brane, which is again similar to the 2+1D YM theory, has been taken into account but in this case the coupling constant is found to vary with the energy scale. At BPS limit, this theory reduces to supersymmetric YM theory. We have considered NG action of a test string and calculate the potential of such confined state located on the boundary. The QCD flux tube tension for large quark-antiquark separation is observed to be a monotonically increasing function of running coupling. The mass spectrum of Pseudoscalar glueball is evaluated numerically from the fluctuations of the axion in the gravity theory using WKB approximation. This produces the mass to be related to the string tension and the levels of the first three energy states. The various results that we obtained quite match with those previously studied through the lattice approach.

1 Introduction

Unlike the four dimensional theory, the three dimensional Yang-Mills theory having the dimension-full coupling $g_{\rm YM}^2$ is not classically scale-invariant. In spite of this difference, the lower dimensional Yang-Mills theory has some interesting similarities with the $3+1$ dimensional counterpart, e.g., ultraviolet freedom, infra-red slavery, energy-scale dependent running effective coupling, confinement, glueball mass spectrum etc. Because of this, the $2+1$ dimensional pure Yang-Mills theory has been studied widely in various approaches [1–5]. So, at low energy scale the color confinement potential has been found to have the well-known linear dependency on the spatial separation between charges, r, i.e., $V(r) \sim \sigma r$. σ represents the QCD flux-tube tension. For large N_c, $2+1$ dimensional pure YM theory is theoretically found to have QCD flux-tube tension [6],

$$\sigma = g_{\rm YM}^4 \frac{N_c^2 - 1}{8\pi}$$

This confinement indicates strong self-coupling dominated regime where the free gluons form the bound states, called glueballs. The presence of such bound state can be observed from

*e-mail: adimanta09@iitkgp.ac.in

its discrete energy spectrum or mass gap in the confined phase. Here we study these two properties of the 2 + 1 dimensional non-perturbative QCD with the holographic approach.

In the planar or large N_c limit of the AdS/CFT duality, the string coupling g_s is so small that the gravity theory is perturbative. This eventually results in very small YM coupling ($g_{YM}^2 N_c \sim g_s \to 0$). But to ensure of the strong coupling nonperturbative behaviour in the gauge theory side the number of D-branes, N_c, living on the boundary of the bulk is taken to be $N_c \to \infty$. This duality has been also proved to be an effective technique to unravel non-perturbative non-conformal QCD-like theories by dragging it to some gravity dual on non-AdS background. Despite the absence of asymptotic freedom due to strong YM coupling, many other non-perturbative behaviours of QCD such as confinement, chiral symmetry breaking, scattering cross-section, glueballs, thermal phase transition, etc., can be studied from holography[1].

Here, we consider non-supersymmetric non-BPS D2 brane solutions of the type-II supergravity [7] as the gravity theory dual to a 2 + 1 dimensional Yang-Mills gauge theory. As the supersymmetry is broken the solutions have more parameters than the usual BPS ones. These parameters are related to the coupling and the UV fixed point, thereby making the dual nonconformal theory QCD-like. Here we consider a stack of N_c number of the non-susy, isotropic D2 branes where the background dilaton field ϕ and consequently the effective coupling of 2+1D QCD-like Yang-Mills theory vary with the length scale of the bulk theory. The low energy decoupling limit is considered to find the near brane geometry of the non-susy D2 brane. Then we configure a probe open string containing the quark-antiquark pair at its extreme endpoints and employed the holographic understandings of the thermal expectation value of minimal Wilson loop to compute the linear $Q\bar{Q}$ confinement potential. The presence of confinement in the low energy scale triggered us to evaluate the mass spectrum of glueball states. We have presented hither, the mass spectrum of Pseudoscalar glueballs in the 2+1D pure YM theory from the Schrödinger-like field equation of the axion's fluctuation using the WKB approximation. The mass spectrum is evaluated numerically but the restrictions in the WKB method constrained us in the lower states only.

2 Gravity Background

In this section we will discuss the non-susy D2 brane solution and its decoupled geometry at low energy limit.

2.1 Non-Susy D2 Brane

Besides the supersymmetric BPS brane, there is a similar sector of the non-supersymmetric Dp brane solutions of the type-II supergravity containing unequal ADM mass and charge[7]. In the BPS or extremal limit, the number of free parameters associated to non-susy branes reduces so that they merge into BPS branes of same dimension.

The non-susy D2 brane solution assumes a form in the String frame as [8].

$$\begin{aligned} ds^2 =& F(\rho)^{-\frac{1}{2}} G(\rho)^{\frac{\beta}{4}+\frac{\delta}{4}} \left(-dt^2 + \sum_{i=1}^{2}(dx^i)^2\right) + F(\rho)^{\frac{1}{2}} G(\rho)^{\frac{1}{5}-\frac{\beta}{4}+\frac{\delta}{4}} \left(\frac{d\rho^2}{G(\rho)} + \rho^2 d\Omega_6^2\right) \\ e^{2\phi} =& e^{2\phi_0} F(\rho)^{\frac{1}{2}} G(\rho)^{\delta-\frac{\beta}{4}}, \quad F_{[6]} = Q\mathrm{Vol}(\Omega_6) \end{aligned} \tag{1}$$

where

$$F(\rho) = G(\rho)^{\gamma} \cosh^2\theta - \sinh^2\theta, \quad G(\rho) = 1 + \frac{\rho_2^5}{\rho^5} \tag{2}$$

are the harmonic functions introduced due to the relaxation in BPS condition. The (2 + 1) dimensional worldvolume is defined with the coordinates (t, x^1, x^2) whereas the seven dimensional transverse space is defined by the spherical coordinates (ρ, Ω_6). ρ is the radial coordinate perpendicular to the worldvolume thereby representing the energy scale of the theory as $\sim 1/\rho$. The brane is taken to be located at the singularty $\rho = 0$. In this geometry, the dilaton field is quite vividly found to be a non-trivial function of the energy scale of the gravity theory. Henceforth, the effective coupling of the theory also varies with the energy scale . Though the geometry contains six parameters only three out of them are independent, namely, mass parameter ρ_2, charge parameter θ and dilatonic parameter δ. The definitions of the harmonic functions $G(\rho)$ and $F(\rho)$ in the range $0 < \rho < \infty$ suggests that the value of the parameter δ should be bounded in the range $-\frac{8}{5} \leq \delta \leq \frac{8}{5}$.

2.2 Decoupled Geometry

Akin to the BPS branes, the non-susy branes also decouple from the bulk theory at low energy limit [8]. However, these geometries are not AdS implying about their non-conformality. From the understanding of gauge/gravity duality, this non-AdS gravity theory corresponds to the non-conformal gauge theory defined in the worldvolume of the brane. As the fundamental string length $\ell_s \to 0$ at the low energy limit, we consider a scaling of the bulk radial coordinate accordingly to get a finite scale in near brane regime and it is given as

$$\rho = \alpha' u, \; \rho_2 = \alpha' u_2, \sinh^2\theta = \frac{L}{\gamma u_2^5 \alpha'^2} \tag{3}$$

where u having mass dimension +1 defines the new radial coordinate of the decoupled geometry as well as the energy scale of the theory. u_2 is a constant parameter that comes into play due to the non-conformal structure of the theory. The other quantity L is defined as $L = 3\pi^2 g_{\mathrm{YM}}^2 N_c$, where g_{YM}^2 is the coupling in 2 + 1 dimensional Yang-Mills theory. Thus using the scaling (3), we can rewrite the decoupled geometry of the background (1) as follows.

$$ds^2 = \alpha' \left[\sqrt{\frac{\gamma u_2^5}{L}} F(u)^{-\frac{1}{2}} G(u)^{\frac{\delta}{4}+\frac{\beta}{4}} \left(-dt^2 + \sum_{i=1}^{2} (dx^i)^2 \right) + \sqrt{\frac{L}{\gamma u_2^5}} F(u)^{\frac{1}{2}} G(u)^{\frac{1}{5}+\frac{\delta}{4}-\frac{\beta}{4}} \left(\frac{du^2}{G(u)} + u^2 d\Omega_6^2 \right) \right],$$

$$e^{2\phi} = \frac{g_{\mathrm{YM}}^4}{4} \sqrt{\frac{L}{\gamma u_2^5}} F(u)^{\frac{1}{2}} G(u)^{\delta-\frac{\beta}{4}} F_{[6]} = 5L\alpha'^3 \mathrm{vol}(\Omega_6), \;\; F(u) = G(u)^{\gamma} - 1, \;\; G(u) = 1 + \frac{u_2^5}{u^5} \tag{4}$$

In these decoupled forms of the harmonic functions, γ can not be zero. So the allowed range of δ is now modified to $-\frac{8}{5} < \delta < \frac{8}{5}$. In this non-conformal decoupled geometry, the curvature of the 7d transverse space is small enough to validate the holographic calculation in the low energy scale. Now the effective gauge coupling λ^2 can be written from the dilaton field following the standard relation $e^{\phi} \sim \frac{\lambda^{\frac{5}{2}}}{N_c}$ [2].

$$\lambda^2 = \frac{1}{(6\pi^2)^{\frac{4}{5}}} \gamma^{\frac{1}{5}} \left(\frac{L}{u_2} \right) F(u)^{\frac{1}{5}} G(u)^{\frac{2}{5}\delta - \frac{\beta}{10}} \tag{5}$$

which evidently varies with the energy scale u and the parameter δ. Thus in the corresponding gauge theory, we have the running coupling similar to QCD. So the theory on the brane is a (2 + 1) dimensional QCD-like theory without flavour.

Like the QCD theory, the effective gauge coupling λ^2 in (5) decays with the increasing energy scale $u = \epsilon u_2$ which is shown in the Figure 1. Again, at the fixed energy scale u, λ^2 is also found to increase monotonically with the increasing value of δ. According to holographic point of view, the type-IIA supergavity describes the non-perturbative gauge theory for the range of $u \sim u_2$, only if

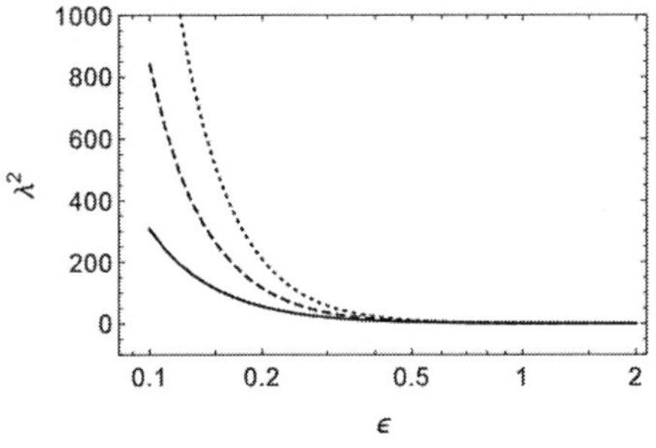

Figure 1: Dependence of λ^2 on the energy scale

$\frac{1}{\gamma^{1/5}} g_{\text{YM}}^2 N_c^{\frac{1}{5}} \leq u_2 \leq g_{\text{YM}}^2 N_c$. At $u \gg u_2$, the geometry (4) reduces to the decoupled geometry of the BPS D2 brane with the effective coupling $\lambda^2 \sim \frac{g_{\text{YM}}^2 N_c}{u}$. On the other hand, at the limit $u \ll u_2$, the effective coupling λ^2 becomes very large for the given range of δ and we have to uplift our theory in the worldvolume of 11d M2 brane to keep supergravity approximation justifiable.

3 Confinement

The establishment of the confinement property has indulged an open string connecting the quark antiquark along its length on the boundary and hanging towards the singularity $u = 0$ in the bulk. Employing the holographic prescription of the minimally swept out Wilson Loop and taking the minimised components of the metric in the NG action, we reach the binding potential of the $Q\bar{Q}$ pair for the maximum separation Δx between them as

$$V = \frac{u_2^{\frac{5}{2}}}{2\pi \sqrt{L}} \left[\frac{5}{8}\delta^2 - \frac{3}{5}\right]^{\frac{1}{4}} \Sigma^{\frac{5\delta}{16\gamma}} \Delta x \tag{6}$$

The $Q\bar{Q}$ potential is linearly proportional to this spatial length and therefore confirms of the confinement property of the theory herein. The coefficient multiplied with Δx has dimension of string tension, which is identified as the QCD string tension σ and is expressed as[9]

$$\frac{\sigma}{g_{\text{YM}}^4 N_c^2} = \frac{9\pi^3}{2} \left(\frac{u_2}{L}\right)^{\frac{5}{2}} \left(\frac{5}{8}\delta^2 - \frac{3}{5}\right)^{\frac{1}{4}} \Sigma^{\frac{5\delta}{16\gamma}} \tag{7}$$

Moreover, the maximum $Q\bar{Q}$ separation Δx is achieved when the vertical distance of the turning point of the probe string from the singularity becomes minimum. This stable configuration can be obtained at

$$u_m = \frac{u_2}{(\Sigma^{1/\gamma} - 1)^{1/5}} \quad \text{where, } \Sigma = \frac{5\delta + 4\gamma}{5\delta - 4\gamma} \tag{8}$$

It is clear from the above turning point that it does not exist for the whole parametric range. Since γ is always positive, it exists only if $\Sigma > 1$ or $5\delta + 4\gamma > 5\delta - 4\gamma$. Then the allowed parametric regime shrinks to $\frac{2\sqrt{6}}{5} < \delta < \frac{8}{5}$. Beyond this range, the test string goes into the singularity without any valid turning point. Here we have taken $u_2 = g_{\text{YM}}^2 N_c$ which makes the ratio $\frac{\sqrt{\sigma}}{g_{\text{YM}}^2 N_c}$ independent of $g_{\text{YM}}^2 N_c$. The variation of the ratio $\frac{u_m}{u_2}$ with the parameter δ in this specified parametric range is depicted in the Figure 2. The nature of the QCD string

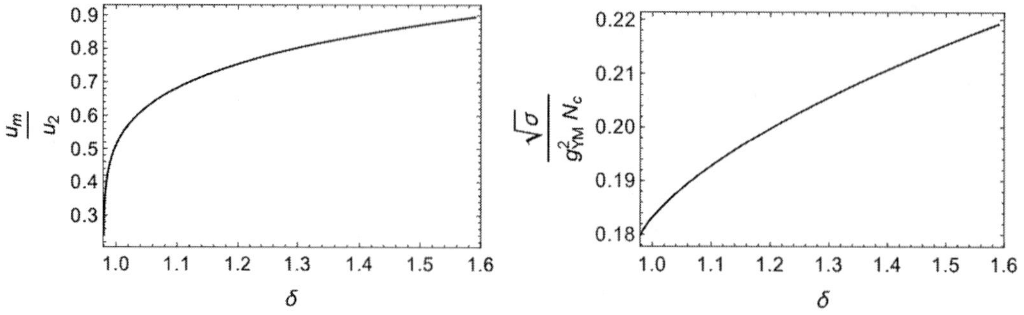

Figure 2: Dependence of the ratio $\frac{u_m}{u_2}$ with δ and the δ-dependence of QCD string tension

Table 1: Values of $\sqrt{\sigma}$ calculated from (7) at $u_2 = g^2_{\rm YM} N_c$.

δ	0.98	1.08	1.18	1.28	1.38	1.48	1.58
$\frac{\sqrt{\sigma}}{g^2_{\rm YM} N_c}$	0.1799	0.1912	0.1984	0.2044	0.2097	0.2144	0.2187

tension σ has been observed in Table 1 [9] by varying the background parameter δ at the entirely nonperturbative regime of the theory that is ensured by taking $u_2 = g^2_{YM} N_c$. Now the theoretical value of $\sqrt{\sigma}$ is 0.1995 for pure Yang-Mills theory. In lattice calculation [5], the value is 0.1975. Both of these values are evaluated at $N_c = \infty$. There are also some recent observations[10], both in lattice and theoretical methods, where the estimated string tension approximately meets these values. Here for various δ, $\sqrt{\sigma}/g^2_{\rm YM} N_c$ varies in the range 0.19 ± 0.02 for large N_c. So our calculation fairly matches with those previous results. It is also clear from Table 1 that the string tension is a monotonically increasing function of δ which is also depicted in Figure 2. As we have seen previously that λ^2 increases with δ, the string tension increases with the increasing effective coupling λ^2. This dependence of the string tension on the effective coupling strengthens its QCD-like nature and indicates towards the confinement property of the pure 2+1D YM theory. The existence of the confinement alludes the theory to be in the self-coupling dominated non-perturbative regime. Now, if we take $u_2 \to 0$, to get a finite turning point in (8) we need to take $\delta \to \frac{8}{5}$ so that the string tension in (7) takes the form $\frac{\sigma}{g^4_{\rm YM} N_c^2} = \frac{9\pi^3}{2} \left(\frac{u_2}{L}\right)^{\frac{5}{2}}$, which goes to zero as $u_2 \to 0$. Thus in the BPS limit, the gravity background (4) reduces to the decoupled geometry of the BPS D2 brane and the dual gauge theory becomes $\mathcal{N} = 8$ deconfined super-Yang-Mills theory in $2 + 1$ dimensions.

4 Mass Spectrum of Pseudoscalar 0^{-+} Glueballs

In the low energy self-coupling dominated regime, the confinement nature of the theory brings the gluons closer and makes the bound states – glueball. Although, the glueball is not observed experimentally till date, the mass spectrum of the glueballs has been calculated theoretically [11] and in lattice QCD [5, 12, 13]. Similar studies have been conducted for the glueballs in the three dimensional QCD by implementing both the lattice [5, 14] and holographic approaches [2]. In this section, we have explored the pseudoscalar glueball spectrum in holographic 3D QCD by incorporating the non-susy brane solutions. The pseudoscalar glueball mass is associated with the axion field of the bulk. As the axion field couples minimally with the background, the fluctuation does not change the background metric. Starting with the linearised field equation for axion fluctuation $\chi = f(u)e^{ik_a x^a}$, where $a = 0, 1, 2$ and $k_a k^a = -M^2$, we have computed the mass spectrum M for the pseudoscalar light glueballs by means of WKB approximation. For the sake of simplicity, we demand the fluctuation field χ to be symmetric on the six dimensional transverse sphere of (4), polarised along the brane worldvolume, and also a function of u. So the potential is almost flat and positive in the whole range except $y_- \leq y \leq y_+$. Thus it forms a small potential well with boundaries at y_- and y_+. The depth of the well is regulated by δ and the mass m. Thus using WKB approximation we can estimate m for given δ. Here due to the technical restrictions of WKB aproximation, we calculate first few energy states of the spectrum.

Table 2: Mass $M_{0^{-+}}$ and $\sqrt{\sigma}$ in unit of $\sqrt{u_2^3/L}$ and $10^{-3}L$ respectively, for various δ

δ	$M_{0^{-+}}$	$M^*_{0^{-+}}$	$M^{**}_{0^{-+}}$	$\sqrt{\sigma}$
0.98	6.15797	9.70152	12.9685	6.07865
1.08	6.25563	9.86602	13.2028	6.45687
1.18	6.35031	10.0252	13.4294	6.70183
1.28	6.44245	10.1799	13.6493	6.90446
1.38	6.53266	10.3310	13.8638	7.08193
1.48	6.62202	10.4803	14.0746	7.24190
1.58	6.71494	10.6346	14.2905	7.38863

In the Table 2, we have listed the dimensionless scaled mass m for different δ, where the actual mass $M_{0^{-+}}$ is $m\sqrt{u_2^3/L}$ [9]. The mass values have been found to increase with the increasing values of δ. Also, we have already seen the effective coupling λ as an increasing function of δ. Therefore combinatorially the increasing δ causes increase in the self-coupling of the gluon fields. Due to this rising of self-coupling, the glueballs become more massive in the sea of gluon. Therefore in this aspect, the calculated mass values show the consistency with the nature of QCD-like theories. We also evaluate the ratio $M_{0^{-+}}/\sqrt{\sigma}$ for $n = 0, 1, 2$ states in the complete non-perturbative region [9] and this put forth an emperical relation of pseudoscalar glueball masses to the QCD flux-tube tension as

$$M_{0^{-+}} \approx \frac{10^3}{6\sqrt{3}\pi^3}(n+2)\sqrt{\sigma} \approx 3(n+2)\sqrt{\sigma} \tag{9}$$

Therefore the mass-to-tension ratio is almost constant for a particular energy state, which is one of the similarities of QCD3 with the four dimensional QCD. From this empirical relation the ratio of the mass of the first excited state to that of the ground state mass is 1.5. To date we have not found the enough data for the full pseudoscalar spectrum. But the complete spectrum is available in four dimensions. So now we first take a comparative look on those results. In QCD4 [12], $M^*_{0^{++}}/M_{0^{++}} = 1.54$ and $M_{0^{-+}}/M_{0^{++}} = 1.50$. On the other hand, in QCD3 these two ratios are respectively 1.5061 and 2.148 in lattice calculation [5]. However in Holographic approach [2, 15], $M^*_{0^{++}}/M_{0^{++}} = 1.72$ in QCD3. So it seems that the glueballs are comparatively more massive in 2 + 1 dimensions. Therefore we can expect the ratio $M^*_{0^{-+}}/M_{0^{-+}}$ which is found to be 1.40 in lattice calculation [12] and 1.46 in holography [15] as QCD4 acquires a higher value in the 2 + 1 dimensional theory. Here we get the ratio in the range 1.575-1.583. Due to the behavioural consistencies of this mass spectrum with the QCD, we can expect that these mass values are also consistent with the pure Yang-Mills theory in 2 + 1 dimensions.

5 Conclusion

Here we have obtained the string tension and the pseudoscalar glueball spectrum in 2 + 1 dimensional strongly coupled gauge theory in the entirely non-perturbative region by assuming $u_2 \sim g^2_{\rm YM} N_c$ from the non-susy D2 brane of type-IIA supergravity. The gravity background has two free parameters – the dimensionless δ and the dimension-full u_2. u_2 controls the nature of the gauge theory and δ measures the zero temperature gluon-condensate in that gauge theory. The string tension σ is found to depend on the dilatonic background parameter δ. It can be found that for a particular value of δ, the string tension matches exactly with the results of the pure Yang-Mills theory in 2 + 1 dimensions. Along with this, the effective gauge coupling λ^2 is monotonically decreasing function of the energy scale u, thereby credibly supporting for the confinement properties achieved in the considered 2 + 1D pure YM-like theory. Here we have elucidated on the numerical computation of the mass spectrum of the pseudoscalar glueball using WKB approximation. The spectrum given in the Table 2 is found to have consistent nature with QCD. The masses increase with the increasing effective coupling of the theory. However the coupling dependence of the spectra has not been shown explicitly. The ratio of the mass to the square root of the string tension has been found to be constant which is also analogous to the four dimensional QCD. All of these similarities stem for the assertion that the holographic dual of the non-susy D2 brane in the considered parametric range is the 2 + 1-dimensional QCD-like theory. But the non-AdS geometrical structure of the gravity background has made it hard to find the complete picture of this QCD-like gauge theory. Due to its complicated mathematical structure, we have evaluated the mass spectrum numerically using WKB approximation. It will be interesting to find the mass spectrum analytically. In this way, we will be able to find the analytic expressions of the masses showing explicit dependence on various parameters. The thermal evolution of the string tension and the glueball spectrum in the 2 + 1 dimensional gauge theory is also an interesting topic to study in non-susy brane background of type-II supergravity.

References

[1] E. Witten, Adv. Theor. Math. Phys. **2**, 505-532 (1998)
[2] O. Aharony, S. S. Gubser, J. M. Maldacena, H. Ooguri and Y. Oz, Phys. Rept. **323**, 183-386 (2000)
[3] D. Karabali and V. P. Nair, Nucl. Phys. B **464**, 135-152 (1996); D. Karabali and V. P. Nair, Int. J. Mod. Phys. A **12**, 1161-1172 (1997); D. Karabali, C. j. Kim and V. P. Nair, Phys. Lett. B **434**, 103-109 (1998)
[4] M. Teper, Phys. Lett. B **311**, 223-229 (1993);
[5] M. J. Teper, Phys. Rev. D **59**, 014512 (1999)
[6] V. P. Nair, Nucl. Phys. B Proc. Suppl. **108**, 194-200 (2002)
O. Philipsen, M. Teper and H. Wittig, Nucl. Phys. B **469**, 445-472 (1996)
[7] J. X. Lu and S. Roy, JHEP **02**, 001 (2005)
[8] K. Nayek and S. Roy, JHEP **03**, 102 (2016)
[9] A. Chakraborty and K. Nayek, Phys. Rev. D **103**, 066001 (2021)
[10] M. Frasca, Eur. Phys. J. C **77**, no.4, 255 (2017); R. Lau and M. Teper, JHEP **10**, 022 (2017)
[11] N. R. Constable and R. C. Myers; JHEP **11**, 020 (1999), C. Csaki and M. Reece; JHEP **05**, 062 (2007) R. de Mello Koch, A. Jevicki, M. Mihailescu and J. P. Nunes, Phys. Rev. D **58**, 105009 (1998)
[12] C. J. Morningstar and M. J. Peardon, Phys. Rev. D **60**, 034509 (1999);

[13] Y. Chen, A. Alexandru, S. J. Dong, T. Draper, I. Horvath, F. X. Lee, K. F. Liu, N. Mathur, C. Morningstar, M. Peardon, S. Tamhankar, B. L. Young and J. B. Zhang, Phys. Rev. D **73**, 014516 (2006)

[14] M. Teper, Phys. Lett. B **397**, 223-228 (1997); O. Philipsen, M. Teper and H. Wittig, Nucl. Phys. B Proc. Suppl. **53**, 626-628 (1997)

[15] C. Csaki, H. Ooguri, Y. Oz and J. Terning, JHEP **01**, 017 (1999)

EPJ Web of Conferences **258**, 10005 (2022) https://doi.org/10.1051/epjconf/202225810005
A Virtual Tribute to Quark Confinement and the Hadron Spectrum (vConf21)

Vortices in $\mathcal{PT}$-symmetric non-Hermitian superfluid

Alexander Begun[1,*], *Maxim* Chernodub[1,2], and *Alexander* Molochkov[1]

[1]Pacific Quantum Center, Far Eastern Federal University, Sukhanova 8, Vladivostok 690950, Russia
[2]Institut Denis Poisson UMR 7013, Université de Tours, 37200 France

Abstract. We discuss the properties of the non-Hermitian $\mathcal{PT}$-symmetric two–scalar fields model. We investigate stability areas of this system and properties of vortices that emerge in the system of two interacting scalar fields. The phase diagram of the model contains stable and unstable regions depending on $\mathcal{PT}$-symmetry breaking, which intercross the regions of $U(1)$-symmetric and $U(1)$-broken phases in a nontrivial way. At non-zero quartic couplings, the non-Hermitian model possesses classical vortex solutions in the $\mathcal{PT}$-symmetric regions. We also consider a close Hermitian analog of the theory and compare the results with the non-Hermitian model.

1 Introduction

Traditional quantum mechanics requires the Hamiltonian to be Hermitian. This condition guarantees that the energy spectrum is real-valued and the time evolution of the system is unitary. However, as some studies show, the requirement of Hermiticity is optional rather than mandatory. The Hermiticity condition can be replaced by the requirement that the Hamiltonian of the system enjoys the invariance under the combined parity $\mathcal{P}$ and time-reversal inversion $\mathcal{T}$ operation ($\mathcal{PT}$-symmetry) [1, 2]:

$$H = H^{\mathcal{PT}}, \tag{1}$$

This symmetry of the Hamiltonian ensures the real-valued energy spectrum and, consequently, the unitary evolution and stability of the system unless the $\mathcal{PT}$-symmetry is broken spontaneously.

Examples of non-Hermitian Hamiltonians with real-valued energy spectrum appeared long ago in theoretical studies [3–5]. Nowadays, the ideas of non-Hermitian systems have found their applications in open systems with balanced gain and loss in different branches of physics, such as optics [6–11], photonics [12–16], superconducting wires [17, 18], $\mathcal{PT}$-symmetric electronic circuits [19] to mention a few.

In this work, we study vortex topological defects in a bosonic non-Hermitian model. We consider a model that consists of a pair of scalar fields associated with interacting condensates. The topological solutions in the multicomponent scalar models appear in various systems ranging from condensed matter to high energy physics. Some of these models can serve as viable extensions of the Standard Model of fundamental particle interactions [20–23].

*e-mail: begun.am@dvfu.ru

2 $\mathcal{PT}$-symmetric scalar field theory

We consider a simplest example of a scalar non-Hermitian $\mathcal{PT}$-symmetric theory which describes dynamics of two complex scalar fields ϕ_1 and ϕ_2 coupled by a mass matrix M. The Lagrangian of theory,

$$\mathcal{L} = \partial_\mu \Phi^\dagger \partial^\mu \Phi - \Phi^\dagger \hat{M}^2 \Phi - U_{int}(\Phi), \tag{2}$$

describes the doublet of two scalar fields $\Phi = \begin{pmatrix}\phi_1 \\ \phi_2\end{pmatrix}$ subjected to the Hermitian self-interaction with the potential

$$U_{int}(\Phi) = \lambda|\phi_1|^4 + \lambda_2|\phi_2|^4 . \tag{3}$$

We consider a repulsive self-interaction with the quartic couplings $\lambda_{1,2} \geq 0$.

The non-Hermiticity emerges from the mass matrix of the Lagrangian (2), namely from the relative sign of its off-diagonal elements:

$$\hat{M}^2_{NH} = \begin{pmatrix} m_1^2 & m_5^2 \\ -m_5^2 & m_2^2 \end{pmatrix}. \tag{4}$$

The opposite signs in front of $m_5^2 \neq 0$ elements correspond to a non-Hermitian theory. The choice of the identical signs of off-diagonal elements (analytically) transforms the theory to its Hermitian counterpart with the mass matrix:

$$\hat{M}^2_{H} = \begin{pmatrix} m_1^2 & m_5^2 \\ m_5^2 & m_2^2 \end{pmatrix}. \tag{5}$$

To highlight the difference between these two types of theories, we rewrite their Lagrangians in terms of individual fields:

$$\mathcal{L}_{\rm NH} = \partial_\nu \phi_1^* \partial^\nu \phi_1 + \partial_\nu \phi_2^* \partial^\nu \phi_2 - m_1^2|\phi_1|^2 - m_2^2|\phi_2|^2 - m_5^2(\phi_1^*\phi_2 - \phi_2^*\phi_1) \\ -\lambda_1|\phi_1|^4 - \lambda_2|\phi_2|^4, \tag{6}$$

$$\mathcal{L}_{\rm H} = \partial_\nu \phi_1^* \partial^\nu \phi_1 + \partial_\nu \phi_2^* \partial^\nu \phi_2 - m_1^2|\phi_1|^2 - m_2^2|\phi_2|^2 - m_5^2(\phi_1^*\phi_2 + \phi_2^*\phi_1) \\ -\lambda_1|\phi_1|^4 - \lambda_2|\phi_2|^4 . \tag{7}$$

Both Hermitian and non-Hermitian models are invariant under the global U(1) transformation, in which the single phase factor (with a real-valued parameter ω) is shared by both complex scalar fields ϕ_1 and ϕ_2:

$$U(1): \qquad \Phi(t,\boldsymbol{x}) \to \Phi'(t,\boldsymbol{x}) = e^{i\omega}\Phi(t,\boldsymbol{x}) \tag{8}$$

The non-Hermitian theory (6) is also invariant under a discrete $\mathcal{PT}$ transformation that consists of the parity inversion $\mathcal{P}$ supplemented by the time conjugation $\mathcal{T}$:

$$\text{NH}: \quad \begin{cases} \mathcal{P}: & \Phi(t,\boldsymbol{x}) \to \Phi'(t,-\boldsymbol{x}) = \sigma_3\Phi(t,\boldsymbol{x}), \\ \mathcal{T}: & \Phi(t,\boldsymbol{x}) \to \Phi'(-t,\boldsymbol{x}) = \Phi^*(t,\boldsymbol{x}). \end{cases} \tag{9}$$

Notice that under the action of $\mathcal{P}$ transformation ϕ_1 transforms as a true scalar, while ϕ_2 transforms as a pseudo-scalar.

In the case of the Hermitian theory, the $\mathcal{P}$ and $\mathcal{T}$ transformations are as follows:

$$\text{H}: \quad \begin{cases} \mathcal{P}: & \Phi(t,\boldsymbol{x}) \to \Phi'(t,-\boldsymbol{x}) = \Phi(t,\boldsymbol{x}), \\ \mathcal{T}: & \Phi(t,\boldsymbol{x}) \to \Phi'(-t,\boldsymbol{x}) = \Phi^*(t,\boldsymbol{x}), \end{cases} \tag{10}$$

Notice that both fields ϕ_1 and ϕ_2 behave as true scalars under the parity inversion.

3 Ground states

In Ref. [24], an analytical study of the ground state of the two-field scalar model (2) has been done for a particular case when only one field was self-interacting ($\lambda_2 = 0$). We extend this analysis to the case of the most general model in which both scalar fields have nonzero self-interacting couplings, $\lambda_{1,2} \neq 0$. Unfortunately, such an extension makes the equations of motion much more complicated, which does not allow us to obtain analytical solutions. In our work, we use numerical analysis to find the ground state of the model as well as its $\mathcal{PT}$ stable regions.

The classical equations of motion can be obtained by the variation of the action corresponding to Lagrangian (6) with respect to the fields ϕ_1^* and ϕ_2^*, respectively:

$$\Box\phi_1 + m_1^2\phi_1 + m_5^2\phi_2 + \frac{\partial V}{\partial\phi_1^*} = 0\,, \tag{11}$$

$$\Box\phi_2 + m_2^2\phi_2 - m_5^2\phi_1 + \frac{\partial V}{\partial\phi_2^*} = 0\,. \tag{12}$$

In the ground state the condensates are coordinate-independent quantities, and Eqs. (12), can be reduced to the non-linear algebraic relations by representing fields ϕ_a in the radial form $\phi_a = v_a e^{i\theta_a}$:

$$m_1^2 v_1 + m_5^2 v_2 + 2\lambda_1 v_1^3 = 0\,, \tag{13a}$$

$$m_2^2 v_2 - m_5^2 v_1 + 2\lambda_2 v_2^3 = 0\,. \tag{13b}$$

In the case of the counterpart Hermitian model, the theory equations (13) take the following form:

$$m_1^2 v_1 + m_5^2 v_2 + 2\lambda_1 v_1^3 = 0\,, \tag{14a}$$

$$m_2^2 v_2 + m_5^2 v_1 + 2\lambda_2 v_2^3 = 0\,. \tag{14b}$$

These systems of equations can be solved numerically. Solutions that correspond to the minima of the energy density of the ground state

$$E_{\text{NH},0} = m_1^2 v_1^2 + m_2^2 v_2^2 + \lambda_1 v_1^4 + \lambda_2 \phi_2^4, \tag{15a}$$

$$E_{\text{H},0} = m_1^2 v_1^2 + m_2^2 v_2^2 + 2m_5^2 v_1 v_2 + \lambda_1 v_1^4 + \lambda_2 v_2^4, \tag{15b}$$

are represented in figure 1.

4 Stability of the ground state of non-Hermitian theory.

To probe the stability of the ground state, we consider weak fluctuations of the scalar fields,

$$\phi_a = v_a + \hat{\phi}_a, \tag{16}$$

with $|\hat{\phi}_a| \ll |v_a|$. Next, we build the quadratic fluctuation matrix that corresponds to the variation of action with respect to the small deviations of the fields about their ground-state values. The configuration is unstable if this matrix contains at least one negative eigenvalue.

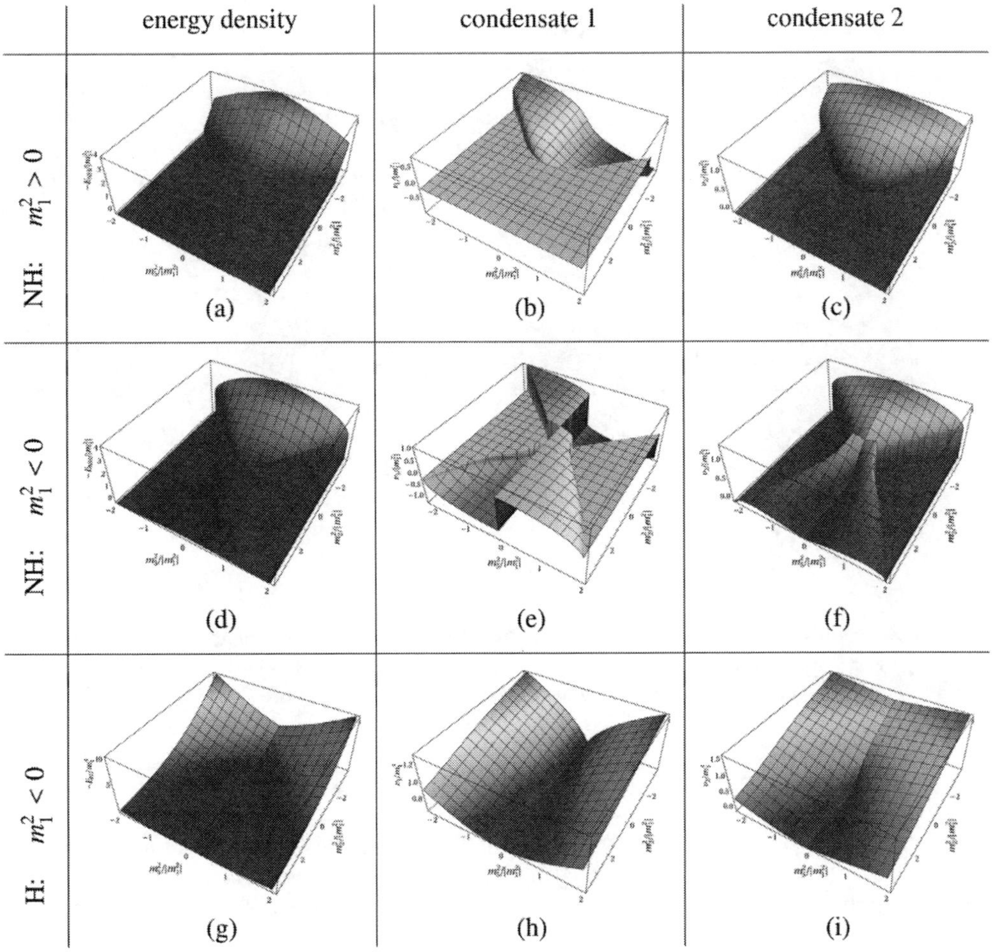

Figure 1. The upper panel depicts (a) the energy density and the condensates (b) v_1 and (c) v_2 in the ground state of the non-Hermitian model (6) with the real-valued mass of the field ϕ_1. The middle panel represents the same quantities obtained for the imaginary mass, $m_1^2 < 0$. The lower panel corresponds to the Hermitian theory (7). All quantities are shown in the units of the absolute value of the mass parameter, $|m_1|$. In all plots, the quartic couplings for both scalar fields are fixed: $\lambda_1 = \lambda_2 = 1$.

In the Hermitian theory, the minimum of energy of the system corresponds to the stable state as the quadratic fluctuation matrix does not contain negative eigenvalues. In the non-Hermitian theory, this criterion does not work in general and we are left with the additional analysis of its fluctuation matrix:

$$\mathcal{M}^2_{\rm NH} = \begin{pmatrix} 4\lambda_1 v_1^2 + m_1^2 & 2v_1^2\lambda_1 & m_5^2 & 0 \\ 2v_1^2\lambda_1 & 4\lambda_1 v_1^2 + m_1^2 & 0 & m_5^2 \\ -m_5^2 & 0 & 4\lambda_2 v_2^2 + m_2^2 & 2v_2^2\lambda_2 \\ 0 & -m_5^2 & 2v_2^2\lambda_2 & 4\lambda_2 v_2^2 + m_2^2 \end{pmatrix}. \tag{17}$$

In the U(1) broken phase, this matrix has one zero eigenvalue which corresponds to the Goldstone mode. In the symmetric U(1) phase, all eigenvalues are generally nonzero.

As it can be seen from figure 2, the ground state of the non-Hermitian model is unstable in some regions of the parameter space. We interpret these new regions as $\mathcal{PT}$-broken

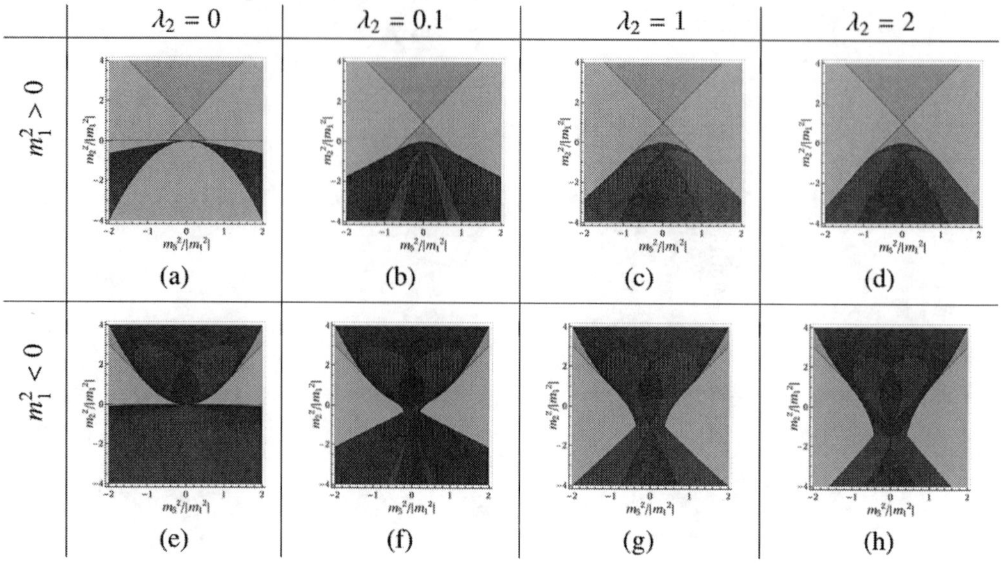

Figure 2. Stability diagrams of the non-Hermitian model in the (m_5^2, m_2^2) plane for the fixed coupling $\lambda_1 = 1$. The stable regions marked by the blue color and the unstable marked by the red color. The upper and panels correspond to $m_1^2 > 0$ and $m_1^2 < 0$, respectively. The dark (light) areas denote regions with non-vanishing (vanishing) condensates.

regions thus providing in our work the extension of the definition of the spontaneously $\mathcal{PT}$ breaking to the case of the interacting model. In the non-interacting limit, $\lambda_{1,2} \to 0$, the $\mathcal{PT}$-instability requirement reduces to the standard criterion of the $\mathcal{PT}$ breaking in the non-interacting model, $4m_5^4 > (m_1^2 - m_2^2)^2$. The system in the $\mathcal{PT}$-broken regions cannot be used as a valid prescription of any steady-state in a physical system. On the contrary, the $\mathcal{PT}$-symmetric regions are stable zones where the steady-state physics can be realized.

5 Vortices at finite couplings

Both Hermitian and non-Hermitian two-field models possess vortex solutions at finite values of the quartic couplings λ_1 and λ_2. We consider examples of the static vortex solutions of the classical equations of motion assuming the form of the scalar fields

$$\phi_a(r,\theta) = v_a f_a(r) e^{in\theta}, \qquad a = 1, 2,$$

where r and θ are the radial coordinates in the (x_1, x_2) plane and $n \in \mathbb{Z}$ is the vorticity of the solution.

The classical equations of motion are given by the following system of differential equations for the profile functions:

$$f_1''(r) + \frac{f_1'(r)}{r} - \frac{n^2}{r^2} f_1(r) - m_1^2 f_1(r) - m_5^2 \frac{v_2}{v_1} f_2(r) - 2\lambda_1 v_1^2 f_1^3 = 0, \tag{18}$$

$$f_2''(r) + \frac{f_2'(r)}{r} - \frac{n^2}{r^2} f_2(r) - m_2^2 f_2(r) + m_5^2 \frac{v_1}{v_2} f_1(r) - 2\lambda_2 v_2^2 f_2^3 = 0. \tag{19}$$

The equations on the profile functions in the counterpart Hermitian model are as follows:

$$f_1''(r) + \frac{f_1'(r)}{r} - \frac{n^2}{r^2} f_1(r) - m_1^2 f_1(r) + m_5^2 \frac{v_2}{v_1} f_2(r) - 2\lambda_1 v_1^2 f_1^3 = 0, \quad (20)$$

$$f_2''(r) + \frac{f_2'(r)}{r} - \frac{n^2}{r^2} f_2(r) - m_2^2 f_2(r) + m_5^2 \frac{v_1}{v_2} f_1(r) - 2\lambda_2 v_2^2 f_2^3 = 0. \quad (21)$$

Using the numerical analysis, we confirm the existence of the stable vortex solutions in the regions characterized by $U(1)$-broken symmetry and, simultaneously, obeying $\mathcal{PT}$-unbroken symmetry. In all other regions the stable vortex solutions do not exist.

An example of the profile functions for a set of coupling constants is shown in figure 3(a). The radial functions feature a linear rise close to the origin which turns into an exponentially slow approach to the corresponding vacuum expectation values at large distances. These properties mark the generic behaviour of all solutions that we have analyzed.

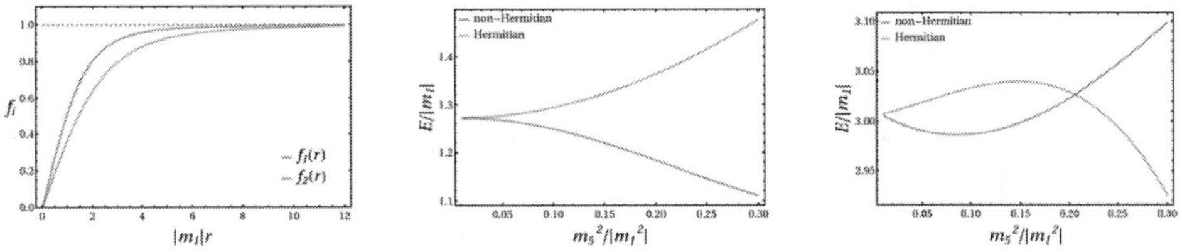

(a) (b) (c)

Figure 3. Panel (a) represents the profile functions of the elementary $n = 1$ vortex solution at the mass parameters $m_2^2 = |m_1^2|$ and $m_5^2 = 0.1|m_1^2|$ with $m_1^2 < 0$ and the equal quartic couplings $\lambda_1 = \lambda_2 = 1$. Panels (b) and (c) represent the Hermitian and non-Hermitian vortex energies respectively vs. the off-diagonal mass squared m_5^2 in different stability areas: (b) $m_2^2 = -m_1^2$ corresponds to the border of the stable and unstable regions and (c) $m_2^2 = 2.5m_1^2 < 0$ resides within the stable region of figure 2(g).

We show the vortex energies in both Hermitian and non-Hermitian theories in, respectively, in figure 3(b) and figure 3(c) in different areas of the phase diagram for the same values of the quartic couplings. While the vortex energies in the Hermitian and non-Hermitian versions trivially coincide with each other at $m_5 = 0$, their behaviour starts to differ from each other with the growth of the off-diagonal mass m_5.

6 Discussion and conclusion

Our work considers the non-Hermitian model of two self-interacting complex scalar fields. Due to the $\mathcal{PT}$-symmetry, this theory can describe open systems that exchange energy with an environment via gain and loss processes. It is the $\mathcal{PT}$-symmetry that provides an exact balance between the gains and losses, thus ensuring the stability of the ground state of the system.

The phase diagram of the non-Hermitian model and its comparison to the Hermitian counterpart are shown in figures 1 and 2, respectively. The Hermitian theory is stable in the whole parameter space. It has the U(1) broken phase with non-vanishing condensates everywhere except the line $m_5^2 = 0$. In contrast, the phase diagram of the non-Hermitian theory has complicated patterns (figure 2) enriched by the intersections of $\mathcal{PT}$-symmetric (stable) and $\mathcal{PT}$-broken (unstable) regions with U(1) symmetric and broken regions (with vanishing and non-vanishing condensates, respectively). This difference is caused by the nontrivial solutions of equations of motion in the case of $\mathcal{PT}$-symmetric self-interacting theory. The notion of the spontaneous $\mathcal{PT}$-breaking is generalized to the interacting model.

Similarly to the Hermitian theory, its non-Hermitian analog possesses the vortex solutions, stable only in the $\mathcal{PT}$-broken regions. We have investigated the vortex profiles and energies and found that the properties of single vortex solutions in the non-Hermitian case are qualitatively similar to the Hermitian ones, apart from the difference in the stability regions imposed by the $\mathcal{PT}$ symmetry in the non-Hermitian model.

Acknowledgement The work is supported by Grant No. 0657-2020-0015 of the Ministry of Science and Higher Education of Russia.

References

[1] C.M. Bender, S. Boettcher, Physical Review Letters **80**, 5243 (1998), `physics/9712001`

[2] C.M. Bender, Contemporary Physics **46**, 277 (2005), `quant-ph/0501052`

[3] E. Caliceti, S. Graffi, M. Maioli, Communications in Mathematical Physics **75**, 51 (1980)

[4] A.A. Andrianov, **140**, 82 (1982)

[5] T. Hollowood, Nuclear Physics B **384**, 523 (1992)

[6] S. Longhi, Physical Review Letters **105**, 013903 (2010)

[7] A. Guo, G.J. Salamo, D. Duchesne, R. Morandotti, M. Volatier-Ravat, V. Aimez, G.A. Siviloglou, D.N. Christodoulides, Physical Review Letters **103**, 093902 (2009)

[8] C.E. Rüter, K.G. Makris, R. El-Ganainy, D.N. Christodoulides, M. Segev, D. Kip, Optics and Photonics News **21**, 47 (2010)

[9] T. Kottos, Nature Physics **6**, 166 (2010)

[10] R. El-Ganainy, K.G. Makris, M. Khajavikhan, Z.H. Musslimani, S. Rotter, D.N. Christodoulides, Nature Physics **14**, 11 (2018)

[11] R. El-Ganainy, M. Khajavikhan, D.N. Christodoulides, S.K. Ozdemir, Communications Physics **2** (2019)

[12] L. Feng, R. El-Ganainy, L. Ge, Nature Photonics **11**, 752 (2017)

[13] R. El-Ganainy, K.G. Makris, D.N. Christodoulides, Z.H. Musslimani, Optics Letters **32**, 2632 (2007)

[14] K.G. Makris, R. El-Ganainy, D.N. Christodoulides, Z.H. Musslimani, Physical Review A **81**, 063807 (2010)

[15] K.G. Makris, R. El-Ganainy, D.N. Christodoulides, Z.H. Musslimani, International Journal of Theoretical Physics **50**, 1019 (2011)

[16] S. Longhi, Laser & Photonics Review **3**, 243 (2009)

[17] J. Rubinstein, P. Sternberg, Q. Ma, Physical Review Letters **99**, 167003 (2007)

[18] N.M. Chtchelkatchev, A.A. Golubov, T.I. Baturina, V.M. Vinokur, Physical Review Letters **109**, 150405 (2012)

[19] J. Schindler, A. Li, M.C. Zheng, F.M. Ellis, T. Kottos, Physical Review A **84**, 040101 (2011)

[20] J. Alexandre, J. Ellis, P. Millington, D. Seynaeve, Physical Review D **99**, 075024 (2019), `1808.00944`

[21] J. Alexandre, J. Ellis, P. Millington, D. Seynaeve, Physical Review D **98**, 045001 (2018), `1805.06380`

[22] A. Fring, T. Taira, Nuclear Physics B **950**, 114834 (2020), `1906.05738`

[23] J. Alexandre, J. Ellis, P. Millington, Physical Review D **102**, 125030 (2020), `2006.06656`

[24] P.D. Mannheim, Physical Review D **99**, 045006 (2019), `1808.00437`

EPJ Web of Conferences **258**, 10006 (2022) https://doi.org/10.1051/epjconf/202225810006
A Virtual Tribute to Quark Confinement and the Hadron Spectrum (vConf21)

Non-Abelian anomalous constitutive relations of a chiral hadronic fluid *

Juan L. Mañes[1,**], *Eugenio* Megías[2,***], *Manuel* Valle[1,****], and *Miguel Á.* Vázquez-Mozo[3,†]

[1]Departamento de Física, Universidad del País Vasco UPV/EHU, Apartado 644, 48080 Bilbao, Spain
[2]Departamento de Física Atómica, Molecular y Nuclear and Instituto Carlos I de Física Teórica y Computacional, Universidad de Granada, Avenida de Fuente Nueva s/n, 18071 Granada, Spain
[3]Departamento de Física Fundamental, Universidad de Salamanca, Plaza de la Merced s/n, 37008 Salamanca, Spain

Abstract. We study the constitutive relations of a chiral hadronic fluid in presence of non-Abelian 't Hooft anomalies. Analytical expressions for the covariant currents are obtained at first order in derivatives in the chiral symmetric phase, for both two and three quark flavors in the presence of chiral imbalance. We also investigate the constitutive relations after chiral symmetry breaking at the leading order.

1 Introduction

Quantum anomalies have important effects in relativistic hydrodynamics, as they induce extra, parity-odd terms in the constitutive relations. These anomalous contributions together with the dissipative transport effects modify the perfect fluid behavior of the constitutive relations in the form of $\langle J^\mu \rangle = \langle J^\mu \rangle_{\rm PF} + \langle J^\mu \rangle_{\rm diss\,\&\,anom}$ for the charge currents, and $\langle T^{\mu\nu} \rangle = \langle T^{\mu\nu} \rangle_{\rm PF} + \langle T^{\mu\nu} \rangle_{\rm diss\,\&\,anom}$ for the energy-momentum tensor [1]. Two well known examples of anomalous transport are the *chiral magnetic effect* (CME), consisting in the generation of an electric current driven by and parallel to an external magnetic field [2], and the *chiral vortical effect* (CVE) where the electric current is triggered by fluid vorticity [3], i.e. $\langle J^\mu \rangle_{\rm anom} = \sigma_{\mathcal{B}} \mathcal{B}^\mu + \sigma_{\mathcal{V}} \omega^\mu$. The two associated susceptibilities, $\sigma_{\mathcal{B}}$ and $\sigma_{\mathcal{V}}$, have been computed using a wide variety of methods, for example Kubo relations [4–6], fluid/gravity correspondence [7–10], as well as the equilibrium partition function formalism (EPFF) [11–13]. The latter technique is especially convenient in systems with spontaneous symmetry breaking (SSB), as in this case the constitutive relations can be obtained directly from the expression of the Bardeen-Zumino (BZ) current [14], thus avoiding the calculation of the, usually cumbersome, Wess-Zumino-Witten (WZW) functional describing the dynamics of Nambu-Goldstone (NG) bosons [15, 16]. An example of anomalous transport phenomenon appearing only in presence of SSB is the *chiral electric effect*, in which the electric current is generated by an external electric field in a perpendicular direction [17, 18].

*Talk given by E. Megías at the vConf21: A Virtual Tribute to Quark Confinement and the Hadron Spectrum, 2-6 August 2021, online.
**e-mail: wmpmapaj@lg.ehu.es
***e-mail: emegias@ugr.es
****e-mail: manuel.valle@ehu.es
†e-mail: Miguel-Vazquez-Mozo@cern.ch

Here, we apply the EPFF to study the anomalous constitutive relations in non-Abelian hadronic fluids with two [14, 18] and three [19] flavors. We will study first the fluid in the unbroken chiral symmetric phase. Later on, we will extend this study to the hydrodynamics in presence of SSB, and describe the dynamics of the NG bosons at the lowest order.

2 Equilibrium partition function formalism

We begin by giving a brief summary of the EPFF, which allows the evaluation of non-dissipative (i.e., time-reversal odd) contributions to the constitutive relations [11, 13, 20, 21]. Let us consider the theory of a time-independent Abelian U(1) gauge connection on the stationary background

$$ds^2 \equiv G_{\mu\nu}(\vec{x})dx^\mu dx^\nu = -e^{2\sigma(\vec{x})}(dt + a_i(\vec{x})dx^i)^2 + g_{ij}(\vec{x})dx^i dx^j \,, \quad (1)$$

$$\mathcal{A} = \mathcal{A}_0(\vec{x})dx^0 + \mathcal{A}_i(\vec{x})dx^i \,. \quad (2)$$

The system partition function is written in terms of its Hamiltonian H and the gauge charge Q as $Z = \text{Tr}\,\exp\left(-\frac{H-\mu_0 Q}{T_0}\right)$, with T_0 and μ_0 the equilibrium temperature and chemical potential. This quantity should be invariant under three-dimensional diffeomorphisms, Kaluza-Klein (KK) transformation $[t \to t + \phi(\vec{x}),\ a_i \to a_i - \partial_i\phi(\vec{x})]$, as well as U(1) time-independent gauge transformations, modulo possible anomalies. Moreover, KK invariance demands that Z depends on the gauge fields only through the invariant combinations $A_0 \equiv \mathcal{A}_0$ and $A_i \equiv \mathcal{A}_i - a_i\mathcal{A}_0$. Given the general form of the partition function in terms of the background metric and field, $\log Z = \mathcal{W}(e^\sigma, A_0, a_i, A_i, g^{ij}, T_0, \mu_0)$, the expectation values of the consistent current and energy-momentum tensor can be obtained from the functional derivatives [11]

$$\langle J_0\rangle_{\text{cons}} = -\frac{T_0 e^\sigma}{\sqrt{g_3}}\frac{\delta\mathcal{W}}{\delta A_0}\,, \quad \langle J^i\rangle_{\text{cons}} = \frac{T_0 e^{-\sigma}}{\sqrt{g_3}}\frac{\delta\mathcal{W}}{\delta A_i}\,, \quad \langle T^{ij}\rangle = -\frac{2T_0 e^{-\sigma}}{\sqrt{g_3}} g^{ik}g^{jl}\frac{\delta\mathcal{W}}{\delta g^{kl}}\,, \quad (3)$$

$$\langle T_{00}\rangle = -\frac{T_0 e^\sigma}{\sqrt{g_3}}\frac{\delta\mathcal{W}}{\delta\sigma}\,, \qquad \langle T_0^{\ i}\rangle = \frac{T_0 e^{-\sigma}}{\sqrt{g_3}}\left(\frac{\delta\mathcal{W}}{\delta a_i} - A_0\frac{\delta\mathcal{W}}{\delta A_i}\right), \quad (4)$$

where $g_3 \equiv \det(g_{ij})$, and thus $\mathcal{W}$ plays the role of a generating functional for the constitutive relations. We study the equilibrium partition function using an expansion in derivatives, which corresponds to a hydrodynamic expansion. The general form of $\mathcal{W}$ at zeroth and first order in derivatives for the system of Eqs. (1)-(2) are respectively given by [11, 13]

$$\mathcal{W}_{(0)} = \frac{1}{T_0}\int d^3x\,\sqrt{g_3}\,e^\sigma P(T,\mu)\,, \quad (5)$$

$$\mathcal{W}_{(1)} = \int d^3x\,\sqrt{g_3}\Big[\alpha_1(T,\mu)\epsilon^{ijk}A_iA_{jk} + \alpha_2(T,\mu)\epsilon^{ijk}A_if_{jk} + \alpha_3(T,\mu)\epsilon^{ijk}a_if_{jk}\Big]\,, \quad (6)$$

with $P(T,\mu)$ the pressure, and $T = e^{-\sigma}T_0$ and $\mu = e^{-\sigma}A_0$ the out-of-equilibrium temperature and chemical potential, respectively. In addition, $A_{ij} \equiv \partial_iA_j - \partial_jA_i$ and $f_{ij} \equiv \partial_ia_j - \partial_ja_i$. The zeroth order $\mathcal{W}_{(0)}$ is responsible for the perfect fluid (PF) contribution to the hydrodynamic constitutive relations, which read

$$\langle J^\mu\rangle_{\text{PF}} = nu^\mu\,, \qquad \langle T^{\mu\nu}\rangle_{\text{PF}} = (\varepsilon + P)u^\mu u^\nu + PG^{\mu\nu}\,, \quad (7)$$

where $\varepsilon = -P + T\frac{\partial P}{\partial T} + \mu\frac{\partial P}{\partial\mu}$ and $n = \frac{\partial P}{\partial\mu}$ are the energy density and charge density, respectively, and $u^\mu = e^{-\sigma}(1, 0, \dots, 0)$ is the local fluid velocity. The coefficients $\alpha_i(T,\mu)$ can be determined for a particular theory by inserting Eq. (6) into Eqs. (3)-(4), and comparing the results with the constitutive relations for this theory. The anomalous contributions to the constitutive relations at first order in derivatives turn out to be

$$\langle J^\mu\rangle_{(1)} = \sigma_{\mathcal{B}}\mathcal{B}^\mu + \sigma_{\mathcal{V}}\omega^\mu\,, \qquad \langle T^{\mu\nu}\rangle_{(1)} = \sigma^{\varepsilon}_{\mathcal{B}}(\mathcal{B}^\mu u^\nu + \mathcal{B}^\nu u^\mu) + \sigma^{\varepsilon}_{\mathcal{V}}(\omega^\mu u^\nu + \omega^\nu u^\mu)\,, \quad (8)$$

where the magnetic field and vorticity are $\mathcal{B}^\mu = \frac{1}{2}\epsilon^{\mu\nu\alpha\beta}u_\nu\mathcal{A}_{\alpha\beta}$ and $\omega^\mu = \frac{1}{2}\epsilon^{\mu\nu\alpha\beta}u_\nu\partial_\alpha u_\beta$, respectively. The transport coefficients in the laboratory rest frame [1] for an ideal gas of Dirac fermions read $\sigma_{\mathcal{B}} = C\mu$, $\sigma_{\mathcal{V}} = \frac{1}{2}C\mu^2 + C_2T^2$, $\sigma^{\varepsilon}_{\mathcal{B}} = \sigma_{\mathcal{V}}$ and $\sigma^{\varepsilon}_{\mathcal{V}} = \frac{1}{3}C\mu^3 + 2C_2\mu T^2$, where $C = 1/(4\pi^2)$ and $C_2 = 1/24$ are related to the axial anomaly [3, 8] and the gauge-gravitational anomaly [5, 9], respectively. Then, one finds $\alpha_1 = -\frac{C}{6}\frac{\mu}{T}$, $\alpha_2 = -\frac{1}{2}\left(\frac{C}{6}\frac{\mu^2}{T^2} - C_2\right)$ and $\alpha_3 = 0$. In the following we will use the EPFF extended to non-Abelian gauge fields.

3 Non-Abelian anomalies

We look next at a theory of chiral fermions coupled to external gauge fields $\mathcal{A}_{\mathrm{L(R)}\mu} \equiv t_a\mathcal{A}^a_{\mathrm{L(R)}\mu}$ with symmetry group $\mathrm{U}(N_f)\times\mathrm{U}(N_f)$, described by the Lagrangian

$$\mathcal{L}_{\mathrm{YM}} = i\bar\psi_{\mathrm{L}}\gamma^\mu(\partial_\mu - it_a\mathcal{A}^a_{\mathrm{L}\mu})\psi_{\mathrm{L}} + i\bar\psi_{\mathrm{R}}\gamma^\mu(\partial_\mu - it_a\mathcal{A}^a_{\mathrm{R}\mu})\psi_{\mathrm{R}}\,, \quad (9)$$

with $t_a = t_a^\dagger$ the Lie algebra generators. The axial anomaly is signaled by the non-invariance of the effective action $\Gamma_0[\mathcal{A}_{\mathrm{L}},\mathcal{A}_{\mathrm{R}}]$ under axial gauge transformations, leading to the anomaly equation $\mathscr{A}_a(x)\Gamma_0[\mathcal{V},\mathcal{A}] = G_a[\mathcal{V},\mathcal{A}]$, where G_a is the consistent anomaly, and $\mathscr{A}_a(x)$ is the local generator of axial gauge transformations[2]. For this theory, the Bardeen form of the chiral anomaly is [23]

$$G_a[\mathcal{V},\mathcal{A}] = \frac{iN_c}{16\pi^2}\epsilon^{\mu\nu\rho\sigma}\mathrm{Tr}\Big\{t_a\big[\mathcal{V}_{\mu\nu}\mathcal{V}_{\rho\sigma} + \tfrac{1}{3}\mathcal{A}_{\mu\nu}\mathcal{A}_{\rho\sigma} - \tfrac{32}{3}\mathcal{A}_\mu\mathcal{A}_\nu\mathcal{A}_\rho\mathcal{A}_\sigma + \tfrac{8}{3}i(\mathcal{A}_\mu\mathcal{A}_\nu\mathcal{V}_{\rho\sigma} + \mathcal{A}_\mu\mathcal{V}_{\rho\sigma}\mathcal{A}_\nu + \mathcal{V}_{\rho\sigma}\mathcal{A}_\mu\mathcal{A}_\nu)\big]\Big\}\,, \quad (10)$$

where N_c is the number of colors, while $\mathcal{V}_{\mu\nu} = \partial_\mu\mathcal{V}_\nu - \partial_\nu\mathcal{V}_\mu - i[\mathcal{V}_\mu,\mathcal{V}_\nu] - i[\mathcal{A}_\mu,\mathcal{A}_\nu]$ and $\mathcal{A}_{\mu\nu} = \partial_\mu\mathcal{A}_\nu - \partial_\nu\mathcal{A}_\mu - i[\mathcal{V}_\mu,\mathcal{A}_\nu] - i[\mathcal{A}_\mu,\mathcal{V}_\nu]$ are the field strengths. Notice that the non-Abelian anomaly G_a includes contributions from the triangle, square and pentagon diagrams, in contrast to the Abelian case in which only triangle diagram contributes. The solution of the anomaly equation can be found using differential geometry methods based on the Chern-Simons effective action and its dimensional reduction, with the result [14, 18, 24–26]

$$\Gamma_0[V,A] = -\frac{N_c}{32\pi^2}\int dt\,d^3x\,\sqrt{g_3}\,\epsilon^{ijk}\,\mathrm{Tr}\Big[\frac{32}{3}i\,V_0A_iA_jA_k + \frac{4}{3}(A_0A_i + A_iA_0)A_{jk} + 4(V_0A_i + A_iV_0)V_{jk} + \frac{8}{3}(A_0^2 + 3V_0^2)A_i\partial_ja_k\Big] + C_2T_0^2\int dt\,d^3x\,\sqrt{g_3}\,\epsilon^{ijk}\mathrm{Tr}A_i\,\partial_ja_k\,, \quad (11)$$

where V_μ and A_μ are KK invariant vector and axial gauge fields. From now on, we neglect the term multiplying C_2, related to the mixed gauge-gravitational anomaly.

4 Constitutive relations in the unbroken chiral symmetric phase

In this section we study the constitutive relations following from the effective action in Eq. (11), corresponding to the chiral symmetric phase of a QCD fluid with two and three flavors in the static geometry (1).

[1] This is the usual frame when computing the transport coefficients by using Kubo formulae, cf. e.g. Refs. [5, 6]. A discussion on the frame dependence of the transport coefficients can be found in e.g. Ref. [22].

[2] We define the vector and axial gauge fields $(\mathcal{V},\mathcal{A})$ in terms of $(\mathcal{A}_{\mathrm{L}},\mathcal{A}_{\mathrm{R}})$ by $\mathcal{A}_{\mathrm{L}} \equiv \mathcal{V} - \mathcal{A}$ and $\mathcal{A}_{\mathrm{R}} \equiv \mathcal{V} + \mathcal{A}$.

4.1 Covariant currents and background

While the consistent currents are computed from the functional derivatives of the effective action, the covariant currents are the ones entering in the constitutive relations. These latter are obtained by adding the BZ polynomials to the corresponding consistent currents, i.e. $J^\mu_{\rm cov} = J^\mu_{\rm cons} + J^\mu_{\rm BZ}$ [27]. Explicit expressions for $J^\mu_{\rm BZ}$ are provided in e.g. Refs. [14, 18, 28]. From $\mathcal{W}_0 = i\Gamma_0$ and using Eqs. (3) and (4), one obtains the following equilibrium expressions [18]

$$\langle J^i_{a\,\rm V}\rangle_{\rm cov} = \frac{N_c}{8\pi^2} e^{-\sigma}\epsilon^{ijk}{\rm Tr}\Big\{t_a\big[(A_0V_{jk}+V_{jk}A_0) + (V_0A_{jk}+A_{jk}V_0) + 2(A_0V_0+V_0A_0)\partial_j a_k\big]\Big\}\,, \quad (12)$$

$$\langle J^i_{a\,\rm A}\rangle_{\rm cov} = \frac{N_c}{8\pi^2} e^{-\sigma}\epsilon^{ijk}{\rm Tr}\Big\{t_a\big[(A_0A_{jk}+A_{jk}A_0) + (V_0V_{jk}+V_{jk}V_0) + 2(A_0^2+V_0^2)\partial_j a_k\big]\Big\}\,, \quad (13)$$

$$\langle {T_0}^i\rangle = -\frac{N_c}{8\pi^2} e^{-\sigma}\epsilon^{ijk}\,{\rm Tr}\Big[(A_0^2+V_0^2)A_{jk} + (A_0V_0+V_0A_0)V_{jk} + \Big(\frac{2}{3}A_0^3+2A_0V_0^2\Big)\partial_j a_k\Big]\,, \quad (14)$$

and vanishing values for the components $\langle J_{a0\rm V}\rangle_{\rm cov} = \langle J_{a0\rm A}\rangle_{\rm cov} = 0$ and $\langle T_{00}\rangle = \langle T^{ij}\rangle = 0$. We particularize these results for a specific case of $N_f = 2, 3$. Due to the non-Abelian character of the charges, the maximal number of chemical potentials that can be introduced consistently equals the dimension of the Cartan subalgebra. Therefore, we take the background

$$V_\mu(\vec{x}) = V_{0\mu}(\vec{x})\,t_0 + V_{3\mu}(\vec{x})\,t_3 + V_{8\mu}(\vec{x})\,t_8\,, \qquad A_0 = A_{00}\,t_0\,, \qquad A_i = 0\,, \quad (15)$$

where A_{00} is constant, and the generators t_a are given by

$$\begin{cases} N_f = 2: & t_0 = \frac{1}{2}\mathbb{1}_{2\times2}\,, & t_3 = \frac{1}{2}\sigma_3\,, & t_8 = 0_{2\times2} \\ N_f = 3: & t_0 = \frac{1}{\sqrt{6}}\mathbb{1}_{3\times3}\,, & t_3 = \frac{1}{2}\lambda_3\,, & t_8 = \frac{1}{2}\lambda_8 \end{cases}\,, \quad (16)$$

with σ_i and λ_i the Pauli and Gell-Mann matrices, respectively. Furthermore, we define the equilibrium velocity field by $u_\mu = -e^{\sigma}(1, a_i)$, while the chemical potentials are $\mu_a = e^{-\sigma}\mathcal{V}_{a0}$ $(a = 0, 3, 8)$ and $\mu_5 = e^{-\sigma}\mathcal{A}_{00}$. Here, μ_5 controls the chiral imbalance of the system [29, 30], whereas $\mu_8 = 0$ for $N_f = 2$. External gauge fields couple through the magnetic non-Abelian vector fields $\mathcal{B}^\mu_a = \frac{1}{2}\epsilon^{\mu\nu\alpha\beta}u_\nu\mathcal{V}_{a\alpha\beta}$. An explicit dependence on u^μ comes through the vorticity vector ω^μ.

While the constitutive relations can be expressed in the basis of the generators of the Cartan subalgebra $\{t_0, t_3, t_8\}$, it would be interesting to write the results also in the alternative Cartan basis formed by the conserved charges. These quantities play a fundamental role in the hydrodynamics of relativistic fluids. In the (uds) flavor sector of QCD, the conserved charges are the baryon number B, electric charge Q, and strangeness S, while $S = 0$ in the (ud) flavor sector. In the $\{B, Q, S\}$ basis, the background defined by Eq. (15) is given by

$$V_\mu(\vec{x}) = V_{B\mu}(\vec{x})B + V_{Q\mu}(\vec{x})Q + V_{S\,\mu}(\vec{x})S\,, \qquad A_0 = A_B B\,, \qquad A_i = 0\,. \quad (17)$$

Let Ψ be a flavor doublet (triplet) of Dirac spinors made out of quarks, i.e.

$$\Psi = \begin{pmatrix} u \\ d \end{pmatrix} \quad \text{for} \quad N_f = 2\,, \qquad \text{and} \qquad \Psi = \begin{pmatrix} u \\ d \\ s \end{pmatrix} \quad \text{for} \quad N_f = 3\,, \quad (18)$$

with each spinor being decomposed into its chiralities according to $q = (q_L\, q_R)^T$. Using that $J^\mu_{a\,V} = \bar\Psi\gamma^\mu t_a\Psi$, we can distinguish between two (three) vector currents for $N_f = 2(3)$, i.e. $J^\mu_{\rm em} = e\bar\Psi\gamma^\mu Q\Psi$, $J^\mu_B = \bar\Psi\gamma^\mu B\Psi$ and $J^\mu_S = \bar\Psi\gamma^\mu S\Psi$, corresponding to the electromagnetic, baryonic and strangeness currents, respectively. These currents, together with the isospin current $J^\mu_I = \bar\Psi\gamma^\mu I_3\Psi$, fulfill the Gell-Mann–Nishijima relation $J^\mu_{\rm em} = eJ^\mu_I + \frac{e}{2}\left(J^\mu_B + J^\mu_S\right)$. By using the relation between the charges $\{B, Q, S\}$ and the generators t_a of the Cartan subalgebra of $U(N_f)$, we will provide explicit expressions for the constitutive relations of $J^\mu_{\rm em}$ and $T^{\mu\nu}$.

4.2 Constitutive relations for $N_f = 2$

The constitutive relations can be computed by particularizing Eqs. (12)-(14) to the background of Eqs. (15)-(16) with $N_f = 2$. While these relations are in equilibrium, the corresponding out-of-equilibrium expressions are obtained by Lorentz covariantization, so that in the end they will be expressed in terms of $\mathcal{B}_a^\mu$ and ω^μ. The result is

$$\langle J^\mu_{a\,\mathrm{V}}\rangle_{\mathrm{cov}} = \frac{N_c}{8\pi^2}\mu_5 \mathcal{B}_0^\mu, \quad (a = 0,3), \tag{19}$$

$$\langle J^\mu_{0\,\mathrm{A}}\rangle_{\mathrm{cov}} = \frac{N_c}{8\pi^2}\left(\mu_0\mathcal{B}_0^\mu + \mu_3\mathcal{B}_3^\mu + (\mu_0^2+\mu_3^2-\mu_5^2)\omega^\mu\right), \tag{20}$$

$$\langle J^\mu_{3\,\mathrm{A}}\rangle_{\mathrm{cov}} = \frac{N_c}{8\pi^2}\left(\mu_3\mathcal{B}_0^\mu + \mu_0\mathcal{B}_3^\mu + 2\mu_0\mu_3\omega^\mu\right), \tag{21}$$

$$q^\mu = \frac{N_c}{8\pi^2}\mu_5\left[\mu_0\mathcal{B}_0^\mu + \mu_3\mathcal{B}_3^\mu + \left(\mu_0^2+\mu_3^2-\frac{1}{3}\mu_5^2\right)\omega^\mu\right], \tag{22}$$

where $\langle T^{\mu\nu}\rangle = u^\mu q^\nu + u^\nu q^\mu$. The charges $\{B, Q\}$ are related to the generators $\{t_0, t_3\}$ by

$$B = \frac{2}{3}t_0, \qquad Q = \frac{1}{3}t_0 + t_3. \tag{23}$$

In the $\{B, Q\}$ basis, the background defined by Eq. (15) is given by Eq. (17) with $V_{B\mu} = \frac{3}{2}V_{0\mu} - \frac{1}{2}V_{3\mu}$, $V_{Q\mu} = V_{3\mu}$, $V_{S\,\mu} = 0$ and $A_{B\mu} = \frac{3}{2}A_{0\mu}$. Besides, the baryon and electromagnetic chemical potentials are defined by $\mu_B = e^{-\sigma}\mathcal{V}_{B0}$ and $\mu_Q = e^{-\sigma}\mathcal{V}_{Q0}$. Let us assume that the electromagnetic field is the only (physical) propagating gauge field. Then, we can define the physical magnetic field by $\mathcal{B}^\mu = \frac{1}{2}\epsilon^{\mu\nu\alpha\beta}u_\nu\mathscr{V}_{\alpha\beta}$ where the physical potential is $\mathscr{V}_\mu$ and its KK invariant form is $\mathbb{V}_\mu$, i.e. $\mathbb{V}_0 = \mathscr{V}_0$ and $\mathbb{V}_i = \mathscr{V}_i - a_i\mathscr{V}_0$. One finds

$$V_{B\mu} = 0, \;\; V_{Q\mu} \equiv e\mathbb{V}_\mu \quad \Longrightarrow \quad V_{0\mu} = \frac{e}{3}\mathbb{V}_\mu, \;\; V_{3\mu} = e\mathbb{V}_\mu, \;\; \mathcal{B}_0^\mu = \frac{e}{3}\mathcal{B}^\mu, \;\; \mathcal{B}_3^\mu = e\mathcal{B}^\mu, \tag{24}$$

a result that implies the following relations between chemical potentials $\mu_0 = \frac{1}{3}\mu_3 = \frac{1}{3}\mu_Q$ with $\mu_Q = e\,\mathbb{V}_0\,e^{-\sigma}$. After using Eq. (24), the constitutive relations of J^μ_{em} and $T^{\mu\nu}$ turn out to be

$$\langle J^\mu_{\mathrm{em}}\rangle_{\mathrm{cov}} = \frac{5e^2N_c}{36\pi^2}\mu_5\mathcal{B}^\mu, \qquad q^\mu = \frac{5N_c}{36\pi^2}\mu_5\left[e\mu_Q\mathcal{B}^\mu + \left(\mu_Q^2 - \frac{3}{10}\mu_5^2\right)\omega^\mu\right]. \tag{25}$$

$\langle J^\mu_{\mathrm{em}}\rangle_{\mathrm{cov}}$ gives the transport coefficient associated with the CME. The absence of a CVE in the vector current for $\mathrm{U(2)_V}\times\mathrm{U(2)_A}$ contrasts with the case $\mathrm{U(1)_V}\times\mathrm{U(1)_A}$, cf. Refs. [6, 12].

4.3 Constitutive relations for $N_f = 3$

Extending the procedure presented in the previous section [19], one finds the following Lorentz covariant expressions of the constitutive relations for $N_f = 3$

$$\langle J^\mu_{a\,\mathrm{V}}\rangle_{\mathrm{cov}} = \frac{N_c}{4\sqrt{6}\pi^2}\mu_5\mathcal{B}_a^\mu, \qquad (a = 0,3,8), \tag{26}$$

$$\langle J^\mu_{0\,\mathrm{A}}\rangle_{\mathrm{cov}} = \frac{N_c}{4\sqrt{6}\pi^2}\left[\mu_0\mathcal{B}_0^\mu + \mu_3\mathcal{B}_3^\mu + \mu_8\mathcal{B}_8^\mu + (\mu_0^2+\mu_3^2+\mu_8^2-\mu_5^2)\omega^\mu\right], \tag{27}$$

$$\langle J^\mu_{3\,\mathrm{A}}\rangle_{\mathrm{cov}} = \frac{N_c}{4\sqrt{6}\pi^2}\left[\mu_3\mathcal{B}_0^\mu + \left(\mu_0 + \frac{1}{\sqrt{2}}\mu_8\right)\mathcal{B}_3^\mu + \frac{1}{\sqrt{2}}\mu_3\mathcal{B}_8^\mu + 2\mu_3\left(\mu_0 + \frac{1}{\sqrt{2}}\mu_8\right)\omega^\mu\right], \tag{28}$$

$$\langle J^\mu_{8\,\mathrm{A}}\rangle_{\mathrm{cov}} = \frac{N_c}{4\sqrt{6}\pi^2}\left[\mu_8\mathcal{B}_0^\mu + \frac{1}{\sqrt{2}}\mu_3\mathcal{B}_3^\mu + \left(\mu_0 - \frac{1}{\sqrt{2}}\mu_8\right)\mathcal{B}_8^\mu + \left(2\mu_0\mu_8 + \frac{1}{\sqrt{2}}\left(\mu_3^2-\mu_8^2\right)\right)\omega^\mu\right], \tag{29}$$

$$q^\mu = \frac{N_c}{4\sqrt{6}\pi^2}\mu_5\left[\mu_0\mathcal{B}_0^\mu + \mu_3\mathcal{B}_3^\mu + \mu_8\mathcal{B}_8^\mu + \left(\mu_0^2+\mu_3^2+\mu_8^2-\frac{1}{3}\mu_5^2\right)\omega^\mu\right]. \tag{30}$$

The relation between the basis of conserved charges $\{B, Q, S\}$ and the basis $\{t_0, t_3, t_8\}$ is

$$B = \sqrt{\frac{2}{3}}t_0\,, \qquad Q = t_3 + \frac{1}{\sqrt{3}}t_8\,, \qquad S = -\sqrt{\frac{2}{3}}t_0 + \frac{2}{\sqrt{3}}t_8\,. \tag{31}$$

In the $\{B, Q, S\}$ basis, the background is given by Eq. (17) with $V_{B\mu} = \sqrt{\frac{3}{2}}V_{0\mu} - \frac{1}{2}V_{3\mu} + \frac{\sqrt{3}}{2}V_{8\mu}$, $V_{Q\mu} = V_{3\mu}$, $V_{S\mu} = -\frac{1}{2}V_{3\mu} + \frac{\sqrt{3}}{2}V_{8\mu}$ and $A_{B\mu} = \sqrt{\frac{3}{2}}A_{0\mu}$, while μ_B and μ_Q are defined as in Sec. 4.2, and the strangeness chemical potential is $\mu_S = e^{-\sigma}\mathcal{V}_{S\,0}$. Assuming, as above, that the electromagnetic field $\mathbb{V}_\mu$ is the only propagating gauge field, one finds

$$\begin{cases} V_{B\mu} = 0\,, \quad V_{S\mu} = 0 \\ V_{Q\mu} \equiv e\mathbb{V}_\mu \end{cases} \Longrightarrow \begin{cases} V_{0\mu} = 0\,, \quad V_{3\mu} = e\mathbb{V}_\mu\,, \quad V_{8\mu} = \frac{e}{\sqrt{3}}\mathbb{V}_\mu \\ \mathcal{B}^\mu_0 = 0\,, \quad \mathcal{B}^\mu_3 = e\mathcal{B}^\mu\,, \quad \mathcal{B}^\mu_8 = \frac{e}{\sqrt{3}}\mathcal{B}^\mu \end{cases}, \tag{32}$$

a result that leads to the relations $\mu_0 = 0$ and $\mu_3 = \sqrt{3}\mu_8 = \mu_Q$, with $\mu_Q = e\,\mathbb{V}_0\,e^{-\sigma}$. Finally, the constitutive relations for the electromagnetic current and energy-momentum tensor writes

$$\langle J^\mu_{\text{em}}\rangle_{\text{cov}} = \frac{e^2 N_c}{3\sqrt{6}\pi^2}\mu_5\mathcal{B}^\mu\,, \qquad q^\mu = \frac{N_c}{3\sqrt{6}\pi^2}\mu_5\left[e\mu_Q\mathcal{B}^\mu + \left(\mu_Q^2 - \frac{1}{4}\mu_5^2\right)\omega^\mu\right]. \tag{33}$$

As in the case of $N_f = 2$, there is no CVE in the vector current for $\text{U(3)}_\text{V} \times \text{U(3)}_\text{A}$.

5 Lowest order constitutive relations in presence of SSB

In this section we study the phase in which chiral symmetry is spontaneously broken. We consider the breaking pattern $U(N_f)_\text{L} \times U(N_f)_\text{R} \to U(N_f)_\text{V}$, with the vector diagonal subgroup as the surviving symmetry. The WZW partition function, which describes the anomaly induced interactions among NG bosons as well as with the external field $\mathcal{A}$ [31–34], is given by $\Gamma^{\text{WZW}}[\xi, \mathcal{A}] = \Gamma_0[\mathcal{A}] - \Gamma_0[\mathcal{A}_g]$ [15, 16, 35, 36], where Γ_0 is the anomalous functional in the symmetric phase, cf. Eq. (11). In this expression $\mathcal{A}_g = g^{-1}\mathcal{A}g + g^{-1}dg$ is the gauge field transformed by the generic element $g \equiv e^{i\xi_a t_a}$, with ξ_a the NG boson fields. In our case, we make the identifications $\mathcal{A} \to (\mathcal{A}_\text{L}, \mathcal{A}_\text{R})$ and $g \to (U, \mathbb{I})$, where $\mathbb{I}$ is the identity element. The matrix U is then written in terms of the conventionally normalized NG boson fields associated with the broken $\text{SU}(N_f)_\text{A}$ symmetry as $U(\xi) = e^{iX(\xi)}$, where

$$X(\xi) \equiv 2\sum_{a=1}^{8}\xi_a t_a = \frac{\sqrt{2}}{f_\pi}\begin{pmatrix} \frac{1}{\sqrt{2}}\pi^0 + \frac{1}{\sqrt{6}}\eta_8 & \pi^+ & K^+ \\ \pi^- & -\frac{1}{\sqrt{2}}\pi^0 + \frac{1}{\sqrt{6}}\eta_8 & K^0 \\ K^- & \bar{K}^0 & -\frac{2}{\sqrt{6}}\eta_8 \end{pmatrix} \quad \text{for} \quad N_f = 3\,, \tag{34}$$

with $f_\pi \approx 92\,\text{MeV}$ the pion decay constant. In the case $N_f = 2$, $X(\xi)$ corresponds to the 2×2 submatrix of pions in Eq. (34) with $\eta_8 = 0$. The NG boson field ξ_0 is absent, as the U(1)_A symmetry is violated in QCD by non-perturbative effects. In the following we will focus on the action at the lowest order in derivatives, that reads

$$\mathcal{W}_{(0)} = \frac{1}{T_0}\int d^3x\,\sqrt{g_3}e^\sigma\left[P(T, \{\mu_a\}) + \mathcal{L}\right], \tag{35}$$

where P is the pressure in absence of NG bosons, and the Lagrangian $\mathcal{L}$ contains all dependence on the NG bosons. The latter has the following expression in terms of KK invariant fields

$$\mathcal{L} = \frac{f_\pi^2}{4}G^{\mu\nu}\text{Tr}\left[D_\mu U\,(D_\nu U)^\dagger\right] = \frac{f_\pi^2}{4}\left[-e^{-2\sigma}\text{Tr}\left(\xi_0\xi_0^\dagger\right) + g^{ij}\text{Tr}\left(\Phi_i\Phi_j^\dagger\right)\right], \tag{36}$$

where the covariant derivative acting on the NG boson field reads

$$\xi_\mu \equiv D_\mu U = \partial_\mu U - i\mathcal{V}_{3\mu}[t_3, U] - i\mathcal{V}_{8\mu}[t_8, U], \tag{37}$$

and we have defined $\Phi_i \equiv \xi_i - a_i\xi_0 = \partial_i U - iV_{3\,i}[t_3, U] - iV_{8\,i}[t_8, U]$. The terms proportional to $[t_0, U]$ are absent as $t_0 \propto 1_{N_f\times N_f}$ commutes with U. By using Eqs. (23) and (31) one can express ξ_0 and Φ_i in terms of the conserved charges as $\xi_0 = -iV_{Q\,0}[Q, U] - iV_{S\,0}[S, U]$ and $\Phi_i = \partial_i U - iV_{Q\,i}[Q, U] - iV_{S\,i}[S, U]$, whereas from Eqs. (24) and (32) we find that $\xi_0 = -ie\mathbb{V}_0[Q, U]$ and $\Phi_i = \partial_i U - ie\mathbb{V}_i[Q, U]$. Then, the leading order Lagrangian of Eq. (36) governs the electromagnetic coupling of the charged mesons $(\pi^\pm, K^\pm)$. Finally, the constitutive relations at this order can be computed from the appropriate functional derivatives of the effective action in (35). The resulting expressions can be brought into a covariant form in terms of the metric $G_{\mu\nu}$ and the four velocity u_μ. The results are

$$\langle J_{0\mu}\rangle_{\rm PF} = n_0 u_\mu, \tag{38}$$

$$\langle J_{a\mu}\rangle_{\rm PF} = n_a u_\mu - \frac{i f_\pi^2}{4}\mathrm{Tr}\left[[t_a, U]\partial_\mu U^\dagger + [t_a, U^\dagger]\left(\partial_\mu U - 2i\sum_{X=Q,S} V_{X\mu}[X, U]\right)\right], \quad (a = 3, 8), \tag{39}$$

$$\langle T^{\mu\nu}\rangle_{\rm PF} = (\varepsilon + P)\, u^\mu u^\nu + P G^{\mu\nu} - \frac{f_\pi^2}{4} G^{\mu\alpha} G^{\nu\beta}\mathrm{Tr}\left[D_\alpha U\left(D_\beta U\right)^\dagger + D_\beta U\left(D_\alpha U\right)^\dagger\right], \tag{40}$$

where $n_a = \frac{\partial P}{\partial n_a}$ $(a = 0, 3, 8)$ are the charge densities. Higher order contributions will be studied in a forthcoming publication [19].

6 Conclusions

We investigated non-dissipative transport induced by chiral imbalance in a hadronic fluid with $N_f = 2, 3$, both at zeroth and first order in the hydrodynamic expansion. By using the EPFF, we have characterized the effects induced by external electromagnetic fields and fluid vorticity. Particularizing to a background with two and three quark flavors and in presence of a chemical potential controlling chiral imbalance, μ_5, we have obtained the constitutive relations for various covariant currents and the energy-momentum tensor in the unbroken chiral symmetric phase. These constitutive relations have been discussed in two different flavor bases: i) the basis of the Hermitian generators of the Lie algebra, t_a, and ii) the basis of conserved charges $\{B, Q, S\}$. We also studied, in the same background and at leading order, the partition function and constitutive relations in the presence of chiral symmetry breaking.

Finally, let us indicate that the general formalism developed in [14, 18] and presented here can be used in a wide variety of physical situations, ranging from the study of other sectors of the Standard Model as e.g. the electroweak sector [37, 38], to the physics of condensed matter systems like superfluids [39, 40] and other systems affected by triangle anomalies [41, 42].

Acknowledgements

The work of J.L.M. and M.V. has been supported by Spanish Science Ministry grant PGC2018-094626-B-C21 (MCIU/AEI/FEDER, EU) and Basque Government grant IT979-16. The research of E.M. is supported by the project PID2020-114767GB-I00 financed by MCIN/AEI/10.13039/501100011033, by the FEDER/Junta de Andalucía-Consejería de Economía y Conocimiento 2014-2020 Operational Program under grant A-FQM-178-UGR18, by Junta de Andalucía under grant FQM-225, and by the Consejería de Conocimiento, Investigación y Universidad of the Junta de Andalucía and European Regional Development Fund (ERDF) under grant SOMM17/6105/UGR. The research of E.M. is also supported by the Ramón y Cajal Program of the Spanish MCIN under grant RYC-2016-20678. M.A.V.-M. acknowledges the financial support from the Spanish Science Ministry through research grant PGC2018-094626-B-C22 (MCIU/AEI/FEDER, EU), as well as from Basque Government grant IT979-16.

References

[1] D.E. Kharzeev, H.U. Yee, Phys. Rev. **D84**, 045025 (2011).
[2] K. Fukushima, D.E. Kharzeev, H.J. Warringa, Phys. Rev. **D78**, 074033 (2008).
[3] D.T. Son, P. Surowka, Phys. Rev. Lett. **103**, 191601 (2009).
[4] D.E. Kharzeev, H.J. Warringa, Phys. Rev. **D80**, 034028 (2009).
[5] K. Landsteiner, E. Megías, F. Pena-Benitez, Phys. Rev. Lett. **107**, 021601 (2011).
[6] K. Landsteiner, E. Megías, F. Pena-Benitez, Lect. Notes Phys. **871**, 433 (2013).
[7] S. Bhattacharyya, V.E. Hubeny, S. Minwalla, M. Rangamani, JHEP **02**, 045 (2008).
[8] J. Erdmenger, M. Haack, M. Kaminski, A. Yarom, JHEP **01**, 055 (2009).
[9] K. Landsteiner, E. Megías, L. Melgar, F. Pena-Benitez, JHEP **09**, 121 (2011).
[10] E. Megías, F. Pena-Benitez, JHEP **05**, 115 (2013).
[11] N. Banerjee, J. Bhattacharya, S. Bhattacharyya, S. Jain, S. Minwalla, T. Sharma, JHEP **09**, 046 (2012).
[12] K. Jensen, P. Kovtun, A. Ritz, JHEP **10**, 186 (2013).
[13] E. Megías, M. Valle, JHEP **11**, 005 (2014).
[14] J.L. Mañes, E. Megías, M. Valle, M.Á. Vázquez-Mozo, JHEP **11**, 076 (2018).
[15] J. Wess, B. Zumino, Phys. Lett. **37B**, 95 (1971).
[16] E. Witten, Nucl. Phys. **B223**, 422 (1983).
[17] Y. Neiman, Y. Oz, JHEP **09**, 011 (2011).
[18] J.L. Mañes, E. Megías, M. Valle, M.Á. Vázquez-Mozo, JHEP **12**, 018 (2019).
[19] E. Megías, M.Á. Vázquez-Mozo, to appear (2021).
[20] K. Jensen, R. Loganayagam, A. Yarom, JHEP **05**, 134 (2014).
[21] S. Bhattacharyya, JHEP **08**, 165 (2014).
[22] E. Megías, K. Landsteiner, F. Pena-Benitez, Acta Phys. Polon. Supp. **6**, 45 (2013).
[23] W.A. Bardeen, Phys. Rev. **184**, 1848 (1969).
[24] V.P. Nair, R. Ray, S. Roy, Phys. Rev. **D86**, 025012 (2012).
[25] K. Jensen, R. Loganayagam, A. Yarom, JHEP **05**, 110 (2014).
[26] J.L. Mañes, M. Valle, M.Á. Vázquez-Mozo, JHEP **05**, 209 (2021).
[27] W.A. Bardeen, B. Zumino, Nucl. Phys. **B244**, 421 (1984).
[28] J.L. Mañes, E. Megías, M. Valle, M.Á. Vázquez-Mozo (2021), 2103.14850.
[29] R. Gatto, M. Ruggieri, Phys. Rev. **D85**, 054013 (2012).
[30] X. Planells, A.A. Andrianov, V.A. Andrianov, D. Espriu, PoS **QFTHEP2013**, 049 (2013).
[31] R. Kaiser, Phys. Rev. **D63**, 076010 (2001).
[32] D.T. Son, M.A. Stephanov, Phys. Rev. **D77**, 014021 (2008).
[33] K. Fukushima, K. Mameda, Phys. Rev. **D86**, 071501 (2012).
[34] T. Brauner, S. Kadam, JHEP **03**, 015 (2017).
[35] J.L. Mañes, Nucl. Phys. **B250**, 369 (1985).
[36] C.S. Chu, P.M. Ho, B. Zumino, Nucl. Phys. **B475**, 484 (1996).
[37] C. García-Recio, L.L. Salcedo, JHEP **07**, 015 (2009).
[38] T. Brauner, O. Taanila, A. Tranberg, A. Vuorinen, JHEP **11**, 076 (2012).
[39] S. Lin, Phys. Rev. **D85**, 045015 (2012).
[40] C. Hoyos, B.S. Kim, Y. Oz, JHEP **10**, 127 (2014).
[41] G. Basar, D.E. Kharzeev, H.U. Yee, Phys. Rev. **B89**, 035142 (2014).
[42] K. Landsteiner, Phys. Rev. **B89**, 075124 (2014).

EPJ Web of Conferences **258**, 10007 (2022) https://doi.org/10.1051/epjconf/202225810007
A Virtual Tribute to Quark Confinement and the Hadron Spectrum (vConf21)

Far from equilibrium Chiral Magnetic Effect in Strong Magnetic Fields from Holography

Sebastian Grieninger[1,2,*] and *Sergio* Morales-Tejera[1,2,**]

[1]Instituto de Física Teórica UAM/CSIC, c/Nicolás Cabrera 13-15, Universidad Autónoma de Madrid, Cantoblanco, 28049 Madrid, Spain
[2]Departamento de Física Teórica, Universidad Autónoma de Madrid, Campus de Cantoblanco, 28049 Madrid, Spain

Abstract. We study the real time evolution of the chiral magnetic effect out-of-equilibrium in strongly coupled anomalous field theories. We match the parameters of our model to QCD parameters and draw lessons of possible relevance for the realization of the chiral magnetic effect in heavy ion collisions. In particular, we find an equilibration time of about $\sim$ 0.35 fm/c in presence of the chiral anomaly for plasma temperatures of order $T \sim 300 - 400$ MeV.
Preprint number: IFT-UAM/CSIC-21-133

1 Introduction

The chiral magnetic effect (CME) is the generation of an electric current in a chirally imbalanced medium by an applied magnetic field [1, 2].[1] Formally it is described by $\vec{J} = 8c\mu_5\vec{B}$, where c is the coefficient of the axial anomaly and μ_5 the axial chemical potential. An experimental observation of the CME in the context of QCD may be possible in heavy-ion collisions. Recent experimental results of the collision of the isotopes ZrZr and RuRu [6] highlight that there are many open questions in our understanding of the underlying physics.

The CME current is formally expressed in terms of a chemical potential (an equilibrium quantity) thus it is specially interesting to understand how the CME is realized far from equilibrium. Answering this question is particularly important in view of a possible realization of the CME in heavy ion collisions. In heavy ion collisions, the magnetic field is only present in the initial stages after the collision and decays quickly whereas the hot QCD matter has a short but non-vanishing equilibration time. The question naturally arising is if the CME builds up fast enough to be measurable before the magnetic field decays. We want to address this question in this work by modeling the strongly coupled quark gluon plasma by means of a holographic model. Previous related studies of out-of-equilibrium chiral transport in holography are [7–15]. In this study [16], we investigate the realtime dynamics of the chiral magnetic effect in strong magnetic fields within holography for the first time in the full backreacted setup.

*e-mail: sebastian.grieninger@uam.es
**e-mail: sergio.moralest@uam.es
[1]For reviews see [3–5].

2 Holographic model

We study a holographic quantum field theory with a $U(1)_A \times U(1)_V$ symmetry which is manifest in the gravity theory in terms of an axial gauge field A^μ and a vector V^μ gauge field, respectively, with the associated field strengths denoted as $F_5 = dA$ and $F = dV$. The anomaly is implemented through a Chern–Simons term which is gauge invariant up to a total derivative. We work with the *consistent* form of the anomaly. Combining these ingredients, the holographic model we consider is the following[2]

$$S = \frac{1}{2\kappa^2}\int_M d^5x\,\sqrt{-g}\left[R+\frac{12}{L^2}-\frac{1}{4}F^2-\frac{1}{4}F_{(5)}^2+\frac{\alpha}{3}\epsilon^{\mu\nu\rho\sigma\tau}A_\mu\left(3F_{\nu\rho}F_{\sigma\tau}+F^{(5)}_{\nu\rho}F^{(5)}_{\sigma\tau}\right)\right]+S_{\rm GHY}+S_{\rm ct} \quad (1)$$

where $S_{\rm GHY}$ is the Gibbons-Hawking-York boundary term to make the variational problem well defined, L is the AdS radius, κ^2 is the Newton constant and α the Chern-Simons coupling. We supplement the action by the appropriate counter-terms $S_{\rm ct}$ to cancel the divergences [17, 18]. The Levi-Civita tensor is defined as $\epsilon^{\mu\nu\rho\sigma\tau} = \epsilon(\mu\nu\rho\sigma\tau)/\sqrt{-g}$. From the action (1), we may derive the equations of motion yielding

$$\nabla_\nu F^{\nu\mu} + 2\alpha\epsilon^{\mu\nu\rho\sigma\tau}F_{\nu\rho}F^{(5)}_{\sigma\tau} = 0, \qquad \nabla_\nu F^{\nu\mu}_{(5)} + \alpha\epsilon^{\mu\nu\rho\sigma\tau}\left(F_{\nu\rho}F_{\sigma\tau}+F^{(5)}_{\nu\rho}F^{(5)}_{\sigma\tau}\right) = 0, \quad (2)$$

$$G_{\mu\nu} - \frac{6}{L^2}g_{\mu\nu} - \frac{1}{2}F_{\mu\rho}F_\nu{}^\rho + \frac{1}{8}F^2 g_{\mu\nu} - \frac{1}{2}F^{(5)}_{\mu\rho}F^{(5)\rho}_\nu + \frac{1}{8}F^2_{(5)}g_{\mu\nu} = 0. \quad (3)$$

We are interested in monitoring the vector current and pressure in the dual field theory. Both one-point functions can be extracted from the full bulk solution through the standard holographic prescription, i.e. varying the renormalized on-shell action with respect to the associated source term. In infalling Eddington-Finkelstein coordinates[3] the metric reads

$$ds^2 = -f(v,u)dv^2 - \tfrac{2L^2}{u^2}dvdu + \Sigma(v,u)^2\left[e^{\xi(v,u)}(dx^2+dy^2) + e^{-2\xi(v,u)}dz^2\right], \quad (4)$$

where v and u denote the time and radial coordinate, respectively. The conformal boundary is located at $u = 0$ where we impose that the metric asymptotes to AdS_5. The minimal ingredients to generate the CME are a finite axial charge density q_5 and a (vector) magnetic field B. Hence, the gauge field ansatz reads (in radial gauge $V_u = A_u = 0$)

$$V = \frac{B}{2}(xdy - ydx) + V_z(v,u)\,dz\,, \qquad A = -Q_5(v,u)\,dv\,. \quad (5)$$

The equations of motion for (4)-(5) may be found in [16]. The asymptotic expansions read

$$Q_5(v,u) = \frac{u^2}{2}q_5 + O(u^3)\,, \qquad V_z(v,u) = u^2\,V_2(v) + O(u^3)\,, \quad (6)$$

$$\Sigma(v,u) = \frac{1}{u} + \lambda(v) + O(u^5)\,, \qquad \xi(v,u) = u^4\left(\xi_4(v) - \frac{B^2}{12}\log(u)\right) + O(u^5)\,, \quad (7)$$

$$f(v,u) = \left(\frac{1}{u}+\lambda(v)\right)^2 + u^2\left(f_2 + \frac{B^2}{6}\log(u)\right) - 2\dot\lambda(v) + O(u^3). \quad (8)$$

In the asymptotic expansions, we introduced $\lambda(v)$ which we use to keep the position of the apparent horizon of the black brane at a fixed radial position $u_h = 1$ throughout the time evolution making use of the diffeomorphism freedom [16, 19]. The coefficient f_2 is related to

[2] In our notation, Greek letters denote the bulk coordinates and small Latin letters the boundary coordinates.

[3] Note that ξ parametrizes the anisotropy of the system, which is induced by the presence of a magnetic field B which breaks the rotational invariance from SO(3) to SO(2) and points in the z-direction.

the energy density of the black brane and the subleading coefficients $V_2(v)$ and $\xi_4(v)$ shall give us the vector current and the pressure anisotropy, respectively. We can extract the one-point functions using the asympotic expansions of the solution

$$2\kappa^2 \langle J_z \rangle = 2V_2(v), \qquad 2\kappa^2 \left\langle J_5^0 \right\rangle = q_5, \tag{9}$$

$$\langle T^v{}_v \rangle = \frac{1}{4\kappa^2}\left[6f_2 - B^2\log(\mu L)\right], \qquad \left\langle T^z{}_z \right\rangle = -\frac{1}{4\kappa^2}\left[2f_2 + 16\xi_4(v) + B^2\log(\mu L)\right],$$
$$\langle T^x{}_x \rangle = \left\langle T^y{}_y \right\rangle = -\frac{1}{8\kappa^2}\left[B^2 + 4f_2 - 16\xi_4(v) - 2B^2\log(\mu L)\right], \tag{10}$$

for the currents and the stress tensor. The presence of the magnetic field induces a trace anomaly thus breaking conformal invariance at the microscopic level. Hence, we have reinstated the AdS radius L in the expressions for the one-point functions since the regularization procedure depends on a renormalization energy scale μ due to the broken conformal invariance. All in all, the problem reduces to solving the full dynamics in the bulk, finding the subleading coefficients of ξ and V_z and substituting them into (9) and (10). The details about the numerical implementation can be found in [16]. Our numerical code is implemented in the programming language *Julia* [20].

The thermodynamic properties of the system are specified by two quantities: temperature and axial chemical potential. At late times, the system equilibrates and we shall label different solutions in terms of their thermodynamical variables in the final equilibrium state. The thermodynamic parameters are also useful to interpret our solutions in terms of results for the QGP produced in heavy-ion collisions. In the dual gravity picture, equilibration implies that the metric becomes stationary at late times and the temperature is given by the black brane $T = -u_h^2/(4\pi)f'(v \to \infty, u_h)$. The chemical potential is the difference of Q_5 at the boundary and horizon, i.e. $\mu_5 = Q_5(v \to \infty, u_h) - Q_5(v \to \infty, 0)$.

We conclude this section by discussing the initial state of the dual quantum field theory. The asymptotic form of the metric ansatz (4) is dual to an infinite-volume non-expanding plasma. By construction the plasma has a uniform charge distribution given by q_5, is immersed in a magnetic field of magnitude B and its uniform energy density is ϵ. All three of them are considered to be homogeneous and constant in time in our model. Finally, we specify the initial conditions of the evolution by giving an initial profile to the fields ξ and V_z. In particular, we choose them to be zero everywhere, which implies that the CME current vanishes in the initial state and the *dynamical* pressure anisotropy, i.e. the anisotropy generated by ξ_4, is zero. Notice that the initial state is anisotropic since the term proportional to B in (10) yields a static contribution to the anisotropy in the pressure (kinematic pressure anisotropy, see [19] for details). Hence, for finite magnetic field, the inital state does not describe an equilibrium solution thus driving the out-of-equilibrium dynamics.

3 Results

To gain an intuition, we monitor the features of our model in terms of the CME current $\langle J \rangle$ and the dynamical pressure anisotropy ξ_4 for different strengths of the magnetic field in section 3.1. The parameter scan is performed at fixed energy density $\epsilon_L = 12$.[4] In section 3.2, we match our model to QCD and give physically relevant values for the parameters.

There is a subtlety related to the definition of the energy density due to the non-trivial renormalization scale dependence in (10): $\epsilon(\mu) = 2\kappa^2\langle T^{vv}\rangle$, which has to be fixed. In this

[4] Working with a different value for ϵ_L only modifies the final equilibrium state for the pressure anisotropy but does not alter the relevant qualitative behavior like the build up time and the presence or absence of oscillations.

subsection, we simply choose $\mu = 1/L$ for computational convenience. Note that the choice $\mu = 1/L$ is not physical since the scale can be changed without changing the values of the physical observables at the boundary due to a scaling symmetry [19]. The physically relevant scale is $\mu = \sqrt{B}$, which we use in section 3.2. Both choices are related through

$$\frac{\epsilon_B}{B^2} = \frac{\epsilon_L}{B^2} + \frac{1}{4}\log(BL^2), \tag{11}$$

where ϵ_L and ϵ_B refer to the energy density at scales $\mu = 1/L$ and $\mu = \sqrt{B}$, respectively.

3.1 B-dependence

In this section, we fix the strength of the anomaly α and consider two qualitatively different values of q_5 for varying strength of B. The results for the larger q_5 are shown in figure 1. We choose to work with dimensionless variables: time, pressure and current are normalized to the energy density ϵ_L, whereas we consider the dimensionless ratio of B to temperature squared. All thermodynamic quantities refer to the final equilibrium state where they are well defined.

As we increase the magnetic field, we observe the appearance of oscillations in the current $\langle J \rangle$ in agreement with the quasinormal modes computation in [8, 9]. As a consequence, the equilibration time increases dramatically for increasing B. The oscillatory behavior of the current indicates that the time-evolution is dominated by the lowest QNM near the real axis. The final equilibrium value matches the equilibrium expression for the CME, i.e. $2\kappa^2 \langle J \rangle_{eq} = 8\alpha\mu_5 B$. Note that oscillatory behavior indicates that we have not reached the final equilibrium state yet. However, we verified that the axial chemical potential μ_5 read off from these states matches the values of the would-be equilibrium state closely.

It is also worth noting that the current builds up faster for larger B. Such behavior is expected in $1+1$ dimensional systems [16]. For the chosen q_5, we clearly observe oscillatory behavior in ξ_4 which is not present for small q_5 (see fig. 1 in [16]).[5]

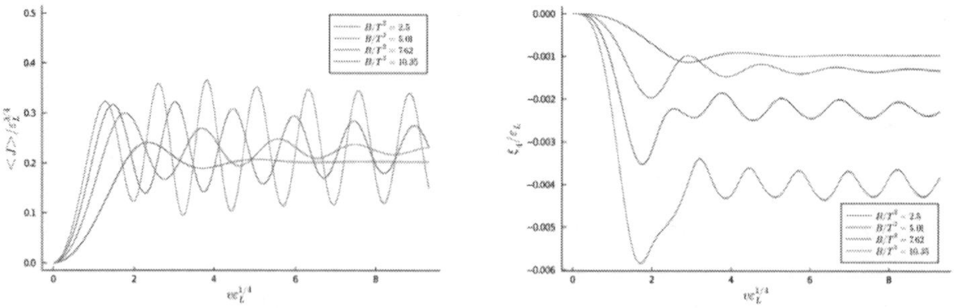

Figure 1: Vector current (left) and dynamical pressure anisotropy (right) for fixed $\alpha = 1.5$ and $q_5 = 1.5$. The magnetic field B is $\{0.5, 1.0, 1.5, 2.0\}$. The axial chemical potential of the would-be final equilibrium state is $\mu_5/T = \{0.11, 0.06, 0.04, 0.02\}$.

3.2 Matching to QCD

Parameters

In this section, we aim to provide simulations in the parameter range experimentally relevant for the Quark-Gluon plasma (QGP). We obtain estimates for the parameters in our model by

[5] At small q_5 oscillations are still there but their amplitude is significantly smaller than the final equilibrium value.

matching to known QCD results, i.e. the entropy and the anomaly. Under an axial gauge transformation, our action has the mixed anomaly $\mathcal{A}_{CS} = \frac{\alpha}{2\kappa^2}$. In order to get an estimate for κ we take the entropy of a black brane $s_{BH} = \frac{A}{4G_N} = \frac{4\pi^4 T^3}{2\kappa^2}$.

In the following, we match these gravity expressions to the entropy of QCD at finite temperature and to the axial anomaly [16, 21]. First, we fix the number of flavors. The up and down quarks are light, whereas the strange quark has a mass of around 95 MeV. The cross over temperature of QCD is at around 175 MeV. Therefore, we include the strange quark in our counting, i.e. we match to three flavor QCD. The Stefan-Boltzmann value of the entropy density is $s_{\rm SB} = 4\left(\nu_b + \frac{7}{4}\nu_f\right)\frac{\pi^2 T^3}{90}$, where $\nu_b = 2(N_c^2 - 1)$ and $\nu_f = 2N_c N_f$ with $N_c = 3$ and $N_f = 3$. Note that the Stefan-Boltzmann value of the entropy is only reached at asymptotically high temperatures and s is usually smaller at the temperatures of interest. As a ballpark value, we take a factor of $3/4$, which yields $\kappa^2 = (24\pi^2)/19 \approx 12.5$. Such a relative factor arises in the strongly coupled $\mathcal{N} = 4$ SYM theory [22]. QCD lattice simulations also indicate a reduction by a factor of around 0.8 at moderate temperatures (see e.g. [23]).

The axial anomaly of three flavor QCD is $\mathcal{A}_{QCD} = 2\,\frac{N_c}{32\pi^2}\left(\frac{4}{9} + \frac{1}{9} + \frac{1}{9}\right) = \frac{1}{8\pi^2}$,where the factor 2 comes from the summation over right- and left-handed fermions and in the bracket we sum over the squares of the electric charges of up, down and strange quarks. The Chern-Simons coupling α is then determined by matching $\mathcal{A}_{CS} = \mathcal{A}_{QCD}$ leading to $\alpha \approx 0.316$.

In typical nucleations of the QGP the temperature, magnetic field and chemical potential (T, μ_5, B) in [MeV] are $(300, 10, m_\pi^2)$ (RHIC) and $(1000, 10, 15m_\pi^2)$ (LHC). The parameter estimates should be viewed as ballpark values representative for RHIC and LHC physics since there are considerable uncertainties in the values. The estimates provide us with two independent dimensionless quantities, which have to be adjusted in the numerical simulations with our two free parameters (ϵ_L, q_5). The dimensionless ratio ϵ_B/B^2 uniquely fixes B/T^2 and hence we work with ϵ_B in this section. In contrast to the previous section, we plot the full pressure anisotropy evaluated at the physical renormalization scale $\mu = \sqrt{B}$ in this section

$$\delta P_i \equiv 2\kappa^2\,\frac{\Delta P_B}{B^2} = 12\,\frac{\xi_4(v)}{B^2} + \frac{1}{2}\,\log\!\left(BL^2\right) - \frac{1}{4}. \tag{12}$$

Fixing ϵ_B/B^2 in eq. (11) does not fix f_2 and B uniquely but rather gives us $B(f_2)$. This means that at fixed ϵ_B/B^2 and vanishing initial dynamical anisotropy $\xi_4(0) = 0$, we are confronted with a one parameter family of relative pressures of the initial state (12) depending on the value of the magnetic field B (for $L = 1$). We shall exploit this feature to study equilibration for several non-equivalent initial states by considering different values of δP_i.

Simulation

In figures 2-3, we present the results for the pressure and the current for RHIC and LHC with the physical parameters estimated in the previous section. We fix our initial state by setting the dynamical pressure anisotropy to zero, fix the ratio ϵ_B/B^2 and q_5 so that we reach the desired final equilibrium state. Neither the vector current nor the pressure anisotropy show oscillatory behavior. The former takes slightly more time to build up than the latter. We observe that in the RHIC simulation in figure 2, the peak in the vector current is reached at $v_{peak} \sim 0.54$ fm/c and the pressure anisotropy reaches its peak at $v_{peak} \sim 0.48$ fm/c.

We display the equilibration times for the simulation with the RHIC parameters in table 1. We use the definition of Chesler and Yaffe to define the equilibration time [24], i.e. the time where the observable is within 10% of its final value. As for the LHC in figure 3, we find the peak in the vector current at $v_{peak} \sim 0.14$ fm/c which is also the time where the pressure anisotropy reaches its peak. We tabulate the equilibration times for the simulation with the LHC parameters, which are about 1/3 shorter than in the RHIC case, in table 2.

A formula to estimate the lifetime of the magnetic field was recently established in [25] as $\tau_B \sim 115$ GeVfm/c/ $\sqrt{s}$, where $\sqrt{s}$ is the collision energy. At RHIC and LHC the collisions take place at around $\sqrt{s} \simeq 200\,\text{GeV}$ and $\sqrt{s} \simeq 5000\,\text{GeV}$, respectively, which yield lifetimes of $\tau_B^{\text{RHIC}} \sim 0.6\,\text{fm/c}$ and $\tau_B^{\text{LHC}} \sim 0.02\,\text{fm/c}$. In this context the equilibration and build up times extracted from our simulations may function as a gauge whether the CME is an observable or not. Our simulations for the RHIC parameters show that the current reaches its equilibrium value before the magnetic field vanishes. On the contrary, for the LHC parameters the magnetic field is too short lived for the current to build up significantly. Hence, we conclude that the chiral magnetic effect should only be observable at RHIC and not at LHC.

In heavy-ion collisions, the magnetic field drops almost instantaneously from its peak value to a base value which is approximately 10% of the original strength at which it stays for most of its remaining lifetime. Since we approximate the magnetic field as static, we did a second simulation for our parameter estimates with 10% of the peak magnetic field. The resulting equilibration times remain effectively unchanged and do not affect the previous results on the observability of the CME at either accelerator. We conclude by remarking that the estimate for the axial chemical potential is the most uncertain one in the literature. To prove that our build up and equilibration times are not influenced by our particular choice of the axial chemical potential $\mu_5 = 10\text{MeV}$, we performed an analogous simulations at a ten times larger axial chemical potential of $\mu_5 = 100\text{MeV}$ [16]. The bottom line is that our results for the build up times and thus the presence of the CME in heavy-ion collisions at RHIC and LHC remain qualitatively unchanged for the larger axial chemical potential.

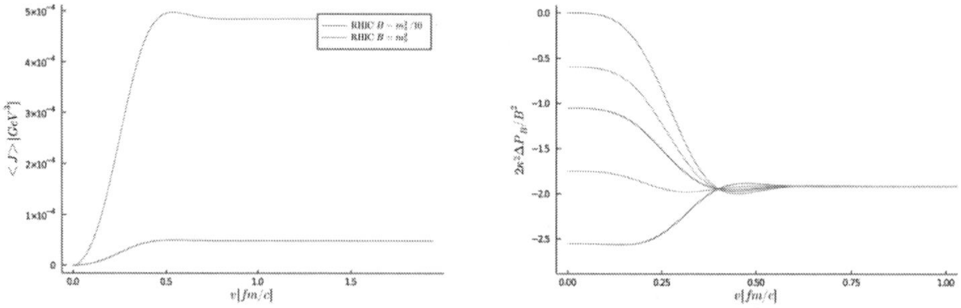

Figure 2: Vector current (left) and pressure anisotropy (right) as a function of time for the physical parameter estimates for RHIC and anomaly $\alpha \simeq 0.316$; for $m_\pi = 140\,\text{MeV}$. The pressure anisotropy is for $B = m_\pi^2$, the results for $B = 0.1\,m_\pi^2$ are qualitatively similar.

4 Conclusions

In this work, we investigated the out-of-equilibrium behavior of the chiral magnetic effect in the presence of strong external magnetic fields. We characterize how the chiral anomaly, the magnetic field and the axial charge density influence the non-equilibrium response of the chiral magnetic vector current and the pressure anisotropy and how they affect their equilibration and build up times. We furthermore provide insights on the build up time of the chiral magnetic current in heavy ion collision experiments at RHIC and LHC. Within our setup, the build up time of the chiral magnetic effect is smaller than the lifetime of the magnetic field and should thus be an observable in heavy ion collisions at RHIC [26]. However, the lifetime of the magnetic field at LHC seems to be so short that the magnetic field already drops to zero before the chiral magnetic current can build up in a meaningful way. Interestingly, for RHIC

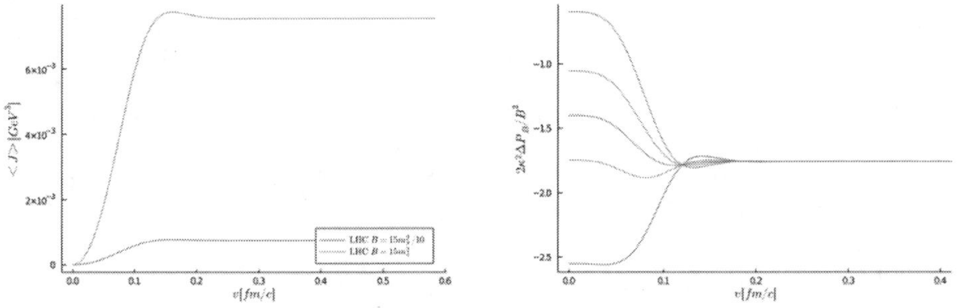

Figure 3: Vector current (left) and pressure anisotropy (right) for the parameter estimates for LHC ($\alpha \simeq 0.316$). $B = 15m_\pi^2$ in the right plot; results for $B = 1.5m_\pi^2$ are qualitatively similar.

RHIC $B = m_\pi^2$					
δP_i	-2.55	-1.75	-1.05	-0.60	0.00
$v_{\text{eq}}^{\langle J \rangle}$ in [fm/c]	0.380	0.380	0.380	0.380	0.380
$v_{\text{eq}}^{\langle \Delta P \rangle}$ in [fm/c]	0.383	0.418	0.334	0.344	0.350
RHIC $B = 0.1m_\pi^2$					
δP_i	-3.70	-2.90	-2.55	-2.21	-1.75
$v_{\text{eq}}^{\langle J \rangle}$ in [fm/c]	0.380	0.380	0.380	0.380	0.380
$v_{\text{eq}}^{\langle \Delta P \rangle}$ in [fm/c]	0.383	0.418	0.310	0.334	0.344

Table 1: Equilibration times v_{eq} for the RHIC simulation at $B = m_\pi^2$ and $B = 0.1\,m_\pi^2$; δP_i labels the different initial conditions for the pressure anisotropy (12).

LHC $B = 15\,m_\pi^2$					
δP_i	-2.55	-1.75	-1.40	-1.05	-0.60
$v_{\text{eq}}^{\langle J \rangle}$ in [fm/c]	0.114	0.114	0.114	0.114	0.114
$v_{\text{eq}}^{\langle \Delta P \rangle}$ in [fm/c]	0.114	0.187	0.085	0.098	0.103
LHC $B = 1.5\,m_\pi^2$					
δP_i	-3.70	-2.90	-2.55	-2.21	-1.75
$v_{\text{eq}}^{\langle J \rangle}$ in [fm/c]	0.114	0.114	0.114	0.114	0.114
$v_{\text{eq}}^{\langle \Delta P \rangle}$ in [fm/c]	0.114	0.187	0.085	0.098	0.103

Table 2: Equilibration times for the LHC simulation at $B = 15\,m_\pi^2$ and $B = 1.5\,m_\pi^2$; δP_i labels the different initial conditions for the pressure anisotropy (12).

we find a shorter equilibration time of $\sim$ 0.35 fm/c (for an initial state with $\delta P_i(0) = 0$) in presence of the chiral anomaly compared to the result of Chesler and Yaffe which estimates the equilibration time as $\sim$ 0.5 fm/c [24]. This is in agreement with the equilibration time estimate of $\sim$ 0.3 fm/c for plasma temperatures of $T \sim 300 - 400$ MeV [27].

Acknowledgments: SG would like to thank the organizers of the (virtual) Quark Confinement and the Hadron Spectrum 2021 conference in Stavanger, Norway, for the opportunity to present this talk. The authors thank J. Ghosh and K. Landsteiner for fruitful collaboration in [16]. SG is supported by the 'Atracción de Talento' program (2017-T1/TIC-5258, Comunidad de Madrid) and through grants SEV-2016-0597 and PGC2018-095976-B-C21. SMT is supported by a FPI-UAM predoctoral fellowship.

References

[1] A. Vilenkin, Phys. Rev. D **22**, 3080 (1980)

[2] K. Fukushima, D.E. Kharzeev, H.J. Warringa, Phys. Rev. D **78**, 074033 (2008), 0808.3382

[3] D.E. Kharzeev, Prog. Part. Nucl. Phys. **75**, 133 (2014), 1312.3348

[4] K. Landsteiner, Acta Phys. Polon. B **47**, 2617 (2016), 1610.04413

[5] D.E. Kharzeev, J. Liao, Nature Rev. Phys. **3**, 55 (2021), 2102.06623

[6] M. Abdallah et al. (STAR) (2021), 2109.00131

[7] S. Lin, H.U. Yee, Phys. Rev. D **88**, 025030 (2013), 1305.3949

[8] M. Ammon, S. Grieninger, A. Jimenez-Alba, R.P. Macedo, L. Melgar, JHEP **09**, 131 (2016), 1607.06817

[9] S. Grieninger, Master's thesis, Jena U., TPI (2016), 1711.08422

[10] M. Haack, D. Sarkar, A. Yarom, JHEP **04**, 034 (2019), 1812.08210

[11] C. Cartwright, JHEP **01**, 041 (2021), 2003.04325

[12] D.E. Kharzeev, Y. Kikuchi, Phys. Rev. Res. **2**, 023342 (2020), 2001.00698

[13] K. Landsteiner, E. Lopez, G. Milans del Bosch, Phys. Rev. Lett. **120**, 071602 (2018), 1709.08384

[14] J. Fernández-Pendás, K. Landsteiner, Phys. Rev. D **100**, 126024 (2019), 1907.09962

[15] S. Morales-Tejera, K. Landsteiner, Phys. Rev. D **102**, 106020 (2020), 2006.16031

[16] J.K. Ghosh, S. Grieninger, K. Landsteiner, S. Morales-Tejera, Phys. Rev. D **104**, 046009 (2021), 2105.05855

[17] S. de Haro, S.N. Solodukhin, K. Skenderis, Commun. Math. Phys. **217**, 595 (2001), hep-th/0002230

[18] M. Ammon, S. Grieninger, J. Hernandez, M. Kaminski, R. Koirala, J. Leiber, J. Wu, JHEP **04**, 078 (2021), 2012.09183

[19] J.F. Fuini, L.G. Yaffe, JHEP **07**, 116 (2015), 1503.07148

[20] J. Bezanson, A. Edelman, S. Karpinski, V.B. Shah, *Julia: A fresh approach to numerical computing* (2015), 1411.1607

[21] S. Grieninger, A. Shukla, JHEP **08**, 108 (2021), 2105.08673

[22] S.S. Gubser, I.R. Klebanov, A.W. Peet, Phys. Rev. D **54**, 3915 (1996), hep-th/9602135

[23] S. Borsanyi, Z. Fodor, C. Hoelbling, S.D. Katz, S. Krieg, K.K. Szabo, Phys. Lett. B **730**, 99 (2014), 1309.5258

[24] P.M. Chesler, L.G. Yaffe, Phys. Rev. Lett. **102**, 211601 (2009), 0812.2053

[25] Y. Guo, S. Shi, S. Feng, J. Liao, Phys. Lett. B **798**, 134929 (2019), 1905.12613

[26] S. Shi, H. Zhang, D. Hou, J. Liao, Phys. Rev. Lett. **125**, 242301 (2020), 1910.14010

[27] U.W. Heinz, AIP Conf. Proc. **739**, 163 (2004), nucl-th/0407067

EPJ Web of Conferences **258**, 10008 (2022) https://doi.org/10.1051/epjconf/202225810008
A Virtual Tribute to Quark Confinement and the Hadron Spectrum (vConf21)

Quantized vortices in pionic superfluid

Oleg Teryaev[1,2,3,*] and *Valentin* Zakharov[2,3,4,**]

[1]Joint Institute for Nuclear Research, 141980, Dubna, Russia
[2]Institute of Theoretical and Experimental Physics, NRC Kurchatov Institute, B. Cheremushkinskaya 25, Moscow 117218, Russia
[3]Dubna State University, 141980, Dubna, Russia
[4]Pacific Quantum Center, Far Eastern Federal University, 10 Ajax Bay, Russky Island, Vladivostok 690950, Russia

Abstract. The interplay between classical vorticity being the main undisputed source of polarization in heavy-ion collisions (HIC) and quantized vortices is considered. The vortex tubes emerging in the rotating pionic (super) fluid polarize the baryons in their cores and explain the emerging global polarization. The appearance of vortices in the region separating participants and spectators in non-central HIC is similar to that for sliding layers of liquid helium. From the other side, it is also the region where the classical vorticity was earlier found to be large forming the vortex sheets. The formation of tubes manifests a threshold at certain critical vorticity implying the vanishing polarization at lower energies. For central HIC the compact jet-like flows may lead to formation of vortex rings related to local polarization. The P-odd momentum correlations for their experimental investigation are suggested. The role of shear and viscosity in the emergence of polarization is discussed.

1 Introduction

Observation of non-vanishing polarization of $\Lambda(\overline{\Lambda})$-hyperons in heavy ion collisions by the STAR Collaboration [1] is considered to be one of most remarkable experimental findings of recent years. The common interpretation nowadays is that spin of the hyperons is aligned with the rotation of plasma produced in peripheral ion-ion collisions. Moreover, such a prediction was made prior to the experimental discovery, (see, e.g. [2–4] and references therein) and explored in some detail after the data appeared [5, 6].

The crucial problem is the transition of rotation to spin which may be realized either through the thermal equilibration of spin with rotating matter [4, 5] or by the making use [2, 6] of axial anomaly in effective theory [7] which naturally led to the predicrion of both polarization decrease with energy [2] and its estimate [3] to be of percent order at several GeV energy.

The latter approach in confined phase may be realized via the quantum vortices in pionic supeerfluid [8]. The transition of the rotation to spin happens in the cores of these vortices via the heavy baryonic degrees of freedom. This process is accompanied by the dissipation which may be considered as a counterpart if absorptive phases. Such phases are necessary

*e-mail: teryaev@jinr.ru
**e-mail: vzakharov@itep.ru

for polarization generation and in non-perturbative QCD describing hadronic processes they are produced by higher twists [9, 10] and Wilson lines [11, 12].

In the present paper we consider the mechanisms of quantized vortices production in the core-corona border region which was earlier found [13] to produce the femto-vortex sheets ("femto-cyclones") corresponding to classical vorticity.

We relate this phenomenon to the sliding layers in liquid helium considered in classical textbook by R. Feynman [14].

Passing from layer to the compact (mini) jet we generalize this picture to vortex rings being the counterpart of classical ones recently suggested to be produced in pA collisions [15] or by thermalized jets [16].

We suggest the specific spin-monentum correlation to explore such quantized vortex rings. We will also discuss the role of shear and viscosity in polarization emergence.

2 Sliding layers and quantized vortices

While both theoretical and exprtimental approaches to liquid helium superfluidity explore the rotating bowl, the situation in HIC is more close to the picture of sliding layers suggested in Feynman's textbook ([14], Chapter 11.9). The relative phase between wave functions of layers results in the extra energy carried by separating surface. It is disappearing in the points where relative phase is proportional to 2π leading to the appearance of the vortex lines residing at this boundary and oriented perpendicular to the sliding velocity.

The critical value of this velocity providing the formation of vortices in liquid helium is [14]:

$$v \geq \frac{1}{md} L, \tag{1}$$

where m is the mass of helium molecule, d is the transverse size of the region occupied by vortices and $L = \ln d/a$ is the logarithmic factor involving this size and "ultraviolate" (UV) scale provided by interatomic distance a.

For HIC the sliding layers correspond to participant and spectator nucleons while d may be related to corona region. L may involve the typical proton radius as a UV scale and is not likely to be large. Assuming the uncertainty relation for $d \sim (< p_T >)^{-1}$ (which will lead to cancellation of $\hbar$ for dimensionfull units) one get the estimate for longitudinal pion momentum $p_L = mv$:

$$p_L \geq \sim < p_T > L. \tag{2}$$

Due to the small L this value defined by beam energy is close to the average transverse momentum $< p_T >$ of order of $1 GeV$. This may explain the observed growth of polarization down to $2.4 GeV$ energy [18]. We may expect that the vortices will not emerge (leading to polarization nullification) only for even lower energies about $1 GeV$ or so.

As the core-corona region is also the main source of classical vorticity[13], it may be interesting to compare it to the quantum one (ω_q) , produced by the vortex. This procedure is possible as soon as pion fluid viscosity is small albeit different from zero. The expression for critical velocity (1) may be than recasted as an expression for critical vorticity

$$\omega_{cl} \geq \omega_q \sim \frac{1}{\mu d^2} L, \tag{3}$$

where we substituted mass by chemical potential (see [8]).

3 Vortex rings, local polarization and handedness

If one pass from consideration of helium flow through long narrow gap (see Fig. 11.32 in [14]), which may correspond to corona region between core and specatators in HIC, to that from the circular one by a sort of (continious) deformation, the vortex lines below and above the gap connect to form the vortex rings.

In HIC this corresponds to the hadronic beams [15] (being most close to the discussed liquid helium example) and jets [16]. One may also consider the rings appearance and evolution in the case of HIC, where 3-fluid hydro [19] or kinetic model [20] can be applied.

While in all these cases the *classical* vorticity in the rings was considered leading to the local Λ polarization, here we suggest to consider the *quantized* vortex rings. While the result is qualitatively the same (local Λ polarization) we offer the another mechanism of its emergence due to excitation of heavy baryonic degrees of freedom in vortex core.

To observe the vortex rings the spin-momentum correlations were suggested in [15, 16]

$$R_\Lambda \sim < \vec{S}_\Lambda \cdot [\vec{t}\vec{p}_\Lambda] >, \tag{4}$$

involving the Λ momentum $\vec{p}_\Lambda$ and (average) spin $\vec{S}_\Lambda$ as well as trigger vector $\vec{t}$ which can be, in particular, jet trust [16].

Let us first point out that these correlations are the particular case of *handedness* first suggested just to measure the polarization of quark and gluons fragmenting to hadronic jets [21]. Later it was also applied [22] in the studies of HIC in kinetic model.

Here we wouild like to suggest some modification of this quantity to probe the quantized vortex rings. Let us note that while correlation (5) may indeed signal the apperance of ring-like structure, it may also appear in inclusive polarization, similar to hadronic case [9–12]. Note that this case was actually a guideline in [15] and that inclusive polarization in HIC was recently studied in the course of preparations for experiments at NICA [23].

For more specific probe of quantized vortex formation, we suggest to consider the *pion* momentum (as forming the ring in the superfluid) instead of that of Λ:

$$R^q_\Lambda \sim < \vec{S}_\Lambda \cdot [\vec{t}\vec{p}_\pi] >, \tag{5}$$

$\vec{t}$ can be either the hadron (or light nucleus, with deuterons being one of the likely first stage beams for NICA [25]) momentum or (mini) jet trust. It is likely that some additional cuts for pion momentum can be applied to select the pions forming the vortex. More studies are definitely required.

4 Conclusions and outlook

We suggested here the that production of quantized vortices at the core-corona (and corona-spectators) border in HIC. This phenomenon may be the quantum counterpart of the other roles of this region relevant for baryon polarization [13, 24].

The critical conditions for vortices emergence are discussed and the rough estimate of the minimal energy required for the global polarization generation is performed leading to rather low value $\sim 1GeV$.

The formation of vortex rings is considered, the role of handedness in their studies is stressed and its new version is suggested.

Let us also note, that vortex (unlike rigid rotation) produces shear besides vorticity. Indeed, if

$$v_i = k\frac{\epsilon_{ijk}x_j s_k}{x^2}, \tag{6}$$

where x_i form the transverse vector w.r.t. polarization s_i direction, its derivative

$$\partial_l v_i = k(\frac{\epsilon_{ilk} s_k}{x^2} - 2\frac{x_l \epsilon_{ijk} x_j s_k}{(x^2)^2}) \tag{7}$$

is not completely antisymmetric in i, l. One may guess is there any connection to the much discussed shear-related polarization [26] and explanation of sign puzzle for local polarization.

The shear *viscosity* plays the crucial role in the transition of rotation to heavy baryon degrees of freedom which is the cornerstone[8] of quantum vortices contribution to polarization. There is also similarity with Kolmogorov cascade with dissipation manifested for smallest vortices.

Finally, let us ask ourselves whether the famous lower bound for viscosity [27] can be related to (upper) bound for polarization?!

4.1 Acknowledgements

We are indebted to Yu. Ivanov, M.Lisa and G. Torrieri for discussions.

The work was supported by Russian Science Foundation Grant 21-12-00237.

References

[1] *Global polarization measurement in Au+Au collisions*, STAR Collaboration (B.I. Abelev et al.), Phys.Rev. C76 (2007) 024915, Erratum: Phys.Rev. C95 (2017) no.3, 039906 STAR-05-11-2007 , e-Print: arXiv:0705.1691 [nucl-ex].

[2] *Chiral vortaic effect and neutron asymmetries in heavy-ion collisions* O. Rogachevsky, A. Sorin, O. Teryaev, Phys.Rev. C82 (2010) 054910, e-Print: arXiv:1006.1331 [hep-ph] ;

[3] *Helicity separation in Heavy-Ion Collisions*, M. Baznat, K. Gudima, A. Sorin, O. Teryaev, Phys.Rev. C88 (2013) no.6, 061901, e-Print: arXiv:1301.7003 [nucl-th];

[4] Λ *polarization in peripheral heavy ion collisions*, F. Becattini, L. Csernai, D.J. Wang, Phys.Rev. C88 (2013) no.3, 034905, Erratum: Phys.Rev. C93 (2016) no.6, 069901, arXiv:1304.4427 [nucl-th];
Angular momentum conservation in heavy ion collisions at very high energy, F. Becattini, F. Piccinini, J. Rizzo, Phys.Rev. C77 (2008) 024906, arXiv:0711.1253 [nucl-th];
F. Becattini, V. Chandra, L. Del Zanna, and E. Grossi, arXiv:1303.3431 [nucl-th].

[5] *Global hyperon polarization at local thermodynamic equilibrium with vorticity, magnetic field and feed-down*, F. Becattini, I. Karpenko, M. Lisa, I. Upsal, S. Voloshin, e-Print: arXiv:1610.02506 [nucl-th] .

[6] *Axial anomaly and energy dependence of hyperon polarization in Heavy-Ion Collisions*, A. Sorin, O. Teryaev, Phys.Rev. C95 (2017) no.1, 011902, e-Print: arXiv:1606.08398 [nucl-th] .

[7] A.V. Sadofyev, V.I. Shevchenko, and V.I. Zakharov, *Notes on chiral hydrodynamics within effective theory approach*, Phys. Rev. D83 (2011) 105025, arXiv:1012.1958 [hep-th].

[8] O. V. Teryaev and V. I. Zakharov, Phys. Rev. D **96**, no.9, 096023 (2017) doi:10.1103/PhysRevD.96.096023

[9] A. V. Efremov and O. V. Teryaev, Phys. Lett. B **150**, 383 (1985) doi:10.1016/0370-2693(85)90999-2

[10] J. w. Qiu and G. F. Sterman, Phys. Rev. Lett. **67**, 2264-2267 (1991) doi:10.1103/PhysRevLett.67.2264

[11] S. J. Brodsky, D. S. Hwang and I. Schmidt, Phys. Lett. B **530**, 99-107 (2002) doi:10.1016/S0370-2693(02)01320-5 [arXiv:hep-ph/0201296 [hep-ph]].

[12] J. C. Collins, Phys. Lett. B **536**, 43-48 (2002) doi:10.1016/S0370-2693(02)01819-1 [arXiv:hep-ph/0204004 [hep-ph]].

[13] M. I. Baznat, K. K. Gudima, A. S. Sorin and O. V. Teryaev, Phys. Rev. C **93**, no.3, 031902 (2016) doi:10.1103/PhysRevC.93.031902 [arXiv:1507.04652 [nucl-th]].

[14] R. Feynman, *Statistical Mechanics: A Set Of Lectures* (W.A. Benjamin, Reading, Mass., 1972)

[15] M. A. Lisa, J. G. P. Barbon, D. D. Chinellato, W. M. Serenone, C. Shen, J. Takahashi and G. Torrieri, Phys. Rev. C **104**, no.1, 011901 (2021) doi:10.1103/PhysRevC.104.L011901 [arXiv:2101.10872 [hep-ph]].

[16] W. M. Serenone, J. G. P. Barbon, D. D. Chinellato, M. A. Lisa, C. Shen, J. Takahashi and G. Torrieri, Phys. Lett. B **820**, 136500 (2021) doi:10.1016/j.physletb.2021.136500 [arXiv:2102.11919 [hep-ph]].

[17]

[18] F. J. Kornas [HADES], Springer Proc. Phys. **250**, 435-439 (2020) doi:10.1007/978-3-030-53448-6_68

[19] Y. B. Ivanov and A. A. Soldatov, Phys. Rev. C **97**, no.4, 044915 (2018) doi:10.1103/PhysRevC.97.044915 [arXiv:1803.01525 [nucl-th]].

[20] A. Zinchenko, A. Sorin, O. Teryaev and M. Baznat, J. Phys. Conf. Ser. **1435**, no.1, 012030 (2020) doi:10.1088/1742-6596/1435/1/012030

[21] A. V. Efremov, L. Mankiewicz and N. A. Tornqvist, Phys. Lett. B **284**, 394-400 (1992) doi:10.1016/0370-2693(92)90451-9

[22] O. Teryaev and R. Usubov, Phys. Rev. C **92**, no.1, 014906 (2015) doi:10.1103/PhysRevC.92.014906

[23] E. Nazarova, R. Akhat, M. Baznat, O. Teryaev and A. Zinchenko, Phys. Part. Nucl. Lett. **18**, no.4, 429-438 (2021) doi:10.1134/S1547477121040142

[24] A. Ayala, M. A. Ayala Torres, E. Cuautle, I. Domínguez, M. A. Fontaine Sanchez, I. Maldonado, E. Moreno-Barbosa, P. A. Nieto-Marín, M. Rodríguez-Cahuantzi and J. Salinas, *et al.* Phys. Lett. B **810**, 135818 (2020) doi:10.1016/j.physletb.2020.135818 [arXiv:2003.13757 [hep-ph]].

[25] A. Arbuzov, A. Bacchetta, M. Butenschoen, F. G. Celiberto, U. D'Alesio, M. Deka, I. Denisenko, M. G. Echevarria, A. Efremov and N. Y. Ivanov, *et al.* Prog. Part. Nucl. Phys. **119**, 103858 (2021) doi:10.1016/j.ppnp.2021.103858 [arXiv:2011.15005 [hep-ex]].

[26] F. Becattini, M. Buzzegoli and A. Palermo, Phys. Lett. B **820**, 136519 (2021) doi:10.1016/j.physletb.2021.136519 [arXiv:2103.10917 [nucl-th]].

[27] G. Policastro, D. T. Son and A. O. Starinets, Phys. Rev. Lett. **87**, 081601 (2001) doi:10.1103/PhysRevLett.87.081601 [arXiv:hep-th/0104066 [hep-th]].

EPJ Web of Conferences **258**, 10009 (2022) https://doi.org/10.1051/epjconf/202225810009
A Virtual Tribute to Quark Confinement and the Hadron Spectrum (vConf21)

Two Schrödinger-like Equations for hadrons

Ruben Sandapen[1,*]

[1]Department of Physics, Acadia University, Wolfville, Nova-Scotia, Canada, B4P 2R6

Abstract. In this talk, based on [1, 2], I argue that the holographic Schrödinger Equation of $(3+1)$-dim, conformal light-front QCD and the 't Hooft Equation of $(1+1)$-dim, large N_c QCD, can be complementary to each other in providing a first approximation to hadron spectroscopy. Together, the two equations play a role in hadronic physics analogous that of the ordinary Schrödinger Equation in atomic physics.

1 Introduction

In $(3+1)$-dim light-front QCD, the internal dynamics of a quark-antiquark meson are governed by the Schrödinger-like equation: [3, 4]

$$\left[\left(-\frac{\nabla^2_{b_\perp}}{x(1-x)} + \frac{m_q^2}{x} + \frac{m_{\bar{q}}^2}{1-x}\right) + U(x, b_\perp)\right]\Psi(x, \mathbf{b}_\perp) = M^2\Psi(x, \mathbf{b}_\perp)\,, \tag{1}$$

where M is the meson mass and $\Psi(x, \mathbf{b}_\perp)$ is the meson light-front wavefunction, with x being the light-front momentum fraction carried by the quark, and $\mathbf{b}_\perp$ the transverse distance between the quark and the antiquark. The quark and antiquark masses, m_q and $m_{\bar{q}}$, are effective masses, and $U(x, b_\perp)$ is an effective confining potential. The advantage of formulating QCD on the light-front is that the structure of Eq. (1) is analogous to that of the ordinary Schrödinger Equation, despite Eq. (1) being fully relativistic and frame-independent. It remains that deriving $U(x, b_\perp)$ from first principles in QCD is an open question. Insights into its analytic form can be gained by introducing a new variable,

$$\zeta = \sqrt{x(1-x)}\mathbf{b}_\perp \tag{2}$$

and factorizing the wavefunction as

$$\Psi(x, \zeta, \varphi) = \frac{\phi(\zeta)}{\sqrt{2\pi\zeta}} e^{iL\varphi} X(x) \tag{3}$$

with $L = |L_z^{\max}|$ being the light-front orbital angular momentum, and $X(x) = \sqrt{x(1-x)}\chi(x)$. Assuming that

$$U(x, \zeta) = U_\parallel(x) + U_\perp(\zeta)\,, \tag{4}$$

Eq. (1) then separates into

$$\left[-\frac{d^2}{d\zeta^2} + \frac{4L^2-1}{4\zeta^2} + U_\perp(\zeta)\right]\phi(\zeta) = M_\perp^2\phi(\zeta) \tag{5}$$

*e-mail: ruben.sandapen@acadiau.ca

and

$$\left[\frac{m_q^2}{x} + \frac{m_{\bar{q}}^2}{1-x} + U_{\parallel}(x)\right]\chi(x) = M_{\parallel}^2\chi(x)\,. \tag{6}$$

with

$$M^2 = M_{\perp}^2 + M_{\parallel}^2 \tag{7}$$

and

$$\int \mathrm{d}x|\chi(x)|^2 = \int \mathrm{d}^2\zeta|\phi(\zeta)|^2 = 1\,. \tag{8}$$

As our notation suggests, Eq. (6) governs the longitudinal dynamics while, for a given x, Eq. (5) governs the transverse dynamics. The x-dependence of ζ means that, in this type of factorization, the so-called transverse and longitudinal dynamics do not decouple completely. However, the variable ζ plays a key role in the search of an analytical form for $U_{\perp}(\zeta)$ [5–7].

2 The holographic Schrödinger Equation

If ζ is identified with the 5th dimension of AdS_5, and L^2 with $(\mu R)^2 + (2-J)^2$, where μ is a mass parameter and R the radius of curvature in AdS_5, then Eq. (5) becomes the wave equation for the freely propagating spin-J string modes in AdS_5, with the confining potential given by [3]

$$U_{\perp}(\zeta) = \frac{1}{2}\varphi''(z) + \frac{1}{4}\varphi'(z)^2 + \frac{2J-3}{2z}\varphi'(z)\,, \tag{9}$$

where $\varphi(z)$ is a dilaton field which distorts the geometry of pure AdS_5.

At this point, it may seem that the choice for the dilaton field is arbitrary. However, if we insist that the underlying action of Eq. (5) remains conformally invariant with a mass scale in Eq. (5), then the latter, which we denote by κ, can only appear via a harmonic confining potential: $U_{\perp}(\zeta) = \kappa^4\zeta^2$. This means that the dilaton field in AdS_5 must be quadratic, $\varphi(z) = \kappa^2 z^2$, and consequently Eq. (9) implies that

$$U_{\perp}(\zeta) = \kappa^4\zeta^2 + 2\kappa^2(J-1)\,. \tag{10}$$

Hence Eq. (10) is fixed simultaneously by the underlying conformal symmetry and holographic mapping. The conformal algebra leading to the uniqueness of the quadratic potential in Eq. (10) can be found in [3, 8]. Solving the holographic Schrödinger Equation, i.e. Eq. (5) with Eq. (10), yields the meson spectrum

$$M_{\perp}^2(n_{\perp}, J, L) = 4\kappa^2\left(n_{\perp} + \frac{J+L}{2}\right) \tag{11}$$

where $n_{\perp}$ is the principal quantum number, and $J = L + S$ with S being the total quark-antiquark spin.

The supersymmetric formulation of Eq. (5) allows baryons, mesons and tetraquarks to be considered in a unified framework [9–12]. This hadronic-level supersymmetry is motivated by an essential feature of colour $SU(N_c)$: a cluster of $N_c - 1$ constituents can be in the same colour representation as the anti-constituent [12]. For $N_c = 3$, it means that a diquark (antidiquark) can be in the same colour representation as an antiquark (quark). It follows that the colour confinement dynamics are the same in a quark-diquark baryon, quark-antiquark meson and a diquark-antiquark tetraquark. The supersymmetrization of Eq. (10) leads to two harmonic-oscillator partner potentials with the same confinement scale, κ, so that Eq. (22) generalizes to:[11]

$$M_{\perp,M}^2 = 4\kappa^2\left(n_{\perp} + L_M + \frac{S_M}{2}\right)\,, \tag{12}$$

$$M^2_{\perp,B} = 4\kappa^2 \left(n_\perp + L_B + \frac{S_D}{2} + 1 \right) , \tag{13}$$

and

$$M^2_{\perp,T} = 4\kappa^2 \left(n_\perp + L_T + \frac{S_T}{2} + 1 \right) , \tag{14}$$

where S_D is the lowest value of the total quark-diquark spin in baryons, S_M is the total quark-antiquark spin in mesons, and S_T is total diquark-antidiquark spin in tetraquarks. Eqs. (12), (13) and (14) show that the masses of baryons with $L_B = L_M - 1$ and $S_D = S_M$ are equal to the masses of mesons with L_M and S_M and tetraquarks $S_T = S_D$ and $L_T = L_B$. Within a family of superpartners, we can thus write $S \equiv S_D = S_M = S_T$. Note that the lowest lying mesons with $n_\perp = L_M = 0$ do not have a baryonic superpartner. Furthermore, pseudoscalar mesons with $n_\perp = L_M = S_M = 0$, like the pion, are predicted to be massless.

In nature, the above superconformal symmetry by broken by non-zero quark masses and longitudinal confinement that generate $M_\parallel$ via Eq. (6). To solve Eq. (6), we need to specify the longitudinal confinement potential. Here we take it to be the 't Hooft potential.

3 The 't Hooft Equation

The 't Hooft Equation is derived from the QCD Lagrangian in $(1 + 1)$-dim, in the large N_c approximation, where only planar diagrams contribute and can be summed to all orders. The result is [13]:

$$\left(\frac{m_q^2}{x} + \frac{m_{\bar{q}}^2}{1-x} \right) \chi(x) + \frac{g^2}{\pi} \mathcal{P} \int \mathrm{d}y \frac{\chi(x) - \chi(y)}{(x-y)^2} = M_\parallel^2 \chi(x) , \tag{15}$$

where $\mathcal{P}$ denotes the principal value prescription and $g = g_s\sqrt{N_c}$ is the (finite) 't Hooft coupling with g_s being the strong coupling and N_c the number of colours. Here we shall refer to the 't Hooft coupling as the longitudinal confinement scale. Eq. (15) coincides with Eq. (6), if we take

$$U_\parallel(x) = \frac{g^2}{\pi} \mathcal{P} \int \mathrm{d}y \frac{\chi(x) - \chi(y)}{(x-y)^2} . \tag{16}$$

Fourier transforming Eq. (16) to position space, we obtain [2]

$$U_L(x^-) = \frac{g^2}{2} P^+ |x^-| = g^2 P^+ b_\parallel \tag{17}$$

where P^+ is the light-front momentum of the meson and $b_\parallel$ is the longitudinal separation between the quark and the antiquark: the 't Hooft potential is a linear confinement potential in position space. Note that $\tilde{z} = P^+ x^-$ is the frame invariant longitudinal coordinate introduced in [14]. To use the 't Hooft Equation for baryons and tetraquarks, we transform the antiquark into a diquark (for baryons), followed by the transformation of the quark into an antidiquark (for tetraquarks). Just as κ, g does not change within a family of superpartners.

We note that the idea of using the 't Hooft potential as a candidate for $U_\parallel(x)$ was first proposed in [15], and that [16, 17] use an alternative model [18, 19] for $U_\parallel(x)$, namely

$$U_\parallel(x) = -\sigma^2 \partial_x (x(1-x)) \partial_x \tag{18}$$

where σ is the longitudinal confinement scale. Eq. (6) then admits analytical solutions: [17]

$$M_\parallel^2 = \sigma(m_q + m_{\bar{q}})(2n_\parallel + 1) + \sigma^2 n_\parallel (n_\parallel + 1) + (m_q + m_{\bar{q}})^2 . \tag{19}$$

On the other hand, the 't Hooft Equation needs to be solved numerically and we follow the matrix method outlined in [15] to do so. We have checked that this numerical method correctly reproduces the analytical solutions, Eq. (19), when using Eq. (18). We draw attention to the recent paper [20] where a comparison between the different models for the longitudinal potential is carried out.

Previously, the Brodsky-de Téramond ansatz [21] was widely used to accommodate light quark masses while neglecting longitudinal confinement. Despite its phenomenological success, the shortcoming of the Brodsky-de Téramond ansatz is that it is not consistent [17] with the Gell-Mann-Oakes-Renner (GMOR) relation [22] . Another ansatz designed to agree with the GMOR relation can be found in [23].

4 Symmetry constraints

Chiral symmetry breaking in QCD is encapsulated in the GMOR relation which predicts that $M_\pi^2 \propto m_{u/d}$ in the limit of small quark masses. Now the holographic Schrödinger Equation predicts that $M_\pi = 0$ while the 't Hooft Equation predicts that, if $m_{u/d} \ll g$, then $M_\pi^2 \propto g m_{u/d}$. Hence, the two equations together correctly satisfy the GMOR relation, provided that g does not scale with $m_{u/d}$ in the chiral limit.

Heavy Quark Effective Theory (HQET) predicts that the mass difference between heavy-light vector and pseudoscalar mesons in their ground states, is suppressed by the heavy quark mass: $M_{qQ}^V - M_{qQ}^P \sim 1/m_Q$. Now, the holographic Schrödinger Equation predicts that, if $M_{qQ}^V \sim \kappa$ and $M_{qQ}^P = 0$, while the 't Hooft Equation predicts that $M_{qQ}^{P/V} \sim m_Q$ if $m_Q \gg g$. Hence, together, two equations satisfy the HQET constraint in the limit $m_Q \gg \kappa$, provided that κ does not scale with m_Q.

Rotational symmetry is restored in the centre-of-mass (CM) frame of a heavy-heavy meson since the system is essentially non-relativistic. In the non-relativistic limit, the light-front holographic potential differs from the 't Hooft potential: $U_\perp \sim (\kappa^4/4)b_\perp^2$ (since $x \sim 1/2$) and $U_\parallel \sim 2m_Q b_\parallel$ (since $P^+ = M \sim 2m_Q$). However, the rotational symmetry constraint applies to instant form potentials in the CM frame. Using the relation between light-front and instant-form potentials given in [24], we find that both light-front potentials correspond to linear instant-form potentials in the CM frame: $V_\perp \sim (\kappa^2/2)b_\perp$ and $V_\parallel \sim (g^2/2)b_\parallel$. Rotational symmetry implies that $V_\perp = V_\parallel$, and is satisfied provided that $g = \kappa$ in the non-relativistic limit.

The above constraints motivate us to consider g to be independent of $m_{u/d}$ in light mesons, κ to be independent of m_Q in heavy-light mesons, and g to coincide with κ in heavy-heavy mesons. These statements also apply to their baryon and tetraquark superpartners since the confinement scales do not change within a family of superpartners.

5 Computing the hadronic spectrum

When solving the 't Hooft Equation, an additional quantum number, $n_\parallel$, emerges. Thus, each hadronic state is identified by four quantum numbers, $n_\perp, L, S, n_\parallel$, with their squared masses given by

$$M_M^2 = M_{\perp,M}^2(n_\perp, L_M, S_M, \kappa) + M_{\parallel,M}^2(n_\parallel, m_q, m_{\bar{q}}, g) , \tag{20}$$

$$M_B^2 = M_{\perp,B}^2(n_\perp, L_B, S_D, \kappa) + M_{\parallel,B}^2(n_\parallel, m_q, m_{[qq]}, g) , \tag{21}$$

$$M_T^2 = M_{\perp,B}^2(n_\perp, L_T, S_T, \kappa) + M_{\parallel,T}^2(n_\parallel, m_{[\bar{q}\bar{q}]}, m_{[qq]}, g) , \tag{22}$$

Table 1. Quark masses and longitudinal confinement scale in GeV. Note that we use $\kappa = 0.523$ GeV for all hadrons, with the constraint $g = \kappa$ for hadrons with two heavy quarks.

Hadron	g	$m_{u/d}$	m_s	m_c	m_b
Light	0.128	0.046	0.357	-	-
Heavy-light	0.410	0.330	0.500	1.370	4.640
Heavy-heavy	0.523	-	-	1.370	4.640

while their parity and charge conjugation are given by

$$P = (-1)^{L_M+1} = (-1)^{L_B} = (-1)^{L_T} . \tag{23}$$

and

$$C = (-1)^{n_\parallel + L_M + S_M} = (-1)^{n_\parallel + L_T + S_T - 1} . \tag{24}$$

À posteriori, we find that

$$n_\parallel \geq n_\perp + L \tag{25}$$

i.e. an orbital and/or radial excitation in the transverse dynamics is always accompanied by an excitation in the longitudinal dynamics.

To calculate the hadronic spectrum, we need to specify the numerical values of the confinement scales and quark masses. We take the diquark mass to be simply twice the quark mass and we use the universal value of the transverse confinement scale, $\kappa = 0.523$ GeV [25]. Our choices for longitudinal confinement scale, g, and the quark masses are shown in Table 1. Note that we find it necessary to allow g to vary with the number of heavy quarks in the hadron, while remaining the same within a family of superpartners and coinciding with κ for hadrons with two heavy quarks. The quark masses in the light hadrons are identical to those used previously in light-front holography [3].

To illustrate qualitatively the agreement with spectroscopic data, we show a sample of Regge trajectories for hadrons with no, one or two charm quarks, taken from [1] where the full sets of numerical results can be found. In general, the agreement for mesons and baryons is quite good. For the tetraquark candidates, which we identify following [11], there is a couple of very notable disagreements. These discrepancies were already present when longitudinal dynamics were neglected and are thus not alleviated by accounting for the latter.

Given that, except for the pion and the kaon, $M_\perp \ll M_\parallel$ for light hadrons and that $M_\parallel \gg M_\perp$ in heavy hadrons, it is legitimate to ask whether the Regge trajectories are sensitive to g for light hadrons and to κ for heavy hadrons. The answer is yes, as illustrated by Fig. 4. This means that the statement that κ remains universal across the full spectrum is not a trivial one.

6 Conclusions

We have argued that, together, the holographic Schrödinger Equation and the 't Hooft Equation describe the main features of the full hadron spectrum. The confinement scale of the holographic Schrödinger Equation is universal across the full spectrum. On the other hand, the confinement scale of the 't Hooft Equation depends on the number of heavy quarks in the hadrons, while remaining the same within a family of superpartners and coinciding with the holographic scale in hadrons with two heavy quarks.

7 Acknowledgements

I thank the organizers of the Virtual Tribute to Confinement and the Hadron Spectrum at the University of Stavanger, Norway, for a successful conference. I am grateful to my coauthors

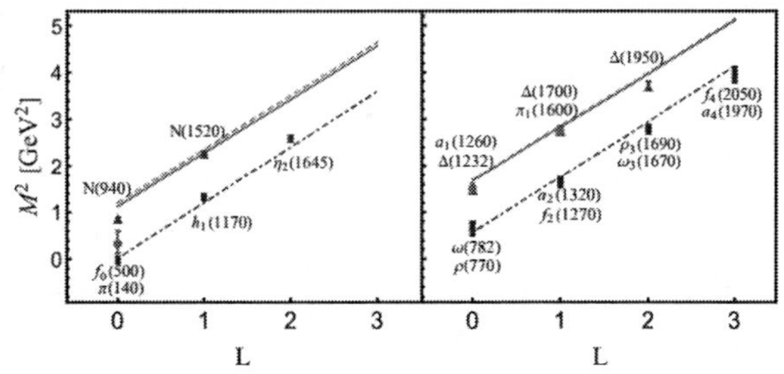

Figure 1. Regge trajectories for the light hadrons. Dot-dashed-blue: mesons, solid-red: baryons, dashed-brown: tetraquarks. Data from PDG [26].

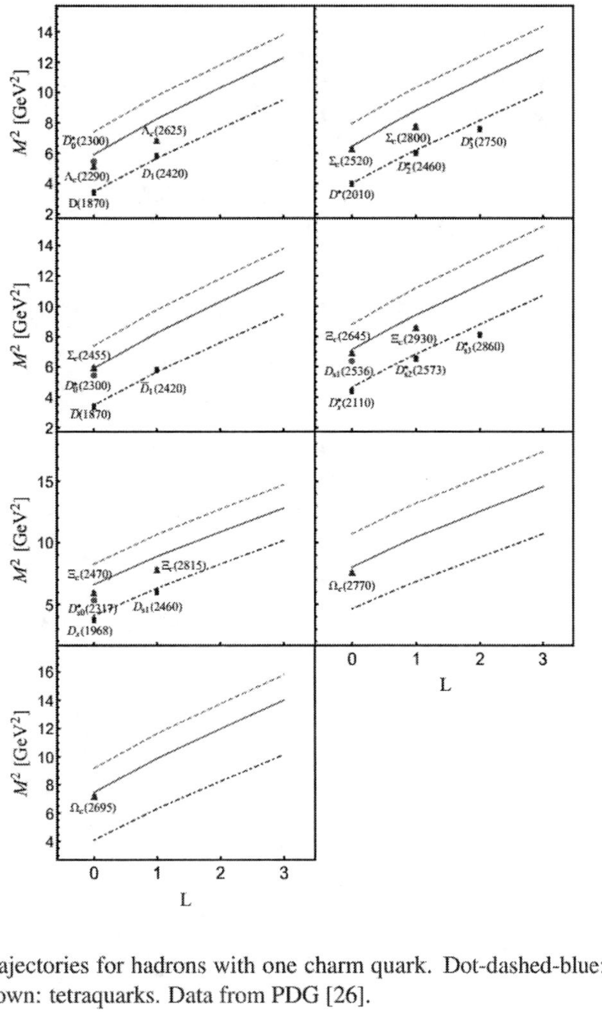

Figure 2. Regge trajectories for hadrons with one charm quark. Dot-dashed-blue: mesons, solid-red: baryons, dashed-brown: tetraquarks. Data from PDG [26].

of [1, 2] on which this talk is based. I also thank Stan Brodsky, Guy de Téramond and Gerald Miller for useful discussions. This research is supported by an Individual Discovery

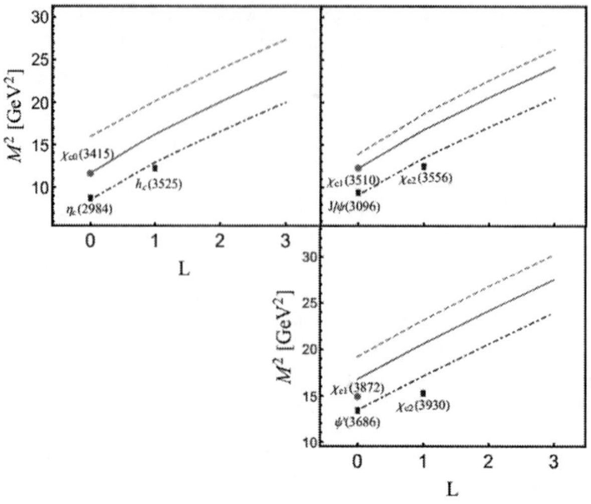

Figure 3. Regge trajectories for hadrons with two charm quarks. Dot-dashed-blue: mesons, solid-red: baryons, dashed-brown: tetraquarks. Data from PDG [26].

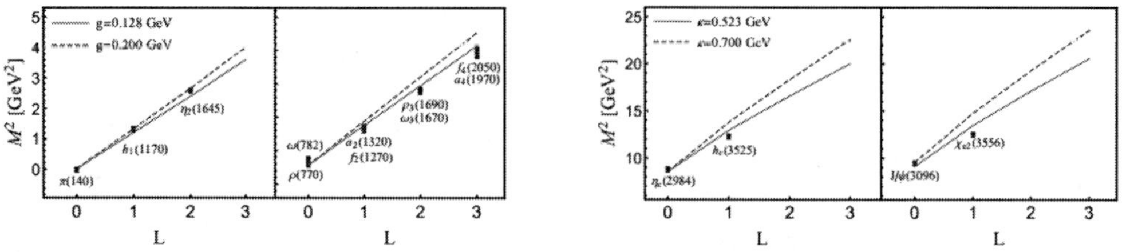

Figure 4. Left: Sensitivity of the Regge trajectories for π and ρ families to g: g = 0.128 GeV (solid-gray) and g = 0.200 GeV (dashed-dark-red). Right: Sensitivity of Regge trajectories for η_c and J/ψ families to κ: κ = 0.523 GeV (solid-gray) and κ = 0.700 GeV (dashed-dark-red).

Grant (SAPIN-2020-00051) from the Natural Sciences and Engineering Research Council of Canada (NSERC).

References

[1] M. Ahmady, S. Kaur, S.L. MacKay, C. Mondal, R. Sandapen, Phys. Rev. D **104**, 074013 (2021), 2108.03482
[2] M. Ahmady, H. Dahiya, S. Kaur, C. Mondal, R. Sandapen, N. Sharma, Phys. Lett. B **823**, 136754 (2021), 2105.01018
[3] S.J. Brodsky, G.F. de Teramond, H.G. Dosch, J. Erlich, Phys. Rept. **584**, 1 (2015), 1407.8131
[4] S.J. Brodsky, H.C. Pauli, S.S. Pinsky, Phys. Rept. **301**, 299 (1998), hep-ph/9705477
[5] G.F. de Teramond, S.J. Brodsky, Phys. Rev. Lett. **94**, 201601 (2005), hep-th/0501022
[6] G.F. de Teramond, S.J. Brodsky, Phys. Rev. Lett. **102**, 081601 (2009), 0809.4899
[7] S.J. Brodsky, G.F. de Teramond, Phys. Rev. Lett. **96**, 201601 (2006), hep-ph/0602252
[8] S.J. Brodsky, G.F. De Téramond, H.G. Dosch, Phys. Lett. **B729**, 3 (2014), 1302.4105
[9] H.G. Dosch, G.F. de Teramond, S.J. Brodsky, Phys. Rev. D **92**, 074010 (2015), 1504.05112
[10] H.G. Dosch, G.F. de Teramond, S.J. Brodsky, Phys. Rev. D **95**, 034016 (2017), 1612.02370
[11] M. Nielsen, S.J. Brodsky, Phys. Rev. D **97**, 114001 (2018), 1802.09652
[12] S.J. Brodsky, G.F. de Téramond, H.G. Dosch, C. Lorcé, Phys. Lett. B **759**, 171 (2016), 1604.06746
[13] G. 't Hooft, Nucl. Phys. B **75**, 461 (1974)
[14] G.A. Miller, S.J. Brodsky, Phys. Rev. C **102**, 022201 (2020), 1912.08911
[15] S.S. Chabysheva, J.R. Hiller, Annals Phys. **337**, 143 (2013), 1207.7128
[16] G.F. De Téramond, S.J. Brodsky (2021), 2103.10950
[17] Y. Li, J.P. Vary (2021), 2103.09993
[18] Y. Li, P. Maris, X. Zhao, J.P. Vary, Phys. Lett. B **758**, 118 (2016), 1509.07212
[19] A.B. Sheckler, G.A. Miller, Phys. Rev. D **103**, 096018 (2021), 2101.00100
[20] C.M. Weller, G.A. Miller (2021), 2111.03194
[21] S.J. Brodsky, G.F. de Teramond, Subnucl. Ser. **45**, 139 (2009), 0802.0514
[22] M. Gell-Mann, R.J. Oakes, B. Renner, Phys. Rev. **175**, 2195 (1968)
[23] T. Gutsche, V.E. Lyubovitskij, I. Schmidt, A. Vega, Phys. Rev. D **87**, 056001 (2013), 1212.5196
[24] A.P. Trawiński, S.D. Głazek, S.J. Brodsky, G.F. de Téramond, H.G. Dosch, Phys. Rev. D **90**, 074017 (2014), 1403.5651
[25] S.J. Brodsky, G.F. de Teramond, H.G. Dosch, C. Lorce, Int. J. Mod. Phys. **A31**, 1630029 (2016), 1606.04638
[26] P. Zyla et al. (Particle Data Group), PTEP **2020**, 083C01 (2020)

AUTHOR INDEX

AUTHOR INDEX